PAPER CHROMATOGRAPHY AND ELECTROPHORESIS

VOLUME II

Paper Chromatography

PAPER CHROMATOGRAPHY AND ELECTROPHORESIS

by Gunter Zweig
DIRECTOR, LIFE SCIENCES DIVISION
SYRACUSE UNIVERSITY RESEARCH CORPORATION
SYRACUSE, NEW YORK

and John R. Whitaker
DEPARTMENT OF FOOD SCIENCE AND TECHNOLOGY
COLLEGE OF AGRICULTURE
UNIVERSITY OF CALIFORNIA
DAVIS, CALIFORNIA

VOLUME II

PAPER CHROMATOGRAPHY

by Joseph Sherma
ASSOCIATE PROFESSOR, CHEMISTRY
LAFAYETTE COLLEGE, EASTON, PENNSYLVANIA

and Gunter Zweig

With contributions by Arthur Bevenue
ASSOCIATE CHEMIST, UNIVERSITY OF HAWAII
HONOLULU, HAWAII

1971

ACADEMIC PRESS New York and London

ACADEMIC PRESS, INC.
111 Fifth Avenue, New York, New York 10003

United Kingdom Edition published by
ACADEMIC PRESS, INC. (LONDON) LTD.
Berkeley Square House, London W1X 6BA

Library of Congress Catalog Card Number: 67-22784

PRINTED IN THE UNITED STATES OF AMERICA

CONTENTS

Preface ix

Contents of Volume I xi

Chapter 1 **History and Introduction**

I. History 1
II. Future Progress in Paper Chromatography 3
References 4

Chapter 2 **Theory, Mechanism, and Fundamentals of Paper Chromatography**

I. Introduction 6
II. Theory and Mechanism 8
III. R_F and R_M Values 14
IV. Artifacts and Anomalous Zones 21
References 27

Chapter 3 **Techniques of Paper Chromatography**

I. The Chromatography Laboratory 30
II. Sample Preparation 32
III. Sample Spotting 32
IV. The Paper 36
V. The Solvent 49
VI. Development 53
VII. Detection Methods 73
VIII. Qualitative Evaluation, Preservation, and Storage of Chromatograms ... 78

IX. Identification of Chromatographic Spots 80
X. Preparative Paper Chromatography 82
References 84

Chapter 4 **Quantitative Paper Chromatography**

I. Formation of the Colored Derivative 90
II. Estimation of the Spots 91
III. Sources of Error in Quantitative Paper Chromatography 105
References 108

Chapter 5 **Amines, Amino Acids, Peptides, and Proteins**

I. Aliphatic and Aromatic Amines 110
II. Amino Acids 120
III. Peptides 140
IV. Proteins 143
References 146

Chapter 6 **Carbohydrates**

I. Introduction 152
II. General Procedure 153
III. Miscellaneous Classes of Carbohydrates 163
References 169

Chapter 7 **Aliphatic Acids**

I. Preparation of Samples 172
II. Filter Paper 176
III. Solvents for Chromatographic Development 177
IV. Detection Reagents 191
V. Quantitative Analysis 195
References 196

Chapter 8 **Steroids, Bile Acids, and Cardiac Glycosides**

I. Steroids 200
II. Bile Acids and Other Steroid Acids 232
III. Cardiac Glycosides and Aglycones (Cardenolides, Bufadienolides) 237
References 244

Chapter 9 **Purines, Pyrimidines, and Related Compounds**

I. Preparation of Samples 248
II. Detection and Identification 250
III. Methods and Techniques 254
IV. Chromatographic Systems 255
V. Quantitative Methods 257
VI. Applications 259
References 281

Chapter 10 **Indoles, Phenols, and Aromatic Acids**

I. Indoles 283
II. Phenols and Aromatic Acids 296
References 315

Chapter 11 **Naturally Occurring Pigments**

I. Fat-Soluble Chloroplast Pigments 319
II. Various Other Porphyrins 334
III. Flavonoids 341
References 350

Chapter 12 **Pesticides**

I. Introduction 355
II. Chlorinated Insecticides 357
III. Organophosphate Insecticides 363
IV. Carbamate Insecticides 375
V. Herbicides 379
VI. Fungicides 386
VII. Miscellaneous Pesticides and Related Products 388
VIII. Food and Feed Additives 391
References 395

Chapter 13 **Antibiotics and Vitamins**

I. Antibiotics 397
II. Vitamins 419
References 433

Chapter 14 **Miscellaneous Organic Compounds**

I. Hydrocarbons 438
II. Alcohols 442

III. Oxo Compounds 447
IV. Organic Peroxides. 454
V. Nitro Compounds 457
VI. Alkaloids 460
VII. Synthetic Dyes 469
VIII. Sulfur Compounds 473
IX. Methylphenylpolysiloxanes 484
X. Phosphorus Compounds 484
XI. Epoxy Resins 494
References 494

Chapter 15. **Inorganic Analysis**

I. Cations 500
II. Common Anions 525
III. Miscellaneous Inorganic Substances........ 530
References 534

Author Index 539

Subject Index 573

PREFACE

Volume II of "Paper Chromatography and Electrophoresis" is presented to the reader with an up-to-date literature search on the subject of paper chromatography. This volume represents a reevaluation of paper chromatography since the last edition of "Paper Chromatography and Paper Electrophoresis" (R. J. Block, E. L. Durrum, and G. Zweig, 2nd Ed., Academic Press, 1958), which was published almost twelve years ago. We have endeavored again, as in previous editions, to give complete experimental procedures so that the researcher does not have to refer to any other source. However, this attempt for complete detail has not been possible at times, and in these cases the reader is referred to earlier experiments found in the second edition of "Paper Chromatography and Paper Electrophoresis" or to original literature.

We have attempted to be selective in our choice of references and to present methods in paper chromatography which in our opinions appear to be most suitable for laboratory work. It is obvious that not all the work published in this field in the past twenty-five years has been covered in this volume, but rather an overall picture of the latest advances as they existed in 1970. The reader should be able to adapt the material presented, with minor modifications, to solve his own experimental problems.

A large number of R_F values have been given, but the reader is cautioned that these values are not intrinsic properties of chemical compounds or inorganic ions, but rather serve to indicate relative migration on paper. As Michael Lederer [*J. Chromatogr.* **33**, 287 (1968)] aptly stated, "R_F values should be approached with the same frame of mind . . . [as uncorrected melting points] . . . and the data of other workers should be considered rather a good hint where one should start when separating a certain mixture rather than as a type of table of chromatograph constants."

Throughout the book, when ammonia, NH_3, NH_4OH, or acetic acid are listed as solvent constituents, the reader may assume that the method requires concentrated ammonia or glacial acetic acid, where applicable. Also the prefix *n* for alcohols or other organic solvents may have been omitted; the reader may assume that *normal* butanol, etc., rather than *sec*-butanol, be used. In all cases, chromatographic solvents, unless stated otherwise, are given in proportion based on volume.

Although some dire predictions have been made about the future of paper chromatography and its replacement as a seperation technique by thin-layer chromatography, it is obvious to us that paper chromatography is still flourishing. The field has become so well established that a smaller number of new techniques may be expected. This book is intended for researchers who want to use an elegant separation technique for complex mixtures—a technique which is inexpensive, convenient, and endowed with a tremendous bibliography.

One of us (G. Z.) wishes to thank Dr. John R. Whitaker, author of Volume I ("Electrophoresis in Stabilizing Media") of this treatise for his patience in extending the deadline for this manuscript a number of times. We also want to thank Dr. Arthur Bevenue for contributing Chapters 7 and 10 to this work.

We wish to dedicate this volume to the late Kurt Jacoby, whose inspiration and constant encouragement made the completion of this project a reality.

GUNTER ZWEIG
JOSEPH SHERMA

CONTENTS OF VOLUME I

Electrophoresis in Stabilizing Media

JOHN R. WHITAKER

I. General Considerations
II. Amines, Amino Acids, and Peptides
III. Proteins
IV. Nucleic Acids, Derivatives, and Related Compounds
V. Carbohydrates
VI. Organic Acids and Derivatives
VII. Electrophoretic Separation of Other Organic Compounds
VIII. Inorganic Ions
AUTHOR INDEX—SUBJECT INDEX

PAPER CHROMATOGRAPHY AND ELECTROPHORESIS

VOLUME II

Paper Chromatography

1

HISTORY AND INTRODUCTION

I. History

Chromatography is defined as "a method of analysis in which the flow of solvent or gas promotes the separation of substances by differential migration from a narrow initial zone in a porous sorptive medium" (*1*). Gas chromatography and solution (liquid) chromatography are major subdivisions of chromatography, and paper chromatography is a subdivision of solution chromatography based on the form of the sorption system. The solvent in paper chromatography is usually organic in nature, and the stationary phase is a sheet of paper containing water or some other polar liquid. In reversed-phase paper chromatography, a fixed, nonpolar sorptive phase, usually prepared by impregnating the paper, is employed in combination with a polar solvent.

The nomenclature of chromatography as devised by M. Tswett is still employed today (*2*). The liquid used to cause the separation of the mixture is the *chromatographic solvent*, *wash liquid*, or *developer*. The resultant series of zones or spots is the *chromatogram*, and the washing of the mixture through the paper to form the chromatogram is termed *formation* or *development* of the chromatogram (*1*).

Paper chromatography may have had its origin with the description by Pliny (23–79 A.D.) of the use of papyrus impregnated with an extract of gallnuts for the detection of ferrous sulfate. During the nineteenth century, vat solutions were tested for purity in the dye industry by spotting on paper or cloth and observing the number of concentric rings formed as the liquid spread (*3*). This method of spot testing colored mixtures on printing, blotting, and filter paper was studied by Runge and Reed who published results independently before 1900 (*4*).

Schönbein and his student Goppelsroeder in the same period (1850–1910) originated capillary analysis, a technique which yields partial separations due to the capillary rise of solutes and solvents through a paper strip dipped into a solution. It was at this stage that the R_F term was introduced (*3*).

Even though Goppelsroeder had mentioned the possibility of employing fresh solvent to improve his separations (*5*), it is M. Tswett who must be given the credit for developing chromatography as defined above (*6a*). Tswett washed a narrow zone of adsorbed pigment mixture through a column with fresh solvent and proved that this washing step was indispensable if the mixture was to be extensively resolved. The earlier methods just described, which Tswett was familiar with and studied, lacked this critical development step.

The next advances were not until the 1930's when Flood (*6b*) investigated chromatography on impregnated papers and Brown (*6c*) developed circular chromatography. However, the great popularity of paper chromatography today is due to the work of Martin, Synge, Consden, and Gordon (*7, 8a*), the first two of whom were recipients of the Nobel Prize in Chemistry in 1952.

In 1938, Neuberger was interested in separating the neutral amino acids and observed that the partition coefficients of acetylated amino acids between water and an immiscible organic solvent differed for various amino acids. This observation was extended by Martin and Synge who attempted in 1941 to separate amino acids by countercurrent distribution (*8b*). This work was tedious and not very successful, so they decided to use silica gel as an inert support to hold the aqueous phase and to pass the immiscible solvent through a bed of the water-containing silica gel in order to conveniently bring about greater contact between the two phases (*8c*).

To eliminate the preparation of silica gel and to reduce the quantity of material needed, the silica gel was later replaced by filter paper as the inert support. With the use of filter paper, the acetylation of the amino acids was no longer necessary, since these compounds could now be detected directly on the paper by treatment with ninhydrin. This idea led to "paper partition chromatography" as we know it today.

The method of paper partition chromatography consists of applying a *small* drop of the solution containing the substances to be separated to a strip of filter paper a short distance from one end. The drop is allowed to dry, and the end of the paper nearest to the spot is placed in the developing solution without immersing the spot itself. The solvent flows down the paper (descending chromatography) by capillarity plus gravity or up by capillarity alone (ascending chromatography), carrying the components of the mixture along at different rates. When the solvent front is near the other end of the paper, the paper is removed and the separated zones are located and observed.

II. Future Progress in Paper Chromatography

In the last decade, published research in paper chromatography has decreased relative to that in thin-layer chromatography. Truter (*9*) has even made the extreme prediction that thin-layer chromatography will ultimately displace column and paper chromatography altogether and relegate gas chromatography to the position of a special technique to be used only for specific problems.

In fact, paper chromatography is today a very important and widely used analytical tool. The technical aspects (sample application, development, zone detection, and quantification, etc.) of paper and thin-layer chromatography are very closely related. With the present commercial availability of cellulose thin-layers on plastic film (e.g., Eastman Chromagram sheets) as well as glass fiber paper (which is resistant to corrosive detection reagents) and glass fiber sheets and cellulose papers loaded with adsorbents such as silica gel, silicic acid, and aluminum hydroxide, the borderline between the two methods is indeed indistinct (*10*).

It is true that thin-layer chromatography may yield better results than paper chromatography. For example, Fig. 1 demonstrates the superior separation of nucleotides on a cellulose layer compared with paper under identical conditions. The zones are clearly resolved on the thin layer but overlap on paper. Another example is the more rapid and sensitive separation of inorganic cations achieved by cellulose thin-layer chromatography as compared with paper chromatography (*11*).

However, results by thin-layer chromatography are not always superior to those by paper chromatography under the same conditions, and procedures worked out on paper cannot always be directly transferred to thin layers (*12*, see Chapter 11). For quantitative analysis, both methods suffer from the same disadvantages in that the quantity of material in the spots is difficult to determine with accuracy (*13*). Besides, paper chromatography is generally less expensive and more convenient, and a huge bibliography of eminently successful procedures has been amassed over the years. And when thin-layer chromatography does have an advantage in speed and resolution, the advantage in many cases is only marginal.

Weaver (*3*) of the Whatman Laboratories has predicted a bright future for paper chromatography based on a greater appreciation of

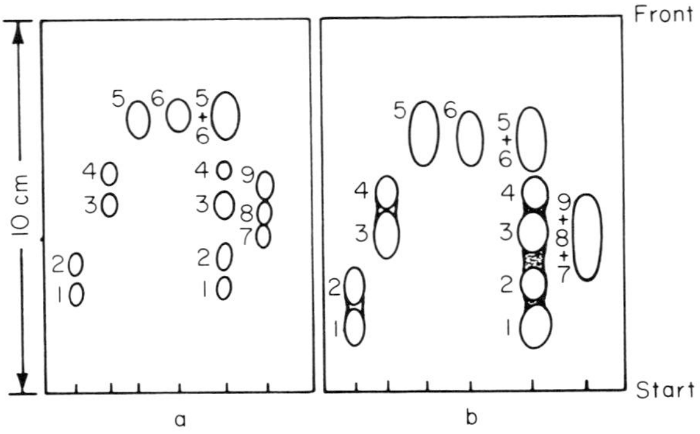

FIG. 1. Comparison of thin-layer chromatography and paper chromatography for the separation of nucleotides on (a) cellulose layers (MN 300G) and (b) on paper (Schleicher and Schuell 2043b) under identical conditions with saturated aqueous NH_4SO_4–1 *M* sodium acetate–isopropanol (80:18:2 v/v). Time: $1\frac{1}{2}$ hours (TLC), 2 hours (PC). Spots visible in UV light. Amounts applied were 0.01–0.03 μmoles each. (1) Adenosine 3′-phosphate; (2) adenosine 2′-phosphate; (3) guanosine 3′-phosphate; (4) guanosine 2′-phosphate; (5) cytidine phosphates; (6) uridine phosphates; (7) adenosine 5′-monophosphate; (8) adenosine diphosphate; (9) adenosine triphosphate [after (*14*)].

the advantages and, at the same time, limitations of paper coupled with an increased understanding of the theory of liquid chromatography. He sees future trends in the following areas: (1) increased popularity of automated techniques for routine uses; (2) greater use of systematic analysis (see Chapter 3) for the characterization and identification of unknown substances; (3) wider use of temperature gradients during development, especially in conjunction with modified papers; (4) development of instrumental methods, such as activation analysis, for solute detection and quantification; (5) based on the present state of knowledge of capillary pore structure in paper, two types of paper are being planned—one offering speed and a fair degree of resolution (comparable to TLC) and the other offering maximum resolution where speed is not of primary importance; (6) production of a wider range of new functional papers giving different interactions (e.g., loaded papers); (7) older techniques will slowly become refined and polished.

REFERENCES

1. H. H. Strain, *in* "Chromatography" (E. Heftmann, ed.),pp. 11–22. Reinhold, New York, 1967.
2. E. Bayer, *Chromatographia* **1**, 153 (1968).
3. V. C. Weaver, *Advan. Chromatog.* **7**, 87 (1968).

4. S. V. Heines, *J. Chem. Educ.* **46**, 315 (1969).
5. I. M. Hais, *in* "Paper Chromatography" (I. M. Hais and K. Macek, eds.), p. 18. Academic Press, New York, 1963.
6a. M. Tswett, *Ber. Deut. Botan. Ges.* **24**, 384 (1906); see H. H. Strain and J. Sherma, *J. Chem. Educ.* **44**, 235 and 238 (1967).
6b. H. Flood, *Tidsskr. Kjemi Bergves.* **17**, 178 (1937).
6c. W. G. Brown, *Nature* **143**, 377 (1939).
7. R. Consden, A. H. Gordon, and A. J. P. Martin, *Biochem. J.* **38**, 224 (1944).
8a. A. H. Gordon, A. J. P. Martin, and R. L. M. Synge, *Biochem. J.* **37**, xiii and 86 (1943).
8b. A. J. P. Martin and R. L. M. Synge, *Biochem. J.* **35**, 91 (1941).
8c. A. J. P. Martin and R. L. M. Synge, *Biochem. J.* **35**, 1358 (1941).
9. E. V. Truter, *Advan. Chromatog.* **1**, 148 (1965).
10. J. Sherma, *Argonne Nat. Lab. Rev.* **5**, 34 (1969).
11. H. A. Foner, *Analyst* **94**, 392 (1969).
12. J. Sherma and G. Zweig, *J. Chromatog.* **31**, 439 (1967).
13. G. Nickless, *Advan. Chromatog.* **7**, 146 (1968).
14. K. Randerath, "Thin-Layer Chromatography," p. 189. Academic Press, New York, 1963.

2

THEORY, MECHANISM, AND FUNDAMENTALS OF PAPER CHROMATOGRAPHY

I. Introduction

Studies of chromatographic theories have provided much information about the mechanism of separations, the interpretation of zone boundaries, and the distribution of substances in the zones (*1*). Theoretical studies usually make use of ideal model systems (e.g., *2*), the problem being, of course, that one very rarely deals with such systems and that eventually complications arise that can only be surmounted by trial and error. Indeed, the experimental approach of chromatography (the selection of the solvent, sorbent, temperature, etc., for the separation of particular substances) remains an empirical one, based primarily on controlled trial and error and the experience and intuition of the researcher.

In chromatographic separations, the solvent, sorbent, and components of the mixture interact with one another (*1*). These elements comprise the *chromatographic system*, the interactions of which determine the degree of separation obtained under any set of conditions. Despite many theoretical studies, the selectivity of chromatographic systems is incompletely understood (*1*). That is, it is seldom possible to predict on theoretical grounds exactly how to obtain a desired separation. The usual approach is to study the literature to see if the same or a similar separation has been achieved and then to adapt the previous procedure to one's own particular problem.

General guidelines for selecting and systematically optimizing chromatographic operating conditions have been published and may be of some use (*3*, *4*). It is recommended, for example, that small loads should be developed on finely divided stationary phases with a rapidly migrating solvent that carries the components of the mixture about one-fourth to one-third way along the support. The rate of solvent flow must be limited to that velocity at which distribution equilibrium is obtained (see below).

Another area that has received attention lately (*5*) is the relationship between structure and chromatographic behavior. The general approach (*6*) is to measure the R_M values (see below) of various molecules and then to assign a value to the contribution of each group or atom. To do this, one measures ΔR_M values of series of closely related compounds in which the necessary group additions, substitutions, and manipulations are made. According to Martin's theory (*7*), one could then calculate the R_M value of any substance from a "ground constant" (which is a function of the chromatographic system) and a summation of the contribution of each atom or group, if these were previously determined for the chromatographic system of interest, and could thereby decide ahead of time whether two substances have R_M values far enough apart to allow their separation.

In practice, the correlation with experimental R_M values is often poor (e.g., *7a*) because of many complex factors involved in the interactions of the chromatographic system: geometrical or positional effects, intramolecular bonding, electronic effects, chain branching, etc. Although accomplishments in this field have been limited, some interesting results have been obtained (*8*, *9*) such as correlation of ΔR_M values with the number of carbon atoms (*10*), degree of hydrogen bonding or position of tautomerism in a given compound.

Needless to say, accurate determinations of R_F and R_M values must be made during such studies so that differences in these values are not due simply to experimental deviations. For this purpose, reversed phase, horizontal tankless chromatography with many replications in systems yielding R_F values between 0.1 and 0.9 is desirable (*6*). Another special method sometimes used to obtain R_M values which can provide structural information is the "moist paper" technique (*10a*): Paper strips are soaked in water, blotted between two sheets of filter paper, spotted with the solutes, and developed only after the weight of the strip has decreased to a value corresponding to a certain amount of water per gram of dry paper (e.g., 0.5 ml/gm).

Strain has pointed out that there is a limited systematic basis for deriving molecular structures from chromatographic behavior because although highly selective with respect to separations, chromatographic methods are nonspecific with respect to structure (*11*). Chromatographic sequences (R values) are useful for *partially* describing or characterizing substances; they provide clues to the kind, location, and number of functional groups present in organic substances. They cannot be used alone to determine identity, nor can structures be used to predict reliably chromatographic behavior or to solve separational problems (*12*). There is, however, considerable value in the systematic use of chromatography in combination with chemical and instrumental analytical methods for the structure elucidation of organic compounds (*13*).

II. Theory and Mechanism

A. Theories of Paper Chromatography

Giddings has published simplified accounts of his general chromatographic theory which is based on a statistical approach to zone spreading caused by random stop-and-go molecular migration during chromatography [the random walk model (*14*)]. Karger has discussed the fundamentals of resolution which are common to all elution chromatographic methods (*15*).

Most workers still rely, however, on the theoretical plate concept developed by Martin and Synge in 1941 for their partition systems (*16*). This theory divides the chromatographic sorbent (column or paper) into a series of theoretical plates which are thought of as little equilibrium chambers equivalent in sense to the cells in a Craig countercurrent distribution apparatus. Since chromatography operates under continuous flow while the Craig apparatus does not, equilibrium is not actually attained at any point in the chromatographic sorbent. However, there is a length of sorbent over which the mixture must pass in order to be subjected to the same degree of separation as in one equilibrium step. This length of sorbent is termed the height equivalent to a theoretical plate (HETP). This theory allows one to conveniently visualize the formation of zones under the influence of nonequilibrium conditions and to describe and compare the performance of chromatographic systems in terms of the number of plates obtained in practice. The reader is referred to the second edition of the Block, Durrum, and Zweig book for a detailed account of this theory (*17*).

B. Mechanism of Paper Chromatography

We will next attempt to describe why, during a chromatographic separation, several substances, all of which start at the same point (the initial zone), move different distances along the paper and end up in separate, discrete zones. Stewart (*18*) has developed a unifying, integrated picture of paper as a chromatographic medium including an equation for the observed plate height (which is the average of all local plate heights over the distance of development) similar to the van Deemter equation for gas chromatography (*19*).

The movement of a solute zone as illustrated by Fig. 1 is explained

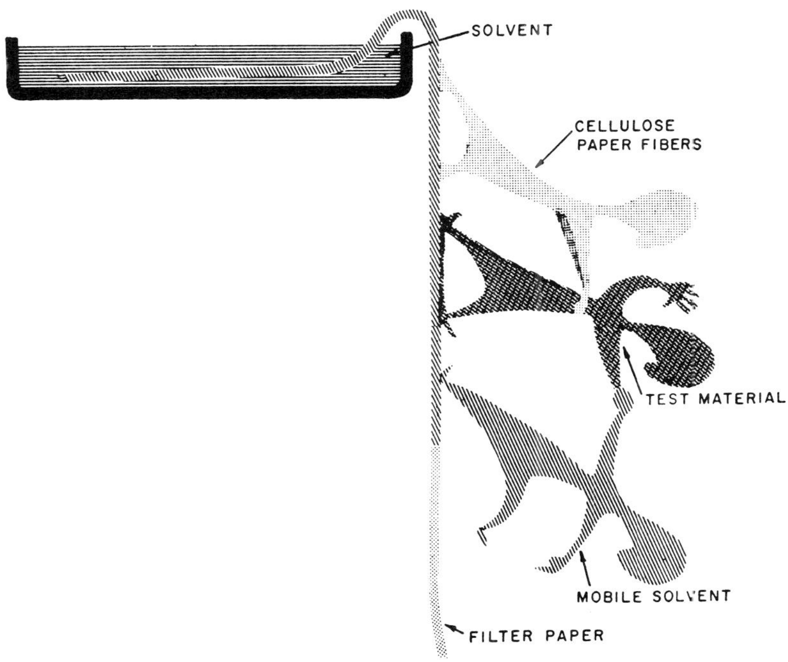

FIG. 1. Pictorial representation of paper chromatography [after (*17*)].

in a simplified and convenient manner as follows: Two opposing forces are in operation as the solvent moves over the mixture spot—the driving force and the resistive action (*1*). The driving force acts to move the substances from the top of the paper in the direction of the solvent flow. The resistive action acts to impede the movement of the substances by dragging them out of the flowing solvent back onto the paper. The resultant of these two forces determines the distance actually traveled by any substance in any given time.

1. *The Role of the Solvent*

Chromatography is a differential migration process. Separations by paper chromatography depend upon the selective retention by the paper of the components of the mixture. The role of the solvent in obtaining differential migration is twofold (*20*). As one component of the chromatographic system, the solvent helps determine the fraction of each solute that remains in solution relative to the fraction sorbed. In this respect it plays a role in the selectivity of the system. This effect of the solvent may be determined in batch experiments without any flow of the solvent. The other role of the solvent is that of a driving force. The movement of the solvent affects all the solutes *in solution* equally, so that the driving force is nonselective. The fractions of the sorbed substances remaining in solution are carried at the same rate until they come into contact with fresh sorbent. At that point, the capacity of the sorbent for the more sorbed constituents will be greater than that for the less sorbed. Consequently, a greater proportion of the more sorbed constituents will be removed and their migration reduced in proportion. As noted before, the solvent will play a role in the system that establishes the relative sorbability of the solutes, but this relative sorbability will not be influenced by the migration of the solvent. Hence, as part of the chromatographic system, the solvent is selective, although the driving force of the solvent is nonselective.

A physicochemical mechanism of cation paper chromatography describing the distribution of the mobile phase on the paper and its effect on the migration of the ions has been recently published (*21*).

2. *The Role of the Paper*

The resistive action in chromatography is the attraction of the components of the mixture to the surface of the sorbent (paper). This sorption is a differential or selective force. As the solvent carries the substances along the paper, each solute will undergo repeated sorption and desorption and migrate at a rate depending upon its mean occupancy time (*14*) on the sorbent relative to that in the mobile phase. Weakly sorbed substances travel more rapidly than those strongly sorbed because they spend relatively more time in the mobile phase. If differences in degree of sorption are sufficiently large, there will be complete separation.

Cellulose paper is a very complex fibrous medium (*22-24*) composed of approximately parallel polysaccharide chains. Hydrogen bonding between the —OH groups of the chains causes some highly ordered regions of crystallinity which account for the strength and insolubility of the cellulose. Other regions of lower order are amorphous, accounting for reactivity, swelling, and sorption of water. Thermodynamic studies indicate that the first water adsorbed by dry cellulose is bonded to two —OH groups of the cellulose while at relatively higher humidities the water molecules are each bonded to —OH and —O groups. These fractions of water are chemically bound and may be considered as part of the fiber and are not free to dissolve the solutes. At still higher humidities, there is a slower uptake of additional water to form a multilayer of loosely bound water which can take part in partition separations. Thus, water (or other hydrophilic solvents) in the amorphous regions is different than bulk water, having a higher degree of organization. The cellulose then consists of a series of connected pools or puddles of water (*19*) in the amorphous regions, the water being more organized near the cellulose fibers but becoming less organized farther from the chains.

Because the exact nature of the paper surface is unknown, the term water–cellulose complex is used (*24*), and paper chromatography can be considered an adsorption process in which the rate of movement of any substance depends upon its overall attraction to this complex surface. This attraction can be due to adsorption on the cellulose molecules or on the bound water as well as to partition between the free (more loosely bound) water in the paper and the flowing organic liquid. In either case, the same types of weak, physical interactions (e.g., van der Waal's forces, dipole–dipole forces, and hydrogen bonds) are involved.

3. *Adsorption versus Partition*

Although it was originally believed that the paper functioned solely as an inert support for the aqueous portion of the developing solvent, hence the original name paper partition chromatography, it is now generally recognized that the paper more commonly acts by a combination of partition, adsorption, and ion exchange. For example, the chromatography of acridine derivatives on wet buffered paper with benzene as the mobile phase has been recently shown to involve partition as well as adsorption (*25*).

Some chromatographers have abandoned making a distinction between adsorption and partition mechanisms, claiming that differences between the two depend only upon definitions of the terms. This view is taken by Weaver (*22*), who states "for adsorption, the solutes must diffuse up to a surface or interface and interact, and, for partition, must pass through the interface," and by Bayer *et al.* (*26*), who say "the terms partition and adsorption chromatography are now out of use, as partition has no specific meaning and adsorption is a partition process."

Many others (e.g., *19*, *27*, *28*) continue, however, to distinguish between the two mechanisms and to study experimentally which mechanism is operative (*29*). Recently, a thermodynamic approach has been used to show the analogies and differences between adsorption and partition chromatography (*30*). In plastid pigment chromatography (*31*), the process is termed adsorption chromatography when dried paper or paper as received from the manufacturer with its natural water content is used. In this case, an increase in the amount of sample applied increases the R_F values of the pigments (see Fig. 1, Chapter 11), a property usually attributed to adsorption systems (*28*). When the paper is made to retain a stationary polar (dimethylformamide or dimethylformamide–H_2O) or nonpolar (Wesson Oil) phase, resolution is obtained by developing with a liquid immiscible with the fixed liquid and the process is termed partition chromatography.

As a compromise, the more general terms sorbent and sorption may be used in discussions of chromatography under all conditions.

4. *The Role of Ion Exchange*

It has been clearly established that ordinary chromatography papers have ion-exchange properties due to carboxyl groups and that under certain conditions this effect can explain the retention of cations and frontal analysis of the developer. The protons liberated by the exchange can be directly detected on the paper by different methods (*32*). The exchange capacity of Whatman No. 1 paper is reported to be about 2.5–7.0 $\times$ 10^{-3} mEq per gram of dry paper (*33*), varying with the pH of the mobile phase (*32*) and decreasing if the paper is prewashed with acetic acid, $HClO_4$, or Na_2H_2 EDTA. Ion exchange is apparently responsible for the formation of multiple zones during the chromatography of sodium orthophosphate on acid-washed Whatman No. 1 paper (*34*); the ascending frontal paper chromatography of cations at low concentrations on Whatman No. 1 paper has been explained by an ion-exchange retention, while a second linear adsorptive mechanism was introduced to account for deviations at higher concentrations (*34a*). The observed relationship between the extent of adsorption and the exchange capacity in another study (*34b*) is explained in terms of the Donnan permeable membrane theory.

5. *Performance of the System* (*14*, *18*, *22*)

In chromatography, the primary purpose is to separate the solute zones without overlap. This separation depends upon two factors—the degree of separation of the zone centers and the compactness of the zones. Obviously, if the zones remain sharp, the relative displacement of the zone centers need not be so great in order to obtain complete resolution. The performance of a paper chromatographic system can be judged, then, on its ability to separate the centers of the solute spots and its ability to keep the spots compact.

Theory indicates that the location of the center of the spot (or in general a chromatographic peak) is determined by the sorptive equilibrium of the solute, which in turn depends upon the chemical nature of the mobile and stationary phases and the resultant interactions involved in the chromatographic system. The spot (or peak) width is determined by the kinetic and mass transport processes related to the equilibration process, which in turn depend upon the physical characteristics of the stationary phase (*14*).

Three factors responsible for spot spreading in paper chromatography are: (a) molecular diffusion of the solute zone in all directions under the influence of its concentration gradient, (b) eddy diffusion due to the migration of the solute through random, multiple paths (which is independent of flow rate), and (c) resistance to mass transfer (non-equilibrium). Molecular diffusion depends upon the viscosity of the solution and the time spent by the solute in solution; eddy diffusion is a function of the size and orientation of the cellulose fibers. It appears, therefore, that minimum diffusion (and the most compact zones) would be obtained by the fast flow of highly viscous solvents through dense paper. Of course, dense paper and viscous solvents do not yield fast flow rates, and anyway, considerations of equilibrium kinetics suggest rather that slow flow rates are desirable so as to obtain as close an approach to equilibrium as possible. Clearly, a compromise must be made in order to obtain the best balance of these factors leading to a minimum average HETP and maximum resolution, and Stewart suggests the use of solvent velocities near 1×10^{-3} cm/sec (*35*).

It must be recognized that in practice one usually accepts whatever flow rate results after the solvent, paper, and other development conditions have been selected and that there is usually a considerable variation in the mobile phase velocity from its initial rapid penetration into the paper to the final, slower arrival of the front some distance down the length of the paper. This decrease in solvent velocity means that molecular diffusion becomes more important the longer the development.

Methods for controlling the velocity of the solvent are usually limited to location of the initial zone farther from the solvent reservoir so as to minimize the effects of the high initial solvent velocity, choice of solvents of different viscosities and papers of different densities, and shaping the paper so as to restrict the flow of solvent—e.g., use of wedge-shaped paper (Fig. 3 in Chapter 6) or a wick to feed the solvent. Other proposed methods include the use of external force fields and centrifugal acceleration (*36*).

DeLigny and co-workers have made extensive studies of peak broadening of amino acids in paper chromatography and reach the following conclusions (*37*): (a) Eddy diffusion and resistance to mass transfer account for more than 50% of the peak broadening at the flow rate obtained in practice and for medium R_F values. (b) Molecular diffusion occurs both in the mobile and stationary phase. (c) The mass transfer term is composed of a contribution of unevenness of flow throughout the paper and a contribution of the slowness of attainment of the partition equilibrium between the mobile and stationary phase. The resistance to attainment of equilibrium resides predominantly in the mobile phase. (d) Peak broadening is a function of the mean flow rate of the solvent which in turn is a function of the distances of the starting point and solvent front from the surface of the solvent in the tank. Equations for the optimum values for these distances with respect to the separation time for a specified separation have been derived.

III. R_F and R_M Values

A. R_F Values

1. *Linear Chromatography*

In order to give a numerical value to the location of a spot on a paper chromatogram developed by linear flow, the term R_F is used.

$$R_F = \frac{\text{Distance traveled by the zone}}{\text{Distance simultaneously traveled by the solvent front}} \tag{1}$$

The distance traveled by the zone can be measured to the geometric center of the zone, its point of highest concentration, or its leading edge, as long as one is consistent and indicates which procedure was followed. Thus, if the solvent front has moved 30 cm and the spot 15 cm, its R_F is 0.50 or $R_F \times 100$ is 50. Since the R_F value indicates nothing as to the size or compactness of the zone, the R_F limits of the leading and trailing zone boundaries are often reported so that separation possibilities become obvious.

Chromatograms of complex mixtures such as inks, which may include zones which are diffuse and incompletely separated, are difficult to represent in terms of R_F values. It is best to record figures or photographs of the results in these cases.

2. *Two-Way and Radial Chromatography*

The results of two-dimensional chromatography are best reported by listing the R_F values in each solvent and also presenting a figure indicating the final locations of the zones.

For radial or circular chromatography the equation

$$R_R \text{ or } rad\ R = \frac{\text{Radial distance to the zone}}{\text{Radial distance to the solvent front}} \tag{2}$$

is used. It has been pointed out that the familiar relationship $R_R{}^2 = R_F$

is not valid because R_F values are greatly reduced in circular paper due to the restricted solvent flow (*19*).

3. *Use of Reference Compounds*

In many instances it is desirable to allow the solvent to run off the paper or to repeat the development with the same or a different solvent in the same direction a number of times. Obviously, in these instances an R_F value cannot be determined. Therefore, a small quantity of a pure compound, such as the dye bromocresol purple (BCP), is used as a marker, and the spot migration values are designated R_{BCP} (or R_F relative to BCP), where

$$R_{BCP} = \frac{\text{Distance traveled by solute zone}}{\text{Distance traveled by BCP}} \tag{3}$$

In this case, R_{BCP} values can, of course, be greater than 1, whereas R_F values as defined above vary only between 0 and 1.0, or between 0 and 100 when reported as percentages.

4. *Identification of Compounds*

As will be seen below, R_F values vary significantly depending upon the experimental conditions. It is therefore generally accepted that R_F values are not sufficient for the identification of a substance (*37a*). The relationship between the accuracy with which chromatographic mobility can be measured and the attempted identification of compounds has been described (*38*).

To establish identity, the technique of mixed chromatography (co-chromatography) may be employed. In this procedure, the solute is mixed with an authentic sample of the compound it is suspected of being. The mixture is chromatographed on paper in several solvents and perhaps also by thin-layer and column chromatography on sorbents other than cellulose. The authentic sample is also run in parallel with the mixture in each system. If the mixture is inseparable under all conditions, and the authentic sample shows the same R_F value, the solute and the authentic known are either identical or very similar (isotopes or isomers).

If an authentic sample of the suspected compound is not available, probable identity can be established using two other reference compounds (*39*).

5. *Factors Influencing R_F Values*

R_F values are not chromatographic constants which define the migration of a substance under all conditions. They are values which vary with the experimental conditions under which they are obtained, and in order for reported R_F values to have any significance, all of these experimental conditions must be recorded. R_F values, however, are, simple to obtain, and if they are reported and used intelligently they can be of great value to the working scientist (*37a*).

Factors which are known to influence R_F values are as follows (*1*, *19*, *39–41*):

a. The composition and aging of the solvent.
b. The nature (brand, grade) of the paper, its method of storage, preparation, and treatment and perhaps its slight heterogeneity. R_F values are lower with thicker papers.
c. R_F values usually increase with increasing concentration in adsorption systems, but there is no effect or a slight decrease in partition systems.
d. The geometry of the paper (circular, rectangular, tapered, etc.).
e. The direction of development (horizontal, ascending, descending). R_F values are often lower by ascending chromatography than by descending due to the effects of gravity on the amount of solvent per unit area of paper.
f. The distance of the origin from the solvent reservoir; zones started farther from the reservoir have lower R_F values; this distance should be kept to a minimum (*19*).
g. The duration and length of the development.
h. The amount (volume) of solvent in the reservoir.
i. Temperature. Increased temperature can lead to faster development and more compact zones, but it is difficult to predict the nature of the change, if any, in R_F values. The influence of temperature is greatest in partition systems.
j. Conditions of equilibration both before and during the run; the humidity of the atmosphere in the tank is especially important.
k. Change in composition of the solvent as it flows over the paper. R_F values may have little meaning if two or more solvent–component fronts are formed on the paper (see section on artifacts below).
l. Irregular flow of the solvent along the paper.
m. The dimensions of the apparatus.
n. Foreign ions and impurities in the solvent or on the paper; for example, organic solvents are often dried or freed of preservatives before use.
o. The pretreatment (e.g., type of desalting) of the sample. In general, spot positions in biological samples, which will likely contain additional components that greatly modify R_F values, are referred to standard compounds which have been co-chromatographed with the sample.

p. The R_F value of component A is influenced by the presence of another component B in the mixture if the distribution coefficient of component A is thereby altered (*41a*).

It is obvious from the above list that the standardization of R_F values is a difficult problem. For this reason, R_F values relative to a standard substance X instead of the solvent front are often reported. These R_X values are in effect corrected R_F values (*40*) which are less affected by such variables. It has been reported (*42*) that under varied conditions, R_X values are 2–3 times more stable than R_F values and that the evaluation of R_X values should be used for comparing the chromatographic results of different laboratories.

6. *Reproducibility of R_F Values*

Reproducibility in paper and thin-layer chromatography was the subject of an international symposium held in October, 1967, the proceedings of which were published in the *Journal of Chromatography*, **33**, pp. 1–293, 1968.

Only rarely in special studies is it really important to sacrifice the inherent simplicity of paper chromatography by rigidly standardizing operating conditions in order to obtain highly reproducible R_F values. In most cases it is the overall pattern or sequence of zones or the position of a zone relative to a standard marker compound, rather than the exact position, that is of interest to the worker. Two areas in which reproducibility is important are when attempting to relate migration behavior with chemical structure (above) and when preparing chromatographic spectra (see Chapters 3 and 13) for systematic qualitative analysis (*6, 40, 43*).

Some steps which may be taken in order to improve reproducibility are as follows (*4, 28*):

a. Maintaining temperature constant to within $\pm 5°$.
b. Mixing the solvent well and maturing it for several days at the temperature of the run.
c. Checking the solvent by developing standard reference compounds.
d. Equilibrating the paper with the atmosphere of the tank for 24 hours after spotting the samples.
e. Sealing the development chamber during the run to insure saturation (*43a*).
f. The solvent should travel 30–35 cm past the origin.
g. Carrying out all manipulations and operations (e.g., drying and detection) following standard procedures.

h. Using paper from a single source and batch, stored under the same conditions. Any pretreatment is carried out in a standard manner.
i. If the paper is impregnated for partition chromatography, the amount and composition of the impregnating liquid and the method of impregnation must be standardized.

A fast method for the statistical evaluation of R_F data has been published (*44*). With conditions such as those listed above standardized, R_F values for the 2,4-dinitrophenylhydrazones of 39 carbonyl compounds showing a standard deviation of $< \pm 0.03$ were obtained by 39 inexperienced operators (*22*).

7. *Publication of Chromatographic Data*

M. Lederer, editor of the *Journal of Chromatography*, has recently presented some thoughts (*45*) concerning the publication of chromatographic data which are here adapted to yield the following guidelines:

a. Since determining an R_F value requires two measurements, each of which cannot be more precise than ± 1 mm, the resultant accuracy is approximately 1–3%, depending upon the distances moved by the spot and the front. Reporting more than two decimal places, therefore, never seems justified.
b. Location of the center of the zone (or the edges or point of maximum concentration if these are used) is done visually and inaccuracies lead to errors of the order mentioned in a.
c. Although spots are ideally round or oval, R_F values may in many cases be usefully recorded for zones of other shapes.
d. As a guide, all experimental conditions should be described as accurately and completely as possible, although exact reproduction of these conditions from one laboratory to another is generally impossible.

Lederer suggests (*45*) that R_F data have about the same value as "uncorrected" melting points and that they should be considered a good hint as to where one should start when separating a certain mixture rather than as chromatographic constants.

B. R_M Values—Mathematical Derivation

The basic definition of R in partition chromatography is either the ratio of the zone velocity to the mobile phase velocity or the fraction of solute in the mobile phase that is moving with the velocity of the mobile phase at any given time (*19*). R therefore reflects the equilibrium distribution of the solute between the two phases and is related to the partition coefficient $\alpha = C_s/C_m$, where C_s and C_m are the respective concentrations of the solute in the stationary and mobile phases.

In any small segment of the chromatographic support, the amount of solute in the mobile phase relative to the amount in the stationary phase is given by

$$\frac{A_m C_m}{A_s C_s} = \frac{A_m}{\alpha A_s} \tag{4}$$

where A_m and A_s are the respective cross-sectional areas (volume fractions) of the phases. Since R is the fraction of solute in the mobile phase, $(1 - R)$ is the fraction in the stationary phase and

$$\frac{R}{1 - R} = \frac{A_m}{\alpha A_s} \tag{5}$$

or

$$R = \frac{A_m}{A_m + \alpha A_s} \tag{6}$$

Under special, seldom-realized conditions, such as the tankless, horizontal, reversed-phase chromatography of Green and co-workers (*6*) mentioned above, $R = R_F$. Therefore, Eq. (6) becomes

$$\alpha = \frac{A_m}{A_s}\left(\frac{1}{R_F} - 1\right) \tag{7}$$

Assuming that for a given system A_m/A_s has a constant value, and taking logarithms and substituting the term

$$R_M = \log\left(\frac{1}{R_F} - 1\right) \tag{8}$$

introduced by Bate-Smith and Westall (*46*), the following is obtained

$$\log \alpha = K + R_M \tag{9}$$

where $K = \log A_m/A_s$.

Martin (*7*) deduced that, for ideal solutions (the behavior of which is approximated by solutions with very low solute concentrations), the partition coefficient for a substance A between two phases is related to the free energy required to transport 1 mole of A from one phase to the other by the expression

$$\ln \alpha = \Delta\mu_A/RT \tag{10}$$

where $\Delta\mu$ is the change in chemical potential.

Thus,

$$\frac{\Delta\mu_{\mathrm{A}}}{2.303\,RT} = K + R_M \tag{11}$$

Equation (7) would apparently allow the calculation of R_F if the distribution coefficient and the cross-sectional areas of the phases were known. This is not practical, however, because $A_{\mathrm{m}}/A_{\mathrm{s}}$ is in fact not constant over the length of a chromatogram (*22*).

It is the $\Delta\mu_{\mathrm{A}}$ term in Eq. (11) that is the basis for the attempts to relate R_M values with molecular structure as mentioned above. This quantity $\Delta\mu_{\mathrm{A}}$ is assumed to be the sum of the contributions of the various groups and atoms comprising the molecule A; that is, the free energy required to transfer a given group between phases is assumed to be independent of the rest of the molecule. It follows that if two compounds differ from each other by only one functional group, then the difference between their R_M values (ΔR_M) is equal to $\Delta\mu/2.303\,RT$ for that group.

Some precautions which must be observed (e.g., adsorption) on application of R_M values in studies of this type have been discussed (*47*).

The theoretical aspects of the correlation between R_M values of cations and the composition of the mobile phase have been studied. The number of chlorine ions with which the cation is complexed in the mobile phase (methanol–10 N HCl–H_2O) and the optimum separation conditions were determined (*48*).

A parallelism has been noted between R_M versus composition and log C_{s} versus composition relationships (C_{s} = solubility) for naphthols and alkaloids in certain partition solvent systems so that in some cases the solubility of substances can be estimated from chromatographic data (*49*). R_M values have also been studied in four-component systems (*50*).

C. Separation Index

A universal separation index, the extent of separation, has been proposed and mathematically described (*51*). It applies to any type of separation system including paper chromatography.

Numerical parameters have also been described by which the ability of any chromatographic system to establish the identity and purity of the members of any class of chemical compounds can be defined (*52*). It was concluded that paper and ion-exchange chromatography rank highest in their ability to determine identity, whereas ion-exchange and adsorption chromatography offer the greatest certainty of chromatographic purity.

IV. Artifacts and Anomalous Zones

Although true equilibrium conditions never exist during the course of a paper chromatography experiment (*24*), one hopes to approach these as closely as possible. One useful criterion for suitable development conditions is that the developed spot, although it will be somewhat larger than the original spot because of diffusion, should be well-defined and have retained the approximate circular or slightly elliptical shape of the original zone (*41*, *53*). Each substance in the mixture should yield one and only one spot. In practice, zones are often diffuse, streaked, flattened, swerved, or otherwise distorted, and a single substance may yield two or more zones. Even if the zones formed are not of ideal size and shape, a chromatographic system may be very useful if it reproducibly yields a desired separation. Some cases of anomalous migration and multiple spot formation will be discussed below.

A. Irregular Zones (*28*, *41*)

1. *Diffuse Zones*

The tendency for diffusion to foil the attempt to obtain sharp, small zones was discussed above (Section II,B,5). To minimize this effect the research worker should make his initial spots as small and sharp as possible and as dilute as the subsequent detection method permits. He should also choose, if possible, a chromatographic system with a flow rate of solvent near optimum (see above). Cellulose paper containing pores of highly irregular shape and size would lead to an increase in HETP and diffuse longitudinal zone spreading.

2. *Trailing or Streaked Zones*

If the system is operating under an adsorption or partition mechanism governed by a downward-curving distribution isotherm, trailing (or tailing) zones are likely to occur unless the solute concentration is very low and within the range of isotherm linearity. A concave isotherm would lead to spots which "head," or have forward elongation. Ionization, dissociation, and association are common causes of nonlinearity of the isotherm and consequent tailing in paper chromatography. Modification of the stationary phase by inclusion of a buffer, an acid, a base, a neutral salt, or a hydrogen-bonding component may help eliminate the nonlinearity of the isotherm (*54*). Tailing can be caused by:

a. An overloaded system; that is, application of an amount of solute exceeding the capacity of the system.
b. A flow of solvent that is too fast, leading to conditions far removed from equilibrium.

c. A gradual, irreversible change in the composition of the solute during migration may result in the formation of an amount of a slower-moving substance which would show up as a trail. This could be due to reaction with the solvent or impurities in the paper. If a faster-running species is formed, the zone may show heading.
d. Strong adsorption to the surface of the sorbent.

In some systems, streaks are formed along the entire length of the chromatogram from the origin to the actual spot. This can be caused by extreme tailing, gradual extraction of a zone precipitated at the origin or gradual conversion of the migrating solute to an insoluble zone or a zone with a low R_F value.

3. *Double-Tailing Zones*

Plastid pigment zones with double tailing portions are often noted when a leaf extract is developed with various solvents on paper or thin layers of cellulose (see Chapter 11, Figs. 1 and 3). The formation of these zones may be attributed to the presence of colorless substances that are much more sorbed than the pigments themselves (*55*). These substances elute the pigments, and even in the initial spot they migrate outward more slowly than the pigments. This would leave a small amount of pigment free of the strongly adsorbed substances at the periphery of the initial spot. As the initial zone is washed, these peripheral portions of the pigment at the sides migrate more slowly than the pigment in the front of the spot and in the center of the spot. The small amount of peripheral pigment at the trailing end of the initial spot migrates rapidly once washed into the region containing the colorless substances. The bulk of the pigment zone, therefore, migrates ahead of the strongly adsorbed polar substance and continues to migrate faster than the trailing regions at the sides until all the colorless material is adsorbed and its rate of migration reduced. A schematic representation of successive stages in the formation of a chromatographic zone with two tails is presented in Fig. 2. The location and distribution of the colorless material depends upon the adsorbability and the amount present.

4. *Projections from the Solvent Front*

Stains or tongues extending back toward the origin often appear after the paper is treated with the detection reagent and perhaps viewed under ultraviolet light. They are due apparently to collections of impurities washed from the paper and can cause problems in detecting zones located near the solvent front.

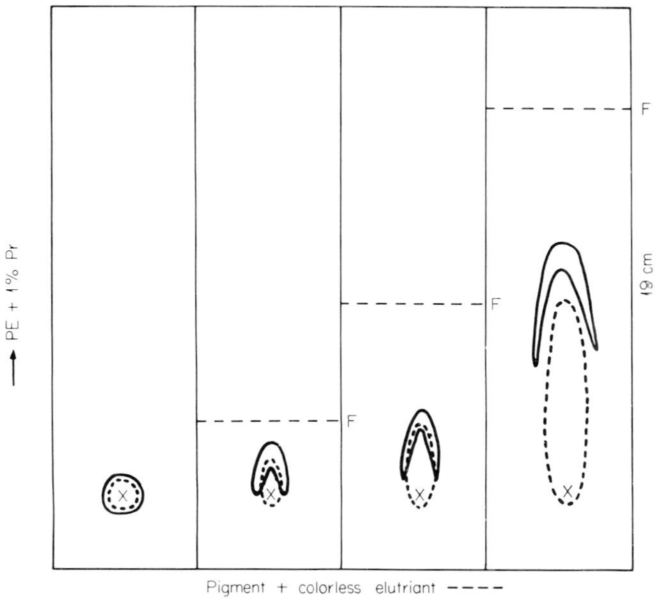

FIG. 2. Schematic representation of successive steps in the formation of a pigment zone with double tails (solid line) formed by a more abundant, more sorbed colorless substance (dashed line). X = origin, PE = petroleum ether, Pr = *n*-propanol, F = front [after (*55*)].

5. *Flattened Zones*

Flattened zones are short zones with the width of the initial spot or an increased width. They usually appear at the solvent front ($R_F = 1$) or at a lower R_F value if the solvent undergoes frontal separation and forms a second front (see Chapter 15). Longitudinal changes in the stationary phase (if the paper is impregnated) and multiple development techniques (Chapter 3) can also produce narrowed zones. The reader is referred to Hais (*54*) for a more detailed discussion of this topic.

6. *Lateral Spreading*

Lateral spreading of zones may occur on paper impregnated with a polar or nonpolar organic phase if the impregnation is too heavy (*28*).

7. *Zones of Other Shapes*

Swerved zones may occur if the paper is irregular or if the sample is placed near the edge of the paper. Heart-shaped, V-shaped, and arrowhead-shaped zones are also encountered and are apparently due to distorted solvent flow through the initial zone.

8. *Spots Remaining at the Origin*

Part of a migrating solute ($R_F \neq 0$) can be left at the origin for a variety of reasons, including low solubility in the mobile phase, formation of an insoluble compound with impurities in the paper, irreversible sorption on the cellulose, and sorption on another solute with $R_F = 0$.

9. *Lost Zones*

It is important to apply enough sample in the initial zone so that the substance can still be detected after migration and concomitant dilution. Zones may be lost if they change during migration—perhaps due to lability in the presence of oxygen and light—to a species which does not react with the proposed detection reagent.

B. Multiple Zoning

Many cases are known in which a single substance forms more than one zone or spot. It is obvious, therefore, that the presence of a certain number of spots on a chromatogram does not necessarily mean that the same number of unique substances were present in the original mixture which was applied at the origin.

A classic example of this effect is the two-way development of the pure xanthophyll lutein mixed with colorless flower extract. For these migrations, 20 μl of a 1:1 mixture of lutein solution and flower extract are spotted on paper, and the development is carried out as indicated in Fig. 3. This results in five yellow zones of lutein (quintuple zoning) as shown, whose formation may be explained in terms of the precipitation, double zoning, and double tailing produced during one-way migration (*55*). The first two of these may be ascribed to the fact that lutein is only slightly soluble in petroleum ether. Consequently, when the initial spot is washed with petroleum ether plus 1% *n*-propanol, there is a separation of a readily soluble fraction from one that is slowly soluble, leaving a "precipitate" at the origin. The double tailing is due to the added flower extract (see Fig. 2), which is similar in nature to the colorless substances normally extracted from leaves along with the pigments.

1. *Causes for the Formation of Multiple Spots*

a. A gradient (e.g., of pH or concentration) in the mobile phase produced by its frontal separation, a temperature change, or a conscious change of solvents, or a gradient in the properties (e.g., activation) of the stationary phase.

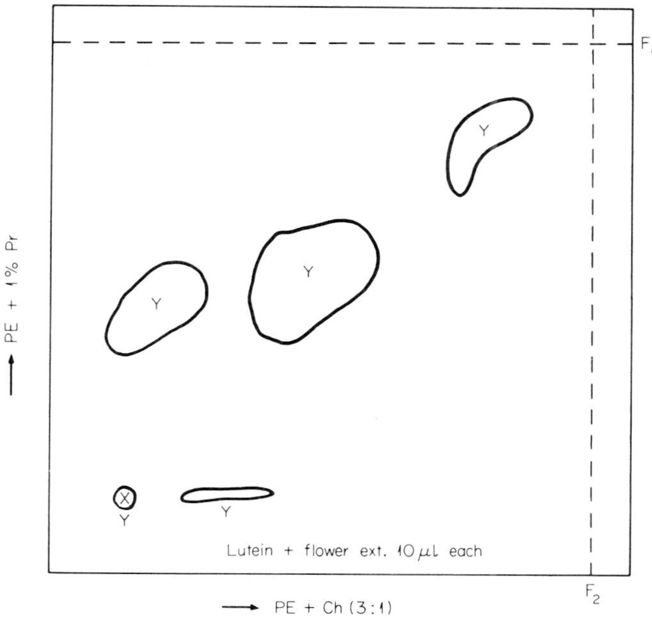

FIG. 3. Pure lutein, approximately 0.007 mg in 10 μl ether–petroleum ether (1:1 v/v), plus 10 μl of the extract of morning glory flowers subjected to two-way migration. F = front, Y = yellow, X = origin, PE = petroleum ether, Pr = *n*-propanol, Ch = chloroform [after (*55*)].

b. Existence of the solute in various ionic forms, e.g.,

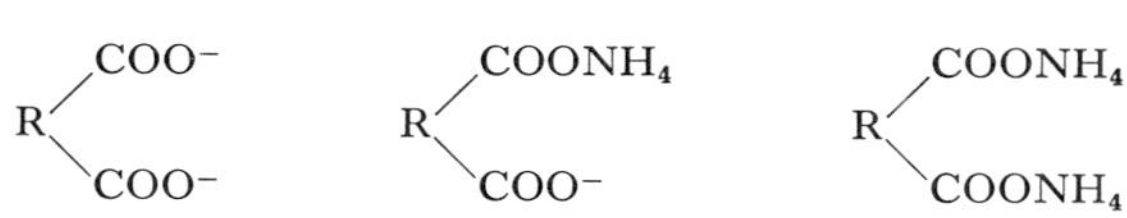

c. Polymerization of the solute during development.

d. Oxidation of the solute before or during development, e.g., methionine → methionine sulfoxide; cysteine → cystine (*56*).

e. Existence of the compound in two isomeric forms, e.g., the cis and trans isomers of the 2,4-dinitrophenylhydrazones of α-keto acids.

f. Overloading of the chromatogram, reaction of one or more of the solutes with the cellulose, "precipitation" of a solute, or any other factor which might leave part of the solute fixed at the origin.

g. Formation of complexes or salts by reaction between the migrating solute and minor impurities in the paper, thus changing the mobility of a portion of the solute. Examples are the effect of alkaline earths on condensed inorganic or ester phosphates and of heavy metals on amino acids. In both cases the addition of complexing agents to the system may be of help (*54*).

h. Heating the initial zone after spotting (see Chapter 3).

2. *Further Examples of Some Artifacts*

Beckett (*57*) has reported that the amino salts of strong organic or inorganic acids form two spots when chromatographed with acidic or neutral solvents, one corresponding to the free base and another to the acid salt. Only a single spot is produced with most ammoniacal solvents or with amines that are salts or weak organic acids (mandelic, propionic, and citric). On the other hand, others (*58, 59*) have reported that if amines such as adrenaline, noradrenaline, histamine, and 5-hydroxytryptamine are chromatographed as salts of CCl_3COOH, or picric acid, double spots are formed. This effect can be prevented by the inclusion of a basic amino acid in the amine solution before chromatography or by use of HCl, CH_3COOH, or oxalic acid salts. Fuller has reported that both hydroxylamine and phenylhydrazine form a complex with an impurity in washed filter paper (Whatman Nos. 1, 3, and 4) which produces a yellow fluorescent spot.

DeVay *et al.* (*60*) and Zweig (*61*) have shown that aspartic and glutamic acids when stored in dilute ethanol undergo partial esterification and form multiple chromatographic spots. These authors recommend that amino acid solutions be stored in 10% v/v 2-propanol as described by Block. Leaback (*62*) has shown that amino sugars give double spots when developed in weakly basic or acidic solvents. Carles (*63*) has pointed out that aconitic acid is easily esterified and consequently an ethanolic extract gives two spots on chromatograms.

Ascending paper chromatography of some sincurarine salts show alterations of R_F values depending on the anion and solvent (*64*). Comets and multiple spots are found particularly with the triiodide, nitrate, and triiodide in 0.9% NaCl. Sincurarine yielded a single spot with an R_F value independent of the salt placed on the paper only in solvents containing HCl.

Schwarz (*65*) has discussed the methods of elucidating the changes that a pure substance may undergo during paper chromatography and has recommended that it be chromatographed two-dimensionally using the same solvent for each direction. Thus, if substance B is in equilibrium with C and decomposes into D and then to A, a two-dimensional chromatogram of a mixture of A, B, C, and D may appear as in Fig. 4. The left vertical column shows the separation of the original mixture in the first direction. The last four columns show the chromatogram

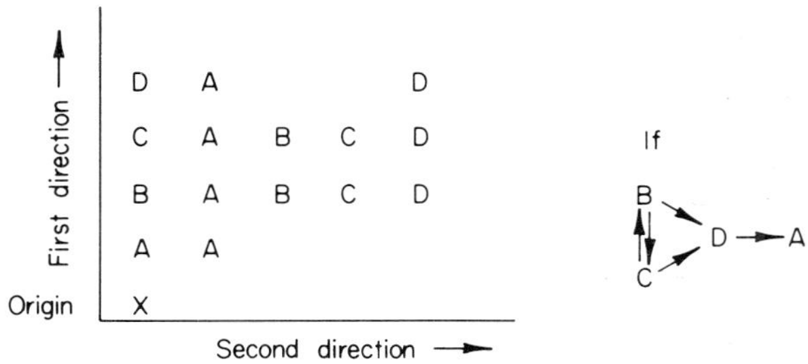

FIG. 4. Theoretical distribution of spots and decomposition products after two-dimensional development with the same solvent in both directions [after (*65*)].

after the second development if decomposition takes place between the runs. The primary spots appear on a diagonal line because the same solvent was used in both runs. All spots above and below the diagonal correspond to substances produced between the first and second developments. This approach has been recently applied to imidazolepropionic acid (*66*).

The subject of multiple zoning in chromatography has been theoretically treated and reviewed (*67*).

REFERENCES

1. H. H. Strain, *in* "Chromatography" (E. Heftmann, ed.), pp. 11–22. Reinhold, New York, 1967.
2. E. Soczewiński, *Advan. Chromatog.* **5**, 3 (1968).
3. J. A. Thoma and C. R. Perisho, *Anal. Chem.* **39**, 745 (1967).
4. J. A. Thoma, *Advan. Chromatog.* **6**, 61 (1968).
5. M. Przyborowska and E. Soczewiński, *J. Chromatog.* **42**, 516 (1969).
6. J. Green and D. McHale, *Advan. Chromatog.* **2**, 99 (1966).
7. A. J. P. Martin, *Biochem. Soc. Symp.* (*Cambridge, Engl.*) **3**, 4 (1949).

7a. J. Becker and C. G. Skinner, *Texas J. Sci.* **20**, 87 (1968).

8. J. Green and S. Marcinkiewicz, *Chromatog. Rev.* **5**, 58–160 (1963).
9. I. E. Bush, "The Chromatography of Steriods." Pergamon Press, Oxford, 1961.
10. E. Soczewiński and M. Bieganowska, *J. Chromatog.* **40**, 431 (1969).

10a. E. Soczewiński and G. Matysik, *J. Chromatog.* **48**, 57 (1970).

11. H. H. Strain, *Anal. Chem.* **33**, 1733 (1961).
12. J. Sherma, *J. Chem. Educ.* **40**, 229 (1963).
13. J. Gasparič, *Advan. Chromatog.* **6**, 3 (1968).
14. J. C. Giddings, *J. Chem. Educ.* **35**, 588 (1958); **44**, 704 (1967); *in* "Chromatography" (E. Heftmann, ed.), pp. 23–42. Reinhold, New York, 1967.
15. B. L. Karger, *J. Chem. Educ.* **43**, 47 (1966).
16. A. J. P. Martin and R. L. M. Synge, *Biochem. J.* **35**, 91 (1941).
17. R. J. Block, E. L. Durrum, and G. Zweig, "A Manual of Paper Chromatography and Paper Electrophoresis," 2nd rev. ed. Academic Press, New York, 1958.

18. G. H. Stewart, *Advan. Chromatog.* **1**, 93 (1965).
19. R. A. Keller and J. C. Giddings, *in* "Chromatography" (E. Heftmann, ed.), pp. 125–138. Reinhold, New York, 1967.
20. H. H. Strain, private communication (1965).
21. P. B. Janardhan and A. Paul, *Indian J. Chem.* **5**, 297 (1967).
22. V. C. Weaver, *Advan Chromatog.* **7**, 87 (1968).
23. C. S. Knight, *Advan. Chromatog.* **4**, 61 (1967).
24. R. Consden, *in* "The Stationary Phase in Paper and Thin-Layer Chromatography" (K. Macek and I. M. Hais, eds.), p. 18. Elsevier, Amsterdam, 1965.
25. D. Ratajewicz, *Chem. Anal.* (*Warsaw*) **11**, 929 (1966).
26. E. Bayer *et al.*, *Chromatographia* **1**, 154 (1968).
27. L. R. Snyder, *J. Chromatog. Sci* **7**, 352 (1969).
28. K. Macek and I. M. Hais, *in* "Paper Chromatography" (I. M. Hais and K. Macek, eds.), pp. 115–169. Academic Press, New York, 1963.
29. J. Oscik, *Bull. Acad. Polon. Sci., Ser. Sci. Chim.* **14**, 879 (1966).
30. E. Soczewiński and M. Przyborowska, *Chem. Anal.* (*Warsaw*) **11**, 849 (1966).
31. H. H. Strain and J. Sherma, *J. Chem. Educ.* **46**, 476 (1969).
32. G. Ackermann and H. P. Frey, *Proc. Conf. Appl. Phys.-Chem. Methods Chem. Anal., Budapest, 1966* **2**, 55 (1966); *Z. Anal. Chem.* **233**, 321 (1968).
33. W. J. Jones and R. A. Keller, *Anal. Letters* **1**, 227 (1968); G. Knudson, L. Ramaley, and R. A. Keller, *J. Chromatog. Sci.* **7**, 500 (1969).
34. M. A. Rommel and R. A. Keller, *J. Chromatog.* **18**, 349 (1965).
34a. G. Knudson, L. Ramaley, and R. A. Keller, *Anal. Letters* **2**, 433 (1969).
34b. R. G. Frew and W. F. Pickering, *J. Chromatog.* **47**, 86 (1970).
35. G. H. Stewart, *Separation Sci.* **1**, 135 (1966).
36. J. G. Herndon, H. E. Appert, J. C. Touchstone, and C. N. Davis, *Anal. Chem.* **34**, 1061 (1962).
37. C. L. deLigny and A. G. Remijnse, *Rec. Trav. Chim.* **86**, 410 (1967); *J. Chromatog.* **35**, 257 and 269 (1968).
37a. G. Hesse, *Chromatographia* **2**, 379 (1969); R. Neher, *J. Chromatog.* **48**, 7 (1970).
38. P. D. Klein and S. A. Tyler, *Anal. Chem.* **37**, 1280 (1965).
39. D. Galanos and V. M. Kapoulas, *J. Chromatog.* **13**, 128 (1964).
40. I. M. Hais, *J. Chromatog.* **33**, 25 (1968).
41. H. G. Cassidy, *Tech. Org. Chem.* **10**, 133–210 (1957).
41a. S. Turina, L. Horvath, and V. Marjanovic, *J. Chromatog.* **37**, 234 (1968).
42. U. Margna, *Eesti NSV Teaduste Akad. Toimetised, Biol. Seer.* **15**, 170 (1966).
43. E. Soczewiński and R. Mańko, *J. Chromatog.* **33**, 40 (1968).
43a. G. H. Stewart and T. D. Gierke, *J. Chromatog. Sci.* **8**, 129 (1970).
44. G. Ledin, Jr., W. R. Gustavson, and A. Furst, *J. Chromatog.* **22**, 376 (1966).
45. M. Lederer, *J. Chromatog.* **33**, 285 (1968).
46. E. C. Bate-Smith and R. G. Westall, *Biochim. Biophys. Acta* **4**, 427 (1950).
47. M. Lederer and A. Pirelli, *Sci. Rept. Ist. Super. Sanita* **1**, 528 (1961).
48. F. W. H. M. Merkus, *Pharm. Weekblad* **103**, 1037 (1968).
49. E. Soczewiński and J. Kuczyński, *Separation Sci.* **3**, 133 (1968); E. Soczewińsky and K. Jurkiewicz, *ibid.* **1**, 387 (1966).
50. E. Soczewiński and W. Maciejewicz, *J. Chromatog.* **22**, 176 (1966).
51. P. R. Rony, *Separation Sci.* **3**, 239 and 519 (1968); **4**, 413 and 493 (1969).
52. P. D. Klein, *Separation Sci.* **1**, 511 (1966).
53. L. Ossicini and M. Lederer, *J. Chromatog.* **17**, 387 (1965).

54. I. M. Hais, *in* "The Stationary Phase in Paper and Thin-Layer Chromatography" (K. Macek and I. M. Hais, eds.), p. 291. Elsevier, Amsterdam, 1965.
55. H. H. Strain, J. Sherma, F. L. Benton, and J. J. Katz, *Biochim. Biophys. Acta* **109**, 1 and 16 (1965).
56. G. Toennis and J. J. Kolb, *Nature* **177**, 281 (1956).
57. A. H. Beckett, M. A. Beaven, and A. E. Robinson, *Nature* **186**, 775 (1960).
58. G. B. West, *J. Pharm. Pharmacol.* **11**, 595 (1959).
59. C. O. Leng, *Far East Med. J.* **2**, 268 (1966)
60. J. E. DeVay, A. R. Weinhold, and G. Zweig, *Anal. Chem.* **31**, 815 (1959).
61. G. Zweig, *Anal. Chem.* **31**, 821 (1959).
62. D. H. Leaback and P. G. Walker, *Biochem. J.* **67**, 22P (1957).
63. J. Carles and S. Lascombes, *J. Chromatog.* **3**, 90 (1960).
64. M. Lederer, R. Marini–Bettólo, and G. Roch, *Ric. Sci.* **36**, 193 (1966).
65. K. Schwarz and A. A. Bitancourt, *Science* **126**, 607 (1957); *Biochem. J.* **75**, 182 (1960).
66. D. H. Hug, L. Wyborny, and D. Roth, *J. Chromatog.* **24**, 285 (1966).
67. R. A. Keller and J. C. Giddings, *J. Chromatog.* **3**, 205 (1960).

3

TECHNIQUES OF PAPER CHROMATOGRAPHY

Among the most attractive features of paper chromatography are its low cost and simple equipment and methods. For example, some qualitative paper chromatographic procedures require only a test tube fitted with a cork and a paper clip. Petri dishes or a set of round glass pie plates might be substituted for the test tube. This is the only equipment necessary besides the filter paper and a few chemical reagents, although more elaborate commercial apparatus is readily available and often preferred.

Detailed information on the techniques and applications of paper chromatography is available in numerous books and articles which are cited throughout this volume. A bibliography of papers published between 1961 and 1965 has recently become available (*1*), and a fairly current continuing listing of references appears regularly in the bibliography section of the *Journal of Chromatography*. A listing of chromatography products, instruments, accessories, supplies, and services and a supplier and manufacturer's directory appeared in the *Journal of Chromatographic Science*, February, 1970.

I. The Chromatography Laboratory

When paper chromatography is used only occasionally in the chemistry laboratory, a small amount of bench space and a chemical hood are usually sufficient for satisfactory work. However, for more precise work it is recommended to set aside a small chromatography room the size of an average darkroom for solvent development and further treatment of chromatograms. This room is especially useful for developing large chromatograms in floor-model chromatographic cabinets. Good ventilation is mandatory for the chromatography room, and it might be convenient to have an automatic switch to start the ventilation system as the door to the room is opened. Some of the commonly used chromatographic solvents, such as phenol, are extremely toxic, and the people exposed to them must be properly protected. Solvents, such as collidine and lutidine, have especially pungent odors which are undesirable in working areas and other solvents, such as petroleum ether, are highly flammable. Some state safety regulations require that a laboratory using large chromatographic cabinets must be provided with an efficient chemical exhaust hood, or the cabinets themselves must be equipped with a self-contained ventilating system. In equipment of the latter type, the papers may be dried after solvent development without opening the lid of the cabinet.

Some means of controlling the temperature of the chromatography room (usually 20–22°C) is desirable. However, if this is impractical, an electric blanket wrapped around the small chromatographic chambers will maintain a fairly constant temperature of 37°C. It is also desirable to keep the cabinets away from outside walls and windows, since fluctuating temperatures within the cabinets will cause uneven chromatograms. Finally, this room should also contain a laboratory shaker for the equilibration of solvents which should be stored in this room after mixing.

The preliminary preparation and final evaluation of chromatograms should be done outside the chromatography room, perhaps in other special rooms set aside for these purposes. Besides the usual equipment found in a well-stocked chemical laboratory, the following items are particularly useful for chromatographic work: rotating evaporator, hair dryer, X-ray viewing box, ultraviolet light (e.g., germicidal lamp), spotting table, constant-temperature oven, and storage cabinet for large sheets of filter paper.

A detailed description of the chromatography laboratory of the Pharmaceutical Division of Ciba Ltd., Basle, designed for processing large numbers of analytical and preparative paper and thin-layer chromatograms comprising a wide range of substances, has been published (*2*). This laboratory unit contains subdivisions for the receiving and allocating of work, preparation and marking of papers, dissolving and spotting of substances, development, drying of chromatograms, photocopying results or recording ultraviolet absorption or fluorescence, chemical or biological detection, qualitative and quantitative evaluation, recording of results, and storage and clean-up.

II. Sample Preparation

No attempt will be made in this section to give a generalized experimental procedure for the preparation of the sample prior to its application to the filter paper since detailed procedures for the individual classes of compounds will be found in later chapters. It is important to have the final sample dissolved in a small volume of a volatile organic solvent such as acetone, ethanol, or $CHCl_3$ at a concentration at which a small spot can be applied to the paper and detected after chromatography.

Material extracted from biological samples must often be concentrated and purified before the extract is subjected to chromatography. The final sample is cleaned up by removing extraneous biological materials such as pigments, proteins, sugars, lipids, and inorganic salts (unless, of course, these substances themselves are being studied), which usually have an unfavorable effect on the chromatographic results. This clean-up may be accomplished by conventional methods such as solvent extraction, precipitation, adsorption chromatography, centrifugation, freezing out, etc. Ion-exchange resins are used for the removal of inorganic ions which interfere with the chromatography of most organic substances. Conversely, ion exchange may be utilized to concentrate ionic substances which are under investigation. Electrolytic desalters are available commercially, for example from Shandon Scientific Co. If organic solvent extracts (e.g., ether, $CHCl_3$, ethyl acetate) of crude biological preparations are used, the salts, peptides, sugars, etc., are essentially eliminated.

Lipids may be removed after the sample is spotted on the paper. The sample is developed with a nonpolar solvent, which carries the lipids with the solvent front. The portion of the paper near the front is removed, and the sample is redeveloped with a more polar solvent (*3*) (see Chapter 11).

In some cases, the test substance is best studied after being converted to a derivative. This is true if the substance is volatile but the derivative is not, or if the derivative is easier to detect, identify, or quantitate than the original substance. One convenient way of preparing many of these derivatives is by injection of a mixture of the sample and reagent into a gas chromatograph. The derivatives are collected at the exit of the chromatograph and subjected to paper chromatography (*3a*).

III. Sample Spotting

Samples are applied to the paper using micropipettes. The sample solutions generally range in concentration from 0.1 to 1%, and a total of 1–1000 μg of sample is applied, depending upon the sensitivity of the subsequent detection procedure. Samples from 1 to 10 μl are conveniently applied with commercial pipettes calibrated in this range. For the application of larger volumes, successive applications of smaller volumes are made with drying in between so as to keep the spot small (about 5 mm diameter). If micropipettes are not available, one dab of a wood or glass rod applicator applies about 5 μl of solution to paper.

The solvent in which the sample is dissolved must not leave a residue at the origin after drying which would interfere with the subsequent chromatography. The nature of the solvent in which the sample is dissolved is important if a system involving paper impregnated with a nonpolar substance is being used. In such a case, the sample should, if possible, be dissolved in a hydrophilic solvent if applied after impregnation and a hydrophobic solvent if applied before impregnation.

There is some evidence from the results of studies with chloroplast pigments on unimpregnated paper that the solvent in which the sample is applied can have an effect on the subsequent chromatographic results. It is possible that solutes are deposited at the origin—after drying the initial zone—in different forms with different solubility characteristics depending upon the solvent employed. Until definitive studies are made, it is best to consider the solvent in which the sample is dissolved as another variable which should be standardized and reported.

A. Procedures for Spotting

A self-filling micropipette, commercially available in sizes ranging from 1 to 10 μl (1 μl = 1 λ = 0.001 ml), is immersed into the solution and filled to the mark by capillarity. Cleanliness is most important for the efficient operation of these micropipettes. They may be cleaned conveniently by suction using in sequence hot soapy water, distilled water, acetone, and air. The tip of the filled pipette is lightly touched to a pencil mark on the paper onto which it empties with an even flow. The solvent should be evaporated quickly with a stream of air. This may be done by holding the pipette in one hand and a hair dryer in the other (*4*) (Fig. 1). For larger amounts of solution (25–50 μl) the use of Kirk transfer pipettes is recommended. The flow of liquid is controlled by an attached microsyringe with a micrometer screw control. By a careful turn of the thumbscrew, as little as 1 μl may be delivered. Inexpensive disposable micropipettes (Kensington Scientific Corp., Oakland, Calif.) are composed of short lengths of precision capillary glass tubing. These pipettes, available in a wide range of sizes, are inserted into a simple holder assembly, filled by capillarity, and emptied onto the paper by squeezing the rubber bulb of the assembly.

FIG. 1. Application of sample with self-filling pipettes (*4*).

Another simple spotting device is the end of a bacteriological transfer loop which may be flamed between applications of different samples. The spots are usually placed about 2–3 cm from the lower edge of the paper, 2–2.5 cm apart from each other. The positions are indicated by light pencil marks.

Samples may be applied semiautomatically if the micropipette is clamped in a vertical position with its tip against the surface of the paper which is supported on a perforated metal disk. A stream of compressed air for drying passes through holes in the metal disk against the lower surface of the paper (*5*).

For the transfer of relatively large amounts of solution to filter paper, small paper disks 6 mm in diameter may be used. These disks are cut with an ordinary paper punch and are then soaked in the solution, dried, and inserted at the origin of the paper chromatogram through two parallel slits 2 mm apart.

Improved resolution of chromatograms, especially for compounds with similar chromatographic properties, may be obtained by a simple variation of spotting (*6*). After the sample has been spotted with a micropipette, three 10 μl portions of water (if an aqueous developing solvent is to be used) are applied in succession to the center of the original dried spot. As the solvent spreads, it carries the solutes to the periphery of the original spot, forming a ring with an empty center. The final chromatogram, especially of slower moving substances, consists of overlapping rings rather than spots.

Devices for the rapid application of radioactive solutions (with screening against beta radiation) (*7*) and the rapid semiautomatic multiple spotting of samples (*8*) on chromatographic paper have been recently reported. A rapid method for checking the calibration of micropipettes for quantitative chromatography has also been described (*9*).

It is normal procedure to heat the initial zone before the run in order to evaporate the solvent. Some substances are not thermostable, and this heating step can lead to the formation of artifacts. For example, α-tripiperidine and isotripiperidine, when dissolved in ethanol–HCl, spotted, dried and developed for 24 hours with butanol–HCl–H_2O (15:1:3 v/v), each gave five spots. When spotted without heating, one and two spots, respectively, were formed (*10*).

B. Procedures for Streaking

Another technique, which is especially recommended for crude biological extracts and preparative chromatography, is streaking, which involves application of the solution to the filter paper as a streak delivered from a pipette at closely controlled rates. For most extracts containing a large amount of extraneous material, the solution should be streaked from one edge of the paper to the other. This streak forms a barrier through which the development solvent must travel. If continuity is not preserved, the solvent may sneak around the applied sample through the least line of resistance, resulting in poorly resolved chromatograms. By applying the sample as a streak rather than a spot, the double-tailing portions (see Chapter 2) characteristic of some chloroplast pigment zones when developed on cellulose can be eliminated (*11*).

A simple, inexpensive pipette, useful for streaking sample solutions in paper chromatography, can be easily constructed (*12*). A survey of available streaking methods and devices along with a description of the construction and performance of a new, versatile, automatic streaking device (the Desaga Autoliner) producing bands of variable length and width has been recently published (*13*). The Shandon chromatography sample streaker, as designed by Bacon (*14*), is shown in Fig. 2. This device uses a glass hypodermic syringe with a right-angled Luer-fitting needle (shown) or can be adapted for use with an Agla micrometer syringe.

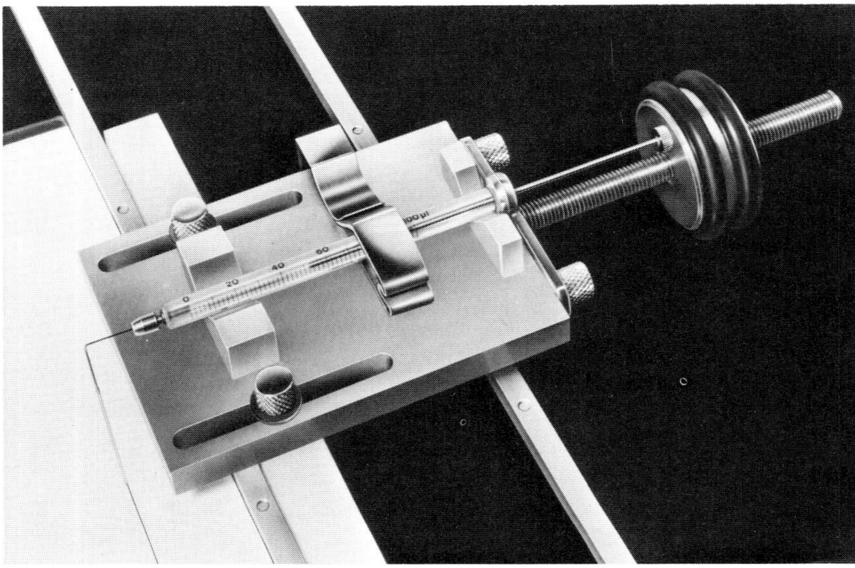

FIG. 2. Chromatography sample streaker. (Courtesy of G. D. Welch, Shandon Scientific Co., Inc., Sewickley, Pa.)

In closing this section, it is most important to stress that for optimum resolution, reproducibility, and qualitative analytical results, the two critical steps in spotting are: (a) the initial zone must be kept small (narrow, if a streak) and (b) the initial zones must be of uniform size. It is important to apply standard solutions and unknowns in the same manner. The application of initial zones for quantitative analysis will be considered in detail in the next chapter.

IV. The Paper

A. Pure Cellulose Papers

Paper has been described (*15*) as a mat of partially bonded fibers with a degree of orientation depending upon the manufacturing process. Commercial papers contain a machine direction, and it is recommended to start the chromatogram in that direction. The machine direction, which is generally along the length of the sheets, is sometimes indicated by an arrow on the wrapping or can be determined by visual inspection of the paper.

Chromatography papers are prepared from cotton linters selected to be especially low in organic and inorganic impurities and uniform in physical characteristics. Whatman papers have approximately the following percentage composition (*16*), varying from grade to grade and batch to batch:

α-Cellulose	98–99
β-Cellulose	0.3–1.0
Pentosans	0.4–0.8
Ether-soluble matter	0.015–0.03
Ammonia	0.001–0.06
Organic nitrogen	$<$0.01
Inorganic material	0.008–0.06

Acid-washed papers with a reduced ash content are available. Troublesome inorganic impurities include copper and iron, which form complexes with amino acids and peptides, leading to tailing zones or multiple spots. Traces of noncarbohydrate organic impurities are also present in paper. For most purposes, however, standard commercial chromatography paper (Whatman 1 and 3 or an equivalent) is satisfactory. In fact, it is reported that chromatography on common filter paper produces generally adequate results, with R_F values comparable to those obtained on Whatman No. 1 paper (*17*).

Many laboratory workers prefer to personally remove interfering metal impurities from commercial paper prior to use by washing with dilute HCl or a metal complexing agent such as EDTA followed by a thorough distilled water wash. The wash solvent is allowed to drip off

TABLE I

WHATMAN CHROMA PAPERS (*16*)

Grade	Basis weight (gm/m^2)	Thickness (mm)	Water diffusion[a] (minutes)	Solvent diffusion[b] A	B	C	Surface
1	87	0.16	9	6	15	17	Smooth
2	97	0.18	12.5	—	—	—	Smooth
3	185	0.38	9	5.5	13.5	16.5	Rough
3 MM	185	0.33	9	5	12.5	15	Smooth
4	92	0.20	5	3	7	8	Smooth
17[c]	440	0.80	3	—	—	—	Smooth
20	93	0.16	25	—	—	—	Smooth
31 ET[c]	190	0.53	2.5	—	—	—	Smooth

[a] Time taken for water to ascend 7.5 cm.

[b] Time in hours for downward diffusion of solvent for 40 cm. Solvents: (A) 60:20:20, MEK–propionic acid–water; (B) 4:1:5, *n*-butanol–acetic acid–water; (C) phenol saturated with water.

[c] Reduced ash, largely sodium sulfate, carbonate and silicate.

TABLE II

COMMERCIAL CELLULOSE CHROMATOGRAPHY PAPERS (*3, 16, 19*)

Whatman Filter Papers ("Selected for Chromatography")

Grade	Basis weight (gm/m^2)	Thickness (mm)	Nearest equivalent[a] chroma grade	Surface	Special properties
Ashless					
40	95	0.21	2	Rough	High purity[b]
41	91	0.21	4	Rough	High purity[b]
42	100	0.20	20	Rough	High purity[b]
Acid hardened					
52	100	0.16	2	Very smooth	High wet strength
54	93	0.17	4	Very smooth	High wet strength
Acid hardened ashless					
540	88	0.14	2	Very smooth	High purity[b] and wet strength
541	82	0.15	4	Very smooth	High purity[b] and wet strength
542	100	0.15	20	Very smooth	High purity[b] and wet strength

Schleicher and Schuell (or Schüll) Papers

Grade no.	Surface	Thickness (inches)	Grade no.	Surface	Thickness (inches)
Ashless			Standard		
589 Green-C	Rough	0.013	470-C	Smooth	0.035
589 Orange-C	Smooth	0.009	470A-C	Extra smooth	0.022
589 Black-C	Smooth	0.008	576-C	Extra smooth	0.004
589 White-C	Smooth	0.007	598-C	Rough	0.012
589 Red-C	Smooth	0.0065	598YD-C	Smooth	0.0085
589 Blue-C	Smooth	0.007	903-C	Smooth	0.017
507-C	Extra smooth	0.004	2040a	Smooth	0.006
			2040b	Smooth	0.0075
			2041	Smooth	0.006
			2043a	Smooth	0.006
			2043b	Smooth	0.0075
			2045a	Smooth	0.006
			2045b	Smooth	0.0075
			2316	Smooth	0.01

TABLE II (*continued*)

Other Commercial Papers[c]

Preparative papers:	FN 8; E 225; ED 320; MN 218
Especially smooth papers:	E 202 S
Acid-washed papers:	E papers designated with the letter A and MN 2214 and 2261
Standard papers:	FN 2, 3 and 4; E 202; MN 214 and 261; M Chr 100 and ELE 130; ED 248 and 613
Paper free of lipid-soluble substances:	E papers designated with the letter F and MN 241ff and 261ff

[a] With respect to diffusion rates and resolution of components.
[b] With respect to inorganic contamination.
[c] FN = Filtrak-Niederschlag; E = Ederol; MN = Machery-Nagel; ED = Eaton-Dikeman.

NOTE: The designations of chromatography papers are often changed so that it may be necessary to consult the various manufacturers in order to obtain a current grade of paper equivalent to one mentioned in this book. For example, Schleicher and Schuell papers designated MGl and Gl formerly referred to smoothed papers and "ausgew" to acid-washed papers. These symbols were nowhere used, however, in the latest S. and S. literature.

the edges of the paper during an overnight descending development. The papers are then air- or oven-dried and stored in a cabinet away from ammonia fumes, since ammonia complexes with cellulose and interferes especially with the separation of amino acids. For this reason, the paper may also be stored over a strong cation exchanger. Organic impurities in filter paper may be removed by washing with ether or with the first solvent to be used for the particular chromatographic development under study. It should be realized that harsh treatment of the paper may oxidize or degrade the cellulose in the course of removing unwanted impurities (*16*).

Table I lists the available grades of Whatman Chroma Papers, their basis weight, speed, thickness, and surface characteristics. These are papers especially manufactured for chromatography. Whatman also supplies various grades of filter papers "selected for chromatography." The user should always specify that he desires such paper when ordering from any manufacturer or supplier. Table II lists the Whatman papers of this type and comparable papers from some other manufacturers (*3*, *16*, *19*). High capacity or thick grades are recommended for larger quantities of materials and preparative use.

The structure of paper gives rise to a system of interconnected capillaries of varying size and shape, and the rate of penetration of any liquid into the paper is governed by the physical and geometrical properties of the capillary channels as well as the physical properties of the penetrating liquid (*15*). Table I also shows the flow rates for some of the papers with various liquids. For a detailed theoretical discussion of liquid absorption in paper, the reader is referred to Weaver (*15*).

Most grades of paper are commercially available in a variety of rectangular sheet, circle, and roll sizes and other special shapes (Table III). These commercial papers can be cut, if necessary, to other

TABLE III

PATTERNS OF PAPER AVAILABLE FROM WHATMAN FOR SPECIAL TECHNIQUES (*16*)

Pattern	Size of paper (mm)	Holes: No.	Holes: Size (mm)	Holes: Center to center (mm)	Other pattern features	Uses
A	254 × 254	4	12	229	—	To fit frames of various 2-dimensional chromatography tanks
H	254 × 254	4	5	229	—	To fit frames of various 2-dimensional chromatography tanks
S	200 × 200	4	7	170	—	To fit frames of various 2-dimensional chromatography tanks
CRL	110 × 213		—	—	Divided by 3 mm wide slots into 12 strips 15 mm wide	For rapid multiple analyses using simple apparatus and ascending chromatography
KCT/14	140 diam.		—	—	Divided by 3 mm wide radial slots into 5 segments	For rapid multiple analyses using simple apparatus by disk chromatography
KCT/26	265 diam.		—	—	Divided by 3 mm wide radial slots into 5 segments	For rapid multiple analyses using simple apparatus by disk chromatography
ERS	457 × 146		—	—	Divided into 6 strips 1.9 cm wide	For rapid multiple analysis using simple ascending development apparatus

sizes and shapes to suit the individual worker. The possibility of contamination of the paper by the scissors during the act of cutting should be borne in mind (*18*).

It is the fibrous structure of paper referred to above which is responsible for some of the disadvantages of paper chromatography relative to thin-layer chromatography (*3*). Papers have been recently produced with very short fibers from cellulose slurry and give a separation of amino acids in about 25% of the time needed for a thin-layer separation with zones which are almost as sharp (*20*). Although these papers have low mechanical strength, it may be possible in the future to produce papers commercially offering superior resolution compared to thin-layer media (*15*).

B. Modified (Altered) Cellulose Chromatography Paper

1. *Chemically Modified Paper*

Whatman No. 1 or S. and S. No. 602 ED paper may be partially acetylated by the following procedure (*21*): A large sheet or filter paper and a glass cloth are rolled into a cylinder which is inserted into a glass cylinder filled with glacial acetic acid containing 5% acetic anhydride. After standing at room temperature overnight, the solution is then replaced by 22.5% acetic anhydride in amyl acetate–1.8% perchloric acid in acetic acid (8:1 v/v). The perchloric acid is slowly added with cooling, so that the temperature remains at 28–29°C. After circulating through the paper for 2 to 3 hours, the paper is removed, washed with cold water, rinsed with methanol and dried at room temperature. The acetyl content is then about 26% by weight (dry). See Chapter 14, Section I,B for another procedure. These papers are useful for separating hydrophobic substances; a mixture of steroids has been resolved on fully acetylated paper using a benzene–methanol–H_2O solvent (4:4:1 v/v) (*22*). Acetylated papers are available commercially from Schleicher and Schuell (No. 2492, 5% acetylated; 2495 and 2496, 21%; and 2497, 45%) and Machery-Nagel (214 AC, 10 and 20%, and 261 AC, 10%).

As mentioned in Chapter 2, ordinary filter paper contains a small number of COOH groups which give the paper a small ion-exchange capacity. Additional carboxyl groups are formed in the paper by oxidation with N_2O_4 to produce paper containing up to 5% COOH groups. Such papers have been used to separate mixtures of amino acids, amines, inorganic ions, and curare alkaloids (*23*).

Copolymerized cellulose has been used as a support for reversed-phase chromatography of dyes and cations (*24*), and wedge-compressed nitrocellulose membrane filters, either intact or impregnated with nonionic detergents or proteins, are a convenient medium for small-scale chromatographic separations or deproteinizations (see Chapter 5) (*25*).

2. *Impregnated Papers*

Many workers impregnate paper sheets with oils to separate hydrophobic substances such as amines (*26*), lipids (*27*), steroids (*28*), vitamins (*29*), and pesticides (see Chapter 11) (reversed-phase chromatography) or with polar liquids to facilitate separations of hydrophilic substances. Such systems will be discussed in the next section. Silicone-treated papers are available commercially—Whatman ST 81 (medium flow rate), 82 (slow), and 84 (fast) and Machery-Nagel 214 WA. Schleicher and Schuell papers No. 2498 and 2499 are impregnated with 5% Dowex 1107.

Paper has also been impregnated with high molecular weight amines and their salts and high molecular weight acids, which act, respectively, as liquid anion and cation exchangers in the reversed-phase chromatography of metal ions (*30*). Papers so impregnated act much like the papers loaded with solid ion-exchange resins (below) except that they cannot be developed with partly organic solvents. Papers impregnated with complexing or precipitating agents are useful for metal ion separations (Chapter 15).

3. *Loaded Papers*

Loaded papers consist of a dispersion of powder, fiber, or resin in a normal cellulosic paper network (*15*).

a. Papers Loaded with Adsorbent Powders. These papers are used to separate less polar and nonpolar substances (such as lipids, sterols, steryl esters, steroids, terpenes, dyes, insecticides, and chloroplast pigments) and inorganic ions. Adsorbent papers generally function in a manner similar to thin layers of the same adsorbent, but in some cases the presence of the cellulose along with the adsorbent can produce a desired separation not attainable on layers or columns of the adsorbent alone (*31*). Silica gel paper has been experimentally compared with silica thin layers and nonimpregnated paper (*32*).

Commercial papers loaded with silica gel (Whatman SG 81, 22% SiO_2 ; S. and S. 966, 25–40%), aluminum hydroxide (Whatman AH 81, 7.5% Al_2O_3 ; and S. and S. 288, 35–40%), and kieselguhr (S. and S. 287) are available. Figure 3 shows the separation of various dyes on SG 81 paper (*33*).

Silica paper can be used for two-way separations by adsorption chromatography in one direction and partition in the other. Twenty cholesteryl esters were so separated by developing with petroleum ether and then with acetic acid–$CHCl_3$–paraffin oil (80:15:5 v/v) after impregnating the paper with paraffin oil between runs (*34*) (cf. Chapter 8).

Besides the separation of hydrophobic compounds by the adsorption mechanism, silica gel-loaded papers are used (*35*) with partition solvents for the rapid ascending separation of polar compounds such as sugars (solvent: methanol–4 *N* NH_4OH + 6% (w/v) boric acid), phenols (nitromethane–formic acid–H_2O, 60:25:15 v/v), and amino acids (phenol–methanol–4 *N* NH_4OH, 1:2:1 v/v).

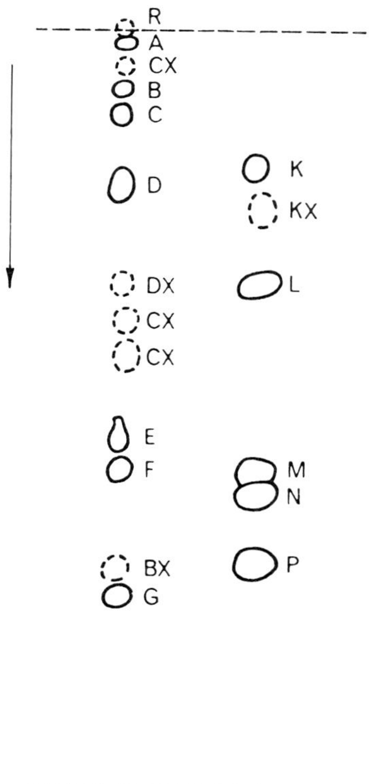

FIG. 3. Separation of 2 μg each of steroid marker dyes and azo dyes on Whatman SG 81 silica paper developed with benzene for 1.5 hours. A = 4-(4-methyl-3′-nitrophenylazo)-3-methyl-5-pyrazolone; B = 1-phenylazo-2-naphthol; C = 1,4-diamino-anthraquinone; D = 4-amino-1-methylaminoanthraquinone; E = 4-(2-hydroxy-1-naphthylazo)-2,3′-dimethylazobenzene; F = 1-(2′-methoxyphenyl)azonaphthol; G = 1,1′-di[*p*-(2′-hydroxy-5′-*tert*-butylphenylazo)-*m*-tolyl]-cyclohexane; R = dye residue; X = impurity in dye sample; K = phenylazoresorcinol; L = *p*-aminoazobenzene; M = Sudan yellow; N = unidentified; P = yellow OB. Paper 5 × 18 cm. [After (*33*).]

The application of adsorbents in paper chromatography has been reviewed (*36*).

b. Papers Loaded with Ion-Exchange Celluloses. The papers available include two cation exchangers and three anion exchangers (*37*):

Cellulose phosphate paper (Whatman P 81) is a bifunctional cation exchanger containing both strongly and weakly acidic groups. It is essentially in the monoammonium form and functions as a strong exchanger over the full pH range. Being a cellulose ester, it is inherently unstable at extreme pH values.

Carboxymethylcellulose paper (Whatman CM 82) is a monofunctional, weakly acidic cation exchanger in the hydrogen form which functions most efficiently above pH 4. It is chemically stable since the ionic group is linked to the cellulose through an ether bond.

Aminoethylcellulose paper (Whatman AE 81) is a monofunctional, weakly basic anion exchanger most effective below pH 9. This and the other anion-exchange celluloses are ether derivatives supplied in the free-base form.

Diethylaminoethylcellulose paper (Whatman DE 81) carries a tertiary amine group and functions best below pH 10.

Ecteola cellulose paper (Whatman ET 81) also carries a tertiary amine group, but it has a more complex structure, being partially cross-linked.

Some modifications of the normal chromatographic technique may be required if these papers are to yield optimum performance. The ionic form of the paper is changed by a descending chromatographic wash with an appropriate aqueous electrolyte. The initial spot is then applied and the run made without drying the paper (the wet-start technique, see Chapter 15). The spot application can be made to the damp paper after blotting away the excess capillary solution or without blotting while the paper hangs in the tank after pretreatment. If a dry start is desired, the sample is applied to the dry, untreated paper as received (*38*). Weak exchange papers cannot be converted with electrolyte and then washed with water to remove the excess and dried before the run because of hydrolysis which occurs as the equilibrating ion is washed away.

Besides the introduction of special interactions in order to increase the ability to resolve ions, an advantage of these papers is that aqueous solvents are usually employed so that the results and reproducibility are less affected by conditions such as temperature, humidity, and chamber equilibration. Also, larger volumes of sample can often be loaded in one application to yield a small initial spot (*15*).

As described in Chapters 5 and 15, two-dimensional separations in which ion-exchange predominates in one direction and partition in the other are easily obtained. Ion-exchange papers developed with solvent gradients (pH and/or concentration) have proved useful for separating mixtures of ionic species with different charges (*37*).

Cellulose ion exchangers have been used for biochemical and inorganic separations. Table IV lists selected applications on the various types of

TABLE IV

APPLICATIONS OF ION-EXCHANGE CELLULOSE PAPERS

Type of paper	Substances separated	Reference
Cellulose phosphate	Lipids	*39*
	Amines	*40–42*
	Amino acids	*43*
	Thiamine and its phosphates	*44*
	Tetracycline and derivatives	*45*
	Inorganic ions	*46, 47*
Carboxymethylcellulose	Lipids	*39*
	Inorganic ions	*48*
Aminoethylcellulose	Lipids	*49*
	Food dyes	*50*
	Inorganic ions	*47*
Diethylaminoethylcellulose	Organic acids	*51*
	Amino acids	*43, 52*
	Proteins	*53*
	Enzymes	*54–56*
	Hormones	*57*
	Nucleotides and Nucleosides	*58*
	Drugs	*59, 60*
	Vitamin C	*61*
	Inorganic ions	*47, 62*
Ecteola cellulose	Carbohydrates	*63*
	Hormones	*64*
	Drugs	*59, 60*

papers (*39-64*). Separations of nucleotides carried out on polyethylene-coated Whatman DE 81 paper in acetic acid–1 *M* trisodium citrate buffer are superior to those on uncoated paper (*65*).

c. Papers Loaded with Synthetic Organic Ion-Exchange Resins. Papers are available from Reeve Angel that contain 45–50% by weight of micropulverized ion-exchange resin incorporated into α-cellulose pulp. These papers are used in a manner analogous to that for ordinary chromatography paper after conversion to the desired ionic form by washing with—or soaking in (*66*)—a suitable electrolyte followed by a distilled water rinse and drying.

The physical properties of the Reeve Angel papers are shown in Table V (*67*), and a partial listing of applications of these papers is given in Table VI (*67*). Both inorganic (*68*) and organic (*69*) ions are separated by ion-exchange chromatography on these papers, and nonionic organic substances are separated by solubilization (partition) chromatography (*70*). Both aqueous and aqueous–organic solvents may be employed (*71*).

TABLE V

PHYSICAL PROPERTIES OF REEVE ANGEL ION-EXCHANGE PAPERS (*67*)

Property	SA-2	WA-2	SB-2	WB-2
Resin[a]	Amberlite IR-120	Amberlite IRC-50	Amberlite IRA-400	Amberlite IR-4B
Resin type	Strong acid	Weak acid	Strong base	Weak base
Resin in paper (%)	45–50	45–50	45–50	45–50
Resin form supplied	Na^+	H^+	Cl^-	OH^-
Paper color	Tan	White	Cream	Yellow
Basis weight (20 × 20/500) (pounds)	37	37	37	37
Thickness (mils)	14	14	14	14
Wet strength	Good	Good	Good	Good
Flow rate	Fast	Fast	Fast	Fast
Approximate exchange capacity (meq/gm dry resin)	4	10	3.3	10

[a] Amberlite is the registered trademark of Rohm & Haas Co., Philadelphia, Pa.

Information previously gained by ion-exchange column experiments is often directly applicable to procedures using these resin-loaded papers (*72*). However, results on these papers are not necessarily equivalent to results obtained in batch equilibrium and column experiments employing the same resin and solvent because of differences in the techniques and the presence of cellulose plus the resin in the ion-exchange paper (*66*).

In order to prevent streaking, it is recommended that the paper be converted into the same ionic form as that used for development. Thus, SA-2 paper is first converted to the hydrogen form with acetic acid and then developed with an acetate buffer for the separation of the basic amino acids (*73*). Amino acids may be detected with ninhydrin (Chapter 5) on SA-2 and WA-2 paper, but this reagent is not suitable for SB-2 and WB-2 papers which contain resins with free amine groups.

By use of ion-exchange papers, inorganic ions from biological samples are removed prior to paper chromatography without unduly diluting the sample, as often occurs when ion-exchange column desalting is employed (*74*). The sample is passed through a combination of WA-2 and WB-2 strips with water, and the water-soluble nonionic compounds running with the solvent front (e.g., phosphorylated carbohydrates) are directly transferred to another paper sheet for chromatography.

Other papers containing ion-exchange resins are available from

TABLE VI

SOME POSSIBLE CHROMATOGRAPHIC SEPARATIONS ON REEVE ANGEL ION-EXCHANGE PAPERS (*67*)

Compound	Paper grade			
	SA-2	WA-2	SB-2	WB-2
Biochemical				
Amines—adrenaline type		×		
Antibiotics	×	×		
Alkaloids	×	×	×	
Acids, fatty	×	×	×	×
Creatine			×	
Cytochrome c		×		
Enzymes		×		
Flavonoids		×		
Hemoglobin		×	×	×
Mucopolysaccharides			×	
Insecticides, hydrolyzed			×	
Vitamins		×		
Inorganic				
Copper, nickel, cobalt		×		
Iodide, iodate, periodate			×	
Indium, rhodium	×			
Scandium, vanadium, titanium			×	
Transuranium			×	
Uranium, vanadium	×			
Zinc, lead			×	

Schleicher and Schuell (Nos. 2493 and 2494, containing 5% resin) and Ederol (208/IK and 208/A) (*3*). Machery-Nagel supplies papers composed of chemically modified cellulose with strong-acid and strong-base exchange groups but a lower ion-exchange capacity than the Reeve Angel papers. The papers from Machery-Nagel, which may be termed dilute exchangers, are similar to papers impregnated with very dilute solutions of liquid ion exchangers; the ion exchangers in the papers do not combine with ions of different charges according to a mass action equation on account of the large distances between the exchange groups (*74a*).

d. Papers Loaded with Inorganic Ion Exchangers. A review of chromatographic separations on paper impregnated with inorganic ion exchangers such as zirconium phosphate, hydrous oxides, and ammonium molybdophosphate has been published (*75*), and various applications are mentioned in later chapters of this volume.

Recently, support-free cerium (IV) phosphate sheets have been prepared and used for metal ion separations by ascending development with $HClO_4$ at various concentrations (Fig. 4). These sheets exhibit porosity, good mechanical and chemical stability, and high ion-exchange capacity and selectivity (*76*).

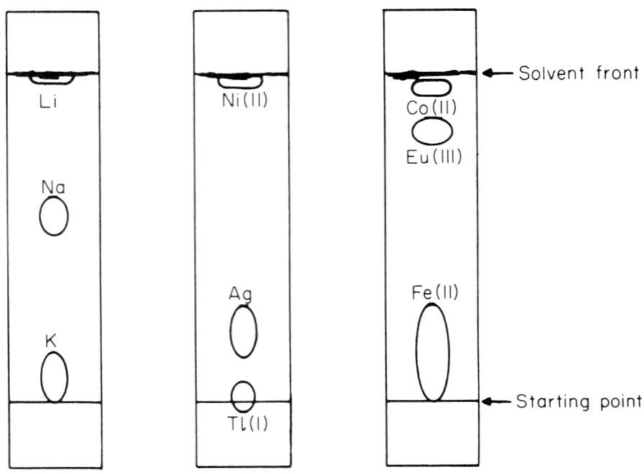

FIG. 4. Representative separations on cerium phosphate paper using 1 *M* $HClO_4$ as the mobile phase. [After (*76*).]

4. *Papers of Glass Fiber*

Glass fiber papers (Schleicher and Schuell 26, 27, and 29 and Whatman GF 81, 82, and 83) are sometimes useful in extreme conditions of temperature and acidity where cellulose papers are unsuitable. These papers are manufactured from fine borosilicated glass fibers and can be employed with detection reagents that are too corrosive for cellulose-based papers (e.g., chromic–sulfuric acid charring solutions).

Although occasionally used untreated (*77*), glass fiber paper is normally impregnated with aqueous salt solutions (*78*), silica gel (*79*), or alumina (*80*, *81*). Chromatography on glass fiber paper has been reviewed (*82*).

An advantage of glass fiber paper is that it can be rendered transparent by immersion in a solvent of the correct refractive index (e.g., methyl salicylate and ethyl benzoate). Transparent sheets are especially suitable for transmission densitometry (Chapter 4) and scintillation counting of separated zones (*83*).

Glass fibers are also less sorptive than cellulose so that the elution of zones prior to estimation may be easier and more complete (*84*). Glass fiber papers are insoluble in organic solvents, they have the ability to

absorb twice as much material and thus larger samples can be applied, and they can be presterilized for the bioautography of antibiotics (Chapter 13) by heating up to 550°C without harming the glass fibers. Glass fiber papers already impregnated with silica gel and silicic acid are commercially available from the Gelman Corporation (ITLC media) (*85*).

V. The Solvent

In paper chromatography, the principal variable is the solvent. Its selection is critical in establishing a chromatographic system with the selectivity necessary to provide the desired separation. And it is during the flow of this solvent—i.e., the chromatographic development—that the solutes undergo differential migration and actually separate as described in Chapter 2.

Historically, the first solvents were composed of two phases: the atmosphere of the chamber was saturated with the aqueous layer, while the immiscible organic layer was used for the development of the chromatogram. Separations obtained in such systems are due mainly to the selective partitioning of the solutes between the two liquids. The phases can be reversed so that the more polar layer is used as the solvent and the paper is impregnated with the less polar layer (perhaps an oil).

Instead of two-phase solvents, so-called direct-phase solvents (*15*), consisting of an organic liquid saturated with a more polar substance (e.g., phenol saturated with water), were later more conveniently used with similar results (*86*). Finally, it was realized that the presence of the stationary, solid support allowed the use of liquid phases that are mutually soluble, for example water-miscible solvents or even pure water as the mobile phase.

A. General Guidelines for Selecting the Solvent

In general, polar solutes are separated in a polar solvent on cellulose which is unmodified or impregnated with a polar liquid. Hydrophobic solutes are separated in reversed-phase systems (*15*). When designing a solvent, the ingredients and their proportions must be decided upon; resort to personal prior experience or results in the literature of others in separating similar mixtures is one's best guide.

As one source of help in selecting a solvent, lists of organic compounds arranged in order of their increasing polarity or strength of attraction to an adsorbent are available. One such series is as follows (*87*):

Petroleum ether
Carbon tetrachloride
Cyclohexane
Carbon disulfide
Diethyl ether
Benzene
Esters
Chloroform
Dichloroethane
Alcohols
Water
Pyridine
Organic acids
Inorganic acids and bases

Another similar classification lists organic liquids in terms of their ability to form hydrogen bonds (*3*). The polarity of the solvent, which for a mixture varies primarily with the concentration of the more polar constituents, determines the degree of mobility of the solutes. The mobility of ionizable substances is influenced as well by the pH of the system.

A simple procedure for the empirical design of a chromatographic solvent has been described by Smith and Feinberg (*88*):

> A representative number of the compounds is made up as a single mixture (e.g., the four cations of the zinc group; a pentose, a hexose and a disaccharide for sugars; an acidic amino acid, a neutral and a basic amino acid) and the mixture is chromatographed in a large range of pure solvents such as water, 2 *N* HCl, 2 *N* ammonia, acetic acid, ethanol, butanol, acetone, chloroform, benzene, ether and petroleum ether. It may be found that all components move in the solvent front in, say, water, acid and ammonia, but remain at the origin in the pure organic solvents.
>
> A suitable mixture based on water: organic solvent must now be designed. However, as water is completely miscible with ethanol and acetone, is partly miscible with butanol and is almost totally immiscible with other organic solvents, a second series of experiments based on water–ethanol mixtures, and butanol saturated with water, could be tried. As the use of neutral solvents often results in elongated spots, it is usual to add various concentrations of hydrochloric acid, ammonia or acetic acid to the water.
>
> A good general idea for a suitable solvent composition should now have emerged and the final composition of solvent can be determined using the complete mixture of substances (e.g., all the amino acids in a protein hydrolysate; all the HCl-soluble cations) and running a number of chromatograms in which the solvent composition is varied only slightly from run to run. Should it be necessary, for other reasons, to base a solvent on water–benzene, which constitute a practically immiscible pair, then, to aid mutual solution, it is necessary to include a third solvent such as glacial acetic acid, which is miscible with both water and benzene.

The preliminary experiments using pure solvents can be carried out quite simply and quickly. The usual procedure is to cut strips of chromatography paper of a size just to slip into large boiling [test] tubes, or 100 ml measuring cylinders. The origin is a single spot about 3 cm up from the bottom of the paper, which dips into 5–10 ml of the solvent in the bottom of the tube or cylinder. The strip of paper can be suspended from a paper clip pushed into the underside of a cork which fits the tube. A solvent run of 10 cm is sufficient to indicate whether the substances move in the solvent front, remain at the origin or move to some intermediate position.

Equations have been published that correlate R_F values for a compound in two solvent systems. They allow prediction of R_F values for a series of compounds in one solvent if their R_F values in a second solvent are known (*89*). An inexpensive device, constructed from a section of an immersion thermometer allows potential solvents to be applied to dried spots of mixture on a paper sheet. The resulting radial development allows the effectiveness of the solvent to be rapidly evaluated (*90*).

It is obvious that the solvent chosen must not decompose the compounds during chromatography as has been found, for example, with glycine (*91*) and tryptophan (*92*). In later chapters, solvents will be given which successfully separate each class of compounds considered.

Many workers no longer bother to carefully equilibrate the paper prior to the run (*46*), although for good reproducibility consistent pretreatment procedures must be followed (see Chapter 2). It has been reported, however, that preequilibration of the paper with the developer substantially reduces the time of development in many cases (*93*). Humidification of a paper by preequilibration influences absorption rates by swelling the fibers and thereby changing the paper structure (*15*). Resolution is also affected by humidity, although it is reported that papers can be equilibrated with certain saturated aqueous salt solutions to eliminate the water content of paper as a practical factor in R_F values (*94*).

B. Types of Solvent Systems (*3*)

1. *Aqueous Stationary Phase*

The paper is saturated with water by soaking and blotting the excess, or the water is taken up from the atmosphere by equilibrating hanging paper in a sealed chamber saturated with water vapor. Aqueous buffer or salt solutions are applied by dipping the paper, drying it, and then allowing it to equilibrate with water as just described. Hydrophilic, medium-polarity solutes are separated on these papers by developing with solvents such as isopropanol–NH_3–H_2O (9:1:2 v/v), *n*-butanol–

acetic acid–H_2O (4:1:5 v/v, organic layer), or phenol saturated with water. When two-phase solvents are employed, the chamber is saturated with both phases.

2. *Stationary Polar (Hydrophilic) Organic Solvent*

For volatile organic liquids (e.g., methanol), the paper is saturated from the atmosphere of the chamber and developed with an immiscible organic solvent. For less volatile liquids (e.g., dimethylformamide), the paper is drawn through the liquid dissolved in a volatile solvent (e.g., ethyl acetate), and the excess liquid is removed by blotting between two thick filter papers, squeezing the paper between rollers (*19*), or by simply holding the paper and letting the excess drip back into the vessel containing the impregnation solution. The paper is then hung in air until the solvent has evaporated (5–10 minutes). Many workers recommend saturating the mobile phase with the stationary liquid (*95*), but others (*19*) suggest that this is not always necessary (if the two phases have very low mutual solubility) or even desirable (because the paper may extract additional impregnating liquid from the mobile phase, becoming overloaded and thereby causing lowered R_F values).

The Zaffaroni systems employ paper impregnated with formamide or propylene glycol and developed with benzene, $CHCl_3$, cyclohexane or mixtures of these (*96*). These types of systems are chosen for the separation of moderately hydrophilic, medium-polarity solutes such as many steroids (*3*).

The reproducibility of R_F values on impregnated papers (*97*) and the influence of the temperature and humdidity during the impregnation upon R_F values (*98*) have been studied. It was found that because of the greater purity and lower volatility of formamide, reproducibility is better with it than with dimethylformamide, and that increases in temperature and humidity during impregnation cause increased R_F values, especially for substances with high R_F values.

3. *Stationary Nonpolar (Hydrophobic) Solvent*

Reversed-phase systems are used for the separation of hydrophobic solutes. The paper is drawn through a solution of petrolatum, kerosene, paraffin oil, silicone oil, etc. dissolved in hexane, petroleum ether, or benzene, hung in air until the solvent evaporates, spotted and developed with an immiscible, more polar solvent such as aqueous isopropanol (70%), dimethylformamide–methanol–H_2O (10:10:1 v/v), or aqueous acetic acid. The mobile phase is usually saturated with the stationary phase prior to the run.

Another technique for impregnating the paper is to dip the top into the impregnating solution which rises up the paper toward the origin.

If the mobile phase to be used for development carries the solutes at its front on nonimpregnated paper, the impregnating solution is not quite allowed to reach the origin line. After evaporating the solvent, the sample is spotted below the impregnated area. The mobile phase will rise through the sample and carry it into the impregnated area, whereupon an ideal thin, transverse origin is produced spontaneously (*99*) (see Fig. 14, Chapter 8).

A recent study has shown that two stationary phases must be considered to be present in reversed-phase systems: the organic non-volatile liquid and the cellulose–water complex. Lipophilic compounds are partitioned between the organic stationary phase and the mobile phase while hydrophilic compounds behave as if the paper did not contain the organic stationary phase. In some separations, both stationary phases are simultaneously involved (*99a*).

As mentioned in Section IV, acetylation makes cellulose less hydrophilic so that systems employing acetylated paper are essentially reversed-phase systems. Such paper preferentially takes up organic components from equilibrating solutions in the chamber.

VI. Development

A. Descending Development

The solvent development step in paper chromatography has been the subject of a great number of variations. The choice of the method may depend on the class of compounds being investigated, or it may be mainly one of personal preference. It is curious that a particular method may predominate in a certain geographic part of the world (e.g., radial chromatography in India). Of historical interest is the technique of capillary analysis (Kapillaranalyse) which was developed by Schönbein, his student Goppelsroeder, and Skraup in the early 1900's for the separation of a large variety of naturally occurring compounds. Water was the solvent of choice, and the chromatographic chamber described by Goppelsroeder over 50 years ago differs little from the equipment which is in use today (*4*).

Consden, Gordon, Martin, and Synge, the fathers of modern paper chromatography, reported in 1944 the successful separation of amino acids by descending chromatography. In this procedure, which is still widely used today, the solvent is permitted to flow down the paper. The obvious advantage of this technique is the relatively constant rate of flow of solvent unimpeded by the force of gravity which restricts the ascending development.

The chromatographic development takes place in a tightly closed chamber which is saturated on the inside with the vapors of all solvent components. This saturation is usually accomplished by placing a small dish containing the solvent to be used at the bottom of the cabinet, or the walls of the chamber may be lined with a fast flow rate filter paper and immersed into the solvent covering the bottom of the chamber. It has been recommended that the chamber be saturated with both phases when two-phase systems are used (*19*). Others say that this is unnecessary since both layers will be in equilibrium with vapor of the same composition (*100*).

The paper may be saturated for a period in the tank after spotting but before the mobile phase is added to the trough. However, most workers now apply the initial zones to the paper and begin development immediately (without preequilibration of the paper) in a tank saturated with the solvent vapors. With this procedure the paper is equilibrating with the solvent (e.g., water is gradually removed by the paper from aqueous solvents) throughout the run.

As was mentioned in Chapter 2, it is preferable to carry out solvent development at a specified temperature without extreme diurnal fluctuations. Thus, it may be advisable to keep the chromatographic cabinet on the unheated floor of a room with minimum window space. Generally, however, extreme measures for solvent saturation and temperature control need not be taken, since in most cases standards and unknowns are chromatographed on the same sheet of paper, which serves to cancel out experimental variations. When reporting chromatographic data, the temperature of development, as well as the method of equilibration and other pertinent conditions (see Chapter 2) should be given, so that results obtained by others may be compared.

At the end of the solvent development, the solvent front is marked with a pencil in order to evaluate R_F values. In another technique, dyes, such as crystal violet and rhodamine, may be incorporated at the origin and travel with the front of most solvent systems (e.g., butanol, ethyl acetate, phenol), thus providing a convenient marker for the solvent front. Glass apparatus is convenient because it allows observation of the solvent front during the run.

For slow-moving substances better resolution may be achieved by permitting the solvent to drip off the serrated lower edge of the paper. It is possible to carry out solvent development for several days and to effect good separation of difficult mixtures (e.g., leucine, isoleucine; and monosaccharides). For lack of a better English word, this technique is called *Durchlauf* chromatography.

Experimental Procedure

In the descending technique, which is the recommended procedure for chromatography on large sheets of paper (18 × 22 inches), the samples are spotted on a line 3 inches from one edge of the paper and at 1 inch intervals from one another. The paper is folded into a Z shape as illustrated in Fig. 5 and transferred to the solvent trough inside the chromatographic cabinet (Fig. 6) where it is fastened to the antisiphon rod with two stainless steel chromatographic clips (Fig. 7). A second

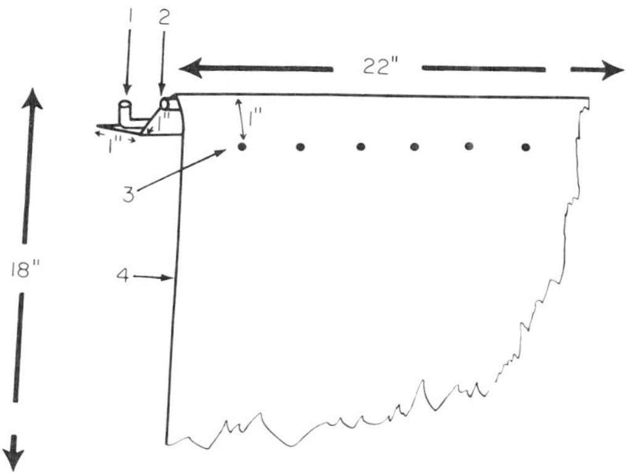

FIG. 5. Front and side views of large paper chromatogram (18 × 22 inches) for descending development. 1 = support rod; 2 = antisiphon rod; 3 = origin; 4 = paper.

chromatogram may be placed on the opposite antisiphon rod so that both papers reach into the same empty trough. A support rod made of heavy glass or a stainless steel rod bent at right angles at the ends is placed into the trough holding the papers in place.

The solvent assembly shown in Fig. 8 consists of a stainless steel cradle (a), glass trough (*101*) (b), antisiphon (c), and support rod (d). The clips are removed, and after equilibrating the papers for some time (2–3 hours), the solvent is added to the troughs through an opening in the lid of the cabinet, and the development proceeds usually overnight. If needed, an automatic solvent addition funnel is connected to a timing device and a solenoid, so that at a preset time a plunger is forced into the liquid, thereby raising the level and actuating a siphon (*102*). At the end of the run, the papers are again clipped to the antisiphon rods and transferred into a drying oven for final analytical evaluation.

Commercial versions of glass desk top chromatographic chambers are available which are simply smaller models of the large cabinet just described. These include a "no-rack" round tube ($2\frac{5}{8}$ × 20 inches)

FIG. 6. Commercial chromatographic cabinet for descending development of large sheets.

for 2 inch wide paper strips, and round jars from 6 × 18 inches to 12 × 24 inches for 4 inch and 7 inch papers, respectively, and a square jar (12 × 12 × 24 inches) for 9 inch paper sheets, all with support racks for glass solvent trough assemblies (available from Fisher Scientific Co.). Shandon Scientific Co. supplies a 13 inch glass tank for descending or ascending chromatography. Figure 9 shows this tank arranged with a solvent trough for descending chromatography.

Recent modifications in the descending technique have been aimed at increased convenience or reproducibility of results. An inexpensive method for descending chromatography in 6 × 1 inch test tubes has been described (*103*), as has a solvent circulating device that pumps solvent from the bottom to the top of extra long tanks (120 cm) (*104*)

FIG. 7. Stainless steel chromatographic clip holding paper on antisiphon rod.

where it is allowed to run down the inner walls and provide proper saturation for the run (*105*). The construction of a trough for descending paper chromatography with an ionic strength, pH, or polarity gradient has also been described (*106*).

Apparatuses have been described for automatically stopping the solvent flow when the liquid front reaches a preselected distance from the origin (*107*). Systems for automatically starting the run by loading solvent into the chamber without disturbing the system have also been suggested (*108*). In one of these systems (*109*), an alarm clock outside the tank is set to go off when solvent equilibration of the paper is reached in the tank. The clock has an arm that rises when the alarm goes off, thus rotating a cylindrical tube cistern to which the arm is attached by a cord and causing solvent to flow from the cistern into the trough to begin the run.

B. Ascending Development

Ascending paper chromatography (i.e., the solvent travels in an upward direction) is especially suitable for the quick analysis of a large number of samples or for exploratory work involving the use of many different solvents. One must bear in mind, however, that the upward flow of solvent is eventually counteracted by gravity, resulting in a definite slowing down after the solvent front has traveled more than 25 cm. The chromatographic development should be stopped at this point. In order to achieve maximum resolution by this method, multiple development should be tried (see below).

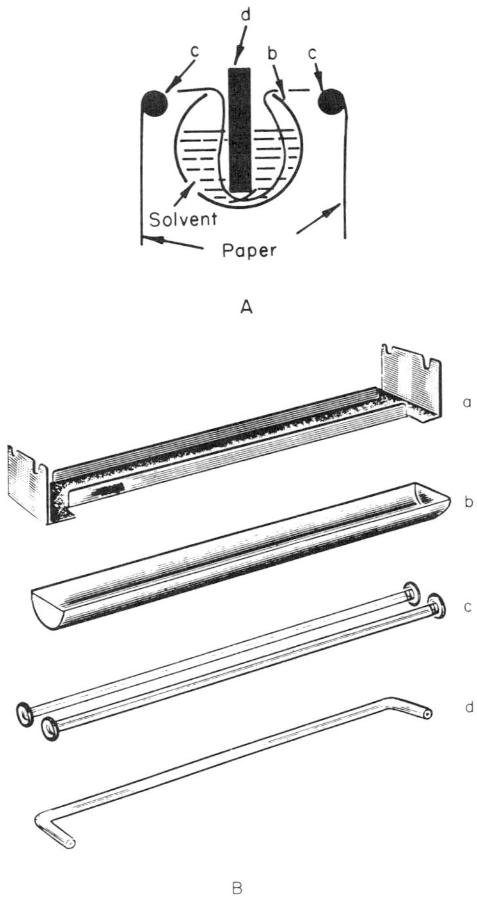

FIG. 8. (A) Cross section of small trough (*101*). (B) Solvent trough assembly: a = stainless steel cradle; b = glass trough; c = antisiphon rods; and d = support rod.

Ascending development is carried out very simply by stapling the spotted paper (e.g., 20 × 20 cm) into the form of a cylinder and standing it in a covered, saturated, paper-lined container (such as an empty 8 pint petroleum ether can or a glass jar) containing solvent on the bottom. The top of the solvent must be well below the level of the origin line on the sheet. Ascending development in test tubes has been described above in Section V.

The ascending–descending technique of Block (*110*) features a modified 10 gallon aquarium tank in which filter paper sheets (18 inches

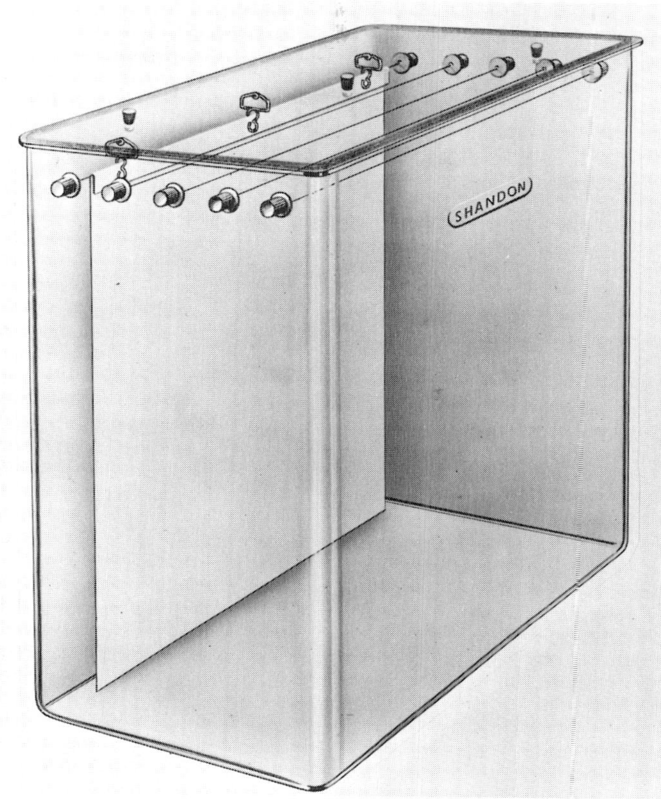

FIG. 9. Universal Sheet Chromatank (13 inches) for developing paper chromatograms. (Courtesy of G. Welch, Shandon Scientific Co., Sewickley, Pa.)

square) are draped over a glass rod, and the solvent is permitted to flow upward past the rod and down on the other side (see Fig. 10).

A stainless steel chromatographic tank for ascending chromatography has been designed by Mitchell and is illustrated in Fig. 11 (*111*). Square sheets of filter paper, 8 × 8 inches, are held in place with two chromatographic clips (see Fig. 7) on a glass or metal stainless steel rod; the solvent troughs may also be made of glass or metal. A glass cover is sealed with masking tape. This tank is especially useful for volatile solvents, and in addition has the advantage of saving space.

Filter paper sheets (10 × 18 inches) may be developed in 10 gallon glass aquarium tanks which are fitted with glass rods running the length of the chamber (see Fig. 10). The sheets are fastened to the rods with stainless steel clips. The tank shown in Fig. 9 is used for ascending

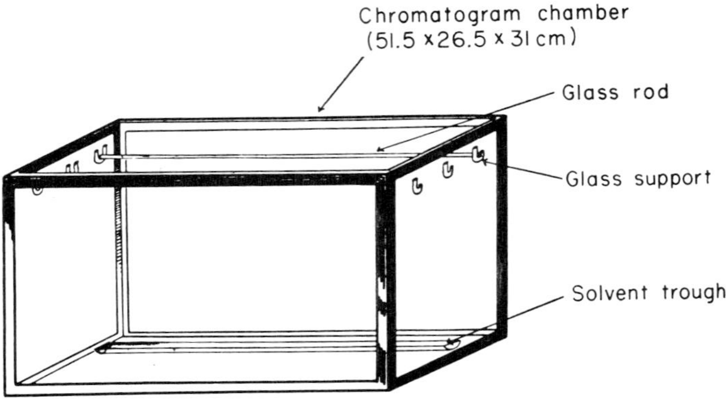

FIG. 10. Chromatographic chamber for ascending–descending technique. [After (*110*).]

chromatography by clipping the paper to the stainless steel support wires at the top.

The Chromatobox is the commercial version of the original design by Barrolier. In this apparatus a paper strip, 40–50 cm long, is sandwiched between a slightly shorter length of Teflon sheeting which has a raised surface to prevent solvent siphoning. This box is made from a solvent-resistant plastic and occupies a space not larger than 3 square inches

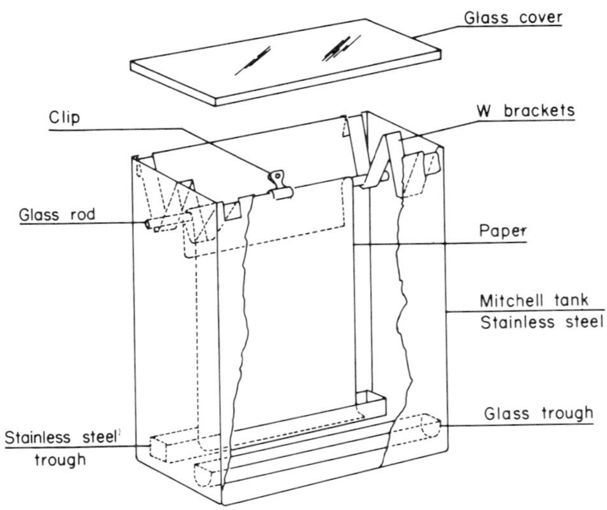

FIG. 11. Stainless steel tank for ascending chromatography. [After Mitchell (*111*).]

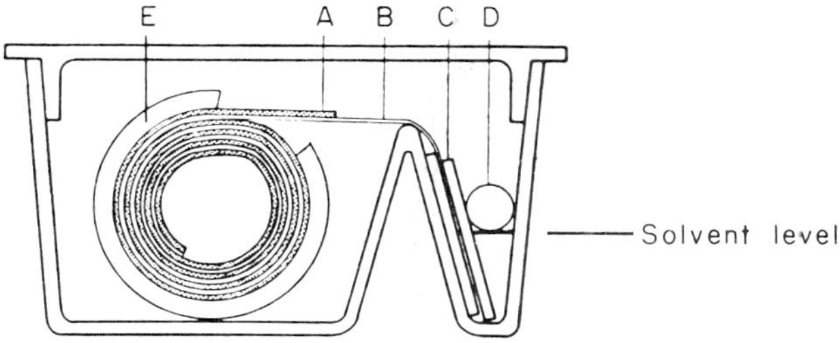

FIG. 12. Diagram of a Chromatobox. A = Teflon band; B = paper strip; C = glass plates; D = glass rod; E = polyethylene support ring for coiled paper. [After Barrollier (*112*).]

(see Fig. 12) (*112*). It is thus possible to achieve a relatively long solvent run in a minimum space. The compactness of the box also assures excellent chamber saturation. However, the solvent front is not visible during the development.

A transistorized solvent front detector for the ascending chromatography of unstable compounds in the dark has been described (*113*). The detector triggers an alarm system when the leading edge of an electrically conductive solvent has reached the end of a predetermined solvent path.

C. HORIZONTAL DEVELOPMENT

In this method, a square or rectangle of paper is laid horizontally on glass rods and the end is dipped into a solvent trough. Volatile solutes are chromatographed on paper placed between glass or metal plates (*114*). Improved reproducibility of R_F values is claimed for the horizontal method (*114*), and the apparatus is compact so that it fits easily in an oven or refrigerator for development at various temperatures.

Horizontal chromatography at elevated temperatures was developed by H. R. Roberts (*115*) and has been applied to the rapid quantitative estimation of sugars, fatty acids, and amino acids. Filter paper sheets (12 × 17 cm) are developed horizontally in a simple stainless steel chamber which is placed in a constant temperature oven at 50° or 60°C. The arrangement of the chromatograph is shown schematically in Fig. 13 and the exact dimensions of the chamber are illustrated in Fig. 14. The chromatograms are spotted and developed in a manner similar to that described before, with the exception that the solvent development may be repeated two to four times for better resolution. Between each run the chromatogram is removed from the tank and dried in the oven at 60° (see also the discussion of multiple development below).

As an example, aspartic and glutamic acids were separated after two

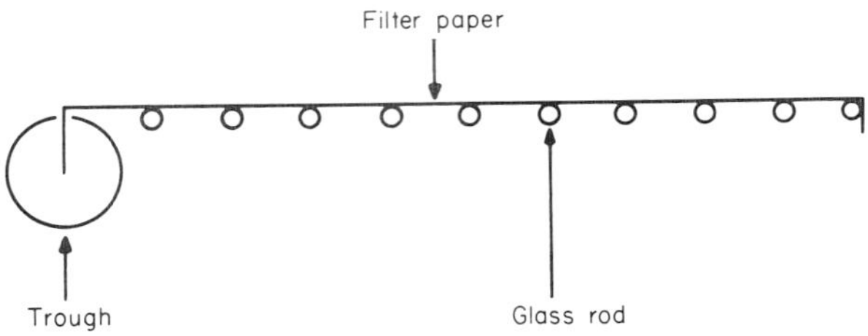

FIG. 13. Position of paper during horizontal solvent development. [After (*115*).]

1 hour developments at 60°C in a solvent consisting of phenol saturated with pH 12.0 buffer (*116*). Similarly, C_2 to C_6 volatile fatty acids could be quantitatively separated after 1 hour in water-saturated *n*-butanol at 50°C and lactose hydrolyzate at 60°C (*115*).

Horizontal development apparatus is now commercially available [e.g., from Schaar and Co., Chicago, Ill. (*3*)].

D. RADIAL DEVELOPMENT

The principle of this technique is that the substances to be chromatographed are resolved into circular zones or arcs instead of the customary round or oval spots. The reader is referred to a comprehensive review on this subject (*117*). Runge in his book "Der Bildungstrieb der Stoffe" (1855) shows numerous illustrations of "radial chromatograms" which

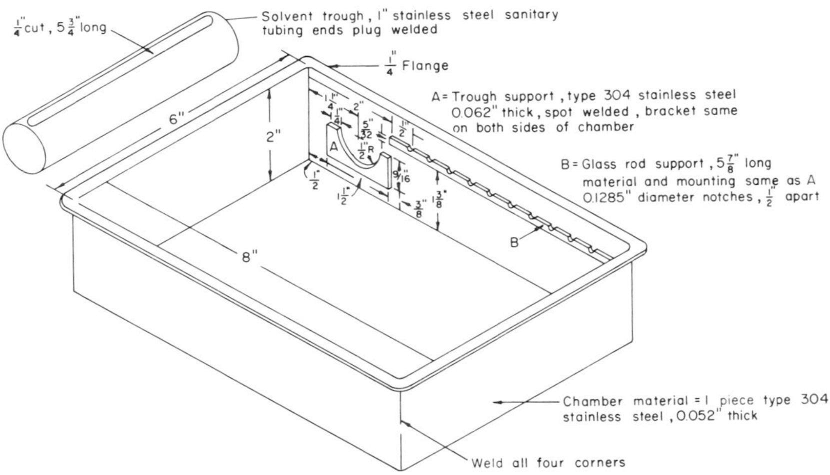

FIG. 14. Stainless steel chamber and solvent trough for horizontal paper chromatography. [After (*115*).]

may have actually been a variation of capillary analysis (Chapter 1). Rutter (1948) is generally credited with originating circular filter paper chromatography, although Brown (1939) described the method earlier (see Chapter 11).

It should be stressed that circular zones are formed during radial development but that circular paper is not required (cf. Fig. 4, Chapter 11). The paper is often in a horizontal position during radial development, but this method differs from horizontal chromatography as described in Section VI,C in the way the mobile phase is fed to the paper (e.g., through a wick).

The apparatus may be fashioned from two equal sized Petri dishes or two circular baking dishes with ground edges (*118*) (Fig. 15), and a

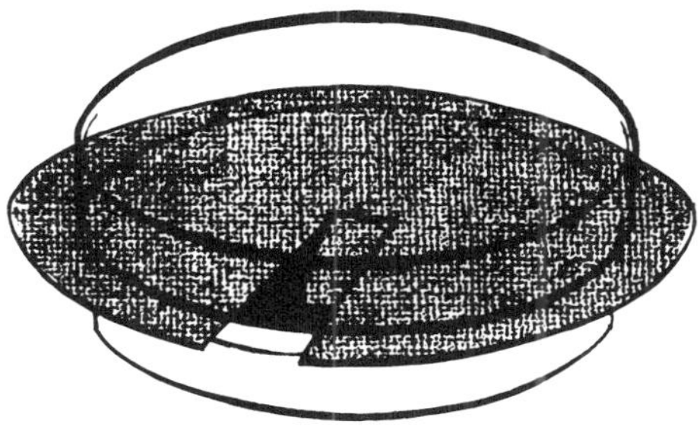

FIG. 15. Petri dish arrangement for circular filter paper chromatography. [After (*118*).]

filter paper disk, slightly larger than the glass dishes. Starting at the center of the paper disk and extending to the periphery, two parallel cuts, 2 mm apart, are cut with a razor blade. This serves as a wick which is bent at the joint perpendicular to the paper and shortened to about 1.5 cm. The wick may also be made from a strip of filter paper rolled into a cylinder (2–3 mm thick) cut at the end into the form of a brush and inserted through a small hole at the center of the disk (*119*), or the solvent may be fed onto the paper disk by means of a glass capillary, 5 mm o.d., that has been ground to a fine taper at the point where contact is made with the underside of the paper (*120*). The size of the capillary regulates the rate of feed and thus the rate of development. A small cone made of filter paper and placed with its apex at the center of the disk can be used as an irrigating wick (*121*).

The drop to be analyzed is placed on the paper disk at the joint of the wick and is air dried. The disk is then placed between the two glass dishes, the lower containing the developing solvent. The solvent rises by capillarity and the rate of development may be readily controlled by the width of the tail and the distance between the liquid surface and the plane of the paper.

For development in Petri dishes with solvents requiring good vapor saturation, the method of DeVries *et al.* (122) is recommended: the spotted paper is equilibrated with solvent in a tilted apparatus which is then leveled to immerse the wick and begin the run (Fig. 16).

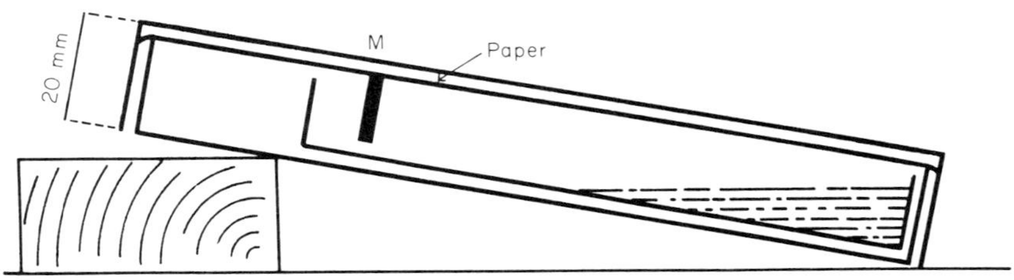

FIG. 16. Equilibration of the paper in Petri dish apparatus prior to radial development; M = the wick. [After (*122*).]

Desiccators can be used for larger tanks with the paper placed between the cover and the bottom part. Solvent is applied to the paper through a wick from a solvent reservoir below. With either Petri dishes or the desiccator, solvent can be fed to the paper by a pipette or burette through a hole in the lid (*123*).

In most cases of chromatographic identification, it is advantageous to analyze both standard and unknown solutions on the same chromatogram. This may be accomplished by applying six to eight equidistant spots on a penciled circle with a radius of 1–2 cm from the center of the paper disk [see Block *et al.* (*4*)]. In this case a series of segments, instead of concentric circles, result after an appropriate color development (*124*) (see Fig. 17).

Sometimes a mixture of compounds can only be resolved by radial chromatography. This may be illustrated by a wedge chromatogram of *o*- and *p*-chloroaniline, diazotized with β-naphthol (*125*) (Fig. 18). No such resolution was achieved by ascending strip chromatography. The separation of carbohydrates by the wedge strip technique of Matthias is illustrated in Chapter 6, Fig. 3. A simple technique for cutting paper for the Matthias method, employing a compass rather than a die, has been described (*126*) and such papers are also commercially available.

An analogous method was used to separate tropane alkaloids on

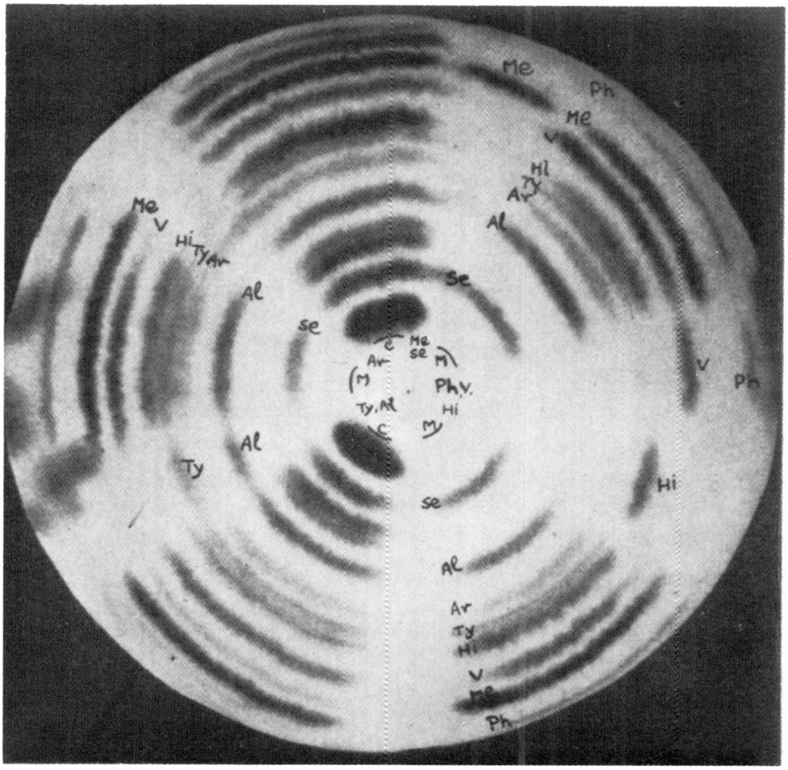

FIG. 17. Circular paper chromatogram of amino acids developed in order to obtain segments rather than rings. [After (*124*).]

triangular paper with a 2 cm × 2 mm tongue at the tip. The alkaloids present in low amounts were concentrated in the narrow lower part of the paper (*127*).

Another variation of circular chromatography is conical paper chromatography (*128*). A filter paper disk is shaped so that it can be folded into a cone. The samples are spotted near the lower edge, and the bottom of the cone is immersed in solvent in a Petri dish placed in a glass vessel with a tight lid. Trapezoidal paper sewn into a truncated cone was used to separate seventeen common amino acids (*129*).

Still another method for circular chromatography has the advantage of speed ($2\frac{1}{2}$ hours for the separation of sugars) and simplicity (*130*). Four to eight pie-shaped segments are cut out of a 1.5 cm diameter circle in the center of a 26 cm filter paper disk. The resulting tips are

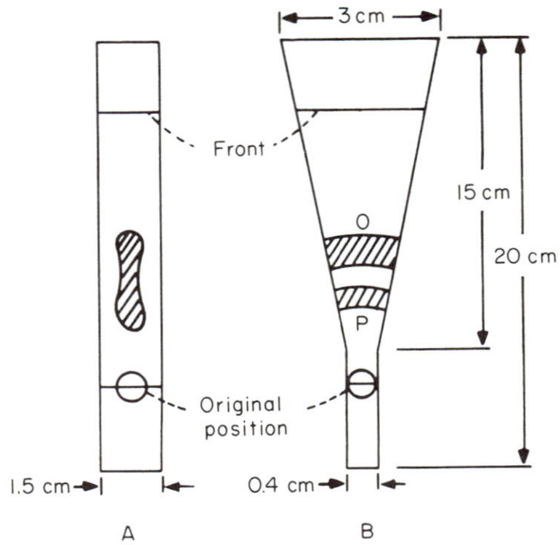

FIG. 18. Chromatography of *o*- and *p*-chloroaniline by the strip (A) and the wedge (B) technique. Development with H_2O–pyridine (4:1 v/v). [After (*125*).]

bent downward, and the samples are applied at the periphery of the circle. The tips of the triangles are immersed into the solvent inside a glass desiccator.

Radial chromatography has received much recent attention. A sandwich apparatus for radial paper chromatography has been described which provides good vapor saturation and speed of development and requires a low solvent volume (*131*). A similar setup affording maximum saturation employs paper placed between glass sheets, one of which has a central hole. An 11 cm diameter disk of Whatman No. 1 paper, folded into a cone and fitted into a solvent container, serves as the wick for irrigation of the chromatogram (*132*).

Accelerated radial paper chromatography has been used for the separation of Cd (R_F 0.82), Bi (0.78), and Hg (0.90) in 15 minutes with butanol saturated with 3 *N* HCl as the solvent (*133*) (see also Section VI,G, below).

In horizontal radial chromatography and the wedge strip method, there is a concentration effect as the annular zones are formed (*100*); the zones increase in length but decrease in width as development occurs, thereby aiding resolution. As described above, the flow of solvent in these methods is restricted through the use of shaped paper or a wick feed. This leads to sharper zones and increased resolution because the solvent, which must fill an increasing number of capillaries as it spreads, fills the finer capillaries first. Since the larger capillaries are not used at an early stage in the solvent flow, diffusion of the solute in these large capillaries is limited (*15*). A theoretical discussion of the analogies and differences between linear and radial chromatography has been published (*134*).

E. Overrun and Multiple Development

The multiple development technique, first proposed by Jeanes (*135*), is used to gain a greater effective distance of solvent flow while retaining a shorter length for the paper itself. After the mixture has been developed by the solvent and the paper dried, the paper is returned to the solvent trough for repeated irrigation of the same or a different solvent. After two to four such developments, an excellent resolution of certain mixtures can be obtained. Another method for effectively lengthening the paper is to allow the solvent to overrun the end and continue flowing, preferably at a constant rate (*136*). This is accomplished by cutting drip points along the bottom of descending chromatograms or allowing the top of ascending chromatograms to protrude out of a slot in the chamber lid.

A simple formula for the multiple development of amino acids for strip and circular chromatography may be found to be generally applicable (*137*). The rule states: The distance between two substances will increase in the second development if $R_{F_1} + R_{F_2} < 1$.

There are a number of developments with the same solvent in the same direction which yield the greatest separation between two substances. Below or above this number, the extent of separation decreases. The equation describing this optimum number of developments is

$$\eta_{\text{opt}} = \frac{\log \dfrac{\log(1 - R_{F_2})}{\log(1 - R_{F_1})}}{\log \dfrac{(1 - R_{F_1})}{(1 - R_{F_2})}}$$

where R_{F_1} and R_{F_2} are the R_F values of the two compounds after one development, $R_{F_1} > R_{F_2}$ (*138*).

Computer-prepared tables relating the R_F values of two solutes to the number of solvent passes required for their separation by a preselected amount have been compiled. These tables aid in the selection of solvents for separations by this technique (*139*).

In fractional chromatography, a solvent of increasing polarity is employed for each subsequent development (*140*).

F. Two-Dimensional Chromatography

For maximum resolution of a complex mixture, two-dimensional paper chromatography is recommended (*141*). The principle of this technique is the successive development of the chromatogram with two different solvents (e.g., a basic and an acidic solvent), whose advancing fronts are at right angles to each other. Fig. 19 is a diagram of a one-dimensional

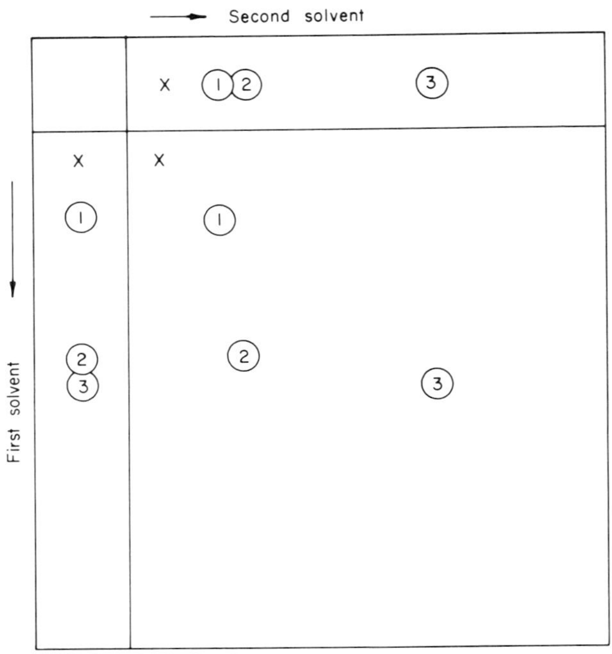

FIG. 19. Diagram of the separation of substances after one- and two-dimensional chromatography.

chromatogram of a three-component mixture, followed by development with another solvent in the second direction. Complete resolution of the mixture is then achieved even though separation in either solvent alone is not possible.

The experimental procedures for one- and two-dimensional chromatography are identical, except that only one sample per paper may be chromatographed at one time in two-dimensional chromatography. Between the first and second solvent development, the chromatogram must be dried thoroughly in a hood or chromatographic oven in order to remove the last traces of the first solvent. When phenol is used for the first direction, it is also recommended to cut off with scissors the leading solvent front which has accumulated paper impurities. If this is not done, an irregular V-shaped second solvent front may result.

A stencil may be fashioned from exposed X-ray film or another suitable rigid transparent plastic, 4 inches square, containing a hole made with a paper punch 1 inch in from two adjoining edges. This stencil is laid along the edges of one corner of a 18 × 22 inch sheet of filter paper, and a circle is drawn with a pencil through the hole as shown in Fig. 20. This mark serves as the origin for a two-dimensional chromatogram.

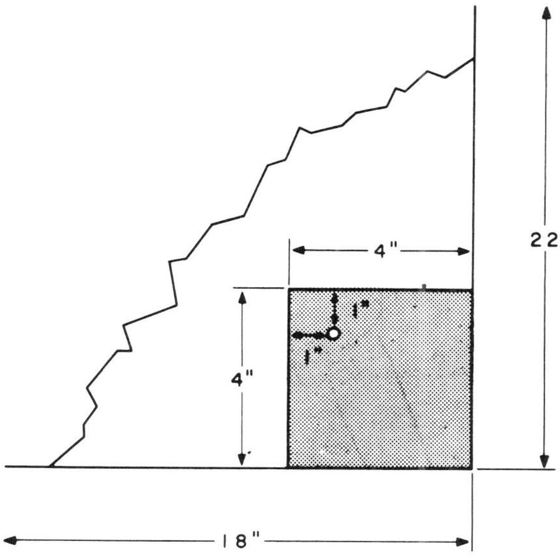

FIG. 20. Stencil to mark origin for two-dimensional chromatograms on large paper sheets.

A filter paper rack accommodating 12 sheets (8 × 8 inches) for use in two-dimensional chromatography is supplied by the Shandon Scientific Co. The rack (Fig. 21) is made of aluminum and consists of two square plates connected by four rods fastened to the plates with nuts. Four holes are punched into the corners of the filter paper sheets (it is suggested that a template be made), and the sheets are placed on the rack by stacking them on the rods through the holes. The sheets are separated from each other by $\frac{3}{4}$ to 1 inch aluminum or stainless steel washers fitting on the rods.

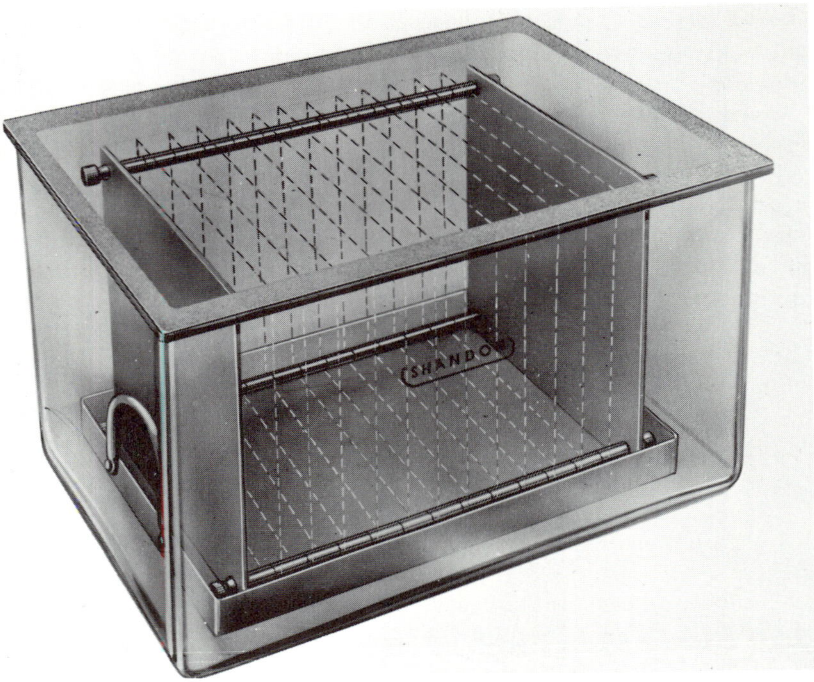

FIG. 21. Chromatank with rack (frame) holding twelve paper sheets for ascending two-dimensional technique. (Courtesy of G. Welch, Shandon Scientific Co., Inc., Sewickley, Pa.)

For optimum results in two-dimensional chromatography, the judicious choice of the solvent pair is important. With three different solvents and three sets of two-dimensional chromatograms, it has been possible to separate the amino acids in protein hydrolyzate (*142*). For example, the following different solvents may be chosen: (a) water-saturated phenol; (b) α-picoline–water–NH_3 (70:28:2 v/v); (c) isopropanol–acetic acid–water (7:2:1 v/v). Three sets of two-dimensional chromatograms may then be run with the solvent combinations shown in the tabulation.

Chromatogram	Solvent (first dimension)	Solvent (second dimension)
I	a	b
II	a	c
III	b	c

These methods are generally applicable to any class of compounds, and a study of the R_F tables will usually suggest a suitable combination of solvents. This scheme has been extended to four solvents and a set of six two-dimensional chromatograms with all possible combinations and has aided in the identification of 40 different components in urine (*143*).

Two-dimensional chromatography with reaction of the solutes on the paper after the first run is used to assess the effect of an enzyme or reagent on the solutes or to aid in their identification (*144*). The same or a different solvent is used in the second direction (see Chapter 6).

Two-dimensional chromatography is often improved if the spots are transferred automatically from one type of paper to another type (*145*) or to a thin layer (*146*) by attaching the papers or the paper and the thin-layer plate together. Phospholipids of yeast were separated using the following combinations of paper types: alumina, formamide-treated, or anion-exchange paper for the first direction with silicic acid paper for the second; or silicic acid paper for the first direction followed by formamide-treated paper (*145*).

Two-dimensional development of labile substances such as plant pigments (see Chapter 11) and tryptophan (*147*) must be carried out with minimum delay and exposure to light between the two runs.

G. Miscellaneous Techniques

Centrifugal chromatography is a method which combines the techniques of chromatography and centrifugal acceleration and allows the rapid separation of mixtures of dyes, amino acids, sugars, alcohols, aldehydes, acids, proteins, vitamins, antibiotics, alkaloids, inorganic compounds, etc. [see, Yao and Chen (*148*) for a review of this topic]. The method was developed mostly by McDonald *et al.* (*149–152*). The apparatus used for this procedure may consist of a balanced, motor-driven disk-shaped head rotating a circular sheet of filter paper (e.g., 18 inches in diameter) or other suitable material in a horizontal plane. The head is made of two round pieces of stainless steel or Plexiglas in the shape of a shallow pie plate. The paper disk is held only at the center and assumes a taut position during the operation. The samples are spotted slightly off-center on the moistened paper, and the disk is slowly set in motion. When the desired speed is attained (300–1500 rpm), the aqueous solvent is fed in a continuous jet stream by means of a capillary tip through an opening of the top plate near the center (e.g., 1.15 ml/minute). The difficulty is the even control of the solvent flow which has a tendency to cause sputtering, thus resulting in rather diffuse bands on the chromatogram. This technique makes possible rapid (5–30 minute) analyses of comparatively simple mixtures of water-soluble compounds, especially substances which are unstable. The resulting chromatograms are often diffuse, irregularly shaped spots with some overlapping; however, for two- to three-component mixtures this is no drawback because it retains the advantage of extreme speed. Figure 22 shows a

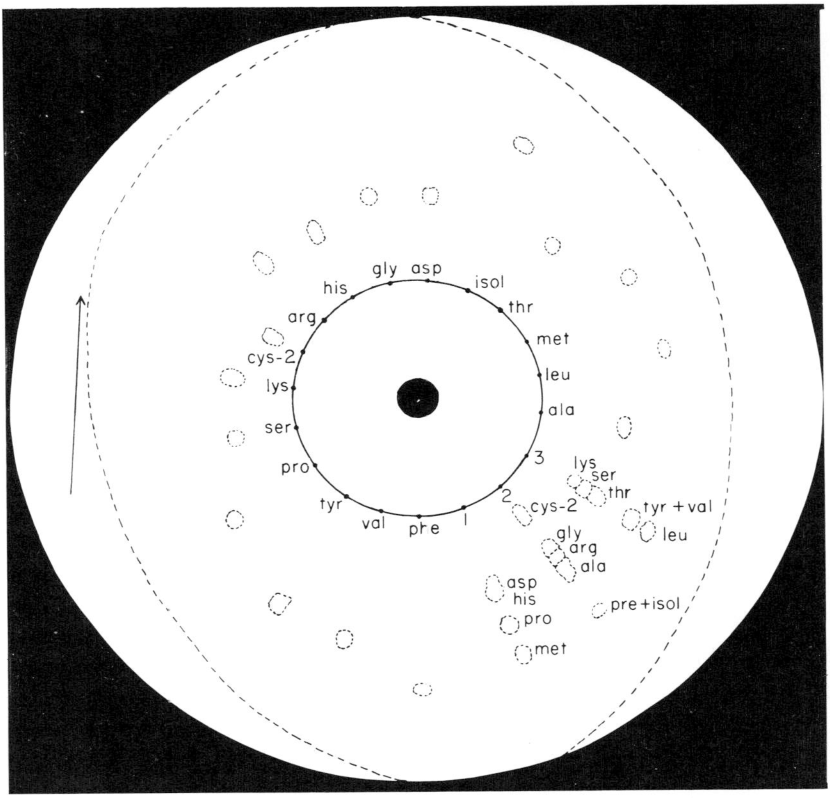

FIG. 22. Relative positions and migration distances of sixteen amino acids and degree of separation of three mixtures following centrifugal chromatography. [After (*149*).]

typical chromatogram. Although circular chromatography is usually used, a device employing paper strips has also been described (*153*). Two-dimensional chromatography on a rotating paper disk separated three dyes in 20 minutes (*154*).

Elatography is a term coined by Becker (*155*) for a method in which the filter paper is partially impregnated with a compound that reacts with the substance to be analyzed either before or during solvent development. This type of chromatography is characterized by the fact that the reaction product rather than the original substance is chromatographed. Speed and simplicity are claimed for this technique. As an example of elatography, aniline and *p*-toluidine may be separated by first reacting the compounds on paper with 2-naphthol-6,8-disulfonic acid (Na_2) and $NaNO_2$ and chromatographing the resultant dyes with 2 *N* Na_2CO_3–concentrated NH_4OH (1:1 v/v).

Separations at *temperatures above and below room temperature* are sometimes required. Horizontal chromatography at elevated temperature was mentioned above, and the reversed-phase chromatography of drugs and alkaloids has been performed at 86°C (*156*). Inorganic ions have been chromatographed at —30°C, at which temperature a good range of solvents can be used with results similar to those obtained at room temperature (*157*). Chromatography in inert atmospheres is necessary for oxygen-unstable solutes.

Rear-phase chromatography is a method used with solvents which form two fronts on the paper (see Chapter 15). The solvent is allowed to flow a certain distance so that the phase boundary reaches the origin and then the sample is applied behind the boundary so that it is developed only with the slower single phase of the solvent. This technique and others, such as *column–disk combination chromatography* and *chromatography with longitudinal barriers* have all been described in detail in the last edition of Block *et al.* (*4*).

Thermal paper chromatography involves chromatography in a temperature gradient. The technique, theory and applications of this method have been described in a series of papers (*158*, *159*).

Gradient development techniques have been described by several workers (*4*, *159a*); these techniques involve the continuous change in composition of solvent during the development.

Paper chromatography has been combined with *paper electrophoresis*, both simultaneously (*159b*) and in sequence (see Chapter 5), and with the *ring oven technique* (*160*) for increased resolution and ease of identification. The *line oven technique* is a variation of the ring oven technique for the rapid analysis of inorganic ions (*161*). *Transfer coefficients* which allow the selection of suitable cellulose columns and the calculation of elution curves based on R_F values on paper have been introduced (*162*). Data derived from paper chromatography also allows the determination of optimum solvent systems for countercurrent distribution procedures (*162a*).

VII. Detection Methods

A. Chemical Detection Methods

After solvent development and air or oven drying (usually around 100°C) of chromatograms, the spots (with the exception of colored compounds) must be visualized. The simplest method is detection by a suitable chemical reaction on paper resulting in a color or fluorescence. If one is dealing with experimental compounds for which no procedures have been described, we suggest consulting Feigl's latest edition of "Spot Tests" or adapting a wet colorimetric method to paper chromatography. The reactions given by twenty-five different substances, representing various chemical classes, with fifteen different color reagents on paper strips have been recently tabulated, and the preparations and methods of use of these reagents are also given (*163*). Some reagents may be used in succession for the detection of several compound types on the same chromatogram. The general utility of iodine vapor (*164*) and π-electron acceptors (*165*) as detection reagents has been recently evaluated. Sulfuric acid spray, followed by heat, is a general detector for most organic compounds, although its use is restricted to glass paper or inorganic thin layers.

1. *Dipping*

For small chromatograms, sheets, strips, or disks, the dipping procedure is recommended. It is necessary, of course, to choose a solvent for the color reagent in which the substances to be detected are insoluble. Acetone is generally the preferred solvent for such a dipping reagent. I. Smith (*86*) lists the following advantages for dipping chromatograms: (a) noxious vapors are not sprayed into the atmosphere; (b) it is economical since reagents may be used over again; (c) since a hood is usually not required, it is space-conserving; (d) the reagent can be applied more uniformly; and (e) colors often appear more rapidly and are more intense and stable.

In one procedure for dipping a chromatogram, a chromatographic trough (Fig. 8) is filled with dipping solution and a sealed glass tube filled with lead shot is laid along the length of the trough. The developed chromatogram is slipped underneath the tube and quickly drawn from end to end through the solution. It is then hung up and air-dried or heated in an oven according to the directions for the particular reaction. Unbreakable metal or polyethylene dip trays can also be employed.

A device for rapidly and uniformly applying controlled amounts of reagent to papers has been designed (*166*). It consists of two porous rollers moving in opposite directions which are continuously moistened with the reagent. Excess reagent is removed from these rollers by two other collecting rollers.

2. *Spraying*

For the application of reagents by spraying, glass atomizers may be used. Atomizers can be made with only moderate skill in glassblowing

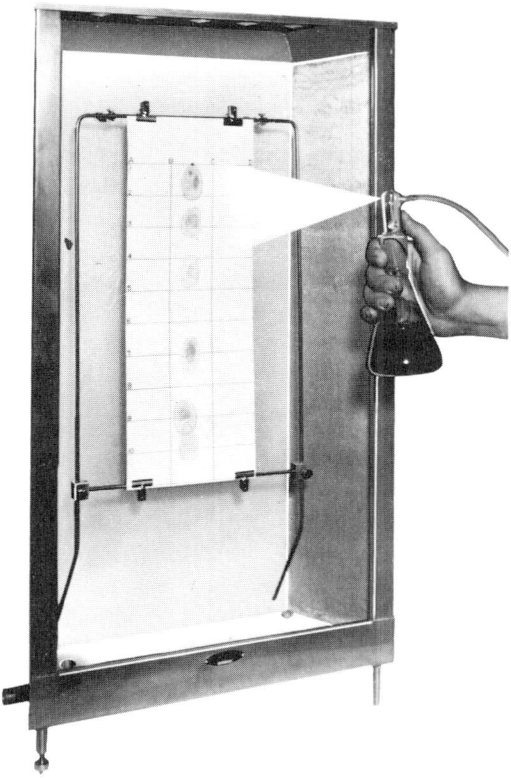

FIG. 23. Water-curtain enclosure for spraying chromatograms. (Courtesy of Microchemical Specialties Co., Berkeley, Calif.)

(*167*) or are available commercially. A commercial 250 ml atomizer is pictured in Fig. 23. A source of compressed air is necessary for operating these atomizers. DeVilbiss atomizers, which are available in most drug stores and do not require compressed air, are suitable for noncorrosive spray reagents. Other sprayers which have been described recently include one giving an extremely fine spray (*168*), one in which the angle of the jet and the size of the drops produced are adjustable (*169*), and an inexpensive sprayer constructed from a side-arm test tube and the spray head from an aerosol can (*170*).

Aerosol spray guns are also commercially available. These comprise a glass bottle holding about 120 ml of reagent, a replaceable propellant unit (usually liquid dichlorodifluoromethane) and a plastic connecting bridge incorporating a release button and a spray nozzle (Fig. 24).

FIG. 24. Chromatographic spray gun. (Courtesy of G. Welch, Shandon Scientific Co., Inc., Sewickley, Pa.)

When spraying for long periods, the propellant container gets very cold and the spray becomes erratic. It is best to alternate several containers under these conditions. Self-contained aerosol bombs containing the reagent mixed with propellant, which were first recommended by Zweig (*171*), are now commercially available for a large variety of reagents.

Chromatograms are sprayed in a chemical hood or a water-curtain enclosure (*172*). The latter, as illustrated in Fig. 23, is constructed of stainless steel, and a running water curtain in the back of the chromatogram washes away the excess spraying reagent. The enclosure is particularly recommended for reagents which have dangerous or noxious residues (such as perchloric acid), and it does not require the use of a chemical hood. The chromatogram is held on a suitable rack by means of chromatographic stainless steel clips. The atomizer is held at a distance of 12–15 inches from the paper and is moved in a slow, but even, left-to-right movement, starting at the top of the paper and working downward. Care must be exercised not to overload the paper with spraying reagent since the spots tend to diffuse and migrate, especially when aqueous reagents are employed.

3. *Other Methods*

As an alternative to either dipping or spraying, the color reagent can be incorporated into the developing solvent, as has been done, for example, with isatin for the detection of imino acids (*172a*). A detection method of "differential charring" may have limited applications for substances for which no suitable spraying reagent is available (*173*). The chromatogram, which must be developed with a solvent that is completely volatile, is heated over a hot plate until the paper gradually begins to darken. Regions in which substances are present tend to darken more rapidly than the rest of the paper, so that the spots appear brown-black on a light brown background.

B. Physical Detection Methods

Many organic substances absorb ultraviolet light between 240–260 mμ and so show up as dark areas which can be circled with a pencil when illuminated with such light. In some cases, the sensitivity of the method is improved if the paper is sprayed with fluorescein in order to increase the fluorescence of the background. Contact printing on blueprint paper may be used to record spots which absorb ultraviolet light. The chromatogram is laid on blueprint paper and held in place with paper clips or plastic rulers (*174*). A quartz germicidal lamp which fits into a fluorescent desk lamp fixture is used for exposure which may last from 1 to 3 minutes. The exact time must be determined experimentally for each procedure. The blueprint paper is then washed with running tap water, and white spots on a blue background indicate the presence of ultraviolet-absorbing compounds. This method is especially useful for purines, pyrimidines, and certain organophosphorus insecticides.

Many organic substances fluoresce upon illumination with ultraviolet light around 360 mμ. With some compounds, it is necessary to heat (*175*) or cool [to liquid-nitrogen temperatures (*176*, *177*)] the chromatogram in order for the fluorescence to occur. Radioactive compounds containing ^{14}C, ^{45}Ca, ^{36}Cl, ^{32}P, ^{60}Co, ^{131}I, or ^{141}Ce can be detected by spraying with 1 ml of a liquid scintillator solution (e.g., 2,5-diphenyloxazole), air drying and examining under ultraviolet light for fluorescence (*177a*).

Certain compounds adsorbed on cellulose exhibit phosphorescence at room temperature when illuminated with 254 mμ light (*178*). Phosphorescence spectra are useful for the sensitive and selective detection of many organic compounds (*179*). Some compounds do not phosphoresce if present on paper as dry spots but do in the wet state. A compound's phosphorescence spectrum may change when its spot is moistened with different solvents. Such differences are helpful in characterizing these substances (*180*).

An assortment of long- and short-wave ultraviolet hand and bench lamps and a portable viewing cabinet designed for examining fluorescent chromatograms under darkened conditions in a brightly illuminated laboratory are available from Black Light Eastern, Westbury, Long Island, N. Y., and other commercial sources. A scanner for fast detection of ultraviolet-absorbing spots on paper chromatograms is also available (Microchemical Specialties Co., Berkeley, Calif.). This instrument scans a 200 mm paper strip in less than 1 minute and punches a small hole at the exact location of the peak of the compound.

X-Ray fluorescence methods permit easy, reliable, and selective detection of chlorine, sulfur, bromine, molybdenum, and other elements in paper chromatograms with detection limits between 0.3 and 10^{-6} gm of the element. With appropriate calibration, semiquantitative determination of these elements is possible. The blank values and the signal/substance quantity ratio depends strongly on the stationary phase, and careful handling and drying of the chromatograms over P_2O_5 before evaluation are absolutely necessary (*180a*).

C. Detection of Radioactive Compounds

The positions of radioactive spots can be determined by autoradiography which will be discussed in detail in Chapter 4. A simple technique for certain isotopes (e.g., alkali metals) is to hand scan the paper with a portable battery-powered radiation survey meter, the detection tube of which is encased in a lead sheet with a narrow slit (*181*). Mechanical scanners are also available which graphically record peaks of radioactivity corresponding to the location of each zone (*181a*).

D. Biological Detection Methods

Bioautographic methods of detection, based on the inhibition of growth of certain organisms, are described in Chapter 13.

VIII. Qualitative Evaluation, Preservation, and Storage of Chromatograms

Various simple devices have been suggested for the rapid measurement of R_F values as defined in Chapter 2. One such aid is a rubber band divided into ten or one hundred equal parts with indelible ink. The rubber band is stretched from the origin (0) to the solvent front (10 or 100), and the R_F values are read off directly where the spots intersect the rubber band. Berbalk's sliding-wire potentiometer and other scales which are mechanically adjustable triangles have been described and reviewed by Barrett (*182*).

A clear cellophane sheet is used for the direct reading of R_F values as shown in Fig. 25 (*183*). A master sheet is constructed as shown in the figure. The whole sheet may be used, and a 250 mm ordinate can be subdivided into 2.5 mm per 0.01 R_F value. Distances in units of 0.10 R_F

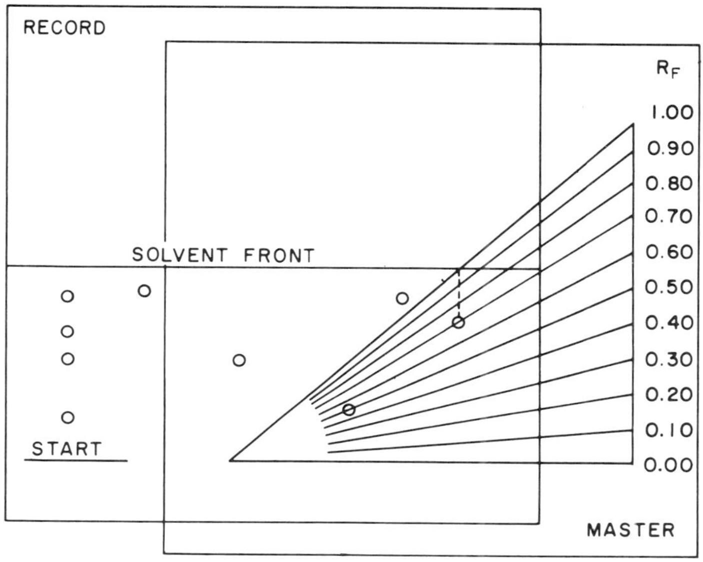

FIG. 25. Use of master cellophane sheet to read R_F values. [After (*183*).]

values are marked off every 25 mm and partially connected to the origin, as shown. In practice, the spaces are further divided into 0.05 R_F units (not shown) by using a different color ink. Thus, for a solvent front that has advanced 15 cm, the difference between 0.01 R_F values is 1.5 mm and the R_F value of a spot can easily be estimated within 0.01 unit. For the actual reading of the R_F value of a spot, the master sheet is placed under the record sheet. The starting positions are aligned and the sheets are moved until the solvent front and the top line ($R_F = 1.00$) intersect directly above the spot. The R_F value is read directly without calculations from the master sheet. This device is very similar to the R_F Measure described by Barrett (*184*). Another similar transparent scale is the Partogrid (*185*), a modification (*186*) of which is applicable to longer distances of solvent flow.

Cellophane sheets are also convenient for recording the results of chromatography (*183*). A sheet of cellophane is superimposed on the developed chromatogram. The outlines of the developed spots are traced with a ballpoint pen. Inks of different colors can be used to mark colors, solvent front, or origin. The sheets are stored in a looseleaf binder, and results can be compared easily by superimposing sheets on one another or on freshly run chromatograms.

Chromatographic results can also be recorded by making Xerox copies or black-and-white or color photographs (*187*) of the actual chromatograms. Ultraviolet photocopies of paper chromatograms of steroids, purines and pyrimidines, and proteins and peptides containing tyrosine or tryptophan may be prepared in lighted rooms if reversal-process photocopying equipment is available. The outline of the spots revealed on the photocopy can be transferred to the original chromatogram by tracing, after which the spot can be eluted for ultraviolet spectrophotometry (*188*).

Alternatively, the chromatograms themselves may be retained. Since the colors on chromatograms tend to fade in time (e.g., ninhydrin color of amino acids), it may be necessary to take steps to ensure their preservation. One way of doing this is to trace the original chromatogram or outline the spots with a soft pencil. Zweig has recommended a copper nitrate–HNO_3 dip, followed by a plastic spray for the amino acids (see Chapter 5). Storage of sprayed (e.g., with diphenylamine urea) papers in the dark can often delay discoloration and increase the useful life for comparative purposes (*189*).

The physical storage of chromatographic sheets is neatly accomplished in a Chromatofile (Applied Science Labs, State College, Pa.). Up to 2000 chromatograms are held by a series of binder clips on brackets in a wall-mounted or mobile file unit.

IX. Identification of Chromatographic Spots

As was explained in Chapter 2, R_F values are not absolute constants and may not be used by themselves for the identification of an unknown compound. Co-chromatography of the unknown in several diverse systems with an authentic sample of the suspected compound can offer tentative identification.

Chromatography used in conjunction with additional confirmatory tests is the best way in which to make identifications. The use of selective color reagents (e.g., dimethylglyoxime for nickel) may be useful, as is the recording of visible, ultraviolet, and infrared spectra after elution of the separated zones from the paper. These spectra are then compared with collections of spectra of authentic compounds. Phosphorescence spectra (see above) and ultraviolet spectra (*190*) can be recorded for substances directly on the paper, although these may not be identical to the spectra obtained for the same substances in solution.

Chemical reactions performed directly on the paper before one-way chromatography or between developments with the same solvent in two-dimensional chromatography can indicate how a substance reacts with various reagents under various conditions and help confirm the identity of the substance (*3; 191*). The later technique, which is termed diagonal chromatography, is also useful for the detection of chromatographic artifacts (Fig. 4, Chapter 2) and for increasing the resolution of the components of the sample. Substances which do not undergo chemical change are situated on the diagonal of the chromatogram after the second run. Interpretation of such results must be carefully made since the diagonal line of spots may not be exactly straight if the conditions differ for the two runs, e.g., the paper may be wetter or drier, or a spot may be displaced by the presence of other sample components in the first dimension but not the second.

A coincidence technique combining ultraviolet absorption photography and autoradiography (see Chapter 4) is useful for the identification of purines, pyrimidines, amino acids, and sugars (*192*).

Substances may also be identified by systematic analysis. In this procedure, R_F values are obtained for the unknown in a series of solvents under carefully controlled conditions. Its chromatographic spectrum is then prepared and identification made by comparison to the spectra of reference compounds. This method is described in detail for antibiotics in Chapter 13. It has also been used for identifying and classifying alkaloids, synthetic dyes, inorganic cations and anions, aromatics, and steroids (*193*). Comparison of data is aided by use of data processing techniques (*194*).

An especially extensive study of the separation and identification of 161 drugs was made (*195*) by (a) separating the drugs into three groups by extraction and ion-exchange procedures, (b) fractionating the groups by paper chromatography in several solvent systems, and (c) identifying the individual drugs by chromatographic comparison with reference substances and the use of several different chemical detection reagents. In addition, correlation studies between chromatographic spectra and chemical structure, and their significance for identification have been made for 1100 bio-organic substances (*195a*).

It will be noted from the above discussion that known authentic reference compounds are usually necessary for identification work. Extensive collections of pure analytical standards of important classes of compounds (e.g., amino acids, pesticides, sugars) are now commercially available in kits (sources include Shandon Scientific Co.; Poly-Science Corp., Evanston, Ill.; and Regis Chemical Co., Chicago, Ill.) for use in identifying compounds, for designing new separation systems, and for quantitative analysis.

A method termed stereochromatography (*196*) has been used for identification and preparative work (see below). The sample is injected into the center of the base of a 25 × 20 × 7.5 cm block of compressed paper pulp. The block is placed on the bottom of a solvent-containing jar and developed until the solvent almost reaches the top of the block. The wet block is then removed and sectioned serially by an electric band saw into 0.5 cm sections. The vertical migration and lateral spread of the solute zones relative to the vertical migration of the solvent (R_F and R_S values) are determined on each section. The R_S value is an additional parameter claimed to be useful for identifying substances.

An overview of methods useful for the identification of substances by paper (and thin-layer) chromatography has been presented by Macek as an introduction to volume 48 of the *Journal of Chromatography* (April, 1970), in which papers presented at the International Symposium on this topic held in Italy in September, 1969 are published. Included are identification methods based on mobility, detection, derivative formation before or during chromatography, decomposition and degradation products, and the combination of paper chromatography with gas chromatography, thin-layer chromatography or other physicochemical methods such as spectroscopy and mass spectrometry.

X. Preparative Paper Chromatography

The recommended procedure for isolating milligram quantities of pure substances from a mixture is scaled-up descending chromatography using high-capacity paper sheets. As much as 0.5 ml of a solution containing the sample is streaked or applied as adjacent spots on Whatman No. 3 MM paper sheets (18 $\times$ 22 inch using repeated applications from a 100 μl pipette). A commercial sample applicator designed for preparative chromatography simplifies the spotting procedure. The streaking pipette (Fig. 26) has a horizontal reservoir of about 0.3 ml and empties only when it touches the paper surface. The sample applicator is a device which holds the streaking pipette in a vertical position where it may be drawn across the paper along a track in a straight line. The paper sheet is held in place by spring clamps, and the solution is dried simultaneously by a stream of warm air. Adjustable stops at each end lift the pipette off the paper to prevent accidental leakage.

After the sample has been applied to several sheets of paper, development in a large-size cabinet proceeds according to directions given in Section VI,A. At the end of the run, after the papers have been dried, guide strips or peel strips (*197*) are developed with suitable color reagents. When one is trying to isolate milligram quantities of a compound, standards should be included in the guide strips and chromatographed along with the sample. Horizontal bands corresponding

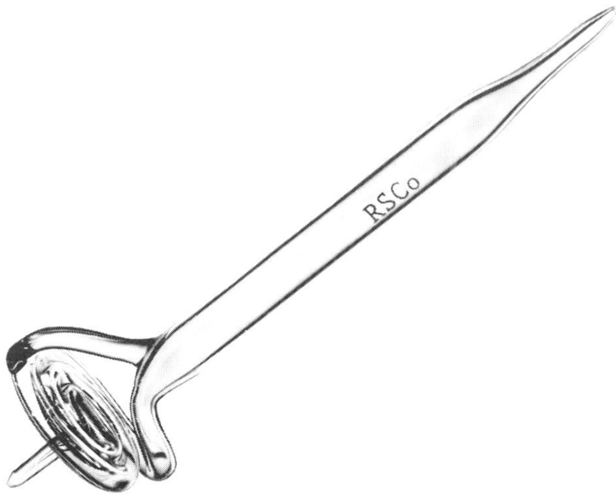

FIG. 26. Streaking pipette for preparative chromatography. [Courtesy of Research Specialties Co. (now American Optical Corp., Analytical Insts. Div., Richmond, Cal.)]

to the markers are cut from the main body of the chromatogram and are eluted by one of the techniques described in Chapter 4. Amounts of amino acids as great as 1.17 gm have been separated from a high-salt medium by four or more successive 15 to 20 hour solvent developments with intermittent drying (*198*).

Using heavier filter paper (about 1.5 mm thick) and descending development, a strip of thin paper is attached above the origin and folded into the trough. In procedure (*199*), a stirrup of Whatman No. 3 MM paper is sewn lengthwise with a sewing machine to the thick paper. In addition, two wicks made from the same paper are also attached. A glass rod through the stirrup serves as support, and the chromatograms are developed in a slightly modified chromatographic cabinet as pictured in Fig. 6. The modification consists of a solvent reservoir fashioned from an inverted round-bottom, long-neck flask resting on a specially built shelf.

In addition to free paper sheets, paper disks tightly compressed into a column, paper pressed into a block, and paper wound into a roll (*200*) can be used for descending preparative chromatography. A paper cylinder rotating at right angles to the solvent flow and fed with sample at a constant rate is used for continuous preparative paper chromatography (*3*). See Block *et al.* (*4*) for detailed descriptions of some of these methods.

References

1. Bibliography of Paper and Thin-Layer Chromatography, 1961–1965, *J. Chromatog.*, Suppl. Vol. (1968).
2. E. von Arx and R. Neher, *Chromatog. Rev.* **5**, 46 (1963).
3. K. Macek, *in* "Chromatography (E. Heftmann, ed.), pp. 139–164. Reinhold, New York, 1967.
3a. H. V. Street, *J. Chromatog.* **41**, 358 (1969); **48**, 238 (1970).
4. R. J. Block, E. L. Durrum, and G. Zweig, "A Manual of Paper Chromatography and Paper Electrophoresis," 2nd rev. ed. Academic Press, New York, 1958.
5. R. E. Bailey, H. P. Pieters, and J. H. Beck, *Nature* **204**, 485 (1964).
6. L. I. Woolf, *J. Chromatog.* **2**, 431 (1959).
7. K. Hruška, *J. Chromatog.* **21**, 350 (1966).
8. C. B. Mankinen and D. Fischer, *J. Chromatog.* **36**, 112 (1968).
9. J. F. Goggins and J. M. Tanzer, *Anal. Chim. Acta* **45**, 526 (1969).
10. J. H. Wisse, H. deKlonia, and B. J. Visser, *Rec. Trav. Chim.* **85**, 865 (1966).
11. J. Sherma and G. Zweig, *J. Chromatog.* **31**, 439 (1967).
12. D. B. Mullick and V. C. Brink, *Chemist-Analyst* **56**, 70 (1967).
13. E. Stahl and E. Dumont, *J. Chromatog.* **39**, 157 (1969).
14. M. F. Bacon, *J. Chromatog.* **16**, 552 (1964); *Chem. & Ind.* (*London*) p. 1692 (1965).
15. V. C. Weaver, *Advan. Chromatog.* **7**, 87 (1968).
16. H. Reeve Angel & Co., Clifton, New Jersey, Techn. Bull. 1012 and C1.
17. F. A. Vega, M. Fernandez, and C. Martin, *Oesterr. Apotheker Ztg.* **21**, 585 (1967).
18. W. E. Lingren, G. Reeck, and R. Marson, *J. Chromatog.* **24**, 269 (1966).
19. K. Macek and I. M. Hais, *in* "Paper Chromatography" (I. M. Hais and K. Macek, eds.), pp. 96–169. Academic Press, New York, 1963.
20. K. Macek and M. Wolf, *J. Chromatog.* **19**, 170 (1965).
21. E. Buras and S. R. Hobart, *Anal. Chem.* **27**, 1507 (1955).
22. F. J. Ritter and J. Hartel, *Nature* **181**, 765 (1958).
23. T. Wieland and A. Berg, *Angew. Chem.* **64**, 418 (1952).
24. J. R. A. Anderson, S. Dilli, J. L. Garnett, and E. C. Martin, *Nature* **201**, 772 (1964).
25. T. I. Přistoupil, *J. Chromatog.* **28**, 124 (1967); *Chem. Listy* **63**, **511** (1969).
26. A. M. Asatoor, *J. Chromatog.* **4**, 144 (1960).
27. J. Schlenk, H. K. Mangold, J. L. Gellerman, W. E. Link, R. A. Morrissette, R. T. Holman, and H. Hayes, *J. Am. Oil. Chemists' Soc.* **37**, 547 (1960).
28. J. A. Labarrere, J. R. Chipault, and W. O. Lundberg, *Anal. Chem.* **30**, 1466 (1958).
29. D. J. Nazir and N. G. Magar, *Biochem. J.* **90**, 268 (1964).
30. U. A. T. Brinkman, *Chem. Tech. Rev.* **21**, 529 (1966).
31. D. Thirkell and R. E. Wuthier, *in* W. and R. Balston Ltd., Maidstone, Kent, England, Tech. Bull. C4, p. 8.
32. Y. Hashimoto, *Japan Analyst* **17**, 1337 (1968).
33. R. C. Longland, *in* W. and R. Balston Ltd., Maidstone, Kent, England, Tech. Bull. C4, p. 12.
34. C. Michalec, Z. Kolman, M. Sulc, and J. Mestan, *J. Chromatog.* **9**, 237 (1962).
35. V. C. Weaver, *J. Chromatog.* **33**, 49 (1968).
36. C. Michalec, *in* "The Stationary Phase in Paper and Thin-Layer Chromatography" (K. Macek and I. M. Hais, eds.), p. 202. Elsevier, Amsterdam, 1965.
37. V. C. Weaver, *in* "Chromatographic and Electrophoretic Techniques" (I. Smith, ed.), 2nd ed., Heinemann, London, 1968.
38. C. S. Knight, *Nature* **194**, 90 (1962).
39. G. V. Marinetti, *J. Lipid Res.* **3**, 1 (1962).

40. K. R. Butterworth, *Brit. J. Pharmacol.* **21**, 378 (1963).
41. M. Roberts, *J. Pharm. Pharmacol.* **14**, 747 (1962).
42. M. Shifrine and G. Zweig, *J. Chromatog.* **14**, 297 (1964).
43. C. S. Knight, *J. Chromatog.* **8**, 205 (1962); *Chromatog. Rev.* **4**, 69 (1962); *Nature* **184**, 1486 (1959).
44. S. David and H. Hirshfield, *Bull. Soc. Chim. France* p. 1011 (1963).
45. E. Addison and R. G. Clark, *J. Pharm. Pharmacol.* **15**, 268 (1963); M. Lederer, *J. Chromatog.* **29**, 306 (1967).
46. P. C. Stein, *Anal. Chem.* **34**, 352 (1962).
47. G. Bagliano, G. Grassini, M. Lederer, and L. Ossicini, *J. Chromatog.* **14**, 238 (1964).
48. C. S. Knight, *Nature* **183**, 165 (1959).
49. R. O. Mumma and A. A. Benson, *Biochem. Biophys. Res. Commun.* **5**, 422 (1961).
50. British Food Manufacturing Research Association, Report of the Trace Materials Committee, *Analyst* **88**, 864 (1963); M. Lederer and L. Ossicini, *J. Chromatog.* **13**, 188 (1964).
51. J. B. Roberts, *J. Inst. Brewing* **68**, 302 (1962).
52. P. J. Peterson and G. W. Butler, *J. Chromatog* **8**, 70 (1962).
53. C. C. Curtain, *Nature*, **191**, 1269 (1961).
54. N. B. Furlong, *Anal. Biochem.* **5**, 515 (1963).
55. J. R. Sherman, *Anal. Biochem.* **5**, 548 (1963).
56. W. G. Ng, W. R. Bergren, and G. N. Donnell, *Clin. Chim. Acta* **15**, 489 (1967).
57. H. K. Sleeman and J. W. Diggs, *Anal. Biochem.* **8**, 532 (1964).
58. J. Hirtz, A. Hatchadourian, and P. Vassort, *Ann. Pharm. Franc.* **26**, 717 (1968); M. Alexos and C. Péaud-Lenoel, *Bull. Soc. Chim. Biol.* **51**, 261 (1969).
59. H. V. Street, *J. Chromatog.* **7**, 64 (1962); *J. Pharm. Pharmacol.* **14**, 56 (1962).
60. H. V. Street and S. K. Niyogi, *J. Pharm. Sci.* **51**, 666 (1962).
61. J. Pawlaczyk, *Bull. Soc. Sci. Lettres Lodz, Classe III* **11**, 91 (1962).
62. N. F. Kember and A. Farmer, *J. Cromatog.* **10**, 106 (1963).
63. J. B. Mooney, *Anal. Chem.* **34**, 1506 (1962).
64. J. M. Loewenstein, F. S. Greenspan, and P. Spilker, *J. Lab. Clin. Med.* **62**, 831 (1963).
65. G. Serlupi-Crescenzi, C. Paolini, and T. Leggio, *Ann. Chim. (Rome)* **57**, 438 (1967).
66. J. Sherma and K. M. Rich, *J. Chromatog.* **26**, 327 (1967).
67. H. Reeve Angel & Co., Clifton, New Jersey, Resin-Loaded Papers Tech. Bull.
68. J. Sherma, *Talanta* **11**, 1373 (1964).
69. T. Yoshino and M. Sugihara, *Kagaku To Kogyo (Osaka)* **40**, 471 (1966).
70. J. Sherma and L. H. Pignolet, *Anal. Chim. Acta* **34**, 185 (1966).
71. J. Sherma, *Anal. Chem.* **39**, 1497 (1967).
72. M. Lederer and L. Ossicini, *J. Chromatog.* **13**, 188 (1964).
73. H. R. Roberts and M. G. Kolor, *Anal. Chem.* **31**, 565 (1959).
74. M. Smillie, *J. Chromatog.* **4**, 494 (1960).
74a. E. Cerrai, G. Ghersini, M. Lederer, and M. Mazzei, *J. Chromatog.* **44**, 161 (1969).
75. G. Alberti, *Chromatog. Rev.* **8**, 246–259 (1966).
76. G. Alberti, M. A. Massucci, and E. Torracca, *J. Chromatog.* **30**, 579 (1967).
77. C. A. Antonopoulos, E. Borelius, S. Gardell, B. Hamnstrom, and J. E. Scott, *Biochim. Biophys. Acta* **54**, 213 (1961).
78. G. Jayme and H. Knolle, *Angew. Chem.* **68**, 243 (1956); *Z. Anal. Chem.* **178**, 84 (1960).
79. C. Michalec, *J. Lipid Res.* **4**, 110 (1963).
80. J. G. Hamilton and J. W. Dieckert, *Arch. Biochem. Biophys.* **82**, 203 (1959).

81. L. H. Klemm, E. P. Antaniades, G. Capp, E. Chiang, and E. Y. K. Mak, *J. Chromatog.* **6**, 420 (1961).
82. F. C. Haer, "An Introduction to Chromatography on Impregnated Glass Fiber." Ann Arbor-Humphrey Sci. Publ., Ann Arbor, Michigan, 1969.
83. R. W. Stoddart and D. H. Northcote, *Biochem. J.* **105**, 61 (1967); Anonymous, "Matters of Fact," Vol. 3, No. 2, p. 5. H. Reeve Angel & Co., Clifton, New Jersey, (1969).
84. G. Jayme and G. Hahn, *Angew. Chem.* **72**, 520 (1960).
85. S. Kruger, *J. Chromatog.* **42**, 148 (1969); J. J. Lech, *ibid.* p. 136.
86. I. Smith, "Chromatographic and Electrophoretic Techniques." Heinemann, London, 1960.
87. E. W. Berg, "Physical and Chemical Methods of Separation." McGraw-Hill, New York, 1963, p. 97.
88. I. Smith and J. G. Feinberg, "Paper and Thin-Layer Chromatography—A Teaching Level Manual." Shandon Sci. Co., Ltd., London, 1965. pp. 18-19.
89. K. A. Connors, *Anal. Chem.* **37**, 261 (1965); C. R. Perisho, *ibid.* **40**, 551 (1968).
90. A. V. DeThomas and F. Pascual, *J. Chem. Educ.* **45**, 303 (1968).
91. A. K. Huggins and V. Moses, *Nature* **191**, 668 (1961).
92. J. Opienska-Blauth, M. Charezinski, M. Sanecka, and H. Brzuszkiewicz, *J. Chromatog.* **7**, 321 (1962).
93. H. Amato, *Kim. Sanayi* **11**, 86 (1963).
94. A. J. Tomisek and P. W. Allan, *J. Chromatog.* **14**, 232 (1964); **33**, 35 (1968).
95. Z. Procházka, *in* "Chromatographic Methods" (O. Mikes, ed.), p. 105. Van Nostrand, Princeton, New Jersey, 1961.
96. A. Zaffaroni, R. B. Burton, and E. H. Keutmann, *Science* **111**, 6 (1950).
97. J. Churáček, *J. Chromatog.* **33**, 45 (1968).
98. T. Beyrich and H. Poser, *J. Chromatog.* **32**, 87 (1968).
99. I. M. Hais, *in* "The Stationary Phase in Paper and Thin-Layer Chromatography" (K. Macek and I. M. Hais, eds.), p. 291. Elsevier, Amsterdam, 1965.
99a. J. Gasparič, *J. Chromatog.* **47**, 51 (1970).
100. R. Stock and C. B. F. Rice, "Chromatographic Methods," 2nd ed. Chapman & Hall, London, 1967.
101. R. Consden, A. H. Gordon, and A. J. P. Martin, *Biochem. J.* **38**, 224 (1944).
102. B. L. Van Duuren, *Anal. Chem.* **32**, 732 (1960).
103. L. O'Donnell, *School Sci. Rev.* **48**, 190 (1966).
104. L. M. Reineke, *Anal Chem.* **28**, 1853 (1956); D. K. Fukushima, H. L. Bradlow, L. Hellman, B. Zumoff, and T. F. Gallagher, *J. Biol. Chem.* **235**, 2246 (1960).
105. H. L. Kimball, N. L. McNiven, and S. Burstein, *J. Chromatog.* **21**, 352 (1966).
106. R. DeWachter, *J. Chromatog.* **36**, 109 (1968).
107. V. N. Chernov and V. M. Bereznikov, *Lab. Delo* p. 503 (1966); T. F. Brodasky and D. Griffith, *J. Chromatog.* **41**, 494 (1969).
108. D. W. Vomhoff and T. C. Tucker, *Chemist-Analyst* **53**, 121 (1964); M. W. Weg, *J. Chromatog.* **28**, 484 (1967); Z. Kučera, *Chem. Zvesti* **22**, 304 (1968).
109. R. H. Powell, Shandon Sci. Co., British Patent 1,029,313 (1966).
110. R. J. Block, *Anal. Chem.* **22**, 1327 (1950).
111. L. C. Mitchell, *J. Assoc. Offic. Agr. Chemists* **40**, 999 (1957).
112. J. Barrollier, *Naturwissenschaften* **42**, 486 (1955).
113. N. Harding, *J. Chromatog.* **24**, 482 (1966).
114. J. Green and S. Marcinkiewicz, *J. Chromatog.* **10**, 35 (1963).
115. H. R. Roberts, *Anal. Chem.* **29**, 1443 and 1447 (1957).

116. H. R. Roberts and M. G. Kolor, *Nature* **183**, 460 (1959).
117. L. Peyron, *Bull. Soc. Chim. France* p. 1494 (1966).
118. L. Rutter, *Nature* **161**, 435 (1948).
119. K. V. Giri, K. Krishnamurthy, and T. A. Venkatasubramanian, *Current Sci. (India)* **21**, 11 (1952).
120. E. Kawerau, *Chromatog. Methods* **1**, No. 2, 7 (1956).
121. S. Berlingozzi and G. Serchi, *Sperimentale, Sez. Chim. Biol.* **3**, 1 (1952).
122. G. DeVries, M. J. Hardonk, and K. H. J. Buschow, *Anal. Chim. Acta* **21**, 568 (1959).
123. I. Irimescu, L. Cociumian, and S. M. Idu, *Z. Med. Labortech.* **8**, 85 (1967).
124. K. Krishnamurthy and M. Swaminathan, *Anal. Chem.* **27**, 1396 (1955).
125. S. Kitahara and H. Hiyama, *Sci. Ind. (Japan)* **28**, 341 (1954).
126. A. Kapri, *Studii Cercetari Biochim.* **8**, 343 (1965).
127. J. Jankulov, *Compt. Rend. Acad. Bulgare Sci.* **17**, 183 (1964).
128. Y. Osawa, *Nature* **180**, 705 (1957).
129. S. Szczepaniak and I. Krzeczkowska, *Chem. Anal. (Warsaw)* **11**, 1197 (1966).
130. A. J. Philippu, *Nature* **182**, 1159 (1958).
131. A. V. DeThomas and F. Pascual, *J. Chem. Educ.* **46**, 319 (1969).
132. M. G. E. El-Hehyawi, *J. Vet. Sci. U.A.R.* **4**, 85 (1967).
133. T. Kwasik and H. Ladzinska-Kulinska, *Zeszyty Nauk. Politech. Lodz., Chem.* **10**, 63 (1961).
134. E. Soczewiński and A. Waksmundzki, *Ann. Univ. Mariae Curie-Sklodowska, Lublin-Polonia* **AA16**, 117 (1961).
135. A. Jeanes, C. S. Wise, and R. J. Dimler, *Anal. Chem.* **23**, 415 (1951).
136. H. Fishbach and J. Levine, *J. Sci.* **121**, 602 (1955); W. Niemierko and M. Krzyzanowska, *J. Chromatog.* **26**, 424 (1967).
137. H. C. Chakrabortty and D. P. Burma, *Anal. Chim. Acta* **15**, 451 (1956).
138. R. Rüdiger and H. Rüdiger, *J. Chromatog.* **17**, 186 (1965).
139. J. A. Thoma, *J. Chromatog.* **12**, 441 (1963); *Advan. Chromatog.* **6**, 106 (1968).
140. K. Macek and J. Večerková, *Pharmazie* **20**, 605 (1965).
141. J. F. Thompson *et al.*, *Botan. Rev.* **25**, 1 (1959).
142. P. Decker, W. Riffart, and G. Oberneder, *Naturwissenschaften* **38**, 288 (1951).
143. P. Decker and K. Kikodse, *Z. Anal. Chem.* **165**, 337 (1959).
144. D. French, A. P. Pulley, M. Abdullah, and J. C. Linden, *J. Chromatog.* **24**, 271 (1966).
145. A. Letters and B. Brown, *Biochim. Biophys. Acta* **116**, 482 (1966).
146. D. G. Irvine and M. E. Anderson, *J. Chromatog.* **20**, 541 (1965); D. G. Irvine, W. Bayne, and J. R. Majer, *J. Chromatog.* **48**, 334 (1970).
147. R. Lipton, T. D. Price, and M. M. Melicow, *J. Chromatog.* **26**, 412 (1967).
148. K.-Y. Yao and H. Chen, *Hua Hsueh Tung Pao* p. 302 (1964).
149. H. J. McDonald, L. P. Ribeiro, and J. Banaszak, *Anal. Chem.* **31**, 825 (1959).
150. H. J. McDonald, L. V. McKendell, and E. W. Bermes, Jr., *J. Chromatog.* **1**, 259 (1958).
151. H. J. McDonald, E. W. Bermes, Jr., and H. G. Shepherd, Jr., *Naturwissenschaften* **44**, 9 (1957); *Chromatog. Methods* **2**, 1 (1957).
152. H. J. McDonald, L. J. Banaszak, and J. Q. Kissane, *Anal. Biochem.* **1**, 44 (1960).
153. K. N. Shlyagin, U.S.S.R. Patent 177,146 (1965).
154. H. J. McDonald and L. V. McKendell, *Naturwissenschaften* **44**, 615 (1957).
155. A. Becker, *Z. Anal. Chem.* **174**, 161 (1960).
156. H. V. Street, *Nature* **193**, 1298 (1962).
157. L. Ossicini and M. Lederer, *J. Chromatog.* **42**, 154 (1969).

158. C. Liteanu and S. Gocan, *Studia Univ. Babes-Bolyai, Ser. 1* **11**, 71 (1966).
159. C. Liteanu and S. Gocan, *Bull. Soc. Chim. France* p. 3836 (1967); p. 1409 and 1416 (1969).
159a. R. DeWachter, *J. Chromatog.* **36**, 109 (1968); F. Franks, *Analyst* **81**, 384 (1956); A. Meniga and N. Muic, *Anal. Biochem.* **1**, 502 (1960); M. Lederer, *Nature* **172**, 727 (1953).
159b. W. E. Lingren, W. E. Reeck, and R. J. Marson, *Anal. Chem.* **40**, 1585 (1968).
160. H. Weisz and G. Schepky, *Mikrochim. Acta* p. 717 (1965).
161. A. Biró, *Magy. Kem. Folyoirat* **65**, 245 (1959).
162. O.-E. Shultz and R. Comberg, *J. Chromatog.* **32**, 346 (1968).
162a. E. Soczewiński, *Advan. Chromatog.* **8**, 91 (1969).
163. E. Hecht, *Clin. Chim. Acta* **13**, 506 (1966).
164. S. Dittrich, *J. Chromatog.* **31**, 628 (1967).
165. L. Fishbein, *J. Chromatog.* **22**, 480 (1966).
166. K. Zwierz, Polish Patent 50,731 (1966).
167. W. J. Wingo, *Anal. Chem.* **25**, 1939 (1953).
168. R. E. Bailey, *Chemist–Analyst* **55**, 56 (1966).
169. R. G. Strickland, *Lab. Pract.* **13**, 523 (1964).
170. A. V. DeThomas and F. Pascual, *J. Chem. Educ.* **45**, 321 (1968).
171. G. Zweig, *Anal. Chem.* **28**, 428 (1956).
172. E. C. Fiebig and H. Siegel, *Anal. Chem.* **30**, 161 (1958).
172a. V. E. Shih and P. M. Madigan, *Clin. Chim. Acta* **24**, 481 (1969).
173. P. C. Caldwell, *Biochem. J.* **60**, xii (1955).
174. H. T. Gordon, *Science* **128**, 414 (1958).
175. G. Weidner and F. Weiss, *Pharmazie* **13**, 395 (1958).
176. A. Szent-Györgi, *Science* **126**, 751 (1957).
177. M. P. Gordon and D. South, *J. Chromatog.* **10**, 513 (1963).
177a. A. A. Noujaim and L. P. Richards, *Can. J. Pharm. Sci.* **3**, 45 (1968).
178. M. Roth, *J. Chromatog.* **30**, 276 (1968).
179. J. D. Pfaff and E. Sawicki, *Chemist-Analyst* **54**, 30 (1965).
180. E. Sawicki and J. D. Pfaff, *Anal. Chim. Acta* **32**, 521 (1965); E. Sawicki, *Talanta* **16**, 1231 (1969).
180a. W. Huber and H. Fricke, *Chromatographia* **3**, 121 (1970).
181. J. Sherma and H. H. Strain, *Anal. Chem.* **34**, 76 (1962).
181a. A. C. Kibrick *et al.*, *Anal. Biochem.* **25**, 40 (1968).
182. G. C. Barrett, *J. Chromatog.* **18**, 435 (1965).
183. O. T. Chortyk and I. Kushnir, *J. Chem. Educ.* **45**, 506 (1968).
184. G. C. Barrett, *British Patent Appl.* 19990/64 (1964).
185. L. B. Rockland and M. S. Dunn, *Science* **111**, 332 (1950).
186. R. M. Nettleton, Jr. and R. B. Mefferd, Jr., *Anal. Chem.* **24**, 1687 (1952).
187. J. Biro, *Herba Hung.* **5**, 275 (1966).
188. S. R. Lerner, *Anal. Biochem.* **26**, 201 (1968).
189. R. Cecil, M. Jack, and J. H. Mack, *Contrib. Boyce Thompson Inst.* **23**, 139 (1965).
190. F. Korte and H. Weitkamp, *Angew. Chem.* **71**, 455 (1959).
191. M. S. J. Dallas, *J. Chromatog.* **48**, 193 (1970); I. M. Hais, *ibid.* **48**, 200 (1970).
192. C. Ponnamperuma, P. Kirk, R. Mariner, and B. Tyson, *Nature* **202**, 393 (1964).
193. V. Betina, *Chromatog. Rev.* **7**, 119 (1965).
194. A. Franck, *Chemiker Ztg.*, **92**, 115 (1968).
195. K. Macek, J. Večerková, and J. Stanislavová, *Pharmazie* **20**, 605 (1965).
195a. L. Reio, *J. Chromatog.* **48**, 11 (1970).

196. C. J. Pfeiffer and J. R. Debro, *Nature* **203**, 1003 (1965).
197. D. J. Maclaurin, J. K. Crossman, and J. W. Green, *Chemist-Analyst* **54**, 118 (1965).
198. A. E. Pasieka and J. E. Logan, *Can. J. Biochem.* **44**, 149 (1966).
199. H. H. Brownell, J. G. Hamilton, and A. A. Casselman, *Anal. Chem.* **29**, 550 (1957).
200. L. B. Rockland, *J. Chromatog.* **16**, 547 (1964).

4

QUANTITATIVE PAPER CHROMATOGRAPHY

Discussed in this chapter will be methods, mostly instrumental, for making quantitative measurements of separated substances on or from chromatograms; for general reviews of this topic, see Jork (*1*) and Boulton (*2*). It will be shown that the production of uniform, reproducible initial spots that contain definite amounts of material is essential in order to obtain accurate results. After this is achieved and the solutes have been separated in a solvent giving zones that are as regular as possible, the solutes are determined by (a) removing the paper plus the spot with subsequent extraction from the paper and determination of the solute by spectrophotometry, measurement of radioactivity, etc., or (b) by direct determination of the solute on the paper by measurement of area, photometric or fluorometric methods, measurement of radioactivity, etc. If the compound of interest is neither radioactive nor originally colored, it must be revealed after separation by treatment with a reagent that produces a colored derivative which is estimated by one of the methods described below.

I. Formation of the Colored Derivative

An excellent discussion of this topic is contained in a recently published paper by Franglen (*3*). The considerations below are taken from this work. The accuracy of quantitative paper chromatography varies with the class of compound and the methods used. Amino acids can be estimated with an accuracy of 0.2–2% and sugars with an accuracy of about 3%, for example.

The concentration of the detection reagent must be high enough to react completely with the greatest concentration of the compound that is likely to occur. This means that more concentrated reagent solutions are needed for quantitative than for qualitative work. For amino acids, for example, a 1% solution of ninhydrin rather than the usual 0.1–0.2% solution is recommended. Use of the more concentrated solutions often leads to darker backgrounds as well, and since the background areas of the paper are used as the blanks, these must be kept as uniform as possible. Dipping the paper through the reagent dissolved in a volatile solvent in which the solute of interest is not soluble usually gives the best results. An apparatus designed to improve the dipping technique for amino acids has been designed (*4*). If sprayed, the detection reagent must be applied as evenly as possible.

The speed of the reaction forming the derivative will vary with different solutes and reagents. Rapid reactions are best to use if possible. For slow reactions, it is recommended that they be allowed to occur for a timed period at room temperature protected from light and laboratory atmosphere (e.g., the chromatogram is stored in a sealed plastic bag in the dark during the reaction period). Heating, if necessary, may increase the background color. The time allowed for the reaction to occur should be chosen so that a maximum difference between the color of the zone and the color of the background results. Zones that fade rapidly may be photographed at the height of their intensity and the quantitation made through photographs.

Plots relating the derivative concentration to the original compound concentration can be of three types. If the compound reacts completely, the color produced is in proportion to the amount of solute applied and the plot is linear with a positive slope. If the reaction is complete at low concentrations of solute but not at higher concentrations, the curve is initially linear but tends to level off. Nonlinear reactions can be used for accurate quantitative results as long as they are reproducible. In some cases an anomalous reaction occurs between the compound and reagent leading to a convex curve with a maximum. Such systems are not desirable for quantitative chromatography.

II. Estimation of the Spots

Once the colored derivative has been formed, the chromatogram is quantitated by direct or elution methods. In either case, the following conditions must be observed (*3*): (a) the volumes of the solutions used to form the initial spots must be accurately measured and the spots of uniform size; (b) appropriate standards are always analyzed along with the test solution; (c) the concentrations of the standard must bracket the concentration of the test solution; and (d) the treatment of the test solution and the standards must be in all ways identical.

A. Visual Estimation

One of the most sensitive photosensitive detectors is the human eye. By comparing the relative intensity of colored spots from standards and and unknowns on paper it may be possible to estimate. concentrations by visual comparison with an accuracy of about 20%. Various workers have claimed that optimum results are obtained when a series of standards of increasing concentration are spotted alternately with a series of dilutions of the unknown decreasing in concentration.

B. Measurement of Spot Area

This is also a visual method wherein the areas of the spots after chromatography are plotted against the logarithm of the amount of substance present in the spots (*5*). If a series of standard spots of the same initial size covering a narrow concentration range are employed, these quantities are linearly related and the concentration of the unknown can be read from a graph. Although perhaps obvious, it might be mentioned that two initial spots of the same size but with one containing twice as much solute as the other can be produced by spotting the same volume (e.g., 10 μl) of two solutions, one of which is twice as concentrated as the other, or by making two 10 μl spots of the more dilute solution and then placing another 10 μl of this solution on top of one of the spots after drying. Application of 10 μl and 20 μl portions of the more dilute solution, on the other hand, will not give spots of the same size.

Spot areas are usually determined by planimetry or by simply measuring the major and minor axes and taking their product as the area. If the zone edges are not sharp, it may help to make a copy of the chromatogram onto high contrast photographic paper (*3*).

The spot can also be photographed with a transparent drawing paper, containing a drawing of a millimeter grid on it, between the chromatogram and the film. The resultant photograph of the spot will have the grid superimposed on it, and the number of squares can be counted to evaluate the area (*6*). Colorless, ultraviolet-absorbing spots are detected and recorded by electrophotography on an office photocopier using copying paper impregnated with zinc oxide (*7*). Spot-area methods are generally accurate to about $\pm 10\%$.

This method, and other direct, comparative methods, require the use of standards which are as similar to the test solution as possible so that the distribution of the compound of interest in all the spots is the same. If the test solution contains impurities which could alter the zone distribution, the use of internal standards (the standard is mixed with the unknown solution so that each is influenced by the same conditions) may help.

If the sample is applied as a streak across the paper strip, the length of the developed zone is proportional to the concentration below a maximum level. This method was applied to antibiotics, sugars, amino acids, and organic acids (see reference *53* in Chapter 6).

C. Methods Involving Elution

The separated solutes can be eluted from the paper and then determined in solution by customary analytical procedures (usually spectrophotometry). A widely used technique consists of developing the color of a test or guide strip. This strip is then placed alongside the rest of the chromatogram, and the unstained regions containing the solutes are cut out with scissors. If the substance is naturally colored, is fluorescent, or absorbs ultraviolet light, the positions of the substances to be cut out can, of course, be located without guide strips. The paper sections are then eluted with a suitable solvent. The following recommended method yields elution of spots from paper in one or two drops of solution. The excised paper spot is stapled to two other strips of filter paper, one of which is shaped to a point (Fig. 1). This strip is folded into a supported trough and placed in a miniature chromatographic chambe ·.g., inverted beaker). It is then developed like a chromatogram by the descending method using a solvent which completely dissolves the substance to be eluted. The compound is thus eluted quantitatively in the first one or two drops of eluate. Such an apparatus for the simultaneous elution of several spots has been described by Skolik (*8*) (see also Chapter 6, Fig. 4).

Another convenient method in which the eluted substance is directly transferred to another chromatogram for further development with another solvent is illustrated in Fig. 2 (*9*). In this procedure the substance moves by ascending development and is directly deposited at the origin of another sheet of filter paper. The spot is kept small by directing a stream of warm air at the top side of the new chromatogram. A similar descending arrangement has also been described (*10*).

In another simple method, the paper containing the spot is attached to a syringe needle which is bent like a hook. When the syringe and paper are held vertically, the solution flows out of the syringe and elutes the substance (*11*). Micro-Soxhlet extractors, which are commercially available in several sizes, can also be used to remove the solute from the paper.

A rapid extraction method employing centrifugation is as follows (*11a*):

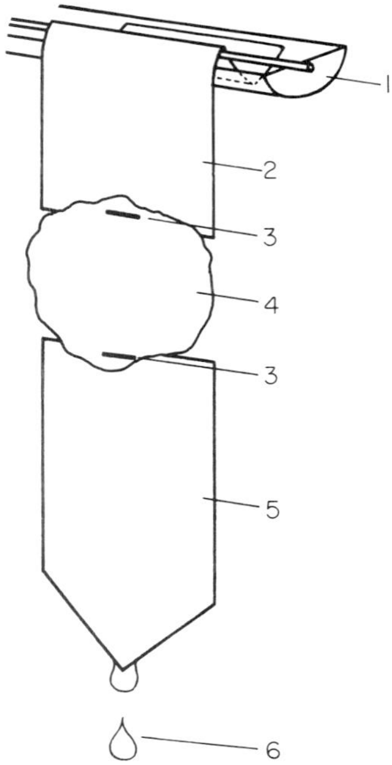

FIG. 1. Technique for elution by descending development.
(1) Solvent trough; (2,5) strips of filter paper; (3) staples; (6) eluted drop.

The section of paper to be extracted is rolled up, saturated with the solvent and placed in a plastic centrifuge tube with a hole in the bottom. The tube is put into a glass conical centrifuge tube, and, after allowing a few minutes for dissolution, the whole assembly is centrifuged. The extraction can be repeated if necessary.

A variation of the method described above is to stain the chromatogram with the detection reagent and then to elute the colored product and determine it by colorimetry. Since the guide strip is not required, this method is faster, and the position of the spots are known more accurately. All the precautions mentioned in Section I for the uniform, reproducible detection of the zones must be strictly followed. It is possible to include excess reagent in the eluent to be sure the compound has completely reacted (*3*).

Errors of 1–2% or below have been claimed for elution methods, although recent work (*12*) indicates that to obtain this level of accuracy, special attention must be paid to the spotting of the samples. The elution step should be ideally quantitative although excellent results can be

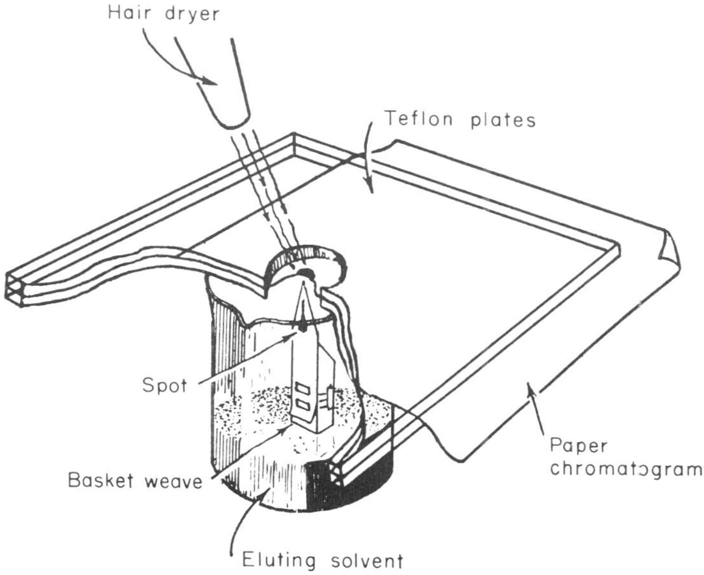

FIG. 2. Method of eluting a substance onto the origin of another chromatogram by ascending development. After (*9*).

obtained by comparison to standards even though the elution is nonquantitative but reproducible.

If cellulose fibers appear in the eluate, they can be removed prior to the analysis, if necessary, by filtering through a small bed of Celite. Unstable substances must be eluted and determined as quickly as possible. If there is a chance that impurities present in the paper (see Chapter 3) may be eluted along with the solute and interfere with the method of analysis chosen, the paper should be washed with the solvent to be used for the actual elution prior to chromatography. A disadvantage of the elution method compared with direct densitometry (below) is that larger quantities of the substances are needed to obtain accurate results (*3*).

D. Direct Photometric Measurement of the Spot on the Paper

1. *Instruments*

Densitometry involves the manual or automatic scanning of the chromatogram with filtered light (visible light up to 500 mμ is usually used, although most commercial instruments permit operation in the near infrared and ultraviolet as well) in order to measure the amount transmitted through or reflected from the paper. The simplest densitometer consists of a filtered light source, an adjustable aperture, a photocell, and a sensitive galvanometer. The chromatogram is placed between the light source and the photocell, and the light is measured with reference to a paper blank. A sensitive, low-cost ultraviolet strip scanner of this type has been designed (*13*). More sophisticated scanning systems record curves corresponding to the chromatographic zones, integrate the areas under the curves, and evaluate the areas by use of a digital computer (*14*).

Bush has described in detail the two principal pieces of equipment required for direct quantitative estimation of chromatograms—a device to treat paper automatically with chemical reagents (*15*) and a photometric scanner for light-absorbing or fluorescent zones (*16*). Electronic systems for the latter have also been described (*17*). Descriptions of various other densitometers appeared in Block *et al.* (18).

A simply designed instrument for direct fluorescent measurements provides for the insertion of the paper chromatogram containing fluorescent spots between a filtered radiation source and a photocell detector (*19*). In contrast to most methods based on the absorption of light, the whole spot is evaluated simultaneously with the use of a shield or mask rather than by scanning the spot with a narrow, transverse slit (see Section II,D,2 below). Another fluorometric scanner has been described which can be equipped with a device to heat the paper to 100°C, thereby increasing the fluorescence emitted by the spots (*20*). Manual fluorometers can be quite easily adapted for continuous strip scanning (*21*).

A commercial densitometry system is shown in Fig. 3. The paper is placed on the stage and a synchronous motor continuously advances it 1 inch per minute. The photometer output is then fed into the recorder which charts the scanned values automatically. Without the recorder, the paper can be manually scanned and the absorbance values read off the meter and plotted on graph paper. Readings on either strips or large sheets of paper can be made. The recorder features a choice of twelve responses; one is linear and is useful when scanning fluorescence, while others are nonlinear to compensate for deviations of Beer's Law. An electronic integrator can be added to the system to provide automatic and continuous integration of the curve areas.

The read-out of the recorder is a series of peaks corresponding to the position and concentration of the spots on the chromatogram. For the accurate quantitative evaluation of these curves, the proper base and separation lines must be drawn. A detailed discussion of the correct procedures for placing these lines has been recently described (*21a*).

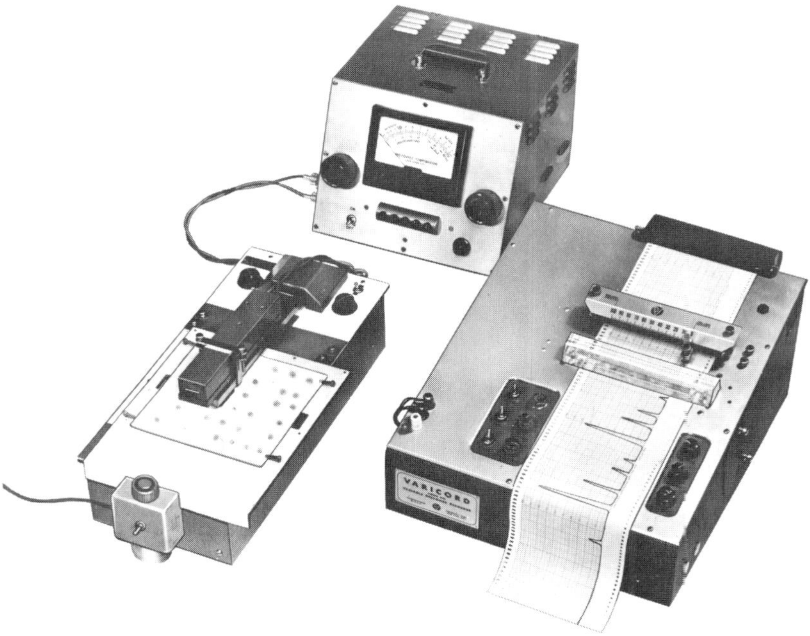

FIG. 3. Densitometry system for the evaluation of chromatograms. Courtesy of Photovolt Corp., New York City.

2. *Methods*

Although densitometry is a form of absorption spectroscopy, Beer's Law is usually not followed (*3*). Densitometers incorporating a Beer's Law response will, therefore, yield a plot which is not a straight line but one which is, nevertheless, usually quite convenient. Many recent instruments, such as that described above, have replacement cams or variable response recorders that allow recording in other parameters which compensate for deviations from Beer's Law.

It is often recommended for reflection and especially transmission densitometry that the paper be made transparent by drawing it through paraffin or some other special mixture. In one study (*22*) it was found that transparent paper has no compelling advantages and that reflection or transmission densitometry on dry paper is generally most suitable.

There are several different ways in which the densitometer can be used. The spot can be scanned, either completely by repeated scans or reproducibly through the center, to give a response curve showing the color variation over whatever path was used. The curve is then integrated to indicate the total spot color (*3*).

Alternatively, the spot is repeatedly scanned over a slightly shifting path (*23*) until the highest recorder curve indicates the maximum spot color. This peak height is proportional, as was spot area (above), to the logarithm of the solute concentration (*3*). This second method is recommended in cases where zones overlap slightly so that the total color in the zone, but not the maximum density, may be affected.

A third method involves determination of the transmittance or reflectance at the center of the spot by a single reading, a procedure first shown by Block (*24*) to give satisfactory results.

Lastly, the whole area of the spot can be measured with a single reading using a relatively large beam of light restricted to the size of the spot by means of a mask.

These four methods have been discussed and compared (*22*). The direct measurement of colored zones on paper has been studied experimentally by many other workers (e.g., *25*), and certain general guidelines have been suggested for obtaining accurate results (*26*). One comparison of direct scanning procedures indicates that fluorometry is most likely to give accurate results (*27*).

3. *Accuracy and Precision*

Most workers report errors of about ±5% for photometric *in situ* methods, although measurement of nanomoles of material with an error of 1–2% has been termed comparatively easily attainable (*3*). It was reported in one study that under optimal conditions a standard deviation of 5–10% and a comparative standard deviation greater than 10% is to be expected (*28*). Densitometry is less accurate than techniques involving elution, but it is more convenient and requires less sample for analysis. See the section on errors below for a summary of conditions required to achieve maximum accuracy in densitometric methods.

E. Radioisotope Techniques

Techniques for the chromatography of radioactive substances are the same as for nonlabeled compounds except that the radioactivity can be exploited for detection, identification, and quantitative estimation. It is best, if possible, to perform the experiments in a separate room which is designed and equipped to minimize the hazards of working with radioisotopes and the chances of spreading the radioactivity into other areas.

1. *Autoradiography*

The simplest and least costly technique for locating radioactive compounds separated by paper chromatography is autoradiography. The developed chromatogram is placed in immediate contact with

X-ray film, and the emitted low-energy particles (beta) react with the silver emulsion of the film. The exposed film, which is then developed by conventional darkroom methods, shows the exact self-image of radioactive spots; thus the term autoradiography. By comparing the position of the radioactive spots on the film with colored spots on the chromatogram treated with a suitable color reagent, tentative identification may be accomplished. The isotopes commonly detected by this method are ^{14}C, ^{35}S, and ^{32}P.

a. Experimental Procedure. The chromatogram is first stamped with radioactive ink to facilitate the matching of chromatogram and film. (Radioactive ink is made by adding a few drops of a ^{14}C-labeled compound or $H_2{}^{35}SO_4$ to a stamp pad.) The chromatogram is next loaded in the dark into a cardboard X-ray film holder together with a sheet of X-ray film. It is recommended that a single chromatogram be placed in each holder, since some ghost images have been observed from adjoining chromatograms which were placed in the same holder. The light-tight holders are stored in subdued light under moderate pressure. A special foam rubber press (*29*) may be helpful, but a set of lead bricks or thick books are satisfactory.

After the film has been exposed for 1 to 7 days or longer, depending on the radioactivity of the tagged compounds, it is developed in the usual manner in trays or tanks. Black or darkened spots on the film, showing the presence of radioactive compounds, may be intensified with Du Pont X-ray film intensifier. The exposed regions then appear blue and may best be viewed through a red filter. Autoradiographs are equally sensitive when thin or thick paper is used (*30*).

Figure 4 is an autoradiogram of ^{14}C-labeled substances derived by exposing an excised bean leaf to $^{14}CO_2$ in the light and chromatographing a hot alcohol extract from the leaf (*31*). Radioactive spots are identified by co-chromatography, in which the radioactive regions are eluted from the chromatogram and rechromatographed together with nonradioactive carrier of a suspected compound. Coincidence of radioactivity and specific color of the same spot is usually taken as final identification. If any doubt persists about the identity of the radioactive compound, it is recommended that repeated isolations be attempted and that a constant specific activity be established (*32*).

In addition to detection and identification, semiquantitative determination of compounds can be obtained by measuring the degree of blackening of the film by densitometry. This method is of lower accuracy than other radiometric methods (below), however.

b. Tritiated Compounds. A major obstacle in obtaining autoradiographs of tritium-labeled compounds is that the very soft β-radiation

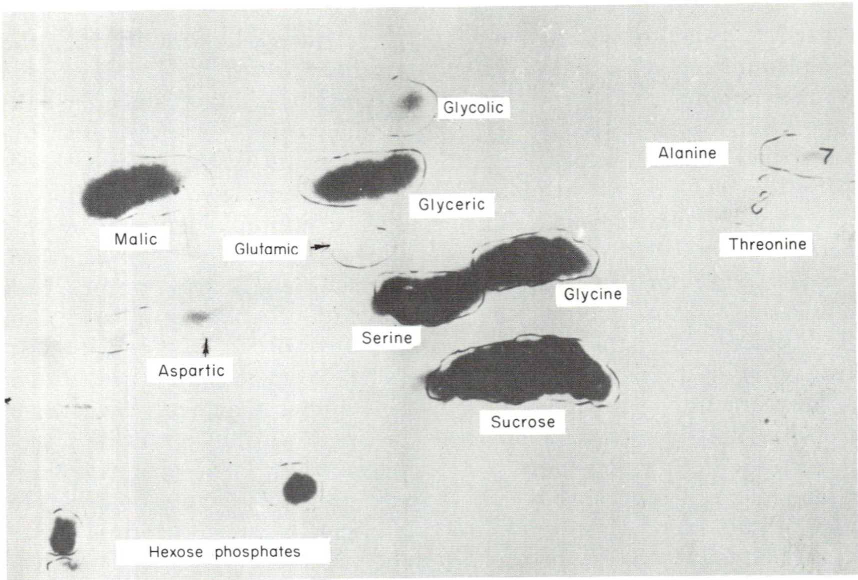

FIG. 4. Autoradiogram of two-dimensional chromatogram of ^{14}C-compounds extracted with 80% ethanol from excised bean leaf. Solvents: (left to right) phenol saturated with water; (down to up) butanol–propionic acid–H_2O (1246:620:874 v/v). After (*31*).

emitted by tritium is adsorbed either by the paper or the protective gelatin layer of the film before it ever reaches the photographic emulsion. For this reason, the chromatogram is first impregnated with a liquid scintillator converting the β-particles into light quanta which in turn cause the exposure of the film (*33*, *34*). The developed chromatograms are dipped into a saturated solution of anthracene (scintillator) in benzene, dried in air, and placed in immediate contact with X-ray film. The compounds on the chromatogram must be insoluble in the scintillator solvent, usually toluene or benzene (*35*).

Tritium can also be detected by directly exposing the chromatogram to highly sensitive X-ray film without the usual protective gelatin coating (e.g., Gevaert Structurix D 10, AGFA SINO) (*36*), or by impregnating the chromatogram itself with photographic emulsion and developing it photographically. This latter method is sensitive for water-insoluble compounds that do not diffuse in the emulsion and is capable of quantitative results (*5*). Sublimation autography involves the sublimation of some of the radioactive material (e.g., tritium-labeled nicotine or steroids) into the emulsion of the film where it produces an image (*35*).

c. *Spark Chambers* (*37*). Spark chamber systems have been designed which yield photographs similar to conventional autoradiographs but with an exposure a thousand or more times shorter. A spark chamber is a radiation detector, consisting of two parallel electrodes (one of which is transparent) placed in a suitable gas with an electric field in between, which is able to produce a visible spark in the position at which charged particles pass through it. A chromatogram containing β-emitting spots is placed against the transparent electrode, and sparks are formed over the spot and photographed. At present, the method is useful for the rapid and convenient location of well separated zones whose level of activity is similar. Semiquantitative analysis is possible by processing electrical signals corresponding to spark positions by digital or analog methods.

2. *Radiometric Methods*

As in chemical detection, radioactive spots can be measured after elution or directly on the paper. The latter method is more convenient and as accurate and is more frequently used. In the elution method, the spots are located (e.g., by autoradiography or scanning), excised from the chromatogram, eluted directly into a counting planchet, and measured by conventional counting techniques with Geiger-Müller tubes, proportional counters, gas-flow counters (for soft β-radiation), or liquid scintillation counters.

Direct measurement involves automatically scanning the chromatogram by drawing it past a slit in front of a Geiger-Müller, proportional, or scintillation counter (*38*). A continuous liquid scintillation scanner for chromatograms carrying soft β emitters has been designed which allows direct analysis of amounts of ^{14}C as low as 10^{-10} curie per 5 cm^2 (*39*). Excised spots may also be counted without elution by liquid scintillation spectrometry.

a. *Scanning Devices*. A simple strip scanner has been described by Frierson (*40*). The chromatogram is attached to the strip chart of a recording Brown potentiometer. As the chart is fed into the apparatus, the chromatogram is brought under a slit arrangement (from the mechanical stage of a microscope) and a Geiger counter. The pulses from the counter are fed to a counting rate meter (Tracer lab. SU-3) in which the 10 ma ammeter is replaced by a 10 ohm resistor. The potential drop across the resistor is proportional to the counting rate and is recorded by the Brown instrument. A curve of distance versus counting rate is obtained for comparison to the chromatogram.

Windowless gas-flow scanners with a 4 π geometry (i.e., both sides of the chromatogram are counted simultaneously) are commercially available (Fig. 5). Details of other equipment may be found in Block *et al.* (*18*).

Two-dimensional chromatogram scanners have been described (*41-44*). A commercial design producing a numerical record consists of an automatic scaler, scanner control, and an automatic typewriter. A minimum print level may be selected that prevents background numbers from being printed. A color print level may also be selected that will print zones of high radioactivity in red rather than black. The time

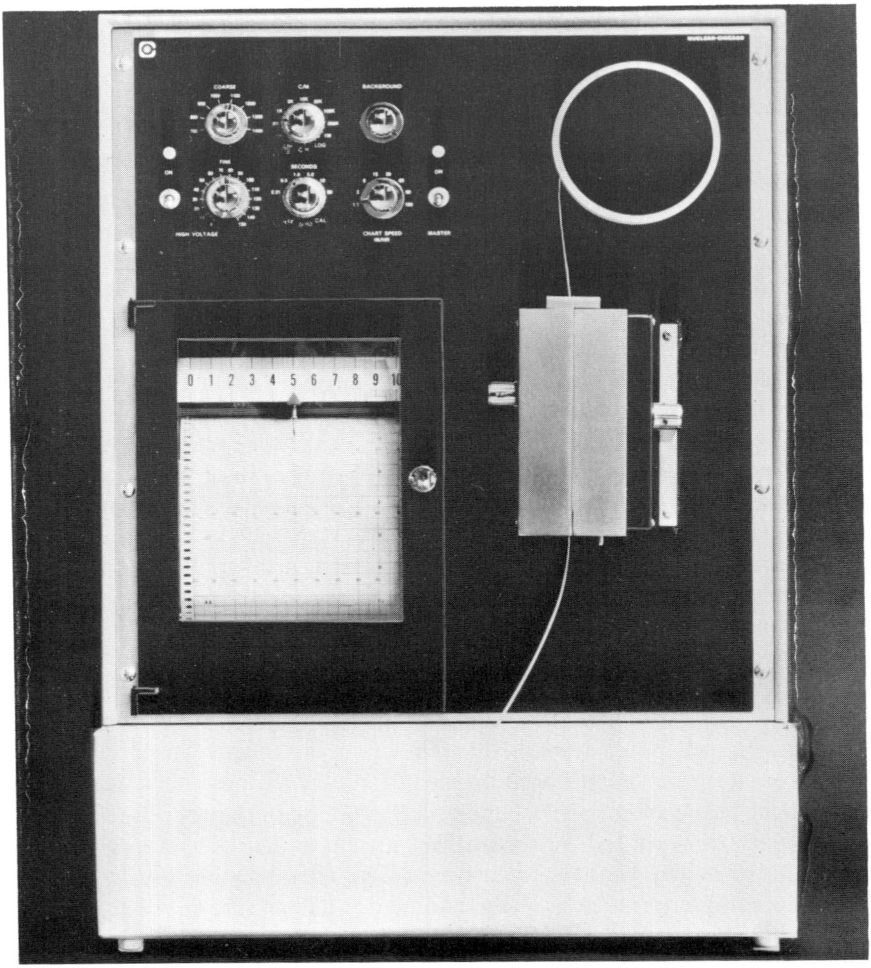

FIG. 5. Biospan Model 1002 Actigraph III paper radiochromatography scanner with dual Geiger-type detectors that scan both sides of the paper in either windowless or ultrathin window operation. Courtesy of Nuclear-Chicago Corp., Cherry Hill, N. J.

involved to scan with high resolution a two-dimensional chromatogram may be as long as the less costly procedure of autoradiography (see Fig. 5 in Chapter 9).

For routine, automatic, bidimensional radiochromatographic scanning (*45*), an assembly of eight scanners with a common printing system to record output data on punched tape for subsequent computer processing has been employed. Each chromatogram is divided into an imaginary matrix of contiguous equal areas, each of which is assayed separately with a pair of Geiger-Müller counters, one on each side of the paper. For greater resolution and statistical accuracy, more areas and a longer counting time per area are required. Another commercial but high-priced β-camera (Baird-Atomic) yields almost instantaneous two-dimensional recording.

b. *Liquid Scintillating Spectrometry.* Liquid scintillation counting is used for tritium as well as for ^{35}S, ^{32}P, and ^{14}C. Pulse height discrimination allows the differentiation of these isotopes if they are mixed together in a zone on a chromatogram. The counting efficiency for ^{14}C is usually between 55 and 85%, while the efficiency for ^{3}H is often as low as 1–2%, although greatly improved with recent instruments.

A paper chromatogram can be examined by cutting it into equal sections (e.g., 4 × 2.5 cm) which are inserted into standard counting vials for liquid scintillation spectrometers. Fifteen milliliters of phosphor solution in toluene are added and the sample is counted at the proper voltage for the specific isotope being measured. The orientation of the paper segment toward the phototube of the spectrometer may have some importance for reproducibility of counting (*46*), so it is recommended that the paper strip be rolled into a cylinder in order to minimize these problems.

As an alternative, the whole spot can be cut out and counted. An example of this technique is taken from a recent paper on the separation of ^{14}C-labeled sugars (*47*). After development and drying of the chromatogram, the spots are located with aniline–phthalate reagent (Chapter 6) and dried with a hot-air drier until the color becomes pronounced. Small circles, which are cut from the paper so that each contains an entire visible spot, are transferred to phosphor and allowed to settle on the floor of the vials. The paper was left in the phosphor at least 2 days before counting. Quenching effects due to the sugar–phthalate complex and to excess reagent must be corrected.

c. *Isotope Indicators.* Amino acids after chromatography may be treated on the paper with ^{131}I-labeled *p*-iodophenylsulfonyl chloride, forming ^{131}I-derivatives (pipsyl derivatives) which can be located by the methods described above (*48*). The amino acid chromatogram may also be exposed to $CH_3{}^{131}I$, resulting in the liberation of ${}^{131}I_2$ in the amino acid zones owing to methylation (*49*).

Detection of metal ions by $H_2{}^{35}S$ (*50*) and reducing sugars and alkaloids with ${}^{110}AgNO_3$ (*51, 52*) has been used. Details of such procedures may be found in Block *et al.* (*18*). A similar technique of double decomposition radiochromatography for the estimation of fatty acids, inorganic phosphates, and sugar phosphates involves the decomposition of silver salts on the chromatograms with $K^{131}I$. After washing the chromatograms with water, the resultant $Ag^{131}I$ spots are counted (*53*).

Organic acids, such as gibberellic acid, may be reacted with diazomethane-${}^{14}C$ generated by alkaline hydrolysis from *N*-methyl-${}^{14}C$-*N*-nitroso-*p*-toluenesulfonamide (Diazald, New England Nuclear). The resultant methyl-${}^{14}C$ ester is chromatographed on paper, and the developed chromatogram is scanned under a windowless counter connected through a rate meter to a recording milliammeter. Areas under the curve are measured with a planimeter and compared with those from known standards (*54*).

Radiometric precipitation techniques combined with paper chromatography [first described by van Erkelens (*50*), see above] have been used to determine submicro amounts of metal ions (*55*). Detection limits ranged from 0.01 μg for lead to 0.0001 μg for beryllium. Two small portions of the mixture are spotted side by side on paper and radioactive tracers of the elements to be analyzed are added to one portion. The chromatogram is developed with a suitable solvent to separate the mixture (ethanol–methanol–2 *N* HCl, 2:2:1 v/v, for calcium, strontium, and barium). The locations of the unknown elements are determined by scanning the paper. A labeled precipitating agent ($H_3{}^{32}PO_4$ for beryllium and zirconium, $H_2{}^{35}SO_4$ for calcium, strontium, barium, and lead) is added 2 cm above the original origin and caused to migrate by developing the paper with another solvent (50% aqueous ethanol for H_2SO_4). The paper is then dried and the zones of precipitated radioactive compounds are cut out and determined by β scintillation counting. For each element, a linear relationship was observed between counting rate and the weight of element in the precipitate.

Nonradioactive sulfate or thiosulfate has been determined on paper chromatograms based on the addition of a constant quantity of the respective radioactive species (*56*).

d. *Neutron Activation Analysis.* The finished chromatogram is placed in an atomic reactor where it undergoes neutron bombardment. Some of the atoms in the separated compounds become radioactive, and the compounds can then be determined by comparing activities to those of similarly treated standards. Depending upon the level of irradiation and the type of paper, time may have to be allowed for the background radiation to drop to a low level before the activities of the compounds of interest can be measured.

This method has been successfully applied to bromine analogs of DDT derivatives where the reaction is $^{81}Br\ (n, \gamma)\ ^{82}Br$ (*57*). Nucleic acid derivatives have been detected by this method, scanning the chromatogram 1 week after irradiation and using an aluminum screen in front of the Geiger tube. This eliminated any stray counts from ^{24}Na, ^{42}K, and ^{35}S [$^{35}Cl\ (n, p)\ ^{35}S$], thus counting ^{32}P exclusively (*49*). Iodine-containing compounds may also be determined quantitatively on chromatograms by short-term neutron activation, resulting in ^{128}I ($t_{1/2}$ = 25 minutes), which must be counted shortly after irradiation with a Geiger-Müller counter or a pulse height analyzer. Treatment of the paper to avoid background disturbances due to sodium and chlorine present in the paper is described (*58*). Trace elements in ores (*59*) and compounds of biochemical interest (*60*) have also been determined by a combination of paper chromatography and activation analysis, and a review of the subject has appeared (*61*). Activation methods are highly sensitive, and in addition, allow the sample preparation, development, drying, etc., to be performed without the precaution associated with handling radioactive compounds.

F. Miscellaneous Methods

Examples are given in later chapters of *in situ* determinations by methods such as titration (*62*), polarography (*63*), high-frequency measurement (*64*), oscillometric (*65*), and amperometric (*66*) methods.

III. Sources of Error in Quantitative Paper Chromatography

One significant source, and perhaps the main source, of error in quantitative paper chromatography has been shown to be the delivery of small volumes of solution to form the initial spots. Errors of individual results up to 20% with coefficients of variation of ±3% to ±10% were found in one study (*12, 67*) with experienced workers using conventional syringes or pipettes. These large errors were due to creep-back on the outside of the needle during repetitive delivery and capillation from within the needle when touching the surface of the paper. If a single delivery is made with a micropipette the errors are significantly less.

A machine has been designed (Fig. 6) for accurate delivery of small replicate volumes by forcible ejection from a syringe that does not actually touch the surface of the paper. An interchangeable printed

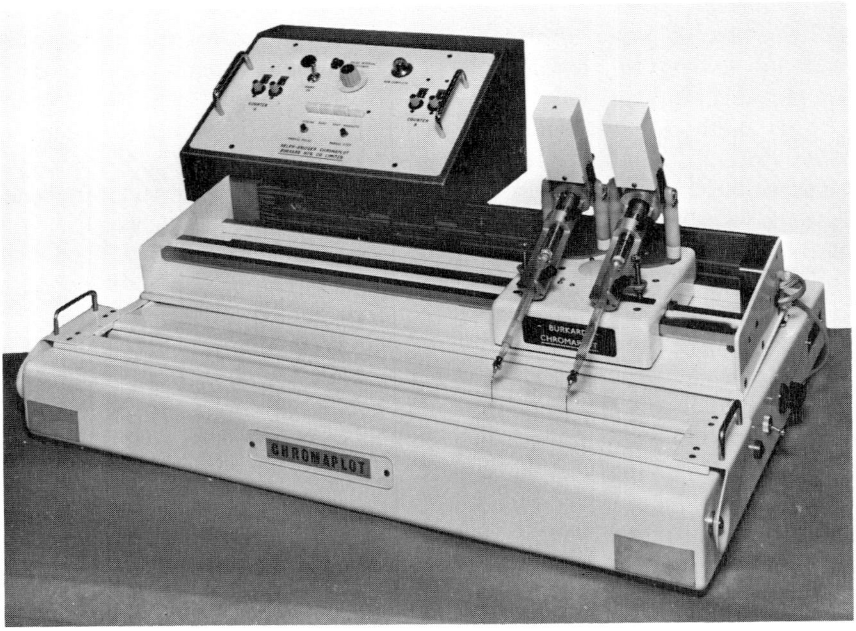

FIG. 6. Chromaplot sample spotting machine. Courtesy Burkard Mfg. Co. Ltd., Rickmansworth, Hertfordshire, England.

circuit key at the rear programs automatic delivery of spots, short bands, or streaks. A series of drops was produced by this machine with a coefficient of variation of $\pm 2.5\%$ (*12*).

For manual spotting, the new Eppendorf pushbutton pipettes (Brinkmann Instruments, Inc.) with disposable tips may prove to be useful for the accurate dispensing of volumes of 5–1000 μl. The throw-away tips are the only part the solution ever contacts, and the fact that they are plastic and are not reusable might reduce the creep-back problem. The operation of the pipette is such that the final bit of solution is blown out of the pipette by the action of a spring-loaded piston so that capillation can also be avoided. Specifications indicate an absolute error of <0.05 and <0.5 μl with a standard deviation of <0.04 and <0.15 μl for the 5 and 50 μl pipettes, respectively (*68*), although as far as is known these pipettes have not been tested for chromatography.

A second major source of error is differences between the slopes of the regression lines—relating the parameter measured (e.g., optical density) to the quantities originally present in the initial spots—from sheet to sheet. These differences can be corrected by using standard and test solutions for each sheet (*47*). The instrument shown in Fig. 6 contains two syringes so that standard and test solutions can be automatically delivered as required. Within a single sheet of paper, new sources of error are minimal during the chromatographic run; under conditions where adjacent spots have very dissimilar quantities of solute, lateral diffusion of the migrating zones may occur leading to slight errors (*47*).

A summary of the chromatographic conditions necessary to achieve minimum error in densitometric quantitative methods follows (*12*):

1. The paper is kept dust free in sealed and resealed containers; it is allowed to equilibrate in the tank overnight (see Chapter 3).
2. The initial zones are delivered by the machine shown in Fig. 6 and dried by a current of cold air (*26*). On each sheet, two quantities of the standard, one a multiple of the other, and two quantities of the unknown, differing by the same factor, are used.
3. The tank is kept at a constant temperature, and a critical solvent volume of 1% is used throughout. The critical solvent volume was discussed in Block *et al.* (*18*, p. 61), where it was recommended that for constant R_F values an amount of solvent approximately 1% of the volume of the cabinet should be used.
4. The initial spots are placed about 11 cm from the upper edge of the paper so that the descending solvent can achieve an even flow rate before it reaches the spots. The solvent front is allowed to reach a constant distance from the origin line in every run.
5. At the end of the run, the paper is removed and rapidly dried in a current of warm air.
6. For color development, the paper is dipped or sprayed evenly on both sides and then dried in a stream of cold air.
7. So that the background color is uniform, densitometry is performed at a set time after color development. If possible, the background reading is reduced by the use of selective light filters. A dual-wave densitometer allowing supression of the background optical density while maintaining high sensitivity and specificity for the zones has been described (*69*).

Using these precautions, the error of individual determinations can be reduced to $\pm 6\%$. With replication, the error will be reduced by $n^{-1/2}$, where n is the number of replicates made (*12*).

For a perspective on the present status of quantitative paper and thin-layer chromatography, the recent remarks of the noted chromatographer L. S. Bark upon reviewing a book on the subject will be quoted (*70*):

> Although there is no doubt that the methods reported [in papers presented at a recent symposium and collected in the book] have potential, it is evident that there is a need for further development before the reproducibility of color reactions on the supports used in these techniques, and the precision of the methods used to measure the amount of the material, are sufficiently reliable to allow the use of any of the direct techniques with confidence for any work involving accuracy and precision. It may be that both these developments and perhaps more symposia and discussion meetings are necessary before any book on this subject can truly include in its title the word "quantitative" with all its analytical implications.

References

1. H. Jork, *J. Chromatog.* **33**, 297 (1968).
2. A. Boulton, *Methods Biochem. Anal.* **16**, 327 (1968).
3. G. Franglen, *in* "Quantitative Paper and Thin Layer Chromatography" (E. J. Shellard, ed.), pp. 17–28. Academic Press, New York, 1968.
4. W. Louwerse, *Analyst* **91**, 56 (1966).
5. V. C. Weaver, *Advan. Chromatog.* **7**, 87–120 (1968).
6. J. S. Knypl and R. Antoszewski, *J. Chem. Educ.* **37**, 49 (1960).
7. H.-E. Sprenger, *Z. Anal. Chem.* **207**, 90 (1965).
8. J. Skolik, *J. Chromatog.* **3**, 273 (1960).
9. G. F. Grogory, *Science* **121**, 169 (1955).
10. A. M. Moore and J. B. Boylen, *Science* **118**, 19 (1953).
11. H. Schriefers, H. Thomas, and F. Pohl, *Clin. Chim. Acta* **8**, 744 (1963).

11a. R. D. Edstrom, *Anal. Biochem.* **26**, 204 (1968).

12. J. W. Fairbairn, *in* "Quantitative Paper and Thin Layer Chromatography" (E. J. Shellard, ed.), pp. 1–15. Academic Press, New York, 1968.
13. D. M. Abelson, *J. Chromatog.* **31**, 630 (1967).
14. A. A. Boulton, R. L. Holder, and H. F. Ross, *Biochem. J.* **96**, 70P (1965).
15. I. E. Bush, *J. Chromatog.* **23**, 94 (1966).
16. I. E. Bush, *J. Chromatog.* **29**, 157 (1967).
17. I. E. Bush and C. P. Hoffman, *J. Chromatog.* **30**, 164 (1967).
18. R. J. Block, E. L. Durrum, and G. Zweig, "A Manual of Paper Chromatography and Paper Electrophoresis," 2nd rev. ed. Academic Press, New York, 1958.
19. G. F. Bailey, *Anal. Chem.* **32**, 1726 (1960).
20. A. A. Boulton, N. E. Chard, and L. Grant, *Biochem. J.* **96**, 82P and 83P (1956).
21. H. Puro, E. Epstein, and B. Zak, *Chemist-Analist* **54**, 54 (1965).

21a. A. Kaiser, *Chromatographia* **2**, 411 (1969).

22. R. B. Ingle and E. Minshall, *J. Chromatog.* **8**, 369 and 386 (1962).
23. G. Franglen, *Joyce-Loebl Rev.* **1**, 6 (1963).
24. R. J. Block, *Science* **108**, 608 (1948); *Anal. Chem.* **23**, 298 (1951).
25. O. Scheltino, *Farmaco (Pavia), Ed. Prat.* **20**, 40 (1965).
26. I. E. Bush, *Methods Biochem. Anal.* **11**, 149 (1963).
27. R. Klans, *J. Chromatog.* **16**, 311 (1964).
28. G. Ackermann and H. J. Wiesner, *J. Chromatog.* **33**, 307 (1968).
29. W. E. Lotz, J. C. Gallimore, and G. A. Boyd, *Nucleonics* **10**, 28 (1952).
30. A. J. Tomisek and B. T. Johnson, *J. Chromatog.* **33**, 329 (1968).
31. G. Zweig and F. M. Ashton, *J. Exptl. Botany* **13**, 5 (1962).

32. S. Aronoff, "Techniques Radiobiochemistry." Iowa State Coll. Press, Ames, Iowa, 1956.
33. A. T. Wilson, *Proc. 2nd U.N. Intern. Conf. Peaceful Uses At. Energy, Geneva,* (1958) Vol. 25, p. 213. United Nations, New York, 1958; *Nature* **182**, 524 (1958); *Biochim. Biophys. Acta* **40**, 522 (1960).
34. E. V. Parups, I. Hoffman, and H. R. Jackson, *Talanta* **5**, 75 (1960).
35. A. T. Wilson and D. J. Spedding, *J. Chromatog.* **18**, 76 (1965).
36. A. Narath and D. Gundlach, *Angew. Chem.* **72**, 707 (1960).
37. B. R. Pullan, *in* "Quantitative Paper and Thin Layer Chromatography" (E. J. Shellard, ed.), pp. 122–133. Academic Press, New York, 1968.
38. R. F. Phillips and W. R. Waterfield, *J. Chromatog.* **40**, 309 (1969).
39. W. E. Spratt, *J. Chromatog.* **17**, 355 (1965).
40. W. J. Frierson and J. W. Jones, *Anal. Chem.* **23**, 1447 (1951).
41. E. A. Piper and H. R. V. Arnstein, *Biochem. J.* **64**, 57P (1956).
42. W. J. Wingo, *Anal. Chem.* **26**, 1527 (1954).
43. F. P. W. Winteringham, *Nature* **172**, 727 (1953).
44. E. B. Chain, M. Frank, F. Pocchiari, and C. Rossi, *Selected Sci. Papers Ist. Super. Sanita* **2**, 75 (1959); **1**, 241 (1956).
45. A. A. Lowe, *in* "Quantitative Paper and Thin Layer Chromatography" (E. J. Shellard, ed.), pp. 119–122. Academic Press, New York, 1968.
46. R. B. Loftfield and E. A. Eigner, *Biochem. Biophys. Res. Commun.* **3**, 72 (1960).
47. J. W. Fairbairn and S. J. Relph, *Chromatographia* **2**, 204 (1969).
48. A. S. Keston, S. Udenfriend, and M. Levy, *J. Am. Chem. Soc.* **69**, 3151 (1947).
49. F. P. W. Winteringham, P. M. Bridges, and G. C. Hellyer, *Biochem. J.* **59**, 13 (1955).
50. P. C. van Erkelens, *Nature* **172**, 357 (1953).
51. M. Jaarma, *Acta Chem. Scand.* **8**, 860 (1954).
52. M. Sarsunova, J. Toelgyessy, and J. Majer, *Cesk. Farm.* **9**, 339 (1960).
53. Z. J. Zubrzycki, A. Z. Budzynski, and I. G. Campbell, *Talanta* **2**, 165 (1959).
54. W. E. Baumgartner, L. S. Lazer, A. M. Dalziel, E. V. Cardinal, and E. L. Varner, *J. Agr. Food Chem.* **7**, 422 (1959).
55. G. A. Welford, E. L. Chiotis, and R. S. Morse, *Anal. Chem.* **36**, 2350 (1964).
56. G. Ackermann and E. Weber, *Talanta* **15**, 63 (1968).
57. F. P. W. Winteringham, A. Harrison, and R. G. Bridges, *Analyst* **77**, 19 (1952).
58. H. J. Born and H. Stärk, *Atomkernenergie* **4**, 286 (1959).
59. R. Coulomb, *Compt. Rend.* **254**, 4328 (1962).
60. J. M. Steim and A. A. Benson, *Anal. Biochem.* **9**, 21 (1964).
61. A. Z. Budzynski and J. Z. Beer, *Anal. Chim. Acta* **46**, 281 (1969).
62. A. Lacourt, G. Sommereyns, J. Hoffmann, A. Stadler-Denis, and G. Wantier, *Compt. Rend.* **234**, 2365 (1952).
63. A. Langer, *Anal. Chem.* **28**, 426 (1956).
64. G. G. Blake, *Anal. Chim. Acta* **13**, 562 (1955).
65. G. Ackermann, *Talanta* **13**, 129 (1966).
66. G. deVries and E. van Dalen, *Rec. Trav. Chim.* **73**, 1028 (1954).
67. J. W. Fairbairn and S. J. Relph, *J. Chromatog.* **33**, 494 (1968).
68. Brinkmann Instruments, Inc., Westbury, New York, 11590, Bull. BR174-B.
69. L. Salganicoff, M. Kraybill, D. Mayer, and V. Legallais, *J. Chromatog.* **26**, 434 (1967).
70. L. S. Bark, *Analyst* 94, 335 (1969).

5

AMINES, AMINO ACIDS, PEPTIDES, AND PROTEINS

As mentioned in Chapter I, the modern era of paper chromatography is considered to have begun with the pioneer work of Consden, Gordon, Martin, and Synge in 1944 on the separation of amino acids by partition methods. Since then, thousands of papers have been published concerning the separation of these acids and related compounds on paper. This large amount of work is due to the fact that the amino acids, along with the flavonoids (Chapter 11), are probably the two classes of compounds most amenable to separation by paper chromatography, due in part to very favorable solubility characteristics in a wide range of convenient developing solvents. Summarized below are many of the important experiments published recently on the separation and detection of these compounds.

I. Aliphatic and Aromatic Amines

Many aliphatic and aromatic amines, derivatives, and related compounds have been separated by paper chromatography. Amines are chromatographed as free bases either in nonaqueous systems or in aqueous systems containing a strong base to surpress dissociation. In aqueous systems containing an acid, the amines are in their salt forms. However, aliphatic amine salts are best separated in neutral systems containing an excess of an electrolyte so that possible interferences (e.g., poor separation, streaking zones) due to the presence of some acids or acid anions are avoided. Alternatively, volatile or labile amines can be converted to nonionic derivatives containing easily detectable functional groups (e.g., 2,4-dinitrophenyl derivatives) or to colored compounds (e.g., azo dyes) which are chromatographed in nonaqueous systems (*1*).

Because the presence of certain anions can disturb separations, samples isolated from biological materials, which are likely to contain various inorganic salts mixed with the amines, must often be given special attention (*2*).

A. Paper

Whatman papers No. 1, 3 MM, or 4 or equivalent grades are most often used. Glass fiber paper is used for the separation of epinephrine, norepinephrine, and their acetyl derivatives (*3*) and acetylated paper [Schleicher and Schuell (S. and S.) No. 2496] for adenine derivatives (*4*).

2,4-Dinitrophenyl derivatives can be separated on paper impregnated with silicone oil (*5*), liquid paraffin (*6*), or formamide (*1*). Benzidine and its isomers are separated on formamide-impregnated paper developed with cyclohexane (*7*). Spraying with 1% *p*-dimethylaminobenzaldehyde in 1 *N* HCl locates spots as follows: benzidine, R_F 0.02, orange; diphenyline, 0.18, yellow; *o*-benzidine, 0.57, yellow; 2-aminodiphenylamine, 0.74, yellow; 4-aminodiphenylamine, 0.33, red. For the separation of toluidine isomers, the best results are obtained by brominating the compounds and separating the bromo derivatives by reversed phase chromatography with ethanol–H_2O–acetic acid (20:14:1 v/v) on paper impregnated with paraffin oil (*1*).

B. Solvents

Tables I–V list selected solvents for amine separations; others can be found in the references mentioned in Section E. For lower aliphatic amines, *n*-butanol saturated with concentrated KCl and paper impregnated with 5% KCl is recommended (*8*).

For higher aliphatic primary amines, ethanol–1 *N* HCl (1:1 v/v) saturated with lauryl alcohol on paper impregnated with lauryl alcohol is suitable (1). Solvents containing *n*-butanol or ethyl acetate plus aqueous hydrochloric, acetic, or formic acid with untreated paper or nonaqueous solvents with impregnated paper are often used for aromatic amines. Tryptamine derivatives, for example, migrate as shown in Table VI on Whatman 1 paper when developed with *n*-butanol–acetic acid (10:4 *v/v*) saturated with water (*9*).

Catecholamines and their metabolites are separated on Whatman No. 1 paper with the upper phase of the following two solvents: (a) *n*-butanol–acetic acid–H_2O (5:1:3 v/v) and (b) ethyl acetate–acetic acid–H_2O (5:1.5:3 v/v). The aqueous phases are used for chamber and paper saturation. Detection is by spraying the chromatograms with a 20% solution of Folin phenol reagent in water (see also Table IV). The R_F values found (*9a*) are shown in Table VII.

TABLE I

R_F VALUES OF AROMATIC AMINES

Compound	R_F in the following solvents[a]						
	1	2	3	4	5	6	7
Aniline	0.90	0.90	0.08	—	—	—	0.30
2-Chloroaniline	0.87	—	—	0.23	0.31	—	—
3-Chloroaniline	0.65	0.81	—	—	—	—	—
4-Chloroaniline	0.80	0.84	—	—	—	—	0.38
4-Bromoaniline	0.66	0.83	—	—	—	—	—
2,4-Dichloroaniline	0.02	0.20	—	—	—	—	—
2,5-Dichloroaniline	0.02	0.10	—	0.33	0.34	—	—
2,4,5-Trichloroaniline	0.00	0.03	—	—	—	—	—
3-Chloro-2-methylaniline	0.45	0.77	—	—	—	—	—
3-Chloro-6-methylaniline	0.35	0.66	—	—	—	—	—
3-Chloro-6-methoxyaniline	0.35	0.66	—	—	—	—	—
3-Chloro-4-methoxyaniline	0.84	0.85	—	—	—	—	—
4-Chloro-2-methylaniline	0.55	0.79	—	0.28	0.23	—	0.64
2-Nitroaniline	0.20	0.63	—	—	—	—	—
2-Nitro-4-chloroaniline	0.04	0.19	—	—	—	—	—
2-Nitro-4-methylaniline	0.06	0.24	—	—	—	—	—
2-Nitro-4-methoxyaniline	0.10	0.37	—	—	—	—	—
3-Nitroaniline	0.51	0.70	—	—	—	—	—
3-Nitro-6-methylaniline	0.19	0.50	—	—	—	—	—
4-Nitroaniline	0.42	0.64	—	—	—	—	—
4-Nitro-2-chloroaniline	0.09	0.31	—	—	—	—	—
4-Nitro-2-methylaniline	0.14	0.46	—	—	—	—	—
4-Nitro-2-methoxyaniline	0.10	0.38	—	—	—	—	—
Benzidine	0.87	0.89	—	—	—	—	—
3,3′-Dichlorobenzidine	0.01	0.19	—	—	—	—	—
3,3′-Dimethoxybenzidine	0.82	0.89	—	—	—	—	—
5,5′-Dichloro-3,3′-dimethoxybenzidine	0.03	0.56	—	—	—	—	—
N-Methylaniline	—	—	0.31	—	—	—	—
N,*N*-Dimethylaniline	—	—	0.72	—	—	—	—
2-Ethoxylaniline	—	—	—	0.34	0.34	—	0.69
2-Methoxyaniline	—	—	—	—	—	—	0.44
2,3-Dimethylaniline	—	—	—	0.29	0.30	—	—
2,4-Dimethylaniline	—	—	—	0.35	0.45	—	—
2,5-Dimethylaniline	—	—	—	0.33	0.42	—	0.77
2,6-Dimethylaniline	—	—	—	0.42	0.54	—	—
2-Methyl-5-chloroaniline	—	—	—	0.27	0.25	—	—
2-Methoxy-5-methylaniline	—	—	—	0.31	0.31	—	—
2-Methoxy-5-chloroaniline	—	—	—	0.16	0.21	—	0.61
2-Aminostilbene	—	—	—	0.16	0.24	—	—
2-Amino-4-chlorophenol	—	—	—	—	—	0.71	—
2-Amino-4,6-dinitrophenol	—	—	—	—	—	0.01	—
2-Amino-5-nitrophenol	—	—	—	—	—	0.53	—

Table continued

TABLE I (*continued*)

Compound	R_F in the following solvents[a]						
	1	2	3	4	5	6	7
1,2-Diamino-4-nitrobenzene	—	—	—	—	—	0.46	—
1,3-Diamino-4-nitrobenzene	—	—	—	—	—	0.41	—
1-Amino-3-acetaminobenzene	—	—	—	—	—	0.33	—
1-Amino-4-acetaminobenzene	—	—	—	—	—	0.24	—
2-Aminophenol	—	—	—	—	—	0.50	—
3-Aminophenol	—	—	—	—	—	0.36	—
4-Aminophenol	—	—	—	—	—	0.29	—
2-Aminophenol-5-sulonamide	—	—	—	—	—	0.07	—
2-Methylaniline	—	—	—	—	—	—	0.57
o-Tolidine	—	—	—	—	—	—	0.06
1-Naphthylamine	—	—	—	—	—	—	0.48
2-Naphthylamine	—	—	—	—	—	—	0.42

[a] (1) 5% acetic acid; Niederslag FN 14 paper (German) impregnated with 0.17 weight (compared to original paper) 5–20% 1-bromonaphthalene in hexane; detection with *p*-dimethylaminobenzaldehyde (*34*). (2) 30% acetic acid; paper as same as (1) with 0.18 weight impregnation (*34*). (3) Cyclohexane; Whatman No. 3 paper impregnated with 50% dimethylformamide in ethanol; detection with diazotized 2,4-dinitroaniline (*35*). (4) Heptane; Whatman No. 3 paper impregnated with 25% dimethylformamide in benzene (*36*). (5) Heptane; Whatman No. 3 paper impregnated with 20% ethylene glycol monoacetate in ethanol (*36*). (6) Chloroform–ethyl acetate (1:1 v/v); Whatman No. 3 paper impregnated with 20% formamide in ethanol (*36*). (7) Heptane; Whatman No. 3 paper impregnated with 20% formamide in ethanol; detection with Ehrlich's reagent and 1-diazo-2-chloro-4-nitrobenzene (*37*).

C. Detection (see also Tables I–VII)

1. *Ninhydrin (1)*

Ninhydrin 0.2% in ethanol; add 2% of acetic acid to the reagent when the paper is buffered to pH 7.5–9, and 5% of acetic acid if buffer is 9–12; paper buffered to pH 1–5 is exposed to ammonia fumes before spraying; heat sprayed paper at 60°C for several minutes. This method is suitable for lower aliphatic amines and diamines.

For higher aliphatic amines and polyamines (*10*), spray with 0.1% ninhydrin in *n*-butanol with 20% pyridine, and heat 5 minutes at 120°C. Secondary amines react poorly with ninhydrin.

2. *Sodium Nitroprusside (1, 10a)*

Spray with a 5% solution in aqueous acetaldehyde (10%) to which Na_2CO_3 (2%) is added before use; reveals secondary amines as blue spots. A 1:1:1 v/v solution of 10% NaOH, 10% sodium nitroprusside, and 10% potassium ferricyanide detects guanidine derivatives. The

TABLE II

R_F VALUES OF SOME ALIPHATIC POLYAMINES ON WHATMAN NO. 3 PAPER (*10*)

Compound	R_F in the following solvents[a]				Color[b]
	1	2	3	4	
Triethylenetetramine	0.42	0.32	0.06	0.01	Brown-blue
Diethylenetriamine	0.45	0.36	0.12	0.02	Brown-blue
Dipropylenetriamine	0.57	0.48	0.15	0.02	Dark gray-blue
Ethylenediamine	0.51	0.44	0.20	0.05	Gray-blue
Trimethylenediamine	0.58	0.49	0.23	0.06	Brown-violet
Tetramethylenediamine	0.66	0.59	0.29	0.08	Blue-violet
Pentamethylenediamine	0.72	0.66	0.36	0.10	Blue-violet
Hexamethylenediamine	0.76	0.73	0.42	0.13	Dark blue-violet
N,N-Diethylethylenediamine	0.78	0.77	0.49	0.21	Green-brown
Methylamine	0.75	0.72	0.50	0.31	Gray-blue-violet
Ethylamine	0.82	0.78	0.63	0.47	Gray-blue-violet
n-Propylamine	0.85	0.83	0.75	0.61	Gray-blue-violet
n-Butylamine	0.88	0.86	0.85	0.73	Gray-blue-violet
Monoethanolamine	0.75	0.67	0.45	0.25	Blue
1-Chloro-2-aminoethane	0.80	0.74	0.59	0.45	Blue-brown
Isopropanolamine	0.82	0.73	0.55	0.35	Blue
3-Aminopropanol	0.82	0.78	0.56	0.35	Blue
N-(2-Hydroxyethyl)-ethylenediamine	0.64	0.60	0.26	0.08	Brown-gray-blue

[a] (1) Methanol–concentrated HCl (1:1 v/v); (2) ethanol–concentrated HCl (1:1 v/v); (3) *n*-propanol–concentrated HCl (2:1 v/v); (4) *n*-butanol–concentrated HCl (4:1 v/v).

[b] Ninhydrin with pyridine (see Section I,C).

TABLE III

R_F VALUES OF SOME AMINO ALCOHOLS ON EDEROL 208 PAPER (SLOW) (*10*)[a]

Compound	R_F in ethyl acetate–dimethylformamide (2:1 v/v)
Monoethanolamine	0.17
Diethanolamine	0.38
Triethanolamine	0.67
3-Aminopropanol	0.19
Monoisopropanolamine	0.31
Diisopropanolamine	0.72
Triisopropanolamine	At front
N-Methyldiethanolamine	0.70
N-(2-Hydroxyethyl)ethylenediamine	0.05

[a] Detection: 0.1% solution of alizarin in ethanol.

TABLE IV

R_F VALUES OF SOME CATHECHOLAMINES AND THEIR DERIVATIVES ON WHATMAN NO. 1 PAPER (*43*)

Compound	R_F [a]	Colors[b] after detection
o-Hydroxyphenylacetic acid	0.84	Purple
3,4-Dihydroxyphenylacetic acid	0.74	Yellow
Vanillic acid	0.84	Purple
Homovanillic acid	0.84	Blue
Vanilmandelic acid	0.68	Purple
3-Hydroxy-4-methoxymandelic acid	0.66	Purple
Ferrulic acid	0.68	Blue-yellow
3-Hydroxy-4-methoxycinnamic acid	0.81	Purple
m-Adrenaline	0.50	Purple
Nor-*m*-adrenaline	0.44	Purple
DOPA	0.14	Pink
α-Methyl-DOPA	0.32	Pink
Tyramine	0.49	Pink
Dopamine	0.38	Pink
L-Phenylephrine	0.49	Pink
Adrenaline	0.26	Green-blue
Noradrenaline	0.19	Green-blue

[a] Solvent: *n*-butanol–acetic acid–H_2O (14:4:50 v/v).

[b] Diazotized *p*-nitraniline.

sensitivity and specificity of this reagent for secondary amines has been questioned recently (*10b*).

3. *Iodine* (*1*)

Dip chromatogram in a 0.3% solution in 5% aqueous potassium iodide; wash several minutes with water to decolorize background; for tertiary amines.

4. *Alizarin*

A 0.1% solution in ethanol detects lower aliphatic amines and amino alcohols (*10*). Bases appear as violet spots on a yellow background.

5. *p-Dimethylaminobenzaldehyde* (*1*)

One gm mixed with 95 ml of ethanol and 5 ml of HCl detects various types of primary aromatic amines via characteristic yellow or red colors, as well as urea, hydrazine, and indole derivatives.

6. *Xanthydrol*

A 1% solution in absolute methanol containing 10% trichloroacetic acid used as a dipping reagent reveals 5-O-substituted tryptamine

TABLE V

R_F VALUES OF THE 3,5-DINITROBENZOYL DERIVATIVES OF ALIPHATIC AMINES (*1, 38*)

Amine	R_F in the following solvents[a]				
	1	2	3	4	5
Ammonia	0.00	0.00	0.00	0.04	0.70
Methylamine	0.00	0.00	0.00	0.28	0.74
Ethylamine	0.01	0.00	0.04	0.49	0.71
n-Propylamine	0.02	0.00	0.09	0.65	0.68
n-Butylamine	0.04	0.03	0.18	0.77	—
Isobutylamine	0.04	0.03	0.17	—	0.66
sec-Butylamine	0.05	0.04	0.15	—	—
tert-Butylamine	0.10	0.10	0.30	—	—
n-Amylamine	0.08	0.09	0.29	0.85	—
Isoamylamine	0.09	0.08	0.26	—	0.62
n-Hexylamine	0.12	0.19	0.47	—	0.56
n-Octylamine	0.23	0.39	0.69	—	0.44
n-Decylamine	0.41	tailing	—	—	0.32
n-Dodecylamine	0.60	tailing	—	—	0.19
n-Tetradecylamine	—	—	—	—	0.11
n-Hexadecylamine	0.82	—	—	—	0.06
n-Octadecylamine	0.89	—	—	—	0.03
Dimethylamine	0.00	0.06	0.18	0.82	0.74
Diethylamine	0.03	0.30	0.57	—	0.60
Di-*n*-propylamine	0.08	0.58	0.77	—	0.49
Di-*n*-butylamine	0.25	0.85	—	—	—

[a] (1) Hexane, paper impregnated with 50% dimethylformamide; (2) hexane–benzene (4:1 v/v), paper impregnated with formamide; (3) hexane–benzene (3:2 v/v), paper as in 2; (4) benzene–$CHCl_3$ (1:1 v/v), paper as in 2; (5) 90% aqueous acetic acid, paper impregnated with 1-bromonaphthalene. Detection: *p*-dimethylaminobenzaldehyde (cf. Section I,C) preceded by spraying with a freshly prepared mixture of 15% $SnCl_2$ (3 ml), concentrated HCl (15 ml), and water (180 ml).

TABLE VI

R_F VALUE OF TRYPTAMINE DERIVATIVES

Tryptamine derivative	R_F
N,*N*-Dimethyl	0.74
N,*N*-Dimethyl oxide	0.81
Gramine	0.73
Bufotenine	0.63
5-Methoxy-*N*-methyl	0.70
5-Methoxy-*N*,*N*-dimethyl	0.72
5-Methoxy-*N*,*N*-dimethyl oxide	0.89

TABLE VII

R_F VALUES OF SOME CATECHOLAMINES

Compound	Solvent a (20 hours)	Solvent b (5 hours)
DOPA	0.27	0.21
Norepinephrine	0.34	0.27
Epinephrine	0.38	0.31
Dopamine	0.45	0.37
Metanephrine	0.51	0.42
Tyramine	0.62	0.51
4-Methoxy-3-hyroxymandelic acid	0.73	0.77
Homovanillic acid	0.87	0.90

derivatives as blue spots on a pale pink background. The color formed is proportional to concentration so that the tryptamines can be quantitatively determined after elution from the paper with 5 ml methanol plus 0.5 ml concentrated HCl (*11*).

7. *Diazotized p-nitraniline*

Twenty-five milliliters of a solution of *p*-nitraniline (0.3%) in HCl (80% w/v) is mixed with 1.5 ml sodium nitrite solution (5% w/v) just before spraying. Spray subsequently with aqueous Na_2CO_3 solution (20% w/v). This is a general reagent giving colored spots with many aliphatic and aromatic amines and amino acids (*12*).

Identification after detection is sometimes complicated by multiple zones (i.e., one substance yields more than one zone), for example, when adrenaline is developed after being spotted in HCl solution (*13*, *14*).

D. QUANTITATIVE ESTIMATION

A few examples of the usual quantitative methods (see also Chapter 4) applied to particular amines will be presented.

1. *Densitometry*

Polyaminocarboxylic acids (e.g., EDTA, nitrilotriacetic acid) on paper are sprayed with 1 *N* Folin–Ciocalteau reagent (phosphomolybdotungstic acid) after drying until the paper is moist and yellow. This is followed by spraying with 2.8 *N* Na_2CO_3 (in a hood) until the yellow color has nearly disappeared. The paper is suspended in a closed container for about 15 minutes until the blue spots of the chelates are fully developed and is then air-dried. The blue spots are scanned by a densitometer equipped with an integrator, and the area (extinction) corresponding to each spot is related to concentration by reference to previously prepared calibration curves (*15*).

A similar procedure is applied to the zone of *p*-aminodiphenylamine ($R_F = 0.1$) separated from other aromatic amines by development with cyclohexane on paper impregnated with a 20% alcoholic formamide solution. The paper is sprayed with *p*-(dimethylamino)benzaldehyde, and the resultant red spot is determined directly by using a recording photometer. The results agree well with the method of photometric determination after elution of the spot with ethanol (*16*).

2. *Fluorescence*

Catecholamine and tryptamine derivatives (0.01–10 μg) on paper are dried and sprayed with 5% (w/v) aqueous glycine adjusted to pH 3 with HCl, and oven-dried at 80°C before being suspended in a sealed Kilner jar (3 liter) containing 20 gm paraformaldehyde moistened with 6 ml water. The jar is heated in an oven at 80°C for 3 hours and the papers are viewed under 360 mμ UV light. The fluorescent compounds are eluted with 0.1 *N* HCl and measured in a spectrophotofluorometer. The colors of the zones combined with the values of the maxima, allowed qualitative identification of some compounds. The fluorescence for two compounds was found to be linear over the range 0.2–4 μg, indicating possible quantitative application (*17*).

E. Detailed Experiments

The usual one-dimensional and two-dimensional development techniques are applied to amine separations (*18*). The paper is employed untreated or occasionally acid-washed (*19*), or it is impregnated as described above. Fan-shaped paper gives improved separations of methylamines (*20*).

Four successful separations of aliphatic amines are represented in Fig. 1 (*1*, *8*, *21*, *22*). The conditions for each run are given in the legend.

Aromatic amines are separated as follows: 0.2% solutions of the amines in acetone are spotted on Whatman No. 1 sheets 46 × 57 cm and air dried. The aqueous phase of *n*-butanol–acetic acid–H_2O (40:10:50 v/v) is placed in trays on the bottom of a tank, and the paper is hung from a tray at the top. The tank is closed and the paper is equilibrated for 2 hours. The organic phase of the solvent is then placed in the tray through openings in the cover, and the chromatograph is developed downward overnight (35–40 cm). The sheet is removed, air dried for 2 hours and sprayed with aqueous potassium persulfate solution (1.0% by weight) containing 1×10^{-3} *M* $AgNO_3$. Most colored spots appear at once, but some require heating in an oven at 45°C. The results in this system are shown in the following tabulation (*23*).

Compound	Color	R_F
o-Toluidine	Blue, changing to yellow	0.78
Benzidine	Brown	0.68
Diphenylamine	Blue-green	0.95
p-Anisidine	Purple	0.71
m-Aminophenol	Brown	0.68
α-Naphthylamine	Purple	0.91
o-Phenylenediamine	Pink tail	0.78
p-Phenylenediamine	Blue-green, changing to violet	0.42
o-Aminophenol	Yellow	0.70
p-Toluidine	Pink	0.93
p-Aminophenol	Violet	0.62

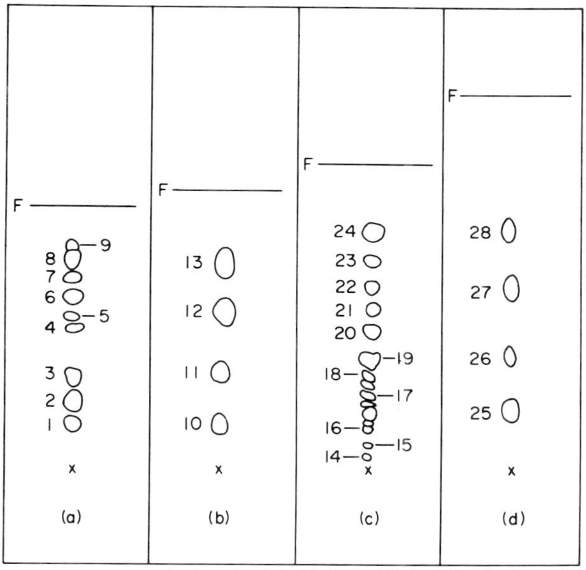

FIG. 1. Separation of mixtures of aliphatic amines. (a) Whatman 3 MM paper; solvent: 2-butanone–propionic acid–H_2O (3:1:1.2 v/v); detection: ninhydrin; (1) lysine, (2) putrescine, (3) histamine, (4) norepinephrine, (5) epinephrine, (6) serotonin, (7) tyramine, (8) tryptamine, and (9) phenylethylamine [after (*21*)]. (b) Paper impregnated with 5% KCl; solvent: *n*-butanol saturated with concentrated KCl solution; (10) methylamine hydrochloride, (11) ethylamine hydrochloride, (12) propylamine hydrochloride, (13) *n*-butylamine hydrochloride [after (*8*)]. (c) Volatile amines (as chlorides) in toasted oat flakes; solvent: *n*-butanol–acetic acid–H_2O (4:1:5 v/v) for 24 hours; detection: ninhydrin; (14) *N,N′*-di-1,4-diaminobutane, (15) 1,3-diaminopropane, (16) 1,4-diaminopropane, (17) methylamine, (18) dimethylamine, (19) ethylamine, (20) isopropylamine, (21) propylamine, (22) isobutylamine, (23) butylamine, (24) amylamine [after (*22*)]. (d) Higher alkyl 3,5-dinitrobenzamides; solvent: hexane; paper impregnated with 50% dimethylformamide; (25) octylamine, (26) decylamine, (27) laurylamine, (28) myristylamine [after (*1*)]. F = solvent front; x = origin.

Amines in natural samples can also be separated by paper chromatography. Urine is either applied directly to the paper or concentrated by adsorption at pH 5.0 on charcoal, from which the amines are eluted with methanol–NH_3 (95:5 v/v) (*18*). Catecholamines (*24*) and tryptophan metabolites (*25*) present in human plasma are separated on ion-exchange resin columns and identified and estimated in the eluate with the aid of paper chromatography and fluorimetry. Volatile amines from plants are extracted with boiling 0.01 *N* HCl, made alkaline, and steam distilled into 0.1–1.0 *N* HCl (*26*). Tryptamines from fresh leaves are extracted with diethyl ether and dried leaves with absolute methanol (*11*).

Tables I–VII contain a selection of R_F values of amines, derivatives and related compounds summarized from the recent literature. Other extensive compilations of R_F values are found in the literature as follows: 2-naphthylamine and derivatives (*18*); aliphatic amines (*27*); 1-alkoxy-2(3)-hydroxy-3(2)-aminocyclohexanes and derivatives (*28*); 6-hydroxy-catecholamines and metabolites (*29*); diaminobenzene derivatives (*30*); primary aromatic amines (*31, 32*); isomeric ureas and thioureas (*33*); β-phenylethylamine derivatives (*14*); 3,5-dinitrobenzoyl derivatives (*38*); arylamines (*39*); phenolic derivatives of phenylethylamine (*40*); guanidine derivatives (*41*); and substituted triazines (*42*).

II. Amino Acids

Many of the considerations mentioned above for amines apply as well to the amino acids which contain both amino and carboxylic acid groups. Because the chromatography of amino acids was covered extensively by Block *et al.* (*44*) and others (*45–47*) and has been reviewed frequently (e.g., *48–50*), it will be the purpose of this section to emphasize important recent work.

Paper chromatography serves mainly for the identification and semi-quantitative estimation of amino acids, while most quantitative amino acid analysis is presently carried out by automated column ion-exchange chromatography. Analyses by paper and column chromatography have been critically compared (*51*). Paper chromatography is also useful for comparing multiple samples, analyzing small volumes of eluates from columns (*52, 53*), and in combination with other separatory methods such as electrophoresis (*54*). Paper chromatography of amino acids has also been compared with thin-layer chromatography (*55*) and microbiological methods (*56*).

The chromatography of small peptides is analogous to that of amino acids, but larger peptides (and most proteins) may not migrate in systems suitable for amino acids . These compounds will be discussed later in this chapter.

A. Preparation of Sample

Paper chromatography is particularly suited to the analysis of protein and peptide hydrolyzates and other amino acid mixtures with low salt content. The most common method for total hydrolysis of proteins or peptides is heating with an *excess* (1:30–1:100 w/v) of redistilled 6 *N* HCl at 105°C for a period of 24–72 hours in evacuated sealed tubes (*57*). The excess HCl is evaporated under reduced pressure over KOH. The hydrolyzate is moistened with water and again evaporated to dryness. This procedure is repeated twice, after which the residue is taken up in a small volume (about 1 ml) of 10% isopropanol for preparation of the sample solution (*44*). Microquantities are treated in sealed capillaries and larger quantities are heated under reflux in thermostatically controlled baths. Partial losses of some amino acids (e.g., tryptophan, cysteine, and tyrosine) occur during hydrolysis. Hydrolysis with 6 *N* HCl–90% formic acid (v/v), 8 *N* sulfuric acid, 80% trifluoroacetic acid, formic acid–acetic acid (10:5.5 v/v), 14% barium hydroxide, or various enzymes has also been suggested (*44*, *57*, *58*).

When biological material is examined, analyses often fail because of the presence of various interferences. Sugars may be eliminated by use of ion-exchange resins (*45*); proteins by ion exchange (*45*), precipitation (*59–61*), dialysis, or ultra-filtration (*62*); and lipids by ion exchange or extraction of protein-free solution with ether or petroleum ether (*45*).

The presence of inorganic salts is especially troublesome. In some cases, good resolution of amino acids is possible even at high loading without preliminary desalting on Whatman 3HR paper eluted with buffered solvent (*63*) or by use of a multiple development technique (see Section D). Desalting procedures employing ion exchange, electrolysis, and electrophoresis have already been described in detail (*44*). A more recent procedure for the purification and desalting of samples of urine for the paper chromatographic determination of amino acids has been described: Shake 20 ml of urine with 2 gm active carbon and 10 drops of $CHCl_3$ and filter through paper. Absorb 10 ml of filtrate on a column of Dowex 50-X4 resin, wash with water, and elute the acids with 1 *N* NH_3 . Evaporate the eluate, dissolve in 0.5 *N* HCl and spot for paper chromatography. If aromatic amino acids are present, run samples through the ion-exchange column only (*64*).

Dry plant material is extracted with ethanol, the extract purified on Dowex-50 resin, and the amino acids separated by paper chromatography–electrophoresis (*65*). Fresh leaves are extracted with 80% ethanol (*66*).

The quantitative paper chromatography of amino acids in foods has been extensively studied and improved (*67*). Proteins are extracted from the foods with appropriate solvents, precipitated with trichloroacetic acid, dissolved in diluted alkali, and reprecipitated. Ten to twenty milligram is then hydrolyzed with 18% HCl for 24 hours as above (*68*).

For the determination of amino acids in seawater, 5 liters is filtered, evaporated by vacuum distillation at 35–50°C to 20–30 ml, and dried at 40–50°C; extract with four portions of acidified acetone (1:100), and evaporate to dryness. Dissolve the residue in 0.5 ml H_2O and spot on the paper (*69*).

The method for analyzing amino acids in musts and wines involves initial removal of proteins by precipitation, and sugar and salts on Dowex 50-X2 resin. The amino acids are finally eluted with 5 *N* NH_3 before preparation of the sample solution (*70*).

These few examples should help the reader to devise appropriate procedures for the extraction, concentration, and purification of amino acids in many types of materials. Standard amino acids dissolved in 10% isopropanol can be stored safely for more than 6 months when kept at less than 20°C (*67*).

The interference due to the presence of salts may also be eliminated by converting the amino acids into derivatives prior to chromatography. The derivatives are soluble in organic solvents and easily separated from the salt. The conversion of amino acids to dinitrophenyl (DNP) derivatives by means of 2,4-dinitrofluorobenzene (DNFB) is most often used. Many procedures for the dinitrophenylation have been published (*44*, *71–73*), and an improved technique suitable for the conversion of amino acids in urine is described below (*74*, *75*).

To 25 ml fresh urine, add 5 *N* NaOH dropwise until phenolphthalein paper turns slightly pink and filter through dry, fluted paper. To 20 ml of filtrate add 5 ml of buffer (8.4 gm $NaHCO_3$ and 2.5 ml 1 *N* NaOH made up to 100 ml with distilled water) and 2 ml freshly prepared 10% (w/v) DNFB solution in ethanol; add 40 ml absolute alcohol and shake or stir the mixture in the dark for 1 hr at 40°C. Cool the mixture to room temperature, adjust the pH to about 12, and transfer to a separatory funnel with a small amount of distilled water and about 10 ml diethyl ether. Shake vigorously, separate the aqueous phase, and extract it twice with 10 ml portions of ether. Discard the ether extracts and add 6 *N* HCl to the alkaline solution until it turns Congo red paper blue. Extract it 6 times with 10 ml portions of ether and then evaporate the combined ether extracts to dryness under vacuum. Immediately before chromatography, take up the residue in 1 ml of acetone or ethyl acetate. To recover the water-soluble DNP compounds, extract the acidic aqueous solution with six 10 ml portions of *n*-butanol–ethyl acetate (1:1 v/v) and evaporate the extract. Immediately before chromatography, take up the residue in 1–2 ml ethyl acetate–acetic acid–*n*–butanol (100:1:99 v/v). All operations should be carried out in the absence of direct illumination.

B. Paper

Unmodified paper such as Whatman Nos. 1, 3, or 4 and S. and S. 589 and 598 have been most often recommended for general use with amino acids and derivatives. However, DeLigney (*76*) has recently determined the order of decreasing efficiency for amino acid separations to be Whatman No. 31 ET > No. 4 > No. 3 > No. 3 MM > No. 1 > No. 2. For quantitative chromatography of amino acids, some workers prefer S. and S. No. 507, and for larger amounts, thicker papers such as Whatman No. 3 or S. and S. No. 470-A (*44*).

Knight (*77*) has examined factors causing lack of reproducibility and suggests specially manufactured papers of high resolving power for optimum results. Control of the initial moisture content of the paper and the flow rate of the solvent, which can be modified by employing a wick-feed device made of another type of paper, are especially important.

Buffered papers are also useful for particular separations which benefit from pH control. For example, the paper may be impregnated with pH 8.4 buffer (50 ml 0.067 *M* boric acid and KCl, and 8.55 ml 0.067 *M* NaOH plus 0.3 gm of EDTA) and developed with *m*-cresol saturated with the same buffer to separate alanine ($R_F = 0.15$), arginine (0.17), hydroxyproline (0.22), tyrosine (0.27), histamine (0.33), valine (0.42), and methionine (0.53); development instead with *n*-butanol–benzyl alcohol–pH 8.4 buffer (45:45:8 v/v) separates valine (0.35), methionine (0.45), isoleucine (0.60), and leucine (0.70). If the chamber is adequately equilibrated and the temperature remains constant, R_F values in such systems are reproducible with a standard deviation of about 0.01 (*45*, *78*).

Glass paper (No. 944-AH, H. Reeve Angel & Co.) chromatography (*79*) is used for the separation of 2,4-dinitrophenylamino acids with solvent systems of isooctane–$CHCl_3$–acetic acid (10:10:1 and 30:20:1 v/v). Development time is 40 minutes. Phenylthiohydantoin amino acids have been identified on starch-impregnated glass paper with the solvent *n*-heptane–$CHCl_3$ (1:1 v/v). The glass paper is first silicated (*80*) and then immersed in 0.5% starch solution and dried at room temperature overnight.

Two types of modified papers with ion-exchange properties have been used for amino acid analyses. Modified cellulose papers have covalently bound ionizable groups, e.g., cellulose phosphate (Whatman P81) and diethylaminoethylcellulose (Whatman DE81). Resin-loaded papers contain particles of conventional polystyrene cation-exchange (Whatman SA-2) or anion-exchange resin (Whatman SB-2) physically incorporated with the cellulose fibers during manufacture. The relative rates of migration of amino acids on both types of papers have been extensively studied (*81*). Most of the common amino acids may be separated on cation-exchange paper (SA-2) with citrate buffer (pH 3.5) having a Na^+ concentration of 0.8 *N* (*82*). The same paper was developed upward with 0.05 *M* potassium phosphate buffer (pH 7.0) to separate radioactive valylhydroxamate and valine (*83*).

The wet-start technique (see Chapter 15, Section I,D,3), in which the sample is applied directly to equilibrated paper, is usually employed with ion-exchange cellulose papers. However, the conventional dry-start method affords higher resolving power for amino acid separations on P81 paper developed with 0.2 *M*, pH 3.1, sodium buffer, probably because of physical solvent gradients along the paper (see Chapter 15, Section I,D,4) (*84*).

Figure 2 illustrates the separation of amino acids by one-dimensional development on the two types of ion-exchange papers. A completely different sequence is obtained on the resin-loaded paper (b) than on the cation-exchange cellulose paper (a). The conclusion reached is that differences in the degree of nonionic interaction between the cellulose–resin matrix and the amino acid side chain combined with differences in the ionizing characteristics of the acidic or basic groups determine the sequence of the compounds within the main groups (i.e., acidic, basic, and neutral) (*84*).

Both types of exchange papers have been used successfully for two-dimensional chromatography of amino acids (*85*). By selecting the pH and composition of the solvents so that an ion-exchange mechanism predominates in one direction and partition–adsorption in the other, separations superior to conventional paper chromatography can result (cf. Chapter 15, Section I,D,2). This technique is illustrated in Fig. 3; with solvent 2, the ion-exchange cellulose paper functions as ordinary cellulose paer.

Results on paper impregnated with the inorganic ion exchanger zirconium phosphate are essentially the same as those obtained on the other ion-exchange papers just mentioned (*86*).

DEAE paper can be used for the separation of amino acids by electrophoresis in one direction and simultaneous centrifugal chromato-

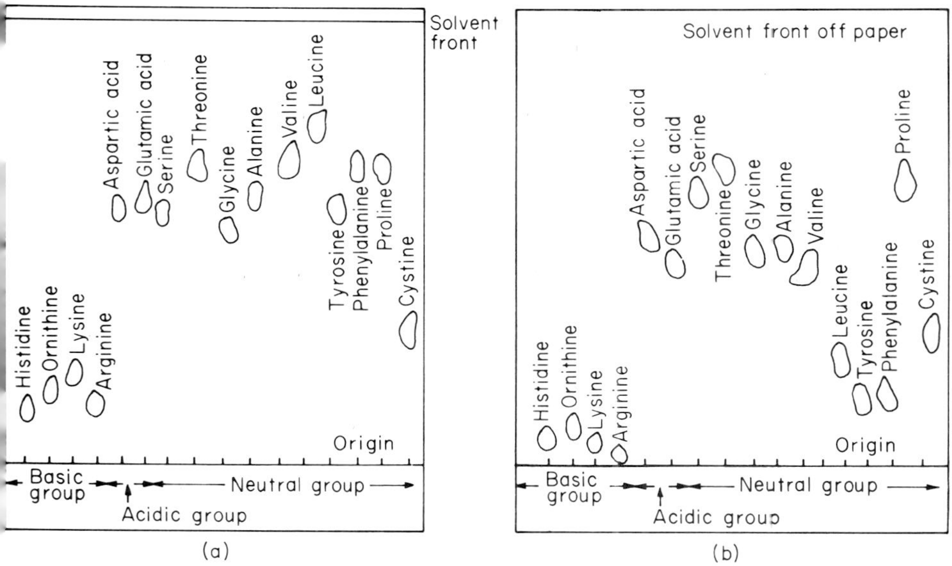

FIG. 2. Effect of exchanger matrix on migration of amino acids on ion-exchange papers. (a) Acids separated in 0.2 *M*, pH 3.1, sodium buffer on cellulose phosphate paper (Whatman P81) and (b) on an experimental sulfonated polystyrene resin-loaded paper (similar to Amberlite SA-2). After (*81*) and (*84*).

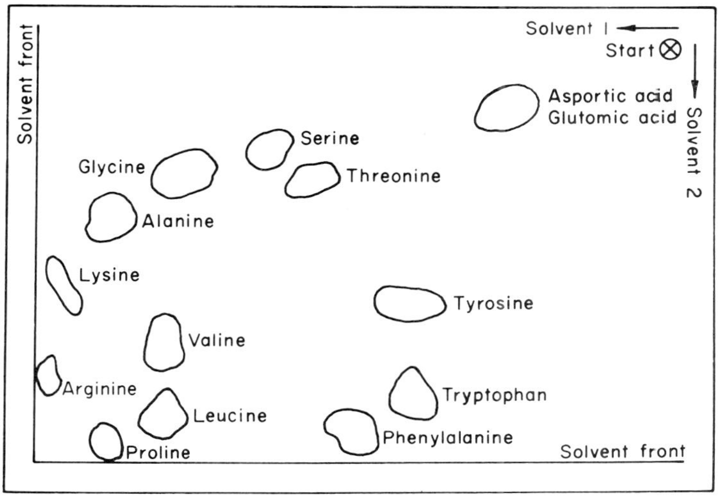

FIG. 3. Two-dimensional separation of amino acids on DEAE anion exchange cellulose paper (Whatman DE81). Solvent 1, 0.001 *M* NaCl (conventional dry-start technique); solvent 2, *m*-cresol–0.3% ammonia. After (*84*) and (*85*).

graphy in the transverse direction (*87*). Twenty-two acids (instead of five when Whatman 3 MM paper is used) in urine, plasma, or red cells can be separated, detected with ninhydrin, eluted with 0.1 *M* NaOH (elution is reproducible but not complete) and estimated by spectrophotometry.

C. Solvents

Hundreds of solvents for the separation of amino acids and their derivatives have already been tabulated and discussed (*44*) in terms of their general effect on R_F values as summarized earlier by Burma (*88*). These solvents are mostly modifications of several types: (a) phenol-based solvents, which are especially useful for separating acidic amino acids; (b) alkaline solvents containing collidine or other pyridine bases, which are often used as the second solvent after phenol for two-dimensional chromatography; (c) and solvents containing alcohols (e.g., *n*-butanol) or ketones (methyl ethyl ketone). Although there is no universal solvent capable of separating all of the usual amino acids in one dimension, solvents such as *n*-butanol–acetic acid–H_2O (4:1:5 v/v) or those mentioned in Section B for buffered paper do separate many of the common amino acids.

In addition, many special solvents have been devised for the separation of specific amino acids or groups. For example, the quantitative separation of asparagine and glutamine is obtained with a freshly prepared mixture of *n*-propanol–acetic acid–H_2O (7:1:1 v/v), but not with the usual butanol solvents (*89*). Acidic amino acids are separated with phenol–ammonia and basic acids with lutidine saturated with 0.022 *M* phosphate buffer (*78*).

t-Amyl alcohol–dioxane–H_2O (5:1:2 v/v) cleanly separates the homologous series of amino acids from glycine to α-amino-*n*-tetradecanoic acid and also valine, leucine, and isoleucine. The paper (Whatman No. 1) is treated with a 0.05% solution of potassium ferrocyanide (0.05 gm salt in 70 ml methanol and 40 ml H_2O) and air dried before use to prevent bearding of spots (*90*).

Iodinated tyrosines are separated by development with *n*-butanol saturated with acetic acid and iodinated thryonines with butanol saturated with 2 *N* NH_4OH (*91*).

Amino acids were chromatographed on moist buffered paper with solvents composed of hydrogen di(2-ethylhexyl)orthophosphoric acid (a liquid ion exchanger) in *n*-hexanol. The migration mechanism is apparently partition plus ion exchange, and separations seem to be feasible although none were attempted (*92*).

With two-dimensional development, large numbers of the common amino acids can be separated on one sheet of paper. Solvent systems for two-dimensional chromatography have been recently tabulated (*75*) and numerous maps of amino acid and DNP–amino acid separations employing various solvent combinations have been published (*44*, *45*). A technique for selecting optimum two-dimensional systems has been suggested (*93*). An improved general system employs *s*-butanol–*t*-butanol–2-butanone–H_2O–diethylamine (40:40:80:50:1 v/v) followed by *s*-butanol–88% formic acid–H_2O (15:3:2 v/v) (*94*). The following solvents are recommended for the separation of neutral sulfur amino acids: *n*-butanol–formic acid–H_2O (75:15:10 v/v) followed by acetone–methyl ethyl ketone–H_2O (40:40:20 v/v) (*95*).

Tables VIII (*96–102*) and IX (*103*, *104*) contain a selection of solvents and R_F values, most of which have been recently reported in the literature. The reader is referred to Block *et al.* (*44*) and Smith (*46*) for earlier collections of R_F values. R_F values have also been tabulated recently for 5-substituted tryptophans (*105*), some new aliphatic 2-amino-3-hydroxy acids (*106*), allenic amino acids (*107*), sulfur amino acids (*108*), and acetylated amino acids (*108a*).

D. Techniques

Hanes has made a detailed study of the resolving power and reproducibility of systems for the one-dimensional separation of the amino acids in protein hydrolyzates (*63*, *109–111*). He finds that a solvent composed of *n*-propanol–ethanol–aqueous pyrophosphate (2:1:1 v/v) on paper buffered to pH 6.5 separates thirteen to fifteen amino acids from a mixture of eighteen; the pairs aspartic acid–cysteic acid and serine–glycine are not resolved but can be in a phenol–aqueous borate system at pH 9.3. The chromatographic apparatus and techniques for this system are quite complicated (involving, for example, a paper feederwick machine-sewn to the top of the paper) and cannot be described here. The reader is referred to Hanes (*109*) and Wade *et al.* (*110*) for details.

A one-dimensional separation of amino acids or peptides from biological mixtures containing high concentrations of proteins and salts is effected, without prior removal of salts or proteins, by redeveloping the chromatogram several times with intermediate drying. Strips of S. and S. 597 or Whatman 3 MM paper are developed by descending flow of *n*-butanol-containing solvents for two to four successive 18 hour periods with drying between developments. Clear separations are obtained from samples in which the NaCl to total amino acid ratio is 7:1; the addition of 20% human serum does not interfere (*112*).

Another method for separating certain amino acids avoiding preliminary desalting utilizes three solvent developments in two directions

TABLE VIII

R_F VALUES OF AMINO ACIDS AND RELATED COMPOUNDS

Amino acid	R_F under the following conditions[a]											
	1	2	3	4	5	6	7	8	9	10	11	12
Cystine	0.13	—	—	—	—	0.09	—	—	—	0.30	0.20	0.33
Cysteine	—	—	—	—	—	0.10	—	—	—	—	0.18	0.58, 0.23
Histidinc	0.16	—	—	—	—	0.12	0.23	—	—	—	0.28	0.35
Lysine	0.19	—	—	—	—	0.12	0.13	—	—	—	0.15	—
Aspartic acid	0.22	—	—	—	—	0.23	0.17	—	—	—	0.40	—
Asparagine	0.22	—	—	—	—	0.12	—	—	—	—	0.21	—
Arginine	0.24	—	—	—	—	0.15	—	—	—	—	0.16	—
Serine	0.27	—	—	—	—	0.22	—	—	—	—	0.42	—
Glycine	0.30	0.16	0.20	0.15	0.17	0.23	0.26	—	—	—	0.37	—
Glutamic acid	0.34	—	—	—	—	0.28	0.19	—	—	—	0.48	—
Glutamine	0.34	—	—	—	—	0.17	—	—	—	—	0.31	—
Threonine	0.39	—	—	—	—	0.26	—	—	—	—	0.50	—
Hydroxyproline	0.43	—	—	—	—	0.22	—	—	—	—	0.45	—
Alanine	0.47	0.22	0.30	0.19	0.23	0.30	0.27	—	—	—	0.52	—
Tyrosine	0.51	—	—	—	—	0.45	0.60	—	—	—	0.50	—
Proline	0.59	—	—	—	—	0.34	—	—	—	—	0.55	—
Tryptophan	0.62	—	—	—	—	0.50	—	—	—	—	0.42	—
Valine	0.62	—	—	0.27	—	0.51	—	—	—	—	0.65	—
Methionine	0.68	—	—	—	—	0.56	—	—	—	—	0.60	0.66
Phenylalanine	0.72	—	—	—	—	0.60	—	—	—	—	0.60	—
Leucine	0.80	—	—	0.38	—	0.70	—	—	—	—	0.70	—
Isoleucine	0.80	—	—	0.35	—	0.67	—	—	—	—	0.68	—
n-Butyric	—	0.31	0.39	0.22	0.29	—	—	—	—	—	—	—
Norvaline	—	0.46	0.52	0.31	0.40	—	—	—	—	—	—	—

Norleucine	—	0.58	0.61	0.41	0.50	—	—	—	—	—	—	—
Heptanoic	—	0.66	0.68	0.49	0.58	—	—	—	—	—	—	—
Octanoic	—	—	0.72	0.55	0.64	—	—	—	—	—	—	—
Nonanoic	—	—	0.74	0.58	0.68	—	—	—	—	—	—	—
S-Methylcysteine *S*-oxide	—	—	—	—	—	0.17	—	—	—	—	—	—
S-Methylcysteine	—	—	—	—	—	0.37	—	—	—	—	—	—
Cysteine methyl ester	—	—	—	—	—	0.16	—	—	—	—	—	—
N-Methylcysteine	—	—	—	—	—	0.12	—	—	—	—	—	—
S-Ethylcysteine	—	—	—	—	—	0.54	—	—	—	—	—	—
Glutathione	—	—	—	—	—	0.24	—	—	—	—	—	—
S-Methylglutathione	—	—	—	—	—	0.28	—	—	—	—	—	—
S-Methylcysteine sulfoxide	—	—	—	—	—	—	—	0.50	0.17	—	—	—
Methionine sulfoxide	—	—	—	—	—	0.22	—	0.40	0.33	—	0.45	0.55
Methionine sulfone	—	—	—	—	—	0.24	—	0.52	0.36	—	0.43	0.56
Taurine	—	—	—	—	—	0.16	—	0.66	0.17	—	0.31	0.60
Cysteic acid	—	—	—	—	—	0.10	—	0.32	0.11	0.71	0.39	0.42
Homocysteic acid	—	—	—	—	—	0.12	—	0.32	0.16	0.27	0.46	0.41
Homocystine	—	—	—	—	—	0.50	—	—	—	0.39	0.31	0.62, 0.34
Homocystine thiolacetone	—	—	—	—	—	—	—	—	—	0.85	0.81	—
Cystathionine	—	—	—	—	—	0.10	—	—	—	0.46	0.19	—
Homoserine	—	—	—	—	—	0.30	—	—	—	—	—	0.58
Adenine	—	—	—	—	—	0.56	—	—	—	—	—	0.58
DL-*threo*-β-Methylaspartic acid	—	—	—	—	—	0.30	—	—	—	—	—	—
DL-β, β-Dimethylaspartic acid	—	—	—	—	—	0.32	—	—	—	—	—	—

[a] (1) Phenol–*n*-butanol–methyl ethyl ketone–propionic acid–acetic acid–H_2O (20:20:50:10:10:20 v/v), circular development on Whatman No. 2 paper (*96*); (2) *n*-butanol–acetic acid–H_2O (4:1:5 v/v), ascending, Whatman No. 1 (*90*); (3) *n*-butyl acetate–*n*-butanol–acetic acid–H_2O (47:9:28:16 v/v) (*90*); (4) *t*-amyl alcohol–dioxane–H_2O (5:1:2 v/v) (*90*); (5) *t*-amyl alcohol–pyridine–H_2O (2:1:1 v/v) (*90*); (6) *n*-butanol–acetic acid–H_2O (12:3:5 v/v) (*46*, *97–99*); (7) *n*-butanol–pyridine–H_2O (1:1:1 v/v), Whatman No. 1 (*100*); (8) pyridine–acetone–NH_3–H_2O (45:30:5:20 v/v) (*101*); (9) isopropanol–formic acid–H_2O (75:12.5:12.5 v/v) (*101*); (10) acetone–0.5% aqueous urea (6:4 v/v), Whatman No. 3 MM (*102*); (11) methanol–pyridine–H_2O (80:4:20 v/v) (*46*, *102*); (12) ethanol–acetic acid–H_2O (65:1:34 v/v), Whatman No. 1 (*99*).

TABLE IX

R_F VALUES OF AMINO ACID DERIVATIVES

Amino Acid	R_F [a] with following solvents				R_F [b]
	1	2	3	4	
Arginine	0.33	0.30	0.82	0.52	0.43
Histidine	0.13	0.21	0.54	0.57	0.40
Aspartic Acid	0.0	0.02	0.02	0.66	0.04
Cysteic Acid	0.0	0.0	0.0	0.68	0.03
Glutamic Acid	0.0	0.02	0.02	0.62	0.04
Alanine	0.25	0.32	0.52	0.48	0.51
Asparagine	0.07	0.10	0.23	0.48	—
Glycine	0.17	0.19	0.34	0.43	0.41
Phenylalanine	0.64	0.51	0.84	0.32	0.60
Glutamine	0.07	0.13	0.29	0.48	—
Isoleucine	0.68	0.56	0.76	0.44	0.70
Lysine	0.76	0.56	0.82	0.02	0.83 (di-)
Leucine	0.68	0.56	0.81	0.45	0.70
Proline	0.20	0.30	0.59	0.55	0.42
Serine	0.12	0.17	0.30	0.53	0.32
Threonine	0.15	0.23	0.35	0.57	0.42
Tryptophan	0.56	0.46	0.74	0.14	—
Valine	0.47	0.45	0.73	0.48	0.65
Cystine	—	—	—	—	0.19 (di-)
Methionine	—	—	—	—	0.64
Tyrosine	—	—	—	—	0.69 (di-)
Tyrisine	—	—	—	—	0.46 (*o*-mono-)

[a] Dinitropyridyl derivatives, on Whatman No. 1 paper (*103*). (1) Toluene-pyridine-ethylene chlorohydrin-0.8 *M* NH_3 (5:1:3:3 v/v). (2) *n*-Butanol-3% NH_3 (1:1 v/v). (3) *iso*-Amyl alcohol-phenol-water (1:1:1 v/v). (4) 1.5 *M* phosphate buffer, pH 6.

[b] Pipsyl derivatives on Whatman No. 1 paper equilibrated 5 hours. Detection: expose paper to Cl_2 gas and dip into saturated solution of *o*-tolidine in acetic acid–H_2O (1:49 v/v) (*104*). Solvent: *n*-butanol–1 *N* NH_3 (5:3 v/v, upper layer).

(*112a*). Untreated urine is applied in one corner of a sheet of Whatman No. 1 paper (35 × 28 cm). The first "desalting" run is overnight with isopropanol–H_2O–formic acid (18:2:2 v/v) along the length of the paper. This is followed by a second run for 4 hours along the breadth of the paper with acetone–pyridine–NH_3 (5:3:2 v/v) and a third run overnight, again along the length of the paper, with isopropanol–*n*-butanol–isoamyl alcohol–formic acid–H_2O (5:2:2:1.5:2 v/v).

For the ascending separation of a homologous series of amino acids on Whatman 1 paper treated with ferrocyanide (see Section C), the paper is cut into tapered strips 28 cm long, 2.3 cm wide at the top, and 1.5 cm wide at the base. The test solution (0.2 μl containing 2–4 μg of each amino acid) is spotted 1.5 cm fron the narrower end, and the unconditioned strips are developed in 30 cm test tubes at room temperature. When the developer reaches the top of the paper, the paper is removed, air dried, and sprayed with 0.25% ninhydrin in acetone. The color develops at room temperature. R_F value deviations are within ± 0.02 (*90*). Straight-chain amino acids developed following this procedure with *t*-amyl alcohol–dioxane–H_2O (9:3:5 v/v) have the following R_F values: hexanoic (C_6), 0.53; C_7, 0.58; C_8, 0.64; C_9, 0.67; C_{10}, 0.68; C_{11}, 0.71; C_{12}, 0.71; C_{13}, 0.72; C_{14}, 0.72.

Two-dimensional chromatography is accomplished by ascending development on 10 inch squares of Ederol 202 paper in saturated tanks with the *s*-butanol solvents described in Section C. Figure 4 illustrates

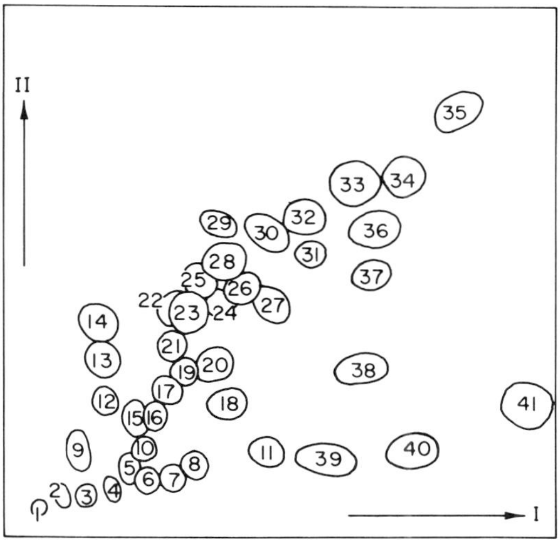

FIG. 4. Two-dimensional separation of various amino acids and related compounds. Solvent I, *s*-butanol–*t*-butanol–2-butanone–H_2O–diethylamine (40:40:80:50:1 v/v); solvent II, *s*-butanol–88% formic acid–H_2O (75:15:10 v/v). Numbered spots: (1) DL-3,4-dihydroxyphenylalanine; (2) L-glutathione; (3) DL-lanthionine; (4) L-cystine; (5) DL-homocysteic acid; (6) L-cysteic acid; (7) L-α-γ-diaminobutyric acid; (8) L-histidine; (9) L-arginine; (10) DL-asparagine; (11) taurine; (12) DL-aspartic acid; (13) L-glutamic acid; (14) α-aminoadipic acid; (15) DL-citrulline; (16) L-glutamine; (17) glycine; (18) DL-serine; (19) L-azetidine-2-carboxylic acid; (20) DL-homoserine; (21) β-alanine; (22) γ-aminobutyric acid; (23) DL-α-alanine; (24) L-proline; (25) β-aminoisobutyric acid; (26) L-tyrosine; (27) *S*-methylcysteine; (28) DL-α-aminobutyric acid; (29) DL-α-aminoisobutyric acid; (30) DL-valine; (31) DL-methionine; (32) DL-norvaline; (33) L-leucine + isoleucine; (34) DL-thyronine; (35) α-aminocaprylic acid; (36) DL-β-phenylalanine; (37) DL-tryptophan; (38) DL-threonine; (39) 1,4-diaminobutane; (40) 1,5-diaminopentane; (41) 1,7-diaminoheptane. After (*94*).

the separation and identification of forty-one amino acids and related compounds by this procedure (*94*).

Solvents are supplied by a small wick in a hole at the center of the paper for the circular chromatography of protein hydrolyzates (*113*). Multiple development is effective for the separation of compounds not resolved by one solvent run (*114*).

Serum amino acids are separated on paper formed into an inverted truncated cone and developed with *n*-butanol–acetic acid–H_2O (4:1:1 v/v). Only three pairs are not separated—aspartic acid–serine, threonine–glutamic acid, and leucine–isoleucine (*115*).

Histidine and aspartic acid are separated by placing the spotted paper strip in a chamber for the establishment of a thermal gradient for 15 minutes before development with *n*-butanol–acetic acid–H_2O (4:1:5 or 9:1:1 v/v). Good resolution is obtained with gradients of 2–3°C per cm (thermal paper chromatography, see Chapter 3) but not with isothermal development at 20° or 60°C (*116*).

The quantities of amino acids applied to the paper depends upon the experimental conditions (method of detection, type and size of the paper, method of development, etc.) and the amino acid itself. In general, the lower limit of detection is between 0.01 and 0.001 μmole, and 0.1 μmole (e.g., 2.5 μl of a .004 *M* solution) is usually a convenient amount for ready detection. The minimum quantities necessary for detection after two-dimensional development have been tabulated for fifty-six amino acids and peptides and range from 0.06 μg for α-alanine to 40 μg for diiodotyrosine (*44*). On thicker paper such as Whatman No. 3, up to 2 mg of each amino acid can be applied for each cm of width (*45*).

E. Detection

Ninhydrin (triketohydrindene hydrate) is the most useful and most widely used spray or dip reagent for detecting amino acids on paper chromatograms. A positive test (usually purple) is obtained with nearly all compounds having a free amino group (peptides, proteins, amino sugars and alcohols, ergot alkaloids, etc.). Many modifications have been proposed for the preparation of the reagent; several of these were listed in Section I,C of this chapter and many others in Block *et al.* (*44*). One of the simplest is a 0.25% solution of ninhydrin in acetone (*90*), which is, however, unstable. Another is composed of 1 gm of ninhydrin in a mixture of ethanol (700 ml), 2,4,6-collidine (29 ml), and acetic acid (210 ml); the paper is dried after spraying at 90°C until spots appear (*75*). We have found that detection limits with this version average 1–2 μg for the acids tested. Still another is 5% of CdI_2 and 1% of ninhydrin in 95% ethanol (*117*).

Improved differentiation of amino acids can be obtained by noting the variety of colors formed with ninhydrin reagent (a 4% solution in isopropyl alcohol containing 5% 2,4,6-collidine), the speed of appearence of these colors at various temperatures and relative humidities, and the fluorescence at 350 mμ after spraying with 1% KOH (*117a*).

The spots formed by reaction with ninhydrin are not stable with time. Zweig (*44*) advises the following procedure for the conservation of chromatograms. The sprayed chromatogram is dipped into dilute copper nitrate [1 ml saturated aqueous $Cu(NO_3)_2$ plus 0.2 ml of 10% v/v HNO_3 diluted to 100 ml with ethanol]. The papers are quickly neutralized with NH_3 vapors, air dried, and sprayed with a clear acrylic spray.

Numerous other reagents, both nonspecific and with a greater degree of specificity than ninhydrin, have been suggested (*44*). These specific reagents, especially when used in sequence on the same chromatogram, are very helpful in establishing the identity of separated amino acids (*118*). Space does not permit relisting these alternate detection methods; interested readers are referred to earlier comprehensive compilations (*44–46*). One example of a specific detection method follows: amides or amino acids dissolved in dilute HCl, or urine, serum, or plasma samples suitably treated, are developed by descending or horizontal circular chromatography on Whatman Nos. 1, 3, 3 MM, 4 or S. and S. 597 filter paper with *n*-butanol–acetic acid–H_2O. Treatment with ninhydrin locates asparagine at $R_F < 0.1$ in the same region as zones of glutamine, arginine, and aspartic acid. For the specific detection of asparagine, the chromatograms are subsequently dipped in a 5% borate solution, washed, treated with 0.05–0.1 *N* HCl and redried. The asparagine spot turns from purple to blue while other amino acids turn red and fade out (*119*).

One recently reported general detection method for amino acids employs the spraying reagent 0.5% *N*-hydroxyphthalonimide in 96% ethanol, followed by air drying and heating for 15 minutes at 140°C; numerous amino acids produce colors in shades of blue, green, and violet (*119a*). In another method, the paper is sprayed with dimethylsulfoxide followed by drying and spraying with 0.3% (w/v) 7,7,8,8-tetracyano-quinodimethan in pyridine–acetone (1:1 v/v). After 15 minutes air drying many amino acids, aliphatic amines, and amino acid derivatives give a brown or green visible color and an orange fluorescence under ultraviolet light (*119b*).

The DNP amino acids are yellow in color and can be detected in ultraviolet light (360 mμ).

Radioactive amino acids containing ^{14}C or ^{35}S are located by scanning the paper with a counting probe (*99*) or by preparing radioautographs (*120*).

F. Quantitative Determination

Quantitative paper chromatography applied to the amino acids is described in great detail in Block *et al.* (*44*). Some of the methods covered are given below.

1. *Quantitative Methods Described Elsewhere* (*44*)

a. Visual Comparison. A series of dilutions of the unknown and a series of known concentrations (equal volumes are always applied) are developed side by side; after detection, the spots are visually compared to find a match with respect to area and color density. The method is tedious and of low accuracy.

b. Elution. The mixture is developed by one- or two-dimensional chromatography; after drying, the zones are detected with dilute ninhydrin or ultraviolet light, cut out, and placed in individual test tubes. Ninhydrin reagent is added, the tubes are shaken, and the extracted color is read in a colorimeter at 550 mμ (440 mμ for proline and hydroxyproline). Accuracy is often better than ±5%.

c. Area of Spot. The procedure is similar to the visual comparison method except that spot areas derived from the standards are plotted on semi-logarithmic paper against concentration, and the concentration of the unknown is read off. If the separated zones have distinct edges, accuracies of ±5% are possible.

d. Total Color of Spot. After one-dimensional development of alternately spotted known and unknown quantities, the color is developed and strips of paper are cut parallel to the direction of solvent flow are scanned using a transmission densitometer. From plots of percentage transmission versus distance, the areas representative of each amino acid are determined and compared with areas of standards. Concentration ratio on Cartesian coordinate paper provides a standard curve for each compound from which the unknwon concentrations can be calculated. Either the entire spot or just the spot centers are read with the densitometer. Accuracy is ±5% with necessary care.

e. Maximum Color Density. This is the procedure recommended and described most fully (*44*). It is the same as method 4 except that the concentration of the unknown is read directly from calibration curves prepared from the maximum color densities of the standard spots. Accuracy of better than ±5% and reproducibility of ±0.5% have been reported.

Seventeen modifications of the *in vitro* procedure (1, b above) and two modifications of the direct paper method (1d and e above) are also described (*44*).

The principal factors affecting the reproducibility of R_F values in amino acid chromatography are changes in the composition of the mobile phase and temperature and nonuniformity of the paper causing uneven solvent migration. The principal source of error in quantitative work is the volumetric error in sample spotting. It was found that butanol–acetic acid–H_2O systems result in reproducible R_F values, but that the error in quantitative determinations is smaller with the phenol–H_2O system (*121*).

Presented below are a few recent modifications and applications of quantitative amino acid determinations.

2. *In Vitro Procedures*

Method 1. A standardized procedure for the quantitative determination of amino acids has been developed. New batches of paper are tested by visual inspection in ultraviolet light for dark gray blotches (reject) or fluorescent spots. If any paper shows a fluorescence, submit a sheet to blank development with *sec*-butyl alcohol–85% formic acid–H_2O (15:3:2 v/v). Dry the paper, draw it through a 0.001 *M* solution of an amino acid and dry it for several hours. Draw the paper through 0.5% ninhydrin solution in ethanol–acetone (1:9 v/v) and after 10 minutes at 20°C, heat the paper for 30 minutes at 60°C. Discard papers that do not give a uniform blue color. Mixtures are developed on the selected paper with the above solvent or with propanol–H_2O (3:1 v/v) or *n*-butanol–acetic acid–H_2O (4:1:1 v/v). Standard solutions of acids are developed alongside the unknown. The air-dried chromatogram is drawn through ninhydrin reagent, air dried, and exposed for 15 hous at ~30°C to an ethanol-saturated atmosphere; a dish containing 50% ethanolic citric acid in the chamber reduces the blank value. Remove the paper and keep it in the dark for 2 hours before examination. Cut out the spots, extract them with 0.1 *M* $NaHCO_3$–96% ethanol (1:1 v/v) and measure the absorbancies of the extracts at 578 mμ. The accuracy and reproducibility is excellent, especially for glycine, alanine, and aspartic acid (*122*).

Method 2. After one- or two-dimensional development, the chromatograms are treated with 0.2% ninhydrin in acetone and kept at 60°C for 10 minutes. The amino acids are eluted and treated for 20–40 minutes at 65°C with 2.5% ninhydrin in methanol. The violet reaction products are mixed with cadmium acetate to give stable orange-red complexes. Upon dilution with methanol, the absorbance is read at 500 mμ. Calibration curves are linear from 25 to 500 γ. Accuracy for amino acids in plant material is $\pm$4–6% (*123*).

A modification of this procedure involves staining the separated amino acids on the chromatogram with a solution prepared by dissolving 0.5 gm cadmium acetate in 50 ml H_2O to which 10 ml glacial acetic acid has been added. Acetone is added to a volume of 500 ml, and solid ninhydrin to a concentration of 0.2% w/v. The colored spots are cut out, sewn to specially cut tongued paper, eluted with methanol, and determined colorimetrically. Amino acids within the range 0.1–40 μg are determined with an accuracy of $\pm$10% (*124*).

Method 3. The dry chromatogram is dipped in 0.5% ninhydrin solution and dried for 30 minutes at 80°C, dipped in H_2O, dried, and redipped in ninhydrin. The dry paper is dipped in $Cu(NO_3)_2$ solution (to 0.1 gm Cu_2O and 0.2 ml 65% HNO_3, 10 ml H_2O is added and the volume is made up to 100 ml with ethanol) and dried at 80°C. Each spot is cut out, extracted for 2 hours with methanol, and the absorbance is read at 500 mμ using methanol as a blank. Standards are similarly treated. The accuracy is 1–2% for amino acids in deslated urine (*125*). Amino acids in feeds are similarly determined by photometry of copper complexes (*126*).

Method 4. The chromatogram is sprayed with diazotized sulfanilic acid and the colored spots eluted with water. The absorbance of the eluate is determined at 496 mμ. Histidine in urine is determined with a sensitivity of 5–40 γ (*127*).

3. *Direct Estimations on Paper*

A photometer has been designed in which errors associated with the measurement of maximum color densities of spots on chromatograms are reduced. Good results are claimed in triplicate assays by the close control of chromatographic conditions and color development (*128*).

Method 1. Quantitative densitometry requires sharply defined, uniform zones for best results. An improved technique for sample application leads to such zones. Five to twenty-five microliter samples of amino acids are saturated into paper disks (4 $\times$ 3 mm, oval) cut out of paper wick (Beckman No. 319329). The disks are allowed to dry and are inserted with pressure into precut holes of the same size in the chromatogram. No cement is required. The chromatogram is then developed by descending solvent flow and sprayed with Sigma ninhydrin (55 parts of 0.3% ninhydrin in ethanol, 45 parts Freon); the densities of each acid are recorded with a scanning densitometer. This method is more accurate and reproducible (relative standard deviation, $\pm$6.1%) than the usual methods with which it was compared (*129*) (Fig. 5).

Method 2. After spraying with ninhydrin, the dried chromatogram is photographed with negative microfilm, and the developed negative

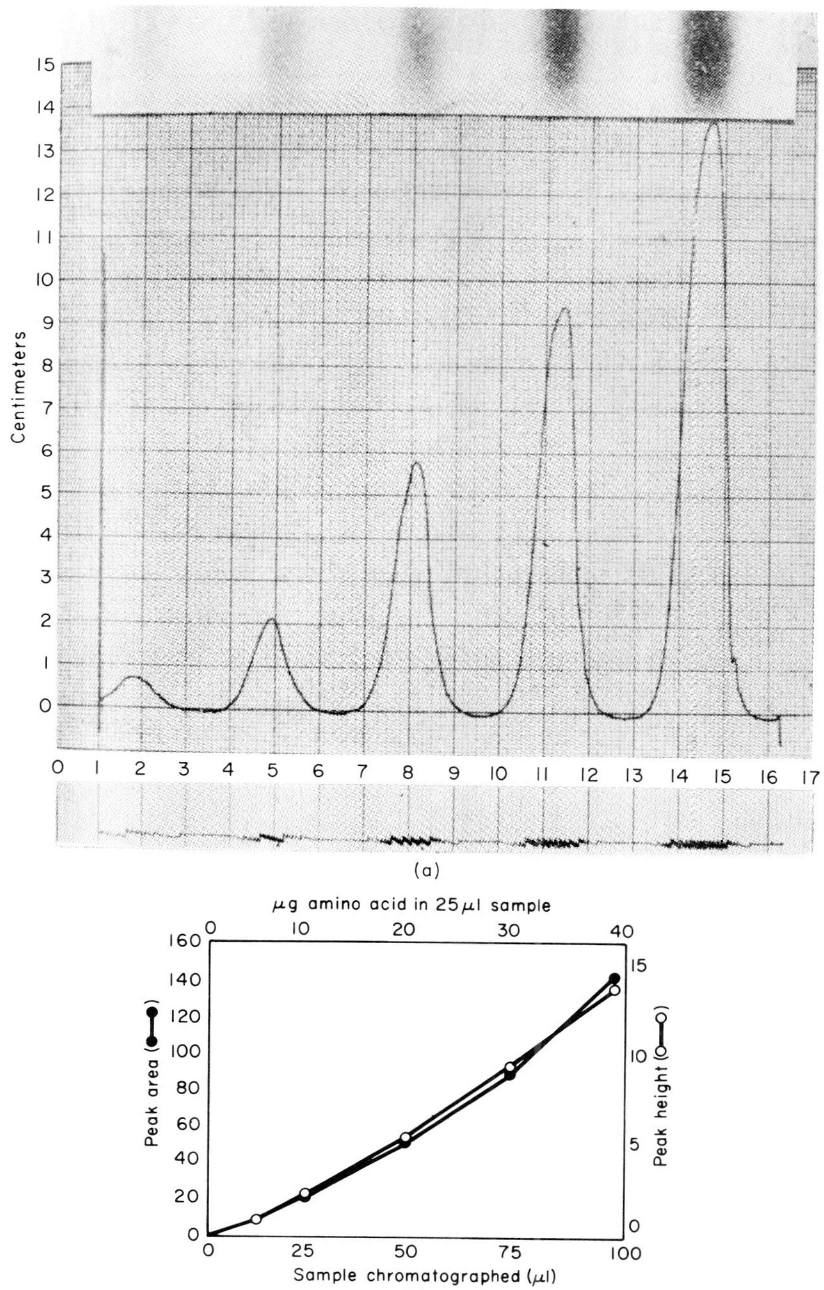

FIG. 5. (a) Scanning of tyrosine spots by a Beckman Spinco Analytrol Model RB. The five peaks represent 5, 10, 20, 30, and 40 μg of tyrosine, respectively. (b) Standard curve of tyrosine separated from an equimolar concentration of twenty-two other amino acids using the solvent *n*-butanol–acetic acid–H_2O (12:3:5 v/v). After (*129*).

is photometrically scanned. The resultant absorbancies are linearly related to the amino acid concentration; optimum results are achieved when analyzing 1–70 μg of amino acid (*129a*).

Method 3. The quantitative fluorimetry of dansyl amino acid derivatives without elution is discussed below (see Section II,G,2).

G. Miscellaneous Detailed Experiments

1. *Amino Acids in Human Materials—Clinical Applications*

Prepare urine samples as described in Section II,A. Spot 15–30 μl of solution on the paper and redevelop three to four times with *n*-butanol–acetic acid–H_2O (4:1:5 and 40:15:5 v/v) and *n*-butanol–formic acid–H_2O (75:15:10 v/v/v) systems. Seventeen spots which develop after ninhydrin treatement can be identified as amino acids when urine samples of healthy people are analyzed by this procedure (*130*).

A scheme with complete details for detection of metabolic disorders utilizing commercial dip tests, spot plate tests, and paper chromatography has been published (*131*). Multiple 50 μl spots (applied in 5 μl increments) of untreated urine are applied to sheets of Whatman No. 1 paper which are developed overnight with various listed solvents; the air-dried chromatograms are then cut into sections to be sprayed and oversprayed with different reagents, the preparation of which is given, in order to reveal certain amino acids, organic and phenolic acids, proteins, and sugars. Two-dimensional chromatography of 50 μl of urine employs development with pyridine–acetone–NH_3–H_2O (10:6:1:4 v/v) overnight, air-drying for 24 hours, and then transverse development for 8 hours with isopropanol–formic acid–H_2O (8:1:1 v/v). After overnight drying, the paper ($11\frac{1}{2} \times 11\frac{1}{2}$ inches) is sprayed with ninhydrin, heated at 85–90°C for 8 minutes, and scanned with a densitometer using a 545 mμ filter and a 4 mm circular opening. A standard curve for each amino acid is prepared by chromatographing known amounts of the pure acids. The quantitative determinations have standard errors of $\pm$ 0.2 μg in amounts below 1–2 μg and 0.5 μg for amounts above 2 μg.

Other determinations of amino acids in human brain tissue (*132*), whole blood samples (*133*), and urine samples of patients (*134*) have also been made.

2. *Separation of Amino Acid Derivatives* (see also Table IX)

Although amino acids are usually chromatographed in the free state, derivatives can be advantageously employed in particular instances. DNP amino acids (the preparation of which was described in Section II,A) are separated [see Block *et al.* (*44*) for a list of twenty suitable solvents], eluted with water and their color measured at 360 mμ by spectrophotometry (*74, 135*). For quantitative analysis, DNP derivatives are probably superior to the free amino acids (*72*).

A micro method for scanning two-dimensional chromatograms has been devised (*136*). Amino acids are reacted with tritiated DNFB, and ^{14}C-labeled DNP amino acids of known specific radioactivity are added to the reaction mixture. The derivatives are chromatographed, and their individual ^{3}H and ^{14}C activities measured separately. The ^{3}H activity, corrected by the decrease in ^{14}C activity, gives the amino acid concentration; it is claimed that this method is as accurate as ion-exchange chromatography (75).

The separation of other derivatives (phenylthiohydantoins, hydrazides, and amino alcohols) has been described in detail earlier (*44*).

2,4-Dinitro-5-aminophenyl derivatives are prepared as follows: 1 μmole of amino acid (or an equivalent amount of peptide) is dissolved in 1 ml of distilled water and 20 mg solid $NaHCO_3$ is added. Two milliliters of 2,4-dinitro-5-fluoroaniline solution (0.75 gm in 100 ml absolute ethanol) is added and the mixture is vigorously stirred for 2 hours. The dinitroaminophenyl peptides are hydrolyzed in 6 *N* HCl for 24 hours. Separation of these derivatives by one- and two-dimensional chromatography is illustrated in Fig. 6 (*137*).

Dansyl derivatives of amino acids are prepared by mixing 1 ml of amino acid (6.5 μmoles in 0.1 *M* $NaHCO_3$) and 1 ml of dansyl reagent (6 mg 5-dimethylamino-1-naphthalene sulfonyl chloride per milliliter in acetone) and leaving reaction mixture overnight. Addition of 8 ml acetone precipitates $NaHCO_3$. The solution is suitable for chromatography after filtering or centrifuging. These compounds are separated by use of nonpolar solvents, such as petroleum ether–acetic acid–H_2O (100:90:10 v/v), which separates glycine, alanine, phenylalanine, leucine, and isoleucine in 12–14 hours on Whatman No.4 paper. The chromatogram can be scanned with a fluorimeter to give quantitative estimation and linear calibration curves up to 10 μg amino acid per zone, each scan needing only 45 seconds (*138*). With an improved scanner, as little as 50–100 ng of amino acids can be analyzed (*139*).

1-Napthylisothiocyanate derivatives (thiohydantoins) are separated on Whatman No. 1 paper by development with aqueous solvents [phosphate buffer (pH 6), 10% pyridine, or 5% acetic acid] or mixtures of $CHCl_3$–ethyl, *n*-butyl or *n*-amyl alcohol-H_2O (8:4:1 v/v). Detection is by fluorescence emitted under ultraviolet light with a sensitivity of 10^{-3} to 10^{-4} μmoles. The luminescent colors of the spots change with the pH of the paper, so that identification of individual amino acids is facilitated. These derivatives are superior to phenylthiohydantoins for

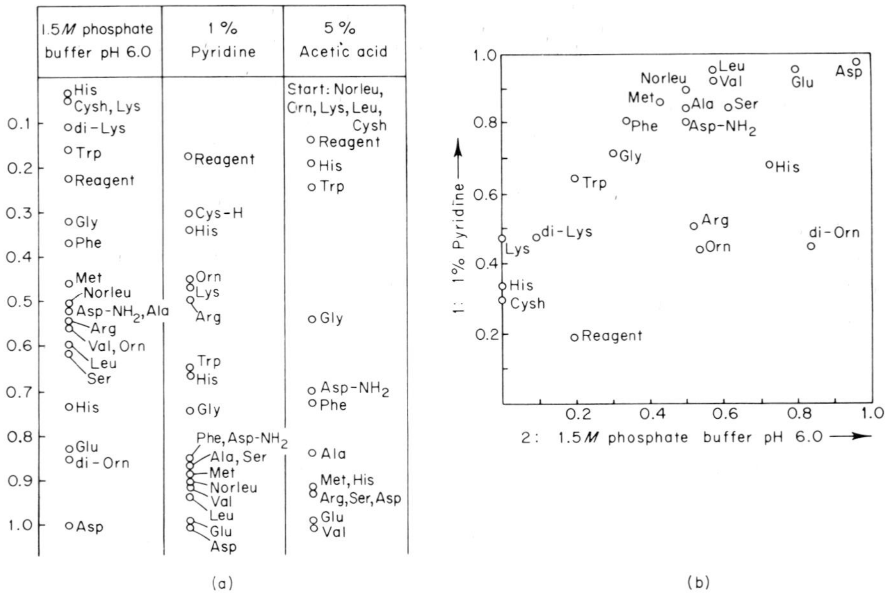

FIG. 6. Chromatographic separation of 2,4-dinitro-5-aminophenyl derivatives on Whatman No. 3 MM paper (a) by one-dimensional development with various solvents and (b) by two-dimensional chromatography. After (*137*).

the identification of *n*-terminal amino acids during sequential protein and peptide analysis (Section III) (*139a*).

III. Peptides

A. Peptide Mapping

Paper chromatography is of practical importance for the characterization of enzymatic digests of proteins by the peptide mapping or "fingerprinting" technique. Fingerprinting is a technique for the separation of peptide mixtures employing high-voltage paper electrophoresis followed by paper chromatography at right angles to the electrophoretic axis (*140*) or vice versa (*141*). This combination separates peptides and amino acids (*141a*) over the entire sheet of paper, after which the peptides are reacted with suitable reagents to produce colored spots. Fingerprinting is used in the determination of protein structures; the protein is partially hydrolyzed under a variety of conditions to yield numerous peptide fragments which are separated and identified by electrophoresis–chromatography (often preceeded by ion-exchange chromatography);

the structure is finally deduced by fitting the fragments back together in the proper sequence to reconstruct the protein molecule. The classic application of such techniques was by Sanger and co-workers (*142, 143*) leading to the elucidation of the complete structure of insulin (*144, 145*).

The fingerprinting method continues to be very widely used (e.g., *146–148*) although some workers report that considerable losses (*149*) and chemical changes (*150*) can occur on paper in the presence of organic solvents [*n*-butanol–acetic acid–H_2O solvents are most used for peptide chromatography (*151*)]. Detailed descriptions of sequence analysis by the fingerprinting technique can be found in Hais (*45*) and Bailey (*152*). An improved version of the technique involves electrophoresis on paper strips supported from a bar along the origin in order to eliminate problems associated with buffer drainage down the paper. This is followed by cutting the electropherogram across the origin, sewing each half to the edge of chromatography paper, and developing by descending chromatography (*152a*). Whatman 3 MM paper is often used for peptide mapping because it is heavy and absorbs sufficient buffer for paper electrophoresis; Whatman Nos. 1 and 3 are also popular.

Detection is usually with some modification of the ninhydrin reagent, for example 0.1% ninhydrin in water-saturated *n*-butanol (*153*). However, higher peptides, cyclic peptides, peptides whose terminal α-amino groups are acetylated or similarly substituted, and some tripeptides (*153a*) produce no color with ninhydrin. Kinins (*154*) and peptides containing arginine are detected with Sakaguchi's reagent [spray with 0.1% α-naphthol in 1 *N* NaOH; after drying, spray with NaClO (commercial Clorox)–ethanol (v/v)] and phosphopeptides by spraying with 0.5 *N* NaOH, heating, spraying with molybdate reagent [5 ml 60% w/w $HClO_4$, 25 ml 4% (w/v) ammonium molybdate, 10 ml 1 *N* HCl, and 60 ml H_2O] and viewing under ultraviolet light (*155*). Higher peptides and cyclic peptides are revealed by chlorination (*153, 156*). *t*-Butyloxycarbonyl derivatives of peptides are heating at 125–130°C for 25 minutes after spraying with ninhydrin (*157*). As an aid in the interpretation of peptide patterns after detection of the zones, densitometry and integrated densitometric scanning of the fingerprints is recommended (*158*). Peptides can also be identified by mass spectrometry (*159*).

The peptide mapping technique can be applied to larger samples by employing paper-strip electrophoresis at pH 4.4 of linearly applied material followed by recycled ascending paper chromatography (*160*).

Peptide molecular weights are estimated by a two-dimensional mapping technique involving gel filtration in a narrow column of Sephadex G-25 as the first stage, followed by transfer of the zones from the column to paper for chromatography as the second dimension (*161*).

A new method for the rapid hydrolysis of N-terminal peptide bonds in dilute aqueous solution at pH 7 employs reaction with β-[Co(trien) $OH(OH_2)]^{+2}$ ion. The resulting amino acid is taken up to form a chelate, $[Co(trien)\ AA]^{+2}$, which can be identified by co-chromatography against marker compounds with *n*-butanol–H_2O–acetic acid (1:1:0.2 v/v) containing 1% pyridine: complex ion of Gly, R_F 0.14; Pro, 0.15; Ala, 0.17; Val, 0.33; Ile, 0.47; Phe, 0.49; Leu, 0.5 (*162*).

B. Peptide Separations by Paper Chromatography

Tables X–XII (*163–165*) contain a listing of recent R_F values and the conditions under which they were obtained. Other collections of data are contained in Bodanszky and Williams (*166*), Moore and Baker (*167*), and Poduska *et al.* (*168*).

TABLE X

R_F Values of Peptides of Tyrosine, Diiodotyrosine, and Thyroxine (*163*)

Compound	R_F with the following solvents[a]	
	1	2
Glycyl-L-tyrosine	0.12	0.11
Glycyl-L-3,5-diiodotyrosine	0.13	0.28
Glycyl-L-thyroxine	0.27	0.57
L-Tyrosylglycine	0.14	0.12
3,5-Diiodo-L-tyrosylglycine	0.09	0.32
L-Thyroxylglycine	0.37	0.58
Glycyl-L-tyrosylglycine	0.09	0.12
Glycyl-3,5-diiodo-L-tyrosylglycine	0.05	0.27
Glycyl-L-thyroxylglycine	0.33	0.51

[a] (1) *n*-Butanol–dioxane–2 *N* NH_3 (4:1:5 v/v); (2) *n*-butanol–acetic acid–H_2O (78:10:12 v/v).

Peptides isolated from bovine liver, brain, kidney, blood, or human placenta can be separated into twelve fractions by repeated descending development with *n*-butanol–acetic acid–H_2O, twice in 4:1:5 v/v and twice in 40:15:5 v/v. Hydrolysis and chromatography of the fractions in the same system separates six to eleven different amino acids in each (*169*).

Alanyl peptides through Ala_6 are separated in 4 hours by descending chromatography on Amberlite SA-2 cation-exchange paper at pH 3.28 employing 0.2 *M* sodium citrate buffer as the solvent. Detection is with

TABLE XI

R_F VALUES OF α AND ε PEPTIDES OF LYSINE[a] (*164*)

Compound	R_F
α-Cbz-Lys-Bz·HCl	0.94
α-Cbz-Gly, ε-Cbz-Lys-Bz	0.93
α-Cbz-Ala, ε-Cbz-Lys-Bz	0.93
α-Cbz-Phe, ε-Cbz-Lys-Bz	0.93
α-Cbz-Leu, ε-Cbz-Lys-Bz	0.93
α-(α-Cbz-Asp-β-Bz), ε-Cbz-Lys-Bz	0.93
α-Gly-Lys·HOAc	0.14
α-Ala-Lys·HOAc	0.18
α-Phe-Lys·HOAc	0.39
α-Leu-Lys·HOAc	0.38
α-(α-Asp)-Lys	0.09
ε-Gly-Lys·HOAc	0.18
ε-Ala-Lys·HOAc	0.21
ε-Phe-Lys·HOAc	0.42
ε-Leu-Lys·HOAc	0.40
ε-(α-Asp)-Lys	0.10

[a] Paper: Whatman No. 1. Solvent: *n*-butanol–acetic acid–pyridine–H_2O (30:6:20:24 v/v).

ninhydrin preceeded by dipping the paper in 10% potassium acetate in ethanol and drying (*170*).

A method for the bioassay of blood oxytocin involves its separation from interfering substances by descending development on prewashed (successively with 3% acetic acid–0.1% HCl, H_2O, and ethanol) Whatman 3 MM paper with *n*-butanol–acetic acid–H_2O (4:1:5 v/v) (R_F oxytocin = 0.75). Detection is with iodine and elution with Tyrode solution (*171*).

Stereoisomeric peptide derivatives and diastereoisomeric dipeptides have also been separated (*172*).

IV. Proteins

The paper chromatography of protein hydrolyzates is described earlier in this chapter and reviewed elsewhere in great detail (*173*). Proteins remain at or near the origin and are separated from their hydrolysis products (amino acids, peptides) and other low molecular weight substances by development with 0.1 *M* acetate buffer, pH 3.7, on nitrocellulose membranes (*174*) (Fig. 7).

The denaturation products of collagen (Schleyer's A and B fractions) are separated by gradient development (acetate buffers, change of ionic

TABLE XII

R_F VALUES OF CYTOACTIVE AMINO ACIDS AND PEPTIDES[a] (*165*)

Compound[b]	R_F
H-L-Glu·OH └─Mel·OEt	0.77
H-D-Glu·OH └─Mel·OEt	0.77
H-L-Glu·Mel·OEt └─OH	0.87
H-D-Glu·Mel·OEt └─OH	0.87
H-L-Glu·OH └─$NHCH_2CH_2C_6H_4M$	0.73
H-D-Glu·OH └─$NHCH_2CH_2C_6H_4M$	0.73
H-L-Glu·OH └─NHC_6H_4M	0.75
H-D-Glu·OH └─NHC_6H_4M	0.75
H-L-Glu·NHC_6H_4M └─OH	0.88
H-D-Glu·NHC_6H_4M └─OH	0.88
H-L-Glu·Mel·OEt └─Mel·OEt	0.97
H-L-Glu-L-Glu·Mel·OEt └─OH └─OH	0.81
H-L-Ser-Mel·OEt	0.78
H-L-Asp·OH └─Mel·OEt	0.65
H-L-Asp·Mel·OEt └─OH	0.77

[a] Paper: Whatman No. 1. Solvent: *n*-butanol–ethanol–propionic acid–H_2O (200:100:40:100 v/v) (ascending).

[b] Mel = L-*p*-bis(2-chloroethyl)aminophenylalanine residue; *M* = *N*,*N*-bis(2-chloroethyl)amino.

strength from 0.01 to 0.5) on sheets of cellulose phosphate ion-exchange paper (Whatman CP 20). Detection and quantitative evaluation is based on the color obtained by dyeing with bromophenol blue. Carboxymethylcellulose sheets are used for the fractionation of gelatin into five zones (R_F 0.0, 0.55, 0.83, 0.87, 1.0); 1% solutions of gelatin are spotted on wetted paper and developed with 0.7% $MgCl_2$ (*175*). Cytochrome *c* has been chromatographed on Amberlite WA-2 carboxylic paper with 0.4 *M* $(NH_4)_2HPO_4$ buffer, pH 7.6 (*176*).

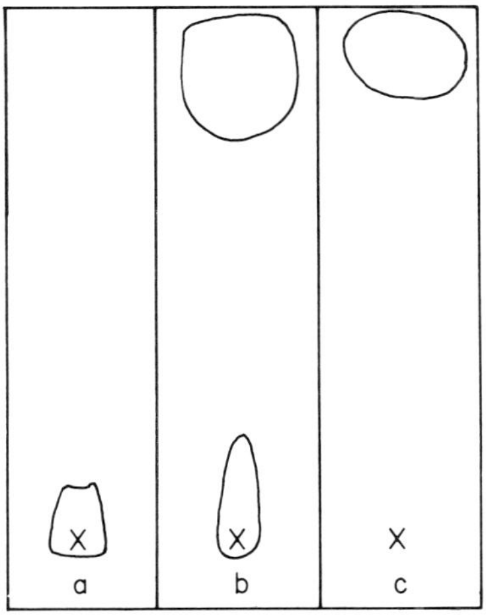

FIG. 7. Group separation of amino acids and proteins on a nitrocellulose membrane in acetate buffer, pH 3.7. (a) bovine serum, (b) bovine serum + protein hydrolyzate, (c) protein hydrolyzate. Sample a was stained by Azocarmine B, samples b and c by ninhydrin. × = origin. After (*174*).

A solvent suggested for the chromatography of protamines on unmodified papers is 18 ml 85% ethanol, 6 ml 20% trichloroacetic acid, and 1 ml 2-butanol (*177*). Insulin from rat pancreas is purified by chromatography overnight on Whatman No. 1 paper with 2-butanol–1% acetic acid (1:1 v/v) (*178*).

Protein particles are separated according to size by a pore-size gradient on wedge-compressed nitrocellulose membranes (Synpor 8, Chemapol Co, Prague) impregnated with 2% Tween 60 and developed with 0.1 *M* Tris hydrochloride and 1 *M* NaCl buffer, pH 7.9. The dyes Amido Black 10B, Ponceau S, and nigrosine are used for detection of the proteins (*179*). Radial development with buffers on several brands of nitrocellulose membranes allows rapid, quantitative determinations of solutions containing 0.001–6% of various proteins in biological samples

(*180*). These membranes also serve for the microchromatographic separation of ribonucleic acids from proteins (*181*).

A rapid chromatographic method, applicable to serum proteins, histones, protamines, and RNA and based on Tiselius's salting out or precipitation chromatography is especially recommended. The sample is applied as a thin line to S. and S. 2040a paper which is developed in a discontinuous salt or acid gradient of decreasing molarity (e.g., 1 *N* HCl for 15 minutes, followed by transfer of the paper to 0.25 *N* HCl, 0.1 *N* HCl, 0.01 *N* HCl, and 50% acetic acid for 15 minutes each). The fractionated proteins are identified with Amido Black B solution. The excess dye is removed with a solution of 10% acetic acid, 45% methanol, and 45% H_2O, and densitometry is used to evaluate the chromatogram (*182*).

The prefered technique for the separation and identification of proteins is undoubtedly electrophoresis, which is the subject of Volume I of this treatise.

References

1. J. Gasparič, *in* "Paper Chromatography" (I. M. Hais and K. Macek, eds.), pp. 414–431. Academic Press, New York, 1963.
2. D. M. Waldon-Edward, *Chem. & Ind.* (*London*) p. 104 (1954).
3. J. S. Stern, M. J. Franklin, and J. Mayer, *J. Chromatog.* **30**, 632 (1967).
4. A. P. Martinez, W. W. Lee, and L. Goodwin, *J. Org. Chem.* **31**, 3263 (1966).
5. I. M. Lockhart, *Nature* **177**, 393 (1956); A. M. Asatoor, *J. Chromatog.* **4**, 144 (1960).
6. M. Katayama and S. Funahashi, *J. Chromatog.* **32**, 777 (1968).
7. J. Gasparič, J. Petránek, and M. Veceřa, *Mikrochim. Acta* p. 1026 (1955).
8. M. Veceřa and J. Gasparič, *Chem. Listy* **52**, 611 (1958); *Collection Czech. Chem. Commun.* **24**, 465 (1959).
9. S. Ghosal and B. Mukherjee, *J. Org. Chem.* **31**, 2285 (1966).
9a. A. Vahidi and D. V. S. Sankar, *J. Chromatog.* **43**, 135 (1969).
10. J. Borecký, *Mikrochim. Acta* pp. 283 and 284 (1966); *J. Chromatog.* **28**, D1 (1967).
10a. E. S.v. Kamienski, *Planta* **50**, 291 (1957).
10b. D. L. van Rheenen and C. B. J. Sipman, *J. Chromatog.* **37**, 341 (1968).
11. E. A. McComb, N. Androulidakis, and V. V. Rendig, *J. Chromatog.* **40**, 125 (1969).
12. V. C. Quesnel, *J. Chromatog.* **24**, 268 (1966).
13. D. J. Roberts and K. J. Broadley, *J. Chromatog.* **27**, 407 (1967).
14. K. J. Broadley and D. J. Roberts, *J. Pharm. Pharmacol.* **18**, 182 (1966).
15. M.F. L'Annunziata and W. H. Fuller, *J. Chromatog.* **34**, 270 (1968).
16. A. Kompisova, V. Kapisinska, and S. Cico, *Chem. Prumysl* **16**, 302 (1966).
17. C. E. Bell and A. R. Somerville, *Biochem. J.* **98**, 1C (1966).
18. E. Boyland and D. Manson, *Biochem. J.* **101**, 84 (1966).
19. N. H. Agnew and J. R. Parrish, *J. Chem. Soc.* p. 203 (1966).
20. I. Toyozawa and S. Ogura, *Nippon Shokuhin Kogyo Gakkaishi* **12**, 46 (1965).
21. J. Awapara, V. E. Davis, and O. Graham, *J. Chromatog.* **3**, 11 (1960).
22. J. Hrdlička and G. Janíček, *Nature* **204**, 1201 (1964).
23. K. C. Khulbe and S. P. Srivastava, *Z. Anal. Chem.* **210**, 282 (1965).

24. W. Tyburczyk, *Acta Physiol. Polon.* **18**, 131 (1967).
25. C. A. Benassi, F. M. Veronese, and A. DeAntoni, *Atti, Ist. Veneto Sci., Lettere Arti, Classe Sci. Mat. Nat.* **122**, 201 (1964).
26. T. Hartmann, *Planta* **65**, 315 (1965).
27. J. M. Bremner and R. H. Kenten, *Biochem. J.* **49**, 651 (1951).
28. A. A. Casselman and R. A. B. Bannard, *J. Chromatog.* **28**, 462 (1967).
29. J. W. Daly, J. Benigni, R. Minnis, Y. Kanaoka, and B. Witkop, *Biochemistry* **4**, 2513 (1965).
30. I. Gemzová and J. Gasparič, *Collection Czech. Chem. Commun.* **32**, 2742 (1967); *J. Chromatog.* **35**, D58 (1968).
31. A. Cee and J. Gasparič, *Mikrochim. Acta* p. 298 (1966); *J. Chromatog.* **33**, D20 (1968).
32. I. Gemzová and J. Gasparič, *Mikrochim. Acta* p. 314 (1966); *J. Chromatog.* **31**, D4 (1967).
33. L. Fishbein, *Rec. Trav. Chim.* **84**, 470 (1965); *J. Chromatog.* **22**, D2 (1966).
34. A. Bassl, H.-J. Heckemann, and E. Baumann, *J. Prakt. Chem.* [4] **36**, 276 (1967); *J. Chromatog.* **36**, D81 (1968).
35. J. Nosek and J. Gasparič, *Collection Czech. Chem. Commun.* **32**, 2030 (1967); *J. Chromatog.* **32**, D11 (1968).
36. J. Gemzová and J. Gasparič, *Mikrochim. Acta* pp. 313 and 318; *J. Chromatog.* **26**, D5, 1967.
37. J. Gasparič and B. Kloucek, *Collection Czech. Chem. Commun.* **31**, 106 (1966); *J. Chromatog.* **22**, D8 (1966).
38. J. Gasparič and J. Borecký, *J. Chromatog.* **5**, 466 (1961).
39. J. Gasparič, *Collection Czech. Chem. Commun.* **29**, 1724 (1964); *J. Chromatog.* **18**, D25 and D26 (1965).
40. R. F. Coward, P. Smith, and O. S. Wilson, *Clin. Chim. Acta* **9**, 385 (1964); *J. Chromatog.* **18**, D24 (1965).
41. R. Pant and H. C. Agrawal, *Z. Physiol. Chem.* **335**, 203 (1964); *J. Chromatog.* **18**, D21 (1965).
42. Chromatographic Data Section, *J. Chromatog.* **23**, D2 and D20 (1966); **34**, D46 (1968).
43. H. Floch, *Biol. Med. (Paris)* **53**, 36 (1964); *J. Chromatog.* **19**, D4 (1965).
44. R. J. Block, E. L. Durrum, and G. Zweig, "A Manual of Paper Chromatography and Paper Electrophoresis," 2nd rev. ed., pp. 85–169. Academic Press, New York, 1958.
45. I. M. Hais, *in* "Paper Chromatography" (I. M. Hais and K. Macek, eds.), pp. 437–494. Academic Press, New York, 1963.
46. I. Smith, ed., "Chromatographic and Electrophoretic Techniques," Vol. I. Wiley (Interscience), New York, 1960.
47. S. Blackburn, "Amino Acid Determination Methods and Techniques." Marcel Dekker, New York, 1968.
48. W. Voegelein, W. A. Vetterli, and S. Napier, *Ciba Rev.* p. 32 (1966).
49. G. Pataki, *Chromatog. Rev.* **9**, 23 (1967).
50. J. E. Eastoe, *Brit. med. Bull.* **22**, 174 (1966).
51. W. Matthias and J. Wagner, *J. Chromatog.* **33**, 316 (1968).
52. P. Wierzchowski, I. Janczarski, and D. Krauze, *Acta Biochim. Polon.* **9**, 343 (1962).
53. J. Gritsch, I. Brad, V. Gheorghiu, and V. Haidu, *Ind. Aliment.* **15**, 585 (1964).
54. S. Blackburn, *Methods Biochem. Anal.* **13**, 1 (1965).
55. C. L. deLigny and A. G. Remijnse, *Rec. Trav. Chim.* **86**, 421 (1967).
56. G. Kovacs and S. Toth, *Agrokem. Talajtan* **15**, 515 (1966).

57. B. Meloun *in* "Paper Chromatography" (I. M. Hais and K. Macek, eds.), pp. 495–533. Academic Press, New York, 1963.
58. K. István, *Magy. Kem. Folyoirat* **59**, 252 (1953).
59. Z. S. Chulkova and J. I. Gumina, *Lab. Delo* 4, **22** (1958).
60. G. C. Müller, G. Bowman, and A. Herranen, *Anal. Chem.* **27**, 1357 (1955).
61. G. Gorbach, G. Dedic, and E. H. Philippi, *Z. Physiol. Chem.* **301**, 185 (1955).
62. C. E. Dent and J. A. Schilling, *Biochem. J.* **44**, 318 (1949).
63. C. S. Hanes, C. K. Harris, M. A. Moscarello, and E. Tigane, *Can. J. Biochem. Physiol.* **39**, 163 (1961).
64. I. I. Shainskii, *Lab. Delo* p. 414 (1966).
65. O. Keerberg, S. Viemer, and J. Viil, *Eesti NSV Teaduste Akad. Toimetised, Biol. Seeria* **16**, 157 (1967).
66. J. M. Das, *Current Sci.* (*India*) **34**, 374 (1965).
67. Y. Kume, *Nippon Nogeikagaku Kaishi* **37**, 196 and 201 (1963).
68. K. Lindner, *Elelmiszervizsgalati Kozlemen* **12**, 185 (1966).
69. N. D. Starikova and O. G. Yablokova, *Tr. Inst. Okeanol., Akad. Nauk SSSR* **79**, 14 (1965).
70. C. J. van Wyk and P. J. Venter, *S. African J. Agr. Sci.* **8**, 57 and 69 (1965).
71. A. Brenner, A. Niederwieser, and G. Pataki, *in* "Thin-Layer Chromatography" (E. Stahl, ed.), p. 391. Academic Press, New York, 1965.
72. G. Biserte, J. W. Hollemann, J. Holleman-Dehove, and P. Sautiere, *Chromatog. Rev.* **2**, 59 (1960).
73. M. Keller and G. Pataki, *Helv. Chim. Acta* **46**, 1687 (1963).
74. G. Pataki, "Duennschichtchromatographie in der Aminosaeure- und Peptidchemie." de Gruyter, Berlin, 1965.
75. G. Pataki, E. Baumann, U. P. Geiger, P. Jenkins, and U. Kupper, *in* "Chromatography" (E. Heftmann, ed.), p. 378. Reinhold, New York, 1967.
76. C. L. DeLigny and E. C. M. Kok, *J. Chromatog.* **38**, 224 (1968).
77. C. S. Knight, *Chromatog. Rev.* **4**, 49 (1962).
78. E. F. McFarren, *Anal. Chem.* **23**, 168 (1951).
79. B. Radhakrishnamurthy and G. S. Berenson, *J. Biol. Chem.* **241**, 2106 (1966).
80. G. S. Berenson and E. R. Dalferes, Jr., *Biochim. Biophys. Acta* **58**, 34 (1962).
81. C. S. Knight, *J. Chromatog.* **8**, 205 (1962).
82. B. A. Svendsen and E. Brochmann-Hanssen, *J. Pharm. Sci.* **51**, 514 (1962).
83. H. George and A. Meister, *Biochim. Biophys. Acta* **132**, 165 (1967).
84. C. S. Knight, *Advan. Chromatog.* **4**, 61 (1967).
85. C. S. Knight, *Nature* **184**, 1486 (1959); **188**, 739 (1960).
86. P. Catelli, *J. Chromatog.* **9**, 534 (1962).
87. S. Bottomley, *Anal. Biochem.* **18**, 472 (1967).
88. D. P. Burma, *Nature* **168**, 565 (1951); *Anal Chem.* **25**, 549 (1953).
89. G. Vogt, *Landwirthsch. Forsch.* **18**, 89 (1965).
90. J. Trzaska and G. N. Kowkabany, *J. Chromatog.* **26**, 141 (1967).
91. P. V. Palacios and F. P. Gil, *Rev. Ecuator. Hig. Med. Trop.* **23**, 39 (1966).
92. E. Soczewiński and M. Rojowska, *J. Chromatog.* **32**, 364 (1968).
93. T. L. Hardy, D. O. Holland, and J. H. C. Nayler, *Anal. Chem.* **27**, 971 (1955).
94. Y. Solberg, *Acta Chem. Scand.* **19**, 1269 (1965).
95. G. E. Gaull and M. K. Gaitonde, *Biochem. J.* **102**, 294 (1967).
96. G. Barbiroli, *Mikrochim. Ichnoanal. Acta* p. 655 (1965).
97. R. Schoental and D. J. Rive, *Biochem. J.* **97**, 466 (1965).
98. I. E. Burrows, G. Shaw, and D. V. Wilson, *J. Chem. Soc.*, C, p. 42 (1968).

99. M. K. Gaitonde and G. E. Gaull, *Biochem. J.* **102**, 959 (1967).
100. A. Zamorani, P. G. Pifferi, G. Lanzarini, and A. Riva, *Gazz. Chim. Ital.* **98**, 472 (1968).
101. T. L. Perry, G. H. Dixon, and S. Hansen, *Nature* **206**, 895 (1965).
102. S. P. Bessman, Z. H. Koppanyi, and R. A. Wapnir, *Anal. Biochem.* **18**, 215 (1967).
103. A. Signor, L. Biondi, M. Terbojevich, and P. Pajetta, *Gazz. Chim. Ital.* **94**, 619 (1964); *J. Chromatog.* **17**, D6 (1965).
104. J. C. Fletcher, *Biochem. J.* **102**, 815 (1967).
105. A. Furst and S. Saxena, *J. Chromatog.* **22**, 493 (1966).
106. H. Mix, *Z. Physiol. Chem.* **337**, 40 (1964).
107. D. K. Black and S. R. Landor, *J. Chem. Soc., C* pp. 281 and 285 (1968).
108. V. R. Villaneuva and M. Barbier, *Bull. Soc. Chim. France* p. 3992 (1967).
108a. J. H. Schmitt and M. H. Zenk, *Anal. Biochem.* **23**, 433 (1968).
109. C. S. Hanes, *Can. J. Biochem. Physiol.* **39**, 119 (1961).
110. E. H. M. Wade, A. T. Matheson, and C. S. Hanes, *Can. J. Biochem. Physiol.* **39**, 141 (1961).
111. E. Tigane, E. H. M. Wade, J. T. Wong, and C. S. Hanes, *Can. J. Biochem. Physiol.* **39**, 427 (1961).
112. A. E. Pasieka, *Can. J. Biochem. Physiol.* **38**, 1137 (1960).
112a. P. I. Singh, V. R. Bajaj, and A. S. Saini, *J. Chromatog.* 43, 554 (1969).
113. K. V. Giri, K. Krishnamurthy, and T. A. Venkatasubramanian, *Current Sci. (India)* **21**, 11 (1952).
114. K. V. Giri, *Current Sci. (India)* **20**, 295 (1951).
115. S. Szczepaniak, I. Krzeczkowska, and E. Nowicka, *Chem. Anal. (Warsaw)* **13**, 155 (1968).
116. C. Liteanu and S. Gocan, *Rev. Roumaine Chim.* p. 1051 (1965).
117. R. Bandklajder Sznelwar and E. de Camargo Fonseca Moraes, *Rev. Fac. Farm. Bioquim. Univ. Sao Paulo* **5**, 431 (1967).
117a. H. Stürchen-Kirchner, *Z. Physiol. Chem.* **349**, 1049 (1968).
118. D. Y. Hsia and T. Inouye, "Inborn Errors of Metabolism," Part 2. Year Book Publ., Chicago, Illinois, 1966.
119. A. E. Pasieka and M. T. Borowiecki, *Biochim. Biophys. Acta* **111**, 553 (1965).
119a. G. Eglite and V. Oskaja, *Latv. PSR Zinat. Akad. Vestis, Kim. Ser.* p. 345 (1968).
119b. M. Guyer, Jr. and E. Sawicki, *Anal. Chim. Acta* **49**, 182 (1970).
120. R. Lipton, T. D. Price, and M. M. Melicow, *J. Chromatog.* **26**, 412 (1967).
121. V. B. Luk'yanov and E. F. Simonov, *Vestn. Mosk. Univ., Ser. II: Khim.* **21**, 36 (1966).
122. S. Ansorge and P. Bohley, *Pharmazie* **22**, 574 (1967).
123. V. Jiracek, *J. Chromatog.* **33**, 312 (1968).
124. J. G. Heathcote and R. J. Washington, *Analyst* **92**, 627 (1967).
125. V. Macek, *Casopis Lekaru Ceskych* **104**, 1402 (1965).
126. R. Lieldiens and Z. Zommers, *Tr. Latv. Nauchn.-Issled. Inst. Zhivotnovodstra i Vet.* **19**, 93 (1966).
127. E. Franzen, *Z. Med. Labortech.* **7**, 174 (1966).
128. J. W. H. Lugg, *J. Chromatog.* **10**, 272 (1963).
129. Y. S. Shin, *J. Chromatog.* **30**, 634 (1967).
129a. I. Krzeczkowska, S. Burzynski, and Z. Czerniak, *Ann. Univ. Mariae Curie-Sklodowska*, Sec. D **1966**, 21, 125-134.
130. I. I. Shainskii, *Uzbesk. Biol. Zh.* **10**, 6 (1966).
131. H. K. Berry, C. Leonard, H. Peters, M. Granger, and N. Chunekamrai, *Clin. Chem.* **14**, 1033 (1968).

132. N. Robinson and C. B. Williams, *Clin. Chim. Acta* **12**, 311 (1965).
133. R. Tanguay, J. Galindo, and M. Gordon, *J. Maine Med. Assoc.* **57**, 237 (1966).
134. D. Lonsdale and W. R. Faulkner, *Cleveland Clin. Quart.* **32**, 5 (1965).
135. J. M. Dellacha and A. V. Fontaine, *Experientia* **21**, 351 (1965).
136. D. Beale and J. K. Whitehead, *Tritium Phys Biol. Sci., Proc. Symp. Detection use, Vienna, 1961* Vol. 1, p. 179. I.A.E.A., Vienna, 1962.
137. Z. Deyl, L. Schinkmannová, and J. Rosmus, *J. Chromatog.* **30**, 614 (1967).
138. A. A. Boulton and I. E. Bush, *Biochem. J.* **92**, 11P (1964).
139. A. A. Boulton, N. E. Chard, and L. Grant, *Biochem. J.* **96**, 69P (1965).
139a. Z. Deyl, *J. Chromatog.* **48**, 231 (1970).
140. V. M. Ingram, *Nature* **178**, 792 (1956); *Biochim. Biophys. Acta* **28**, 539 (1958).
141. A. M. Katz, W. J. Dreyer, and C. B. Anfinsen, *J. Biol. Chem.* **324**, 2897 (1959).
141a. P. J. Peterson, *J. Chromatog.* **48**, 301 (1968); P. J. Peterson and L. Fowden, *J. Chromatog.* **48**, 575 (1970).
142. F. Sanger, *Biochem. J.* **39**, 507 (1945).
143. F. Sanger and E. O. P. Thompson, *Biochim. Biophys. Acta* **9**, 225 (1952); *Biochem. J.* **53**, 353 (1952).
144. A. P. Ryle, F. Sanger, L. F. Smith, and R. Kitai, *Biochem. J.* **60**, 541 (1955).
145. E. O. P. Thompson, *in* "Readings from *Scientific American, Bio-Organic Chemistry*," p. 34. Freeman, San Francisco, California, 1968.
146. S. K. Chan and E. Margoliash, *J. Biol. Chem.* **241**, 335 (1967).
147. K. Bell, H. A. McKenzie, and D. C. Shaw, *Biochim. Biophys. Acta* **154**, 284 (1968).
148. F. C. Stevens, A. N. Glazer, and E. L. Smith, *J. Biol. Chem.* **242**, 2764 (1967).
149. W. A. Schroeder, R. T. Jones, J. Cormick, and K. McCalla, *Anal. Chem.* **34**, 1570 (1962).
150. R. J. Ambler, *Biochem. J.* **89**, 349 (1963).
151. C. J. O. R. Morris, *Brit. med. Bull.* **22**, 168 (1966).
152. J. L. Bailey, "Techniques in Protein Chemistry," 2nd ed. Elsevier, Amsterdam, 1967.
152a. G. Custer, III, *J. Chromatog.* **42**, 429 (1969).
153. M. Kimura, K. Murayama, M. Nomoto, and Y. Fujita, *J. Chromatog.* **41**, 458 (1969).
153a. K. Kito and T. Murachi, *J. Chromatog.* **44**, 205 (1969).
154. E. Gapanhuk and O. B. Henriques, *J. Chromatog.* **32**, 782 (1968).
155. B. Blombäck, M. Blombäck, P. Edman, and B. Hessel, *Biochim. Biophys. Acta* **115**, 371 (1966).
156. F. Reindel and W. Hoppe, *Chem. Ber.* **87**, 1103 (1954).
157. Y. Wolman and Y. S. Klausner, *J. Chromatog.* **24**, 277 (1966).
158. S. C. Glauser, H. Wagner, Jr., R. Arbuckle, and E. M. Glauser, *Biochim. Biophys. Acta* **130**, 525 (1966).
159. E. Bricas, J. van Heijenoort, M. Barber, W. A. Wolstenholme, B. C. Das, and E. Lederer, *Biochemistry*, **4**, 2254 (1965).
160. K. C. Hoerman, and K. Kamel, *Anal Biochem.* **21**, 107 (1967).
161. P. R. Carnegie, *Nature* **206**, 1128 (1965).
162. D. A. Buckingham, J. P. Collman, D. A. Happer, and L. G. Marzilli, *J. Am Chem. Soc.* **89**, 1082 (1967).
163. T. Shiba and H. J. Cahnmann, *J. Org. Chem.* **29**, 3063 (1964).
164. J. D. Padayatty and H. van Kley, *J. Org. Chem.* **31**, 1934 (1966).
165. M. Szekerke, R. Wade, and F. Bergel, *J. Chem. Soc.* p. 1910 (1965).
166. M. Bodanszky and N. J. Williams, *J. Am. Chem. Soc.* **89**, 685 (1967).
167. T. B. Moore and C. G. Baker, *J. Chromatog* **1**, 513 (1958).

168. K. Poduska, G. S. Katrukha, A. B. Silaev, and J. Rudinger, *Collection Czech Chem. Commun.* **30**, 2410 (1965).
169. N. V. Pokidova and T. P. Buzina, *Antibiotiki* **11**, 594 (1966).
170. S. Ellis and J. M. Nuenke, *J. Biol. Chem.* **242**, 4623 (1967).
171. M. N. Gioia de Coch, H. Coda, C. Fielitz, H. M. Cabot, A. Fraga, and J. A. Coch, *J. Endocrinol.* **38**, 355 (1967).
172. G. Losse, H. Raue, and K. Koehler, *Z. Chem.* **7**, 105 (1967); T. Sokolowska, and J. F. Biernat, *J. Chromatog.* **13**, 269 (1964).
173. T. E. Wheeler, *Serum Proteins Dysproteinemias, 1964* pp. 159 and 162. Lippincott, Philadelphia, Pennsylvania, 1964.
174. T. I. Přistoupil, *J. Chromatog.* **26**, 121 (1967); T. I. Přistoupil, M. Kramlová, and J. Štěrbíková, *J. Chromatog.* **42**, 367 (1969).
175. Z. Deyl and J. Rosmus, *Chromatog. Rev.* **8**, 225 (1966).
176. K. Motonaga and K. Nakanishi, *J. Biochem.* (*Tokyo*) **58**, 98 (1965).
177. N. Muic and A. Meniga, *Bull. Sci., Conseil Acad. RSF Yougoslavie* **11**, 185 (1966).
178. K. W. Taylor and G. H. Smith, *Biochem. J.* **91**, 491 (1964).
179. T. I. Přistoupil, *J. Chromatog.* **28**, 89 (1967).
180. T. I. Přistoupil, M. Kramlová, and J. C. Štěrbíková, *J. Chromatog.* **34**, 370 (1968); M. Kramlová and T. I. Přistoupil, *ibid.* **37**, 264 (1968).
181. T. I. Přistoupil and M. Kramlová, *J. Chromatog.* **32**, 769 (1968).
182. K. Bauer, *J. Chromatog.* **35**, 538 (1968).

6

CARBOHYDRATES

I. Introduction

As with the amino acids, paper chromatography of carbohydrates has been covered in great depth (*1*). Therefore, only recent applications and deveolopments will be discussed below. The reader is referred also to other reviews of this topic which have been published (*2–6*).

Most authors consider the mechanism of separation of sugars on paper to be a partition process (*3*), although adsorption phenomena may also be important in some cases (*7*). Relationships between R_F values and sugar structure have been studied by many workers (*2, 3*). The mobility of sugars depends greatly on their particular ring form; within a certain ring form, the R_F values depend on the configuration of the hydroxyl groups. In series with the same configuration of hydroxyls, the nature of the substituents attached to the ring is important.

It was noted early (*8*) that the influence of structure on the behavior of sugars is in some cases additive. For example, in the oligosaccharide series, the successive addition of hexose units brings about a corresponding decrease in migration. This principle has been extensively reviewed (*9*) and found to hold in some, but not all, solvents (*10*). Recently, mathematical equations have been derived in order to predict, with the knowledge of the R_F values of six aldohexose or aldomethylose sugars, the R_F values of two others (*11*).

In addition to effects of molecular dimension, the type of bond is important; for example, 1,4 disaccharides migrate faster than 1,6 disaccharides. Paper chromatography is capable of separating isomeric sugars differing only in stearic configuration (*3*), but optical isomers of the D and L series have not been resolved (*2*).

II. General Procedure

A. Preparation of Sample

Initial zones formed from 3–5 μl of 1–2 % solutions of sugars in water or 50–90 % ethanol are preferable for paper chromatography. The optimum sample size for the separation and detection of most carbohydrates is about 50 μg, although quantities as low as 1 μg are detected with ^{14}C-tagged sugars.

Extracts from biological samples should be freed of substances other than carbohydrates. Excessive acidity or alkalinity is to be avoided during preliminary purification so that sugars are not hydrolyzed or epimerized (*2*). Desalting is achieved by precipitation methods, ion exchange (on weak base and strong acid resins), or pyridine extraction (*1*). Hot pyridine causes chemical transformations in some sugars (*12*), and an improved procedure has been suggested (*13*) in which 2 ml pyridine is added to 0.5 ml sample (urine) to precipitate the salts. After mixing and allowing to stand for 10 minutes, the tube is centrifuged at 2000 rpm for 5 minutes. Supernatant is spotted over the marked area "CDFGE" of a strip of Whatman No. 1 chromatography paper (Fig. 1), pyridine

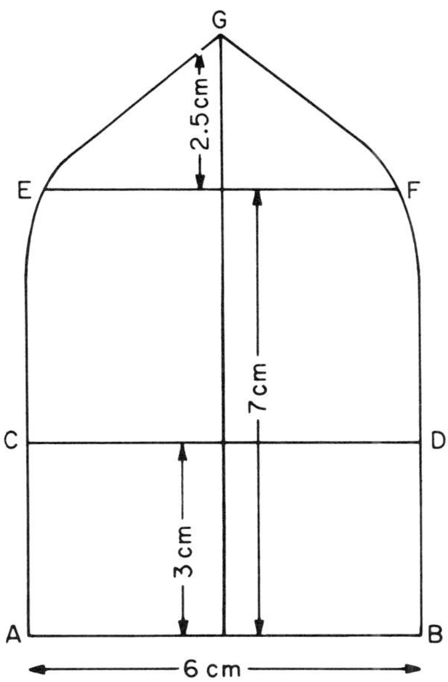

FIG. 1. Paper strip used in the described method of sample desalting. After (*13*).

being evaporated off with a blast of cold air. Two washings, each of 0.5 ml pyridine, are similarly spotted. Next the strip is immersed in diethyl ether in a jar for 10 minutes, after which the strip is taken out and kept under a fan for 10 minutes to completely remove the pyridine. The sugars are washed in 0.3–0.5 ml distilled water by the descending technique (running time 1 hour at 70°C) in a water-saturated incubator adjusted to this temperature. The sugar solution is received in a graduated tube and is adjusted to a definite volume before spotting. Recovery of sugars is 100%, and the sample may be preserved on the paper before elution for any length of time.

Deionization has been avoided in the identification of urinary sugars by employing standards prepared in a solution similar in composition to urine. Use of ion exchangers can lead to the retention and loss of acidic and basic sugar derivatives (e.g., uronic acids and amino sugars) which may be present in complex samples (*2*).

Macromolecular substances (proteins) are precipitated with mercuric acetate (*14*), lead acetate, Somogyi's reagent [0.3 N $Ba(OH)_2$–5% aqueous $ZnSO_4 \cdot 7H_2O$], or 70% ethanol (*15*).

Water extracts of various products (fruits and vegetables) are purified on a dextran column of the Sephadex type, which does not retain sugars. The eluate is directly subjected to paper chromatography (*16*).

Sugars have been determined in such other diverse samples as fruit juices and wines (*17*, *18*), yeast (*19*), leaves of virus-infected plants (*20*), urine of infants with chronic disorders (*21*), Siberian silkworms (*22*), plant (*23*) and fungal (*24*) cell walls, and blood (*15*).

B. Hydrolysis

Polysaccharides are hydrolyzed chemically with sulfuric (*3*), hydrochloric (*3*), or nitric (*23*) acid. 0.5 N H_2SO_4 is employed for partial hydrolysis and 1–4 N H_2SO_4 for complete hydrolysis. The temperature and heating period vary with the experimenter, 1–6 hours at 100°C being typical. Neutralization with $NaHCO_3$ (*25*) or $Ba(OH)_2$ (*26*) follows hydrolysis. Acid hydrolysis may lead to the formation of oligosaccharides as well as to monosaccharides (*3*).

Hydrolysis, in solution or on the paper, with enzymes such as β-amylase, α-glycosidase, or invertase is also employed (*27*). A novel method (*28*, *29*) involves chromatography of oligosaccharides in one direction, treatment with an enzyme directly on the paper, followed by separation of the hydrolytic products in a second direction. Multiple (two to four) ascents with the same solvent (*n*-butanol–pyridine–H_2O, 9:5:7, 6:4:3, or 6:4:4 v/v) in both directions are recommended.

C. Paper

Untreated chromatography paper is most often used, although prewashing with acetic and hydrochloric acids and then water [for phosphoric acid esters which are strongly sorbed on unwashed paper (*2*)], or with a mixture of 2.5 gm hydroxyquinoline, 81 ml acetic acid, and 1000 ml H_2O (*30*) has been recommended. In addition, the following stationary phases have been employed for sugar separations:

a. Paper impregnated with *dimethylsulfoxide*; developing solvent: diisopropyl ether or benzene–dimethyl sulfoxide (19:1 v/v) (*31*).
b. Strongly basic *anion-exchange paper* (Amberlite SB-2), developed with a NaCl solution of steadily increasing molarity, separates mucopolysaccharides; R_F : hyaluronic acid, 0.63, heparin sulfate, 0.49; chondroitin sulfate, 0.32; heparin, 0.24 (*32*).
c. Paper impregnated with *sodium molybdate* (0.5% w/v aqueous solution, adjusted to pH 5); solvent: butanol–pyridine–H_2O (6:4:3 v/v); R_F : 0.50 and 0.57 for 2-*C*-methyl-*L*-ribose and -arabinose, respectively (*33*).
d. Paper containing *35% H_2O* was used in the elucidation of the structure of three disaccharides isolated from *Asclepiadaceae* (*34*).
e. Paper impregnated with *3% KCl* or *0.5% borax* is used to separate D-mannitol, D-glucitol, and galactitol (*35*).
f. *Glass fiber sheets* (Gelman ITLC SG) impregnated with *1% NaH_2PO_4* (w/v), dried at room temperature 1–3 days and stored in aluminum foil; detection: char by spraying with concentrated H_2SO_4 on both sides of the paper and hang vertically in an oven at 230°C for 10 minutes; yields brown or dark gray spots which can be quantified by densitometry (*36*) (see Fig. 2); R_F values (*24*) are given in the tabulation.

	R_F in the following solvents[a]		
	1	2	3
Galactose	0.36	0.60	0.13[b]
Glucose	0.45	0.70	0.21[b]
Mannose	0.56	0.76	0.29[b]
Galactosamine	0.00	0.00	0.37
Glucosamine	0.00	0.06	0.49
Glucuronic acid	0.00	0.32	0.00
No. of developments	3	1	1
Time for 15 cm run (minutes)	15	35	25

[a] 1. $CHCl_3$–methanol–pyridine–H_2O (130:40:1:5 v/v); 2. ethyl acetate–methanol–acetic acid–H_2O (65:20:7.5:10 v/v); 3. $CHCl_3$–methanol–NH_3–H_2O (130:63:7:4 v/v).

[b] R_F value of major zone.

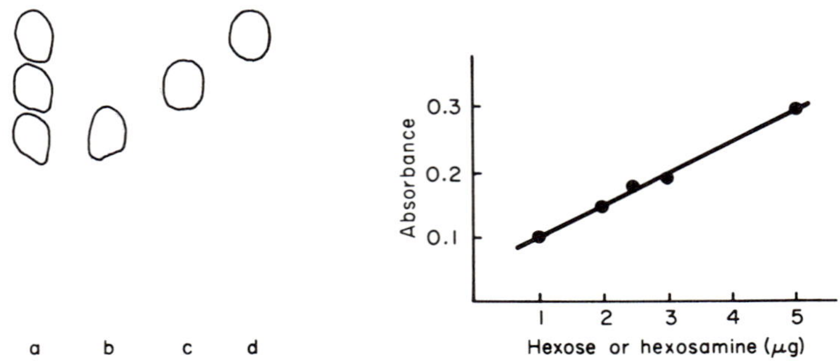

FIG. 2. (*left*) A chromatogram showing the separation of a mixture of glucose, galactose, and mannose (a); galactose (b); glucose (c); and mannose (d) on a glass fiber sheet developed for three 15 minute periods with $CHCl_3$–methanol–pyridine–H_2O (130:40:1:5 v/v). (*right*) Densitometer readings of charred spots on glass fiber paper chromatograms demonstrating linearity among varying quantities of standards. After (*24*).

D. TECHNIQUES

All the classical procedures described in Chapter 3 are useful for carbohydrate derivatives. One-dimensional multiple development is often used, or the solvent can be allowed to overrun the end of the paper, in which case mobilities are compared with that of pure known standards.

Centrifugal chromatography (*37*) on Whatman No. 3 paper with 4:1:5 v/v butanol–acetic acid–H_2O is rapid and yields the following R_F values: raffinose, 0.32; maltose, 0.37; sucrose, 0.45; glucose, 0.49; fructose, 0.53; and rhamnose, 0.75.

Circular chromatography of 10–100 γ of sugars can be carried out in a chamber consisting of two watch glasses (45 cm diameter) between which is inserted the sheet of paper (45 $\times$ 45 cm) provided with a wick which dips into a vessel containg the solvent (butanol–acetone–H_2O, 4:5:1 v/v). After 30 hours the front reaches the edge of the two glasses, and spots are detected with the following R_F values (*38*): lactose, 0.15; sucrose, 0.21; galactose, 0.28; glucose, 0.31; fructose, 0.38; arabinose, 0.40; and xylose, 0.45.

Chromatography on wedge-shaped tongues of paper (the cone–stripe method), which in effect is a type of circular chromatography, has been used to separate nine sugars (R_F 0.48–0.86) in 32 hours with benzene–butanol–pyridine–H_2O (3:10:5:4 v/v) as the solvent (*39*). Matthias had earlier employed this same system with triple development for the separation of ten sugars (*3, 39a*) (Fig. 3).

Paper electrophoresis in the first dimension combined with paper chromatography in the second separates various hexosamines (*40*).

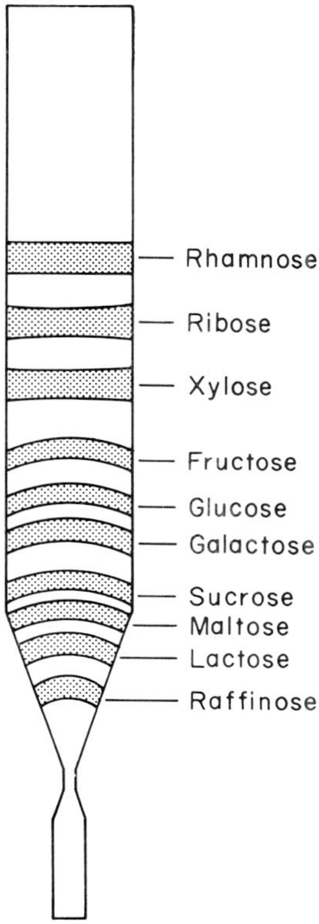

FIG. 3. Sugar separation by triple development on wedge-shaped strips; solvent: benzene–*n*-butanol–pyridine–H_2O (3:10:5:4 v/v). After (*3*, *39a*).

E. SOLVENTS

Water soluble mono- and oligosaccharides are chromatographed in water-containing mobile phases on paper equilibrated for several hours with the solvent vapor. Biphasic solvents (e.g., butanol–acetic acid–H_2O, 4:1:5 v/v) must be used with careful temperature control. Monophasic solvents containing more or less water have become increasingly popular of late. Recommended solvent systems are (*2*): (a) for mono- and disaccharides: *n*-butanol–pyridine–H_2O (9:5:4 v/v), and isopropanol–pyridine–H_2O (7:7:2 v/v); (b) for oligosaccharides: isoamyl alcohol–pyridine–H_2O (7:7:6 v/v), and ethyl acetate–*n*–butanol–acetic acid–H_2O (6:8:5:8 v/v). Solvents containing an acid and an alcohol should be prepared several days before use.

Numerous other solvent systems for various carbohydrates have been listed earlier (*1*). Some which have been recently employed (*41*) are: (a) butanol–ethanol–H_2O (5:1:4 or 80:22:38 v/v) (*14, 23*); (b) collidine–ethanol–H_2O (1:1:1 v/v) (*16*); (c) ethyl acetate–pyridine–H_2O (8:2:1, 2:2:1, or 10:4:3 v/v) (*6, 16, 23*); (d) butanol–ethanol–H_2O–NH_3 (120:23:80:2 v/v) (*21*); (e) benzene–butanol–pyridine–H_2O (1:5:3:3 v/v) (*30*); (f) acetone–butanol–H_2O (7:2:1 v/v) (*15*); (g) methyl ethyl ketone–H_2O (10:1 v/v) (*31*); (h) ethyl acetate–acetic acid–piridine–H_2O (5:1:5:3 v/v), tank equilibrated with ethyl acetate–pyridine–H_2O (40:11:6 v/v), for uronic acids (*23*); (i) ethyl acetate–acetic acid–formic acid–H_2O (18:3:1:4 v/v), for uronic acids and acidic oligosaccharides (*23*). Other solvent systems are given in the separate sections and R_F tables later in the chapter.

Some workers incorporate the detecting reagent with the solvent. For the identification of sucrose (*42*), the following has been used: ethyl acetate (water–saturated)–acetic acid (6:3 v/v) or butanol (water–saturated)–acetic acid (4:1 v/v), to either was added 0.3% 2-thiobarbituric acid (w/v) and 5% trichloroacetic acid (w/v).

F. Detecting Reagents

The reactions of various types of sugars with fourteen classic detecting reagents have been tabulated earlier (*1, 3*). The more important of these as well as some new detection methods are described below. Most of these reagents may be applied by spraying or dipping techniques.

1. *Ammoniacal* $AgNO_3$ (*2, 3, 15*)

The chromatogram is drawn through a solution prepared by mixing 0.1 ml of H_2O saturated with $AgNO_3$, 20 ml acetone, and a drop of water. After drying, spray with a solution prepared by dissolving 2 gm NaOH in a minimum amount of water and made up to 100 ml with methanol. Silver-brown spots of reducing carbohydrate derivatives appear at once; nonreducing compounds require short heating; sensitivity is 1 μg for reducing sugars. Reagents containing 33.3% $AgNO_3$ and 66.7% ammoniacal $AgNO_3$ have also been recommended (*43*).

2. *Aniline hydrogen phthalate* (*15, 21, 23, 30*)

Spray with a solution composed of 0.93 gm reagent, 1.66 gm phthalic acid, and 100 ml H_2O–saturated butanol, and heat the paper for 5 minutes at 105°C (130° for sucrose); observe fluorescence under ultraviolet light; very sensitive for aldoses, weak reactions given by ketoses and oligosaccharides; aldohexoses and methylpentoses yield brown spots, aldopentoses red spots. This reaction lends itself well to quantitative estimation by elution and spectrophotometry at 480–490 mμ or by photographing the chromatograms and photometry of the negatives.

Other amine detecting reagents suggested in the literature are benzidine (*44*) and β-naphthylamine (*38*). The latter, after heating the sprayed paper for 10 minutes at 150°C, yields yellow–brown spots with aldoses and ketoses. Urea in HCl solution reveals ketoses and their methylated derivatives and oligosaccharides; densitometry can be applied with an accuracy of ±5% (*17*).

3. *o- or p-Aminophenol*

Dissolve 0.15 gm reagent in 20 ml ethanol and 10 ml 50% H_3PO_4; spray paper and heat several minutes at 105°C; 1–5 μg of aldoses and ketoses give a positive test (*22*).

4. *Orcinol*

Dissolve 0.5 gm of reagent and 15 gm trichloroacetic acid in 100 ml H_2O–saturated *n*-butanol; spray paper and dry for 20 minutes at 105°C; ketohexoses give yellow spots, ketoheptoses blue-green spots. Non-reducing carbohydrate spots are eluted centrifugally from the chromatogram with water, and the eluate reacted with orcinol (100 μg per ml) in 50% (v/v) H_3PO_4 on a spot tile (*45*). Heating 30 minutes at 105°C forms pale yellow to deep reddish brown colors.

5. *β-Naphthylamine* (0.10 *gm*)–*Thymol* (1 *gm*)–80% *Ethanol* (150 *ml*)–H_3PO_4 (2 *ml*) (*2*)

Spray paper and heat at 110°C for a few minutes. Colors: fructose and fructose oligosaccharides, yellow; aldohexoses, brown; aldopentoses, pink; uronic acids, red.

6. *Thermal Detection* (*46*)

Sugars separated in the solvent ethanol–pyridine–benzene–0.5% sodium versenate are detected by heating at 115–195°C. Arabinose and xylose appear readily and sucrose and mannose only slowly.

7. *Malonic Acid*

A developed chromatogram is drawn quickly through the solution (1 gm malonic acid and 1 ml aniline in 100 ml absolute methanol), air-dried and then dried at 105°C for 10 minutes (longer for sucrose). Sugar zones are brown, yellow or gray, and fluoresce in ultraviolet light (*47*).

8. *Glucose Oxidase Reagent*

This reagent, containing reduced *o*-dianisidine, detects glucose (R_F 0.36 in *n*-butanol–pyridine–H_2O, 9:5:8 v/v) and deoxyglucose (R_F 0.50) as green spots changing to reddish brown (*48*).

9. *p-Aminobenzoic Acid*

Three grams of reagent is added gradually with stirring to 5.0 ml hot H_3PO_4; when dissolved, 300 ml of *n*-butanol–acetone–H_2O (10:5:2 v/v) is added. After dipping the paper and heating at 105°C for a few minutes, pentoses give a dark red color, while other sugars are stained dark brown. The solution is stable below 25°C for long periods (*13*).

10. *Diphenylamine–Aniline*–H_3PO_4

Diphenylamine (0.15 gm) is dissolved in 20–25 ml of ethyl acetate from a total volume of 100 ml. Then 0.8 ml of aniline, the remaining ethyl acetate, and 11 ml of 80% (w/v) H_3PO_4 are added. The chromatograms are dipped, dried at room temperature, and finally heated at 95–100°C until the background is faintly gray. The reagent gives specific color reactions that can be used to distinguish between sugars of different structure and to detect various nonreducing sugar derivatives with acid-labile blocking groups (*49*).

11. *Acids*

Five percent H_2SO_4 and 5% $HClO_4$ (by volume) (light, even spraying followed by heating at 105°C for 90 minutes) detects D-mannose, D-galactose, lactose, D-glucose, and maltose. Five percent H_3PO_4 detects D-fructose, sucrose, D-xylose, D-mannose, D-galactose, and lactose (*50*).

12. *p-Anisidine–Periodate*

Spray the chromatogram first with *p*-anisidine solution (1 gm in 100 ml 70% ethanol), heat 5–10 minutes at 105°C and then dip in a solution of 10 ml sodium periodate in 100 ml acetone. Depending upon the compound assayed and the mobile phase employed, spots of different colors appear on a brownish background. The sensitivity ranges from 1 to 20 μg; polyols (white), sugar acids (uronic acids red or brown, aldonic acids white), 2-deoxy sugars (yellow), amino sugars (brown or yellow) and their acetyl derivatives (gray or yellow), and pentoses and hexoses (red fluorescence in ultraviolet light) are detected and differentiated (*51*).

13. *Enzymatic Hydrolysis*

Enzymatic hydrolysis followed by a modification of the colorimetric method for cyanide detects cyanoglycosides (*52*).

14. 2,3,5-*Triphenyltetrazolium Chloride*

Thoroughly dried chromatograms are dipped into a 0.5% TTC solution in methanol to which is added NaOH to a final concentration of 0.25 *N*. The final color is developed by steaming the papers for 2 minutes in a chamber with a steam flow of 120 ml condensed water per minute. The papers are then washed with tap water and dried; uronic acids, neutral sugars, and hexosamines are detected with a sensitivity of 1 to 2 μg or less. For quantitative determinations with an accuracy of 10%, the colored spots on the paper (formazan dye) are extracted with methanol–acetic acid (8:1 v/v), and optical densities measured at 482 mμ (*52a*).

15. *Pyridinium Barbiturate*

The chromatogram is treated with an aqueous solution of pyridium barbiturate and heated. Reducing sugars produce brown spots; pyruvic acid and some other carbonyl compounds form pink spots and glyoxal deep red (*52b*).

A study has been reported on the relative sensitivity and specificity of twelve color reagents, including many of the above, for the detection of eleven sugars and derivatives found in the carbohydrate moiety of glycoproteins, indicating the general applicability of specific sugar color reagents for the analyses of complex molecules (*52c*).

G. Quantitative Analysis

Two general methods are used for quantitation of nonradioactive sugars separated by paper chromatography. In the first, sugars are sprayed with a suitable reagent and the quantity present is then estimated while still on the paper by spot length (*53*), absorbance (*54–56*), reflectance (*57*), or by eluting the colored spot from the paper and determining the absorbance of the eluate in a photometer (*58*, *59*). In the second method, marker strips are sprayed and used to locate sugars on the unsprayed portion of the chromatogram. The unsprayed areas containing the sugars are removed and the sugars are eluted and estimated by appropriate reactions. Elution may be effected with water, 50% ethanol, or ethanol–HCl–H_2O (250:12:35 v/v) (*30*). A Soxhlet apparatus is used for continuous extractions.

One test used with the second method employs ferricyanide (*60*). One milliliter of eluate (containing 0–0.06 μmoles of sugar) is mixed with 1 ml potassium ferricyanide solution (0.5 gm per liter H_2O) and 1 ml carbonate–cyanide solution (5.3 gm Na_2CO_3 and 0.65 gm KCN per liter H_2O). The tubes are capped, heated 15 minutes in boiling water and cooled in tap water for 15 minutes. Five milliliters of ferric solution (1.5 gm ferric ammonium sulfate plus 1 gm sodium lauryl sulfate per liter of 0.05 *N* H_2SO_4) is added to each tube with mixing. With reducing sugars, a blue color develops, the absorbance of which is determined at 680 mμ 15 minutes after adding the ferric solution. Substances without a free reducing group (sucrose, inositol, etc.) fail to react; deoxy and phosphorylated sugars give less color. Beer's law is followed, cellulose fibers in the eluate do not interfere, and sensitivity is of the order of 0.01 μmole of sugar. Alternatively, the Somogyi copper reagent can be used after elution from the paper in a microelution tube (*61*) (Fig. 4).

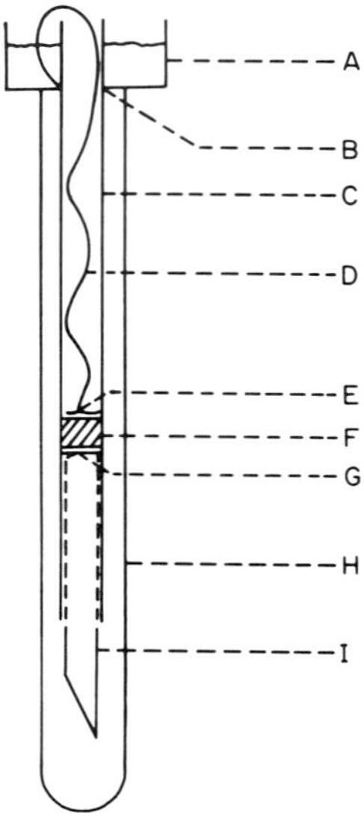

FIG. 4. Specifications for the assembly of the microelution tube. (A) a 1 cm in height by 3 cm in diameter plastic cap holding 3 ml water; (B) paraffin seal; (C) 10 cm of glass tube with 0.5 cm bore; (D) a strand from a glass wool wick 10 cm long; (E) loose cellulose powder; (F) 0.4 cm filter paper plug; (G) small plug of glass wool; (H) 15 × 150 mm test tube; (I) a rolled-up cut out square of a paper chromatogram. After (*61*).

Separated sugars can also be determined by indirect titration. A solution of $AgNO_3$ is sprayed on the chromatogram, the excess is removed by hyposulfite, and the metallic silver formed after reduction of Ag^+ by the sugars is dissolved in 4 *N* HNO_3 and titrated with standard 0.004 *N* NH_4SCN. For 0.02–0.250 mg of sugar, the error is ±5% (*62*).

Chromatography combined with polarography is also used to analyze sugar mixtures (*63*). An improved method for the detection and radioassay of monosaccharides has been described (*64*). The separated radioactive monosaccharides are located by spraying with one of the following modified reagents which eliminate possible dilution of the zones: (a) A solution of $AgNO_3$ (1 gm in 4 ml H_2O) is added to 125 ml acetone; spray the paper, air-dry, and spray with a solution of 1.2 gm

NaOH in 1.2 ml H_2O to which 99.8 ml absolute ethanol is added; monosaccharides appear as dark spots on a yellow background. (b) 2, 3, 5–Triphenyl–2*H*–tetrazolium chloride, 0.2% (w/v) in 100 ml $CHCl_3$ to which 5 ml triethylamine is added; heat for 10 minutes at 110°C after spraying.

The zones are cut out, placed in low-background scintillation vials and decolorized with 10 μl of 4 *N* HNO_3 to avoid possible quenching in the subsequent radioassay. After about 1 hour, 0.5 ml H_2O is added to the decolorized sample, and the suspension is allowed to stand 16 hours. The aqueous extract is mixed with 10 ml scintillation solvent (e.g., a PPO–POPOP mixture) and counted in a scintillation spectrometer.

III. Miscellaneous Classes of Carbohydrates

Some important classes of carbohydrates will now be considered separately, and recently obtained R_F values will be presented. Other compilations of R_F values have been published earlier (*1–3*). Sugar phosphates are covered in Chapter 14.

A. Aldoses and Ketoses

Systems for the separation of the simple sugars have been mentioned above. Another useful solvent is phenol (saturated with water) plus 1% NH_3. Most of the reagents mentioned above detect these sugars; urea–HCl detects ketoses in the presence of aldoses.

B. Oligosaccharides and Polysaccharides

Polysaccharides are not suitably separated by paper chromatography because of their limited solubility and migration in organic solvents. However, paper chromatographic analysis of sugar components after partial and total polysaccharide hydrolysis is of great value in the structure determination of polysaccharides (*3*). Also useful in this regard are color reactions specific for certain types of linkages in the oligosaccharides resulting from hydrolysis (*2*).

Disaccharides are separated in the solvents used for monosaccharides, while higher oligosaccharides require solvents with an increased water content (*2*). Continuous and multiple development techniques are very useful. The chromatography of oligosaccharides in general has been reviewed (*65*).

Oligosaccharides from chitin have been separated into six zones (hexasaccharide to acetylglucosamine, in ascending order) on Whatman No. 1 paper by ascending development with isoamyl alcohol–pyridine–H_2O (1:1:0.8 v/v) (*66*). The chromatogram is dried, humidified for

2 hours in a cylinder with a water-saturated atmosphere, rolled up, and soaked in chlorine reagent (Cl_2 gas absorbed in 1 liter CCl_4 to which is added 5 gm $BaCO_3$ and 5 gm anhydrous $CaCl_2$) for 20 minutes. After aeration for 30 minutes to remove excess Cl_2, spray with a fresh solution prepared by boiling 1 gm starch, 0.25 g KI and 1 ml 5 *N* HCl in 100 ml H_2O. A stable blue color is formed by spots containing $<1\ \mu g$ of oligosaccharide. This test is much more sensitive than $AgNO_3$ for these compounds.

Separations of starch (*67*) and xylan (*68*) hydrolyzates have been performed by paper chromatography in solvents such as those described in Section II,E. Separation of acidic mucopolysaccharides is possible in systems of acetate buffer and ethanol (*69*).

C. Methylated Sugars

These derivatives are separated in *n*-butanol–ethanol–H_2O (5:1:4 or 4:1:5 v/v), methyl ethyl ketone–H_2O–NH_3 (100:50:3 v/v) or *n*-butanol–water (azotropic mixture) (see Table I) (*68, 70, 71*). Paper impregnated with dimethylsulfoxide (DMSO) (isopropyl ether saturated with DMSO as the mobile phase) or 2% NH_4Cl [octane–isopropanol–10% NH_3 (50:25:2 v/v) or isooctane–isopropanol–10% NH_3 (65:25:2 v/v) as the mobile phase] have also been used (*72, 73*). Detection is with the aniline–phthalate reagent.

D. Glucosides

These are separated on Whatman No. 1 or 3 paper with the solvents shown in Table II (*74, 75*). Detection is with $AgNO_3$–NaOH or aniline–hydrogen phthalate.

E. Sugar Acids and Lactones

These compounds are separated on Whatman No. 1 paper with the solvents shown in Table III (*76*). Detection is with alkaline $AgNO_3$. A separation on glass fiber paper has also been reported (*77*). For uronic acids, solvents ethyl acetate–acetic acid–pyridine–H_2O (5:1:5:3 v/v, ethyl acetate–acetic acid–formic acid–H_2O (18:3:1:4 v/v) and phenol–H_2O–formic acid (100 gm:100 ml:1 ml 88%) are recommended (*6*).

F. Sugar Esters

Glucose, mannose, and rhamnose formates are separated using phenol saturated with H_2O, 2-butanol saturated with water, or *n*-butanol–acetic acid–H_2O (2:1:1 v/v) as the developing solvent with aniline–phthalate detection. Solvents containing NH_4OH, aliphatic amines, or HCl cause hydrolysis of the esters (*78*).

TABLE I

R_g VALUES[a] OF METHYLATED SUGARS (*68, 70, 71*)

Compound	R_g in the following solvents[b] 1	2
Tetramethylmannose	1.00	
2,3,5-Trimethylarabinose	0.98 (0.95)	0.82
Tetramethylgalactose	0.91 (0.86)	
2,3,4-Trimethylarabinose	0.84	
2,5-Dimethylarabinose	0.84 (0.86)	
3,4,6-Trimethylmannose	0.77	
2,3,4-Trimethylarabinose	0.84	
2,3,4-Trimethylgalactose	0.79 (0.73)	
2,3,6-Trimethylgalactose	0.79	
2,4,6-Trimethylgalactose	0.79	
2,4-Dimethylarabinose	0.62	
Dimethylmannose	0.66	
2,5-Dimethylgalactose	0.54	
2,3-Dimethylgalactose	0.55 (0.48, 0.43)	
2-Methylarabinose	0.55	
2,4-Dimethylgalactose	0.50 (0.48)	
4-Methylmannose	0.40	
2-Methygalactose	0.30	
3,4-Dimethylrhamnose	0.86	
2,3,5-Trimethylgalactose	0.73	
3-Methylrhamnose	0.66	
3,4-Dimethylgalactose	0.43	
Rhamnose	0.35	
Galactose	0.14	
2,3,4-Trimethylglucose	0.86	
2,3-Dimethylarabinose	0.61	
2,3,4-Trimethylxylose	0.94	0.80
2,3-Dimethylxylose	0.73	0.58
2-Methylxylose	0.38	0.20
Oligosaccharides	0.04	

[a] R_g = R_F relative to that of 2,3,4,6-tetramethyl-D-glucose. Figures in parentheses indicate different R_F values reported in separate studies.

[b] (1) *n*-Butanol–ethanol–H_2O (4:1:5 v/v); (2) *n*-butanol–H_2O (azeotropic mixture).

Acetylated sugars are detected with $AgNO_3$ after chromatography on acetylated paper (*3*) (solvent: ethyl acetate–tetrahydrofuran–H_2O, 6:35:47 v/v) or on paper impregnated with dimethylsulfoxide (*79*) (solvents: petroleum ether, isopropyl aclohol, or diethyl ether–dimethylsulfoxide, 25:1 v/v).

TABLE II

R_F VALUES OF SOME SUGARS AND GLUCOSIDES (*74, 75*)

Compound	R_F in following solvents[a]			
	1	2	3	4
Methyl-α-L-fucopyranoside	0.61			
Methyl-α-L-rhamnopyranoside	0.78	0.65		
6-Deoxy-L-talose		0.60		
L-Fucose		0.43		
L-Rhamnose		0.48		
L-Arabinose			0.18	0.13
D-Ribose			0.27	0.54
2-*C*-Methyl-L-arabinose			0.37	0.34
2-*C*-Methyl-L-ribose			0.39	0.71
Methyl-β-L-arabinopyranoside			0.40	0.31
Methyl-β-D-ribopyranoside			0.55	0.86
Methyl-2-*C*-methyl-β-L-arabinopyranoside			0.58	0.52
Methyl-2-*C*-methyl-β-L-ribopyranoside			0.74	0.92

[a] (1) Ethyl acetate–*n*-propanol–H_2O (5:3:2 v/v); (2) ethyl acetate–pyridine–H_2O (2:1:2 v/v); (3) *n*-butanol–ethanol–H_2O (4:1:5 v/v); (4) 100 ml solvent 3 plus 5 gm phenylboronic acid and 2.5 ml ethanol.

TABLE III

R_g VALUES[a] OF SOME SUGAR ACIDS AND LACTONES (*76*)

Compound	R_g in the following solvents[b]		
	1	2	3
Glucuronic acid	0.00	0.42	0.46
Galacturonic acid	0.00	0.33	0.85
Tagaturonic acid	0.00	0.59	0.62
Fructuronic acid	0.00	0.64	0.62
Glucuronolactone	—	1.10	1.25
Galactonolactone	1.00	1.00	1.00
Mannonolactone	0.85	0.91	1.12
Gulonolactone	0.77	0.89	0.84
Altronolactone	1.12	1.09	1.32
Talonolactone	1.03	0.98	1.07
Allonolactone	1.12	1.02	1.09
Idonolactone	1.12	1.03	1.11
Gluconolactone	1.77	1.08	1.22

[a] Relative to galactonolactone.

[b] (1) Pyridine–ethyl acetate–H_2O (20:72:23 v/v); (2) pyridine–ethyl acetate–acetic acid–H_2O (5:5:1:3 v/v); (3) ethyl acetate–acetic acid–H_2O (3:1:3 v/v).

Sugar sulfates are chromatographed with *n*-butanol–ethanol–H_2O (3:1:1 v/v) plus 3% cetylpyridinium chloride (*80*). R_F values in this system are: glucose, 0.21; glucose 3-sulfate, 0.67; glucose 6-sulfate, 0.41; and glucose disulfates, 1.00. For location of the zones, the $AgNO_3$–NaOH reagent is used after the paper is first rinsed with $CHCl_3$ to remove the detergent.

Table IV (*81*) shows R_F values for various sugar esters.

TABLE IV

R_F VALUES OF SOME SUGAR ESTERS[a] (*81*)

Compound	R_F in the following solvents[b]			
	1	2	3	4
4-*p*-Cumaroylrhamnose	0.80	0.78	0.68	0.79
1-*p*-Cumaroyl-β-D-glucose	0.66	0.61	0.65	—
2-*p*-Cumaroylglucose	0.69	—	0.64	0.68
3-*p*-Cumaroylglucose	0.66	0.63	0.68	0.67
6-*p*-Cumaroylglucose	0.59	0.56	0.55	0.59
1-Feruloyl-β-D-glucose	0.60	—	0.57	—
2-Feruloylglucose	0.63	0.60	0.57	0.59
3-Feruloylglucose	0.62	0.57	0.65	0.58
4-Feruloylglucose	0.60	0.55	0.67	0.54
6-Feruloylglucose	0.55	0.47	0.49	0.48
Caffeylglucose	—	0.31	0.32	—
Glucosyl-(1 → 2)-6-feruloyl-β-D-glucose from *Raphanus*	0.39	0.20	0.67	—
Glucosyl-(1 → 2)-6-*p*-cumaroyl-β-D-glucose from *Raphanus*	0.42	0.26	0.73	—
Glucose	0.22	—	0.37	0.10
Ferulic acid	0.84	0.87	0.27	0.80

[a] Paper: Schleicher & Schüell 2043b Mgl. Wedge strip form according to Matthias (Fig. 3).

[b] (1) *n*-Butanol–acetic acid–H_2O (4:1:2 v/v); (2) *n*-pentanol–acetic acid–H_2O (4:1:5 v/v); (3) H_2O–concentrated HCl (97:3 v/v); (4) *n*-butanol–ethanol–H_2O (4:1:5 v/v).

G. SUGAR HYDRAZONES AND AZINES

(a) Dialkyl and aralkyl hydrazones—Schleicher and Shuell (S. and S.) No. 2043bM paper; 96% ethanol–benzene (3:1 v/v) as solvent (*82*).

(b). Monoalkylhydrazones—FN 12 paper; *n*-butanol–acetone–H_2O (2:7:1 v/v) solvent (*82*).

(c) Azines—S. and S. 2043b paper; acetone–H_2O (4:1 v/v) as solvent; R_F values (*83*): azine of D-ribose, 0.19; L-arabinose, 0.23; D-xylose, 0.18;

D-galactose, 0.11; D-mannose, 0.08; D-glucose, 0.06; L-rhamnose, 0.25; D-fructose, 0.27; L-sorbose, 0.16; maltose, 0.14; and lactose, 0.12.

H. Deoxy Sugars (*84, 85*)

Paper: Whatman No. 1. Solvent: (a) *n*-butanol–ethanol–H_2O (4:1:5 v/v); (b) ethyl acetate–*n*-propanol–H_2O (7:1:2 v/v); (c) *n*-butanol–acetic acid–H_2O (70:7:23 v/v). Technique: deoxysugars applied as 2% solutions in methanol; descending development.

Detection: (a) Air-dried chromatograms are dipped in a solution composed of 2 gm diphenylamine and 2 ml aniline dissolved in 100 ml acetone to which 10 ml H_3PO_4 (80%) is added, and they are heated at 80–85°C for 10 minutes; all 3-, 4-, and 6-deoxyhexoses give yellow spots which change color on storage so that the three classes can be differentiated; other deoxy sugars also react.

(b) Dip the chromatogram in reagent A (1 volume of 0.1 *M* periodic acid mixed with 20 volumes of acetone right before use) and allow the acetone to evaporate; dip in reagent B (0.01 *M* ethylene glycol in ethanolic solution saturated with thiobarbituric acid), dry again and heat at 60°C for 10–20 minutes. Compounds yielding malondialdehyde on periodate oxidation (2-deoxysugars and 3-deoxypolyols) give brilliant red spots on a pale yellow background which fluoresce orange on a violet background under ultraviolet light.

R_F values solvent (a):

Monodeoxyglucose	R_g (relative to glucose)
2-	2.48
3-	2.41
4-	1.86
5-	2.88
6-	2.66

By a combination of paper and thin-layer chromatography, all possible stereoisomeric 6-deoxyhexoses, 3-*O*-methyl-6-deoxyhexoses and 6-deoxyhexuloses have been differentiated (*86*).

I. Sugar Alcohols

Sugar alcohols are separated with a two-phase system composed of acetone–sodium acetate–H_2O (*86a*).

J. Amino Sugars

Solvents for amino sugars and *N*-acetyl derivatives (*6, 87*): (a) butanol–acetic acid–H_2O (4:1:5 or 12:3:5 v/v); (b) butanol–pyridine–H_2O (6:4:3 v/v); (c) aqueous phenol saturated with NH_3.

Detection: ninhydrin (cf. Chapter 5), the Morgan–Elson (*p*-aminobenzaldehyde and acetylacetone) reagent (*2, 3*), or by chlorination (*66*).

Comments: Artifacts (multiple zoning, streaking) are detected after radioactive amino sugars are developed in various recommended solvents when radioactive methods (scanning, radioautography), but not color methods, are employed (*88*).

Detailed Experiment (*89*)

A serum or tissue sample is hydrolyzed with 6 *N* HCl for 8 hours in a boiling H_2O bath. The neutralized hydrolyzate is passed through a Dowex 50 column and the hexosamines are eluted with 2 *N* HCl. The eluates are reacted with ninhydrin in a capillary tube in the presence of pyridine, and the pentoses formed are separated by paper chromatography using butanol–benzene–H_2O–pyridine–ethanol (4:1:1:1:1 v/v). The pentoses are eluted with water; a portion is mixed with 0.1 ml 80% phenol and 5% H_2SO_4, cooled, and the color measured in a colorimeter. An alternate solvent for the separation of glucosamine and galactosamine is ethyl acetate–pyridine–butanol–butyric acid–H_2O (10:5:5:1:5 v/v) (*90*).

References

1. R. J. Block, E. L. Durrum, and G. Zweig, "A Manual of Paper Chromatography and Paper Electrophoresis," 2nd rev. ed., pp. 170–214. Academic Press, New York, 1958.
2. F. Percheron, *in* "Chromatography" (E. Heftmann, ed.), pp. 573–605. Reinhold, New York, 1967.
3. K. Macek, *in* "Paper Chromatography" (I. M. Hais and K. Macek, eds.), pp. 289–330. Academic Press, New York, 1963.
4. M. Toth, *Elelmiszervizsgalati Kozlemen.* **12**, 201 (1966).
5. R. W. Wheat, *Methods Enzymol.* **8**, 60 (1966).
6. D. H. Northcote, *Brit. Med. Bull* **22**, 180 (1966).
7. C. K. Hordis and G. N. Kowkabany, *Anal. Chem.* **30**, 1210 (1958).
8. A. J. P. Martin, *Biochem. Soc. Symp.* (*Cambridge, Engl.*) **3**, 4 (1948).
9. I. E. Bush, *Methods Biochem. Anal.* **13**, 357 (1965).
10. M. Lederer and A. Pirelli, *Sci. Rept. Inst. Super. Sanita* **1**, 582 (1961).
11. H. Amato, *J. Chromatog.* **33**, 500 (1968).
12. M. Fedoroňko and K. Linek, *Collection Czech. Chem. Commun.* **32**, 2177 (1967).
13. A. S. Saini, *J. Chromatog.* **24**, 484 (1966).
14. P. Esau and M. A. Amerine, *Am. J. Enol. Viticult.* **17**, 265 (1966).
15. V. K. Gorodetskii, *Vopr. Med. Khim.* **13**, 627 (1967).
16. I. G. Mokhnachev and V. P. Leshchenko, *Konserv. i Ovoshchesushil. Prom.* **19**, 29 (1964).
17. A. A. Pinto, C. H. Godoy, and C. L. Rayo, *Bol. Soc. Quim. Peru* **27**, 211 (1961).
18. S. Ullah and F. M. Khan, *Pakistan J. Sci. Ind. Res.* **9**, 178 (1966).
19. T. A. Tonchev and G. K. Bambalov, *Nauchni Tr., Vissh Inst. Khranitelna Vkusova Prom.-Plovdiv* **12**, 339 (1965).
20. P. J. Goodman, M. A. Watson, and A. R. C. Hill, *Ann. Appl. Biol.* **56**, 56 (1965).

21. M. Giovannini, V. Consonni, and A. Calleri, *Minerva Pediat.* **17**, 1702 (1965).
22. V. B. Kupressova, *Izv. Sibirsk. Otd. Akad. Nauk SSSR, Ser. Biol. Med. Nauk*, p. 141 (1965).
23. R. W. Stoddart, A. J. Barrett, and D. H. Northcote, *Biochem. J.* **102**, 194 (1967).
24. S. J. Kraeger and J. G. Hamilton, *J. Chromatog.* **41**, 113 (1969).
25. D. F. Durso and J. C. Paulson, *Anal. Chem.* **30**, 919 (1958).
26. E. Gulbaran, *J. Am. Leather Chemists' Assoc.* **59**, 619 (1964).
27. J. Wohnlich, *Intern. Etude Methodes Separation Immediate Chromatog., Paris, 1961* p. 203.
28. J. H. Pazur and S. Okada, *J. Biol. Chem.* **241**, 4146 (1966).
29. D. French, A. P. Pulley, M. Abdullah, and J. C. Linden, *J. Chromatog.* **24**, 271 (1966).
30. G. S. D'yachkov, V. A. Lepeshkina, and V. A. Rostimeshin, *Tr. Vses. Nauchn.-Issled. Inst. Torf. Prom.* p. 71 (1967).
31. T. G. Bonner, E. J. Bourne, S. E. Harwood, and D. Lewis, *J. Chem. Soc., C* p. 2229 (1966).
32. G. Manley, *Nature* **206**, 1253 (1965).
33. A. A. J. Feast, W. G. Overend, and N. R. Williams, *J. Chem. Soc., C* p. 303 (1966).
34. H. Allgeier, *Helv. Chim. Acta* **51**, 668 (1968).
35. M. Kulhanek and Z. Sevcikova, *Folia Micribiol. (Prague)* **11**, 62 (1966).
36. J. R. Swartwout, J. W. Dieckert, O. N. Miller, and J. G. Hamilton, *J. Lipid Res.* **1**, 281 (1960).
37. J. Davidek and J. Hrdlička, *Sb. Vysoke Skoly Chem.-Technol. Praze, Potravinarska Technol.* **9**, 113 (1966).
38. G. Barbiroli, *Rass. Chim.* **17**, 62 (1965).
39. N. Jako, *Elelmiszervizsgalati Kozlemen.* **13**, 167 (1967).
39a. W. Matthias, *Naturwissenschaften* **41**, 17 (1954).
40. S. Ohkuma and T. Shinohara, *Nature* **202**, 593 (1964).
41. L. Hough and J. K. N. Jones, *Methods Carbohydrate Chem.* **1**, 21 (1962).
42. C. Stella and L. Niccolai, *Riv. Viticolt. Enol.* **19**, 104 (1966).
43. J. Z. Beer, *Carbohydrate Res.* **1**, 297 (1965).
44. B. G. Chan and J. C. Cain, *Nature* **192**, 69 (1961).
45. F. R. Jevons, *J. Chromatog* **16**, 252 (1964).
46. I. Krzeczkowska, *Ann. Univ. Mariae Curie-Skladowska, Lublin-Polonia* **D19**, 343 (1964).
47. H. Zentner, *Chem.-Ind. (London)* p. 1836 (1968).
48. A. Martinsson, *J. Chromatog.* **24**, 487 (1966).
49. J. Kocourek, M. Tichá, and J. V. Koštíř, *J. Chromatog.* **24**, 117 (1966).
50. K. Horitsu, *Hirosaki Daigaku Nogakubu Gakujutsu Hokoku* **11**, 51 (1965).
51. L. A. Veiga and E. L. Chandlier, *Anal. Biochem.* **20**, 419 (1967).
52. W. D. Bennett and B. A. Tapper, *J. Chromatog.* **34**, 428 (1968).
52a. J. Mes and L. Kamm, *J. Chromatog.* **43**, 480 (1969).
52b. K. Shozaburo and S. Kiroku, *Kenkyu Hokoku* **15**, 25 (1967).
52c. J. Mes and L. Kamm, *J. Chromatog.* **38**, 120 (1968).
53. I. R. Shimi, M. S. N. El Dein, and G. M. Iman, *Analyst* **86**, 535 and 744 (1961); **90**, 564 (1965).
54. T. V. Belen'kaya, A. I. Skrygan, and N. M. Syalitskaya, *Vestsi Akad. Navuk Belarusk SSR, Ser. Fiz.-Tekhn. Navuk*, p. 76 (1961).
55. S. A. Marutyan and A. D. Dogramadzhyan, *Vinodelie i Vinograd. SSSR* **23**, 13 (1963).
56. T.-J. Chou, *Hua Hsueh Tung Pao* p. 372 (1965).
57. J. E. Jeffrey, E. V. Partlow, and W. J. Polglase, *Anal. Chem.* **32**, 1774 (1960).

58. C. M. Wilson, *Anal. Chem.* **31**, 1199 (1959).
59. J. P. Thornber and J. P. Northcote, *Biochem. J.* **81**, 455 (1961).
60. G. Guinn, *J. Chromatog.* **30**, 178 (1967).
61. B. G. Chan and J. C. Cain, *J. Chromatog.* **22**, 95 (1966).
62. R. G. Katkevich and Yu. Yu. Katkevich, *Latvijas PSR Zinatnu Akad. Vestis, Kim. Ser.* p. 102 (1967).
63. N. A. Arkhipovich and L. M. Zelenina, *Sakharn. Prom.* **39**, 366 (1965).
64. A. E. Gal, *J. Chromatog.* **34**, 266 (1968).
65. R. W. Bailey and J. B. Pridham, *Chromatog. Rev.* **4**, 114 (1962); J. E. Courtois, *Pharm. Biol.* **5**, 485 (1968).
66. R. F. Powning and H. Irzykiewicz, *J. Chromatog.* **17**, 621 (1965).
67. M. Richter and H. H. Stroh, *Staerke* **18**, 115 (1966).
68. M. Alam and R. J. McIlroy, *J. Chem. Soc., C* p. 1577 (1967).
69. T. A. Good, *Anal. Biochem.* **19**, 109 (1967).
70. G. O. Aspinall, V. P. Bhavanandan, and T. B. Christensen, *J. Chem. Soc.* p. 2682 (1965).
71. G. O. Aspinall and V. P. Bhavanandin, *J. Chem. Soc.* p. 2696 (1965).
72. K. Wallenfelds, G. Bechtler, R. Kuhn, H. Trischmann, and H. Egge, *Angew. Chem. Intern. Ed. Engl.* **2**, 515 (1963).
73. F. Petek, *Bull. Soc. Chim. France Engl.* p. 263 (1965).
74. P. M. Collins and W. G. Overend, *J. Chem. Soc.* p. 1912 (1965).
75. R. J. Ferrier, W. G. Overend, G. A. Rafferty, H. M. Wall, and N. R. Williams, *J. Chem. Soc., C* p. 1092 (1968).
76. J. Hickman and G. Ashwell, *J. Biol. Chem.* **241**, 1426 (1966).
77. B. Radhakrishnamurthy, *Arch. Biochem. Biophys.* **99**, 534 (1962).
78. T. McCullough, E. DeJong, and C. Caster, *J. Chromatog.* **37**, 545 (1968).
79. B. Wickberg, *Carbohydrate Chem.* **1**, 31 (1962).
80. D. A. Rees, *Nature* **185**, 309 (1960).
81. L. Birkofer, C. Kaiser, M. Donike, and W. Koch, *Z. Naturforsch.* **20**, 426 (1965); *J. Chromatog.* **20**, D5 (1965).
82. H. H. Stroh and H.-G. Scharnow, *Chem. Ber.* **98**, 1594, 1595, and 1597 (1965); *J. Chromatog.* **23**, D4 and D5 (1966).
83. H. H. Stroh, A. Arnold, and H.-G. Scharnow, *Chem. Ber.* **98**, 1409 (1965); *J. Chromatog.* **23**, D5 (1966).
84. S. McNally and W. G. Overend, *J. Chromatog.* **21**, 160 (1966).
85. P. J. Anderson, *J. Chromatog.* **21**, 163 (1966).
86. H. Kaufmann, P. Muehlradt, and T. Reichstein, *Helv. Chim. Acta* **50**, 2287 (1967).
86a. D. Scheffer and R. Kickuth, *Z. Anal. Chem.* **230**, 254 (1967).
87. A. Yamamoto, C. Miyashita, and H. Tsukamoto, *Chem. & Pharm. Bull. (Tokyo)* **13**, 1036 (1965).
88. G. Sheppard, *J. Chromatog.* **40**, 312 (1969).
89. R. M. Nair and P. A. Kurup, *Indian J. Biochem.* **4**, 117 (1967).
90. H. Mukerjee and J. Sri-Ram, *Anal. Biochem.* **8**, 393 (1964).

7

ALIPHATIC ACIDS

Aliphatic acids have been separated by paper chromatography either as the free acids, the acid salts, or some form of derivative, as discussed later in this chapter. Solvent mixtures for chromatographic development are many and varied, and representative types are discussed herein. Spray reagents for detection of acids on the paper are generally acid–base indicators; metallic salts may be used with acid derivatives; the 2,4-dinitrophenylhydrazones of the keto acids are visible as yellow spots which may be enhanced by treatment with an alcoholic basic solution. Inspection of some acid spots may be accentuated when viewed under ultraviolet light. The chromatographic technique is limited only by the degree of ingenuity of the investigator, as evidenced by several detailed studies on this subject (*1–4*). Lipid chromatographic analysis has been recently reviewed thoroughly (*5*) and therefore will not be included in the following discussion.

I. Preparation of Samples

A. Salts

Sodium, ammonium, and diethylamine salts of the lower carbon fatty acids are usually used for the separation of these acids by paper chromatography (*6*, *7*). A water extract of the unknown is made alkaline with one of the aforementioned reagents and the solution is applied to the paper; or the extract may be applied to an area of the paper previously dampened with diethylamine; or an alcoholic mixture of the acids may be neutralized with 0.1 *N* ethylamine, using bromocresol green as an indicator and applied to the chromatogram (*8*). Quantitative determinations may require preliminary purification of the acids by steam distillation, silica gel columns, or anion exchange resins.

B. Hydroxamates

Conversion of the volatile aliphatic acids to their respective hydroxamate derivatives permits good resolution of some of the straight-chain acids. Formic and acetic acid, not readily resolved by chromatography as acids or salts, are readily separated as the hydroxamate derivatives. The solution containing 1 mg of the acid is made basic with NH_4OH, evaporated to dryness, and extracted twice with 1.0 ml portions of 0.4 *N* HCl in ethanol. Ethereal diazomethane is added until the yellow color persists, and esterification is allowed to proceed for 10 minutes. Excess diazomethane is destroyed by adding several drops of 0.4 *N* alcoholic HCl. Two milliliters of 2 *N* NH_2OH sulfate–3.5 *N* NaOH (1:1 v/v) is added, and the solution is allowed to stand at room temperature for 15 minutes. The hydroxamate solution is concentrated on a steam bath and spotted on the chromatographic paper (*9*). A modification of the procedure consists of a reflux period of 30 minutes at 90°C in a water bath after the addition of a neutralized solution of the hydroxylamine reagent (*10*).

C. Dinitrophenyl Derivatives

Fatty acid 2,4-dinitrophenylhydrazines may be prepared as follows by refluxing fatty acids (10 mg) with thionyl chloride (0.3 ml) for 30 minutes and vacuum-distilling the mixture. A dry pyridine solution (0.1 ml) of 5% 2,4-dinitrophenylhydrazine is added to the residue, the mixture is warmed and then cooled to room temperature and allowed to stand for 20 minutes. Benzene (2 ml) is added and the excess pyridine is extracted with 10–20 ml of warm water. The benzene layer is applied to the chromatographic paper (*11*).

D. Acids

Nonvolatile water-soluble aliphatic acids may be extracted from plant material with hot 80% aqueous ethanol (*12*) or from foodstuffs with 2% sulfuric acid (*13*).

E. Keto Acids

Although the α-keto acids have been chromatographed as the free acids (*14*), the use of derivatives of the acids are preferred because of ease of extraction from biological materials, the increased stability over the parent acid, and because the 2,4-dinitrophenylhydrazones are colored and can be detected on the paper chromatogram without further reactions with color reagents. Two milliliters of 0.5% 2,4-dinitrophenylhydrazine (DNPH) in 6 *N* HCl is added to 50 ml of deproteinized sample (diluted tenfold with water) of blood or urine (*15*, *16*), and the mixture is allowed to react at room temperature for 30 minutes. The aqueous solution is extracted three times with 15 ml portions of chloroform containing 20% ethanol or ethyl acetate. The combined solvent fraction is extracted with 15 ml of 1 *N* Na_2CO_3. The retained Na_2CO_3 fraction is washed with 10 ml of the chloroform–ethyl solvent mixture; the carbonate solution is acidified in the cold (0–4°C) with 5 ml 6 *N* HCl. The hydrazones are extracted from the solution with three portions of chloroform–ethanol (or ethyl acetate); the combined extracts containing the keto acid and derivatives are concentrated by evaporation under a gentle air stream. It has been suggested that deproteinization of biological materials—for the preparation of DNPH derivatives—can be best accomplished by the addition of 0.05–0.1 volumes of ice-cold 60% $HClO_4$ to the homogenate which is kept in an ice bath (*17*). After 10 minutes, the mixture is filtered directly into an excess of freshly prepared DNPH solution (1 gm DNPH in 1 liter of 2 *N* HCl); the mixture is stored at room temperature for 2 hours and then overnight at 4°C. Using these conditions, all keto acids investigated were quantitatively converted to the derivative form with the exception of β-mercaptopyruvate, which required several days of incubation. Some decomposition of acetoacetate in dilute $HClO_4$ was noted.

For clinical screening purposes, 2.5 ml urine is combined with 2.5 ml of 0.3% DNPH (in 1 *N* HCl) in a stoppered tube and allowed to stand for 10 minutes. Ethyl acetate (10 ml) is added and the tube is shaken for 2 minutes. The ethyl acetate layer is removed and transferred to a beaker. The urine mixture is reextracted with 10 ml ethyl acetate, the ethyl acetate extracts are combined and concentrated at room temperature or under reduced pressure. The residue is dissolved in 0.25 ml of 1:1 ethanol–ethyl acetate and applied to the chromatographic paper for keto acid analysis (*18*).

When a keto acid hydrazone produces several isomers on the finished chromatogram, such as *cis*- and *trans*-pyruvic acid (*19*), the nitroquinoxaline derivative of the acid may be preferred. With this technique, no isomerism has been reported. The preparation of this derivative is essentially the same as that for the hydrazones (*20*, *21*). Sufficient 1,2-diamino-4-nitrobenzene in 2 *N* HCl is added to a deproteinized solution of α-keto acids; the mixture is warmed to 60°C and allowed to stand for 16 hours at room temperature. The nitroquinoxalinol derivatives are extracted and purified in the same manner as described for the DNPH derivatives, except that a dilute solution of NaOH is substituted for the Na_2CO_3 solution.

F. Higher Fatty Acids

Natural fatty acid mixtures may contain both saturated and unsaturated components which are not readily separated on the paper chromatogram. Hydrogenation in acetic acid in the presence of a platinum oxide catalyst (*1*) may be used to determine the total amount of acid of each chain length. Supplementary comparative data are obtained by destroying the unsaturated components with strong alkaline permanganate solution at room temperature for 16–18 hours and chromatographing the remaining saturated acids. Better separation of the higher fatty acids are possible with chromatography of a derivative of the acid.

1. *Hydroxamates* (*22*)

An ether solution of the acid is esterified by the addition of diazomethane solution. The ether is evaporated and the residue is dissolved in an alkaline solution of hydroxylamine in methanol (1 mole hydroxylamine hydrochloride–2 moles KOH–1 mole ester). The mixture is heated 2–3 minutes in a water bath; KCl is filtered off; excess KOH is neutralized with tetrahydrofuran–acetic acid (4:1); the solution is applied to the paper chromatogram for analysis.

2. *2,4-Dinitrophenylhydrazones of Acetol Esters* (*23*)

An alcoholic solution of the fatty acids is acidified with 0.05 *N* methanolic HCl. Bromoacetone is added and the mixture is refluxed 30 minutes at 80–90°C. The mixture is cooled to 50°C, and excess DNPH in 2 *N* methanolic HCl is added and allowed to react for 30 minutes. Benzene layer is used for paper chromatography.

3. *Mercury Compounds* (*24*)

Unsaturated fatty acid methyl esters and 20% excess of mercuric acetate in absolute methanol (1–2 ml methanol; 1 gm mercuric acetate) is heated 30 minutes at 80°C. An excess of ether or benzene and water are added and the mixture is shaken vigorously. The organic layer is used for paper chromatography.

4. *Peracid Treatment* (*25*)

A 1-gm mixture of fatty acids (saturated and unsaturated) is dissolved in 14 ml acetic acid followed by the addition of a 1:1 mixture of 88% formic acid and 30% hydrogen peroxide. The mixture is allowed to stand overnight; it is then diluted with 15 ml water and extracted with 2 ml chloroform; the extract is used for paper chromatography.

5. *Cholesteryl Esters of Long-Chain Fatty Acids*

The esters of lauric, myristic, palmitic, stearic, oleic, linoleic, and linolenic acids are prepared by the method of Swell and Treadwell (*26*).

A mixture of 0.182 mole thionyl chloride and 0.058 mole fatty acid is heated near the boiling point, with constant stirring, in a 250 ml Erlenmeyer flask. Excess thionyl chloride is removed by suction. The acid chloride is warmed and 0.040 mole cholesterol is added gradually. When reaction has subsided, 0.040 mole anhydrous pyridine is added and the mixture is heated at 80°C for 20 minutes. Absolute alcohol is added to dissolve the ester; acetone is used if the fatty acid contains more than ten carbons. Labarrere *et al.* (*27*) substituted oxalyl chloride for the thionyl chloride reagent because its use with unsaturated fatty acids was more efficient and because it was more easily removed from the reaction mixtures. The derivatives are useful in cholesterol metabolism studies.

II. Filter Paper

Whatman papers Nos. 1 and 4 have been used for the separation of the fatty acid salts. Prewashing the paper with an organic acid followed by a water rinse or an ammoniacal rinse, depending on the pH of the indicator spray reagent subsequently used, will eliminate ghost spots, minimize tailing of the spots, and will permit better resolution of a mixture of the acids.

Whatman papers Nos. 1, 2, 3 MM, and 4 have been used for the separation of the keto acid 2,4-dinitrophenylhydrazones. Prewashing the papers with buffer solutions (which should be soluble in the mobile phase of the chromatographic solvent) will improve resolution of the derivatives. Treatments of the paper have included the use of glycine–0.1 *M* NaOH, pH 8.2–8.4 (*28*); 1 *N* sodium bicarbonate, pH 8.2 (*15*); and 0.1 *M* phosphate buffer, pH 7.4 (*29*). A critical factor is the stability of the acid derivatives; organic solvents may decarboxylate the hydrazone (*28*). The use of phosphate buffers may prevent the problem, but even then the solutions may degrade if kept for more than one day. The oxalacetic acid derivative is stable at room temperature even when dissolved in 10% Na_2CO_3, but it will decompose to the pyruvic acid derivative if it is exposed to moderate heat (*29*).

Pretreatment of filter papers have included the application of liquid paraffins, castor oil, olive oil, petroleum jelly, and silicones for use in the chromatographic separation of the long-chain fatty acids. In this reversed phase partition system, the treated paper repels the water and retains the less polar component of the solvent system. Examples of the process are as follows:

1. The paper is passed through a solution composed of 7 gm water-washed USP heavy mineral oil in 100 ml of ether. After a few minutes of air drying, the paper is ready for use (*25*).

2. Whatman Nos. 1 or 3 papers are immersed in liquid paraffin (10% solution, v/v, in benzene or ether) for 5–10 minutes. The treated papers are then air dried (*1*, *30*). Excess paraffin may result in loss of acids if the cupric acetate detection procedure is used (*1*). Papers may also be impregnated with paraffin by overnight descending chromatography. Skipski *et al.* (*31*) states that optimal acid separations are obtained with papers containing 19–20% paraffin oil, and Ashley (32) has noted that a concentration of 7% paraffin in the paper is preferable for fatty acids of carbon chain length C_{12}–C_{18}, and a 12% concentration for carbon chain lengths of C_{18}–C_{24}. The treated papers are preferably stored over a desiccant and paraffin shavings in a vacuum desiccator.

3. Chromatographic papers have been impregnated with silicones, using the ascending chromatographic technique (*33*), or passing the paper through a silicone solution and blotting the paper between sheets of adsorbent paper, then drying in an oven at 110°C for 1 hour. The concentration of the solution may range from 5% to 10% silicones dissolved in benzene or cyclohexane. The paper acquires hydrophobic properties with this treatment (*34*).

4. A mixture of redistilled methylcellosolve and distilled water (9:1 v/v) is mixed thoroughly with an equal volume of *n*-butyl ether. Whatman No. 1 paper is passed through the separated methylcellosolve–water layer and allowed to drain for 5 minutes at room temperature (20°C). The treated paper, when developed in the *n*-butyl ether phase of the mixture, will give improved separation of the C_6–C_{12} dibasic acids in the presence of monobasic acids (*35*).

III. Solvents for Chromatographic Development

1. *Lower Fatty Acids*

n-Butanol–1.5 *N* aqueous NH_3 (1:1)—The butanol layer separates the C_2–C_7 straight-chain fatty acids more efficiently than their corresponding sodium salts (*36–38*).

n-Butanol–ethanol–3 *N* NH_4OH (4:1:5)—The alcohol phase is used for the separation of the fatty acid sodium salts (*36*).

95% ethanol–NH_4OH (100:1) (*39*)

n-Propanol–concentrated aqueous NH_3 (7:3) (*6*, *40*)

n-Butanol–H_2O–diethylamine (100:15:1) (*39*, *41*, *42*)

n-Hexanol–1.5 *N* aqueous NH_3 (1:1)—The alcohol layer separates alkoxy acids (*43*).

Isooctane–acetone–95%ethanol–NH_4OH (40:30:30:1)—Solvent separates alkoxy acids (*43*).

See Tables I and II.

TABLE I

R_F VALUES OF SOME LOWER FATTY ACIDS

Fatty acid	Solvent[a]				
	1	2	3	4	5
Formic[b]	0.10	0.31	0.37	—	—
Acetic[b]	0.11	0.33	0.37	0.28	0.08
Propionic[b]	0.19	0.44	0.48	0.41	0.08
n-Butyric[b]	0.29	0.54	0.57	0.51	—
n-Valeric[b]	0.41	0.60	—	0.56	—
n-Caproic[b]	0.53	0.68	0.69	—	—
n-Heptanoic[b]	0.62	0.72	—	—	—
n-Octanoic[b]	0.65	0.76	0.78	—	—
n-Nonanoic[b]	0.67	—	—	—	—
Bromoacetic	—	—	0.45	—	0.02
Dibromoacetic	—	—	—	—	0.10
Tribromoacetic	—	—	—	—	0.31
Chloroacetic	—	—	—	—	0.14
Dichloroacetic	—	—	—	—	0.27
Trichloroacetic	—	—	—	—	0.45
2-Bromopropionic	—	—	0.52	—	0.25
3-Bromopropionic	—	—	—	—	0.20
2,3-Dibromopropionic	—	—	—	—	0.27
2-Chloropropionic	—	—	—	—	0.23
3-Chloropropionic	—	—	—	—	0.19
Bromobutyric	—	—	0.60	—	—
Iodoacetic	—	—	0.50	—	—
Iodopropionic	—	—	0.55	—	—
Iodobutyric	—	—	0.62	—	—

[a] The solvents for the above columns are: (1) *n*-butanol–1.5 *N* NH_4OH (1:1 v/v); (2) 95% ethanol–concentrated NH_4OH (100:1); (3) *n*-propanol–concentrated NH_4OH (7:3 v/v); (4) *n*-butanol–H_2O–diethylamine (100:15:1 v/v); (5) *n*-butanol–1.5 *N* aqueous NH_3 (1:1 v/v).

[b] Data from Block *et al.* (*104*).

2. *Hydroxamate Derivatives of Fatty Acids*

Water-saturated butanol
Water-saturated isobutyric acid
Water-saturated phenol
n-Butanol–acetic acid–H_2O (4:1:5)
n-Butanol–acetic acid–H_2O (4:1:1)
Amyl alcohol–acetic acid–H_2O (4:1:5)
Octyl alcohol–formic acid–H_2O (3:1:3)
Benzene–formic acid (1:1)
Benzene–formic acid–H_2O (1:1:1)
Benzene–acetic acid–H_2O (4:3:4)
n-Butanol–benzene–H_2O (1:1:1)

See references (*10, 44–47*).

TABLE II

R_F VALUES OF SOME SEMIVOLATILE ALKOXY ACIDS[a]

Alkoxy acid	R_F in the following solvents[a] 1[b]	2[c]	3[c]
Methoxyacetic	0.10	0.18	0.05
Ethoxyacetic	0.19	0.17	0.09
n-Propoxyacetic	0.28	0.09	0.13
n-Butoxyacetic	0.42	0.18	0.25
3-Ethoxypropionic	0.22	0.04	0.17
4-Methoxy-*n*-butyric	0.19	0.04	0.14
n-Hexanoic	0.58	0.39	0.43
2-Ethyl-*n*-hexanoic	0.70	0.55	0.71

[a] (1) *n*-Butanol–1.5 *N* NH_4OH (1:1 v/v); (2) *n*-hexanol–1.5 *N* NH_4OH (1:1); (3) isooctane–acetone–95% ethanol–concentrated NH_4OH (40:30:30:1).

[b] Data from Block *et al.* (*104*).

[c] Data from Nair (*43*).

The acid component in some of the above solvent mixtures should repress the ionization of the hydroxamic acids, thus preventing the tailing effect of the spots on the chromatogram; also, the presence of the acid in the mixture will usually increase the solubility of the water-immiscible solvent phase.

Two-dimensional chromatography may improve separation of the acids; a representative solvent pair that has proved useful is as follows:

First dimension: 95% ethanol plus 5% NH_3–pyridine–H_2O (3:1:1)
Second dimension: 95% ethanol–dioxane–H_2O–acetic acid (60:20:19:1)

See Lugg and Overell (*48*) and Table III.

3. *2,4-Dinitrophenylhydrazides of Fatty Acids*

90% Methanol–acetic acid–tetrahydronaphthalene (10:2:1)
80% Ethanol–acetic acid–tetrahydronaphthalene (10:2:1)
90% Methanol–tetrahydronaphthalene (10:1)

Prior to chromatography, the spotted papers are sprayed uniformly with tetrahydronaphthalene (Tetralin) (*11*) (See Table IV).

4. *Nonvolatile Water-Soluble Aliphatic Acids*

n-Butanol–acetic acid. Water-saturated *n*-butanol plus 1–2 moles of acetic acid; a good solvent for the separation of malic, tartaric, citric, and succinic acids.

Isooctane–95% ethanol–acetone–90% formic acid (44:44:11:1)

Chloroform–95% ethanol–90% formic acid (65:33:2)

TABLE III

R_F VALUES OF HYDROXAMIC ACID DERIVATIVES

Hydroxamic acid derivative	R_F in the following solvents[a]										
	1[b]	2[b]	3[b]	4[b]	5[b]	6[b]	7[c]	8[c]	9[d]	10	11[e]
Formic	.42	.26	—	.008	0	0	.45	.57	—	0.23	0.02
Acetic	.50	.34	.30	.11	0	.01	.57	.70	—	0.49	0.04
Propionic	.62	.51	.50	.25	.01	.05	.68	.78	—	0.70	0.10
Butyric	.72	.67	.68	.45	.04	.13	.74	.80	—	0.89	0.18
Valeric	.79	.78	.81	.67	.11	.32	.83	.84	—	0.83	0.35
Isovaleric	—	—	—	—	—	—	—	—	—	0.83	0.23
Caprylic	—	—	—	—	—	—	—	—	—	0.86	0.90
Caproic (hexanoic)	.84	.86	.83	.81	.26	.61	.87	.90	—	0.84	0.52
Heptanoic	.86	.89	.85	.88	.51	.87	—	—	—	—	—
Octanoic	.88	.89	—	.90	.77	.95	—	—	—	—	—
Nonanoic	.89	.89	.90	.90	.88	.97	—	—	—	—	—
Decanoic	.90	.90	—	.92	.92	.97	—	—	—	—	—
Oxalic	—	—	—	—	—	—	.14, .40	.22, .28, .32	—	—	—
Malonic	—	—	—	—	—	—	.11, .23	.19, .32	—	—	—
Succinic	—	—	—	—	—	—	.40, .72	.45, .52	—	—	—

Glutaric	—	—	—	—	—	—	.47	.36, .52	—	—	—
Adipic	—	—	—	—	—	—	.54, .67	.44, .60	—	—	—
Pimelic	—	—	—	—	—	—	.60, .73	.52, .69	—	—	—
Azelaic	—	—	—	—	—	—	.63, .74	.66 streaked	—	—	—
Sebacic	—	—	—	—	—	—	.89	.74, .89	—	—	—
Citric	—	—	—	—	—	—	.90, .23	.20, .29	—	—	—
Tartaric	—	—	—	—	—	—	.10	.19	—	—	—
Pyruvic	—	—	—	—	—	—	.59, .86	.54, .62, .73	—	—	—
Pelargonic	—	—	—	—	—	—	—	—	—	0.87	0.94
Acetyl phosphate	—	—	—	—	—	—	—	—	.52		
Propionyl phosphate	—	—	—	—	—	—	—	—	.65	—	—
Butyryl phosphate	—	—	—	—	—	—	—	—	.76	—	—
Valeryl phosphate	—	—	—	—	—	—	—	—	.81	—	—
Caproyl phosphare	—	—	—	—	—	—	—	—	.86	—	—

[a] (1) 1-Butanol–acetic acid–H_2O (4:1:5 v/v); (2) amyl alcohol–acetic acid–H_2O (4:1:5 v/v); (3) amyl alcohol–formic acid–H_2O (75:25:75v/v); (4) octyl alcohol–formic acid–H_2O (75:25:75 v/v); (5) benzene–formic acid–H_2O (75:75:75 v/v); (6) benzene–acetic acid–H_2O (100:75:100v/v); (7) water-saturated isobutyric acid; (8) water-saturated phenol; (9) water-saturated butanol; (10) butanol–acetic acid–H_2O (4:1:1 v/v); (11) benzene–formic acid (1:1 v/v).

[b] Data from Thompson (*44*).

[c] Data from Fink (*46*).

[d] Data from Stadtmann (*47*).

[e] Data from Ueno (*10*).

TABLE IV

R_F VALUES OF 2,4-DINITROPHENYLHYDRAZIDES OF FATTY ACIDS[a]

Acid	No. of C atoms in acid	R_F in the following solvents[b] 1	2	3
Acetic	2	0.85	0.93	0.77
Butyric	4	0.88	0.97	0.82
Caproic	6	0.83	0.93	0.81
Heptanoic	7	0.78	0.90	0.78
Caprylic	8	0.74	0.87	0.74
Nonanoic	9	0.69	0.81	0.68
Capric	10	0.64	0.78	0.64
Undecanoic	11	0.57	0.72	0.56
Lauric	12	0.54	0.70	0.51
Tridecanoic	13	0.45	0.62	0.43
Myristic	14	0.42	0.60	0.38
Pentadecanoic	15	0.34	0.52	0.30
Palmitic	16	0.30	0.49	0.25
Heptadecanoic	17	0.25	0.41	0.19
Stearic	18	0.21	0.38	0.15
Nonadecanoic	19	0.16	0.30	0.12
Arachidic	20	0.13	0.27	0.09
Behenic	22	0.08	0.18	0.05

[a] Ascending chromatography at 30°C.

[b] (1) 90% methanol–acetic acid–Tetralin (10:2:1 v/v); (2) 80% ethanol–acetic acid–Tetralin (10:2:1 v/v); (3) 90% methanol–Tetralin (10:1 v/v).

Chloroform–95% ethanol–90% formic acid (50:50:1)
n-Butanol–benzyl alcohol–H_2O–formic acid (45:45:9:1)
tert-Butyl alcohol–benzyl alcohol–H_2O–formic acid (22:68:9:1)
Isopropyl alcohol–*tert*-butyl alcohol–benzyl alcohol–H_2O–formic acid (16:16:48:16:1)
Phenol–H_2O–formic acid (75:25:1)
n-Pentanol–5 *M* formic acid (1:1)—*n*-Pentanol–H_2O–formic acid (20:12:1) is refluxed for 1 hour; water layer is decanted and 0.05% 8-quinolinol is added to the organic phase.
Mesityl oxide–H_2O–85% formic acid (75:75:36)
n-Butanol–formic acid–H_2O (10:2:15)
Ether–acetic acid–H_2O (13:3:1)
Ethyl acetate–acetic acid–H_2O (3:1:1)
2-Ethyl-1-butanol–5 *M* formic acid (2:3)
95% Ethanol–H_2O–concentrated NH_4OH (8:1:1)
Methylcellosolve–eucalyptol–88% formic acid (5:5:2)
Ethylcellosolve–eucalyptol–88% formic acid (5:5:2)
n-Propanol–eucalyptol–88% formic acid (5:5:2)
Ethylcellosolve–concentrated NH_4OH–H_2O (80:5:15)

Ethanol–concentrated NH_4OH–H_2O (80:5:15)
Carbon tetrachloride–acetic acid–H_2O (4:1:1)
Ethyl ether–88% formic acid–H_2O (5:2:1)
n-Amyl formate–H_2O–formic acid (7:1:2)

See references (*35–64*).

Aqueous alcohols containing a "swamp" acid are most frequently used for chromatographic development, and the addition of water to the mixture is apparently necessary for efficient separation of the aliphatic acid components. Acetone or alcohol should be added to mixtures containing isooctane or chloroform as an aid for incorporating small amounts of water in the mixture. Mesityl oxide or ethyl acetate, which are relatively rapid movers on the chromatographic paper, may be used to replace the alcohol of the mobile phase. The high volatility of these latter solvents frees the paper from the swamp acid in less time after development of the chromatogram. The formic acid component of a mixture may be removed more rapidly from a chromatogram by including eucalyptol in the solvent system. The large number of acids that may be present in a biological system may require some chromatographic experimentation by the investigator; it may be necessary to develop the chromatogram in two or more sequential solvent mixtures, either by several one-dimensional procedures and/or by a two-dimensional procedure. In such combinations, the use of an alkaline solvent in one direction and an acid solvent in the other should prove most efficient in separations of the unknown acids because of their specific effects on the acids either in terms of the anion or molecular characteristics of the acids. (See Table V).

5. *Free α-Keto Acids*

n-Butanol–H_2O–propionic acid (10:7:5)
α-*n*-Dimethylaminoisobutyric nitrile–*n*-butanol–H_2O (4:6:3)
Water-saturated *n*-butanol–formic acid (95:5)
Water-saturated 2-butanol–propionic acid (95:5)—The solvent mixture is aged 3–4 weeks to permit esterification equilibrium. The butanol must be peroxide free to avoid keto acid losses during chromatography
Toluene–acetic acid–H_2O (100:5:60)—An additional 4 ml acetic acid is added to the separated toluene phase of the mixture.

6. *α-Keto Acid 2,4-Dinitrophenylhydrazones*

20% Ethanol in *n*-butanol
Ethanol–H_2O (83:17). The chromatographic paper is impregnated with 0.1 *M* potassium phosphate buffer at pH 7.3. The concentration and the pH of the phosphate buffer as well as the ethanol–water ratio of the solvent have a decisive influence on the separation of the DNPH derivatives.
n-Butanol equilibrated with an equal volume of 3% aqueous ammonia—More concentrated ammonia decomposes some hydrazones to brown products and clean chromatographic separations will not be obtained.

TABLE V

R_F OF ORGANIC ACIDS (*104*)

Acid	R_F in the following solvents[a]												
	1	2	3	4	5	6	7	8	9	10[b]	11	12	13
Aconitic	0.36	0.81	0.78	—	0.70	0.73	0.02	0.09	—	—	—	—	—
Adipic	0.86	—	—	—	—	0.79	0.20	0.38	—	.575	—	0.10	—
Ascorbic	—	—	—	—	0.44	—	—	—	0.32	—	—	—	0.28
Azelaic	—	—	—	—	—	—	—	—	—	.940	—	0.23	—
Caffeic	—	—	—	—	—	—	—	—	—	—	0.63	—	—
Chlorogenic	—	—	—	—	—	—	—	—	—	—	0.29	—	—
Citric	0.26	0.31	0.37	0.15	0.38	0.23	0.02	0.06	0.46	.055	0.22	—	—
Fumaric	0.63	—	0.86	0.90	—	0.82	0.17	0.35	—	.485	0.80	—	0.95
Glutaric	0.78	—	0.80	—	—	—	—	—	—	.465	0.66	—	0.83
Glyceric	—	—	—	—	—	—	—	—	.049	—	—	—	—
Glycerate (Ca)	—	—	—	—	—	—	—	—	0.48	—	—	—	—
Glycolic	0.59	0.51	—	0.45	—	0.42	0.34	0.37	—	—	—	—	0.67
Glycolate (K)	—	—	—	—	—	—	—	—	0.64	—	—	—	—
Isocitric	—	—	—	—	—	—	—	—	0.46	—	—	—	—
Itaconic	—	—	—	—	—	—	—	—	0.82	—	—	—	—
α-Ketoglutaric	—	0.63	—	0.36	—	0.63	0.08	0.28	0.69	—	0.25	—	—
Lactic	0.72	0.66	0.77	0.72	0.77	0.59	0.48	0.61	0.80	.460[c]	—	—	—
Levulinic	0.91	—	—	—	0.94	—	—	—	—	—	—	—	—
Maleic	—	—	0.46	—	0.46	0.38	0.13	0.25	0.78	.335	—	—	—
Malic	0.42	0.39	0.44	0.25	0.49	0.32	0.09	0.23	0.55	.115	0.30	—	—
Malonic	0.48	0.63	0.60	0.52	0.61	0.51	0.07	0.17	0.74	.250	—	—	—
Oxalic	0.18	—	0.05	0.12	—	0.02	0.02	0.05	0.18	.135	0.11	—	0.71
Pimelic	—	—	—	—	—	—	—	—	—	.730	—	0.14	—
Pyruvate (K)	—	—	—	—	—	—	—	—	0.83	—	—	—	—
Sebacic	—	—	—	—	—	—	—	—	—	1.000	—	0.33	—
Suberic	—	—	—	—	—	—	—	—	—	.845	—	0.17	—
Succinic	0.66	0.70	0.72	0.76	0.78	0.66	0.14	0.30	0.79	.340	0.60	—	0.84
Syringic	0.95	—	—	—	—	—	—	—	—	—	—	—	—
Tartaric	0.19	0.22	0.23	0.09	0.26	0.12	0.06	0.15	0.33	.045	0.10	—	—
Tricarballylic	0.52	0.63	0.67	—	—	—	—	—	—	—	—	—	—

[a] (1) Phenol–H_2O–formic acid (75:25:1); (2) mesityl oxide–H_2O–formic acid (75:75:36); (3) butanol–formic acid–H_2O (10:2:15); (4) ether–acetic acid–H_2O (13:3:1); (5) ethyl acetate–acetic acid–H_2O (3:1:1); (6) ethylcellosolve–NH_4OH–H_2O (80:5:15); (7) ethanol–NH_4OH–H_2O (80:5:15); (8) CCl_4–acetic acid–H_2O (4:1:1); (9) ether–formic acid–H_2O (5:2:1); (10) *n*-amyl formate–H_2O–formic acid (7:1:2); (11) 1-pentanol–H_2O–formic acid (20:12:1); (12) *n*-butyl ether layer of methylcellosolve–H_2O–*n*-butyl ether (9:1:10); (13) ethyl ether–formic acid–H_2O (20:5:3).

[b] R_F values are relative to migration of sebacic acid.

[c] Lactic acid produces two spots; R_F 0.460, 0.600.

n-Butanol–ethanol–0.5 *N* NH_4OH (70:10:20)
n-Butanol–1 *N* $NaHCO_3$ (1:2)
n-Butanol–amyl alcohol–2-octanol (1:1:1)
n-Hexanol saturated with 0.8 *N* NH_4OH
n-Butanol–ethanol–H_2O (40:10:50)
Glycine–0.1 *M* NaOH buffer, pH 8.4. Separations with this solvent will depend on the affinity (adsorption) of the hydrazones for wet cellulose, and different grades of paper will probably give different R_F values. Chromatographic efficiency is best obtained in an atmosphere containing phenol vapors and traces of HCN.
Benzene–*tert*-pentanol–ethanol–H_2O (5:3:1:2). A useful solvent mixture for pyruvic acid, using filter paper impregnated with 0.2 *M* borate buffer, pH 8.2.
tert-Pentanol–ethanol–H_2O (5:1:4). A useful solvent for α-ketoglutaric acid using paper impregnated with 0.2 *M* sodium phosphate buffer, pH 6.3.

See references (*15*,*28*, *69–76*) and Table VI.

7. *α-Keto Acid 4-Nitro-2-Alkyl-Quinoxalinols*

Methanol
Methanol–H_2O (9:1)
Ethanol
Ethanol–H_2O (9:1)
n-Butanol–5 *N* NH_4OH (2:1)
Amyl alcohol–5 *N* NH_4OH (2:1)
Benzyl alcohol–5 *N* NH_4OH (2:1)
1% aqueous Na_2CO_3

Separations will depend on the adsorption rate of the derivatives with the wet cellulose and will vary with the grade of chromatographic paper. See references (*77–79*) and Table VI.

8. *Higher Fatty Acids*

The water-insolubility of the higher fatty acids and their derivatives requires the use of reversed phase chromatography. Two solvent phases, the stationary (or immobile) and the mobile phase are employed. The stationary phase is usually a nonpolar substance, and the cellulosic chromatographic paper must be modified by impregnation with such materials as paraffins, oils, and silicones. Equilibration of the paper with the stationary phase is necessary prior to development of the chromatogram. Usually, the mobile solvent phase is saturated with the stationary phase. The storage of impregnated paper for more than several days prior to use is not wholly desirable, because it may result in poor separation of the acids. The temperature of development may be critical; below 30°C may not be satisfactory for the C_{20}–C_{24} saturated acids with certain solvents, whereas 37°C may readily separate the C_{12}–C_{24} acids (*25*). Solvents and chromatographic data are listed in Tables VII (*1*, *22*, *24*, *25*, *80–86*), VIII, and IX (*33*).

TABLE VI

R_F VALUES OF α-KETO ACIDS AND DERIVATIVES[a] (*104*)

Compound	Free acids			2,4-Dinitrophenylhydrazones						Nitro-2-*R*-quinox-alinols	
	1	2	3	4	5	6	7	8	9	10	11
Oxalacetic acid	0.16	0.04	0.08	0.08	0.24	0.06	0.78	0.32; 0.47	0.63; 0.76	0.12	0.13
α-Ketoglutaric acid	0.21	0.07	0.51	0.16	0.12	0.04	0.69	0.33	0.62	0.14	—
Pyruvic acid	0.29	0.46	0.64	0.36	0.41	0.62; 0.80	0.45	0.51	0.75; 0.81	0.65	0.57
α-Ketobutyric acid	—	—	0.76	—	—	—	—	0.63	0.80	0.73	0.74
α-Ketoisovaleric acid	—	—	0.83	—	—	—	—	0.80	0.86	—	—
Dimethylpyruvic acid	0.60	0.78	—	—	—	—	—	—	—	0.86	0.85
Cinnamoylformic acid	0.75	0.95	—	—	—	—	—	—	—	—	—
Acetoacetic acid	—	—	—	0.50	0.55	—	—	—	—	—	—
p-Hydroxyphenylpyruvic acid	—	—	—	0.70	—	—	—	0.75	0.83	—	—
Phenylpyruvic acid	—	—	—	0.78	—	0.91	—	0.87	0.86	—	—
o-Hydroxyphenylpyruvic acid	—	—	—	0.81	—	—	—	—	—	—	—
Acetone	—	—	—	—	0.93	—	—	—	—	—	—
Glyoxylic acid	—	—	—	—	—	—	—	0.39	0.71; 0.77	—	—
β-Sulfopyruvic acid	—	—	—	—	—	—	—	0.18; 0.22	0.62; 0.83	—	—
β-Mercaptopyruvic acid	—	—	—	—	—	—	—	0.47	0.77	—	—
D-α-Keto-β-methylvaleric acid	—	—	—	—	—	—	—	0.77	0.86	—	—
α-Ketoisocaproic acid	—	—	—	—	—	—	—	0.83	0.88	—	—
α-Keto-ε-aminocaproic acid	—	—	—	—	—	—	—	0.72	0.95	—	—
α-Keto-γ-methiolbutyric acid	—	—	—	—	—	—	—	0.43	0.82	—	—
β-Indopyruvic acid	—	—	—	—	—	—	—	0.87	0.83	—	—

[a] Solvents for the columns: (1) *n*-butanol–H_2O–propionic acid (10:7:5); (2) α-*n*-dimethyl-aminoisobutyric nitrile–*n*-butanol–H_2O (4:6:3); (3) water-saturated *n*-butanol–formic acid (95:5); (4) *tert*-amyl alcohol–ethanol–H_2O) (5:1:4); (5) *n*-butanol–ethanol–0.5 *N* NH_4OH (7:1:2); (6) *n*-butanol–1 *N* $NaHCO_3$ (1:2); (7) glycine-0.1 *M* NaOH buffer, pH 8.4; (8) *n*-butanol–H_2O–ethanol (5:4:1); (9) methanol–benzene–*n*-butanol–H_2O (4:2:2:2); (10) *n*-butanol–5 *N* NH_4OH (2:1); (11) benzyl alcohol–5 *N* NH_4OH (2:1).

TABLE VII

SOLVENTS FOR THE SEPARATION OF HIGHER FATTY ACIDS AND DERIVATIVES

Mobile phase	Stationary phase	Paper treatment	Compounds resolved	Reference
Methanol–acetone (1:1)	Rubber	Impregnated with latex	Esters	*(80)*
Ethanol–H_2O (75:25)	Toluene	Impregnated with toluene and olive oil	C_{10}–C_{22} Acids	*(81)*
Methanol–H_2O (85:15) sat. with paraffin oil + 0.001 *N* HCl	Paraffin	Impregnated with paraffin	C_{12}–C_{16} Acids	*(82)*
Methanol–H_2O (90:10) sat. with paraffin oil + 0.001 *N* HCl	Paraffin	Impregnated with paraffin	C_{14}–C_{20} Acids	*(82)*
Methanol–H_2O (95:5) sat. with paraffin oil + 0.001 *N* HCl	Paraffin	Impregnated with paraffin	C_{18}–C_{24} Acids	*(82)*
90%, 95%, 70%, or 50% aq. Acetic acid	Paraffin	Impregnated with paraffin	C_{10}–C_{20}	*(1)*
90% aq. Acetone	Paraffin	Impregnated with paraffin	C_{10}–C_{20}	*(1)*
70% or 50% aq. Acetic acid	Castor oil	Impregnated with castor oil	C_{10}–C_{20}	*(1)*
50% aq. Acetone	Castor oil	Impregnated with castor oil	C_{10}–C_{20}	*(1)*
Acetic acid–88% formic acid–30% hydrogen peroxide (6:1:1) at 37°C	Mineral oil	Impregnated with mineral oil	C_{12}–C_{24}	*(25)*
Methanol–H_2O (80:20) sat. with Decalin	Decalin	Octadecyloxymethyl cellulose sprayed with Decalin	Acids	*(83)*
Acetic acid–H_2O (9:1) sat. with hydrocarbon (190–220°C)	Hydrocarbon (190–220°C)	Sprayed with kerosene	C_{14}–C_{16} Acids	*(84, 85)*
Methanol–Decalin (8:1)	Decalin	Sprayed with Decalin	Dinitrophenylhydrazones	*(86)*
Methanol–Decalin–ethylacetate (40:7:3)	Decalin	Sprayed with Decalin	Dinitrophenylhydrazones	*(86)*
90% Methanol–acetic acid–Tetralin (30:1:3)	Tetralin	Equilibrated with Tetralin	Mercury compds. of unsat. esters	*(24)*
Diethylene glycol–acetic acid–Tetralin (60:20:11)	Tetralin	Equilibrated with Tetralin	Mercury compds. of unsat. esters	*(24)*
Methanol–acetic acid–petroleum ether (140–170°C) (30:1:7)	Petroleum ether	Equilibrated with petroleum ether	Mercury compds. of unsat. esters	*(24)*
Ethyl acetate–tetrahydrofuran–H_2O (0.6:3.5:4.7)	Tetrahydrofuran	Acetylated paper	Hydroxamates	*(22)*

TABLE VIII

R_F VALUES OF HIGHER FATTY ACIDS AND DERIVATIVES[a] (*104*)

No. of C atoms	Fatty acid	Hydrox-amates	2,4-DPNH	Free acids					Mercury acetate addition compounds—unsaturated acids		
		1	2	3	4[b]	5	6	7	8	9	10
5	Valeric acid	0.84	—	—	—	—	—	—	—	—	—
6	Caproic acid	0.72	0.67	—	—	—	—	—	—	—	—
7	Enanthic acid	0.64	—	—	—	—	—	—	—	—	—
8	Caprylic acid	0.57	0.60	—	—	—	—	—	—	—	—
9	Pelargonic acid	0.51	—	—	—	—	—	—	—	—	—
10	Capric acid	0.46	0.51	0.80	1.00	—	—	—	—	—	—
11	Undecylic acid	0.40	—	0.80	0.77	—	—	—	—	—	—
12	Lauric acid	0.38	0.43	0.70	0.63	—	—	0.62	—	—	—
14	Myristic acid	0.34	0.33	0.56	0.48	0.80	—	0.51	—	—	—
16	Palmitic acid	0.30	0.25	0.39	0.35	0.72	—	0.33	—	—	—
18	Oleic acid	0.30	—	0.46	—	0.72	—	—	0.17	0.20	0.62
18	Stearic acid	0.24	0.17	0.29	0.26	0.59	—	0.19	—	—	—
20	Arachidic	—	0.12	0.16	0.19	0.44	—	0.09	—	—	—
10	Sebacic	—	—	0.93	—	—	—	—	—	—	—
9	Azelaic	—	—	0.87	—	—	—	—	—	—	—
18	Elaidic	—	—	0.34	—	—	—	—	—	—	—
18	Linolenic	—	—	0.45	—	—	—	—	0.78	0.87	0.84
22	Erucic	0.22	—	0.14	—	—	—	—	0.07	0.10	0.40
22	Behenic	—	—	—	0.14	0.31	—	0.04	—	—	—

C	Acid	(1)	(2)	(3)	(4)	(5)	(6)	(7)	(8)	(9)	(10)
24	Lignoceric	—	—	—	0.11	0.21	—	0.01	—	—	—
	2-Hydroxymyristic	—	—	—	—	0.94	0.89	—	—	—	—
	2-Hydroxypalmitic	—	—	—	—	0.93	0.82	—	—	—	—
	2-Hydroxystearic	—	—	—	—	0.93	0.64	—	—	—	—
	2-Hydroxylignoceric	—	—	—	—	0.92	0.00	—	—	—	—
26	Cerotic	—	—	—	0.09	—	—	—	—	—	—
10	9-Decenoic	—	—	—	—	—	—	—	0.67	0.88	0.85
11	10-Undecenoic	—	—	—	—	—	—	—	0.59	0.71	0.84
12	9-Dodecenoic acid	—	—	—	—	—	—	—	0.56	0.65	0.84
14	9-Tetradecenoic and	—	—	—	—	—	—	—	0.42	0.46	0.80
16	9-Hexadecenoic acid	—	—	—	—	—	—	—	0.27	0.29	0.74
12	2-Dodecenoic acid	—	—	—	—	—	—	—	0.57	0.66	0.85
14	2-Tetradecenoic acid	—	—	—	—	—	—	—	0.43	0.47	0.81
16	2-Hexadecenoic acid	—	—	—	—	—	—	—	0.29	0.30	0.74
18	2-Octadecenoic acid	—	—	—	—	—	—	—	0.17	0.21	0.62
22	Brassidic acid	—	—	—	—	—	—	—	0.08	0.10	0.44
18	Linoleic acid	—	—	—	—	—	—	—	0.57	0.62	0.82
18	β-Eleostearic acid	—	—	—	—	—	—	—	0.77	0.86	0.81
18	Stearolic acid	—	—	—	—	—	—	—	0.75	0.84	0.82
22	Behenolic acid	—	—	—	—	—	—	—	0.50	0.64	0.77
18	Ricinoleic acid	—	—	—	—	—	—	—	0.64	0.53	0.81

[a] Conditions for the columns: (1) ethyl acetate–tetrahydrofuran–H_2O (0.6:3.5:4.7), acetylated Whatman No. 1 (25% acetyl); (2) methanol–Decalin (8:1); (3) 75% ethanol, paper impregnated with 12% olive oil in toluene; (4) acetic acid–H_2O (9:1), paper saturated with paraffin (190–220°C), circular chromatography; (5) 95% acetic acid saturated with paraffin oil; (6) 65% acetic acid saturated with paraffin oil nos. 5 and 6 developed at 40°C using Whatman No. 1 paper impregnated with paraffin oil (12–15%); (7) acetic acid–88% formic acid–30% hydrogen peroxide (6:1:1) at 37°C, Whatman No. 1 paper drawn through 7% mineral oil solution; (8) 90% methanol–acetic acid–Tetralin (30:1:3); (9) diethylene glycol–acetic acid–Tetralin (60:20:11); (10) methanol–acetic acid–petroleum ether (30:1:7).

[b] Given as R_c (migration relative to capric acid).

TABLE IX

R_F VALUES OF HIGHER FATTY ACIDS USING REVERSED PHASE CIRCULAR PAPER CHROMATOGRAPHY[a]

Acid	R_F[b]						
	1	2	3	4	5	6	7
Lauric	0.85	0.76	0.81	0.70	0.85	0.87	0.96
Myristic	0.76	0.65	0.73	0.57	0.73	0.72	0.81
Palmitic	0.66	0.55	0.62	0.45	0.52	0.63	0.60
Stearic	0.55	0.48	0.50	0.38	0.41	0.53	0.48
Oleic	0.65	—	0.63	0.46	0.51	0.60	0.60
Linoleic	0.74	—	0.72	0.56	0.71	0.71	0.79
Linolenic	0.83	—	0.80	0.69	0.84	0.85	0.94

[a] From (*33*).

[b] Whatman No. 3 filter paper; in above columns 1, 2, and 3, paper was impregnated with 10 % liquid paraffin in benzene; papers for remaining columns were treated as follows: (4) 10 % Dow-Corning stopcock grease; (5) 5 % Dow-Corning High Vacuum grease; (6) 10 % Vaseline; (7) 5 % Vaseline. Solvents for above columns were: (1) 90 % acetic acid; (2) glacial acetic acid–formic acid (88 %)–hydrogen peroxide (30 %) (6:1:1); (3) aqueous methanol 95 %; (4) aqueous methanol 90 %; (5) aqueous methanol 65 %; (6) aqueous methanol 90 %; (7) aqueous methanol 80 %.

9. *Cholesteryl Esters of Higher Fatty Acids*

The solvent mixtures for the chromatography of the cholesteryl esters on siliconized paper are listed in Table X. The resolution of a mixture of saturated and unsaturated materials is not wholly satisfactory with the

TABLE X

R_F VALUES OF CHOLESTERYL ESTERS OF LONG-CHAIN FATTY ACIDS[a]

Ester	R_F[b]		
	1	2	3
Cholesterol	0.87	0.90	0.92
Cholesteryl ester			
Laurate	0.42	0.00	0.50
Myristate	0.38	0.00	0.43
Palmitate	0.33	0.00	0.36
Stearate	0.28	0.00	0.00
Oleate	0.32	0.17	0.41
Linoleate	0.37	0.24	0.47
Linolenate	0.41	0.30	0.53

[a] Using siliconized Whatman No. 1 Filter paper.

[b] Solvents for the above columns were: (1) chloroform–methanol–formic acid–H_2O (40:60:1:5); (2) acetone–ethanol–formic acid–H_2O (50:50:15:5); (3) acetone–ethanol–H_2O (50:50:5).

chloroform solvent, but separation of the components of such a mixture is improved by development with the acetone solvent (Table X). The R_F values of the esters will be influenced by the type of paper (Whatman Nos. 1 and 3 are preferable) and the amount of silicone on the paper. A high silicone content will repress the migration rate of the esters (*27*).

IV. Detection Reagents

Detection reagents and the preparatory conditions for their use, not necessarily complete, for fatty acids and their derivatives are given in Table XI (*11, 24, 36, 39, 40, 48, 51, 65, 66, 68, 80, 87–97*). Additional data on the detection technique are discussed below.

Pretreatment of the chromatographic paper by washing with 2 *N* acetic acid, distilled water, and 10 *N* NH_4OH, respectively, will improve resolution and minimize tailing of the acids and will improve detectability limits when acid–base indicator reagents are used for locating the acid spots. Amounts of 0.01–0.004 μM of the acid salts may be detected with this type of reagent.

The acid hydroxamate derivatives, when sprayed with the ferric chloride reagent (Table XI), produce relatively stable spots in a sensitivity range of 0.01–0.20 μM of the derivative. Prewashed filter paper is not necessary in this procedure.

The reagents listed in Table XI for the water-soluble aliphatic acids are applicable to the free acids. If the developing solvents include an acid component, the paper should be free of this acid prior to the use of an acid–base indicator reagent.

Organic acids in fruit juices have been detected on the chromatogram with a dip spray composed of 0.3 ml each of aniline and furfural in 100 ml of acetone (*64*); the acids will produce red spots on a pink background. If the background is not pink, the chromatogram should be treated with steam for about 15 minutes to remove remnants of the acid component of the development mixture.

Improvement in the use of acid–base indicators for the detection of organic acids may be obtained by the incorporation of the indicator reagent with the developing solvent (*2, 3, 99*). For example, 0.008% methyl red is suitable in a solvent mix of ethyl acetate–acetic acid–H_2O (2:1:1). Bromophenol blue (0.015%) is preferable in a solvent mixture of ethyl formate–98% formic acid–H_2O (2:1:1). The additional incorporation of sodium acetate or sodium formate (0.05%) will suppress the formation of acidic bands on the chromatograms. Stable red spots on a pink background are formed. Phenol violet or thymol blue are suitable with basic solvents, such as ethanol–H_2O–NH_3 (70:26:4).

TABLE XI

DETECTION REAGENTS FOR FATTY ACIDS

Fatty acid	Color reagent	Reagent composition	Specific treatment conditions	Remarks	References
Lower fatty acids	Bromophenol blue	50 mg in 100 ml H_2O + 200 mg citric acid	—	H_4N + ethylamine salts give blue spots, yellow background	*39, 48, 51*
	Bromocresol purple	40 mg in 100 ml formalin–ethanol (1:5) to pH5 with 0.1 *N* NaOH	Expose sprayed sheet to NH_3 vapors	Yellow spots, purple background	*87*
	Bromothymol blue	0.04% aq. solution to pH 7.5 with NaOH	—	Yellow spots-anions; blue spots-cations	*36*
	Methyl red and bromothymol blue	200 mg each in formalin–ethanol (100 ml:400 ml) to pH 5.2 with 0.1 *N* NaOH	Expose sprayed sheet to NH_3 vapors	Red spots, green background	*88, 89*
	Chlorine gas	—	Expose iodo acids to gas	Yellow spots, white background	*40*
	Fluorescein	0.1% in 75% aq. ethanol	Expose sprayed bromo acids to chlorine	Red spots	*40*
	Silver nitrate	0.01 *N* $AgNO_3$ and 5 *N* NH_4OH (1:1 v/v)	—	Differentiates formic acid and acetic acid	*90*
	Ninhydrin	0.1% ninhydrin + 0.1% collidine in ethanol	Sprayed paper heated 10 minutes at 40–50°C	Differentiates formic acid and acetic acid	*91*
Hydroxamate derivatives	Ferric chloride	1% ferric chloride in acidic ethanol	—	Purple spots, yellow background	
2,4-dinitro-phenyl-hydrazide derivatives	Potassium hydroxide	0.5 *N* KOH in ethanol	—	Brown spots	*11*
Water-soluble aliphatic acids	Chlorophenol red, bromocresol green, bromophenol blue	0.04% of selected indicator in ethanol, to pH 7 with NaOH	—	Yellow spots, mauve background	*92*
	Silver nitrate	0.1 *N* $AgNO_3$ + 0.1 *N* NH_4OH (1:1 v/v)	Dry sprayed sheet 4 hours at room temperature; protect from light	Black, yellow, white spots for different acids, tan background	*51*

	Aniline xylose	1 gm aniline + 1 gm xylose in 100 ml methanol	Reagent-treated paper air dried, then heated 30 minutes at 105°C	Brown spots	*93*
	p-Dimethylamino-benzaldehyde	4% in acetic anhydride + sodium acetate crystals	Sprayed paper heated at 140°C 1–2 minutes	Red-orange spots	*94*
	Bromocresol green, bromophenol blue, $KMnO_4$	(a) 0.075% bromocresol green mixed with 0.025% bromophenol blue; (b) 0.5% (b) $KMnO_4$ + 1% Na_2CO_3	Spray paper with 1:1 mix of solutions (a) and (b)	Specific colors with different acids	*95*
	p-Nitroaniline	1 gm reagent + 2 ml HCl + 20 ml H_2O; heated and diluted with H_2O (to 180 ml) and 20 ml 5% sodium nitrite	Spray damp ammoniacal paper with reagent; respray immediately with 15% aq. Na_2CO_3	Red spots	*96*
Free α-keto acids	*o*-Phenylenediamine	0.05% in 10% CCl_3COOH	Spray paper and heat 2 minutes at 100°C	Yellow-green spots fluoresce under UV light	*65*
	Semicarbazide	0.1% in 0.15% sodium acetate	Use Eaton & Dikeman No. 613 paper	Dark spots under UV light	*66*
	Nessler reagent		Spray ammonia-exposed paper with Nessler reagent	Orange spots	*68*
Higher fatty acids	Sudan IV	Reagent in 50% aq. methanol	For fatty acid esters	—	*80*
	Bromothymol blue	0.4% reagent in methylcellosolve	—	—	*97*
	Cupric acetate	—	(a) Heat chromatogram to 80–100°C; (b) immerse paper in 2% cupric acetate solution; (c) wash paper in 0.01% HOAc; (d) dip paper in 1.5% $K_4Fe(CN)_6$	Brown spots	*24*
	Diphenylcarbazone	0.2% in ethanol	Reagent spray followed by 0.05 *N* nitric acid in ethanol	Purple spots for mercuric acetate compounds of fatty acid esters	*24*

Chromatograms sprayed with a 10% solution of potassium ferrocyanide, dried at room temperature and then resprayed with a 0.5% solution of ferric ammonium sulfate solution in 70% ethanol and heated for 5 minutes at 100°C will produce blue and blue-violet colors for spots of oxalic, malonic, malic, and citric acids (*4*). If the R_F values of oxalic acid and malic acids are similar, a bright blue color is produced by oxalic and a blue-violet color with malic acid.

A saturated aqueous ammonium vanadate solution is highly sensitive for the detection of glycolic acid, giving a white spot (*4*).

A solution of pyridine–acetic anhydride (7:3 v/v) will detect citric and aconitic acid spots on heating the sprayed chromatogram 3 minutes at 80–90°C (*4*). On the assumption that acids form stable pyridine salts, Cerbulis and Taylor (*100*) recommended the use of pyridine in the solvent mix, such as *n*-butanol–pyridine–H_2O (3:2:1.5); the developed chromatogram is air dried 4–6 hours and then viewed under ultraviolet light; the acids appear as dark blue spots.

A method which produces stable spots (*101*) involves treating the chromatogram by dipping in a solution of dimethylglyoxime (1% in 95% ethanol) and, after partially drying, spray the sheet with alkaline nickel biuret solution (1 gm nickel sulfate heptahydrate plus 1 gm biuret in 50 ml distilled water; warm to dissolve biuret and filter to remove precipitated nickel hydroxide). The acids form pink to red spots. Washing the finished chromatogram in a solution of ethanol–NH_3 (1:1) will stabilize the spots indefinitely.

The α-keto acid 2,4-dinitrophenylhydrazones are colored and can be detected on the chromatogram without any further treatment and are visible as yellow spots; a secondary spray with 2% ethanolic KOH will produce intense red-brown spots. The use of a nickel salt will differentiate the cis and trans isomers of the 2,4-dinitrophenylhydrazones of pyruvic, glyoxylic, phenylglyoxylic, phenylpyruvic, and *p*-hydroxyphenylpyruvic acids. The cis isomers will produce red spots, the trans spots will not produce any color (*19*). The developed chromatogram is first sprayed with a 5% (w/v) solution of nickel chloride in 70% aqueous ethanol, followed by a spray of 0.3% suspended zinc powder in aqueous ethanol. The paper is heated at 100°C for several minutes. Trans spots may give a positive color response if they are isomerized by heat. Detection of the nitro quinoxaline derivatives are visualized in the same manner as the 2,4-dinitrophenylhydrazones.

In addition to the reagents listed in Table XI, the higher fatty acids may be detected by treatment with phenylhydrazine, Rhodamine B, and aqueous silver nitrate, respectively (*98*), producing the following colors:

Acid	Color
Linoleic	Reddish brown
Oleic	Red
Elaidic	Red
Erucic	Dark red
Stearic	Reddish brown
Undecylenic	Red
Pelargonic	Faint red-brown
Enanthic	Faint red-brown
Caproic	Black
Butyric	Black
Valeric	Black

The use of quinine, chlorophyll, anthracene, or anthranol (*98*) will produce intense green fluorescence with linoleic acid and blue fluorescence with oleic acid.

A spray reagent composed of Nile blue sulfate (0.2% in 50 ml ethanol), trethanolamine (about 15 ml), and ethanol (50 ml) may be applied to fatty acid spots; the spots are blue, the background is pale red. Triethanolamine is added to the Nile blue sulfate solution until the blue color changes to a clear red, then the solution is diluted with the ethanol (*30*).

The unsaturated cholesteryl esters of long-chain fatty acids (*27*) may be detected by staining the chromatogram with iodine vapor; dark brown spots will appear. The paper may then be subjected to steam treatment for 5 minutes to remove the iodine color, followed by a spray of 10% phosphotungstic acid in ethanol (*102*), which will reveal both the saturated and unsaturated compounds.

V. Quantitative Analysis

Techniques for the quantitative analysis of many groups of compounds developed by paper chromatography are discussed in general terms in Chapter 4, some of which may be applicable to the aliphatic acids. Semiquantitative procedures include the visual comparison of standard spots of known amounts of the acids with the sample spots; or the area of the spot may be related to the amount of sample applied to the chromatogram. Maximum color density measurements may be obtained with a densitometer (See Chapter 4) and compared to a calibration curve similarly obtained from standard spots of known amounts of acids.

With the aid of guide spots, the unstained spot areas of aliphatic acid hydroxamates may be extracted with 5% sodium acetate (*9*, *10*). To the extract is added 0.5 ml of 1% sulfanilic acid in 25% aqueous acetic acid, followed by 5 drops of iodine solution (1.3 gm iodine in 100 ml acetic acid); let stand for 4 minutes. Excess iodine is destroyed by the addition of 0.1 *N* sodium thiosulfate. Two drop of α-naphthylamine (0.6% of reagent in 30% acetic acid) are added. The solution is diluted to a given volume with water and measured with a photometer at 520 mμ. Minor modifications of the method have been made by Ueno (*10*), including the use of 0.1 *M* acetate–HCl buffer (pH 3) to dilute the solution prior to color measurement. Maximum intensity may be obtained 10 minutes after the addition of the α-naphthylamine and the intensity will remain constant 30 minutes at room temperature. The lower limit of the method is 0.5 μg per milliliter of fatty acid.

Free keto acids sprayed with the semicarbazide reagent (see Table XI) may be quantitatively measured by cutting out the spots on the paper and placing the cut paper in test tubes. Three milliliters of 0.025% 2,4-dinitrophenylhydrazine in 0.5 *N* HCl are added to each tube, followed by 1 ml of 40% KOH. The tubes are allowed to stand 10 minutes at room temperature and then shaken for an additional 10 minutes to extract the hydrazone from the paper. Colors are measured at 435 mμ (*67*).

Chromatographic spots, or streaks, of the keto acid 2,4-dinitrophenylhydrazones may be cut from the paper, placed in test tubes, and extracted with 10 ml of 1 *N* NaOH. The filtered extract is measured at 455 mμ (*15*). Another technique consists of extracting the excised spots with 10 ml of 0.1 *M* phosphate buffer solution (pH 7.4) for 5 hours with occasional shaking. The filtered yellow solutions are measured in a suitable colorimeter (*29*).

Paper chromatograms of the higher fatty acids may be treated with mercuric acetate (0.1% plus 0.5 ml acetic acid per liter), the treated paper is washed thorougly with tap water, dried, and sprayed with 0.2% *s*-diphenylcarbazide in 95% ethanol. The acids are estimated by extracting the spots of the mercury color complex with methanol–toluene (1:1) and measuring the extract with a photocolorimeter at 540 mμ. Measurements are linear between 5 μg and 60 μg of acid. The color is stable for at least 4 hours (*25*, *33*, *103*).

References

1. P. E. Ballance and W. M. Crombie, *Biochem. J.* **69**, 632 (1958).
2. R. D. Hartley and G. J. Lawson, *J. Chromatog.* **4**, 410 (1960).
3. R. D. Hartley and G. J. Lawson, *J. Chromatog.* **7**, 69 (1962).

4. F. Stern, G. Grumet, F. Trabal, A. Mennis, and H. H. Zinsser, *J. Chromatog.* **19**, 130 (1965).
5. G. V. Marinetti, ed. "Lipid Chromatographic Analysis," Vol. 1. Marcel Dekker, New York, 1967.
6. F. A. Isherwood and C. S. Hanes, *Biochem. J.* **55**, 824 (1953).
7. A. R. Jones, E. J. Dowling, and W. J. Skraba, *Anal. Chem.* **25**, 394 (1953).
8. S. David and J. Monnier, *Bull. Soc. Chim. France* p. 797 (1955).
9. F. Bergmann and R. Segal, *Biochem. J.* **62**, 542 (1956).
10. Y. Ueno, *J. Biochem. (Tokyo)* **48**, 161 (1960).
11. Y. Inoue and M. Noda, *Bull. Agr. Chem. Soc. Japan* **19**, 214 (1955).
12. F. R. Bharucha and G. V. Joshi, *Naturwissenschaften* **43**, 327 (1956).
13. E. Becker, *Z. Lebensm.-Untersuch.-Forsch.* **98**, 249 (1954).
14. H. E. Umbarger and B. Magasanik, *J. Am. Chem. Soc.* **74**, 4253 (1952).
15. D. Seligson and B. Shapiro, *Anal. Chem.* **24**, 754 (1952).
16. E. Kvamme and L. Hellman, *Anal. Chem.* **26**, 1995 (1954).
17. E. Kun and M. Garcia-Hernandez, *Biochim. Biophys. Acta* **23**, 181 (1957).
18. H. K. Berry, C. Leonard, H. Peters, M. Granger, and N. Chunekamrai, *Clin. Chem.* **14**, 1033 (1968).
19. A. M. Tokushige and E. S. Cook, *Nature* **191**, 797 (1961).
20. D. J. D. Hockenhull and D. G. Floodgate, *Biochem. J.* **52**, 38 (1952).
21. I. Smith, *Nature* **171**, 43 (1953).
22. F. Michael and H. Schweppe, *Angew. Chem.* **66**, 136 (1954).
23. O. Hirayama and M. Noda, *Abura Kagaku* **5**, 16 (1956).
24. Y. Inoue, M. Noda, and O. Hirayama, *J. Am. Oil Chemists' Soc.* **32**, 132 (1955).
25. M. A. Buchanan, *Anal. Chem.* **31**, 1616 (1959).
26. L. Swell and C. R. Treadwell, *J. Biol. Chem.* **212**, 141 (1955).
27. J. A. Labarrere, J. R. Chipault, and W. O. Lundberg, *Anal. Chem.* **30**, 1466 (1958).
28. S. M. Altmann, E. M. Crook, and S. P. Datta, *Biochem. J.* **49**, lxiii (1951).
29. G. A. Abdel-Tawab, E. Broda, and G. Kellner, *J. Chromatog.* **2**, 99 (1959).
30. R. Chayen and E. Margery Linday, *J. Chromatog.* **3**, 503 (1960).
31. V. P. Skipski, S. M. Arfin, and M. M. Rapport, *Arch. Biochem. Biophys.* **87**, 259 (1960).
32. B. D. Ashley and V. Westphal, *Arch. Biochem. Biophys.* **56**, 1 (1955).
33. C. V. Viswanathan and B. Meera Bai, *J. Chromatog.* **7**, 507 (1962).
34. T. H. Kritchevsky and A. Tiselius, *Science* **114**, 299 (1951).
35. J. L. Occolowitz, *J. Chromatog.* **5**, 373 (1961).
36. F. Brown, *Biochem. J.* **47**, 598 (1950).
37. R. L. Reid and M. Lederer, *Biochem. J.* **50**, 60 (1951).
38. J. W. Chittum, T. A. Gustin, R. L. McGuire, and J. T. Sweeney, *Anal. Chem.* **30**, 1213 (1958).
39. E. P. Kennedy and H. A. Barker, *Anal. Chem.* **23**, 1033 (1951).
40. M. H. Hashmi and C. F. Cullis, *Anal. Chim. Acta* **14**, 336 (1956).
41. A. R. Jones, E. J. Dowling, and W. J. Skraba, *Anal. Chem.* **25**, 394 (1953).
42. E. R. Hiscox and N. J. Berridge, *Nature* **166**, 522 (1950).
43. J. H. Nair, *Anal. Chem.* **25**, 1912 (1953).
44. A. R. Thompson, *Australian J. Sci. Res.* **B4**, 180 (1951).
45. O. Perila, *Acta Chem. Scand.* **10**, 143 (1956).
46. K. Fink and R. M. Fink, *Proc. Soc. Exptl. Biol. Med.* **70**, 654 (1949).
47. E. R. Stadtman and H. A. Barker, *J. Biol. Chem.* **184**, 769 (1950).
48. J. W. H. Lugg and B. T. Overell, *Nature* **160**, 87 (1947).

49. J. W. H. Lugg and B. T. Overell, *Australian J. Sci. Res.* **A1**, 98 (1948).
50. J. B. Stark, A. E. Goodban, and H. S. Owens, *Anal. Chem.* **23**, 413 (1951).
51. M. L. Buch, R. Montgomery, and W. L. Porter, *Anal. Chem.* **24**, 489 (1952).
52. F. E. Resnik, L. A. Lee, and W. A. Powell, *Anal. Chem.* **27**, 928 (1955).
53. F. Bryant and B. T. Overell, *Nature* **168**, 167 (1951).
54. G. D. Kalyankar, P. R. Krishnaswamy, and M. Sreenivasaya, *Current Sci.* (*India*) **21**, 220 (1952).
55. R. Y. Shkolnik, *Dokl. Akad. Nauk SSSR* **90**, 847 (1953).
56. F. W. Denison and E. F. Phares, *Anal. Chem.* **24**, 1628 (1952).
57. J. E. Loffler and E. H. Reichl, *Mikrochim. Acta* p. 79 (1953).
58. R. I. Cheftel, R. Munier, and M. Macheboeuf, *Bull. Soc. Chim. Biol.* **33**, 840 (1951).
59. R. I. Cheftel, R. Munier, and M. Macheboeuf, *Bull, Soc. Chim. Biol.* **34**, 380 (1952).
60. R. I. Cheftel, R. Munier, and M. Macheboeuf, *Bull. Soc. Chim. Biol.* **35**, 1085 (1953).
61. M. Ohara and Y. Suzuki, *Science* (*Tokyo*) **21**, 362 (1951).
62. J. K. Palmer, *Conn. Agr. Expt. Sta., New Haven, Bull.* **589**, 12 (1955).
63. H. Kalbe, *Z. Physiol. Chem.* **297**, 19 (1954).
64. J. Fitelson, *J. Assoc. Offic. Anal. Chemists* **52**, 646 (1969).
65. T. Wieland and E. Fischer, *Naturwissenschaften* **36**, 219 (1949).
66. B. Magasanik and H. E. Umbarger, *J. Am. Chem. Soc.* **72**, 2308 (1950).
67. H. E. Umbarger and B. Magasanik, *J. Am. Chem. Soc.* **74**, 4253 (1952).
68. L. A. Liberman, A. Zaffaroni, and E. Stotz, *J. Am. Chem. Soc.* **73**, 1387 (1951).
69. D. Cavallini, N. Frontali, and G. Toschi, *Nature* **163**, 568 (1949).
70. D. G. Walker and F. L. Warren, *Biochem. J.* **49**, xxi (1951).
71. M. S. F. El Hawary and R. H. S. Thompson, *Biochem. J.* **53**, 340 (1951).
72. F. A. Isherwood, *Brit. Med. Bull.* **10**, 202 (1954).
73. D. J. D. Hockenhull, M. Herbert, and A. D. Walter, *Biochem. J.* **56**, 73 (1954).
74. D. Cavallini and N. Frontali, *Ric. Sci. Suppl.* **23**, 807 (1953).
75. A. I. Virtanen, J. K. Miettinen, and H. Kuntto, *Acta Chem. Scand.* **7**, 38 (1953).
76. F. A. Isherwood and C. A. Niavis, *Biochem. J.* **64**, 549 (1956).
77. D. J. D. Hockenhull and D. G. Floodgate, *Biochem. J.* **52**, 38 (1952).
78. D. J. D. Hockenhull, *Nature* **171**, 982 (1953).
79. D. J. D. Hockenhull, G. D. Hunter, and M. Herbert, *Chem. & Ind.* (*London*) p. 127 (1953).
80. J. Boldingh, *Experientia* **4**, 270 (1948).
81. V. Kobrde and R. Zahradnik, *Chem. Listy* **48**, 1189 (1954).
82. B. D. Ashley and V. Westphal, *Arch. Biochem. Biophys.* **56**, 1 (1955).
83. R. G. Baker, *Biochem. J.* **54**, xxxix (1953).
84. H. Wagner, L. Abish, and K. Bernhard, *Helv. Chim. Acta* **38**, 1536 (1955).
85. O. Perila, *Acta Chem. Scand.* **9**, 864 (1955).
86. O. Hirayama and M. Noda *Aburu Kagaku* **5**, 16 (1956).
87. R. L. Reid and M. Lederer, *Biochem. J.* **50**, 60 (1951).
88. R. E. B. Duncan and J. W. Porteous, *Analyst* **78**, 641 (1953).
89. O. Priori, *Boll. Lab. Chim. Provinciali* (*Bologna*) **6**, No. 2, 45 (1955).
90. S. M. Partridge, *Biochem. J.* **42**, 238 (1948).
91. A. T. H. Burness and H. K. King, *Biochem. J.* **68**, 32P (1958).
92. F. Brown, *Nature* **167**, 441 (1951).
93. J. Saarnio, E. Niskasaari, and C. Gustafsson, *Suomen Kemistilehti* **25B**, 25 (1952).
94. K. Schreier and K. Gaedke, *Naturwissenschaften* **43**, 130 (1956).
95. J. Paskova and V. Munk, *J. Chromatog.* **4**, 241 (1960).
96. A. E. Whitfield, *J. Chromatog.* **4**, 350 (1960).

97. W. Baker, J. B. Harbone, and W. D. Ollis, *J. Chem. Soc.* p. 3215 (1952).
98. H. P. Kaufmann and J. Budwig, *Fette Seifen* **54**, 156 (1952).
99. R. D. Hartley, *Biochem. J.* **69**, 3P (1958).
100. J. Cerbulis and M. W. Taylor, *Anal. Chem.* **35**, 114 (1963).
101. E. Savory, *J. Chromatog.* **14**, 549 (1964).
102. C. Michalec, *Naturwissenschaften* **42**, 509 (1955).
103. C. V. Viswanathan and B. Meera Bai, *J. Chromatog.* **6**, 264 (1961).
104. R. J. Block, E. L. Durrum, and G. Zweig. "A Manual of Paper Chromatography and Electrophoresis," 2nd rev. ed. Academic Press, New York, 1958.

8

STEROIDS, BILE ACIDS, AND CARDIAC GLYCOSIDES

I. Steroids

The successful and widespread application of paper chromatography to the lipophilic steroids and related compounds began in the early 1950's with the introduction of the Zaffaroni and Bush solvent systems. Reversed-phase chromatography has also been successfully employed. Paper chromatography is now used for the qualitative and quantitative analysis of steroids, preparative work, and for structural and stereochemical studies.

Reviews of this topic have been written by Neher (*1*), Bush (*2*), and others (*3*, *4*).

A. Sample Preparation and Application

1. *Sample Application*

Depending on the detection method, 0.1–20 μg of sample dissolved in 1–10 μl of solvent can be used. Solvents for the sample include acetone, methanol, $CHCl_3$, or methanol–$CHCl_3$ (1:1 v/v). Standard steroids are stored as 0.2–1 % solutions in stoppered containers in the cold. Application is with a micropipette as a spot or short streak, or, for preparative work, a long streak containing an amount per centimeter of paper known to be consistent with good separation. Air can be blown on the underside of the origin to hasten drying, but this can lead to anomalous adsorption at the origin.

A large initial zone can be applied past the origin and washed back and concentrated with ethyl acetate–methanol (2:1 v/v) (*5*). In another method for concentrating steroid extracts prior to chromatography (*6*), a starting line is drawn several centimeters from one end of the paper and lanes 5–10 mm wide are marked parallel to the direction of flow of solvent. The solutions to be chromatographed (40–100 μl) are then pipetted along their respective lanes and the solvent is allowed to dry. The sheet is then hung a small tank specially kept for this purpose with the lower edge dipping into a mixture of $CHCl_3$-ethyl acetate (1:1 v/v) containing a small addition of methanol. An ascending chromatogram is thus run with a solvent in which all the steroids move with the front. The solvent is allowed to ascend to the starting line. The result, when dried, is that all the steroids are compressed into small rectangles at the starting line. This preliminary chromatogram takes 5–10 minutes.

A similar procedure may be used for the preliminary concentration of steroids extracted from urine. The dried extract is dissolved in 0.1–1.0 ml methanol–acetic acid (v/v) to which one drop of Sudan red is added. The total solution is spotted on a strip of Schleicher and Schuell Paper No. 2043b shaped to a point at the opposite end. The strip is developed by ascending chromatography with methanol–acetic acid (v/v). The migration of the steroids may be followed by the movement of the dye. After 25–30 minutes the steroids have reached the top of the strip, which is dried *in vacuo*. The tip containing the concentratied steroids is cut off and inserted into a slit of another strip of paper, which is developed by any of the techniques described below.

A method for the deposition of 1–2 ml of sample involving concentration onto 2–3 mm of a wick, painting onto the paper, and treatment as above is also described in Block *et al.* (*6*).

Authentic standard steroids are usually run along with the mixture to be analyzed. A marker dye that migrates with the solvent front or near the steroids for overrun development can be added to the sample.

2. *Extraction of the Sample*

a. Extraction of Estrogens from Urine (*7*). After hydrolysis in boiling HCl for 7–10 minutes, the estrogens are extracted into benzene. The extract is washed with 9% $NaHCO_3$, and the estrogens are extracted as phenolates into 2 *N* aqueous NaOH. This extract is acidified with HCl and the liberated estrogens are extracted back into benzene. The benzene solution is concentrated and applied to the paper. Fluorescent and colored impurities accompany the estrogens; these can be removed by a preliminary 24-hour development with methylcyclohexane on formamide-impregnated paper (*8*).

b. Extraction of Urinary 17-*Ketosteroids* (*7*). After extracting the estrogens with NaOH as above, the benzene layer contains the 17-ketosteroids. It is washed until neutral and applied to the paper.

c. Extraction of Aldosterone from Urine. The urine is adjusted to pH 1.0 by adding concentrated HCl. During hydrolysis the urine is kept in the dark at room temperature for 24 hours. Extraction is carried out in a glass-stoppered separatory funnel with 3, 2, and 2 volumes of dichloromethane. The combined dichloromethane extracts are washed successively with a 0.1 volume of 1 *N* NaOH, 0.1 *N* NaOH, 0.1 *N* acetic acid and H_2O, dried over Na_2SO_4 and evaporated under reduced pressure. The residue is dissolved in 8 ml 70% methanol and washed with 2 ml toluene–petroleum ether (1:1 v/v). The upper layer is discarded and the methanol is evaporated from the aqueous methanol extract. The aqueous residue is extracted three times with four volumes of dichloromethane and the extract is then evaporated to dryness. The residue is dissolved in a small volume of $CHCl_3$ and applied to the paper (*9*).

d. Extraction of Labeled Cortisol Metabolites. These substances are extracted with ten volumes of ethyl acetate from an enzymatic reaction medium in which radioactive cortisol is incubated with rat liver homogenate (*10, 11*). Procedures for the extraction of corticosteroid fractions from the organs of guinea pigs (*12*), cortical steroids from blood, and estrogens from human placenta (*6*) have also been described.

3. *Chromatography of Derivatives*

Reactions can be performed on the mixture *in vitro* and the resultant derivatives chromatographed in parallel with standards treated in the same way. Useful reactions include oxidation (e.g., with CrO_3, MnO_2, or OsO_4), reduction (Zn), esterification (acetic anhydride), ether formation (diazomethane), saponification (KOH), and hydrazone formation (Girard's reagent) (*1*). For details of applications and reaction conditions, consult work by Bush (*2, 13, 14*). The mobilities of the products aid in the identification of the original steroids.

The assay of aldosterone by the double isotope dilution derivative method developed by Kliman and Peterson (*14a*) and later modified (*9*) involves acetylation of dried extracts of plasma or urine with ^3H-labeled acetic anhydride in benzene in the presence of pyridine to convert aldosterone quantitatively to the ^3H-labeled diacetate. A measured amount of authentic aldosterone ^{14}C-diacetate is added to each sample, and the double-labeled steroid is identified and purified by paper or thin-layer chromatography. The samples are treated with 0.5% chromic acid in glacial acetic acid to form a monoacetate oxidation product, which is also purified by chromatography. The ^{14}C-content of the purified steroid is assayed by liquid scintillation counting. The tritiated aldosterone serves as the indicator of steroid loss during the procedure.

B. Detection and Identification Methods

1. *Color Tests*

A large number of steroid color test have been described, some of which are selective, and some of which are general. A scheme for the tentative identification of steroids requiring 50–100 μg of sample is based on profiles of chromatographic data and the results of four generally applicable color test indicated in Table I (*15*). The details of these tests are as follows:

a. 15% ethanolic *phosphoric acid.*
b. *Carr–Price*—saturated solution of $SbCl_3$ in $CHCl_3$ (the bottle of reagent grade solid $SbCl_3$ is stored in a dessicator after being opened).
c. *Kagi–Miescher*—1 gm *p*-anisaldehyde in 100 ml glacial acetic acid containing 2 ml concentrated H_2SO_4 .
d. *AACS*—1 ml *p*-anisaldehyde is added to 100 ml $CHCl_3$ saturated with $SbCl_3$ followed by 2 ml concentrated H_2SO_4 to form a two-phase system. Keep the solution in the dark at room temperature for 1.5 hours before use. The chromatograms, freed from solvent by suspension in a dark, well-ventilated chamber for 24 hours, are dipped in the upper $CHCl_3$ phase (free from any H_2SO_4 droplets) and dried in a darkened fume hood prior to heating at 90°C for 3 minutes. After visual inspection, heating is repeated for 3 minutes. Any fluorescence produced persists for several weeks, but visible colors may fade.

Color reactions listed in Table II are shown by some sterols after the dried chromatogram is inspected under ultraviolet light or is sprayed with (*16*):

e. A solution of *antimony (III) chloride* (50% in ethanol). Heat 5–10 minutes at 70°C. The colors produced in daylight (and in ultraviolet light of 365 mμ) are noted.
f. A solution of *antimony (V) chloride* (20% in $CHCl_3$).
g. A solution of *bismuth (III) chloride* (33% in ethanol). Heat some seconds at 60°C. Colors are produced in daylight and ultraviolet light.
h. A solution of *$CaCl_2$* (50% in ethanol–H_2O) and heated 10–15 minutes at 80–90°C. In ultraviolet light of 365 mμ fluorescent spots are visible.
i. Moisten the chromatogram with *trichloroacetic acid* dissolved in two drops of water.
j. A solution of *phosphotungstic acid* (15% in ethanol). Heat 3–4 minutes at 60°C.

TABLE I

A COMPARATIVE STUDY OF THE REACTIONS OF STEROIDS WITH FOUR GENERAL STEROIDS TESTS[a] (*15*)

Steroid	Ethanolic Phosphoric Acid		Carr-Price		Kagi-Miescher		AACS	
	Visible	UV (366 mμ)	Visible	UV (336 mμ)	Visible	UV (366 mμ)	Visible	UV (366 mμ)
Androsterone	—	Violet white 1	Purple 2	Red orange 4	Violet 1	Blue 2	Blue 4	Purple 4
Epiandrosterone	Violet 0	—	Purple 2	—	—	—	Blue 2	Purple 2
Dehydroepiandrosterone	Brown violet 1	Gray white 2	Orange brown 4	Wine 4	Blue 1	Blue gray 2	Blue 4	Blue black 2
4-Androstene-3,17-dione	Violet 0	Orange pink 1	Tan ±	White yellow 2	Wine purple 2	Rust 4	Orange 3	Gold 4
5β-Androstane-3,11,17-one	—	—	—	White yellow 1	Yellow 0	Brown 2	—	—
Androsterone benzoate	—	Blue white 2	Purple gray 2	Wine 2	—	—	Blue 3	Purple 2
17β-Hydroxy-5β-androstane-3-one	Violet 1	—	Pink 1	Pink orange 4	Blue 4	Brown 1	Wine 2	—
Testosterone	Violet 0	Gray white 2	Red violet 2	Orange 4	Purple 2	Pink 3	Purple 2	Wine 3
Testosterone propionate	—	White 0	Tan 1	Pink orange 1	Violet 1	Brown 3	Pink 2	Orange 2
Testosterone benzoate	—	White 0	Purple 1	Yellow white 0	Blue black 2	Brown 3	Wine 3	Wine 3
17α-Ethyl-19-nortestosterone	Orange 0	Pink 1	Orange pink 4	Gold 4	Blue 1	Blue gray 2	Purple 4	Wine 4
19-Norandrost-4-ene-17α-ethyl-17β-ol 3β-propionate	Blue 0	Light blue 1	Gray 2	Violet 1	Blue 1	Blue gray 1	Blue 2	Violet 4
Pregnenelone	—	Pink 1	Red brown 4	Wine 4	Yellow 1	Blue gray 2	Blue 3	Blue violet 4
5α-Pregnane-3β,20α-diol	—	—	Brown 2	Dark orange 4	—	—	Blue 4	Violet 4
5α-Pregnane-3β,20β-diol	—	White 0	Brown 2	Dark orange 4	—	—	Blue 3	Violet 4
3β-Hydroxy-5α-pregnane-20-one	—	Blue white 2	Red gray 2	Yellow 0	Yellow 0	Rust 0	Blue brown 3	Brown 3

5β-Pregnane-3,20-dione	Violet 0	Blue white 2	Gray 1	Yellow 2	Yellow 1	Brown 4	Brown 3	Red brown 4
Progesterone	—	White 1	Tan 0	Yellow white 2	Purple 2	Pink 4	Orange 3	Gold 4
11α-Hydrocyprogesterone	Orange 0	Violet white 0	—	Orange yellow 2	Violet 2	Brown 3	Orange 3	Gold 4
3β,21-Dihydroxy-5α-pregnane-20-one	Violet 0	Blue 2	Brown orange 4	Orange 4	Blue 0	Blue 1	Blue gray 2	Tan 0
Deoxycorticosterone	Violet 0	Gray white 2	Tan 1	Pink orange 1	Blue black 4	Dark brown 4	Brown 4	Olive 3
Corticosterone	Yellow 0	Yellow white 4	Tan yellow 1	Blue white 4	Brown 2	Brown 1	Gold 3	Olive 3
21-Hydroxy-4-pregnane-3,11,20-trione	—	White 1	Purple 1	Pink orange 1	Brown 3	Orange brown 3	Amber 3	Orange 3
3β,17α,21-Trihydroxy-5-pregnene-20-one	Violet 1	Orange white 2	Brown orange 4	Orange 4	Yellow 2	Blue 2	Blue 2	Blue black 4
17α,21-Dihydroxy-5β-pregnane-3,11,20-trione	Yellow 0	Blue 3	Brown 1	Purple 3	Brown 3	Brown 3	Wine 3	Wine 3
17α,21-Dihydroxy-5α-pregnane-3,11,20-trione	Yellow 0	—	Brown 1	Purple 3	Brown 3	Yellow 4	Yellow 4	Orange 3
11β,17α,21-Trihydroxy-5β-pregnane-3,20-dione	Yellow orange 2	Yellow white 3	Yellow 0	Blue white 3	Yellow 1	Brown 2	Brown 2	Orange 4
Reichstein S	Violet 3	Brown orange 4	Purple 1	Orange 1	Orange brown 3	Orange 4	Wine 3	Brown 3
Cortisol	Yellow 3	Chartreuse 4	Brown 1	Yellow 1	Yellow black 3	Brown 3	Purple 4	Purple 3
Cortisone	Yellow 0	Light blue 4	Tan 1	Violet 1	Brown 2	Brown orange 3	Light wine 3	Orange 2
Δ^1-Cortisone	Yellow 1	White 3	Tan 0	Violet 0	Yellow 3	Tan 1	Tan 2	Purple 3
Δ^1-Cortisol	Brown orange 1	Orange white 2	Tan 2	Blue violet 1	Yellow 1	Tan ±	Blue brown 3	Blue violet 4
3α,17α-Dihydroxy-21-acetoxy-5β-pregnane-11,20-dione	—	—	—	White yellow 1	—	—	Tan 1	—

[a] These data are based on 10 $\mu g/cm^2$ steroid spotted on chromatographic paper. Dash = no color response. The numbers following the colors indicate their relative intensity rating: 4 = approximately 2 μg of steroid/cm^2 is readily distinguishable; 3 = 4 $\mu g/cm^2$; 2 = 7 $\mu g/cm^2$; 1 = 10 $\mu g/cm^2$; 0 = 15–25 $\mu g/cm^2$.

TABL

COLOR REACTIONS SHOW

Compound	Color reaction				
	Ultraviolet light 365 mμ	e. $SbCl_3$	f. Sb_2Cl_5	g. $BiCl_3$	h. $CaCl_2$ in ultraviolet light
Cholesterol	—	Violet	Yellow-brown	Violet	—
β-Sitosterol	—	Violet	Very faint yellow	Violet	—
Stigmasterol	—	Faint violet	Gray-purple	Rose-violet	—
Lanosterol	—	Orange	Yellow	Orange	—
Ergosterol	+	Very faint purple	Pink	—	++
Zymosterol	—	Pink-brown	Orange-yellow	Faint yellow	Faint
7-Dehydrocholesterol	+	Purple-brown	Brown-blue	Faint purple-brown	+
Vitamin D_2	+	Brown	Brown-blue	Gray-brown	+
Lumisterol	+	Faint brown	Purple	—	+

[a] Letters correspond to references in text.

k. A solution of *silicotungstic acid* (25% in ethanol). Heat 3–4 minutes at 60°C.

l. A mixture of *dimethyl-p-phenylenediamine* and *m*-tolylenediamine (1:1) (1% in water).

m. *Millon's reagent* (one part of mercury dissolved in two parts of concentrated HNO_3). Heat 2–4 minutes at 40–50°C; spray again and heat.

n. A solution of *urea* in water (50%). Heat 10–15 minutes at 80°C. In ultraviolet light of 365 mμ and 256 mμ, fluorescent spots appear.

o. A solution of *$NaIO_4$* (1% in water). Spray after 5 minutes with 1% $KMnO_4$.

y Some Sterols (*16*)

ie following reagents[a]						
i. Trichloroacetic acid	j. Phosphotungstic acid	k. Silicotungstic acid	l. Dimethyl-*p*-phenylenediamine	m. Millon's reagent	Urea in ultraviolet light 365 mμ	o. $NaIO_4$–$KMnO_4$
—	Pink-violet	Pink-orange	—	Very faint yellow	—	—
—	Pink-violet	Pink-orange	—	—	—	—
—	Violet-brown	Faint purple brown	—	—	—	—
—	Orange-yellow	Orange-yellow	—	—	−(365 mμ) +(256 mμ)	Yellow
Faint purple	Faint purple	Faint purple-brown	Very faint blue	Yellow	++	—
—	Yellow-brown	Yellow	Blue	Yellow	Faint (365 mμ) +(256 mμ)	Yellow
Green-purple	Purple-brown	Purple-brown	—	Yellow	+	Yellow
—	Brown	Brown	Blue	Faint yellow	—	Yellow
Purple	Purple-brown	Purple-brown	Blue	Yellow	+	Yellow

Six tests have been devised for the detection of estrogens on filter paper, three of them are applicable to the detection of sterols and two to other steroids. The sensitivities range in general from 2–8 μg of compound per square centermiter of paper, and the colors and fluorescences for each compound are helpful in identification work (*17*). Results of these test, described below, are given in Tables III and IV:

p. (A) 0.5% solution of sulfanilic acid in 2% HCl. (B) 0.5% solution of $NaNO_2$. (C) 10% solution of Na_2CO_3. All reagents are stable separately. A mixture of equal volumes of reagents A and B is freshly prepared. The test strip is dipped into the reagent and then held in the air for 3 seconds. It is then passed through the carbonate

TABLE

DETECTION OF ESTROGENS

Compound	Test p Color	Test q Color	Test q Fluorescence	Test r Color
Estradiol-17α	or.	y.	gr. y.	oc. br.
Estradiol-17β	or. t.	gr. y.	gr.	ol. → br.
Estrone	t. or.	pi.	pe.	oc. br.
Equilin	y. t.	gr. y.	gr. y.	pu. r. → ro.
Equilenin	or. r. → ver.	pi.	pe.	sal. r.
6-Dehydroestrone	—	pi.	pi.	pi. → ro.
17α-Ethinylestradiol	oc. y. → or. br.	pi. cor.	r. or.	ol. r. br.
7-Ketoestrone	or. br.	pi.	or. cor.	y.
1α-Hydroxyestrone	y.	or. pi.	gr. bl.	y. → y. or.
6-Ketoestradiol	r. or. → r. br.	t.	or. pi.	y.
Estriol-3,16α,17β	or. → or. br.	pi.	pe. pi.	oc. → or. br.
Estriol-3,6β, 17β	or. r. → r. br.	ol. y.	bl.	r. br.
Diethylstilbestrol	pi. br. → vi.	—	—	gr. → ol. gr.
Methoxydoisynolic acid	—	—	bl.	lav.
2-Hydroxyestradiol-17β	br.	pi.	pi.	br.
2-Methoxyestrone	or. br. → br.	or. t.	pe.	br. r.
2-Methoxyestriol	br. or.	—	gr. y.	br. y.
10-Hydroxy-$\Delta^{1,4}$-estradiene-3,17-dione	or. br.	y.	g. y.	or.

[a] Key: r. = red, or. = orange, oc. = ochre, y. = yellow, br. = brown, ol. = olive, ro. = rose, sal. = salmon, lav. = lavender, vi. = violet, ver. = vermillion, cor. = coral.

solution (reagent C). The color develops immediately and continues to intensify or change color for 15 minutes. Sensitivity of the test for the estrogens is 1–3 μg per square centimeter of filter paper.

q. Solution of 4.0 gm urea and 0.3 gm stannous chloride in 10 ml of 40% H_2SO_4 (gentle heating is necessary). This reagent should be prepared fresh each day. The test strip is dipped into the freshly prepared reagent, placed on a glass plate, and heated until maximum color is produced (surface temperature of hot plate is about 80°C). This strip is then viewed under ultraviolet light. Sensitivity of the test for estrogens is 2–5 μg per square centimeter of paper.

r. (A) 2% solution of *p*-nitrobenzenediazonium chloride in 50% acetic acid. The reagent is stable for at least 1 month. The test is carried

III

ON FILTER PAPER[a] (17)

Test s		Test t		Test u	
Color	Fluor-escence	Color	Fluor-escence	Color 1st dip	Color 2nd dip
r. or.	or. pi.	or. → br.	or. y.	cor. → or.	or. r.
y. oc.	or. pi.	or. pi. → or. br.	gr. y. → pi.	y. or.	or. br.
y. → y. br.	gr. y.	gr. y.	or. y. → pi. or.	y.	t. or.
br. or.	gr. y.	oc. → or. br.	pu.	y. → or. y.	or.
ol. gr.	or. r.	br.	r. cor.	or.	or.
ol. br.	pi.	br.	or. pi.	or.	or. br.
wi.	cor.	wi.	cor. pi.	or. → wi.	wi. br.
ol. br.	pi.	y. br.	pi. cor.	y. oc.	br.
or. y.	bl.	y.	gr. bl.	or. pi.	pi. br.
y.	gr. y.	y.	gr. y.	—	—
y. → r.	or. pi.	ol. → pi. t.	pi.	pi.	or.
ol. gr. br.	gr.	ol. → br.	pi.	gr.	oc.
or.	—	—	—	y. or. → or.	br.
ol.	gr. y.	y.	pi.	gr. y.	pi. or.
br.	gr. y.	br.	gr. bl.	or. pi.	or. → br.
t.	r. or.	t.	or. pi.	or. br. → r. br.	r. br.
y.	pi. or.	y.	y. gr.	or. → y. br.	br.
or.	gr. y.	y.	bl.	or.	or.

gr. = green, wi. = wine, t. = tan, pi. = pink, bl. = blue, pe. = peach, pu. = purple,
Tests are described in text.

out by dipping the strip into the reagent and heating it on a glass plate at about 100°C until maximum color intensity develops. The *p*-nitrobenzenediazonium chloride solution is stable for at least 2 months if kept in a brown bottle. The test varies in sensitivity from 2 to 5 μg per square centimeter of paper for the estrogens.

s. (A) 50:1 mixture (v/v) of concentrated H_2SO_4 and formalin. This reagent is stable for about 2 weeks. (B) Mixture of equal volumes of chloroform and acetic anhydride. The strip is passed through reagent (A) and placed into a puddle of reagent (B) which has been pipetted onto a glass plate. The strip is then immersed completely by means of a glass rod. After observing the developed color, the strip is viewed under the ultraviolet lamp for fluorescences. If reagent (B) evaporates too rapidly, a little may be pipetted directly

TABLE IV

THE DETECTION OF STEROLS ON FILTER PAPER[a] (*17*)

Compound	Test s		Test t		Test u	
	Color	Fluorescence	Color	Fluorescence	Color 1st dip	Color 2nd dip
Cholesterol	Brown → brown purple	Orange pink	Blue → purple	Pink	Salmon pink	Salmon orange
Cholestanol	Yellow ochre	Blue yellow	—	Blue	—	—
Ergosterol	Orange → red brown	Orange pink	Pink → wine	Pink Orange → brown orange	Red	Red → brown
7-Ketocholesterol	Yellow	Green yellow	Yellow	Blue	Orange	Orange
Cholestan-3-one	—	Blue	—	Blue	—	—
Δ^4-Cholesten-3-one	—	Blue green	—	Blue yellow	—	—
Lanosterol	Tan → ochre	Green yellow	Yellow	Yellow blue	Orange	Orange
Zymosterol	Yellow brown	Blue → yellow	Yellow ochre	Yellow blue	Ochre	Ochre
Calciferol (Vitamin D_2)	Orange brown	Orange	Orange → brown	Blue yellow → orange	Orange → red brown → brown	Brown
Vitamin D_3	Orange brown	Salmon orange	Yellow brown	Blue pink → orange pink	Yellow → ochre brown	Brown
Coprostanol	Yellow	Yellow → pink yellow	—	Blue	Violet	Yellow → orange
7-Dehydrocholesterol	Red → brown red → purple brown	Salmon	Orange → pink purple	Peach	Orange red → brown red	Red brown
Squalene	Ochre	Blue	—	Blue	Olive pink	Pink purple

Test q (10 $\mu g/cm^2$)

Steroid	Color	Fluorescence
Dehydroepiandrosterone	Red → purple	Light blue
Reichstein's Substance A	Orange → red	Rose-coral
Δ^4-Pregnene-2α, 17α, 21-triol-20-one	Red → brown	Coral
3α, 11β, 20α, 21-Tetrahydroxypregnane	Red	Yellow-white
Cortisol	Red	Coral
Corticosterone	Tan → green	Yellow-green
Tetrahydrocortisol	Red	Coral
$\Delta^{1,4}$-Androstadien-10-ol-3,17-dione	Yellow	Yellow
Cholic acid	Pink → pink-orange	Brilliant blue

Test u (10 $\mu g/cm^2$)

Steroid	Color 1st dip	Color 2nd dip
Dehydroepiandrosterone	Red purple → brown	Brown → purple → gray
Δ^4-Androstene-3,17-dione	Pink → purple	Pink → purple
Prednisolone	Yellow	Brown
Δ^9-Etiocholanolone	Brown	Brown
Progesterone	Purple → red	Wine
Pregnanolone	Ochre	Purple → brown
$\Delta^{1,4}$-Androstadien-10-ol-3,17-dione	Yellow → orange	Pink → orange
Δ^4-Androsten-1α-ol-3,17-dione	Purple	Brown → purple
3β, 11β, 17α, 20β, 21-Pentahydroxyallopregnane	Yellow → brown	Yellow → brown

[a] Tests are described in text.

onto the strip. Sensitivity of the test varies for the estrogens from 1 to 4 μg per square centimeter and for the sterols from 4 to 8 μg per square centimeter of paper.

t. (A) 2:1 mixture (v/v) of concentrated H_2SO_4 and formalin. The mixing is done in cracked-ice bath to prevent overheating of the solution. This reagent is stable for about 1 week. (B) Mixture of equal volumes of chloroform and acetic anhydride. This procedure is exactly the same as that in s. Sensitivity of the test varies from 2 to 5 μg per square centimeter for the estrogens and from 4 to 10 μg per square centimeter for the sterols.

u. Four grams of $ZnCl_2$ is mixed with 0.5 ml of water, and 25 ml acetyl chloride is slowly added with continuous stirring. This reagent is usable for about 1 hour from time of preparation. The strip is passed through the boiling reagent and placed on a preheated glass plate. Heating (90–100°C) is continued until maximum color intensity develops. Redipping and reheating of the strip will further change the color in most cases. This test should be done in a hood because of the noxious effects of the acetyl chloride vapors. Sensitivity of the test is from 2 to 5 μg per square centimeter for the estrogens and from 4 to 8 μg per square centimeter for the sterols.

Other test are described below:

v. *Zimmerman reagent*—The chromatogram is sprayed with 2% *m*-dinitrobenzene in absolute ethanol, dried at room temperature, and sprayed with 2.5 *N* KOH. After drying at 70–100°C, red-violet spots appear. This test is used mostly for ketosteriods.

w. *t-Butyl chromate*—Stock solution: *t*-butanol (10 ml) is diluted with CCl_4 (10 ml) and cooled in ice; CrO_3 (5 gm) is added in small portions with stirring; after standing 10 minutes and diluting with CCl_4 (65 ml), the solution is percolated through 5 gm anhydrous Na_2SO_4 supported on a small filter; 10 ml more of CCl_4 is used to wash the filter and apparatus; 1 gm $NaSO_4$ is added before storage at 4°C. Shortly before use, 1 ml of stock solution is added to a mixture of xylene (80 ml) and pyridine (20 ml) with shaking. The paper is evenly sprayed and heated at 100°C $\pm 5°$ for 5 minutes. If the paper is now dipped through 1 *N* NaOH, dried at 100°C and examined under ultraviolet (360 mμ) light, 3-hydroxy- or 3-keto-, 4- or 5-unsaturated steroids in the androstane, pregnane, cholane, cholestane, and spirostane series are detected. If instead, the alkali-treated paper is dipped through Zimmerman's reagent, purple or blue ketone spots (produced from 17-hydroxy-20-hydroxy- or 17-hydroxy-20-ketone pregnane derivatives) appear upon warming with a hair drier (*18*). A similar method for the detection of non-ketonic steroids by oxidation of hydroxyl groups to ketone groups which are detected with the Zimmerman reagent is as follows (*19*):

1. The dried paper chromatograms are passed quickly through a freshly prepared solution of 0.25% CrO_3 in 90% acetic acid. The strips are blotted lightly and placed in an oven at 90°C for 15 minutes.

2. The strips are then developed with a modified Zimmerman reagent as follows. The strips are first passed through a solution of 2.5 *N* KOH in absolute ethanol, blotted thoroughly and passed through a solution of 2% *m*-dinitrobenzene in ethanol, again thorougly blotted and placed in an oven at 60°C for 1–2 minutes to develop the color. The colored zones should be marked as soon as possible since some of the color products tend to fade very rapidly. The compounds tabulated below were spotted on paper and treated in this way.

Compound	Color after oxidation and Zimmerman
Androstane-3α,17β-diol	Purple
Androstane-3β,17β-diol	Purple
Etiocholane-3α,17β-diol	Purple
Δ^{16}-Androsten-3β-ol	Blue fading to tan
Δ^{16}-Androsten-3α-ol	Blue fading to tan
19-Nor-5α-androstane-3β.17β-diol	Purple
19-Nor-5α-androstane-3α,17β-diol	Purple
19-Nor-5β-androstane-3α,17β-diol	Purple
19-Nor-5β-androstane-3β,17β-diol	Purple
Allopregnane-3α,20β-diol	Blue fading to tan
Pregnane-3α,20α-diol	Blue fading to tan
Pregnane-3α,17α,20α-triol	Purple
β-Allocortol	Purple
Pregnanolone	Blue fading to tan
Pregnane-3α,20α-diol diacetate	Negative

x. *2,4-Dinitrophenylhydrazine*—One of the most general methods for detecting ketonic steroids is as follows (*20*). The chromatograms are drawn through the DNPH solution and held in the air for 10 minutes. The wet strips are then washed for 1 minute in the neutralizing fluid, 10 minutes in the decolorizing fluid, 5–10 minutes in the reducing fluid, and finally in tap water. If the decolorization is not complete, subsequent washings repeated once or twice will be helpful. After the last washing in water, the strips are air dried and can be stored for further use. The procedure described increases the sensitivity of the DNPH qualitative test about five times as compared with the data of Neher (*21*) where nonimpregnated papers are used. Reagents: (a) 0.1% solution of DNPH in 2 *N* HCl; stable for 2–3 weeks if stored in a cool and dark place; (b) 1% $KMnO_4$; (c) 10% $Na_2CO_3 \cdot 10H_2O$; (d) neutralizing fluid: dilute 10 ml of (c) to 100 ml with water. (e) decolorizing fluid: mix 2 ml of (b) and 1 ml of (c); dilute to 100 ml with water; stable for 1 day; (f) reducing fluid: 0.2% solution of ascorbic acid; add one drop of concentrated hydrochloric or sulfuric acid per 100 ml of the fluid; stable for one day. All operations should be carried out at room temperature. The results for various steroids with this method are given in the tabulation below.

	PGS	TSN	ADN	PGN	DHA	17-OH-PGS	PDN
Final color of the spot	Orange	Red-orange	Purple-orange	Lemon yellow	Light lemon yellow	Red-orange	Yellow
Amount visually detectable $\mu g/cm^2$ (approximate)	0.5	0.5	0.5	1	5	0.5	1

Abbreviations: PGS = progesterone; TSN = testosterone; ADN = Δ^4-androstene-3,17-dione; PGN = Δ^5-pregnen-3β-ol-20-one; DHA = dehydroepiandrosterone; 17-OH-PGS = 17α-hydroxy-progesterone; PDN = pregnane-3,20-dione.

In another study, methods were developed for detecting: (a) Steroid ketone spots on a paper chromatogram by applying a *2,4-dinitrophenylhydrazine* reagent with subsequent washing with 2 *N* NaOH and then H_2O followed by 2 *N* HCl and then H_2O. (b) Steroid alcohols by direct chromic acid oxidation on the paper, permitting the ketone produced to migrate out of the area sprayed with *chromic acid* and detecting the ketone formed with *2,4-dinitrophenylhydrazine*. (c) Methyl ketones with a *sodium nitroprusside* reagent. (d) Secondary methyl carbinols by direct *chromic acid* oxidation and detection of the methyl ketone formed (*22*).

y. *Methods for detecting steroids containing α-ketol or vic-glycol groups* (*23*). Steroids (C_{21}) with a 21-ol-20-one or a 20,21-diol structure produce formaldehyde on periodate oxidation. The dried papergram is sprayed with a reagent consisting of 1 volume saturated aqueous potassium periodate solution plus 3 volumes 95% ethanol. Four to eight minutes later, the partially dried paper is sprayed with a freshly prepared reagent containing 15 gm ammonium acetate, 0.3 ml glacial acetic acid and 1 ml 2,4-pentanedione in 100 ml methanol. Ten minutes after the second spray, a formaldehydogenic steriod appears as a yellow-green fluorescent spot on a dark background under ultraviolet (Mineral Light). The spot becomes yellow under ordinary light on a colorless background in 30–60 minutes. The background gradually turns yellow on longer standing.

Only the 21-unsubstituted C_{21} steroids with a 17,20-diol structure yield acetaldehyde on periodate oxidation. The dried papergram is sprayed with a reagent prepared by dissolving 2.5 gm of periodic acid in 10 ml of water and diluting the resulting solution with 90 ml of *t*-butanol. Three to seven minutes later, the partially dried paper is sprayed with a freshly prepared reagent containing 2 gm sodium nitroprusside and 15 ml piperidine in 100 ml methanol. Ten to fifteen minutes after the second spray, an acetaldehydogenic steroid appears as a blue spot on a straw-colored background. The whole background turns blue after 20 minutes. Exposing the sprayed paper strip momentarily to NH_3 vapor intensifies the blue color.

Periodate oxidation of a steroidal α-ketol generates a nonvolatile acid as one of the products. It is feasible to detect the acid formation from readily oxidizable neutral steroids directly on a papergram. The dried papergram is sprayed with a reagent containing 1 volume saturated aqueous KIO_4 solution and 3 volumes 95% ethanol. One hour later, the paper strip is sprayed with the phenol red–Tris buffer reagent developed for steroid acids. The acidogenic steroids appear immediately as yellow spots on a light pink background. The spots become more conspicuous on drying.

A steriod containing a secondary α-ketol or a *vic*-glycol structure, except a 17,20-diol, generates a nonvolatile aldehyde on periodate oxidation. The part of the papergram bearing the spot to be tested is sprayed with a reagent prepared by dissolving 2.5 gm periodic acid in 10 ml water and diluting the resulting solution with 90 ml 95% ethanol. The other part of the papergram is covered with a piece of cardboard at the time of spraying. Immediately after spraying, the paper strip is rolled in the form of a hollow cylinder and kept for 4 hours in an atmosphere saturated with 95% ethanol. Upon removal, it is air dried and then transferred, still in the form of a cylinder, to another jar which contains a shallow layer of 1:1 methanol–chloroform mixture on the bottom. An ascending type chromatogram is developed until the solvent front moves at least 2 inches beyond the area which has been wetted by the periodic acid solution. The strip is again air dried, the part which has been wetted by the periodic acid solution is cut off and the rest is sprayed with Schiff's reagent (*6, 7*). One hour later, the air-dried strip is placed in a closed container to prevent the evaporation of sulfur dioxide and left at room temperature overnight. The aldehyde spot now situated in the front becomes deep blue on an almost colorless background. Traces of red or purple areas along the front are not considered as a positive result. This test is applicable to papergrams developed with both Zaffaroni and Bush systems. The former are dried in a 100°C oven having a forced-air draft system.

z. *Molybdophosphoric acid.* After development, the paper is drawn through 8% alcoholic molybdophosphoric acid and then through 2.5% aqueous KOH. Steroids ($\simeq$10 μg) with a free phenolic group appear as dark blue spots; 2-hydroxy- and 2-methoxy-estrone give a positive reaction before the alkali test, the color being intensified by the alkali (*23a*).

For further general and selective color reagents, consult references (*1, 6, 24, 25*). See also Fig. 1 and Tables V–XIII.

2. *Physical Detection Methods*

Steroids with α,β-unsaturated oxo groups absorb ultraviolet light of 240 mμ. Fluorescent screens and photocopying equipment for use with steroids have been described (*26, 27*). Estrogens absorb at 280 mμ (*28*) and bufadienolides (see Section III) at 300 mμ (*29*). Some steroids fluoresce in ultraviolet light (*30*), and the spots can be photographed (*31*). Steroids sensitive to ultraviolet light may be decomposed upon exposure (*32, 33*). Radioactivity methods (Chapter 4) are applicable to the detection of labeled steroids (*34–36*).

3. *Identification of Steroids*

The following are helpful in establishing the identity of steroids after chromatography: (a) the reaction of the substance to detection reagents of the type described above; (b) R_F values or R values relative to a standard compound migrating with about the same velocity; (c) ultraviolet or infrared absorption spectra of the steroids after elution and perhaps preliminary purification; (d) ultraviolet and visible absorption spectra of sulfuric acid chromogens of steroids (*6, 37*); (e) chromatography of derivatives formed before spotting (Section A,3) or on the chromatogram itself (*1*); (f) gas and column chromatography of the compounds separated on paper (*38*); (g) ΔR_M values in the presence and

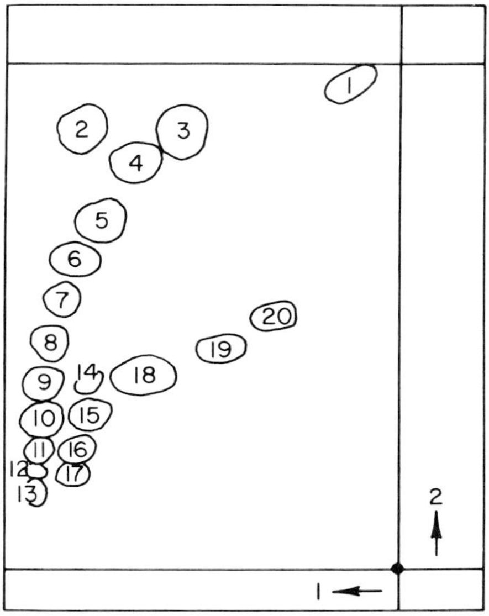

FIG. 1. Two-dimensional separation of a mixture of synthetic cholesteryl esters by a combination of adsorption and reversed-phase partition chromatography. Silica gel-impregnated paper spotted with the mixture (2–5 μg of each compound) is developed twice with petroleum ether (60–90°C). After drying, the paper is immersed in 0.5% paraffin oil in ether and then developed for 4–5 hours with acetic acid–$CHCl_3$–paraffin oil (80:15:5 v/v). Detection: Dry paper at 80–90°C and spray with phosphomolybdic acid (10% in ethanol) or 50% $SbCl_3$ in acetic acid. Compounds: (1) Cholesterol; (2) cholesteryl formate; (3) cholesteryl acetate; (4) cholesteryl butyrate; (5) cholesteryl caproate; (6) cholesteryl caprylate; (7) cholesteryl caprinate; (8) cholesteryl laurate; (9) cholesteryl myristate; (10) cholesteryl palmitate; (11) cholesteryl stearate; (12) cholesteryl arachidate; (13) cholesteryl behenate; (14) cholesteryl palmitoleate; (15) cholesteryl oleate; (16) cholesteryl eicosenate; (17) cholesteryl erucate; (18) cholesteryl linoleate; (19) cholesteryl linolenate and arachidonate; (20) cholesteryl $C_{20:5}$ and $C_{22:6}$. After Č. Michalec *et al.* (*54*).

absence of formaldehyde indicate the number of polar functional groups and to a lesser degree the nature of the steroid skeleton (*39*).

C. TECHNIQUES

Descending development is most often employed, but ascending, horizontal (*40*), circular (*41*), and centrifugal (*42*) methods have also been applied. Two-dimensional development, both conventional and with esterification between the runs (*43*), multiple development (*2*), and overrun development on serrated paper have all proved useful.

TABLE V

DATA CONCERNING THE PAPER CHROMATOGRAPHIC SEPARATION OF SOME STEROLS, VITAMINS, AND PROVITAMINS (*16*)

Compound	R_F [a]																
	1	2	3	4	5	6	7	8	9	10	11[b]	12	13	14	15	16	17[b]
Cholesterol	0.52	0.56	0.49	0.73	0.56	0.30	0.77	0.0	—	0.63	0.33	—	0.29	0.49	0.50	0.35	0.42
Cholesterol acetate	—	—	—	—	—	0.11	—	—	—	—	0.09	—	—	—	—	—	—
β-Sitosterol	0.54	0.65	0.43	0.65	0.47–0.58	—	0.80	—	—	—	0.25	—	0.00	0.51	0.42	0.26	0.38–0.33
Phytosterol from rapeseed or olive oil	—	—	—	—	0.33–0.41	—	—	—	—	—	—	—	—	—	—	0.33–0.40	0.42
Stigmasterol	0.53	0.52	—	—	—	—	0.80	—	—	0.30	0.27	—	0.00	—	—	—	0.38
7-Dehydrocholesterol	0.94	0.88	0.48	0.67	—	—	—	0.9	—	—	0.40	—	0.39	0.57	0.56	—	—
β-Cholestanol	0.56	0.63	—	—	—	—	—	—	—	—	0.27	—	—	—	—	—	—
Ergosterol	0.95	0.84	0.43	0.68	—	0.47	0.60	0.0	—	0.37	0.42	—	0.00	0.56	0.00	—	0.89
Zymosterol	—	—	—	—	—	—	—	—	—	—	0.43	—	0.41	0.61	0.48	—	—
Vitamin D_3	—	—	0.76	0.91	—	—	—	—	—	—	0.36	—	0.46	0.68	0.68	—	—
Vitamin D_2	—	—	0.80	0.91	—	—	—	—	—	—	0.29	—	0.44	0.66	0.69	—	—
Lumisterol	—	—	—	—	—	—	—	—	—	—	0.33	—	0.27	0.59	0.52	—	—
Suprasterol II	—	—	—	—	—	—	—	—	—	—	—	—	0.36	0.52	0.57	—	—
Tachysterol	—	—	—	—	—	—	—	—	—	—	—	—	0.00 0.37 0.98–	0.00 0.61 0.98	0.00 0.58 0.98	—	—
Vitamin A	—	—	—	—	—	—	—	—	~0.9	—	0.30	—	0.93 0.98	—	—	—	—
Vitamin A acetate	—	—	—	—	—	—	—	—	~0.5	—	0.01	0.32	—	—	—	—	—
α-Tocopherol	—	—	—	—	—	—	—	—	—	—	0.13	0.07	0.15	0.58	—	—	—

[a] Systems: (1) quilon/ethanol–H_2O (8:2); (2) quilon/methanol; (3) quilon/methanol–H_2O–ethylene glycol monomethyl ether (65:20:20); (4) quilon/methanol–H_2O (95:5); (5) quilon/ethanol–H_2O (8:2); (6) sodium stearate or palmitate/methanol–carbon tetrachloride–H_2O (18:5:2); (7) aluminium oxide/hexane–ether (3:1); (8) H_2O/phenol–methanol–H_2O (13:5:30:56.5); (9) H_2O/isopropanol–H_2O (1:1); (10) phenylcellosolve/heptane; (11) paraffin/acetic acid–H_2O (84:16); (12) silicone grease/acetonitrile–H_2O (6:4); (13) paraffin/ethylene glycol monoethyl ether–*n*-propanol–methanol–H_2O (35:10:30:25); (14) paraffin/*n*-propanol–methanol–H_2O (15:82:3); (15) paraffin/methanol–H_2O (95:5); (16) petroleum (b.p. 220–240°C/pyridine–H_2O (85:15); (17) paraffin/acetic acid–H_2O (84:16). Solvents are given in volume proportions. The first mentioned component of each system is the stationary phase. Quilon is a chromium sterato complex used to render the paper hydrophobic. For detection methods, see Table II.

[b] Values reported independently by different workers.

TABLE VI

R_F VALUES OF ESTROGENS IN MODIFIED ZAFFARONI SYSTEMS (55)

Mobile phase	Stationary phase [formamide–methanol–formic acid (v/v)]	R_F value			
		Estriol	Estradiol	Estrone	Equilenin
Skellysolve B–benzene	1:1:0	—	—	0.30	0.22
Benzene	1:1:0	—	0.3	0.65	—
	1:1:0	—	—	0.53	0.50
	1:1:0.5	0	0.13	0.35	—
	1:1:2	0	0.22	0.50	—
Benzene–chloroform	1:1:1	0	0.74	0.95	—
	1:1:1.5	0	0.74	0.95	—
	0.5:1:1	0.03	0.43	0.75	—
Chloroform	1:1:0.5	0.1	0.56	0.67	—
	1:1:1	0.02	0.60[a]	0.60[a]	—
	1:1:1 + 1 part water	0.03	0.67[a]	0.67[a]	—

[a] R_F values are not precise because a definite separation could not be detected.

TABLE VII

RELATIVE MOBILITIES OF 17-KETOSTEROIDS IN TWO ZAFFARONI TYPE SYSTEMS (56)[a]

Compounds	Propylene glycol/petroleum ether	Dimethylsulfoxide/heptane	
	R_{AND}	R_{DOC}	R_{CORT}
3α-Hydroxy-5α-androstan-17-one (androsterone)	1.00	2.28	—
3β-Hydroxy-5β-androstan-17-one	1.00	2.25	—
4-Androstene-3,17-dione	1.70	2.05	—
3β-Hydroxy-5α-androstan-17-one (epiandrosterone)	0.70	1.85	—
3α-Hydroxy-5β-androst-9-en-17-one	0.70	1.80	—
3α-Hydroxy-5β-androstan-17-one	0.70	1.75	—
3β-Hydroxy-5-androsten-17-one (dehydroepiandrosterone)	0.70	1.35	—
5β-Androstane-3,11,17-trione	0.75	0.75	—
11β-Hydroxy-5α-androstane-3,17-dione	0.17	0.27	10.3
3α-Hydroxy-5β-androstane-11,17-dione	0.26	0.20	6.7
3α,11β-Dihydroxy-5α-androstan-17-one	0.10	0.12	3.5
3β,11β-Dihydroxy-5β-androstan-17-one	0.048	0.09	2.25
3β,11β-Dihydroxy-5α-androstan-17-one	—	0.075	2.15

[a] R_{AND} = mobility relative to androsterone as reference steroid; R_{DOC} = mobility relative to deoxycorticosterone as reference steroid; R_{CORT} = mobility relative to cortisone as reference steroid. Ambient temperature was 24 ± 1.5° for the DMSO system. The DMSO was employed as a 50% v/v solution in methanol. Detection is by ultraviolet light and the Zimmerman, tetrazolium, and anisaldehyde tests.

TABLE VIII

R_F AND R_F VALUES OF PROGESTERONE ANALOGS IN ZAFFARONI TYPE SOLVENTS (*57*)[a]

R_F value of monohydroxy congeners

Substituent position	Substituent orientation[b]	R_F [a] 1	R_F [a] 2	R_F [a] 3
2α	e	0.08	0.58	0.87
4	—	0.28	0.60	—
6β	a	—	0.16	0.63
7α	a	—	0.08	0.41
9α	a	—	0.23	0.61
11α	e	—	0.14	0.38
11β	a	—	0.29	0.57
12α	a	—	0.16	0.63
14α	a	—	0.23	0.59
15α	q	—	0.06	0.37
15β	q	—	0.10	0.29
16α	q	—	0.06	0.30
17α	q	0.02	0.40	0.78
18	—	—	0.34	0.64
19	—	—	—	0.31
21	—	—	0.33	0.83
Progesterone	—	0.3	0.9	1.0

R_P value of monomethyl congeners

Substituent position	Substituent orientation[b]	R_P [a] 1
2α	e	1.64
2β	a	1.78
4	—	1.42
6α	e	1.31
6β	a	1.18
7α	a	1.04
7β	e	1.22
16α	q	1.33
16β	q	1.03
17α	q	1.43

R_P value of monohalo congeners

Substituent position	Substituent orientation[b]	R_P [a] 1	R_P [a] 2
2α-Fluoro	e	0.17	0.86
2α-Iodo	e	1.25	1.05
4-Chloro	—	0.2	0.7
4-Bromo	—	0.18	0.76
6α-Fluoro	e	0.21	0.53
6β-Fluoro	a	0.63	0.56
6α-Chloro	e	0.77	0.68
6β-Chloro	a	0.63	0.61
16α-Chloro	q	0.53	—
17α-Fluoro	q	1.65	1
17α-Bromo	q	1.65	1

R_P value of dehydro congeners

Double bond position	R_P [a] 1	R_P [a] 2
1	0.38	0.97
6	0.71	1.07
8	0.75	0.85
9 (11)	0.96	0.96
11	0.81	0.92
14	0.75	0.8
16	0.81	0.63

[a] Solvents: (1) formamide/skellysolve B; (2) formamide/skellysolve B–benzene (1:1 v/v); (3) formamide/benzene. R_P = R_F relative to progesterone. Detection: ultraviolet light, and $FeCl_3$–ferricyanide reagent for 4-hydroxyprogesterone.

[b] e = equitorial; a = axial; q = quasi.

TABLE IX

MEAN R_F VALUES OF Δ^4- AND $\Delta^{1,4}$-3-KETOSTEROIDS AFTER CHROMATOGRAPHY ON FULLY ACETYLATED PAPER (*58*)

Steroid	Mean R_F value[a]		
	1	2	3
C_{27} steroid			
Δ^4-Cholesten-3-one	0.28	0.09	—
C_{21} steroids			
Progesterone	0.48	0.35	0.88
11-Deoxycorticosterone	0.54	0.45	0.77
17α-Hydroxyprogesterone	0.54	0.47	0.75
Reichstein's Substance S	—	0.57	0.57
Cortisone	0.61	0.60	0.43
Hydrocortisone	0.68	0.68	0.36
Prednisone	—	0.65	—
Predinisolone	—	0.73	—
17α-Ethynyltestosterone	—	0.45	—
C_{21} steroid acetates			
11-Deoxycorticosterone acetate	—	0.30	—
Cortisone acetate	—	0.46	—
Hydrocortisone acetate	—	0.56	—
C_{20} steroid			
17-Methyltestosterone	—	0.51	—
C_{19} steroids			
Δ^4-Androstene-3,17-dione	0.52	0.41	—
$\Delta^{1,4}$-Androstadiene-3,17-dione	—	0.47	—
Δ^4-Androsten-17α-ol-3-one	—	0.52	—
Testosterone	0.60	0.57	—
C_{19} steroid acetate			
Testosterone acetate	0.27	0.32	—
C_{18} steroid			
19-Nortestosterone	—	0.55	—

[a] Solvents: (1) benzene–methanol–H_2O (1:10:3); (2) benzene–methanol–H_2O (4:4:1); upper (most polar) layer as mobile phase, lower layer used to impregnate the paper; (3) same as 2 with lower layer as mobile phase and upper layer used to impregnate the paper. Detection: ultraviolet light and 2,4-dinitrophenylhydrazine.

TABLE X

$R_F \times 100$ VALUES[a] OF STEROIDS AND STEROID ESTERS IN FORMIC ACID–PETROLEUM ETHER MIXTURES (*59*)

Steroid	R_F 1 Free	1 Formate	2 Free	2 Formate	2 Acetate	2 Propionate	3 Free	3 Formate
Cholestenone	96	—	98	—	—	—	—	—
Ostrone	—	—	—	—	—	—	—	22
Androstane derivatives								
5α-Androstan-17-one	95	—	97	—	—	—	—	—
Androst-5-en-3β-ol	—	93	—	93	—	—	—	—
5α-Androstane-3α, 17β-diol	—	77	—	84	—	—	—	87
5β-Androstane-3α, 17β-diol	—	78	—	88	—	—	—	89
Androst-5-ene-3β, 17α-diol	—	76	—	86	—	—	—	89
Androst-5-ene-3β, 17β-diol	—	78	—	85	—	—	—	89
17α-Methyl-5α-androstane-3β, 17β-diol	11	48	—	66	—	—	—	82
17α-Methyl-androst-5-ene-3β, 17β-diol	—	48	—	55	—	—	—	79
5α-Androstan-3β-ol-17-one	—	25	—	54	62	77	—	68
5α-Androstan-3α-ol-17-one	7.5	32	11	51	59	78	55	68
5β-Androstan-3α-ol-17-one	5.5	29	8	52	57	73	31	74
5β-Androstan-3β-ol-17-one	6	30	—	—	—	—	47	82
5β-Androst-9(11)-en-3α-ol-17-one	7.5	32	11	58	64	81	51	76
Androst-5-en-3β-ol-17-one	5	28	9	51	56	74	29	72
5β-Androstan-17β-ol-3-one	5	25	9	43	—	—	—	75
Androst-4-en-17β-ol-3-one	—	8	3	20	23	42	16	49
Androst-4-en-17α-ol-3-one	—	8	3	19	20	39	16	49
Androsta-1, 4-dien-17β-ol-3-one	—	—	—	—	—	—	—	29
17α-Methylandrost-4-en-17β-ol-3-one	0	—	6	—	32	—	25	—
19-Norandrost-4-en-17β-ol-3-one	—	8	—	16	—	—	—	48

5α-Androstane-3, 17-dione	5	—	13	—	—	—	46	—
5β-Androstane-3, 17-dione	5	—	13	—	—	—	45	—
Androst-4-ene-3, 17-dione	3	—	3	—	—	—	20	
Androsta-1, 4-diene-3, 17-dione	—	—	—	—	—	—	10	—
Androst-4-ene-3, 11, 17-trione	—	—	2	—	—	—	24	—
Androst-4-en-11β-ol-3, 17-dione	—	—	0	—	—	—	2	
5β-Androstane-3α, 11β-diol-17-one	—	—	—	3	—	—	—	27
Androst-4-ene-11β, 17β-diol-3-one	—	—	—	2	—	—	—	13
Pregnane derivatives								
5α-Pregnan-3β-ol	—	91	—	93	—	—	—	—
5β-Pregnan-3β-ol	—	83	—	93	—	—	—	—
5α-Pregnane-3α, 20α-diol	—	86	—	88	—	—	—	93
5β-Pregnane-3α, 20α-diol	—	86	—	88	—	—	—	94
Pregn-5-ene-3β, 20α-diol	—	85	—	88	—	—	—	92
Pregn-5-ene-3β, 20β-diol	—	84	—	89	—	—	—	93
5α-Pregnane-3α, 17α, 20α-triol	—	23	—	36	—	—	—	70
5β-Pregnane-3β, 16β, 20α-triol	—	22	—	—	—	—	—	73
5α-Pregnan-3α-ol-20-one	—	70	35	76	—	—	—	86
5β-Pregnan-3α-ol-20-one	—	65	28	77	—	—	—	89
Pregn-4-en-20β-ol-3-one	—	19	—	28	33	52	—	72
Pregn-5-en-3β-ol-20-one	—	61	23	76	80	86	60	86
Pregna-5, 16-dien-3β-ol-20-one	—	—	—	—	—	—	58	86
5α-Pregnane-3, 20-dione	19	—	36	—	—	—	68	—
5β-Pregnane-3, 20-dione	19	—	36	—	—	—	68	—
Pregn-4-ene-3, 20-dione	4	—	11	—	—	—	48	—
Pregna-4, 16-diene-3, 20-dione	—	—	—	—	—	—	49	
16α-Methyl-pregn-4-ene-3, 20-dione	—	—	—	—	—	—	62	—
Pregn-4-ene-3, 11, 20-trione	—	—	0	—	—	—	8	—

[a] Solvents: (1) formic acid–petroleum ether (80–100°C) (1:1 v/v); (2) formic acid–petroleum ether–methanol (10:9:1 v/v); (3) formic acid–petroleum ether–benzene (2:1:1 v/v). The top layer of these mixtures is the mobile phase. The paper (Whatman No. 1, 2, 4, and 3 MM) is soaked with and then equilibrated for 4 hours in the vapors of the bottom layer before development. Detection: ultraviolet light, phosphomolybdic acid and alkaline *m*-dimtrobenzene. Up to 0.5 mg of steroid or derivative per square centimeter can be chromatographed.

TABLE XI

R_F VALUES OF SOME 16-DEHYDROSTEROIDS AND THEIR ACETATES ON SILICIC ACID-IMPREGNATED PAPER (*60*)

Compound[a]	R_F[b]											Colors with visualizing reagents[c]			
	1	2	3	4	5	6	7	8	9	10	11	PMA	PTA	RA	Allen
An (α)	0.17	0.47	0.52	0.38	0.33	0.61	0.59	0.63	0.42	0.30	0.91	Bluish green	Pink	Mauve	Mauve
Ae	0.10	0.36	0.42	0.30	0.25	0.53	0.47	0.55	0.34	0.21	0.88	Bluish green	Pink	Mauve	Mauve
An (β)	0.08	0.29	0.34	0.26	0.18	0.46	0.41	0.48	0.28	0.17	0.84	Bluish green	Pink	Mauve	Mauve
Andien	0.09	0.31	0.37	0.28	0.20	0.47	0.43	0.50	0.31	0.19	0.85	Purple	Rose	Blue	Mauve
Oe	0.24	0.64	0.69	0.51	0.46	0.77	0.71	0.74	0.55	0.40	0.95	Mauve	Orange	Orange-red	Rose
An (α) acetate	0.55	0.84	0.86	0.95	0.68	1.0	0.91	0.85	0.72	0.60	1.0	Bluish green	Pink	Mauve	Mauve
Ae acetate	0.61	0.85	0.87	0.96	0.69	1.0	0.92	0.87	0.72	0.64	1.0	Bluish green	Pink	Mauve	Mauve
An (β) acetate	0.60	0.85	0.88	0.95	0.70	1.0	0.92	0.87	0.71	0.65	1.0	Bluish green	Pink	Mauve	Yellow
Andien acetate	0.60	0.86	0.87	0.94	0.69	1.0	0.92	0.87	0.70	0.64	1.0	Purple	Rose	Blue	Mauve
Oe acetate	0.60	0.86	0.88	0.96	0.71	1.0	0.94	0.90	0.72	0.65	1.0	Mauve	Orange	Red	Purple

[a] Abbreviations: 5α-androst-16-en-3α-ol (An (α), 5α-androst-16-en-3β-ol (An (β), 5β-androst-16-en-3α-ol (aetiochol-16-en-3α-ol, Ae), androsta-5, 16-dien-3β-ol (Andien), oestra-1, 3, 5(10), 16-tetraen-3-ol (Oe). Ascending development without equilibration was employed.

[b] Solvents: (1) petroleum ether (b.p. 83–97°C)–benzene (1:1 v/v); (2) benzene; (3) benzene–ether (99:1 v/v); (4) toluene–ether (99:1 v/v); (5) toluene–ethyl acetate (99:1 v/v); (6) ethylene chloride; (7) benzene–methylene chloride (50:50 v/v); (8) benzene–ethylene chloride (50:50 v/v); (9) benzene–methyl ethyl ketone (99:1 v/v); (10) toluene; (11) methylene chloride.

[c] PMA = phosphomolybdic acid reaction; PTA = phosphotungstic acid reaction; RA = resorcyaldehyde in glacial acetic acid (0.5% w/v)–concutrated H_2SO_4 in glacial acetic acid (5% v/v) (1:1 v/v); heat at 100°C after spraying; Allen = concentrated H_2SO_4–ethanol–H_2O (40:9:1 v/v) (*61*).

TABLE XII

R_F VALUES OF SOME CORTICOSTEROIDS, ESTROGENS, AND KETOSTEROIDS IN ETHYLENE GLYCOL SYSTEMS (*62*)[a]

Steroid	R_F		
	Toluene solvent	CCl_4 solvent	Petroleum ether solvent
Cortisol	0.03	0.01	—
Aldosterone	0.07	0.08	—
Cortisone	0.10	0.10	—
11-Deoxycortisone	0.43	0.15	—
Corticosterone	0.51	0.20	—
11-Deoxycorticosterone	0.88	0.88	—
Cortisone acetate	0.70	0.26	—
11-Deoxycortisone acetate	0.85	0.40	—
11-Deoxycorticosterone acetate	0.96	0.98	—
Estrone	—	0.50	—
Estradiol	—	0.12	—
Androstane-3α,11β-diol-17-one	—	—	0.06
Etiocholan-3α-ol-11,17-dione	—	—	0.13
Δ^4-Androsten-11β-ol-3,17-dione	—	—	0.20
Δ^4-Androstene-3,11,17-trione	—	—	0.24
Δ^5-Androsten-3β-ol-17-one	—	—	0.47
Androstan-3β-ol-17-one	—	—	0.56
Etiocholan-3α-ol-17-one	—	—	0.54
Androstan-3β-ol-17-one	—	—	0.67
3,5-Cycloandrostan-6β-ol-17-one	—	—	0.76
Androstane-3,17-dione	—	—	0.85
3β-Chloro-Δ^5-androsten-17-one	—	—	0.94
Testosterone	—	—	0.22
17-Methyltestosterone	—	—	0.43
Pregnane-3,20-dione	—	—	0.92
Progesterone	—	—	0.84

[a] Method: Steroids in chloroform solution were applied to sheets of Whatman No. 1 paper (15 × 48 cm). After the application of the solution the strips were soaked in a 30% solution of ethylene glycol in methanol, blotted between filter paper and exposed to air at room temperature for 20 minutes. The chromatograms were then transferred to the chromatography tanks, saturated with vapors of the mobile phase, and developed by the descending technique at 17–20°C. For chromatographic separation of corticosteroids, toluene or carbon tetrachloride saturated with ethylene glycol, and for 17-ketosteroids petroleum ether (b.p. 45–65°C) were used. Detection: ultraviolet light, tetrazolium reagent or *m*-dinitrobenzene.

TABLE XIII

R_F VALUES OF SOME STEROIDS IN SYSTEMS WITH STATIONARY TRIETHYLENE GLYCOL (*63*)

Steroid	R_F				
	Cyclo-hexane	Petroleum ether	Tetrachlo-romethane	Benzene	Toluene
Cortisol	—	—	0.01	0.04	0.02
Aldosterone	—	—	—	0.08	0.05
Cortisone	—	—	0.01	0.10	0.06
Corticosterone	—	—	0.04	0.25	0.21
11-Desoxycortisone	—	—	0.05	0.23	0.20
11β-Hydroxyetiocholanolone	0.01	0.02	0.05	0.16	0.22
11β-Hydroxyandrosterone	0.02	0.03	0.11	0.26	0.27
7-Ketodehydroepiandrosterone	0.02	0.03	0.12	0.36	0.33
Estrone	0.03	0.04	0.16	0.37	0.37
$\Delta^{4,6}$-Androstadien-17β-ol-3-one	0.04	0.05	0.28	0.50	0.48
Testosterone	0.09	0.12	0.30	0.54	0.55
Adrenosterone	0.07	0.11	—	—	—
11-Ketoetiocholanolone	0.04	0.06	0.24	0.50	0.62
11-Ketoandrosterone	0.08	0.12	0.35	—	—
17α-Hydroxyprogesterone	—	—	0.33	0.80	0.62
Dehydroepiandrosterone	0.16	0.22	0.56	0.60	0.69
Epiandrosterone	0.22	0.31	0.62	0.64	0.68
Etiocholanolone	0.25	0.32	0.63	0.66	0.71
Androsterone	0.34	0.41	0.68	0.68	0.74
Methyltestosterone	0.22	0.26	0.74	0.84	0.90
$\Delta^{4,6}$-Androstadiene-3,17-dione	0.31	0.30	0.73	0.85	0.93
Δ^{4}-Androstene-3,17-dione	0.40	0.50	0.75	0.85	0.93
Androstane-3,17-dione	0.68	0.70	0.85	0.93	0.96
Progesterone	0.64	0.82	0.93	—	—
3β-Chloro-Δ^{5}-androsten-17-one	0.93	0.95	—	—	—
Pregnane-3,20-dione	0.84	0.90	—	—	—

[a] Method: Same as Table XII. Detection: Ketosteroids detected by Zimmerman reagent.

Sealed glass development tanks (*2, 44*) are satisfactory, and the temperature should be as constant as possible, especially when using Bush solvents (see next section). Some workers suggest that the temperature for equilibration and the run should be the same (*1*), but others employ elevated temperature for saturation of the chamber and paper and room temperature for development (*45*). Before the application of detection reagents, developed papers are thoroughly dried (30–60 minutes in air if unimpregnated and 12–14 hours in air or 20–40 minutes at 85°C if impregnated). Elution of spots from impregnated paper is easier if the paper is not thoroughly dry; impregnating agent can be removed from the eluate *in vacuo* (*1*).

The novel method of paper strip elution chromatography has been applied to the separation of Δ^4-3-ketosteroids. The descending solvent drips off the end of the paper into a collecting funnel, the stem of which passes through a small hole in a polyethylene stopper inserted through the bottom of the glass chamber. The solvent is collected in test tubes held in a rotating fraction collector, and an elution profile is prepared by plotting the fluorescence of each fraction against the fraction (tube) number (*46*).

Extra-long tanks with a circulation device for proper solvent saturation have been described for steroid separations (*47*), as has a device for the automatic start of development after equilibration of the paper for an appropriate time when using biphasic Bush type systems (*48*).

D. Chromatographic Systems

After the earliest attempts to separate steroids as derivatives (*49, 50*), introduction of the Bush and Zaffaroni solvent systems allowed excellent separations of uncombined steroids. These and other types of systems for steroid separations will now be described.

1. *Aqueous Solvents*

In the Bush systems, the aqueous phase is usually aqueous methanol absorbed by the spotted, dry paper during equilibration for 4–12 hours in a chamber saturated with both phases. For example, benzene, methanol, and H_2O are thoroughly mixed in 2:1:1 volume proportions and the layers are allowed to separate. The lower layer is used to saturate the chamber and the upper is used as the mobile phase. Saturation and chromatography at a controlled, elevated temperature (e.g., $35 \pm 1°C$) maintains the correct degree of saturation, increases the solubility of the steroids, and prevents streaking. Development can also be at room temperature (*45*). Other stationary phases are formic acid–methanol, formic acid, aqueous ethanol, *t*-butanol, or acetic acid.

Alternatively, the paper can be impregnated with the stationary phase, hung at room temperature for 20 minutes, spotted, and developed at room temperature without further equilibration of the paper.

Mobile phases are weakly polar and can be composed of hexane, isooctane, petroleum ether, cyclohexane, toluene, $CHCl_3$, ethyl acetate, etc. Such systems are useful for the separation of androgens, estrogens, gestagens, and corticoids, i.e., for the C_{19} and C_{21} steroids containing two to five oxygen atoms in the molecule (*24*). Systems containing acetic acid separate steroidal alkaloids and saponins. Acidic and basic systems separate sulfates and glucosiduronates of steroid hormones and their metabolites, and basic systems separate estrogens and steroid acids (*25*).

2. *Nonaqueous Solvents*

In the Zaffaroni systems, the paper is impregnated with formamide, propylene or ethylene glycol, or dimethylformamide, and the mobile phase is a solvent such as hexane, benzene, $CHCl_3$, or cyclohexane, mixtures of these, or three-component mixtures (e.g., ethyl acetate–butyl acetate–H_2O). The mobile phase is usually saturated with the stationary phase, or it can be saturated with a polar component similar to the stationary phase (e.g., formamide as the stationary phase, cyclohexane saturated with dimethylformamide as the mobile phase) (*1*).

The paper is drawn through a solution of the stationary phase in a solvent (e.g., 30% formamide in ethyl acetate or methanol); the excess is removed by blotting between thick papers; the paper is hung is a hood 5–10 minutes to dry; the sample in a solvent such as $CHCl_3$ or acetone is applied; and development is begun at once in a tank saturated with the vapors of all components of the system. Development is quite rapid and the spots are usually round. For multiple development, the paper is dried 5 minutes between runs (*24*).

Steroids separated by systems of the Zaffaroni type include corticosteroids, 17-ketosteroids, estrogens, sterols, sapogenins, and steroid alkaloids. Phenylcellosolve is a stationary phase suited to the separation of nonpolar steroids, and carbitol and ethylene glycol are useful for polar steroids (e.g., aldosterone) (*25*).

3. *Classical Aqueous System* (*1*)

Normal aqueous systems can be used without special impregnating or equilibrating procedures for the chromatography of polar and ionic steroids. These solvents can be neutral, acidic, or alkaline in nature, and can be mono- or biphasic.

4. *Reversed-Phase Systems*

For the separation of low-polarity steroids, the paper is made hydrophobic by treatment with Vaseline, paraffin oil, kerosene, silicone oil, olive oil, etc. The impregnating agent acts both as the stationary phase and as a support for the less polar component of the solvent. The paper (thicker papers such as Whatman No. 3 are often used) is dipped into a solution of paraffin oil (e.g., 10–20%) in ether or benzene, the surplus is removed by pressing between two thick papers, and the solvent is allowed to evaporate. The samples are applied and developed with acetic acid–$CHCl_3$–paraffin oil (65:25:10 v/v). The R_F values of cholesterol and ergosterol are 0.65 and 0.70 in this system (*24*).

The paper is also rendered hydrophobic by acetylation (see Chapter 3). Weakly polar steroids may be separated by development with the more polar phase of benzene–methanol–H_2O (4:4:1 v/v) after equilibrating the acetylated paper with the less polar phase. The more polar corticosteroids are separated better if the phases are reversed, the less polar being used for the development (*25*).

5. *Monophasic, Weakly Polar, Anhydrous Solvents*

Development with heptane, isooctane, petroleum ether, etc. on untreated paper separates epimeric steroid acetates (*1*, *51*).

6. *Adsorbent-Loaded Papers*

Paper impregnated with silica gel, alumina, or other absorbents have been used for steroid separations. For example, four major classes of cholesterol esters (saturated esters, oleate, linoleate, and linoleate plus arachidonate) are separated on slilicic acid paper using 8–10% benzene in hexane (*38*). Aqueous solvents can also be employed for various separations.

Glass-fiber paper developed with isooctane serves for the separation of cholesterol (R_F = 0.34), 7-dehydrocholesterol (0.03), and ergosterol (0.15). Detection may be by charring (concentrated H_2SO_4 plus heat) (*52*). Glass papers impregnated with silicic acid, potassium silicate, monopotassium phosphate, and alumina have been used to separate steroids as closely related as an alcohol and its corresponding ketone and cis–trans isomers (*53*).

Glass-fiber sheets impregnated with silica gel and previously treated with 10% $AgNO_3$ solution provide the separation of C_{19} steroids extracted from biological tissue. Development is carried out twice with $CHCl_3$ followed by cyclohexane–ethyl acetate (7:1 v/v). Extracted steroid sulfates are separated on glass fiber sheets impregnated with silica gel and developed twice with $CHCl_3$–acetone–acetic acid (110:35:6 v/v). The spots are detected by spraying with methylene blue reagent or with H_2SO_4–ethanol (1:1) and heating at 110° for 15 minutes (*53a*).

E. Chromatographic Data

Figures 1 and 2 (*53*, *54*) and Tables V–XIII (*16*, *55–63*) present mobility data for specific steroids in chromatographic systems of the type just listed. Other solvent systems, methods for treatment of paper, and migration data are given in Block *et al.* (*6*). Comprehensive tables of R_F values will also be found in Neher (*1*) and Hais and Macek (*7*). R_F values for thirty-seven 1-oxygenated steroids in various solvent systems from among fourteen which are suggested, have been recently published (*63a*).

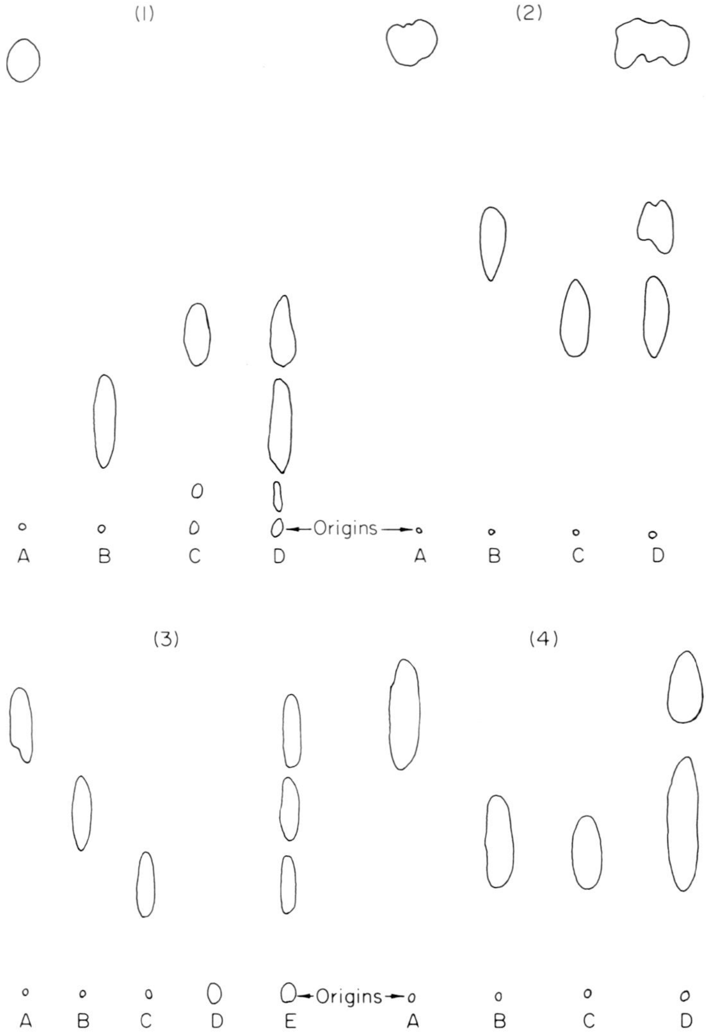

FIG. 2. Separation of Steroids on Treated Glass Paper. (1) Chromatogram showing the separation on monopotassium phosphate-treated glass paper. Developing solution, benzene–ethyl alcohol (400:1 v/v); developing time, 7 minutes. (A) 11-Deoxycorticosterone, (B) corticosterone, (C) 11-dehydrocorticosterone, (D) mixture of B and C. (2) Chromatogram showing the separation on monopotassium phosphate-treated glass paper. Developing solution, benzene–ethyl alcohol (200:3); developing time, 7 minutes. (A) Cortisone acetate, (B) cortisone, (C) hydrocortisone, (D) mixture of A, B, and C. (3) Chromatogram showing the separation of some estrogens on monopotassium phosphate-treated glass paper. Developing solution, isooctane-benzene (1:2); developing time, 7 minutes. (A) Estrone, (B) estradiol-17β, (C) estradiol-17α, (D) estriol, (E) mixture of A, B, C, and D. (4) Chromatogram showing the separation on potassium silicate-treated glass paper. Developing solution, isooctane; developing time, 7 minutes. (A) Pregnan-3β-ol-20-one, (B) 5-pregnen-3β-ol-20-one, (C) allopregnan-3β-ol-20-one, (D) mixture of A, B, and C. After J. G. Hamilton and J. W. Dieckert (*53*).

F. QUANTITATIVE ANALYSIS

The usual quantitative methods (see Chapter 4) have been applied to steroid analyses. A machine for the treatment of chromatograms with chemical reagents improves quantitative results obtained by densitometry (*64*). Addition of 2% v/v water to the Zimmerman reagent reduces the background color without any loss of sensitivity when determining 17-ketosteroids (*65*).

Steroids may also be determined by ultraviolet spectrophotometry after elution from the chromatogram. An elutor has been described for removal of highly polar conjugated steroids with a small volume of solvent without eluting any paper fibers (*66*). Ketosteroids are eluted with methanol from acetylated paper held between two inclined glass slides (*67*). The eluted steroid is determined with the aid of a calibration curve and the corrected extinction (a) of the steroid:

$$a = \frac{c(e - d)}{b - d}$$

where b = measured extinction of the solution of the steroid chromatographed and eluted from the standard solution; c = measured extinction of solution of a similar quantity of steroid not chromatographed; d = measured extinction of solution obtained after elution of the paper blank from the control strip; e = measured extinction of solution of the steriod to be determined after chromatography and elution.

Fluorimetry after elution has been used for the analysis of ketosteroids (*46*) and fluorimetry on the paper for coumestrol (*68*). Radioactivity measurement is used for the quantitation of labeled steroids (*69*).

G. MOBILITY–STRUCTURE RELATIONSHIPS

Detailed discussions of this topic appear in Neher (*1*), Oakey (*3*), and Hais and Macek (*7*). Some important considerations are as follows: (a) The nature and position of the functional groups are significant. (b) Oxygen-containing functional groups lower R_F values—carboxyl and primary alcohol groups have the greatest effect and ester groups the least; the ratio of carbon to oxygen atoms is important. (c) An isolated double bond has little effect upon mobility.

A study of a series of seven Zaffaroni-type solvents revealed that the mobility of a steroid in each system could be estimated by inspection of the steroid's structure, thus permitting the choice of a suitable solvent without experimentation (*70*).

H. Applications

Some selected applications of paper chromatography to steroids will now be mentioned: soybean sapogenols were separated and identified by horizontal, circular chromatography with hexane–$CHCl_3$-acetic acid (100:10:2.5 v/v) (*41*); 20-deoxy steroids in the 3α-hydroxy-5β-pregnane series were purified by paper, thin-layer, and column chromatography after preparation (*71*); the contribution to polarity of hydroxyl and other groups at the C-17, C-20, and C-21 positions of the pregnane side chain was evaluated by descending chromatography on untreated Whatman No. 2 paper with toluene–isooctane–aqueous methanol solvents (*72*); after preliminary purification by Girard separation and silica gel column chromatography, cortisol and cortisone were identified in concentrated hog urine extracts by paper chromatography using Bush and Zaffaroni systems (*73*); for its detection in plant extracts, coumesterol is separated from masking substances by chromatography in acetone–H_2O (3:7 v/v), elution from the paper, and redevelopment with acetone–acetic acid–H_2O (3:1:6 v/v) (*74*); and a comprehensive procedure for urinary steroid analysis employing paper chromatography has been recently published (*75*).

II. Bile Acids and Other Steroid Acids

Bile acids have been separated as anions and as unionized molecules by the use of acidic and alkaline systems (*76–78*). Neutral systems of the Bush and Zaffaroni type are suitable for the free acid esters. Zaffaroni's method of developing ketosteroids as Girard's reagent T hydrazones has been applied to the methyl esters of bile keto acids (*79*). Bile acids have also been separated as hydroxamic acid derivatives (*80, 81*).

A. Preparation of Sample

If the concentration of bile acids in a biological sample is high, it can be applied directly to the paper. The extraction method of Haslewood and Wooten yields the acids as conjugates or as free acids after basic hydrolysis (*82*): Bile liberated from gall bladders is extracted with successive quantities of ethanol until fresh extracts are colorless. Further ethanol is added to precipitate mucin, etc. The resulting bile salt solution is filtered, evaporated to dryness under reduced pressure, transferred to a dish with fresh ethanol and water, again evaporated, and finally dried over H_2SO_4 in a vacuum desiccator. The crude bile salts are purified by dissolving them in water in a dish, adding decolorizing charcoal and 5 *N* NaOH until a pink color is obtained with phenolphthalein, evaporating to dryness at 70°C on a water bath, and extracting with ethanol in a

Soxhlet apparatus. The extract is filtered and evaporated, and the bile salts obtained as before. For hydrolysis, 1.0 gm of bile salts is dissolved at room temperature in 10 ml of 2.5 *N* NaOH and heated in a small bomb at 110–120°C for 5–6 hours. When cool, the contents are washed into a flask, acidified with HCl, excess NaCl is added, and the precipitate is stirred and left for some hours. The precipitate is then collected, washed with water, and dried *in vacuo*. Free bile acids are extracted with boiling ethyl acetate.

Cholyl hydroxamic acid (as well as other bile acid hydroxamic acids) is prepared as follows (*81*): To a methanolic solution of methyl cholate (2 gm; melting point 155°C) is added a mixture of equal volume (25 ml) of 12.5% methanolic KOH solution and 12.5% methanolic hydroxylamine solution, and refluxed for 5 minutes. After being cooled, the mixture is diluted with an equal quantity of water, acidified (litmus) by adding acetic acid, and evaporated on a water bath until no bubbling is observed and a crystalline substance begins to separate. After standing overnight at room temperature, the crystals are collected. (yield 1.9 gm). Recrystallizations from aqueous methanol afford a pure sample (needles) of cholylhydroxamic acid.

B. Systems (see Table XIV)

1. *Basic Solvents* (*1*)

Examples of basic solvents : *n*–propanol–NH_4OH–H_2O (90:2:8 v/v), *n*-propanol–ethanolamine–H_2O (90:5:5 v/v), collidine–H_2O (10:3.5 v/v, NH_3 atmosphere), and isoamyl alcohol saturated with 3 *M* $(NH_4)_2CO_3$ or 2% NH_3 . For medium polarity acids, two modified Bush solvents are recommended: hexane–*tert*-butanol–4 *N* NH_4OH (10:3:10 v/v) and $CHCl_3$-methanol–1 *N* NH_4OH (1:1:1 v/v). For weakly polar acids, Zaffaroni systems are better; the lower phase of ethylene glycol–butyl ether–piperidine (10:20:1 v/v) is diluted with 1.5 volumes of methanol and used to impregnate the paper; after equilibration for 1 hour, the ether phase is used for development (*83*).

2. *Acidic Solvents* (*1*)

For the separation of free and conjugated bile acids, the paper is impregnated with a 10–20% solution of 70–95% aqueous acetic acid in ether or methylene chloride followed by 3–5 minutes drying. Suitable solvents include heptane–isopropyl ether (4:1, 2:3, or 1.5:8.5 v/v), heptane–amyl acetate (2:3 or 1:4 v/v), or butanol saturated with 3% aqueous acetic acid.

TABLE XIV

PAPER CHROMATOGRAPHIC SEPARATION OF STEROID ACIDS (*83*)

Acids	R_F values in solvent system[a] 1	2	3	4	Recommended detection method[b]
Acetyleburicoic acid	0.77	0.95	0.93	0.95	PMA
3-Keto-$\Delta^{8,24(28)}$-ergostadien-21-oic acid	0.42	0.85	0.92	0.95	PMA
Eburicoic acid	0.23	0.66	0.73	0.93	PMA
Damarenolic acid	0.22	0.66	—[c]	—	PMA
Oleanolic acid	0.20	0.60	0.67	—	PMA
Polyporenic acid-A	0.00	0.14	0.30	0.20	PMA
Dehydrocholic acid	0.00	0.00	0.06	0.35	DNPH
3-Keto-Δ^4-etienic acid	0.00	0.00	—	0.47	UV
17α-Hydroxy-3-keto-Δ^4-etienic acid	0.00	0.00	—	0.39	UV
11β,17α-Dihydroxy-3-keto-Δ^4-etienic acid	0.00	0.00	—	0.17	UV
Deoxycholic acid	0.00	0.00	0.15[d]	0.20	PMA
Cholic acid	0.00	0.00	—	0.07	PMA

[a] Solvent: (1) ethylene glycol–piperidine/*n*-butyl ether–piperidine; (2) ethylene glycol–piperidine/*n*-amyl acetate–piperidine; (3) *n*-hexane–butanol–4 *N* NH_4OH (10:3:10 v/v); (4) chloroform–methanol–1 *N* NH_4OH (1:1:1 v/v). Chromatography performed in 4.5 × 3.5 × 9 inch jars.

[b] PMA = phosphomolybdic acid reagent; UV = ultraviolet scanning; DNPH = dinitrophenylhydrazine reagent.

[c] Spot appeared as a smear.

[d] Spot appeared with tailing.

3. *Separation of Hydroxamic Acids* (*81*)

Hydroxamic acid derivatives of bile acids are separated by ascending development with toluene–acetic acid–H_2O (5:5:1 v/v). R_F values are as follows: lithocholyl, 0.57; deoxycholyl, 0.34; chenodeoxycholyl, 0.22; hydeoxycholyl, 0.09; cholyl, 0.06.

4. *Separation of Girard's Hydrazones T* (*79*)

These derivatives of methyl esters of bile keto acids are separated by *n*-butanol–isopropanol–H_2O (6:2:1 v/v).

5. *Separations on Glass Paper*

Table XV shows R_F values on glass paper treated with monopotassium phosphate, potassium silicate, and silicic acid with three developing solvents. Bile acids and some of their ketonic oxidation products are

TABLE XV

MOBILITIES OF BILE ACIDS ON GLASS PAPER TREATED WITH VARIOUS SOLUTIONS AND ON GELMAN ITLC-SG SILICA GEL GLASS FIBER PAPER (*53, 84*)

Bile acid	R_F value[a]		
Glass Paper Impregnated with Silicic Acid			
	1	2	3
Lithocholic	0.68	1.00	1.00
3,12-Diketocholanic	0.35	1.00	1.00
Dehydrocholic	0.25	0.92	1.00
3α-Hydroxy-12-ketocholanic	0.13	0.78	1.00
3α-Hydroxy-7,12-diketocholanic	0.00	0.57	1.00
Deoxycholic	0.00	0.45	0.84
3,12-Dihydroxy-7-ketocholanic	0.00	0.19	0.63
3α,6β,7β-Trihydroxycholanic	—	—	0.44
3α,6α,7β-Trihydroxycholanic	—	—	0.40
3α,6β,7α-Trihydroxycholanic	—	—	0.32
Cholic	0.00	0.00	0.28

Bile acid	R_F value[a]			
Glass Paper Treated with Monopotassium Phosphate				
	4	5	6	7
3,12-Diketocholanic	0.67	1.00	1.00	0.47
Lithocholic	0.55	1.00	1.00	0.55
Dehydrocholic	0.46	1.00	1.00	0.30
3α-Hydroxy-12-ketocholanic	0.27	0.83	1.00	0.20
3α-Hydroxy-7,12-diketocholanic	0.21	0.67	1.00	0.15
Deoxycholic	0.09	0.41	0.84	0.05
3α,6β,7β-Trihydroxycholanic	—	—	0.76	—
3α,12α-Dihydroxy-7-ketocholanic	0.00	0.15	0.62	1.00
3α,6β,7α-Trihydroxycholanic	—	—	0.56	—
3α,6α,7β-Trihydroxycholanic	—	—	0.42	—
Cholic	0.00	0.15	0.26	0.00

Bile acid	R_F value[a]		
Glass Paper Treated with Dilute Potassium Silicate			
	8	9	10
3,12-Diketocholanic	0.54	0.83	1.00
Dehydrocholic	0.47	0.74	1.00
Lithocholic	0.36	—	—
3α-Hydroxy-12-ketocholanic	0.26	0.67	1.00
3α-Hydroxy-7,12-diketocholanic	0.20	0.55	1.00
Deoxycholic	0.13	0.45	0.84
3α,6β,7β-Trihydroxycholanic	—	0.12	0.80
3α,12α-Dihydroxy-7-ketocholanic	0.00	0.24	0.60
3α,6β,7α-Trihydroxycholanic	—	0.08	0.38
3α,6α,7β-Trihydroxycholanic	—	0.00	0.20
Cholic	0.00	0.08	0.18

Table continued

TABLE XV (*continued*)

Bile acid	R_F value[a]		
Silica Gel-Impregnated Glass Fiber Paper (ITLC-SG)	11	12	13
Free bile acids			
Lithocholic[b]	0.98	Solvent Front	0.92
Chenodeoxycholic	0.96	Solvent Front	0.80
Deoxycholic	0.98	Solvent Front	0.78
Cholic	0.98	Solvent Front	0.33
Conjugated bile acids			
Glycolithocholic	0.94	Solvent Front	0.82
Glycochenodeoxycholic	0.68	Solvent Front	0.61
Glycodeoxycholic	0.73	Solvent Front	0.56
Glycocholic	0.43	0.89	0.24
Taurolithocholic	0.24	0.86	0.88
Taurochenodeoxycholic	0.09	0.55	0.76
Taurodeoxycholic	0.09	0.56	0.75
Taurocholic	0.03	0.19	0.43
Lecithin[c]	0.03	0.20	0.12

[a] Solvent: (1) Benzene–ethyl alcohol (200:1); (2) benzene–ethyl alcohol (200:3); (3) benzene–ethyl alcohol (200:5); (4) benzene–ethyl alcohol (100:1); (5) benzene–ethyl alcohol (200:7); (6) benzene–ethyl alcohol (100:5); (7) chloroform; (8) benzene–ethyl alcohol (100:5); (9) benzene–ethyl alcohol (100:9); (10) benzene–ethyl alcohol (100:11); (11) isooctane–isopropyl ether–glacial acetic acid–isopropyl alcohol (1:1:1:1 v/v). This system is suitable for the separation of glycine-conjugated bile acids. The free bile acids can be separated with the same solvent system in the proportions of 10:1:1:1; (12) chloroform–isopropyl ether–glacial acetic acid–isopropyl alcohol (3:2:2:2). This system is suitable for the separation of taurine-conjugated bile acids; (13) chloroform–isopropyl alcohol–ammonium hydroxide (20:25:1). This basic solvent system is suitable for the separation of both, glycine- and taurine-conjugated bile acids and lecithin.

[b] The corresponding cholanoic acid derivatives are: chenodeoxycholic, 3α, 7 dihydroxy; cholic, 3α, 7α, 12α-trihydroxy; deoxycholic, 3α, 12α-dihydroxy; lithocholic, 3α-hydroxy.

[c] Lecithin, i.e., phosphatidyl choline.

separated, as are cis–trans isomers of hyocholic acid. The table also shows data for bile acids developed on Gelman ITLC-SG silica gel glass fiber paper with three new solvent systems: an acidic solvent suitable for separating glycine-conjugated bile acids, a polar modification of this solvent for taurine conjugates, and a basic solvent for glycine and taurine-conjugated bile acids, and lecithin (*53*, *84*). R_F values in various systems are listed in Block *et al.* (*6*).

C. DETECTION (see also Table XIV)

1. *Phosphomolybdic acid*—10% solution in ethanol; heat at 90°C for a few minutes after spraying.
2. *$SbCl_3$*—50% solution in glacial acetic acid or a saturated solution in $CHCl_3$.
3. *H_2SO_4*—10% solution in acetic anhydride (Lieberman's reaction); or spray with concentrated H_2SO_4 and heat gently for detection on glass paper.
4. *H_3PO_4*—85% H_3PO_4–H_2O–95% ethanol (85:25:25 v/v); heat at 90°C for 20 minutes after spraying.
5. *Aqueous I_2*—cholic acid is detected as a blue zone.
6. *$FeCl_3$*—1% solution in isopropyl ether detects hydroxamate derivatives of bile acids; a 10% solution in 0.2 *N* HCl can also be used.

D. QUANTITATIVE ESTIMATION (*85*)

After chromatographing the bile acids on Whatman 3 MM paper by the method of Sjövall (*86*), guide strips are visualized with phosphomolybdic acid (see above). The undeveloped areas containing the bile acids are eluted with ethanol, and the extracts are evaporated. To the residues, 65% H_2SO_4 is added, and the solutions are heated for various lengths of time. Optical densities are determined at the maxima shown in the tabulation below. Labeled bile acids have been determined by autoradiography (*87*).

Bile acid	Treatment (minutes at 60°C)	Maximum absorption (mμ)
Tauro- and glycodeoxycholic	60	390
Deoxycholic	15	385
Cholic, taurocholic, and glycocholic	15 or 60	320, 389

III. Cardiac Glycosides and Aglycones (Cardenolides, Bufadienolides)

A. PREPARATION OF SAMPLE

As an illustration, the extraction of cardiac aglycones from *Strophanthus* seeds is described (*6*). One gram of the seed is ground in a coffee mill and then extracted in a Soxhlet apparatus with 150 ml methanol for 2 hours. The extract is then evaporated to 25 ml, diluted with 5 ml water, and extracted three times with 25 ml light petroleum. The aqueous solution is then diluted with 30 ml 0.1 *N* H_2SO_4 and refluxed for 30 minutes. After cooling, the solution is extracted twice with 50 ml portions of $CHCl_3$ and the extract is evaporated to 5 ml. One drop of this extract is usually sufficient for paper chromatography. Other plant juice material may be extracted with a mixture of chloroform–ethanol (3:1), and portions spotted on Schleicher and Schuell Paper No. 2043b (*88*).

For the extraction of bufadienolides from the venom of the parotid glands of *Bufo paracnemis* Lutz 1925, the viscous mass collected by pressure is dried in a desiccator over P_2O_5. The aqueous suspension of this material is repeatedly extracted with $CHCl_3$–methanol (9:1 v/v), and the organic layers are concentrated and dried (*89*).

B. PAPER

Whatman Paper No. 3 MM may be impregnated with ethylene glycol by passing it through a 30% methanolic solution. The paper may also be saturated with propylene glycol as the stationary phase (*90*).

Whatman Paper No. 1 is saturated with formamide. The paper may also be impregnated with silica gel by passing it first through a 20% waterglass solution, then rinsing it in 3 *N* HCl for 24 hours, and drying the paper for 5 hours at 110°C (*91*).

The paper may be dipped into the aqueous phase of the solvent used for development (e.g., water saturated with butanol), dried for 10 minutes, and blotted between two sheets of filter paper with the aid of a rubber roller (*92*).

C. SOLVENTS (see also Tables XVI–XX)

1. *n-Butanol saturated with water* has been used for the separation of highly polar cardiac glycosides and aglycones (*92–94*). The butanol is refluxed for 2 hours with CaO and is then distilled.
2. *n-Butanol–acetic acid–H_2O* (4:1:5 v/v, upper phase) is used for the separation of saponins (*95*).
3. *Petroleum ether (35–60°C)–toluene–ethanol–H_2O* of different composition has been used to separate sapogenins from sapogenin acetates (*96*).
4. *Chloroform–methanol–H_2O* (10:2:5) is used for the separation of glycosides and aglycones (*97*).
5. *Benzene–chloroform* (75:25) (formamide paper) has been used for derivatives of digitoxigenin (*91*).
6. *Chloroform–tetrahydrofuran–methylglycol–formamide* (30:45:1.5:9) has been used for the glycosides from *Digitalis purpurea* (*98*).
7. *Xylene–methyl ethyl ketone* (v/v) has been used for the glycosides (*88*).

TABLE XVI

R_F VALUES OF CARDIAC GLYCOSIDES AND AGLYCONES (*1*)

Compound	R_F [a]								
	1	2	3	4	5	6	7	8	9
Digitalinum verum	—	—	—	—	0.95	F	—	—	0.16
Gitorin	—	—	—	—	0.78	F	—	—	—
Digilanid C	0.08	0.12	0.15	—	0.86	F	0.25	—	0.33
Digilanid B	0.14	0.18	0.25	—	0.66	F	0.10	—	0.47
Digilanid A	0.37	0.46	0.49	—	0.28	F	0.03	—	0.85
Deacetyldigilanid C	0.01	0.01	0.02	—	—	—	—	—	0.13
Deacetyldigilanid B	0.02	0.03	0.05	—	0.73	F	—	—	0.19
Deacetyldigilanid A	0.07	0.10	0.15	—	0.39	F	—	—	0.50
Strospeside	—	—	—	—	—	—	—	0.75	0.61
Digoxin	0.68	0.75	0.76	—	0.29	0.72	—	0.16	—
Digoxigenin	0.58	0.64	0.61	—	—	—	—	0.32	—
Gitoxin	0.76	0.81	0.82	0.40	0.18	0.55	—	0.29	—
Gitoxigenin	0.82	0.84	0.84	0.80	0.17	0.52	—	0.42	—
Acetyldigoxin-α	—	—	—	—	—	—	—	0.39	—
Acetyldigoxin-β	—	—	—	—	—	—	—	0.45	—
Acetylgitoxin-α	—	—	—	—	0.08	0.41	—	0.59	—
Acetylgitoxin-β	—	—	—	—	—	—	—	0.64	—
Digitoxin	0.88	0.91	0.90	0.68	0.03	0.27	—	0.69	—
Digitoxigenin	0.92	0.93	0.92	0.91	0.03	—	—	0.81	—
Acetyldigitoxin-α and -β	—	—	—	—	0.02	0.21	—	0.90–0.95	—
Oleandrin	—	—	—	—	—	—	—	0.93	—
Oleandrigenin	—	—	—	—	—	—	—	0.72	—
Nervifolin	—	—	—	—	—	—	—	0.55	—
Cymarin	—	—	—	—	—	—	—	0.55	—
Allocymarin	—	—	—	—	—	—	—	0.48	—
Strophanthidin	—	—	—	—	—	—	—	0.40	—
Allostrophanthidin	—	—	—	—	—	—	—	0.32	—
Isostrophanthidin	—	—	—	—	—	—	—	0.76	—
Strophanthidol	—	—	—	—	—	—	—	0.26	—
Cymarol	—	—	—	—	—	—	—	0.36	—
Emicymarin	—	—	—	—	—	—	—	0.11	—
Alloemicymarin	—	—	—	—	—	—	—	0.07	—

[a] Solvents: (1–3) mixtures of $CHCl_3$–methanol–H_2O; (4) benzene–$CHCl_3$ (2:3 v/v), formanide-treated paper; (5) octanol–pentanol–H_2O–formanide (5:6:8:2 v/v), paper impregnated with the organic phase and developed with the aqueous phase; (6) same as (5), 6:2:1:4 v/v; (7) same as (5), pentanol–H_2O (1:1 v/v); (8) isobutanol–H_2O; (9) butanol–pyridine–H_2O (3:1:3 v/v).

TABLE XVII

DISTANCES OF MIGRATION OF CARDIAC GLYCOSIDES, AGLYCONES, AND THEIR ACETATES (*96*)

Substance	Distance of migration (cm)			
	Benzene[a]		Toluene[b]	
	24 hours	48 hours	24 hours	48 hours
Periplogenin	2.8	4.2	0.7	2.7
Periplogenin acetate	21.6	43.6	6.9	16.9
Alloperiplogenin acetate	15.0	32.0	6.6	—
Alloperiplocymarin	11.8	28.9	2.5	7.1
Periplocymarin	16.2	38.9	3.8	10.0
Emicymarin	0.8	1.1	0.7	3.0
Emicymarin acetate	—	—	13.6	23.4
Alloemicymarin	1.0	1.4	1.0	2.5
Cymarol	2.5	3.6	1.0	3.0
Strophanthidol	2.0	2.6	0.6	1.4
Strophanthidol acetate	8.3	15.8	4.7	12.0
Strophanthidin acetate	4.1	6.1	2.8	7.1
Cymarin	4.0	6.2	2.0	5.1
Cymarin acetate	—	—	15.1	—
Sarmentocymarin	3.3	4.1	0.8	3.1
Sarmentogenin	2.5	2.5	1.0	2.0
Sarmentogenin acetate	—	—	15.5	25.3
Sarmentoside A	2.3	2.3	0.7	1.3

[a] Paper treated with formamide; solvent saturated with the stationary phase.
[b] Paper treated with propylene glycol; solvent saturated with the stationary phase.

8. *Chloroform saturated with ethylene glycol* is another useful solvent for the glycosides. The papers, previously impregnated with ethylene glycol (see above) are saturated with chloroform overnight before development (*99*). The tendency of this solvent to separate into two phases is overcome by placing a dry sheet of Whatman paper No. 3 MM on top of the chromatogram. *Toluene*, sautrated with propylene glycol, has been used for the glycosides and aglycones (*90*).
9. *Formamide–benzene, formamide–*$CHCl_3$, *and propylene glycol–*H_2O *(4:1)/benzene–*$CHCl_3$ *(1:1)* (paper impregnated with the stationary phase previously diluted with acetone 1:5 and partially dried at 70°C before development) have been used for the chromatography of bufadienolides (*89*).

TABLE XVIII

R_F VALUES AND COLOR REACTIONS OF SOME CARDIAC GLYCOSIDES IN THE SYSTEM METHANOL–CHLOROFORM–WATER (10:10:6) (*101*)

Substance	R_F	Color reaction: Anisaldehyde reagent[a]: Daylight	Anisaldehyde reagent[a]: Ultraviolet light	Zimmerman reagent (daylight)	Legal reagent[b] (daylight)
Diginin	0.24	Green blue	Dark blue	Blue	Red
Digitoxin	0.51	Green blue	Dark blue	Blue	Red
Gitoxin	0.70	Green blue	Light blue	Blue	Red
Saponin	0.77	Yellow	Dark violet	—	—
Gitoxigenin	0.84	—	Light blue	Blue	Red
Purpurea glycoside A	0.84	Green blue	Light blue[c]	Blue	Red
Purpurea glycoside B	0.93	Green blue	Light blue	Blue	Red

[a] Anisaldehyde reagent = Anisaldehyde–H_2SO_4–glacial acetic acid (1:2:100) followed by heating 5 minutes at 90°C.

[b] Legal reagent = sodium nitroprusside.

[c] The dark blue fluorescence of purpurea glycoside A is masked by the light blue fluorescence of gitoxigenin, which has the same R_F value.

TABLE XIX

R_F VALUES OF SOME AGLYCONES AND DERIVATIVES IN THE BUSH SYSTEM B_5 (*102*)[a]

Substance	R_F
"β"-Anhydrodigitoxigenin	0.91
"α"-Anhydrodigitoxigenin	0.63
"β"-Anhydrodigoxigenin	0.73
"β"-Anhydrodigoxigenone	0.59
"β"-Anhydrosarmentogenin	0.69
"β"-Anhydrouzarigenin	0.94
Digitoxigenin	0.76
Digitoxigenin acetate	0.89
Digoxigenin	0.075
Digoxigenin diacetate	0.70
Digoxigenone	0.68
Gitoxigenin	0.26
Gitoxigenin diacetate	0.88
Periplogenin	0.43
Sarmentogenin	0.10
Sarmentogenin diacetate	0.85
Sarnovide mono(?)acetate	0.72
Sarverogenin	0.40
Strophanthidin	0.16
Strophanthidol	0.05
Uzarigenin	0.83

[a] Solvent B_5 = benzene–methanol–H_2O (10:5:5 v/v).

TABLE XX

MIGRATION OF CARDIAC GLYCOSIDES AND AGLYCONES (*6*)[a]

	Migration (cm)			R_F			
Compound	1	2	3	4	5	6	7
Strophanthidin	—	—	13.6	—	—	—	—
Sarverogenin	—	—	32.4	—	—	—	—
Sarmentogenin	—	—	8.0	—	—	—	—
Strophanthidol	—	—	14.0	—	—	—	—
Sarmentocymarin	4.1	7.5	>37	—	—	—	—
Sarverosid	8.5	17	—	—	—	—	—
Cymarol	3.7	8	—	—	—	—	—
Cymarin	5.5	11	—	—	—	—	—
Emicymarin	1.5	1.5	—	—	—	—	—
Digitalinum verum	—	—	—	0.17	—	—	—
Gitoxigenin glycoside	—	—	—	0.24	—	—	—
Gitoxin	—	—	—	0.29	0.29	—	0.80
Glucogitaloxin	—	—	—	0.53	—	—	—
Strospesid	—	—	—	0.71	—	—	—
Verodoxin	—	—	—	0.81	—	—	—
Digitoxin	—	—	—	—	0.69	0.22	0.95
Gitaloxin	—	—	—	—	0.55	—	—
Digitoxigenin	—	—	—	—	0.81	0.50	0.90
Gitoxigenin	—	—	—	—	0.42	—	0.55
Digitoxigenin glycoside	—	—	—	—	—	0.42	—
Digitoxigenin acetate	—	—	—	—	—	0.91	—

[a] Solvents: (1) benzene, saturated with formamide (formamide paper); (2) benzene–chloroform (9:1), saturated with formamide; (3) chloroform, saturated with formamide (*97*); (4) chloroform–tetrahydrofuran–methylglycol–formamide (30:45:1.5:9) (*95*); (5) xylene–methyl ethyl ketone (v/v), saturated with formamide (ascending development) (*85*); (6) benzene–chloroform (75:25) (formamide paper) (*88*); (7) chloroform, saturated with ethylene glycol (*96*).

10. *Benzene–ethyl acetate–formamide* (30:69:1 v/v) (formamide–impregnated paer) was used to separate cardenolides containing 2-deoxy sugars which were extracted with methanol from seeds of *C. acutangulus* (*89a*).

D. TECHNIQUES USED (*92*)

The extracted glycosides are dissolved in ethanol and spotted on Whatman No. 1 paper which has been treated with water saturated with *n*-butanol. The paper is equilibrated in the tank with both separated phases overnight. The chromatogram is then developed by the descending technique. Since the solvent front is not easily visible, a small amount of dye (1-amino-2-methylanthraquinone) is added as marker, having an R_F of 1. After 11–16 hours at 14°C, the development has been completed.

E. Detection of Spots (see also Table XVIII)

1. *Alkaline m-Dinitrobenzene*

After the separation of cardiac glycosides, developed with a benzene solvent on formamide-treated paper at 15°C for 24 hours, the paper is dried at 60°C. It is then sprayed with a 10% solution of *m*-dinitrobenzene in benzene and placed in an oven at 60°C. After drying, the paper is sprayed with methanolic NaOH (6 gm NaOH dissolved in 25 ml water, diluted with 45 ml methanol). Purple spots, changing to blue, denote the position of the cardiac glycosides. Since the spots fade within 10 minutes, the positions must be marked immediately (*106*).

2. *3,5-Dinitrobenzoic Acid*

For the detection of cardiac aglycones, a 1% solution of 3,5-dinitrobenzoic acid in 0.5 *N* 50% v/v aqueous methanolic KOH is very suitable. This reagent gives a purple color which has the advantage of being reasonably permament (*6*).

3. *Trichloroacetic Acid*

After separation of cardiac glycosides, aglycones, and acetates, developed with benzene or toluene on propylene glycol- or formamide-treated paper, the dried chromatograms are sprayed with a 25% solution of trichloroacetic acid in chloroform. The sprayed sheets are hung in a drying oven at 120°C for 20 minutes and then inspected in the dark under an ultraviolet lamp (*96*).

4. *Blood*

For the detection of steroid sapogenins the chromatogram is sprayed evenly with whole blood or a blood cell suspension in saline. The sapogenins appears as lighter or darker spots (depending on their concentration) on a tan background (*96*).

5. *Triphenyltetrazoliumchloride*

This reagent (0.2%) in 1 *N* NaOH detects ketotriol cardiac glycosides (*91*).

6. *Anisaldehyde–acetic acid–H_2SO_4 (0.5:50:1 v/v)*

This solution is sprayed on the dried paper which is heated at 90°C for several minutes (*101*).

7. *Sodium periodate* (2%)–$KMnO_4$ (*1*%) (*4:1*)

The solution is made up freshly in 2% Na_2CO_3 (pH adjusted to 7.2), sprayed on the paper, left for 5–6 minutes, and washed with running water. Brown spots on a white background indicate the location of saponins (*95*).

8. *Ultraviolet Absorption*

Glycosides and aglycones possessing a coumarin ring (e.g., bufotalin, hellebrigenin) exhibit a maximum absorption at 300 mμ. Using a mercury high-pressure lamp and suitable filters, the dry chromatogram may be placed on top of photocopy paper, covered with 3 mm thick Plexiglas and given an exposure time of 35 seconds at a focal distance of 100 cm. White spots on a black background result. The following filters have been used: (a) Plexiglas and Pyrex glass to filter out light below 300 mμ; (b) 300 gm $CoSO_4 \cdot 7H_2O$ and 500 gm of $NiSO_4 \cdot 7H_2O$ dissolved in 2 liters of water and poured into a 20 mm thick filter; (c) CG (Corning Glass #9863) red filter. Combination of all three filters gives good transmission between 300 and 350 mμ with a maximum at 325 mμ (*29*).

F. Mobility Data

Tables XVI–XX (*1*, *6*, *96*, *101*, *102*) give migration data for specific cardiac glycosides and aglycones in various systems. Other data is listed in Hais and Macek (*7*).

References

1. R. Neher, "Steroid Chromatography." Elsevier, Amsterdam, 1964.
2. I. E. Bush, "The Chromatography of Steroids." Pergamon Press, Oxford, 1961.
3. R. E. Oakey, *J. Chromatog.* **8**, 2 (1962).
4. F. M. Ganis, *in* "Lipid Chromatographic Analysis" (G. V. Martinetti, ed.), Vol. 2. Marcel Dekker, New York, 1969.
5. I. E. Bush, *Biochem. J.* **50**, 370 (1952).
6. R. J. Block, E. L. Durrum, and G. Zweig, "A Manual of Paper Chromatography and Paper Electrophoresis," 2nd. rev. ed. Academic Press, New York, 1958.
7. O. Siblíková, *in* "Paper Chromatography," (I. M. Hais and K. Macek, eds.), pp. 363–408. Academic Press, New York, 1963.
8. L. R. Axelrod, *J. Biol. Chem.* **208**, 579 (1954).
9. T. J. Benraad and P. W. C. Kloppenborg, *Clin. Chim. Acta* **12**, 565 (1965).
10. P. de Moor and C. Denef, *Arch. Ges. Physiol.* **295**, 56 (1967).
11. C. Denef, Thesis, Universitaire Boekhandel, Leuven (1966).
12. I. Gy. Fazekas and A. T. Fazekas, *Folia Endocrinol.* (*Pisa*) **18**, 94 (1965).
13. I. E. Bush and M. Willoughby, *Biochem. J.* **67**, 689 (1957).
14. I. E. Bush and V. B. Mahesh, *Biochem. J.* **71**, 705 (1959).

14a. B. Kliman and R. E. Peterson, *J. Biol. Chem.* **235**, 1639 (1960).

15. S. Katz, *Arch. Biochem. Biophys.* **91**, 54 (1960).
16. J. W. Copius Peereboom, J. B. Roos, and H. W. Bukes, *J. Chromatog.* **5**, 500 (1961).
17. L. R. Axelrod and J. E. Pulliam, *Arch. Biochem. Biophys.* **89**, 105 (1960).
18. R. W. H. Edwards and D. J. H. Trafford, *J. Chromatog.* **21**, 275 (1966).
19. D. Kupfer, E. Forchielli, M. Stylianov, and R. I. Dorfman, *J. Chromatog.* **4**, 500 (1960).
20. R. Stupnicki and E. Stupnicka, *J. Chromatog.* **9**, 235 (1962).
21. R. Neher, *J. Chromatog.* **1**, 205 (1958).
22. S. C. Pan, *J. Chromatog.* **8**, 449 (1962).
23. S. C. Pan, *J. Chromatog.* **9**, 81 (1962).
23a. B. Hoffmann, *J. Chromatog.* **34**, 269 (1968).
24. Z. Procházka, *in* "Laboratory Handbook of Chromatographic Methods" (O. Mikes, ed.), pp. 101–107. Van Nostrand, Princeton, New Jersey, 1961.
25. E. Heftmann, *in* "Chromatography" (E. Heftmann, ed.), 1st ed. pp. 488–491. Reinhold, New York, 1961.
26. E. von Arx and R. Neher, *Helv. Chim. Acta* **39**, 1664 (1956).
27. N. A. Drake, W. J. Haines, R. E. Knauff, and E. D. Nielson, *Anal. Chem.* **28**, 2036 (1956).
28. D. D. Perrin, *Nature* **178**, 1244 (1956).
29. R. Bernasconi, H. P. Sigg, and T. Reichstein, *Helv. Chim. Acta* **38**, 1767 (1955).
30. P. J. Ayres, S. A. Simpson, and J. F. Tait, *Biochem. J.* **65**, 647 (1957).
31. H. Silberman and R. H. Thorp, *J. Pharm. Pharmacol.* **5**, 438 (1953).
32. S. A. Simpson *et al.*, *Helv. Chim. Acta* **37**, 1163 (1954).
33. H. Silberman and R. H. Thorp, *J. Pharm. Pharmacol.* **9**, 476 (1957).
34. D. L. Berliner, O. V. Dominguez, and G. Westenskov, *Anal. Chem.* **29**, 1797 (1957).
35. E. Bojesen, *Scand. J. Clin & Lab. Invest.* **8**, 55 (1956).
36. M. N. Cayen and P. A. Anastassiadis, *Anal. Biochem.* **15**, 84 (1966).
37. A. Zaffaroni, *J. Am. Chem. Soc.* **72**, 3828 (1950).
38. K. H. Gabbay and C. Waterhouse, *J. Chromatog.* **11**, 241 (1963).
39. D. J. H. Trafford and R. W. H. Edwards, *Analyst* **93**, 453 (1968).
40. H. R. Roberts and R. S. Robison, *Nature* **194**, 88 (1962).
41. B. Gestetner, *J. Chromatog.* **13**, 259 (1964).
42. J. S. Matthews and M. de los Angeles Cervantes, *J. Chromatog.* **9**, 195 (1962).
43. R. W. H. Edwards, *Biochem. J.* **82**, 48P (1962); *J. Endocrinol.* **22**, XXVI (1961).
44. E. von Arx and R. Neher, *J. Chromatog.* **8**, 145 (1962).
45. W. R. Eberlein and A. M. Bongiovanni, *Arch. Biochem. Biophys.* **59**, 9 (1955).
46. D. M. Abelson, P. K. Bondy, and J. Piskorski, *J. Chromatog.* **5**, 332 (1961).
47. H. L. Kimball, N. L. McNiven, and S. Burstein, *J. Chromatog.* **21**, 532 (1966).
48. H. K. Kley, *J. Chromatog.* **42**, 563 (1969).
49. A. Zaffaroni, R. B. Burton, and E. H. Keutmann, *J. Biol. Chem.* **177**, 109 (1949).
50. E. Heftmann, *J. Am. Chem. Soc.* **73**, 851 (1951).
51. J. R. Pasqualini and M. F. Jayle, *J. Chromatog.* **5**, 336 (1961).
52. C. H. Edwards *et al.*, *J. Chromatog.* **11**, 349 (1963).
53. J. G. Hamilton and J. W. Dieckert, *Arch. Biochem. Biophys.* **82**, 203 and 212 (1959).
53a. A. H. Payne and M. Mason, *Anal. Biochem.* **26**, 460 and 463 (1968).
54. Č. Michalec, Z. Kolman, M. Šulc, and J. Meštan, *J. Chromatog.* **9** 237 (1962).
55. B. Kadis, C. Harris, and H. A. Salhanick, *J. Chromatog.* **7**, 430 (1962).
56. S. Katz and J. R. Broich, *J. Chromatog.* **6**, 514 (1961).
57. J. L. Swanson, J. C. Warren, C. Harris, and H. A. Salhanick, *J. Chromatog.* **8**, 359 (1962).

58. F. J. Ritter and J. Hartel, *J. Chromatog.* **1**, 461 (1958).
59. R. W. H. Edwards, *J. Chromatog.* **12**, 212 (1963).
60. D. B. Gower, *J. Chromatog.* **14**, 424 (1964).
61. W. M. Allen, S. J. Hayward, and A. Pinto, *J. Clin. Endocrinol.* **10**, 54 (1950).
62. L. Stárka and M. Prusíková, *J. Chromatog.* **2**, 304 (1959).
63. L. Stárka, *J. Chromatog.* **4**, 334 (1960).
63a. J. J. Schneider, *J. Chromatog.* **37**, 89 (1968).
64. I. E. Bush, *J. Chromatog.* **23**, 94 (1966).
65. K. Boyce and F. A. Jenner, *Clin. Chim. Acta* **16**, 451 (1967).
66. L. Kornel, *Anal. Chem.* **36**, 443 (1964).
67. J. Hartel, A. Bol Raap, and F. J. Ritter, *J. Chromatog.* **3**, 482 (1960).
68. A. L. Livinston, E. M. Bickoff, J. Guggolz, and C. R. Thompson, *Anal. Chem.* **32**, 1620 (1960).
69. H. J. M. Hansen, *J. Chromatog.* **35**, 129 (1968).
70. P. Kabasakalian and A. Basch, *Anal. Chem.* **32**, 458 (1960).
71. M. L. Lewbart and J. J. Schneider, *J. Org. Chem.* **33**, 1707 (1968).
72. J. J. Schneider and M. L. Lewbart, *J. Chromatog.* **35**, 287 (1968).
73. G. Tegeler and B. Schülke, *Arch. Exptl. Veterinäermed.* **21**, 777 (1967).
74. J. Churý, *J. Chromatog.* **27**, 329 (1967).
75. J. B. Roberts, I. E. Bush, and N. B. Gibree, *Anal. Biochem.* **23**, 378 (1968).
76. J. Beyreder and H. Rettenbacher-Daübner, *Monatsh. Chem.* **84**, 99 (1953); N. Watanabe, *J. Biochem.* (*Tokyo*) **46**, 681 (1959).
77. D. Kritchevsky and M. R. Kirk, *Arch. Biochem. Biophys.* **35**, 346 (1952); *J. Am. Chem. Soc.* **74**, 4484 and 4713 (1952).
78. J. Sjövall, *Acta Chem. Scand.* **6**, 1552 (1952).
79. K. Tanaka and K. Takeda, *J. Biochem.* (*Tokyo*) **39**, 333 (1952).
80. W. H. Elliott, *Biochem. J.* **62**, 427 and 433 (1956).
81. K. Shimizu, T. Usui, T. Fujioka, and K. Yamasaki, *J. Biochem.* (*Tokyo*) **45**, 13 (1958).
82. G. A. D. Haslewood and V. Wooten, *Biochem. J.* **47**, 584 (1950).
83. S. C. Pan, A. I. Laskin, and P. Principe, *J. Chromatog.* **8**, 32 (1962).
84. H. E. Gallo-Torres and J. G. Hamilton, *J. Chromatog. Sci* **7**, 513 (1969).
85. S. Ericksson and J. Sjövall, *Acta Chem. Scand.* **8**, 1099 (1954).
86. J. Sjövall, *Acta Chem. Scand.* **8**, 339 (1954).
87. M. D. Siperstein, F. M. Harold, I. L. Chaikoff, and W. G. Dauben, *J. Biol. Chem.* **210**, 181 (1954).
88. M. Wichtl, *Naturwissenschaften* **43**, 158 (1956).
89. R. Zelnik, L. M. Ziti, and C. V. Guimaraes, *J. Chromatog.* **15**, 9 (1964).
89a. E. V. Rao and D. V. Rao, *Indian J. Pharm.* **30**, 214 (1968).
90. E. Heftmann, P. Berner, A. L. Hayden, H. K. Miller, and E. Mosettig, *Arch. Biochem. Biophys.* **51**, 329 (1954).
91. F. Turba and C. Scholtissek, *Z. Physiol. Chem.* **296**, 87 (1954).
92. E. Schenker, A. Hunger, and T. Reichstein, *Helv. Chim. Acta* **37**, 680 (1954).
93. E. Habermann, W. Müeller, and A. Schreglmann, *Arzneimittel-Forsch.* **15**, 578 (1954).
94. C. H. Hassall and S. L. Martin, *J. Chem. Soc.* p. 2766 (1951).
95. N. L. Dutta, *Nature* **175**, 85 (1955).
96. E. Heftmann and A. J. Levant, *J. Biol. Chem.* **194**, 703 (1952); E. Heftmann and A. L. Hayden, *ibid.* **197**, 47 (1952).
97. M. Okada, A. Yamada, and K. Kometani, *J. Pharm. Soc. Japan* **72**, 930 (1952).

98. E. Haack, F. Kaiser, M. Grube, and H. Springler, *Naturwissenschaften* **43**, 301 (1956).
99. G. J. Rigby and D. M. Bellis, *Nature* **178**, 415 (1956).
100. O. Schindler and T. Reichstein, *Helv. Chim. Acta* **34**, 108 (1951).
101. C. Günzel and F. Weiss, *Z. Anal. Chem.*, **148**, 250 (1955); *Pharmazie* **10**, 725 (1955).
102. I. E. Bush and D. A. H. Taylor, *Biochem. J.* **52**, 643 (1952).

9

PURINES, PYRIMIDINES, AND RELATED COMPOUNDS

The separation and identification of purine and pyrimidine bases, nucleotides, and nucleosides have aided greatly in obtaining the proof of structure of nucleic acids. The nucleic acids are hydrolyzed (generally enzymatically), and the hydrolysis products are resolved by one- or two-dimensional chromatography. Once the individual degradation products are identified, attempts are made to deduce what the sequence of the nucleotide units was in the parent nucleic acid. The method is similar to that used to determine the sequence of amino acids in peptides for the determination of protein structures (cf. Chapter 5) but is not so successful. Nucleic acid constituents are also separated by ion-exchange column chromatography (*1*) and electrophoreisis. (*Nucleotide* is the term used for the monomer of which nucleic acids are composed. A nucleotide consists of a pyrimidine or purine base and a sugar esterified with phosphoric acid. A *nucleoside* is the phosphorus-free combination of the base and sugar.)

I. Preparation of Samples

A. Hydrolysis of Nucleic Acids

1. *Acid Hydrolysis*

Nucleic acids are conveniently hydrolyzed with 1 ml of 72% perchloric acid (for 50–100 mg dry sample) at 100°C for 60 minutes. The solution is diluted with an equal volume of water and applied directly to the paper. The bases of RNA and DNA are not quantitatively liberated by perchloric acid, some destruction of purines occurs and excessively charred extraneous material which interferes with the chromatography often results (*2*).

Hydrolysis with formic acid at 175°C for 30 minutes has been used (*3*) and is satisfactory for purified DNA (*2*). Formic acid causes interfering extraneous material to be formed when applied to tissue samples (*2*).

Purines are liberated from DNA by heating the sample for 1 hour at 100°C in enough HCl to obtain a pH of 2.8. Acids heated in 6 *N* HCl at 120°C for 2 hours yield pyrimidine bases with partial destruction of purines. Heating ribonucleic acids for 1 hour at 100°C in 1 *N* HCl liberates adenine, guanine, cytidylic acid, and uridylic acid (*3*) and splits part of the pyrimidine nucleotides to nucleosides.

Methods for the determination of RNA and DNA after the hydrolysis of small amounts of dried, defatted tissue employing trichloro- and trifluoro-acetic acids have been described in detail (*2*).

2. *Alkaline Hydrolysis*

The nucleic acids are hydrolyzed to nucleotides by boiling with 1 *N* NaOH for 18 hours. The hydrolyzate is applied to the paper but must be neutralized if saturated $(NH_4)_2SO_4$–H_2O-isopropanol (79:10:2 v/v) is to be employed as the solvent (*3*). An alternative temperature of 37°C has been used (*2*), as has 0.33 *M* KOH to avoid deamination of CMP to UMP (*4*). Hydrolysis with 1 *N* NaOH at 22°C for 24 hours yields mono-and polynucleotides (*5*). Under some weakly alkaline conditions, nucleosides are also formed, but enzymatic hydrolysis is recommended for the quantitative formation of nucleosides (*6*). $Cd(OH)_2$ has been used to hydrolyze yeast nucleic acid (see below) to nucleotides (*3*).

3. *Acid and Base Hydrolysis*

A sample of ribonucleic acid may be hydrolyzed by three methods, each yielding different decomposition products (*3*).

1. Ten milligrams of ribonucleic acid is digested with 0.1 ml of 72% w/w $HClO_4$ at 100°C for 1 hour. The hydrolyzate is made up to a volume of 1.0 ml and is chromatographed with 65% isopropanol–HCl (2 *N* with respect to HCl). The following R_F values are obtained: guanine, 0.33; adenine, 0.48; cytosine, 0.57; uracil, 0.73.

2. Twenty milligrams of the sample of ribonucleic acid is hydrolyzed for 1 hour with 1.0 ml of *N* HCl, and the hydrolyzate is again chromagraphed with acidic isopropanol. The following R_F values are obtained: guanine, 0.33; adenine, 0.48; cytidylic acid, 0.76; uridylic acid, 0.91.

3. Ten milligrams of ribonucleic acid is hydrolyzed with 1.0 ml of 0.3 *N* KOH. The digest is adjusted to pH 3.6 with 60% w/w $HClO_4$, and the precipitated $KClO_4$ is removed by centrifugation. The super-

natant solution containing ribonucleotides is analyzed by paper electrophoresis.

The isolation of deoxyribonucleic acids, the preparation and purification of yeast ribonucleic acid, and the extraction of purines from urine have been described in Block *et al.* (*3*).

B. Application of Samples

As mentioned above, hydrolyzates may be applied directly to the paper or after appropriate concentration or dilution. Up to 1 ml of extract has been applied as a streak (*7*). Reference compounds are generally dissolved in water, dilute acid, or buffer at a concentration of about 2 mg/ml, and test spots contain 1–10 μg(*7*). In one extensive study, 3 μl of 0.01 *M* solutions were applied (*8*), and in another, each spot contained 0.15 μmoles (*9*). Deoxyribonucleotides have been applied in the solid state by trapping 2–8 μg in a capillary melting-point tube, tapping it out onto the origin and adding 1–2 μl of water to impregnate the paper (*10*).

II. Detection and Identification

A. Ultraviolet Light

The major method for detecting purines, pyrimidines, and related compounds is based on their absorption of short-wave ultraviolet light. A fluorescent screen or a 0.005% fluorescein spray is useful for increasing the visual contrast and therefore the sensitivity of detection. Contact prints can be made on photographic paper for a permanent record (*3*).

Pyrimidine nucleosides and nucleotides, in particular compounds containing uracil or thymine (*10a*) as a base component, are extremely unstable to ultraviolet irradiation while on filter paper. Guanine derivatives decompose slowly, while adenine-containing compounds are most resistant (*9*).

B. Chemical Detection Methods

1. Hg(II) *Salts*

The dried chromatogram is immersed for 5 minutes in 0.2% eosin in absolute ethanol saturated with $HgCl_2$, washed with 95% ethanol until only faint pink, and dried. Under 254 or 365 mμ ultraviolet light, purine bases appear as dark spots (due to a purine–Hg complex) on a brightly fluorescent background. Quantities of 0.02–0.5 γ of purines are detected (*11*).

Treatment with 0.25 *M* $Hg(NO_3)_2$ in 0.5 *M* HNO_3 followed by washing out the excess with dilute HNO_3 and water and treatment with aqueous ammonium sulfide solution forms black spots of mercuric sulfide at the site of the purine–Hg complexes (*12*). For compounds whose mercury complexes are water soluble (e.g., xanthine, uric acid), the washing step is eliminated; the chromatograms are sprayed with 0.25% Hg(II) acetate in 95% ethanol acidified with a few drops of glacial acetic acid, then sprayed with 0.05% diphenylcarbazone in 95% ethanol and heated a short time at 90°C until the background color fades (*3*).

2. $AgNO_3$

Paper sprayed with 0.1% $AgNO_3$ and exposed to light turns brown. If the $AgNO_3$ solution is adjusted to pH 1.8 with formic acid (4 ml per 100 ml solution), purines are revealed as yellow, orange-yellow, or brownish yellow spots, while pyrimidines are not detected. At neutral pH (0.1% $AgNO_3$ solution, pH not adjusted), purines give white or yellow spots and pyrimidines mauve or gray-mauve spots. Amounts below 1 μg can be detected in many systems, and the bases can be recovered from the sprayed chromatograms for determination of absorption spectra by eluting with 0.1 *N* HCl overnight (*13*).

Nucleoside polyphosphates are also revealed with $AgNO_3$ (*14*) or $AgNO_3$–$Na_2Cr_2O_7$ (*15*) solutions.

3. *Diphenylamine*

The spray solution is composed of 1% diphenylamine in glacial acetic acid to which 2.75% (v/v) concentrated H_2SO_4 is added, and the whole is diluted with one-half of its volume of H_2O. After spraying, the paper is clamped between two glass plates and placed in an oven at 90°. Purine deoxyribosides and deoxyribotides give purple-blue colors in 5–10 minutes (*16*).

For pyrimidine nucleotides and nucleosides, the paper is first treated with Br_2 : The dried chromatogram is put on a glass plate and dabbed carefully with a piece of cotton-wool soaked in a solution of bromine–H_2O–acetic acid (1:50:10 v/v). The paper must not be excessively wetted or the zones will diffuse. The paper is placed between two glass plates at 100°C for 5 minutes and then treated with diphenylamine–H_2SO_4 reagent to produce blue spots. The sensitivity of the diphenylamine spray is increased about four times by this prior bromination (7–20 μg can be detected) (*17*).

4. Cl_2 *Oxidation*

Purines can be detected with a sensitivity of 0.2 μg by placing the chromatogram saturated with 50–100% water in a Cl_2 atmosphere for

3 minutes. After drying at 80–90°C, it is placed in an NH_3 atmosphere. Xanthine derivatives give a violet color and other purines a green color under ultraviolet light (*18*).

5. *Molybdic Acid Reagent*

Nucleotides are detected via phosphoric acid with this reagent, which is also useful for sugar phosphates (cf. Chapter 6)—5.0 ml of perchloric acid (specific gravity 1.61), 10.0 ml of 1 *N* HCl, 25 ml of 4% $(NH_4)_2MoO_4$, and 60 ml of water. The dried chromatogram is sprayed with the reagent, and after a preliminary drying in a current of warm air, it is heated for 7 minutes at 85°C. To develop the color of the spots more fully, the paper is allowed to regain moisture from the air and is exposed to hydrogen sulfide gas for 5–10 minutes. The recommended method for the formation of molybdenum blue is the exposure of the sprayed and dried chromatogram to ultraviolet illumination.

6. *Periodate and Aniline Phthalate*

Nucleotides and nucleosides with α-glycol groups are detected with periodate. Sugar components of nucleosides are detected with aniline hydrogen phthalate after hydrolysis on the paper (cf. Chapter 6).

7. *Cysteine–H_2SO_4 Reagent* (*Modified Dische Reagent*)

The reagent consists of a solution of 0.5 gm of cysteine hydrochloride in 100 ml of 3 *N* H_2SO_4. The air-dried chromatogram is sprayed with this solution, and after 5–10 minutes' heating at 85°C pink spots appear, indicating the position of deoxyribosides and deoxyribotides (*3*).

8. *Specific Reagent for Adenine Compounds* (*19*)

The dried chromatogram is sprayed with 0.07% aqueous $KMnO_4$, placed in an atmosphere of chlorine gas for 15 seconds, and dried at 100°C for 5 minutes. Yellow-orange spots result. When the paper is next sprayed with 3 *N* KOH, the spots turn red. The sensitivity for this reagent is 0.2 γ for adenine and 0.3 γ for bound adenine, e.g., AMP, ADP, ATP.

9. *1,2-Naphthoquinone-4-sulfonic Acid* (*Folin's Reagent*)

Two-tenths percent in half-saturated Na_2CO_3 forms violet spots with adenine and blue spots with guanine (*3*).

10. *Detection of Pyridine Nucleotides*

Oxidized pyridine nucleotides (DPN, TPN) are detected by treating the chromatogram with 1 *M* aqueous KCN. Ultraviolet illumination produces fluorescent spots of the pyridine nucleotides; adenine-containing compounds exhibit a quenching effect (*3*).

11. *Detection of Pyrimidine Reduction Products* (*3*)

The chromatogram is first sprayed with 0.5 *N* NaOH (for hydrolysis) and then with *p*-dimethyl-aminobenzaldehyde (DMAB) reagent (1 gm DMAB, 10 ml concentrated HCl, 100 ml ethanol). After 2–6 hours at room temperature, yellow spots are formed.

12. *Detection of Thiouracils*

Thiouracils are detected by one of two methods: (a) The chromatogram is exposed to iodine vapors and dark brown spots on a yellow background appear. (b) The chromatogram is sprayed first with an alcoholic solution of dichlorobenzoquinonechlorimide, followed by 0.1 *N* alkali (*3*). Orange-yellow spots are formed.

Additional detection methods are listed in the tables and figures in this chapter.

C. Enzymatic Detection

DPN is reduced to DPNH by applying a mixture of alcohol, alcohol dehydrogenase, and buffer (pH 9), to the paper. Strong fluorescence indicates the presence of DPNH. The presence of DPNH can be confirmed by oxidizing it enzymatically as above, but using pH 7.0 buffer. Blue quenching indicates the formation of DPN (*3*).

D. Bioautography

Adenine and guanine have been determined microbiologically (*20*). Deoxyribosides and ribotides have been determined against *Lactobacillus leichmannii* (*21*, *22*). See Chapter 13 for procedural details.

E. Methods Based on Radioactivity (see Chapter 4)

Compounds labeled with ^{32}P or ^{14}C can be detected by autoradiography. An isotope derivative method has been described in which the unknown sample, containing pyrimidines, is treated with pipsyl-^{131}I chloride (pipsyl = *p*-iodophenylsulfonyl), and the reaction mixture is diluted with 50% acetone. Another portion of the unknown sample is reacted with pipsyl-^{35}S chloride. Both solutions are extracted with ether after acidification, the extracts are concentrated to a small volume and are chromatographed. The dried strips are counted with and without Al filters. The ratio of the two counts should be constant for a pure spot or band (*3*).

F. Identification by Instrumental Analytical Methods

Spots can be eluted after detection and identified by their infrared or ultraviolet (220–300 mμ) absorption spectra. Pyrimidine and purine bases in amounts as low as 0.1 μg have been identified by mass spectrometry after chromatographic separation, location by ultraviolet light, and elution (*23*) (see Fig. 1).

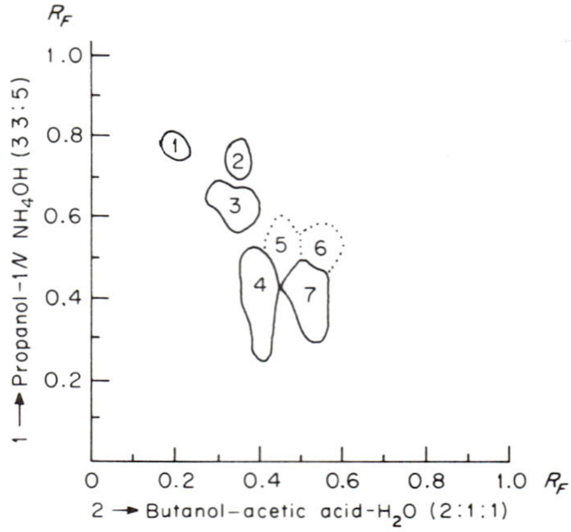

FIG. 1. Chromatogram of a mixture of eight bases. Detection in ultraviolet light and identification by mass analysis: (1) thymine, (2) adenine, (3) cytosine, 5-methylcytosine, uracil, hypoxanthine, (4) xanthine, (5) no significant mass spectra, (6) trace guanine, (7) guanine. After M. H. Studier *et al.* (*23*).

III. Methods and Techniques

One- and two-dimensional (*24*) descending chromatography in all-glass tanks is usually employed. Thermostated development has been recommended for some strongly acid solvents. Circular chromatography is only occasionally used (*25*).

The amount of sample applied is determined by the sensitivity of the detection method. Unless a strongly acidic solvent is to be used, HCl applied as part of the sample solution must be removed by a stream of air prior to development.

Multiple and continuous (overrun) development methods are also useful. Continuous ascending chromatography with *n*-butanol–acetone–5% NH_3–acetic acid–H_2O (9:3:2:2:4 v/v) separates microgram amounts of AMP, ADP, ATP, and uric acid (*26*).

Paper chromatography and electrophoresis were employed in combination for the isolation of eleven pure, labeled trinucleotides from the enzymatic hydrolyzate of RNA obtained from *Escherichia coli* and baker's yeast grown in a ^{32}P-phosphate containing medium (*27*).

Anion-exchange column chromatography, paper chromatography, and anion-exchange thin-layer chromatography were applied in combination to isolate a new nucleotide, adenisine 5′-tetraphosphate, in the acid-soluble fraction from horse skeletal muscle (*28*).

Free nucleotides from natural sources (plant materials) are purified and concentrated prior to paper chromatography by adsorption onto charcoal from acid solution, washing, and eluting with 50% aqueous ethanol containing 0.1% NH_3. The charcoal (e.g., Norit A, carboraffin, or Desorex) is boiled with dilute HCl and washed with water until chloride free before use (*29*).

Procedures for the removal of inorganic phosphate from inosine and adenosine phosphates prior to chromatography and for the separation of purines from pyrimidines are as follows (*3*). Whatman No. 1 filter paper (20 × 60 cm), washed with Versene and water prior to chromatography, is folded 38 cm from one end and the edges are fastened with paper clips. The samples to which phosphate-^{32}P is added as a marker are spotted 2–3 cm from the fold. The chromatogram is first developed for 3–4 hours by the ascending technique with isopropyl ether–90% HCOOH (9:6 v/v). The phosphates which have moved to the front are located by scanning with a radiation monitor and are cut off the paper horizontally. The remaining chromatogram is now developed with the second solvent, isobutyric acid–*N* NH_4OH–0.1 *M* Versene (100:60:1.6) by descending chromatography for 16–18 hours. The paper is dried at 80°C for 20 minutes, and the spots are detected by molybdate spray or ultraviolet illumination. The following R_F values are observed by this technique: AMP, 1.00; ADP, 0.80; ATP, 0.66; IMP, 0.53; IDP, 0.39; ITP, 0.26.

In a similar fashion, the pyrimidines can be separated from the purines by a preliminary 3-hour chromatography with water or 0.01 *M* phosphate buffer (pH 7.1) as solvent. The pyrimidines travel about twice as fast as the purines. The purines are located with an ultraviolet lamp and are cut off in a line parallel to the points of application. The filter paper is unrolled, and the pyrimidines are chromatographed in the usual manner.

The separation between pyrimidines and their deoxyribonucleosides can be improved if the mixture is acetylated prior to spotting and development. The alcoholic hydroxyl groups of the nucleosides react while the phenolic hydroxyls of the pyrimidines do not, resulting in more widely separated R_F values (*30*).

IV. Chromatographic Systems

Untreated standard chromatography paper (e.g., Whatman No. 1) is usually used with a wide variety of solvents [many were listed and described earlier in Block *et al.*, (*3*)] (see also Chapter 1, Fig. 1). For the separation of bases in nucleic acid hydrolyzate and nucleotides, the solvents listed in the following tabulation, have been recommended (*29*, *31*).

Solvent system[a]	Hours for run
Descending	
1. Morpholine–*n*-butanol–H_2O (1:1:1)	48
2. Isobutyric acid–H_2O–acetic acid (100:50:1)	22
3. *n*-Propyl alcohol–concentrated NH_4OH–H_2O (60:30:10)	15
4. Isopropyl alcohol–concentrated HCl (170:41)	8
5. Isopropyl alcohol–H_2O–28% NH_4OH (85:15:1.3)	15
6. Isopropyl alcohol–concentrated HCl–H_2O (170:40:39)	20
7. *n*-Butanol–H_2O (86:14)	15
8. *n*-Butanol–H_2O (86:14) in NH_3 atmosphere supplied by the addition of 5% NH_4OH inside the chamber	20
9. *n*-Butanol–ethanol–H_2O (50:15:35)	14
10. *n*-Butanol–formic acid–H_2O (77:10:13)	15
11. Isobutyric acid–0.5 *M* NH_4OH (5:3)	24
Ascending	
12. *n*-Butanol saturated with 10% aqueous solution of urea	15
13. 5% Na_2HPO_4 saturated with isoamyl alcohol	16
14. 5% ammonium citrate saturated with isoamyl alcohol at pH 9.3 and 3.6	27
15. 5% KH_2PO_4 saturated with isoamyl alcohol	16
16. Saturated $(NH_4)_2SO_4$–H_2O–isopropyl alcohol–0.1 *M* Versenate (58:39:2:1)	20
17. Isopropyl alcohol–acetic acid–H_2O (60:30:2)	20
18. *n*-Butanol saturated with H_2O in an NH_3 and CO_2 buffered vapor supplied by the addition of excess solid NH_4HCO_3 inside the bell jar	

[a] The numbers in parenthesis represent the composition in volume percentages. The pH is adjusted where necessary with concentrated NH_4OH.

Solvent 7 is very widely used for base separations. With ammonia in the atmosphere or the solvent, the sequence of various pairs is reversed. Solvent 6 carries various impurities found in hydrolyzate with the solvent front. Solvent 11 is good for the separation of higher phosphorylated nucleotides, and in combination with solvent 3, it gives excellent two-dimensional nucleotide separations. Ethanol–1 *M* ammonium acetate (pH 3.7) followed by isobutyric acid–0.5 *M* aqueous NH_3 (3.7) gives two-dimensional separations of diribonucleoside monophosphates of all possible base pairings (*31a*).

Nucleic acids have been applied to Schleicher and Schuell 2043b Mgl paper strips as a streak from edge to edge and development effected using a series of aqueous solvents for 15 minutes each (e.g., 1.0 *N*, 0.25 *N*, 0.1 *N*, and 0.01 *N* HCl and then 50% acetic acid) (*32*).

Ion-exchange papers have also been used and would seem to be ideal, considering the success of column ion-exchange separations (*1*). Paper loaded with weak acid cation-exchange resin (Amberlite WA-2) and developed with 0.1 *M* acetic acid adjusted to pH 4.8 with NH_4OH separates thymine, thymine dimer, and cytosine (*33*) and gives the following R_F values: 5′-UMP, 0.98; 5′-TMP, 0.88; 5′-GMP, 0.81; 5′-CMP, 0.77; uridine, 0.73; thymine dimer, 0.63–0.70; uracil, 0.56–0.59; thymidine, 0.53; thymine, 0.40–0.48; 5′-AMP, 0.37; guanosine, 0.36; cytidine, 0.33; cytosine, 0.15–0.17; guanine, 0.15; adenosine, 0.10; adenine, 0.05 (streaking zone).

Nucleoside phosphates are separated on diethylaminocellulose paper which is either untreated or coated with polyethylene (*33a*). Conventional filter paper impregnated with polyethylene (PEI paper) separates nucleoside diphosphate sugars and related nucleotides (*34*) (Fig. 2). Adenine nucleotides are separated on paper impregnated with $(NH_4)_2SO_4$, with 70% ethanol as the solvent (*34a*).

Figures 1–6 (*23*, *34–38*) and Tables I–XII (*7*, *8*, *10*, *36*, *37*, *39–46a*) indicate specific conditions for the separation and detection of various purines, pyrimidines, and related compounds. Tables of R_F values determined in the early 1950's were published in Block *et al.* (*3*). Further mobility data for nucleoside phosphates is given in Chapter 14 in the section on phosphorus compounds, and R_F values for thirty-one 2- and 6-substituted purines in six solvent systems along with eight reagents for spot detection are tabulated in Macek and Stanislarova (*47*), and R_F values for sixty-five nucleotides in isopropyl alcohol–concentrated NH_3–H_2O (7:1:2 v/v), ethanol–1 *M* ammonium acetate, pH 7.5 (7:3 v/v), and isobutyric acid–concentrated NH_3–H_2O, pH 3.7 (66:1:33 v/v) in Narang *et al.* (*47a*). A recent report indicates that chromatography of sugar nucleotides in solvents containing ammonium acetate may lead to degradation and artefact formation (*47b*).

V. Quantitative Methods

In almost every case, the spots are revealed in ultraviolet light after chromatography and then are marked, eluted, and the eluate determined by ultraviolet spectrophotometry (*2*, *7*, *26*, *31*). Blanks are obtained by eluting equal areas of paper free from analyzed substances near the spot. Molar extinction coefficients for use in subsequent calculations [the maximum extinction technique (*3*)] have been published (*3*). Direct densitometry or planimetry on contact photographs (*48*), or densitometry on chromatograms treated with molybdenum blue reagent (*3*) have also been used for quantitative analysis.

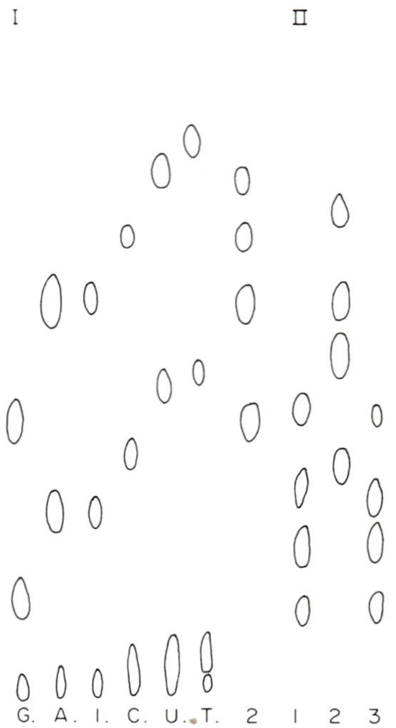

FIG. 2. Contact print photograph of nucleoside mono-, di-, and triphosphates and nucleoside diphosphate sugars after chromatography on PEI paper for 5 hours at 20°C with 0.3 *M* LiCl. Abbreviations: G = guanosine, A = adenosine, I = inosine, C = cytidine, U = uridine, and T = thymidine. (Part I) Separation of nucleoside diphosphate sugars from nucleoside mono-, di-, and triphosphates. Nucleoside diphosphate sugars have the greatest mobility. Nucleoside di- and triphosphates remain at the origin and monophosphates move between them and the nucleoside diphosphate sugars. Lane I_2 illustrates the separation of a mixture of UDP–Glc, CDP–Glc, ADP–Glc, and GDP–Glc. (Part II) Chromatography was performed with 0.3 *M* LiCl for 4 hours at 20°C. Reading down the spots are: lane II_1—TMP, dCMP, dAMP, and dGMP; lane II_2—UDP–Glc, CDP–Glc, ADP–Glc, and GDP–Glc; lane II_3—UMP, CMP, AMP, and GMP. Amounts spotted were always between 0.05 and 0.1 μmole. Mobilities increase in the order: guanosine, adenosine, cytidine, uridine, and thymidine nucleotides. The solvent front is at 35 cm from the origin. After H. Verachtert *et al.* (*34*).

Elution is usually performed with <1 ml of water or 0.1–1 *N* HCl. Phosphate buffer (*31*), pH 7, and 0.002% Li_2CO_3 (*26*) have also been employed. The amount of compound in an eluate is often determined by the differential extinction method of Vischer and Chargaff (*49*), described in detail in Block *et al.* (*3*). This consists of taking two optical density readings, one at a wavelength giving the maximum reading and another (e.g., 290 mμ) yielding a low value. Readings are made on the eluate and a standard solution containing 10 μg/ml. To calculate the quantity, the difference between the maximal and lower readings of the

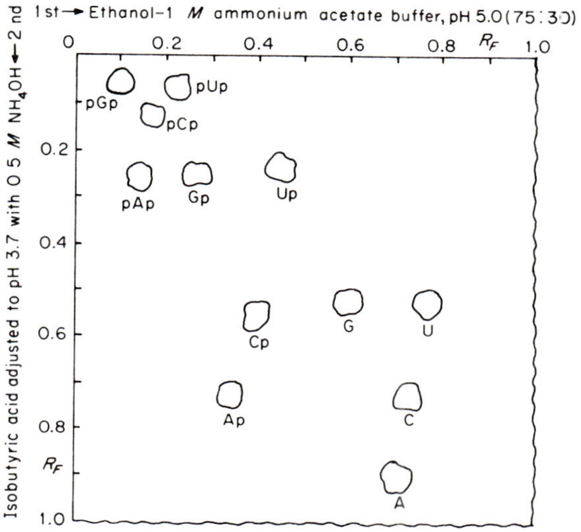

FIG. 3. Separation of ribonucleosides, ribonucleoside 3′(2′)-phosphates, and ribonucleoside 3′(2′)-diphosphates. Descending development; detection by ultraviolet light. After J.-I. Mukai (*35*).

eluate is multiplied by 10 (the value of the standard), and this product is divided by the difference in extinction for the standard at the two wavelengths.

Neutron activation of mononucleotide phosphorus with subsequent β-counting of the induced ^{32}P activity has been used to determine base ratios of very small samples of RNA (*4*).

VI. Applications

As examples, paper chromatography has been used recently in conjunction with the following studies: the depolymerization of RNA by Pb(II) ion (*50*), investigation of the metabolism of adenosine-$^{14}C_{10}$ and AMP-$^{14}C_{10}$ blood platelets (*51*), comparison of the *in vivo* methylated components of ribosomal RNA from *E. coli* and *Saccharomyces cerevisiae* (*52*), study of the chemical reactions between DNA and H_2O_2 or NH_2OH (*53*), synthesis of ^{32}P-labeled ribonucleoside 5′-monophosphates (*54*), discovery of the first specific inhibitor of dihydroorotate dehydrogenase (*55*), the occurrence of UDP-apiose in parsley (*56*), synthesis of 5-azacytosine and its methyl derivatives (*57*), analysis of nucleoside- and nucleotide-containing pharmaceutical preparations (*58*), the large-scale fractionation of di- and trinucleotides obtained from pancreatic ribonuclease digests (*59*), and a rapid assay for UDP-glucose-glycogen glucosyltransferase (*60*).

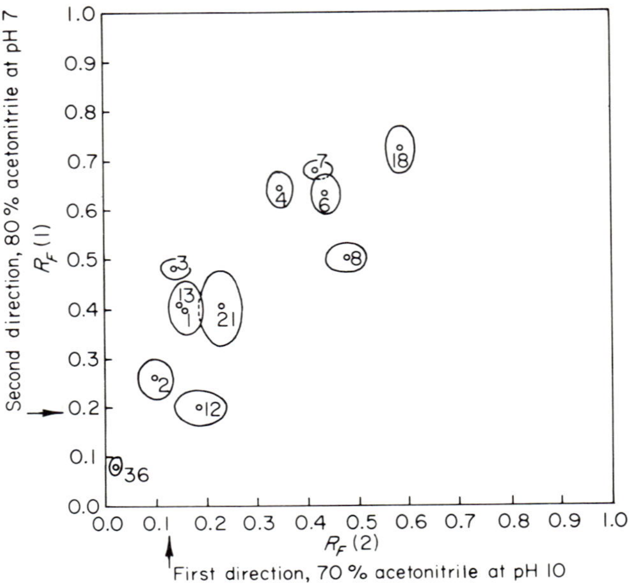

FIG. 4. Two-dimensional chromatogram of selected purines and pyrimidines using 70% acetonitrile–water at pH 10 for the first dimension, and 80% acetonitrile–pH 6.86 sodium phosphate buffer as the second dimension. Paper sprayed with water adjusted to pH 10 and dried in air before the first run. Detection in ultraviolet light (purple spots on a light blue background); (1) adenine, (2) guanine, (3) cytosine, (4) uracil, (6) thymine, (7) 6-methyluracil, (8) 5-nitrouracil, (12) xanthine, (13) hypoxanthine, (18) thiouracil, (21) isocytosine, (36) deoxycytocytidylic acid. After B. A. Berger and C. E. Hedrick (*36*).

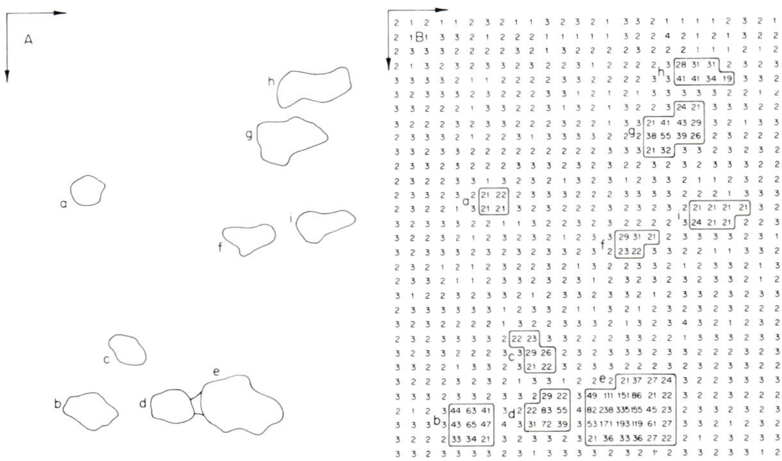

FIG. 5. Autoradiograph (A) and numerical map (B) of the activity of a two-dimensional radiochromatogram of an aqueous alcoholic extract of rat diaphragm incubated with adenine-8-^{14}C (the chromatogram was developed: horizontally in solvent 6 (8 hours), vertically in solvent 4 (30 hours, solvent allowed to run off paper). (a) Uric acid, (b) adenine, (c) xanthine, (d) hypoxanthine, (e) inosine, (f) AMP, (g) ADP, (h) ATP, (i) IMP. See Table IX for definition of the solvents. After L. Rossi and C. Rossi (*37*).

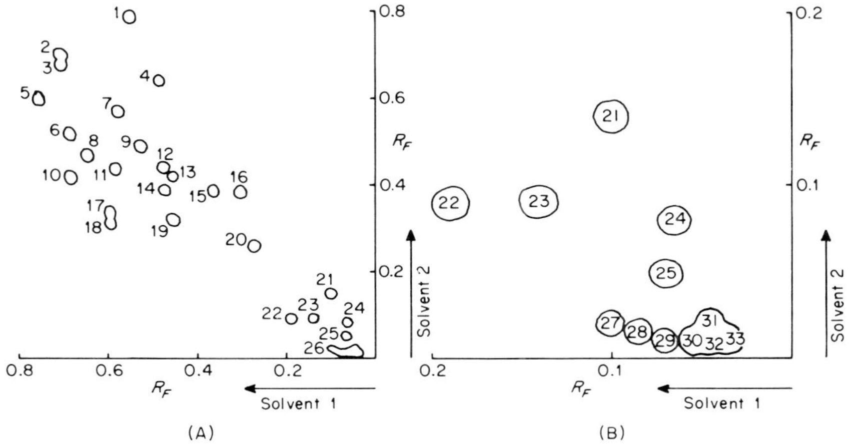

FIG. 6. (A) Separation obtained on a two-dimensional descending chromatogram with the solvents allowed to reach the edge of the chromatogram; (1) β-ureidoisobutyrate, (2) thymine, (3) thymidine, (4) β-ureidopropionate, (5) dihydrothymine, (6) dihydrouracil, (7) uracil, (8) cytosine, (9) β-aminoisobutyrate, (10) deoxyadenosine, (11) adenine, (12) hypoxanthine, (13) β-alanine, (14) uridine, (15) xanthine, (16) orotic acid, (17) cytidine, (18) adenosine, (19) inosine, (20) thymidine monophosphate, (21) uridine monophosphate, (22) adenosine monophosphate, (23) cytidine monophosphate, (24) guanosine monophosphate, (25) uridine diphosphate, (26) adenosine diphosphate, cytidine diphosphate, adenosine triphosphate, cytidine triphosphate, guanosine diphosphate, uridine triphosphate, and guanosine triphosphate. Solvent 1: *n*-propanol–NH_3 (specific gravity 0.88)–H_2O (3:1:1 v/v); solvent 2: upper phase of *tert*-amyl alcohol (technical)–90% formic acid–H_2O (3:1:3 v/v). Whatman No. 1 paper (42 × 42 cm). Knowns applied in 2 μl spots of 0.05 *M* solutions. Paper equilibrated for 12 hours before the run. Development times 15 and 18 hours in solvents 1 and 2, respectively. Detection by ultraviolet light. (B) Distribution of nucleotides on a two-dimensional chromatogram after prolonged separation (48 hours) in each solvent. Spots 21–25 are as in (A); (27) adenosine diphosphate, (28) cytidine diphosphate, (29) adenosine triphosphate, (30) cytidine triphosphate, (31) guanosine diphosphate, (32) uridine triphosphate, and (33) guanosine triphosphate. The solvent front reaches the bottom of the sheet where it is absorbed by a pad of filter paper stapled to the bottom of the paper. After J. W. Bradbeer and B. C. Jarvis (*38*).

TABLE I

CHROMATOGRAPHIC DATA FOR PURINES, PYRIMIDINES, AND DERIVATIVES ARRANGED ACCORDING TO INCREASING R_F VALUE IN THE FIRST SOLVENT SYSTEM[a] (*8*)

Compound	R_F values ($\times$ 100) in various solvent systems[b]								
	1	2	3	4	5	6	7	8	9
Uridine 5′-diphosphate	0	0	7	12	66	0	0	0	0
Adenosine 5′-triphosphate	0	4	6	13	39	0	0	0	0
Uridine diphosphate glucose	0	6	8	16	69	2	0	0	0
Adenosine 5′-diphosphate	2	4	6	12	40	0	0	0	0
Uridine 5′-triphosphate	3	3	5	15	63	0	0	0	0
2,5-Diamino-4,6-deoxypyrimidine	3	3	8	14	—	2	0	0	0
Guanosine 5′-phosphate	5	4	17	22	39	2	0	0	0
S-Adenosylmethionine	5	35	17	14	7	8	1	0	0
Cytidine 5′-phosphate	7	6	22	20	50	4	2	0	0
Orotidine monophosphate	8	2	12	12	74	1	1	0	0
Adenosine 5′-phosphate	8	8	27	25	42	5	3	0	0
Cytosine 5-carboxylic acid	9	23	—	22	26	8	19	0	3
Uridine 5′-phosphate	12	4	23	24	70	6	5	0	0
2,6-Diamino-5-formylamino-4-hydroxypyrimidine	12	10	30	20	19	11	9	0	1
Uric acid	14	18	30	27	22	5	7	4	12
5-Aminouridine	17	23	35	42	24	25	7	0	14
Deoxycytidine 5′-phosphate	15	7	29	22	64	4	6	0	0
6-Amino-2,8-dihydroxypurine	15	9	31	23	33	—	4,17	0	5
6-Amino-5-formylaminouracil	15	16	31	26	22	16	12	0	6
Orotidine	17	17	30	20	66	7	12	0	0
Xanthosine	18	32	34	28	26	13	12	0	10
5-Methyldeoxycytidine 5′-phosphate	19	7	36	27	65	8	10	0	0
2-Amino-8-hydroxypurine	19	24	28	37	21	24	8	0	6

Guanosine	20	27	36	44	29	32	11	0	11
Deoxyuridine 5′-phosphate	21	4	38	17	83	6	12	0	0
6-Amino-5-formylamino-4-hydroxypyrimidine	21	22	40	30	19	18	21	0	7
5-Ribosyluracil	21	25	35	43	53	34	12	2	0
8-Hydroxy-7-methylguanine	22	15	46	43	43	35	23	4	11
6-Amino-5-formylamino-1-methyluracil	22	27	42	35	27	19	22	1	7
3-Methyluric acid	22	28	41	38	36	24	19	3	18
6,8-Dihydroxypurine	23	18	38	42	41	36	20	2	19
Guanine	23	19	45	42	22	36	13	0	11
Thymidine 5′-phosphate	24	5	36	27	88	6	13	0	0
Xanthine	24	27	45	46	25	41	21	0	22
5-Hydroxymethyluridine	24	31	38	48	59	32	13	0	19
Inosine	24	36	39	48	30	35	14	0	14
1-Methyluric acid	25	28	42 ±	42	—	27	20	3	30
5-Aminouracil	25	31	41	46	21	36	12	5	22
7-Methylinosine	26	Unstable	38	29	31	15	13	0	4
5-Hydroxymethylorotic acid	26	22	32	30	53	15	19	0	5
7-Methyluric acid	26	25	49	37	54	22	20	3	22
5-Hydroxymethyldeoxycytidine	26	56	53	44	54	34	—	—	—
2,6-Diamino-7-methylpurine	28	30	42,52	33	14	16	15	4,24	0
7-Methylguanosine	28	Unstable	45	28	40	16	15	0	5
5-Amino-6-carboxyuracil	29	16	43	25	24	10	27	11	10
6,8-Dihydroxy-2-methylpurine	29	23	54	50	43	36	29	5	18
1,7-Dimethylguanine	29	60	49	39	25	23	17	0	7
6-Carboxypurine	30	16	43	22	35	10	27	4	4
6-Amino-2-hydroxypurine	30	22	52	34	32	24	19	0	6
1-Methylguanine sulfate	30	36	54	48	20	35	18	2	14
5-Formyluridine	30	43	38	45	56	34	20	4	31
1-Methylguanosine	30	49	50	47	38	32	16	1	13
6-Methylaminopurine deoxyribonucleoside	30	52	53	26	38	12	18	0	6
6-Amino-8-hydroxypurine	31	27	51	55	44	52	10	0	19

Table continued

TABLE I (*continued*)

Compound	R_F values ($\times$ 100) in various solvent systems[b]								
	1	2	3	4	5	6	7	8	9
5-Hydroxydeoxyuridine	32	18	47	57	68	46	25	6	40
5-Amino-6-methyluracil	32	31	54	46	25	40	16	1	28
2-Aminopurine	32	37	57	60	17	48	23	1	23
Cytidine	32	47	47	45	45	29	19	0	12
1-Methyladenosine	32	51	53	28	42	15	22	0	5
4-Amino-5-imidazolecarboxamide ribonucleoside	32	57	50	48	48	36	21	0	19
Adenosine	32	64	56	63	33	51	20	3	25
Deoxyguanosine	33	36	50	55	24	39	18	1	20
Isobarbituric acid	34	4,14	44	53	57	42	31	9	36
6-Aminouracil	34	22	47	46	Pink str.	38	32	6	22
2,6-Diaminopurine hemisulfate	34	31	56	39	16	28	29	0	11
5-Methylaminouracil	34	40	61	61	35	53	10,21	11	45
2-Amino-4,6-dioxypyrimidine	35	21	51	48	Pink str.	39	33	6	22
7-Methylguanine	35	24	55	46	26	35	19	2	11
Uridine	35	33	46	55	67	42	25	6	35
3-Methylguanine	36	22	53	34	33	25	21	2	7
2-Hydroxy-6-methylpurine	36	30	55	45	40	33	27	2	11
5-Hydroxymethyldeoxyuridine	36	39	49	59	73	47	25	4	36
6-Azauridine	36	42	49	53	63	42	27	9	39
5-Methylcytidine	36	47	53	41	49	30	20	0	12
5-Hydroxymethyluracil	37	36	48	56	65	46	29	8	33
4,5,6-Triaminopyrimidine sulfate	37	48	57	40	22	32	29	0	2,14
1-Methylinosine	37	56	54	45	49	37	30	3	19
2-Amino-4,6-dioxy-5-methylpyrimidine	38	17	56	45	68	33	38	5	14
Orotic acid	38	23	39	31	57	15	34	9	6
6-Amino-5-formylamino-3-methyluracil	38	26	54	50	50	38	31	5	25

2,4-Diamino-6-oxypyrimidine	38	28	54	36	51 pink	23	29	3	9
6-Methylamino-5-(*N*-formylmethylamino) uracil	38	29	57	42	60	31	45	7	19
3-Methylhypoxanthine	38	30	52	40	30	31	27	4	11
Hypoxanthine	38	35	53	57	28	49	25	6	26
5-Hydroxymethylcytosine	38	42	49	40	44	30	23	0	10
Deoxyinosine	38	42	49	58	31	45	26	4	24
3-Methylxanthine	38	43	55	60	35	51	38	16	31
6-Amino-5-formylamino-3-methyl-4-oxypyrimidine	38	45	53	35	30	25	40	6	13
4,6-Diamino-5-formylamino-2-methylpyrimidine	38	48	57	37	40	20	31	0	0
6-Amino-5-formylamino-1-methyl-4-oxypyrimidine	39	30	55	42	39	33	41	6	17
3,7-Dimethyluric acid	39	38	60	50	57	40 ±	41	9	34
1-Methylguanine	40	25	50	45	32	30	23	0	7
7-Methylxanthine	40	32	56	57	32	47	42	15	28
2-Methyladenine	41	46	65	71	41	64	33	0	12
8-Hydroxypurine	42	40	63	76	31	66	26	13	51
2-Amino-6-methylpurine	42	43	62	64	27	59	32	2	26
5-Formylcytosine	42	48	55	52	35	40	33	7	26
5-Formyldeoxyuridine	42	63	55	64	73	53	26	5	44
6-Methylaminopurine ribonucleoside	42	77	68	70	51	62	25	7	43
Uracil-6-acetic acid	43	17	51	26	65	15	34	14	33
5-Methyl-5-hydroxybarbituric acid	43	25	53	51	63	45	43	25	52
4,6-Diamino-2-oxypyrimidine	43	29	42	19	52	13	49	35	10
Adenine-*N*′-oxide	43	29	61	41	32	37	37	5	8
Barbituric acid	43	37	45	28	51	12	52	31	21
5-Methoxymethyluridine	43	37	55	58	74	47	35	7	39
Deoxyadenosine	43	74	66	70	35	62	31	5	35
Uracil-5-carboxylic acid	44	9	46	23	54	10	38	26	44
5-Acetylamino-6-amino-3-methyluracil	44	26	59	48	78	39	40	0	25
5-Hydroxymethyl-6-methyluracil	44	39	55	55	75	49	40	9	34
5-Formyluracil	44	56	50	56	58	47	39	24	54
2-Methylhypoxanthine	45	32	62	60	39	50	35	8	28

Table continued

TABLE I (*continued*)

Compound	R_F values ($\times$ 100) in various solvent systems[b]								
	1	2	3	4	5	6	7	8	9
5-Hydroxymethylorotic acid lactone	45	37	49	46	49	38	47	37	49
5-Methyluridine	45	46	57	64	74	55	35	8	47
Cytosine	45	49	53	48	45	37	32	0	15
2-Methyl-4,5,6-triaminopyrimidine sulfate	45	55	64	43	35	32	36	0	15 ±
7-Methyladenine	45	57	60	58	30	42	35	0	15
4-Amino-5-imidazolecarboxamide	45	58	58	56	50	48	36	1	25
5-Bromouridine	46	42	58	69	72	61	39	17	65
1-Methyladenine	46	47	59	35	31	18	41	1	8
6-Succinoaminopurine	47	7	61	17	55	6	27	4	45
2-Thioorotic acid	47	25	44	—	69	—	44	21	13
8-Methylhypoxanthine	47	33	63	62	46	57	30	5	33
1,7-Dimethyluric acid	47	35 ±	65	63	82	48	40	17	43
6-Hydroxy-2-methylaminopurine	47	34	58	60 ±	49	50 ±	30	0	25
1,3-Dimethyluric acid	47	43	64	60	67	49	42	16	42
5-Formylamino-3-methyl-6-methylaminouracil	48	30	63	52	54	39	49	7	22
Adenine	48	40	65	71	33	63	37	6	33
5′-Thiomethylinosine	48	53	61	62	45	53	38	11	44
Deoxycytidine	48	61	57	55	58	40	29	0	18
7-Methylhypoxanthine	49	35	62	64	36	50	42	12	28
6-Amino-1-methyluracil	49	57	58	53	Pink str.	44	51	11	27
2-Amino-4-hydroxypyrimidine	50	35	66	66	55	57	40	4	33
9-Methylhypoxanthine	50	37	59	57	36	47	42	11	25
6-Amino-2-thiouracil	50	38	57	str.	65	57	45	18	53
Uracil	50	41	56	65	69	55	46	21	47
Deoxyuridine	50	45	58	67	80	59	41	13	52
1-Methyl-6-methylamino-5-(*N*-formylmethylamino)-uracil	50	46	64	52	75	44	49	12	32

2,6-Diamino-5-formylamino-4-methoxypyrimidine	50	52	61	47	51	26	40	0	5
6-Amino-1,3-dimethyl-5-formylaminouracil	50	58	61	51	51	43	48	8	27
2-Dimethylamino-6-hydroxypurine	51	36	68	61	37	52	42	4	28
1-Methylhypoxanthine	51	37	63	62	36	51	42	13	30
1-Methylxanthine	52	35	67	69	47	65	49	25	51
5-Methylcytosine	53	55	60	55	52	44	38	0	15
3-Methyladenine	53	55	61	48	46	43	45	4	12
4,6-Diaminopyrimidine	53	60	63	56	47	46	49	3	15
6-Dimethylaminopurine ribonucleoside	53	83	74	80	58	74	36	13	61
5-Fluorouracil	54	42	62	70	75	63	55	38	64
2-Hydroxy-4-methylpyrimidine	54	48	66	60	49	55	41	7	38
2-Thiobarbituric acid	54	50	46	41	72	21	58	39	15
Spongothymidine	55	63	61	79	81	66	42	12	59
3,8-Dimethylxanthine	56	38	70	69	48	64	56	24	45
5-Ethoxymethyluridine	56	51	66	70	85	63	—	—	—
5-Methyldeoxycytidine	56	67	65	59	63	46	35	1	21
5-Methylorotic acid	57	27	44	—	78	18	51	8	16
2-Methyladenine sulfate	57	51	69	76	44	68	46	3	30
5-Acetyluracil	57	54	66	70	71	64	52	36	63
1-Methylcytosine	57	61	63	55	55	45	44	2	18
9-Methyladenine	57	67	71	66	40	59	47	12	30
5′-Thiomethyladenosine	57	87	75	82	54	77	45	14	54
2-Thiocytosine	58	34,93	59	89	55	87	45	5	33
2-Amino-4-hydroxy-6-methylpyrimidine	58	38	72	70	63	63	44	6	36
4-Hydroxypyrimidine	58	46	68	70	45	65	54	25	51
3,7-Dimethylxanthine	58	46	70	64	52	56	62	30	42
6-Methylpurine	58	52	76	81	46	77	46	22	57
Purine	58	53	73	73	35	69	45	22	52
2,4-Dihydroxypyrimidine-6-methylsulfone	58	60	61	39	66	23	62	50	61
6-Amino-3-methyl-5-(*N*-formylmethylamino) uracil	59	35	64	64	83	58	52	3	52
5-Fluorodeoxyuridine	59	42	65	72	86	66	50	29	73

Table continued

TABLE I (*continued*)

Compound	R_F values ($\times$ 100) in various solvent systems[b]								
	1	2	3	4	5	6	7	8	9
6-Methyluracil	59	50	68	71	77	66	59	29	55
4,6-Diamino-2-thiopyrimidine	59	80	62	68	56	52	44	2	25
5-Methylbarbituric acid	60	33	54	24 ±	63	10	65	55	46 ±
2-Amino-5-methyl-4-oxypyrimidine	61	47	72	71	67	64	48	7	38
5-Methoxymethyluracil	61	50	68	72	81	65	57	29	58
6-Methylaminopurine	61	61	77	85	49	80	47	8	53
Thymine	62	56	68	77	80	70	61	34	61
Thymidine	62	59	68	77	87	71	48	19	61
2-Aminopyrimidine	62	80	78	77	52	72	58	35	60
6-Methoxypurine ribonucleoside	62	81	78	77	59	70	60	29	58
3-Methyluridine	62	83	68	70	92	65	50	18	59
4,5-Diamino-6-methylpyrimidine	63	69	64	60	49	55	52	0	13
6-Dimethylaminopurine	63	71	77	88	52	82	55	12	65
5-Bromodeoxyuridine	64	53	68	81	85	75	56	32	77
1,7-Dimethylxanthine	66	42	76	77	58	68	70	42	56
4-Hydroxy-6-methylpyrimidine	66	47	76	75	54	72	58	33	59
8-Mercapto-6-methylpurine	66	49	73	77	49	77	58	45	74
1-Methyluracil	66	50	69	70	81	60	64	38	57
1,7-Dimethylhypoxanthine	66	70	73	64	52	57	60	29	40
3-Methyl-6-methylamino-5-(*N*-formylmethylamino)-uracil	67	41	76	64	89	56	70	27	51
5,6-Dimethyluracil	69	70	75	80	85	77	69	40	66
2-Amino-4-methylpyrimidine	69	88	82	87	63	83	65	39	72
1,3-Dimethylxanthine	70	55	78	80	61	75	67	50	65
6-Azathymine	71	46	73	82	78	83	70	66	77
6-Methoxypurine	72	54	82	84	58	84	68	52	69

6-Methyl-2-thiouracil	72	56	76	79	78	73	77	64	75
6-Amino-1,3-dimethyluracil	72	83	75	76	Pink str.	75	74	38	61
5-Ethoxymethyluracil	73	59	78	86	89	79	70	49	75
6-Amino-2,4-dimethylpyrimidine	73	87	72	69	75	70 ±	69	4	16,30
2-Amino-4,6-dimethylpyrimidine	74	92	82	89	78	86	70	31	67
1-Methylthymine	75	67	78	72	85	69	75	56	68
3-Methyl-6-methylaminouracil	76	43	78	84	Pink str.	77	73	38	67
Dithiouracil	76	73	77	88	88	86	83	90	89
Theophylline-7-acetic acid	77	45	74	36	88	17	72	54	54
3-Methyluracil	77	61	78	81	91	77	75	58	70
2-Thiothymine	77	63	77	92	87	83	75	70	84
5-Ethoxymethy-6-methyluracil	77	69	81	88	81	84	78	53	76
6-Furfurylaminopurine	78	85	87	94	79	89	77	58	85
6-Methylmercaptopurine	79	68	82	90	55	82	78	71	77
1,3,7,9-Tetramethyluric acid	79	87	80	80	93	71	79	58	65
1,3,8-Trimethylxanthine	80	61	84	86	75	83	80	63	76
1,3,7-Trimethylxanthine	82	88	84	81	79	79	85	74	73
4-Hydroxy-2-methylmercaptopyrimidine	83	59	82	88	76	85	86	78	81
4,6-Dimethylpyrimidine	84	—	89	—	74	—	81	80	77
1,3-Dimethyluracil	86	91	85	87	97	84	88	82	83
Propyl-2-thiouracil	88	76	88	93	95	95	93	92	92
5-Butoxymethyluracil	88	85	86	96	95	93	89	87	87
2,4-Dimethoxy-5-methylpyrimidine	90	95	93	93	89	92	91	97	93

[a] Whatman No. 1 paper; ascending development at 28°C; tank equilibrated with the solvent or the aqueous phase for two-phase solvents. Detection: 253.7 mμ ultraviolet light. ± following a figure indicates an abnormally high variability in replicate determinations of the R_f values.

[b] Solvents: (1) *tert*-butanol–methyl ethyl ketone–formic acid–H_2O (40:30:15:15); (2) *tert*-butanol–methyl ethyl ketone–H_2O–NH_4OH (40:30:20:10); (3) *n*-butanol–glacial acetic acid–H_2O (50:25:25); (4) upper phase from a mixture of water–*sec*-butanol–*tert*-butanol (48.4:43:8.6); (5) isopropyl alcohol–H_2O–concentrated HCl (65:18.4:16.6); (6) upper phase from a mixture of *sec*-butanol and H_2O; (7) Ethyl acetate–formic acid–H_2O (70:20:10); (8) upper phase from a mixture of ethyl acetate–H_2O–formic acid (60:35:5); (9) *tert*-butanol–methyl ethyl ketone–H_2O–formic acid (44:44:11:0.26).

TABLE II

R_{p_i} AND R_F VALUES OF DEOXYRIBONUCLEOTIDES AND RIBONUCLEOTIDES[a] (*10*)

Compound	R_{p_i} Solvent I	R_{p_i} Solvent II	R_F Solvent III	Compound	R_{p_i} Solvent I	R_{p_i} Solvent II	R_F Solvent III
d-AMP	1.58	0.93	0.44	AMP	1.36	0.75	0.17
d-ADP	1.33	0.46	0.32	ADP	1.11	0.34	0.11
d-ATP	1.18	0.26	0.23	ATP	0.94	0.19	0.07
d-CMP	1.45	0.92	0.46	CMP	1.21	0.68	0.17
d-CDP	1.13	0.45	0.33	CDP	0.96	0.29	0.09
d-CTP	0.79	0.23	0.23	CTP	0.77	0.16	0.06
d-GMP	1.19	0.68	0.29	GMP	0.86	0.54	0.08
d-GDP	0.91	0.31	0.20	GDP	0.64	0.24	0.06
d-GTP	0.74	0.15	0.13	GTP	0.50	0.11	0.04
d-UMP	1.07	1.12	0.46	UMP	0.83	0.86	0.17
d-UDP	0.77	0.55	0.34	UDP	0.62	0.38	0.10
d-UTP	0.60	0.26	0.24	UTP	0.50	0.19	0.06
d-TMP	1.27	1.17	0.48	IMP	0.79	0.68	0.15
d-TDP	0.94	0.69	0.38	IDP	0.54	0.28	0.10
d-TTP	0.76	0.41	0.28	ITP	0.39	0.15	0.07
P_i	1.00	1.00	0.26	XMP	0.60	0.48	0.09
				XDP	0.32	0.20	0.06
				XTP	0.21	0.10	0.04

[a] The R_{p_i} of the nucleotides in solvents I and II may vary if the time of development varies. Solvents: (I) isobutyric acid–1 *N* NH_4OH–0.1 *M* Na EDTA (pH 8.2) (50:30:0.8), used in descending chromatography, and developed from 18–48 hours depending on the nucleotides; (II) absolute ethanol–1 *M* ammonium acetate (pH 3.8)–0.1 *M* Na EDTA (pH 8.2) (75:29:1), used in descending chromatography and developed 24–48 hours; (III) 95% ethanol–0.10 *M* sodium borate buffer (pH 10) (60:40), used in ascending chromatography and developed 15–18 hours. Paper: acid washed (1 *N* HCl followed by distilled water) Whatman No. 1; for Solvent III, the acid-washed paper is dipped in 0.10 *M* borate buffer, pH 10, and dried in air. Method: descending or ascending development, tanks and sheets equilibrated with the solvent prior to development. Detection: ultraviolet light and molybdate reagent for orthophosphate (P_i).

TABLE III

R_F VALUES OF NUCLEOTIDES, PURINES, PYRIMIDINES, AND RELATED COMPOUNDS[a] (*7*)

Compounds	R_F				
	1	2	3	4	5
Adenine	0.34	0.46	0.74	0.55	0.48
Adenosine	0.17	0.54	0.59	0.54	0.55
AMP	0.17	0.16	0.38	0.17	0.81

Table continued

TABLE III (*continued*)

Compounds	R_F 1	2	3	4	5
Deoxyadenosine	0.31	0.63	0.66	0.64	0.58
Deoxy-AMP	0.26	0.18	0.50	0.22	—
Hypoxanthine	0.23	0.43	0.52	0.62	—
Inosine	0.12	0.42	0.36	0.66	0.79
IMP	0.14	0.10	0.24	0.10	0.89
Deoxyinosine	0.20	0.49	0.47	0.69	0.82
Guanine	0.12	0.29	0.45	0.43	—
Guanosine	0.10	0.30	0.41	0.53	0.76
GMP	0.12	0.08	0.26	0.10	—
Deoxyguanosine	0.18	0.41	0.48	0.55	0.79
Deoxy-GMP	0.17	0.09	0.36	0.13	—
Xanthine	0.19	0.33	0.38	0.51	0.64
Xanthosine	0.35	0.31	0.33	0.62	0.75
Uracil	0.38	0.48	0.60	0.70	0.80
Uridine	0.18	0.40	0.46	0.72	0.86
UMP	0.22	0.15	0.36	0.26	0.92
Deoxyuridine	0.29	0.45	0.51	0.73	0.88
Cytosine	0.31	0.51	0.73	0.64	—
Cytidine	0.14	0.46	0.58	0.65	0.75
CMP	0.15	0.13	0.36	0.22	0.88
Deoxycytidine	0.22	0.56	0.66	0.73	0.77
Deoxy-CMP	0.22	0.15	0.48	0.28	—
Thymine	0.52	0.56	0.71	0.72	—
Thymidine	0.42	0.54	0.67	0.77	0.85
TMP	0.36	0.17	0.50	0.32	—
Orotic acid	0.22	0.31	0.54	0.49	0.88
Uric acid	0.19	Dec.	0.19	Dec.	—
Allantoin	0.27	0.32	0.28	0.62	—

[a] Solvents: (1) isopropyl ether–*n*-butanol–98% formic acid (30:30:20), run in downward direction at 20–24°C for 5 hours; (2) *n*-butanol–*n*-propanol–ethanol–25% NH_4OH–H_2O (40:40:10:45:15 v/v), run in downward direction at 26°C for 12 hours in solvent-saturated atmosphere; (3) 80% (v/v) formic acid–*n*-butanol–*n*-propanol–acetone–30% (w/v) trichloroacetic acid (25:40:20:25:15 v/v), run in ascending direction at 20–24°C for 18 hours in solvent-saturated atmosphere; (4) methanol–isopropanol–25% NH_4OH–H_2O (45:30:15:10 v/v), run in downward direction at 20–24°C for 12 hours in partially water-saturated atmosphere; (5) H_2O–*n*-propanol–25% NH_4OH (90:10:1 v/v), run in upward direction at 20–24°C for 3–4 hours. Dec. = partial decomposition. Chromatography is carried out in glass jars; Whatman No. 1 papers (20 × 50 cm) are washed by immersion in 1 *N* HCl for 60 minutes and then in distilled water until neutral. This method of washing gives reproducible and low blanks for ultraviolet absorption. Detection: 253.7 mμ ultraviolet light. Allantoin was detected by spraying with 0.25% mercuric acetate in 95% ethanol, drying, and then spraying with 0.05% diphenylcarbazone in 95% ethanol.

TABLE IV

R_F VALUES OF VARIOUS HYDROXYPURINES, AMINOHYDROXYPYRIMIDINES, AND IMIDAZOLES[a] (*39*)

Compounds	R_F (× 100) 1	2	3	4	5	6	7	Absorption or fluorescence in ultraviolet light
Purine	60	75	75	60	66	60	42	Absorption
2-Hydroxypurine	16	37	48	37	40	41	80	Absorption
6-Hydroxypurine (hypoxanthine)	25	54	59	50	41	41	60	Absorption
8-Hydroxypurine	33	63	50	54	51	46	56	Absorption
2,6-Dihydroxypurine (xanthine)	22	60	22	32	37	43	45	Absorption
2,8-Dihydroxypurine	14	27	26	30	33	40	60	Fluorescence
6,8-Dihydroxypurine	30	33	31	26	35	38	40	Absorption
2,6,8-Trihydroxypurine (uric acid)	5	17	12	14	23	22	40	—
4,5-Diamino-2-hydroxypyrimidine	7	31	45	27	7	6	4	Fluorescence
3,4-Diamino-6-hydroxypyrimidine	6	20	60	32	40	30	60	Fluorescence
4,5-Diamino-2,6-dihydroxypyrimidine	0	5	30	6	5	6	—	Fluorescence
2,4,5-Triaminopyrimidine	10	33	58	23	36	30	55	Fluorescence
4,5,6-Triaminopyrimidine	11	35	—	37	53	38	64	Fluorescence
4-Amino-5-imidazole carboxamidine	16	22	55	42	35	41	62	Absorption
4-Amino-5-imidazole carboxamide	23	34	66	57	58	58	70	Absorption

[a] Solvents for above columns: (1) butanol–H_2O (86:14); (2) propanol–H_2O (77:23); (3) butanol–propionic acid–H_2O (43:27:30); (4) *t*-butanol–ethyl methyl ketone–formic acid–H_2O (40:30:15:15); (5) *t*-butanol–ethyl methyl ketone–NH_3 (15 *N*)–H_2O (40:30:10:20); (6) isopropanol–HCl (10 *N*)–H_2O (65:18.4:16.6); (7) ammonium bicarbonate (16%). Descending development on 10 × 50 cm strips of Whatman No. 1 paper. Detection: ultraviolet light except for uric acid which was detected by spraying with Folin–Ciocalteu reagent.

TABLE V

R_F VALUES OF PURINES AND PYRIMIDINES IN BUTANOL AND ACETONITRILE SOLVENTS[a] (*36*)

Compound	R_F Solvent A	Solvent B	Solvent C	Solvent B 2nd Dimension[b]
Adenine	0.35	0.24	0.42	0.16
Guanine	0.32	—[c]	0.23	0.10
Cytosine	0.14	0.18	0.45	0.14

Table continued

TABLE V (*continued*)

Compound	R_F Solvent A	Solvent B	Solvent C	Solvent B 2nd Dimension[b]
Uracil	0.30	0.40	0.65	0.35
1,3-Dimethyluracil	0.75	1.00	—[c]	1.00
Thymine	0.49	0.48	0.70	0.44
6-Methyluracil	0.46	0.48	0.66	0.42
5-Nitrouracil	0.10	0.17	0.47	0.48
5-Aminouracil[d]	0.70	—[c]	0.08	0.02
6-Aminouracil	0.14	0.11	0.27	0.12
Caffeine	0.66	—[c]	0.73	0.54
Xanthine	—[c]	1.00	0.20	0.19
Hypoxanthine	0.18	—[c]	0.37	0.15
Purine	0.45	0.33	0.54	0.19
Uric Acid	0.25	—[c]	0.09	0.03
Theobromine	0.30	—[c]	0.73	0.48
Theophylline	0.45	0.72	0.71	0.48
Thiouracil	0.47	—[c]	0.65	0.59
Dithiouracil[d]	0.64	—[c]	0.58	0.77
6-Methylthiouracil	0.59	—[c]	0.69	0.66
Isocytosine	0.20	0.23	0.43	0.23
6-Mercaptopurine[d]	0.29	—[c]	0.34	0.31
Adenosine	—[e]	—[e]	0.43	0.12
Thymidine	—[e]	—[e]	0.73	0.44
Uridine	—[e]	—[e]	0.40	0.12
Deoxyadenosine	—[e]	—[e]	0.44	0.12
Deoxyguanosine	—[e]	—[e]	0.36	0.16
Deoxyuridine	—[e]	—[e]	0.51	0.22
Deoxycytidylic acid	—[e]	—[e]	0.06	0.02

[a] Solvent A: *n*-butanol equilibrated with 0.4 *M* sodium phosphate buffer, pH 6.86 (prepared by dissolving 0.2 mole of sodium dihydrogen phosphate and 0.2 mole of disodium hydrogen phosphate in distilled water and diluting to 1 liter); Whatman No. 1 sheets are sprayed with the aqueous buffer solutions and dried in air before spotting the compounds; the mobile phase is the equilibrated butanol. Solvent B: prepared by mixing 8 volumes of acetonitrile with 2 volumes 0.4 *M* sodium phosphate buffer, pH 6.86; paper is sprayed with pH 6.86 sodium phosphate buffer and dried before spotting the compounds. Solvent C: 70% acetonitrile–H_2O (v/v), adjusted to pH 10 (apparent pH) with concentrated NH_4OH; paper is pretreated by spraying with water adjusted to pH 10 (ammoniacal), and drying in air.

[b] The paper could not be pretreated as suggested above when Solvent B was used as the second solvent for two-dimensional chromatography (see Fig. 4).

[c] No spot.

[d] Fluorescent.

[e] Not determined.

TABLE VI

R_F VALUES OF HYDROPEROXIDIC DERIVATIVES OF THYMINE[a] (*40*)

Thymine derivative	R_F
cis-4-Hydroxy-5-hydroperoxythymine	0.51
trans-4-Hydroxy-5-hydroperoxythymine	0.33
cis-4-Hydroperoxy-5-hydroxythymine	0.62
trans-4-Hydroperoxy-5-hydroxythymine	0.43
trans-4-Hydroperoxy-5-bromothymine	0.83
Hydrogen peroxide	0.68
Thymine	0.55

[a] Paper: Whatman No. 1. Solvent: *n*-Propanol–1 *N* HCl (85:15). Temperature: 2°C.

TABLE VII

R_F VALUES OF SOME METHYLATED GUANINES AND THEIR RIBOSIDES (*41*)

Compound	R_F [a]			
	1	2	3	4
Guanine	0.26	0.49	0.09	0.08
1-Methylguanine	0.26	0.62	0.19	0.13
6-Hydroxy-2-methylaminopurine	0.50	0.66	0.25	0.18
2-Dimethylamino-6-hydroxypurine	0.43	0.69	0.28	0.20
2-Ethylamino-6-hydroxypurine	0.60	0.82	—	—
Guanosine	0.37	0.42	0.10	0.02
1-Methylguanosine	0.45	0.64	0.16	0.11
6-Hydroxy-2-methylaminopurine riboside	0.55	0.63	0.18	0.08
2-Dimethylamino-6-hydroxypurine riboside	0.49	0.66	0.23	0.10
Uracil	0.69	0.72	0.38	0.20
Uridine	0.70	0.70	0.21	0.08

[a] Solvents: (1) isopropanol (680 ml), 11.6 *N* HCl (176 ml), H_2O to 1 liter; (2) isopropanol (700 ml), H_2O (300 ml) with NH_3 in vapor phase; (3) *n*-butanol (770 ml), H_2O (130 ml), 98% formic acid (100 ml); (4) *n*-butanol (860 ml), H_2O (140 ml) with NH_3 in vapor phase.

TABLE VIII

R_F VALUES OF NUCLEIC ACID DERIVATIVES IN ACETONITRILE-BUFFER SOLVENTS[a] (*42*)

Compound	R_F								
	A	B	C	D	E	F	G	H	I
Deoxyribosides									
Thymidine	0.57	0.66	0.65	0.61	0.35	0.83	0.74	0.73	0.70
Deoxycytidine	0.34	0.63	0.59	0.57	0.30	0.67	0.74	0.53	0.49

Table continued

TABLE VIII (*continued*)

Compound	R_F								
	A	B	C	D	E	F	G	H	I
Deoxyadenosine	0.37	0.40	0.38	0.38	0.13	0.67	0.59	0.48	0.48
Deoxyguanosine	0.25	0.35	0.28	0.30	0.09	0.69	0.48	0.45	0.42
Deoxyribo-5′-nucleotides									
Thymidylic acid	0.08	0.07	0.07	0.055	0	0.43	0.23	0.16	0.16
Deoxycytidylic acid	0	0.05	0.05	0.035	0	0.35	0.21	0.08	0.11
Deoxyadenylic acid	0	0.06	0.06	0.035	0	0.34	0.19	0.07	0.09
Deoxyguanylic acid	0	0.03	0.01	0.028	0	0.31	0.11	0.07	0.08
Acyl derivatives[b]									
TrThymidine	0.96	—	—	—	—	—	—	—	—
MMTrThymidine	1.0	—	—	—	—	—	—	—	—
DMTrThymidine	1.0	—	—	—	—	—	—	—	—
T-OAc	0.93	—	—	—	0.77	—	—	0.93	0.92
pT-OAc	0.08	—	—	—	—	—	—	—	0.50
Cyanoethyl pT	—	—	—	—	0.22	—	—	0.83	0.52
Trichloroethyl pT	0.42	—	—	—	—	—	—	—	0.61
Trichloroethyl pT-OAc	0.57	—	—	—	—	—	—	—	—
TrTpT	0.62	—	—	—	—	—	—	—	—
MMTrTpT-OAc	0.57	—	—	—	—	—	—	—	—
d-pC^{An}-OAc	0.23	—	—	—	—	—	—	—	0.73
d-pC^{An}	—	—	—	—	—	—	—	—	0.43
d-pA^{Bz}-OAc	0.10	—	—	—	—	—	—	0.63	—
d-pA^{Bz}	—	—	—	—	—	—	—	0.25	—
d-pA^{Bz}-OBz	—	—	—	—	—	—	—	0.90	—
d-pG^{Ac}-OAc	—	—	—	—	—	—	—	—	0.55
d-pGpGpG	—	—	—	—	—	—	—	—	0.18
MMTrTpC^{An}-OAc	0.67	—	—	—	—	—	—	—	—
d-TpC^{An}-OAc	0.41	—	—	—	—	—	—	—	—
dTpA	—	—	—	—	—	—	—	0.15	—
DMTrTpA^{Bz}-OAc	0.75	—	—	—	—	—	—	—	—
DMTrTpA^{Bz}	0.60	—	—	—	—	—	—	—	0.77
DMTrTpA^{Bz}pG^{Ac}-OAc	0.50	—	—	—	—	—	—	—	—
MMTrTpG^{Ac}-OAc	0.56	—	—	—	—	—	—	—	—

[a] Solvents (by volume): A = acetonitrile–ammonium acetate (80:20); B = acetonitrile–ammonium acetate–28% NH_4OH (70:20:10); C = acetonitrile–ammonium acetate–28% NH_4OH–isopropanol (60:20:10:10); D = acetonitrile–ammonium acetate–28% NH_4OH–*n*-butanol (60:20:10:10); E = acetonitrile–ammonium acetate–28% NH_4OH (80:10:10); F = acetonitrile–NH_4Cl, pH 5 (60:40); G = acetonitrile–ammonium acetate–28% NH_4OH (60:30:10); H = acetonitrile–ammonium acetate (70:30); I = acetonitrile–NH_4Cl, pH 5 (70:30). Buffers are 0.1 *M*, pH 7, unless otherwise noted. Ascending development without prior equilibration; detection with short-wave ultraviolet light.

[b] The abbreviations used in this table are derived from those used by Khorana *et al.* (*43, 44*).

TABLE IX

R_{Hy} VALUES[a] OF PURINE, PYRIMIDINE, PURINE NUCLEOSIDES, AND PURINE NUCLEOTIDES[b] (*37*)

	R_{Hy} [c]					
Compounds	1	2	3	4	5	6
Adenine	1.20	1.06	0.85	1.02	1.34	0.67
Guanine	0.90	0.58	0.56	0.72 (dec)	0.63	0.71
Cytosine	0.93	1.14	0.74	1.05	1.35	1.25
Uracil	1.07	1.09	1.41	1.11	1.30	1.29
Hypoxanthine	1.00	1.00	1.00	1.00	1.00	1.00
Xanthine	0.87	0.73	0.75	0.88 (dec)	0.61	0.84
Uric acid	0.66	0.39	0.43	0.62	0.36	0.67
Adenosine	1.06	1.19	0.57	1.00	1.40	0.89
Guanosine	0.80	0.73	0.39	0.89	0.63	1.11
Inosine	0.82	0.93	0.50	0.96	0.70	1.28
Xanthosine	0.77	0.77	0.38	0.83	0.48	1.17
AMP	0.51	0.32	0.07	0.70	0.12	1.23
ADP	0.27	0.23	0.00	0.44	0.09	1.38
ATP	0.20	0.19	0.00	0.30	0.07	1.45
GMP	0.39	0.14	0.06	0.55	0.05	1.39
IMP	0.43	0.20	0.07	0.67	0.06	1.56

[a] $R_{Hy} = R_F$ relative to hypoxanthine.

[b] All the values (the average of five determinations) were obtained from one-dimensional descending chromatograms on Whatman No. 1 paper at 18 ± 2°C. The partial decomposition in a solvent is indicated by "dec." Tank equilibrated with the solvent before the run. Detection was with short wavelength ultraviolet light.

[c] Solvent systems used: (1) *n*-butanol–acetic acid–H_2O (40:11:25 v/v)—hypoxanthine runs about 30 cm in 40 hours of development with overflow, its R_F is 0.46; (2) *n*-propanol–NH_4OH (d = 0.90)–H_2O (3:1:1 v/v)—hypoxanthine runs 30 cm in 40 hours of development with overflow, its R_F is 0.53; (3) *n*-butanol–99% formic acid–H_2O (77:10:13 v/v)—hypoxanthine runs 40 cm in 10 hours of development with overflow, its R_F is 0.27; (4) 95% ethanol–1 *M* ammonium acetate (pH 3.8) (75:30 v/v)—hypoxanthine runs 32 cm in 27 hours of development with overflow, its R_F is 0.54; (5) *n*-butanol–diethyl glycol–H_2O (4:1:1 v/v) (NH_3 atmosphere)—hypoxanthine runs 35 cm in 70 hours of development with overflow, its R_F is 0.31; (6) 5% (w/v) Na_2HPO_4 in isoamyl alcohol (100:50 v/v)—hypoxanthine runs 25 cm in 10 hours of development with overflow, its R_F is 0.60.

TABLE X

MEAN R_F VALUES (20–40 MEASUREMENTS) OF AUTHENTIC PURINE AND PYRIMIDINE COMPOUNDS IN ISOBUTYRIC ACID–ACETIC ACID–H_2O (19:0.4:11 v/v)[a] (*45*)

Authentic substances	R_F Wedge strips 3 × 60 cm	R_F Normal strips 4 × 60 cm
ITP	0.13	0.08
CTP	0.14	0.08
UTP	0.15	0.08
GTP	0.15	0.08
IDP	0.17	0.09
GDP	0.18	0.09
UDP	0.19	0.11
ATP	0.19	0.11
ADP	0.24	0.17
I-5′-MP	0.30	0.27
G-5′-MP	0.30	0.27
U-5′-MP	0.30	0.28
C-5′-MP	0.32	0.31
A-5′-MP	0.41	0.40
Guanosine	0.46	0.46
Inosine	0.49	0.47
Uridine	0.48	0.48
Cytidine	0.53	0.52
Guanine	0.54	0.57
Hypoxanthine	0.58	0.57
Uracil	0.60	0.57
Cytosine	0.60	0.59
Adenosine	0.76	0.74
Adenine	0.82	0.83

[a] Schleicher and Schuell 2043b mgl or VEB Neiderschlag FN 3 paper developed for 18–20 hours by descending chromatography.

TABLE XI

APPROXIMATE R_F VALUES (ASCENDING DEVELOPMENT) OF NUCLEOTIDES, NUCLEOSIDES, AND BASES[a] (*46*)

Compound	R_F													
	1	2	3	4	5	6	7	8	9	10	11	12	13	14
Deoxyadenosine 5′-triphosphate	—	.51	—	—	—	—	—	.35	.24	.33	.42	—	—	—
Deoxyguanosine 5′-triphosphate	—	.28	—	—	—	—	—	—	.05	—	—	—	—	—
Thymidine 5′-triphosphate	—	.30	—	—	—	—	—	.25	.06	—	—	—	—	—
Deoxyadenosine 5′-diphosphate	—	.67	—	—	—	—	—	.52	.45	.37	.55	—	—	—
Deoxyguanosine 5′-diphosphate	—	.38	—	—	—	—	—	—	.06	—	—	—	—	—
Thymidine 5′-diphosphate	—	.39	—	—	—	—	—	.37	.06	—	—	—	—	—
Deoxyadenosine 5′-monophosphate	—	.73	—	.70	—	—	—	.60	.53	.40	.72	—	—	—
Deoxycytidine 5′-monophosphate	—	—	—	.90	—	—	—	.59	.21	.43	—	—	—	—
Deoxyguanosine 5′-monophosphate	—	.45	—	.80	—	—	—	.42	.08	.35	—	—	—	—
Thymidine 5′-monophosphate	—	.51	—	—	.83	—	—	.59	.20	.44	—	—	—	—
Adenosine 5′-triphosphate	—	.40	—	.85	.44	—	—	.29	<.05	.13	.42	.07	—	—
Cytidine 5′-triphosphate	—	.32	—	—	.50	—	—	.21	<.05	—	.33	—	—	—
Guanosine 5′-triphosphate	—	.21	—	—	.40	—	—	.11	<.05	—	.19	—	—	—
Inosine 5′-triphosphate	—	—	—	—	—	—	—	.10	<.05	—	—	.06	—	—
Uridine 5′-triphosphate	—	.19	—	—	.70	—	—	.16	<.05	—	.20	.05	—	—
Adenosine 5′-diphosphate	—	.45	—	.78	.47	—	—	.43	<.05	.22	.51	.11	—	—
Cytidine 5′-diphosphate	—	.40	—	—	.54	—	—	.33	<.05	—	.47	—	—	—
Guanosine 5′-diphosphate	—	.23	—	—	.44	—	—	.16	<.05	.08	.26	.05	—	—
Inosine 5′-diphosphate	—	—	—	—	—	—	—	.17	<.05	—	—	.10	—	—
Uridine 5′-diphosphate	—	.28	—	—	.72	—	—	.18	<.05	—	.26	.15	—	—
Adenosine 2′-monophosphate	—	.70	.77	—	—	—	.52	—	—	.39	.65	—	—	—
Adenosine 3′-monophosphate	—	—	—	—	—	—	.52	—	—	.39	.65	—	—	—
Adenosine 3′:5′-cyclic phosphate	—	.74	.67	—	—	—	.27	—	—	.65	—	—	—	—
Adenosine 5′-monophosphate	—	.60	.85	.72	.50	—	.52	.56	<.05	.26	.65	.27	—	.29
Cytidine 2′- and 3′-monophosphates	.86	—	—	—	—	—	.84	—	—	.42	—	—	—	—
Cytidine 2′:3′-monophosphate	.63	.60	—	—	—	—	.72	—	—	.66	—	—	—	—
Cytidine 5′-monophosphate	.84	.55	.94	.90	.57	—	.87	.45	<.05	.25	.59	—	—	.82
Guanosine 2′- and 3′-monophosphates	—	.43	—	—	—	—	.62	.22	—	.42	—	—	—	—
Guanosine 5′-monophosphate	—	.38	—	.82	.47	—	.73	.30	<.05	.08	.36	.13	—	.53
Inosine 5′-monophosphate	—	.39	—	—	—	—	—	.31	<.05	—	—	.25	—	.65

Uridine 2′- and 3′-monophosphates	.78	—	—	—	—	—	—	—	—	.36	—	—	—	—
Uridine 2′:3′-monophosphate	.53	—	—	—	—	—	—	—	—	—	—	—	—	—
Uridine 5′-monophosphate	.76	.43	—	.90	.74	—	—	.32	<.05	.24	.34	.33	—	.80
Deoxynucleosides														
Deoxyadenosine	—	—	—	—	.41	.40	—	.94	.71	.76	—	—	—	—
Deoxycytidine	—	—	—	—	—	.26	.71	.82	.67	.75	—	—	—	—
Deoxyguanosine	—	—	.66	—	.30	.14	.50	—	.45	.57	—	—	—	—
Deoxyinosine	—	.69	.68	—	.34	.09	—	—	—	—	—	—	—	—
Deoxyuridine	—	—	—	—	.79	—	—	—	.25	—	—	—	—	—
Thymidine	—	—	—	—	.87	.49	—	.83	—	—	—	—	.36	—
Ribonucleosides														
Adenosine	—	.87	.59	.59	.40	.34	.36	.89	.59	.65	.90	—	—	.14
Cytidine	—	.80	.89	.83	.51	.13	.76	.76	.40	.62	.76	—	—	.70
Guanosine	—	.50	—	.47	.29	.13	.36	.52	.23	.42	—	.46	—	.43
Inosine	—	—	.73	—	.34	.07	—	.61	—	—	—	—	—	—
Uridine	—	.60	.90	.84	.75	.14	—	.60	—	—	—	.70	—	.70
Xanthosine	—	.51	—	—	.26	<.02	—	—	—	—	—	—	—	—
Purine and pyrimidine bases														
Adenine	—	.95	.44	.43	.40	.40	—	.94	.55	.65	.94	—	—	.11
Cytosine	—	.95	.83	.75	.51	.27	.74	.85	.58	.66	.88	—	—	.72
Guanine	—	.65	—	—	.29	.13	—	—	—	—	—	.46	—	.05
Hypoxanthine	—	.70	.60	—	.28	.18	—	—	—	—	—	—	—	—
Thymine	—	—	.74	—	.85	.50	—	.83	—	—	—	—	.50	—
Uracil	—	.70	.80	.75	.75	.33	—	.72	—	—	—	.70	—	.63
Xanthine	—	.54	—	—	.25	—	—	—	—	—	—	—	—	—

[a] Sample Size: 5 μl of solution, 4 mg/ml, superimposed with 1 μc of a ^{14}C-labeled compound or 100 μc of a ^{3}H-labeled compound. Development: Paper fastened into a cylinder, taped to the inner wall of a chamber containing 20 ml of solvent to equilibrate 3 hours at 30°C, and then lowered into the solvent to develop 16 hours at 30°C. Detection: Under ultraviolet light and by autoradiography. Solvents: (by volume) (1) water-saturated $(NH_4)_2SO_4$[70 gm $(NH_4)_2SO_4$ plus 100 gm H_2O]–0.1 *M* phosphate buffer, pH 6 isopropanol (79:19:2); (2) isobutyric acid–2.0 *N* NH_4OH (66:34); (3) isoamyl alcohol–5% aqueous $Na_2HPO_4 \cdot H_2O$ (1:1); (4) isoamyl alcohol–5% aqueous sodium citrate · $2H_2O$ (1:1); (5) isopropanol–concentrated HCl (65:16.7), dilute to 100 ml with H_2O; (6) water-saturated *n*-butanol–15 *N* NH_4OH (100:1); (7) saturated $(NH_4)_2SO_4$ in H_2O–1 *M* sodium acetate–isopropanol (80:18:2); (8) isobutyric acid–15 *N* NH_4OH–H_2O (66:1:33); (9) isopropanol–15 *N* NH_4OH–0.1 *M* H_3BO_3 (7:1:2); (10) isopropanol–15 *N* NH_4OH–0.1 *M* H_3BO_3 (6:1:3); (11) isobutyric acid–1 *N* NH_4OH–0.2 *M* EDTA (100:60:0.8); (12) 95% ethanol–1 *M* ammonium acetate, pH 5.0 (adjusted with acetic acid) (7:3); (13) ethy acetate–H_2O–formic acid (60:35:5)—equilibrate for $16\frac{1}{2}$ hours in lower phase, run for 3 hours in upper phase; (14) 13.8 gm $NaH_2PO_4 \cdot H_2O$ in 900 ml H_2O, adjust pH to 6.8 with H_3PO_4 and dilute to 1 liter with H_2O, then add 600 gm $(NH_4)_2SO_4$ and 20 ml *n*-propanol.

TABLE XII

R_F VALUES OF NUCLEOTIDES, NUCLEOSIDES AND DINUCLEOSIDE PHOSPHATES[a] (*46a*)

Compound	R_F		
	1	2	3
Uridine	0.40	0.15	—
5-Bromouridine	0.40	0.34	—
5-Iodouridine	0.40	0.34	—
3-Methyl-5-bromouridine	0.50	0.72	—
3-Methyl-5-iodouridine	0.50	0.72	—
2′,3′-*O*-Ethoxymethyleneuridine	0.68	0.67	—
2′,3′-*O*-Ethoxymethylene-5-bromouridine	0.67	0.65	—
2′,3′-*O*-Ethoxymethylene-5-iodouridine	0.70	0.70	—
2′,3′-*O*-Ethoxymethylene-3-methyl-5-bromouridine	0.89	0.78	—
2′,3′-*O*-Ethoxymethylene-3-methyl-5-iodouridine	0.89	0.82	—
Uridine 2′(3′)-phosphate	0.10	—	0.30
Uridine 2′,3′-cyclic phosphate	0.30	—	0.42
5-Bromouridine 2′(3′)-phosphite	0.32	—	0.58
5-Iodouridine 2′(3′)-phosphite	0.34	—	0.60
5-Bromouridine 2′,3′-cyclic phosphate	0.32	—	0.58
5-Iodouridine 2′,3′-cyclic phosphate	0.32	—	0.60
5-Bromouridine 3′-phosphate	0.11	—	0.38
2′,5′-Di-*O*-(1-ethoxyethyl)-5-bromouridine 3′-phosphate	0.28	—	—
2′,5′-Di-*O*-(1-ethoxyethyl)-5-iodouridine 3′-phosphate	0.30	—	—
Uridylyl-(3′ → 5′)-uridine	0.19	—	0.52
Uridylyl-(3′ → 5′)-5-bromouridine	0.18	—	0.60
Uridylyl-(3′ → 5′)-5-iodouridine	0.19	—	0.60
Uridylyl-(3′ → 5′)-3-methyl-5-bromouridine	0.31	—	—
Uridylyl-(3′ →5′)-3-methyl-5-iodouridine	0.36	—	—
5-Bromouridylyl -(3′ → 5′)-uridine	0.19	—	0.58
5-Iodouridylyl-(3′ → 5′)-uridine	0.20	—	0.60
5-Bromouridylyl-(3′ → 5′)-adenosine	0.19	—	—
2′-*O*-Tetrahydropyranyluridylyl-(3′ → 5′)-2′,3′-*O*-ethoxymethylene-5-bromouridine	0.62	—	—
2′-*O*-Tetrahydropyranyluridylyl-(3′ → 5′)-2′,3′-*O*-ethoxymethylene-5-iodouridine	0.61	—	—
2′-*O*-Tetrahydropyranyluridylyl-(3′ → 5′)-2′,3′-*O*-ethoxymethylene-3-methyl-5-bromouridine	0.73	—	—
2′-*O*-Tetrahydropyranyluridylyl-(3′ → 5′)-2′,3′-*O*-ethoxymethylene-3-methyl-5-iodouridine	0.72	—	—

[a] Solvents: (1) isopropanol–concentrated NH_3–H_2O (7:1:2 v/v/v); (2) *n*-butanol saturated with H_2O; (3) ethanol–1 *M* ammonium acetate. Paper: Whatman No. 1. Detection: ultraviolet light and the Isherwood reagent in the case of phosphorus-containing compounds.

References

1. W. E. Cohn, *in* "Chromatography" (E. Heftmann, ed.), pp. 627–660. Reinhold, New York, 1967.
2. W. J. Kleinschmidt and J. A. Manthey, *Arch. Biochem. Biophys.* **73**, 52 (1958).
3. R. J. Block, E. L. Durrum, and G. Zweig, "A Manual of Paper Chromatography and Paper Electrophoresis," 2nd rev. ed. Academic Press, New York, 1958.
4. M. Flikke and E. Steinnes, *Arch. Biochem. Biophys.* **118**, 82 (1967).
5. K. C. Smith and F. W. Allen, *J. Am. Chem. Soc.* **75**, 2131 (1953).
6. D. T. Elmore, *J. Chem. Soc.* p. 2084 (1950).
7. E. Gerlach, R. H. Dreisbach, and B. Deuticke, *J. Chromatog.* **18**, 81 (1965).
8. K. Fink and W. S. Adams, *J. Chromatog.* **22**, 118 (1966).
9. H. Ishihara, *J. Chromatog* **4**, 376 (1960).
10. B. S. Vanderheiden, *Anal. Biochem.* **22**, 231 (1968).

10a. R. Teoule and J. Cadet, *J. Chromatog.* **43**, 368 (1969).

11. H. Descimon, *Bull. Soc. Chim. Biol.* **49**, 452 (1967).
12. E. Vischer and E. Chargaff, *J. Biol. Chem.* **176**, 703 (1948).
13. D. S. Letham, *J. Chromatog.* **20**, 184 (1965).
14. Z. Körös, *Magy. Kem. Lapja* **7**, 23 (1952).
15. R. M. Regura and I. Asimov, *J. Am. Chem. Soc.* **72**, 5781 (1950).
16. J. G. Buchanan, C. A. Dekker, and A. G. Long, *J. Chem. Soc.* p. 3162 (1950).
17. J. Jonsen, L. Haavaldsen, and S. Laland, *J. Chromatog.* **1**, 291 (1958).
18. H. Michl, *Naturwissenschaften* **40**, 390 (1953).
19. E. Gerlach and H. J. Döring, *Naturwissenschaften* **42**, 344 (1955).
20. N. Fries, *Physiol. Plantarum* **2**, 212 (1949); **4**, 410 (1951).
21. V. Kocher, R. Karrer, and H. R. Müller, *Intern. Z. Vitaminforsch.* **21**, 403 (1950).
22. W. A. Winsten and E. Eigen, *J. Biol. Chem.* **177**, 989 (1949).
23. M. H. Studier, R. Hayatsu, and K. Fuse, *Anal. Biochem.* **26**, 320 (1968).
24. J. Picard, *J. Chromatog.* **7**, 223 (1962).
25. K. V. Giri, P. R. Krishnaswamy, G. D. Kalyankar, and P. L. N. Rao, *Experientia* **9**, 296 (1953).
26. W. Niemierko and M. Krzyzanowska, *J. Chromatog.* **26**, 424 (1967).
27. G. Högenaver, *Monatsh. Chem.* **98**, 494 (1967).
28. K. Jaroszewicz, *Bull. Acad. Polon. Sci. Ser. Sci. Biol.* **15**, 1 (1967).
29. K. Šebesta, *in* "Laboratory Handbook of Chromatographic Methods" (Ŏ. Mikes, ed.), pp. 118–122. Van Nostrand, Princeton, New Jersey, 1961.
30. G. Kaldor and C. Heidelberger, *Biochim. Biophys. Acta* **36**, 249 (1959).
31. C. L. Washington and D. Ludwig, *J. N. Y. Entomol. Soc.* **73**, 168 (1965).

31a. J.-I. Mukai and S. Akune, *J. Chromatog.* **33**, 557 (1968).

32. K. Bauer, *J. Chromatog.* **32**, 529 (1968).
33. R. D. Brown and C. E. Holt, *Anal. Biochem.* **20**, 358 (1967).

33a. J. F. Morrison, *Anal. Biochem.* **24**, 106 (1968); G. Serlupi-Crescenzi, C. Paolini, and T. Leggio, *Anal. Biochem.* **23**, 263 (1968).

34. H. Verachtert, S. T. Bass, J. Wilder, and R. G. Hansen, *Anal. Biochem.* **11**, 497 (1965).

34a. E. D. Duee, *Bull. Soc. Chim. Biol.* **50**, 1215 (1968).

35. J.-I. Mukai, *J. Chromatog.* **21**, 498 (1966).
36. B. A. Berger and C. E. Hedrick, *Anal. Biochem.* **16**, 260 (1966).
37. L. Rossi and C. Rossi, *J. Chromatog.* **30**, 278 (1967).
38. J. W. Bradbeer and B. C. Jarvis, *J. Chromatog.* **20**, 624 (1965).
39. M. N. Khattak, N. T. Barker, and J. H. Green, *Analyst* **91**, 526 (1966).

40. B. Ekert and R. Monier, *J. Chromatog.* **3**, D20 (1960).
41. J. D. Smith and D. B. Dunn, *Biochem. J.* **72**, 294 (1959).
42. T. F. Gabriel, *J. Chromatog.* **36**, 518 (1968).
43. E. Ohtsuka, M. W. Moon, and H. G. Khorana, *J. Am. Chem. Soc.* **87**, 2956 (1965).
44. T. M. Jacob and H. G. Khorana, *J. Am. Chem. Soc.* **87**, 2971 (1965).
45. R. Rammelt and H. Bergmann, *J. Chromatog.* **31**, 279 (1967).
46. Schwarz BioResearch Radiochemical Cataloge, pp. 6–7, 48, and 52 (1969). Courtesy of Dr. S. S. Rennert, Schwarz BioResearch, Orangeburg, N. Y.
46a. A. Holý, J. Smrt, and F. Šorm, *Collection Czech. Chem. Commun.* **33**, 3816 (1968).
47. K. Macek and J. Stanislavova, *J. Chromatog.* **4**, 156 (1960).
47a. S. A. Narang, T. M. Jacob, and H. G. Khorana, *J. Am. Chem. Soc.* **89**, 2158 (1967).
47b. K. C. Tovey and R. M. Roberts, *J. Chromatog.* **47**, 287 (1970).
48. J. E. Edström, *Biochim. Biophys. Acta* **9**, 528 (1952); B. M. Johnstone and G. P. Briner, *Nature* **183**, 681 (1959).
49. E. Vischer and E. Chargaff, *J. Biol. Chem.* **163**, 781 (1947).
50. W. R. Farkas, *Biochim. Biophys. Acta* **155**, 401 (1968).
51. H. Holmsen and M. C. Rozenberg, *Biochim. Biophys. Acta* **155**, 326 (1968).
52. L. A. Isaksson and J. H. Phillips, *Biochim. Biophys. Acta* **155**, 63 (1968).
53. H.-J. Rhaese, E. Freese, and M. S. Melzer, *Biochim. Biophys. Acta* **155**, 476 and 491 (1968).
54. R. H. Symons, *Biochim. Biophys. Acta* **155**, 609 (1968).
55. V. Santilli, J. Škoda, J. Gutand, and F. Šorm, *Biochim. Biophys. Acta* **155**, 623 (1968).
56. H. Sandermann, Jr. and H. Grisebach, *Biochim. Biophys. Acta* **156**, 435 (1968).
57. A. Piśkala, *Collection Czech. Chem. Cummun.* **32**, 3966 and 4271 (1967).
58. T.-I. Yang and J.-Ch. Su, *Chung Kuo Nung Yeh Hua Hsueh Hui Chih* **5**, 71 (1967).
59. S. Aoyagi and Y. Inoue, *J. Biochem. (Tokyo)* **64**, 603 (1968).
60. J. A. Thomas, K. K. Schlender, and J. Larner, *Anal. Biochem.* **25**, 486 (1968).

10

INDOLES, PHENOLS, AND AROMATIC ACIDS

I. Indoles

The indole derivatives have retained the attention of investigators because of their important role in the regulation of growth and maturation of plants and also because of their possible contribution to problems in human mental diseases and neurological problems in livestock. The preparation of samples and methods for paper chromatographic detection are discussed below.

A. Preparation of Samples

1. *Isolation from Plant Material* (*1–5*)

The plant sample is frozen to $-10°C$ with Dry Ice and macerated at this temperature with small amounts of ethanol. After 24 hours of storage at $-10°C$, the crude extract is centrifuged, the residue is washed several times with ethanol, and the ethanol is evaporated at reduced pressure. The residue is adjusted to pH 3 with H_3PO_4 and extracted for $1\frac{1}{2}$ hours with ether which is shaken immediately before use with acidic $FeSO_4$. The ether extract is shaken three times with 5% $NaHCO_3$, the bicarbonate extracts are adjusted to pH 3 with H_3PO_4, reextracted with ether, and washed with a small amount of H_3PO_4 at pH 3. The final extract is evaporated under a stream of nitrogen and redissolved in anhydrous ether.

In an alternate procedure, the sample (e.g., ground sweet corn) is extracted with 50% acetone under low temperature conditions (cold room). The extract is salted out with NaCl (250 gm NaCl per 1.5 liters solution), and the acetone layer is concentrated under reduced pressure to a small volume. The indoles are extracted with ether, concentrated, and the residue is chromatographed.

Note: Indoles isolated from plant sources should be kept under nitrogen during the evaporation steps and stored at low temperatures in dark containers. The compounds are extremely labile and lose their biological activity.

In another procedure for the extraction of tryptamine derivatives, fresh plant leaves are exposed to diethyl ether for 3–5 minutes. Plant juices are rapidly extracted from a small amount of sample by crushing the sample within a piece of Tygon tubing closed at one end with a Mohr pinch cock clamp; the juice may be withdrawn from the tubing by means of a micropipette (*6*). The expressed juice, containing 10–100 μg tryptamine derivatives, is spotted on Whatman No. 1 chromatographic paper. To isolate the tryptamines from dried leaf material, the ground sample (40 mesh) is Soxhlet-extracted for 7 hours with absolute methanol. The methanol is evaporated and 3–4 volumes of water are added to the liquid residue. The mixture is acidified with 0.1 *N* HCl, extracted twice with equal volumes of toluene, then made alkaline with NH_4OH and extracted three times with diethyl ether. The ether is evaporated and the residue is made to volume with methanol; samples are then spotted on the chromatographic paper (*7*).

2. *Isolation from Urine, Blood, and Spinal Fluid*

a. Urine. A 5 ml sample of urine in a 25 ml test tube is adjusted to pH 1 (about 5 drops HCl); 10 ml ethyl acetate is added and the tube is stoppered and shaken for 2 minutes. The ethyl acetate layer (top layer) is removed and placed in a beaker. The ethyl acetate extraction procedure is repeated. The extracts are combined, evaporated, and 0.5 ml 50% ethanol is added to the residue (*8*). A variation of this procedure includes treatment of the sample after the first ethyl acetate extraction with 10% $NaHCO_3$, then acidified and reextracted with ethyl acetate. Further purification of the extracts is accomplished by passage through a deactivated charcoal column, followed by elution of the indoles with phenol (*9, 10*). Preliminary purification of urine has been considered unnecessary for the examination of tryptophan metabolites (*11*).

The use of sequential solvents of graded polarity under variable pH conditions permits the fractionation of urinary indoles without too much overlapping in the multiple extracts; it also makes possible the analysis of more highly concentrated urine extracts by paper chromatography. In this procedure (*12*), urine (200–300 ml) is concentrated 10–20-fold *in vacuo* at 37–40°C. The concentrate is centrifuged and filtered. The filtrate is saturated with crystalline $(NH_4)_2SO_4$, again filtered, and divided into two equal parts. One of the portions is adjusted to pH 10 with 40% NaOH and then extracted four times with equal volumes of ethyl ether and four times with equal volumes of 2-butanone. The second portion is adjusted to pH 2 with HCl and subjected to the same extraction procedure. The four extracts are evaporated to dryness *in vacuo* and made up to a volume (equivalent to 100 mg creatinine per milliliter of extract) with 95% ethanol. For chromatography, 0.3 ml of each ethanol extract is applied to Whatman No. 1 chromatographic paper and developed by two-dimensional chromatography.

A modified procedure for group separation and purification of indoleamines has been proposed by Tanimukai (*13*). The concentrated urine (one fourth of a 24 hour urine volume) is adjusted to pH 4.5 with NH_4OH. Dowex 50-X2 (100–200 mesh) resin, H^+ form (5 gm wet weight) is added, the mixture is stirred for 45 minutes and then transferred to a glass column (2 × 30 cm). Prior to use, the resin is prepared as follows. The H^+ form of the resin is successively treated with acetone, 2 *N* NaOH, 2 *N* HCl, 1 *N* NH_4OH in 65% ethanol, and 2 *N* HCl, with water washing of the resin between each treatment. The urine effluent and the combined 30 ml distilled water wash of the column resin is retained for the detection of conjugated amines. The water-washed resin is subsequently eluted with 30 ml 0.1 *N* sodium acetate, 30 ml distilled water, and 30 ml 50% ethanol, respectively, to remove the neutral and aliphatic base substances. Sixty milliliters 0.1 *N* sodium acetate followed by 50 ml distilled water are substituted for the 30 ml washes if it is necessary to remove such compounds as tryptophan or 5-hydroxytryptophan from the resin. The indoleamine fraction is eluted with 30 ml of 1 *N* NH_4OH in 65% ethanol; the eluate is concentrated *in vacuo* at 36°C to a volume of 0.5 ml. Absolute ethanol (3 ml) is added to the concentrated extract and again concentrated to 1 ml. After centrifugation, the supernatant liquid is transferred to a conical-bottomed tube and evaporated to dryness *in vacuo* over sulfuric acid. The dried residue should contain any indoleamines, indoleamino acids, and phenolic amines. The above-mentioned fraction which contains any conjugated amines is adjusted to pH 1 with HCl and hydrolyzed for 30 minutes on a steam bath. The hydrolyzed solution is cooled, filtered, the pH is adjusted to 4.5, followed by the same procedure as above for the free amines. The samples should be stored *in vacuo* until they are used for chromatographic analysis.

b. Blood Serum. The sample (0.5 ml) is added to 5 ml of a mixture of absolute ethanol–acetone (1:1) in a 10 ml centrifuge tube (*11*). By means of either a magnetic stirrer or wrist action shaker, the contents of the tube are mixed for 10 minutes. The mixture is centrifuged and the supernatant liquid is decanted into a flask which has a previously calibrated 10 ml mark. The precipitate is washed with 2-ml portions of the solvent mixture, and the washings are added to the liquid until the 10 ml volume is reached; 0.1 ml of this solution is directly chromatographed or concentrated if necessary.

c. Spinal Fluid. The sample (1 ml) is gently heated to boiling, a drop of 10% trichloroacetic acid is added and heated in a water bath for 5 minutes at 50°C, then cooled and centrifuged; 0.1 ml of the supernate is used for chromatography (*11*).

The above procedures for blood serum and spinal fluid are applicable to the analysis of tryptophan metabolites.

B. Solvents for Chromatographic Development (see also Tables I and II)

1. *One-dimensional Development*

Isopropanol–NH_4OH–H_2O (10:1:1)—for separation of plant indoles (*14*)
n-Propanol–NH_4OH–H_2O (6:3:1)—for separation of 5-hydroxytryptamine, *N*-methyl-5-hydroxytryptamine, bufotenin, and tryptamine (*15*)
n-Butanol–acetic acid–H_2O (4:1:5) (*16*), organic phase—for separation of anthranilic acid and kynurenine (*17*)
Water-saturated *n*-butanol–NH_4OH (7:1) (*18*)
n-Butanol–pyridine–H_2O (1:1:1) (*19*)
n-Butanol–formic acid–H_2O (81:7:12)—for separation of L-kynurenine, kynurenic acid, and anthranilic acid (*20*)
Methanol–*n*-butanol–benzene–H_2O (2:1:1:1) (*19*)
Methanol–*n*-butanol (4:2) (*21*)
Ethyl acetate–pyridine–H_2O (4:2:4)—for separation of plant indoles (*18*)
Methyl ethyl ketone–pyridine–H_2O (70:15:15)—for separation of indole-acetic acid and tryptophan (*15*)
Formamide-saturated benzene–ethyl acetate (3:1)—using formamide-treated chromatographic paper (30% solution of formamide in acetone); paper is air-dried 20 minutes prior to use (*22*)
Dichloromethane containing 1.4% formamide by volume—also using formamide-treated paper (*22*).

The solvent mixtures containing formamide are useful for the separation of indole acids and their derivatives.

2. *Two-dimensional Development*

First direction: isopropanol–NH_4OH–H_2O (80:5:15)
Second direction: *n*-butanol–acetic acid–H_2O (4:1:5)
For the separation of indole, skatole, indole aldehyde (*23*)

First direction: isopropanol–acetic acid–H_2O (7:2:1)
Second direction: *n*-butanol–NH_4OH–H_2O (75:5:20) (*24*)

First direction: *n*-butanol–acetic acid–H_2O (4:1:5)—develop at least 15 hours
Second direction: aqueous KCl, 20%—3–4 hour development
For the separation of urinary indoles (*10*)

First direction: *n*-butanol–acetic acid–H_2O (4:1:5)
Second direction: Distilled H_2O (*11*)

TABLE I

VALUES OF INDOLE COMPOUNDS

Compound	R_F [a]						
	1	2	3	4	5	6	7
Indican	—	0.34	—	—	—	—	—
Tryptophan	0.50	0.44	0.62	0.42	0.24	0.18	0.34
N-Methyltryptophan	0.56	—	0.66	—	0.39	—	—
N-Acetyltryptophan	0.88	—	0.68	0.58	0.46	—	—
Tryptamine	0.71	—	0.78	0.82	0.82	0.65	0.72
1-Methyltryptamine	0.70	—	0.75	0.85	—	—	—
2-Methyltryptamine	0.72	—	0.78	0.85	—	—	—
α-Methyltryptamine	0.78	—	0.82	0.89	0.87	—	—
α-Ethyltryptamine	0.84	—	0.86	0.89	—	0.79	—
α,α-Dimethyltryptamine	0.83	—	0.87	0.90	0.89	—	—
N,*N*-Dimethyltryptamine	0.76	—	0.80	0.92	—	—	—
5-Hydroxytryptamine	0.47	0.38	0.70	0.65	0.57	0.52	—
5-Methoxytryptamine	0.60	—	0.73	—	0.76	—	—
6-Methoxytryptamine	0.62	—	0.75	—	0.76	—	—
5,6-Dimethoxytryptamine	0.50	—	0.66	—	0.63	—	—
Gramine	0.74	—	0.76	—	—	0.80	—
Indole	0.97	—	0.97	0.93	0.93	0.79	0.78
Skatole	0.97	—	0.97	0.95	0.95	0.9	—
Indole-3-carboxylic acid	—	—	—	—	—	0.20	0.62
Indole-3-acetic acid	0.90	—	0.80	0.52	0.38	0.25	0.43
Indole-3-propionic acid	0.92	—	0.86	0.59	0.45	0.35	0.49
Indole-3-butyric acid	0.95	—	0.91	0.65	0.55	0.44	0.53
Indolylacetonitrile	0.93	—	—	0.92	—	0.75	—
Indole-3-aldehyde	0.90	—	—	—	—	0.72	0.46
5-Hydroxyindoleacetic acid	0.77	—	0.70	0.31	0.20	0.17	—
Indoxyl sulfate	0.45	0.25	0.78	0.71	0.61	—	—
Indolyl-3-pyruvic acid	—	—	—	—	—	0.12	—
Anthranilic acid	—	0.90	—	—	—	—	—
3-Hydroxyanthranilic acid	—	0.84	—	—	—	—	—
5-Hydroxyanthranilic acid	—	0.54	—	—	—	—	—
L-Kynurenine	—	0.37	—	—	—	—	—
D-Kynurenine	—	0.31	—	—	—	—	—
L-3-Hydroxykynurenine	—	0.29	—	—	—	—	—
D-3-Hydroxykynurenine	—	0.24	—	—	—	—	—
L-5-Hydroxykynurenine	—	0.17	—	—	—	—	—
D-5-Hydroxykynurenine	—	0.14	—	—	—	—	—
Indole-3-butyronitrile	—	—	—	—	—	—	0.79
Indole-3-lactic acid	—	—	—	—	—	0.30	0.33
Indole-3-acrylic acid	—	—	—	—	—	—	0.80
5-Hydroxytryptophan	—	—	—	—	—	0.11	—
7-Hydroxyindoleacetic acid	—	—	—	—	—	0.09	—

[a] Solvents and references: (1) *n*-butanol–acetic acid–H_2O (60:15:25 v/v) (*19*); (2) *n*-butanol–acetic acid–H_2O (4:1:5 v/v) (*16*); (3) *n*-butanol–pyridine–H_2O (1:1:1 v/v) (*19*); (4) isopropanol–NH_4OH–H_2O (8:1:1 v/v) (*19*); (5) isopropanol–NH_4OH–H_2O (20:1:2 v/v) (*19*); (6) isopropanol–NH_4OH–H_2O (10:1:1 v/v) (*4*); (7) isopropanol–NH_4OH–H_2O (80:5:15 v/v) (*2*).

TABLE II

R_F VALUES OF INDOLEAMINES AND RELATED COMPOUNDS

Compound and reference	R_F [a]							Color		
	1	2	3	4a	4b	5a	5b	Ehrlich Reagent	DMCA[b]	DPNA[c]
Indole	0.91	0.92	—	—	—	—	—	Pink-red	Green	—
Skatole	0.92	0.92	—	—	—	—	—	Purple	Purple-blue	—
Indole-3-acrylic acid	0.43	0.86	—	—	—	—	—	Green	Blue	—
Indole-3-carboxylic acid	0.35	0.90	—	—	—	—	—	Blue-purple	—	—
5-Benzyloxyindole-3-acetic acid	0.52	0.87	—	—	—	—	—	Blue	Blue	—
Indole-3-aldehyde	0.90	0.86	—	—	—	—	—	—	—	—
Indole-3-acetonitrile	0.90	0.92	—	—	—	—	—	Pink	Purple-red	—
5-Hydroxytryptophan (*25*)	0.18	0.15	0.25	0.36	0.42	—	—	Purple	Blue	Purple-red
Serotonin	—	—	0.70	0.49	0.37	—	—	Blue-purple	Blue	Red
N-Methylserotonin	—	—	0.85	0.50	0.42	—	—	Blue-purple	Blue	Red
Bufotenin	—	—	0.90	0.60	0.51	—	—	Blue-purple	Blue	Red
5-Hydroxyindoleacetic acid (*26*)	0.24	0.70	0.25	—	—	—	—	Purple	Blue	Purple-red
Tryptophan	—	—	—	0.50	0.62	—	—	Red-purple	Purple	—
Tryptamine	—	—	—	0.70	0.54	—	—	Red-purple	Purple	—
N-Methyltryptamine	—	—	—	0.73	0.58	—	—	Red-purple	Purple	—

N-Dimethyltryptamine	—	—	—	0.75	0.64	—	—	Red-purple	Purple	—
N-Diethyltryptamine	—	—	—	0.83	0.66	—	—	Red-purple	Purple	—
N-Ethyltryptamine	—	—	—	0.84	0.58	—	—	Red-purple	Purple	—
Psilocin	—	—	—	0.75	0.52	—	—	Green-blue	Green	—
5-Methoxytryptamine	—	—	—	0.64	0.34	—	—	Blue-purple	Blue	—
5-Methoxy-*N*-dimethyltryptamine	—	—	—	0.70	0.47	—	—	Blue-purple	Blue	—
Melatonin	—	—	—	0.87	0.39	—	—	Blue	Blue-green	—
6-Hydroxytryptamine	—	—	—	0.46	0.35	—	—	Green	Blue-green	—
Kynuramine (*13*)	—	—	—	0.66	0.69	—	—	Red	Red-purple	—
Kynurenine	—	—	—	—	—	0.37	0.67	—	—	—
3-Hydroxykynurenine	—	—	—	—	—	0.30	0.56	—	—	—
Sodium-acetylkynurenine	—	—	—	—	—	0.82	0.81	—	—	—
Anthranilic acid	—	—	—	—	—	0.88	0.66	—	—	—
3-Hydroxyanthranilic acid	—	—	—	—	—	0.85	0.60	—	—	—
Kynurenic acid	—	—	—	—	—	0.56	0.55	—	—	—
Xanthurenic acid	—	—	—	—	—	0.56	0.45	—	—	—
Xanthurenic acid 8-methyl ether (*11*)	—	—	—	—	—	0.55	0.51	—	—	—

[a] Solvents: (1) and (3) isopropanol–NH_3–H_2O (8:1:1); (2) *n*-butanol–acetic acid–H_2O (4:1:1); (4a) *n*-butanol–acetic acid–H_2O (12:3:5), first direction; (4b) 20% aqueous KCl, second direction; (5a) *n*-butanol–acetic acid–H_2O (4:1:5), first direction; (5b) distilled water, second direction.

[b] *p*-Dimethylaminocinnamaldehyde.

[c] *p*-Nitroaniline (see Table III).

The potentially large number of indole compounds that may be present in biological materials makes it difficult to obtain a complete assay by one-dimensional chromatography, and therefore the two-dimensional technique will minimize the problem. Multiple extraction techniques applied to the sample (Section I), using variable conditions of pH and sample extraction solvents of different polarities (*12*) may also clarify questionable analytical data. The possibility of loss or change in chemical composition of the compounds, because of air oxidation and/or photochemical effects, are minimized by conducting the experimental work in areas of subdued light and preparing and storing the samples under a blanket of nitrogen. For precise work, the developing tank should be shielded from light, and preferably, the tank should be filled with nitrogen and chromatography conducted at reduced temperature (about 15°C). As an example, it has been noted that smog conditions, a common occurrence in some cities, will affect the chromatographic results of indole and phenol acids (*27*); the indole spots on the chromatogram may be destroyed by the ozone component of the atmosphere. Shaw and Trevarthen (*27*) obtained normal indole spots with chromatograms developed in isopropanol–NH_4OH–H_2O (8:1:1), but was unable to do so with a development mixture of benzene–propionic acid–H_2O (100:70:5), and this differential effect was attributed to ozonization under acidic conditions with the latter solvent mixture when the experiments were conducted on a smoggy day.

C. Detection Reagents

There are many chromogenic reagents that react with indole compounds. Some of them are given in Tables III (*2*, *7*, *12*, *24*, *26*, *28–33*) and IV (*33*). One of the more commonly used detection reagents, Ehrlich's (*p*-dimethylaminobenzaldehyde), which requires lengthy times for color development and produces unstable indole colored spots, has been improved by superimposing on the Ehrlich spray an additional spray of aqueous sodium nitrite (*34*); the spots appear immediately and are stable for at least 1 month.

p-Dimethylaminocinnamaldehyde (DMCA) is considered to be ten times more sensitive than the Ehrlich reagent (*31*); indole-3-acetic acid can be detected at the 0.5 μg level (*25*). When indole and skatole may have the same R_F values, DMCA will produce a green color with indole and a purple color with skatole.

Durkee and Sirois (*25*) compared the reaction results of indoles with Ehrlich's and the DMCA reagents and made the observation that 3-substituted indoles in which a carbonyl or nitrile group is attached to the β-carbon atom of the pyrrole ring; react incompletely; the reaction is greatly improved if either group is separated by a methylene group.

Because of the increasing importance of *N*-methylated indoleamines in the study of the biochemical aspects of schizophrenia, Tanimukai (*13*) has developed a sensitive chromatographic procedure for these compounds (see Section I). The purified urine sample (equivalent to one-fourth of a 24 hour urine volume) is dissolved in 100–200 μl of 95% ethanol and spotted on Whatman No. 1 paper. Two-dimensional chromatography is employed; first, with *n*-butanol–acetic acid–H_2O (12:3:5), then with 20% aqueous potassium chloride. *p*-Dimethylamino-benzaldehyde or *p*-dimethylaminocinnamaldehyde reagent is used to detect the indoles. The latter reagent is more sensitive for the detection of *N*-methylated indoleamines. Indoleamine content as low as 1 μg per 1000 ml of urine can be detected.

D. Quantitative Methods

The spots obtained on chromatograms sprayed with a solution of *p*-dimethylaminobenzaldehyde (1% reagent in 1 *N* HCl) are measured at 560 mμ. (See Chapter 4). Plant indole spots, located with the aid of guide strips, may be eluted with water and measured at 280 mμ (*18*). Colored spots produced by reaction with methanolic xanthydrol (10% methanolic xanthydrol–acetic acid, v/v, and exposure to HCl vapors) are eluted with methanol–acetic acid (1:2) and measured at 580 mμ (*35*). The blue color developed by tryptamine derivatives and xanthydrol reagent (1% xanthydrol and 10% trichloroacetic acid in absolute methanol) is eluted from the paper with 5 ml methanol; after the addition of 0.5 ml HCl, the colors are measured at 600 mμ (*7*).

The following procedures have been used for the measurement of tryptophan metabolites in biological fluids (*11*). The compounds on the developed chromatograms are observed under ultraviolet light (3655 Å) and marked with a pencil. The pieces of paper containing 3-hydroxykynurenine and xanthurenic acid are each placed in an Erlenmeyer flask fitted with a ground glass stopper; 3.8 ml distilled water is added. After 15–16 hours, 1 ml of diazotized sulfanilic acid (0.5% in 2% HCl mixed before using with an equal volume of 0.5% $NaNO_2$ aqueous solution) and 0.2 ml of pyridine are added; temperature should be maintained at 15°C. Absorbance due to xanthurenic acid is read immediately at 510 mμ. After 60–80 minutes, absorbance due to 3-hydroxykynurenine is read at 450 mμ. The pieces of paper containing kynurenine are eluted with 5 ml of a 1% solution of *p*-dimethylamino-benzaldehyde in 50% acetic acid; the colored solution is read at 450 mμ.

TABLE III

DETECTION REAGENTS FOR INDOLE COMPOUNDS

Detection reagent	Reagent composition	Reactive conditions with paper chromatograms	Remarks	References
p-Dimethylaminobenzaldehyde (Ehrlich's reagent)	(a) 0.5% in 1% alcohol HCl or 1% in 1 *N* aqueous HCl	Spray	Secondary spray with NH_4OH causes brown indican spot to change to red	*2, 24, 28*
	(b) 2 gm in 100 ml HCl–acetone (10:90)	Dip and heat to 40°C	Detects 0.05 μg serotinin and other indoles	*29*
	(c) 2% in concentrated HCl	Spray	Secondary spray with 1% $NaNO_2$ causes blue spots; indican = orange	*12*
	(d) 0.125 gm reagent + 0.1 ml ferric chloride in 100 ml 65% H_2SO_4	Spray	Hydroxyskatoles (HSK) causes yellow spots; 5-HSK, blue spot	*30*
p-Dimethylaminocinnamaldehyde (DMCA)	2 gm in 100 ml 95% ethanol + 100 ml 6 *N* HCl	Spray	Analysis of schizophrenic urines; indole = green; skatole = purple ten times more sensitive than Ehrlich reagent	*25, 30, 31*
4-*N*,*N*-bis(2 chloroethyl)-aminobenzaldehyde or 4-*N*,*N*-bis(2-chloroethyl)-amino-2-tolualdehyde	1 gm either reagent in HCl–methanol (25 ml:75 ml)	Spray	Hydroxyskatoles	*30*
Terephthalaldehyde	0.2 gm in acetone–acetic acid (90 ml:10 ml)	Spray	6-Hydroxyskatole = blue spot	*30*

Formaldehyde	10 ml 35% HCHO + 10 ml 25% HCl + 20 ml 95% ethanol	Spray and heat to 105°C 3 minutes	6-Hydroxyskatole = blue spot; other skatole isomers = brown	*30*
Paraformaldehyde	—	Paper heated in presence of paraformaldehyde 100–155°C 30 minutes	Primary and secondary indoleamines give fluorescent spots with UV light	*32*
p-Nitroaniline	1 ml reagent (0.6 mg/ml 1.2 *M* HCl) + 1 drop $NaNO_2$ solution (1.2 gm/100 ml H_2O) + 5 ml 1/15 *M* phosphate buffer (pH 7) + 1 *N* NaOH to pH 6.5, and diluted to 10 ml	Spray twice; dry sheet between each spray	5-Hydroxyindoles = red spots	*26*
Xanthydrol	(a) 0.2 gm in ethanol–HCl (90 ml:10 ml)	Spray	4-,6-Hydroxyskatoles = yellow spots; 5-,7-hydroxyskatoles = blue-gray spots	*30*
	(b) 1% in absolute methanol containing 10% trichloroacetic acid	Dip	5-*O*-substituted tryptamine derivatives	*7*
1-Nitroso-2-naphthol (N2N)	(a) 1% N2N in 95% ethanol (b) 2 *N* HCl in 2 volumes 5% $NaNO_2$ solution	Spray with (a) and dry; then spray with (b)	5-Hydroxyindoles = violet spots	*30*
$FeCl_3$–$HClO_4$	2 ml 0.05 *M* $FeCl_3$ + 100 ml 5% $HClO_4$	Spray	Indoles = red spots; blue-yellow after 24 hours	*33*
Gold chloride	0.5% aqueous solution	Spray	7-Hydroxyskatole = violet spot; 4-, 5-, 6-isomers = gray-green spots	*30*
KNO_2	1 gm in 20 ml concentrated HNO_3 and 80 ml 95% ethanol	Spray	—	*33*

TABLE IV

DETECTION AND SENSITIVITY OF INDOLE COMPOUNDS[a] (*33*)

Compound	Fluorescence in ultraviolet light	Color reactions with							
		$FeCl_3$–$HClO_4$		*p*-Dimethylamino benzaldehyde		KNO_2–HNO_3		Cinnamaldehyde HCl	
		Color	(μg)	Color	(μg)	Color	(μg)	Color	(μg)
Indole	Pale green	Orange red	3	Light red	3	Red	3	Pinkish brown	3
2-Methylindole	Green	Pinkish yellow	—	Scarlet	—	—	—	—	—
3-Methylindole	Light blue	Pinkish brown	3	Blue	1	Yellow	3	Light brown	3
Indole-3-aldehyde	Pale yellow	Light brown	3	Light brown	3	Yellow	10	Yellow	3
Indole-3-carboxylic acid	Blue	Orange	1	Pink	1	Red	3	Orange	3
Indole-3-acetic acid	Ash	Pink	1	Bluish pink	1	Red	3	Yellowish brown	1
Indole-3-propionic acid	Light blue	Light brown	1	Bluish green	1	Yellow	3	Light brown	1
Indole-3-butyric acid	Light blue	Brown	1	Bluish green	1	Yellow	3	Light brown	1
Indole-3-acetamide	Yellowish brown	Pink	1	Pinkish brown	1	Pinkish brown	1	Yellowish brown	3

Indole-3-acetonitrile	Greenish blue	Green	3	Yellow	1	Ashy brown	3	Yellow	5
Indole-3-butyronitrile	Violet	Orange brown	3	Bluish light brown	1	Orange	10	Yellowish brown	10
Tryptophan	Yellowish green (after treatment with $HClO_4$)	Light brown	1	Pink	1	Yellow	1	Light brown	3
Tryptamine	—	Dull brown	—	—	—	Yellow	3	—	—
Gramine	Gray	Pinkish blue	5	Yellow	1	—	—	—	—
Acetyl tryptophan	Light blue	Light brown	—	Ash	—	—	—	—	—
Isatin	Brown	Yellow	5	Yellow	3	Yellow	3	Light yellow	3
Dihydroxyindole	Brown	—	—	Yellow	—	—	—	—	—
N-Acetylisatin	Brown	—	—	Light yellow	—	Yellow	50	Light yellow	5
N-Acetylindoxyl	Blue	Light brown	10	Orange	3	Light brown	10	Brown light	3
Indoxyl acetate	Green	Blue	1	Grayish brown	3	Light ash	10	Brown	1
Indican	Light violet	Blue	3	Brownish orange	3	Yellow	10	Yellowish pink	1
Indican glucoside	Yellow	—	—	Pale yellow	—	—	—	—	—
Indigotin	Light blue	Blue	—	Black	—	—	—	—	—
Indigo disulfonate	Light blue	Blue	—	Black	—	—	—	—	—
Indigo tetrasulfonate	Light blue	Blue	—	Black	—	—	—	—	—

[a] See also Table III.

The kynurenic acid spots are eluted with 5 ml ethanol (spectroscopic grade) and the solution is read at 243 mμ. *N*-α-acetylkynurenine (read at 227 mμ) and xanthurenic acid 8-methyl ether (read at 237 mμ) spots are eluted from the paper with ethanol. Amounts of the metabolites as small as 1–5 μg can be measured.

The red dye produced on the chromatogram by the reaction of *p*-nitroaniline and the 5-hydroxyindoles (see Table III) is extracted from the paper areas with 2.5 ml 70% aqueous *n*-propanol in a test tube (*26*). The tubes are placed in a boiling water bath and allowed to remain for 2 minutes after the alcohol begins to boil. The cooled tubes are centrifuged and the decanted solution is measured at 490 mμ; 1–5 μg amounts of the compounds can be measured.

II. Phenols and Aromatic Acids

The literature abounds with paper chromatography data on the phenols, phenolic derivatives, and related aromatic compounds (*36–47*). One recent publication contains data on more than seven hundred compounds belonging to the series of phenol derivatives and related compounds (*38*). Obviously, this chapter can only include selective, although typical, examples of the paper chromatographic technique applicable to this group of compounds.

A. Preparation of Samples

Standard solutions of phenols and organic acids are dissolved in volatile organic solvents (ethyl ether, ethanol, methanol, acetone, chloroform, and others) at concentrations of 0.1–1.0%. About 5–20 μg of each substance is deposited on the chromatographic paper, although as little as 1 μg of some compounds may be detectable.

1. *Extraction from Urine* (*48*, *49*)

A measured volume of urine is treated with 0.1 volume of concentrated HCl and heated (refluxed) for 20 minutes to hydrolyze the ethereal sulfates. The urine is cooled and extracted continuously with ether for 6 hours. Unconjugated phenolic acids may also be extracted with ethyl acetate from acidified urine (pH 1) which has been saturated with NaCl; the phenols are reextracted with 10% $NaHCO_3$, the pH is adjusted to 1–2, and the phenols are again extracted with ethyl acetate. Aminophenols are removed from the extract with 2 *N* HCl and the residual nitrophenols are transferred to 2 *N* NaOH. The latter is acidified, and the nitrophenols are transferred to 10 ml ether which is then used for paper chromatography.

The metabolic products of hydroxybenzoic acids in urine have been identified by Bray (*50*). The urine is acidified with 0.2 volume of 2 *N* H_2SO_4 and extracted continuously for 6 hours to concentrate the unconjugated phenols. The acid content of the urine is then adjusted to 2 *N*, and the urine is hydrolyzed by boiling under reflux for 1 hour and again continuously extracted with ether for 20 hours to remove phenols originally conjugated. The residues after removal of the solvent from the two extracts are separately dissolved in 5 ml acetone. Hydroxy acids may be concentrated by extracting the ether solutions with saturated $NaHCO_3$ solutions (*51*). Biological samples may often be analyzed without previous concentration. Kelemen (*52*) applied 2.5–20 μl urine to the paper to analyze for aromatic amino acids.

Estrogen glucuronides in human urine have been hydrolyzed with β-glucuronidase (*53*). A portion of a 24 hour urine sample is acidified to pH 4.5. Five milliliters 1.0 *M* sodium acetate buffer (pH 4.5) is added to each 100 ml urine, 300 units per 1 ml of urine, β-glucuronidase are added and the urine is incubated at 30°C for 48 hours. The estrogens are extracted with ether, the ether extract is extracted four times with 20 ml portions of 2 *N* NaOH; the NaOH extract is immediately acidified with concentrated H_2SO_4 to pH 1.5–2.0 and reextracted with ether. The final ether extract is washed with $NaHCO_3$ and water and dried with Na_2SO_4 and evaporated to dryness *in vacuo* below 45°C. The residue is dissolved in 1–2 ml methanol prior to paper chromatography. Acid hydrolysis of the glucuronides may release unwanted chromogens (*54*) and enzyme hydrolysis is therefore preferable; however, where rapid routine clinical examinations are necessary, acid hydrolysis combined with absorption of the chromogens by activated charcoal during the washing procedure may be feasible.

To determine the suspected presence of pentachlorophenol in urine (*55*), 25 ml portions of urine, 5 ml concentrated H_2SO_4, and 15 ml petroleum ether are mixed together and shaken for 2 minutes. By means of a separatory funnel, the ether phase is removed; the aqueous layer is reextracted with 15 ml petroleum ether. The two ether layers are combined, washed twice with distilled water, and concentrated with the aid of a stream of nitrogen. The ether concentrate of the pentachlorophenol is applied to the paper chromatogram as such or first converted to the ether derivative with diazomethane and then chromatographed.

2. *Isolation of Pentachlorophenol from Blood* (*56*)

One to 5 ml blood, 20 ml 0.1 *N* H_2SO_4, and 12 ml benzene are placed in a 125 ml glass-stoppered Erlenmeyer flask which contains a Teflon magnetic stirring bar. The flask is stoppered and the contents heated at

50°C with constant stirring on a combination hot plate-magnetic stirrer for 20 minutes. The flask is removed from the heater and immediately cooled in an ice bath. The contents of the flask are transferred to a 40 ml centrifuge tube and centrifuged for 10 minutes; the upper benzene layer is removed by pipette and placed in a test tube. The benzene fraction is concentrated to about 1 ml on a steam bath (40°C) with the aid of a stream of filtered air or nitrogen. The concentrated extract may be used as is for chromatography or it may be converted to the ether derivative with diazomethane and then chromatographed.

3. *Isolation from Processed Foods* (*57*)

The material to be analyzed (25–50 gm) is acidified with H_2SO_4 and deproteinized by the addition of 2–3 ml of a mixture of 15% $K_4Fe(CN)_6$ and 30% $ZnSO_4$. The mixture is heated to 100°C and filtered. The filtrate is extracted with ether–petroleum ether (v/v) and the extract is dried over anhydrous Na_2SO_4 and evaporated *in vacuo*. The residue is dissolved in 75% methanol and used for chromatography.

4. *Isolation of Plant Phenolic Acids* (*58*)

Leaves or whole plants are chopped up and placed into boiling 80% ethanol followed by a reflux period. The filtered ethanol extract is evaporated to dryness *in vacuo*, and the residue is refluxed on a steam bath for 1 hour with 2 *N* HCl. Alkaline hydrolyzates are prepared with 2 *N* NaOH, and the mixture is allowed to stand for 5 hours in a nitrogen atmosphere. The acid hydrolyzates are extracted for 6 hours with ether; the alkaline hydrolyzates are acidified to pH 2 with HCl prior to extraction. The ether extracts are evaporated to dryness, and the residues are dissolved in ethanol for chromatography.

5. *Isolation of Monohydroxyphenols from the Vanilla Bean* (*59*)

Shredded fresh beans are soaked for a period of 3 weeks, and then extracted with equal volumes of 95% ethanol and water. The extract is concentrated and extracted with ethyl ether. The ether extract is washed once with 10% sodium bisulfite solution.

6. *Phenols in Hair Dyes* (*60*)

If a hair dye preparation is completely miscible with water, a 1–2 ml portion is diluted with an equal volume of water and spotted directly on the chromatogram. A preparation not completely miscible with water is diluted with *n*-hexane (H_2O to hexane, 2:20). The mixture is shaken gently in a separatory funnel in the presence of 10 ml aqueous acetic acid (1:9). The lower aqueous layer is used for chromatography. If the hair dye preparation is a dry solid or paste, 1 gm is mixed thoroughly with 5 ml aqueous acetic acid (1:19). The mixture is filtered through glass wool, and the filtrate is used for chromatography.

7. *Phenolic Acids in Cigarette Smoke and Tobacco*

Yang and Wender (*61*) extract cigarette smoke samples, previously trapped in a 1:1 mixture of anhydrous acetone–absolute ethanol at low temperature, with ether. The ether extract is shaken with 5% $NaHCO_3$ solution, and the bicarbonate solution is removed from the mixture and chilled. The aqueous layer is acidified with concentrated HCl, saturated with NaCl, and shaken several times with ether to reextract the phenolic acids. The ether extract is concentrated to a small volume. Tobacco powder is extracted with *n*-pentane to remove coloring material; the powder is then ether extracted, and the remainder of the procedure is similar to that applied to tobacco smoke solutions.

8. *Phenols in Automobile Exhaust* (*62*)

Samples of automobile exhaust are collected in impingers, each containing 250 ml NaOH in a series of three impingers ranging from 0.1 *N* to 1.0 *N* NaOH at ice-water temperature. The phenols are released by adding HCl to the solutions. Each solution is extracted twice with 100 ml portions of chloroform. About 50 ml of the chloroform extract from each impinger is combined and reacted with 25 ml 0.0105 *M* diazotized *o*-nitraniline. The mixture is shaken, allowed to stand 2 minutes, and made alkaline by the addition of 25 ml Na_2CO_3 (20% by weight). The mixture is then acidified with dilute HCl and the azo dyes are extracted with ether. The solvent phase is dried over anhydrous magnesium sulfate and evaporated to dryness with a stream of cool air. The azo dye extract is dissolved in chloroform, made to a specific volume, and used for chromatography.

9. *Isolation of Chlorogenic Acid* (*63*)

Plant material is boiled in 95% ethanol for 1 minute to destroy phenol oxidase activity. The plant material is then crushed with sand and extracted three times with 70% ethanol (ten times the weight of the sample) at 50–55°C. The extracts are concentrated to a total volume of 10 ml and 10 μl portions are spotted on the paper chromatogram.

10. *Preparation of Colored Derivatives*

Hossfeld (*64*) made satisfactory qualitative separations of many simple phenols by paper chromatography using a mixture of the sodium salts of phenylazo dyes derived by coupling phenols with diazotized sulfanilic acid. The dyes were prepared by the method of Cheronis (*65*): A solution is prepared consisting of sulfanilic acid (2.6 gm), 20 ml water, and 3 ml

6 *N* NaOH. $NaNO_2$ (1 gm) is added with stirring, and the mixture is poured slowly into a beaker which contains 30 ml water, 40 gm crushed ice, and 2 ml concentrated H_2SO_4. After 30 minutes at a temperature of about 5°C, a slight excess of the phenol is added, and the diazotization is allowed to proceed for 15 minutes. Twelve milliliters 6 *N* NaOH and 50 ml saturated NaCl solution are added, the mixture is cooled for 10 minutes, and the crystallized dye is filtered by suction.

B. Solvents (see Tables V–IX)

Petroleum ether–ethyl ether (1:1)—separation of naturally occurring alkyl-substituted phenols (*66*)

Methyl ethyl ketone–H_2O (v/v)—separation of phenols coupled with diazotized sulfanilic acid (*64*)

Isoamyl alcohol–xylene–H_2O (40:60:50 v/v plus 16.25% acetic acid)—for di- and trihydroxyphenols (*67*)

2% Acetic acid

Sodium formate–90% formic acid–H_2O (10 gm:1:200)

Isopropyl alcohol–NH_4OH–H_2O (20:1:2)

8% NaCl–acetic acid (100:1)—above combinations for separation of phenols and phenolic acids (*68*)

n-Butanol–benzene–formic acid–H_2O (100:19:10:25)—aged over the aqueous phase at 73°C for 3 days; aqueous layer in bottom of tank for descending chromatography; for separation of phenols and phenolic acids (*69*)

Petroleum ether (40–60°C)–formic acid (2:1)—separation of *tert*-butyl methoxyphenol, -quinol, *p*-benzoquinone, *n*- and *p*-methoxyphenol (*70*)

m-Cresol–acetic acid–H_2O (50:2:48 v/v), organic phase (*71*)

n-Butanol–acetic acid–H_2O (4:1:5) (*71, 72, 72a*)

Benzene–acetic acid–H_2O (125:72:3)—useful in estimating free hydroxyl groups of phenols, phenolic acids, and phenolic aldehydes (*73*)

Butanol–acetic acid–H_2O (60:15:25) or benzene–acetic acid–H_2O (2:2:1), organic phase—separation of phenols and phenolic acids (*74*)

sec-Butanol–2% aqueous Na_2CO_3 (v/v)—separation of phenols diazotized with sulfanilic acid (*75*)

n-Butanol saturated with 5 *N* NH_4OH (*76*)

n-Butanol–2% aqueous ammonia (1:1)—for phenolic compounds (*77*)

n-Butanol–concentrated NH_4OH (4:1 v/v)—separation of methylol phenols (*78*)

n-Butanol–benzene–concentrated NH_4OH (50:2:48 v/v)—separation of phenols and aminophenols from their respective glycosides (*79*)

n-Butanol–pyridine–saturated NaCl (1:1:2), organic phase—separation of phenols and phenolic acids (*80*)

sec-Butanol–2% Na_2CO_3 (v/v), first direction; *n*-butanol–ethanol–H_2O (20:6:20), second direction—separation of *o*- and *m*-cresol (*75*)

Benzene–acetic acid–H_2O (2:2:1), first direction; HCl (20%) or $KHCO_3$ (22%), second direction (*51, 81*)

Universal buffer solutions (Britton-Robinson) of different pH as mobile phase, using Whatman No. 3 paper impregnated with 10% olive oil in benzene—for separation of chlorinated phenols (*82*)

Petroleum ether (80–100°C) saturated with formic acid or xylene saturated with formamide or benzene–acetic acid–H_2O (2:2:1), organic phase, using Whatman No. 1 paper impregnated with methanol–formamide solution (30:100)—separation of chlorinated cresols and chlorinated xylenols (*83*)

Benzene–acetic acid–H_2O (2:2:1), first direction; HCl (20%) (w/v), second direction—separation of halogenated phenols (*84*)

Benzene–diethylamine–H_2O (3:2:5), lower aqueous phase used to equilibrate chromatographic chamber—separation of isomeric mononitrophenols (*85*)

Benzene–cyclohexane–dipropylene glycol (30:70:3), chromatographic paper pretreated with formamide—separation of azo dyes derivatives of phenols (*86*)

Acetic acid in concentrations ranging from 2.0 *M* to 8.0 *M* for separation of phenols on ion-exchange resin paper (45% Amberlite IR 120 cation exchange resin) (*87*)

Cyclohexane–ethyl acetate–acetic acid (5:1:1)—separation of mono- and dihydric phenols (*88*) or butanol–H_2O–acetic acid (6:2:1)—separation of phenolic aldehydes and hydroxybenzoic acids (*88*) using chromatographic paper impregnated with Amberlite IRA-400 Cl^- form ion-exchange resin

Acetone–H_2O (40:60, 45:55, 50:50)—separation of alkyl-substituted 1-phenyl-azo-2-naphthols on Whatman No. 3MM paper impregnated with mineral oil (5 gm oil/100 ml ether) (*89*)

Acetone–H_2O (70:30)—separation of pentachlorophenol and/or its ether derivative using Whatman No. 1 paper impregnated with 10% paraffin oil (*90*)

Isoamyl alcohol–ethanol–NH_3 (40:40:25) or isoamyl alcohol–ethanol–H_2O (40:40:25)*; or methyl isobutyl ketone–ethanol (70:30) saturated with NH_3—separation of α- and β-naphthols (*91*)

Benzene–cyclohexane (1:12 v/v plus 1% methanol), using chromatographic paper previously impregnated with 4% NaOH—separation of simple phenols (*92*)

n-Butanol–ethanol–3.0 *N* ammonium carbonate buffer (40:11:19) or *n*-butanol–benzene–3.0 *N* ammonium carbonate buffer (80:5:15) (*93*)

sec-Butanol–water (4:1), using Whatman No. 1 paper washed with acetic acetic acid, followed by a wash with phosphate buffer (pH 7.4)—separation of phenolic acids (*94*)

Cyclohexane–ethyl acetate (5:1) saturated with a mixture of dimethylformamide–H_2O (5:2)—separation of phenols on paper previously treated with a mixture of dimethylformamide–acetone (75:25) (*95*)

n-Propanol–kerosene (45:55), papers developed in presence of vapors from a 25% NH_4OH solution

Na_3PO_4 (0.125 *M*), using Whatman No. 1 paper impregnated with tricresyl phosphate (9.1% by volume in benzene)—for qualitative and quantitative analysis of Bisphenol A and its impurities (*96–98*)

* Whatman No. 1 paper impregnated with 3% NaOH before use.

Butanol–ethanol–pyridine–H_2O (3:1:1:1) or butanol–pyridine–H_2O (3:1:1)—separation of aromatic acids (*99*)

Butanol–ethanol–pyridine–H_2O (3:1:1:1)—separation of mono- and dicarboxylic acids (*100*)

Formamide–H_2O (60:40 v/v), pH 3.5 buffered with formic acid—separation of ellagic and flavellagic acids (*101*)

TABLE V

R_F VALUES OF AROMATIC ACIDS, PHENOLS, AND RELATED COMPOUNDS

Compound	R_F [a]										
	1	2	3	4	5	6	7	8	9	10	11
Aromatic Acids											
o-Hydroxy											
benzoic (salicylic acid)	0.78	0.88	0.66	0.90	0.43	0.65	—	0.95	0.84	—	—
hippuric	0.22	0.43	0.68	0.86	0.30	—	—	—	—	—	—
phenylacetic	0.76	0.57	0.86	0.91	0.49	—	—	—	—	—	—
mandelic	0.58	0.13	0.82	0.54	0.30	—	—	—	—	—	—
	(0.68)	—	(0.88)	(0.68)	—	—	—	—	—	—	—
phenylpropionic	0.70	0.74	0.77	0.91	0.72	—	—	—	—	—	—
cinnamic	0.35	0.70	0.44	0.93	0.75	—	—	—	—	—	—
phenyllactic	0.59	0.29	0.84	0.85	0.28	—	—	—	—	—	—
phenylpyruvic	dec.	0.21	0.85	0.70	0.43	—	—	—	—	—	—
phenylpyruvic lactone	0.54	0.88	0.44	0.85	0.87	—	—	—	—	—	—
m-Hydroxy											
benzoic	0.39	0.53	0.70	0.88	0.50	—	—	—	—	—	—
hippuric	0.33	0.09	0.75	0.78	0.19	—	—	—	—	—	—
phenylacetic	0.46	0.49	0.83	0.92	0.46	—	—	—	—	—	—
mandelic	0.37	0.07	0.87	0.68	0.19	—	—	—	—	—	—
phenylpropionic	0.54	0.61	0.77	0.92	0.68	—	—	—	—	—	—
cinnamic	0.44	0.58	0.39	0.93	0.60	—	—	—	—	—	—
phenyllactic	0.51	0.12	0.86	0.82	0.21	—	—	—	—	—	—
phenylpryuvic	dec.	0.13	streaks	0.74	dec.	—	—	—	—	—	—
p-Hydroxy											
benzoic	0.23	0.55	0.54	0.87	0.68	—	—	—	—	—	—
hippuric	0.15	0.07	0.72	0.78	0.18	—	—	—	—	—	—
phenylacetic	0.42	0.49	0.82	0.92	0.50	—	—	—	—	—	—

mandelic	0.33	0.07	0.87	0.68	0.16	—	—	—	—	—	—
phenylpropionic	0.51	0.61	0.73	0.93	0.69	—	—	—	—	—	—
cinnamic	0.28	0.58	0.31	0.93	0.67	—	—	—	—	—	—
phenyllactic	0.45	0.13	0.85	0.82	0.19	—	—	—	—	—	—
3-Methoxy-4-hydroxy											
benzoic (vanillic acid)	0.22	0.80	0.66	0.89	0.60	—	—	0.92	0.81	—	—
hippuric	0.16	0.18	0.66	0.74	0.16	—	—	—	—	—	—
phenylacetic	0.39	0.66	0.82	0.88	streaks	—	—	—	—	—	—
mandelic	0.28	0.16	0.84	0.71	0.20	—	—	—	—	—	—
phenylpropionic	0.43	0.81	0.74	0.91	0.65	—	—	—	—	—	—
cinnamic	0.27	0.80	0.35	0.88	0.63	—	—	—	—	—	—
Protocatechuic acid	0.06	0.16	0.73	0.83	streaks	—	—	0.85	0.35	—	—
Gentisic	0.68	0.26	0.53	0.89	0.34	0.51	—	—	—	—	—
α-Resorcylic	0.24	0.09	0.64	0.78	0.39	—	—	—	—	—	—
β-Resorcylic	0.39	0.38	0.48	0.92	0.34	0.55	—	0.93	0.54	—	—
γ-Resorcylic	0.77	0.11	0.50	0.54	0.48	—	—	—	—	—	—
Gallic	dec.	0	streaks	0.60	dec.	0.43	—	0.68	0.08	—	—
Syringic	0.18	0.79	0.57	0.87	dec.	—	—	—	—	—	—
Homogentisic	—	0.09	0.90	0.75	0.34	—	—	—	—	—	—
Caffeic	—	0.19	0.36	0.78	0.53	—	—	—	—	—	—
Kynurenic	0.52	0.16	0.37	0.55	0.21	—	—	0.64	—	—	—
Xanthurenic	0.08	0.03	0.24	0.53	0.23	—	—	0.69	—	—	—
3-Hydroxy-anthranilic	—	0.38	streaks	0.85	0.65	—	—	0.90	—	—	—
5-Hydroxy-indoleacetic	0.32	0.16	0.58	0.76	0.45	—	—	—	—	—	—
Pyrogallol carboxylic	—	—	—	—	—	0.71	—	—	—	—	—
Tannic acid	—	—	—	—	—	0.99	—	—	—	—	—
Quinolinic	—	—	—	—	—	—	—	—	0.50	—	—
Anthranilic	—	—	—	—	—	—	0.26	—	—	—	—
Mandelic	—	—	—	—	—	—	0.28	—	—	—	—
p-Aminobenzoic	—	—	—	—	—	—	0.09	—	—	—	—
Hippuric	—	—	—	—	—	—	0.31	—	—	—	—

Table continued

TABLE V (*continued*)

Compound	R_F[a]										
	1	2	3	4	5	6	7	8	9	10	11
o-Coumaric	—	—	—	—	—	—	0.27	—	—	—	—
Sulfanilic	—	—	—	—	—	—	0.11	—	—	—	—
Cinnamic	—	—	—	—	—	—	0.62	—	—	—	—
Phenylpropionic	—	—	—	—	—	—	0.57	—	—	—	—
Phenylacetic	—	—	—	—	—	—	0.44	—	—	—	—
Benzoic	—	—	—	—	—	—	0.41	—	—	—	—
Phenols											
Phenol	—	—	—	—	—	0.97	—	—	—	0.93	—
Catechol	—	—	—	—	—	0.96	—	0.91	0.74	—	—
Resorcinol	—	—	—	—	—	0.97	—	0.91	0.63	—	—
Hydroquinone	—	—	—	—	—	0.96	—	0.88	0.69	—	—
Pyrogallol	—	—	—	—	—	0.94	—	0.77	0.34	—	—
Hydroxyquinol	—	—	—	—	—	0.48	—	—	—	—	—
Phloroglucinol	—	—	—	—	—	0.96	—	0.76	0.16	—	—
Orcinol	—	—	—	—	—	—	—	0.91	0.75	—	—
o-Hydroxybenzyl alcohol	—	—	—	—	—	—	—	—	—	0.84	—
m-Hydroxybenzyl alcohol	—	—	—	—	—	—	—	—	—	0.81	—
p-Hydroxybenzyl alcohol	—	—	—	—	—	—	—	—	—	0.79	—
2,6-Dimethylolphenol	—	—	—	—	—	—	—	—	—	0.67	—
2,4-Dimethylolphenol	—	—	—	—	—	—	—	—	—	0.61	—
2,4,6-Trimethylolphenol	—	—	—	—	—	—	—	—	—	0.34	—
p-Cresol	—	—	—	—	—	—	—	—	—	0.92	—
p-Cresolmonomethylol	—	—	—	—	—	—	—	—	—	0.89	—
2,6-Dimethylol-*p*-cresol	—	—	—	—	—	—	—	—	—	0.82	—
o-Cresol	—	—	—	—	—	—	—	—	—	0.93	—
2,4-Dimethylol-*o*-cresol	—	—	—	—	—	—	—	—	—	0.79	—
m-Cresol	—	—	—	—	—	—	—	—	—	0.94	—

Conjugated Phenols											
m-Aminophenyl sulfuric acid	—	—	—	—	—	—	—	0.33	—	—	0.55
m-Aminophenyl glucoside	—	—	—	—	—	—	—	0.33	—	—	0.23
m-Aminophenol	—	—	—	—	—	—	—	0.75	—	—	0.90
m-Hydroxyphenylsulfamic acid	—	—	—	—	—	—	—	0.22	—	—	0.15
p-Aminophenyl sulfuric acid	—	—	—	—	—	—	—	0.33	—	—	0.10
p-Aminophenol	—	—	—	—	—	—	—	0.64	—	—	0.81
p-(2,4-Dihydroxyphenyl) azophenyl sulfuric acid	—	—	—	—	—	—	—	0.78	—	—	0.11
p-(2,4-Dihydroxyphenyl) azophenol	—	—	—	—	—	—	—	0.90	—	—	0.25
3-Quinolyl sulfuric acid	—	—	—	—	—	—	—	0.74	—	—	0.82
3-Hydroxyquinoline	—	—	—	—	—	—	—	0.88	—	—	0.70
6-Quinolyl sulfuric acid	—	—	—	—	—	—	—	0.54	—	—	0.62
6-Hydroxyquinoline	—	—	—	—	—	—	—	0.77	—	—	0.90
7-Quinolyl sulfuric acid	—	—	—	—	—	—	—	0.58	—	—	0.62
7-Hydroxyquinoline	—	—	—	—	—	—		0.74		—	0.90
8-Quinolyl sulfuric acid	—	—	—	—	—	—	—	0.56	—	—	0.61
8-Quinolyl glucoside	—	—	—	—	—	—	—	0.55	—	—	0.50
8-Hydroxyquinoline	—	—	—	—	—	—	—	0.70	—	—	0.90
7-Coumarinyl sulfuric acid	—	—	—	—	—	—	—	0.53	—	—	0.78
7-Coumarinylglucoside	—	—	—	—	—	—	—	0.67	—	—	0.46
7-Hydroxycoumarin (umbelliferone)	—	—	—	—	—	—	—	0.95	—	—	0.64
4-Methylcoumarin-7-yl sulfuric acid	—	—	—	—	—	—	—	0.66	—	—	0.90
7-Hydroxy-4-methylcoumarin	—	—	—	—	—	—	—	0.90	—	—	0.77

[a] Solvents and references: (1) isopropanol–concentrated NH_4OH–H_2O (8:1:1 v/v) (*49*); (2) benzene–propionic acid–H_2O (2:2:1 v/v) (*49*); (3) 20% KCl (*49*); (4) *n*-butanol–acetic acid–H_2O (8:2:2 v/v) (*49*); (5) *n*-butanol–pyridine–dioxane–H_2O (70:20:5:5 v/v) (*49*); (6) *n*-butanol–pyridine–saturated NaCl (1:1:2 v/v) (*80*); (7) benzene–acetic acid–H_2O (2:2:1 v/v) (*48*); (8) *n*-butanol–acetic acid–H_2O (4:1:5 v/v) (*71, 72, 72a*); (9) *m*-cresol–acetic acid–H_2O (50:2:48 v/v) (*71*); (10) *n*-butanol–NH_4OH (4:1 v/v) (*78*); (11) ethyl methyl ketone–2 *N* NH_4OH (2:1 v/v) (*72a*).

TABLE VI

R_F VALUES OF PHENYLAZOBENZENESULFONIC ACID DYES (*75*)

Phenol	R_F [*sec*-butanol–2% Na_2CO_3 (v/v) solvent]
Acetoguaiacone	0.11
2,6-Dimethoxyphenol	0.12
Guaiacol	0.19
Phenol	0.30
o-Chlorophenol	0.37
o-Hydroxybenzaldehyde	0.39
α-Naphthol	0.43
o-Cresol	0.43
m-Cresol	0.43
2,6-Dimethylphenol	0.48
2,4-Dichlorophenol	0.63
p-Methylguaiacol	0.64
2,5-Dimethylphenol	0.66
o-Ethylphenol	0.75
m-Ethylphenol	0.76
p-Ethylguaiacol	0.80
Eugenol	0.81
3,5-Dimethylphenol	0.83
o-Phenylphenol	0.84
m-Phenylphenol	0.86
p-Phenylphenol	0.89
p-Cresol	0.89
2,3,5-Trimethylphenol	0.90
Dihydroeugenol	0.93
3,4-Dimethylphenol	0.94
β-Naphthol	0.95
p-Ethylphenol	0.95
Thymol	0.96
4-(α-Methylbenzyl)-2-phenylphenol	0.98
2,4-Dimethylphenol	0.99

Phenol–H_2O (3:1 w/v), plus 0.9% formic acid—separation of dihydroshikimic, shikimic, and L-quinic acids (*102*)

Benzene–acetic acid–H_2O (2:2:1 v/v), organic phase—separation of hydroxybenzoic acids (*50*)

Benzene–acetic acid–H_2O (2:2:1), organic phase—separation of dihydroxybenzoic acids (*103*)

Ethyl acetate–acetic acid–H_2O (2:1:2), organic phase—separation of gallic acid, caffeic acid, chlorogenic acid (*103*)

Chloroform–acetic acid (99:1), using Whatman 3MM paper impregnated with 20% formamide—separation of benzoic, *p*-hydroxybenzoic, salicylic, and dehydroacetic acids (*104*)

Triethylamine (1%) in cumene, using Whatman 3MM paper impregnated with 20% formamide—separation of methyl through butyl and benzyl *p*-hydroxybenzoates (*104*)

TABLE VII

R_F VALUES OF CHLORINATED PHENOLS AND RELATED COMPOUNDS (*136, 137*)

Compound	R_F [a]				
	1	2	3	4	5
2,3-Dichlorophenol	—	0.84	—	—	—
3,4-Dichlorophenol	—	0.95	—	—	—
3,4-Dichlorocatechol	0.54	0.62	—	—	—
4,5-Dichlorocatechol	0.42	0.52	—	—	—
2,3-Dichloroquinol	0.38	—	—	—	—
2,5-Dichlorophenol	0.93	—	—	—	—
2,5-Dichloroquinol	0.33	—	—	—	—
2,3-Dichlorophenylmercapturic acid	—	0.70	—	—	—
3,4-Dichlorophenylmercapturic acid	—	0.75	—	—	—
2,4-Dichlorophenol	—	0.85	—	—	—
2,6-Dichlorophenol	—	0.73	—	—	—
3,5-Dichlorophenol	—	0.94	—	—	—
4-Chlorocatechol	0.33	—	—	—	—
3-Chlorocatechol	0.41	—	—	—	—
4-Chlororesorcinol	0.19	—	—	—	—
Chloroquinol	0.17	—	—	—	—
3,5-Dichlorocatechol	0.51	—	—	—	—
2,6-Dichloroquinol	0.39	—	—	—	—
L-2-Chlorophenylmercapturic acid	—	0.56	—	—	—
L-3-Chlorophenylmercapturic acid	—	0.61	—	—	—
L-4-Chlorophenylmercapturic acid	—	0.60	—	—	—
L-2,4-Dichlorophenylmercapturic acid	—	0.61	—	—	—
L-2,6-Dichlorophenylmercapturic acid	—	0.53	—	—	—
L-3,5-Dichlorophenylmercapturic acid	—	0.65	—	—	—
2-Chlorothiophenol	—	0.50	—	—	—
3-Chlorothiophenol	—	0.61	—	—	—
4-Chlorothiophenol	—	0.59	—	—	—
2,4-Dichlorothiophenol	—	0.02	—	—	—
2,6-Dichlorothiophenol	—	0.56	—	—	—
3,5-Dichlorothiophenol	—	0.39	—	—	—
2-Chloro-3-nitrophenol	—	—	0.75	—	0.69
3-Chloro-2-nitrophenol	—	—	1.0	—	0.69
3-Chloro-4-nitrophenol	—	—	0.07	—	0.69
4-Chloro-3-nitrophenol	—	—	0.14	—	0.87
o-Nitrophenol	—	—	1.0	—	0.57
2-Amino-3-chlorophenol	—	—	—	0.95	—
3-Amino-2-chlorophenol	—	—	—	0.91	—
3-Amino-4-chlorophenol	—	—	—	0.83	—
4-Amino-3-chlorophenol	—	—	—	0.67	—
o-Chloroaniline	—	—	—	1.0	—
o-Aminophenol	—	—	—	0.17	—

Table continued

TABLE VII (*continued*)

Compound	R_F [a]				
	1	2	3	4	5
N-Acetyl-*S*-(2-nitrophenyl)-L-cysteine	—	—	0.0	—	0.39
2-Chloro-4-nitrophenol	—	—	0.53	—	0.66
2-Chloro-6-nitrophenol	—	—	1.0	—	0.60
3-Chloro-5-nitrophenol	—	—	0.16	—	0.84
4-Chloro-2-nitrophenol	—	—	1.0	—	0.66
m-Nitrophenol	—	—	0.09	—	0.81
2-Amino-4-chlorophenol	—	—	—	0.62	—
2-Amino-6-chlorophenol	—	—	—	0.93	—
3-Amino-5-chlorophenol	—	—	—	0.51	—
4-Amino-2-chlorophenol	—	—	—	0.45	—
m-Chloroaniline	—	—	—	1.0	—
m-Aminophenol	—	—	—	0.4	—
2-Chloro-5-nitrophenol	—	—	0.50	—	0.69
5-Chloro-2-nitrophenol	—	—	1.0	—	0.66
p-Nitrophenol	—	—	0.02	—	0.59
2-Amino-5-chlorophenol	—	—	—	0.51	—
5-Amino-2-chlorophenol	—	—	—	0.70	—
p-Chloroaniline	—	—	—	1.0	—
p-Aminophenol	—	—	—	0.2	—
N-Acetyl-*S*-(4-nitrophenyl)-L-cysteine	—	—	0.0	—	0.42

[a] Solvents: (1) benzene–acetic acid–H_2O (1:1:2 v/v); (2) *n*-butanol–ethanol–3 *N* $(NH_4)_2CO_3$ buffer (40:11:19 v/v); (3) petroleum ether (90°C) saturated with 98% formic acid; (4) chloroform–acetic acid–H_2O (2:1:1 v/v); (5) *n*-butanol–3 *N* $(NH_4)_2CO_3$–3 *N* NH_4OH (4:3:3 v/v).

Methyl ethyl ketone–cineole–formic acid–H_2O (50:50:20:16 v/v) (*105*)

Ethyl acetate–formic acid–H_2O (10:2:3), use organic phase after phases are allowed to separate for a period of 6 hours—separation of chlorogenic and caffeic acids (*106*)

90 ml acetone and 3 ml NH_4OH diluted to 100 ml with water—separation of benzoic, *p*-hydroxybenzoic, and salicylic acids (*107*)

Propanol–3.5% NH_4OH–H_2O (350:12:105)—separation of isomeric phthalic acids (*108*)

n-Butanol–saturated with 0.005 *N* NH_4OH—separation of hydroxybenzoic acids, *m*- and *p*-hydroxyhippuric acids

n-Butanol saturated with ammonium carbonate (3.0 *N*)—separation of number of aromatic acids (*109*)

n-Butanol–dioxane–2 *N* NH_4OH (4:1:5 v/v)—separation of thyroacetic acids (*110*)

Acetic acid (2%)—separation of cinnamic acid derivatives (*112*)

Acetic acid–deionized H_2O (0.01, 0.1, 0.2, 0.4, 1, 2, 5, 10, 20, and 50% acetic acid, v/v), chromatography performed in darkness on Whatman No. 1 paper—separation of *cis*- and *trans*-cinnamic acid derivatives (*111*)

TABLE VIII

ADDITIONAL R_F VALUES OF PHENOLS (*83–85*)

Compound	R_F [a]			
	1	2	3	4
o-Chlorophenol	0.69	0.78	—	—
p-Chlorophenol	0.12	0.77	—	—
o-Bromophenol	0.47	0.88	—	—
p-Bromophenol	0.32	0.81	—	—
o-Iodophenol	0.93	0.89	—	—
p-Iodophenol	0.92	0.56	—	—
o-Cresol	—	—	0.50	—
6-Chloro-*o*-cresol	—	—	0.89	—
4-Chloro-*o*-cresol	—	—	0.66	—
4,6-Dichloro-*o*-cresol	—	—	0.91	—
m-Cresol	—	—	0.34	1.00
6-Chloro-*m*-cresol	—	—	0.73	—
4-Chloro-*m*-cresol	—	—	0.55	—
4,6-Dichloro-*m*-cresol	—	—	0.84	—
2,4,6-Trichloro-*m*-cresol	—	—	0.90	—
2,3-Dimethylphenol	—	—	0.62	—
4-Chloro-2,3-dimethylphenol	—	—	0.73	—
2,5-Dimethylphenol	—	—	0.60	—
4-Chloro-2,5-dimethylphenol	—	—	0.76	—
3,4-Dimethylphenol	—	—	0.44	—
6-Chloro-3,4-dimethylphenol	—	—	0.79	—
2,6-Dichloro-3,4-dimethylphenol	—	—	0.92	—
3,5-Dimethylphenol	—	—	0.59	—
4-Chloro-3,5-dimethylphenol	—	—	0.72	—
2,4-Dichloro-3,5-dimethylphenol	—	—	0.91	—
5-Ethyl-3-methylphenol	—	—	0.70	—
4-Chloro-5-ethyl-3-methylphenol	—	—	0.81	—
2,4-Dichloro-5-ethyl-3-methylphenol	—	—	0.92	—
o-Nitrophenol	—	—	—	0.44
m-Nitrophenol	—	—	—	0.65
p-Nitrophenol	—	—	—	0.10
2,4,6-Trinitrophenol	—	—	—	0.89
o-Hydroxybenzoic acid	—	—	—	0.27
m-Hydroxybenzoic acid	—	—	—	0.00
p-Hydroxybenzoic acid	—	—	—	0.00
α-Naphthol	—	—	—	1.00
β-Naphthol	—	—	—	1.00
o-Hydroxydiphenyl	—	—	—	1.00
2,2′-Dihydroxydiphenyl	—	—	—	0.77
Phenol	—	—	—	1.00
m-Cyanophenol	—	—	—	0.61
p-Cyanophenol	—	—	—	0.20

[a] (1) Benzene–acetic acid–H_2O (2:2:1, v/v); (2) HCl (20% w/v); (3) xylene saturated with formamide; (4) benzene–diethylamine–H_2O (3:2:5, v/v).

TABLE IX

R_F VALUES OF AROMATIC ACIDS AND RELATED COMPOUNDS

Compound	R_F [a]		
	1	2	3
o-Sulfamidobenzoic acid	0.28	—	—
m-Sulfamidobenzoic acid	0.23	—	—
p-Sulfamidobenzoic acid	0.15	—	—
Phloroglucinol	—	0.46	—
Salicylic acid	—	0.48	—
m-Hydroxybenzoic acid	—	0.39	0.47
p-Hydroxybenzoic acid	—	0.23	0.50
Protocatechuic acid	—	0.02	0.24
p-Hydroxyphenylacetic acid	—	0.42	0.46
3,5-Dihydroxybenzoic acid	—	0.20	—
β-Resorcylic acid	—	—	0.28
Gentisic acid	—	—	0.24
Vanillic acid	—	—	0.58
Syringic acid	—	—	0.65
o-Hydroxycinnamic acid	—	—	0.44
p-Hydroxycinnamic acid	—	—	0.46
Caffeic acid	—	—	0.21
Ferulic acid	—	—	0.52
Sinapic acid	—	—	0.53
3,4-Dihydroxyphenylacetic acid	—	—	0.20
3-(*p*-Hydroxyphenyl)propionic acid	—	—	0.57
3-(3,4-Dihydroxyphenyl)propionic acid	—	—	0.29

[a] (1) Whatman No. 4 paper, *n*-butanol–5% NH_3 (1:1) (*138*); (2) Whatman No. 1 paper, isopropanol–NH_4OH–H_2O (20:1:2) (*68*); (3) Reeve Angel Grade SB-2 Amberlite ion-exchange paper impregnated with Amberlite IRA-400 base resin in Cl^- form, *n*-butanol–H_2O–acetic acid (6:2:1) (*88*).

C. Techniques

Separations of phenols and aromatic acids have generally been carried out with Whatman No. 1 or Whatman No. 4 paper. Whatman No. 3 MM paper, pretreated with formamide or mineral oil, has also been used. Phenols and related compounds have been separated on filter papers impregnated with ion-exchange resins by development with aqueous solutions of acetic acid (*87*) or an acidic organic solvent mixture (*88*).

Treatment of the paper with Na_2CO_3 or Na_2HPO_4 solutions facilitates the separation of acidic compounds, i.e., phenol and carboxylic acids, from nonacid substance (*113*). Sodium salt dyes, formed by the reaction of phenols and diazotized sulfanilic acid, are separated on Whatman No. 1

paper which was previously treated with 4% Na_2CO_3 and dried prior to solvent development (*64*). Other modifications of the paper, such as treatment with NaOH or tricresyl phosphate, or the use of buffering media in the developing solvent, have proved to be useful (see Section II,B).

Circular paper chromatography has been employed for a mixture of phenols and organic acids using the solvent 1-pentanol–glacial acetic acid–water (4:1:5 v/v) (*114*). Circular papers, impregnated with tricresyl phosphate, have been developed with an aqueous solution of trisodium phosphate to determine the principal impurities in Bisphenol A. The various phenolic derivatives are separated as concentric rings on the paper and are detected by treating the dried paper with *p*-nitrobenzene-fluoborate (*98*).

An interesting technique of two-dimensional chromatography of phenols involves both $ZnCO_3$ impregnation and reversed-phase development on the same chromatogram (*115, 116*). Whatman No. 1 paper is immersed in a solution containing 40 gm $ZnCO_3$, 30 gm $(NH_4)_2CO_3$, 500 ml H_2O, and 200 ml NH_4OH. The paper is air dried and then heated at 95°C for 3 hours. The solvent for the first direction is dry benzene using the ascending development. The dry paper is next immersed in 5% ethyl oleate in petroleum ether (40–60°C), care being taken to impregnate only the area which has not been traversed by the substances during the first development. The second direction is developed with 40% ethanol for 16 hours.

Whatman No. 7 paper is impregnated with a saturated solution of boric acid, and the paper is developed with the organic phase of *n*-butanol saturated with a 2% NH_4OH solution for the separation of phenolic products derived from degraded lignin substances (*77*). Jurd (*117*) has noted that Na_2HPO_4 ionizes only strongly acidic phenolic hydroxyl groups, whereas sodium borate ionizes most phenolic groups; but a mixture of boric acid and sodium acetate will not ionize phenolic hydroxyl groups.

D. Detection

p-Dimethylaminobenzaldehyde, 500 mg in 1 ml of concentrated HCl and 100 ml of absolute ethanol (*30, 118*). Another mixture consists of 4% *p*-dimethylaminobenzaldehyde in acetic anhydride containing a few crystals of sodium acetate. After the paper is sprayed, the chromatogram is heated for 1–2 minutes at 130–150°C. The reagent is very sensitive for the detection of hippuric acid, which gives an orange spot. It will detect amounts as low as 1 μg of such compounds as sesamol, *D*-catechin, and phloretin (*119, 120*). Quantitative estimations can be made by elution of the azlactone with methanol and determination of the color density at 460 mμ (*72, 121*).

2,6-Dichloroquinonechlorimide (2% alcoholic solution), followed by one of the following sprays: saturated $NaHCO_3$ (pH 8), borax (pH 9), or Na_2CO_3 (pH 10). The reagent will differentiate halogenated benzenes (*136*).

Nitrobenzenediazonium fluoborate. A 1% solution of the reagent in acetone is painted on the chromatogram with a brush. After drying, the chromatogram is painted with a 0.1 *N* solution of KOH in methanol, causing colored spots to appear for phenols. The reagent is prepared by heating 14 gm of *p*-nitroaniline in a mixture of 30 ml concentrated HCl and 30 ml water. The mixture is cooled to 5°C and a solution of 8 gm $NaNO_2$ in 20 ml water is added. Sixty milliliters of 40% fluoboric acid is added, and the yellow precipitate is separated by filtration, washed successively with fluoboric acid, alcohol, and ether, and dried in a vacuum desiccator (*78*). The reagent has also been applied as a spray (*96, 98*).

N-(1-naphthyl)ethylenediamine (*122, 137*). The chromatogram is sprayed with a reagent composed of 2 *N* HCl, 0.1% $NaNO_2$, 0.5% ammonium sulfamate, and 0.1% *N*-(1-naphthyl)ethylenediamine · 2HCl.

β-Napthol (*79*). For the detection of aminophenols, the chromatogram is first sprayed with a 1% (w/v) solution of HNO_2, followed 5 minutes later with a 1% solution of β-naphthol in 1 *N* NaOH; red colors will result.

p-Nitraniline (*50*). Twenty-five milliliters of a solution of *p*-nitraniline (0.5%) in HCl (80% w/v) is mixed with 1.5 ml of $NaNO_2$ solution (5% w/v) just before spraying. Subsequently, the strip is sprayed with Na_2CO_3 solution (20% w/v). Reagent will detect hydroxybenzoic and aminohydroxybenzoic acids.

Fast Red Salt GG (a stabilized water-soluble diazo salt of *p*-nitroaniline); 0.05% spray solution, followed by spray of Na_2CO_3 solution. For phenolic compounds (*123*).

Diazotized Sulfanilic Acid (*124*). Twenty-five milliliters of freshly prepared 5% $NaNO_2$ is slowly added at 0°C to 5.0 ml sulfanilic acid solution (0.9 gm sulfanilic acid and 9 ml concentrated HCl diluted to 100 ml with water). The dried chromatogram is lightly sprayed with the reagent, followed by spraying the damp paper with 20% Na_2CO_3.

$FeCl_3$ Solution (0.1% w/v). A general reagent for phenols (*50, 125, 126*). A 2% ethanolic solution of $FeCl_3$ produces green spots with chlorogenic and caffeic acids with a sensitivity of 10 μg (*106*). Another mixture of 5% $FeCl_3$–1% potassium ferricyanide (1:1 v/v), has been applied to ion-exchange resin chromatographic papers, followed by steaming the paper for 5 minutes. Blue color will mark location of the phenols (*87*).

Phloroglucinol (5 gm in 300 ml methanol) (*139*). Reagent is sprayed on the chromatogram, dried, and followed by a secondary spray of KOH solution (0.1 *N*). For detection of *o*-dihydroxy compounds.

Amino-J acid (disodium salt of 2-naphthylamine-5,7-disulfonic acid) (*126a*). The spots of the phenolic compounds are coupled *in situ* with a cold alkaline solution of diazotized Amino-J (about 0.02 *M*). A dye is formed; α- and β-naphthols may be differentiated.

$AgNO_3$ (0.1 *N*). A convenient reagent for any easily oxidizable compound: reagent is in ammoniacal form. Many complex compounds, such as the pyrogallol tannins and hydroxybenzoic acids produce black spots on a white background (*50, 127*). A modification of the reagent consists of a mixture of saturated aqueous $AgNO_3$ (50 ml) diluted to 1 liter with acetone, plus enough water to dissolve the precipitate; can be used either as a spray or as a dip reagent (*128*). A third variation consists of a mixture of $AgNO_3$, 0.005 *M* (170 mg $AgNO_3$ in 1 ml water plus 5 ml NH_4OH, diluted to 200 ml with ethanol); reagent No. 2 consists of pyrogallol 0.0005 *M* (13 mg pyrogallol in 200 ml ethanol). The paper is sprayed with the $AgNO_3$ solution, air dried, then sprayed with the pyrogallol solution to detect benzoic, *p*-hydroxybenzoic, and salicylic acids (*107*).

Sodium 1,2-naphthoquinone-4-sulfonate (0.5 gm of compound in 95 ml distilled water is mixed with 5 ml glacial acetic acid) (*60*). For the detection of phenolic materials used as coloring agent in hair dyes.

Millon's reagent (modified) (*59*). Ten milliliters mercury are treated with 50 ml nitric acid on a steam bath, then diluted to 250 ml with water; allow to stand several hours, then filter; for detection of monohydroxyphenols in vanilla extracts.

Ceric ammonium nitrate (100 gm in 250 ml 2 *N* nitric acid); for detection of phenylphenols (*39*).

Antimony trichloride (34 gm $SbCl_3$ in 100 ml chloroform). The paper is sprayed, then heated 5 minutes at 100°C. Examine under both visible and ultraviolet (3660 Å) light for plant phenols (*129*).

Hydrogen peroxide, 0.3% solution. For the detection of some aromatic acids. The treated, wet paper, is illuminated with a Mineralight ultraviolet lamp equipped with filter to absorb most of visible radiation above 400 mμ. Blue fluorescent spots become visible within a few minutes (*130*).

NaOH solution (1 *N*). When various nitrophenols are sprayed with this reagent, characteristic colors are produced (*48*).

$NaNO_2$; 2% in 50% ethanol or 1% in 10% acetic acid is used to detect chlorogenic and caffeic acids (*63*, *106*). On exposure to ammonia, red-yellow spots appear; sensitivity, 5 μg.

Copper sulfate, 2% solution in water (*36*). The paper is first sprayed with $CuSO_4 \cdot 5H_2O$ solution and then with dilute ammonia. Aromatic and aliphatic acids produce blue, green, or brown spots.

Methyl red indicator. Five parts of 0.1% alcoholic methyl red solution is mixed with 10 parts of 0.667 *M* phosphate buffer solution of pH 7.0. Aromatic acids give pink spots on a yellow background. Other indicators (e.g., bromocresol green, bromophenol blue) may also be used (*93*, *136*).

3,5-Dinitrobenzoyl chloride, 2% solution in toluene (*131*). The sheets are sprayed, followed by dipping in a 0.5% solution of NaOH, then a water wash, and finally the sheets are dipped in an aqueous 0.002% solution of rhodamine and dried. In this procedure, invisible hydroxy compounds are converted to 3,5-dinitrobenzoate esters. The rhodamine-treated areas show dark spots on a bright orange background under ultraviolet light.

Sucrose (*132*). Two grams are mixed with 10 ml of concentrated HCl and 90 ml absolute ethanol. Complete solution is not achieved, but the suspension is sprayed on the chromatogram. The paper is then heated for 1 minute at 85–95°C. Characteristic colors are produced with many di- and trihydroxyphenols.

Ultraviolet absorption and fluorescence (*133*). When chromatograms are viewed with a Hanovia ultraviolet light (2536 Å) provided with a Chance 0 × 7 Filter, 1,2-dihydroxynaphthalenes appear as dark spots, while 1-naphthol, 2-naphthol, and derivatives cause fluorescence after being exposed to NH_3 vapors. Isomers of cinnamic acid may also be detected (*112*).

Antimicrobial agents used in food products, such as benzoic acid, *p*-hydroxybenzoic acid, salicylic acid, and dehydroacetic acid (labeled *System I* for what follows), and methyl- through butyl- and benzyl-*p*-hydroxybenzoates (labeled System II) are detected and identified by paper chromatography by the following procedures (*134*):

System I: Chromogenic agents; 1% potassium permanganate in a 2% solution of Na_2CO_3, 2% aqueous solution of ferric chloride; $AgNO_3$ (170 mg dissolved in 1 ml water), plus 5 ml NH_4OH, and diluted to 200 ml with ethanol. Separate papers are sprayed with one of these three reagents.

System II: The paper is sprayed with a reagent which consists of 1 part by weight mercury with 2 parts fuming HNO_3 diluted with 2 volumes of water.

Pearl and McCoy (*135*) have used water solutions of 30 different stabilized, commercially available, diazo salts for locating and identifying phenolic compounds on paper chromatograms. The developed chromatograms were dried, the dried papers were exposed to NH_3 vapors and sprayed immediately with a 0.05% water solution of the diazo salt; the paper was air dried and the colors of the spots, if any, were recorded; after 30 minutes the papers were sprayed with a saturated solution of sodium carbonate in water and allowed to dry; the colors were again recorded.

References

1. T. A. Bennet-Clark and N. P. Kefford, *Nature* **171**, 645 (1953).
2. A. J. Vlitos and W. Meudt, *Contrib. Boyce Thompson Inst.* **17**, 197 (1953).
3. L. E. Weller, S. H. Wittwer, and H. M. Sell, *J. Am. Chem. Soc.* **76**, 629 (1954).
4. B. B. Stowe and K. V. Thimann, *Arch. Biochem. Biophys.* **51**, 499 (1954).
5. N. P. Kefford, *J. Exptl. Botany* **6**, 129 (1955).
6. E. A. McComb and V. V. Rendig, *Chemist-Analyst* **49**, 55 (1960).
7. E. A. McComb, N. Androulidakis, and V. V. Rendig, *J. Chromatog.* **40**, 125 (1969).
8. H. K. Berry, C. Leonard, H. Peters, M. Granger, and N. Chunekamrai, *Clin. Chem.* **14**, 1033 (1968).
9. M. D. Armstrong and K. S. Robinson, *Arch. Biochem. Biophys.* **2**, 287 (1954).
10. C. E. Dalgliesh, *Biochem. J.* **64**, 481 (1956).
11. D. Coppini, C. A. Benassi, and M. Montorsi, *Clin. Chem.* **5**, 391 (1959).
12. H. Sprince, C. Parker, J. T. Dawson, Jr., D. Jameson, and F. C. Dohan, *J. Chromatog.* **8**, 457 (1962).
13. H. Tanimukai, *J. Chromatog.* **30**, 155 (1967).
14. T. A. Bennet-Clark, M. S. Tambiah, and N. P. Kefford, *Nature* **169**, 452 (1952).
15. D. Jerchel and R. Muller, *Naturwissenschaften* **38**, 561 (1951).
16. C. E. Dalgliesh, *Arch. Biochem. Biophys.* **58**, 214 (1955).
17. C. Mitoma, H. Weissbach, and S. Udenfriend, *Arch. Biochem. Biophys.* **63**, 122 (1956).
18. R. Muller, *Beitr. Biol. Pflanz.* **30**, 1 (1953).
19. J. B. Jepson, I. Smith, and B. J. Stevens, private communication (1954).
20. W. B. Jacoby and D. M. Bonner, *J. Biol. Chem.* **221**, 689 (1956).
21. C. A. Benassi, *Boll. Soc. Ital. Biol. Sper.* **27**, 420 (1951).
22. R. A. Heacock and M. E. Mahon, *Can. J. Biochem. Physiol.* **38**, 1500 (1960).
23. A. Fischer, *Planta* **43**, 288 (1954).

24. P. Decker and I. Sano, *Z. Physiol. Chem.* **300**, 245 (1955).
25. A. B. Durkee and J. C. Sirois, *J. Chromatog.* **13**, 173 (1964).
26. E. J. Cowles and A. C. Hilding, *J. Chromatog.* **19**, 326 (1965).
27. K. N. F. Shaw and J. Trevarthen, *Nature* **182**, 664 (1958).
28. K. K. Reddi and E. Kodicek, *Biochem. J.* **53**, 286 (1953).
29. B. Rodnight, *Biochem. J.* **64**, 621 (1956).
30. R. A. Heacock and M. E. Mahon, *J. Chromatog.* **17**, 338 (1965).
31. J. Harley-Mason and A. A. P. G. Archer, *Biochem. J.* **69**, 60 (1958).
32. E. J. Cowles, G. M. Christensen, and A. C. Hilding, *J. Chromatog.* **35**, 389 (1968).
33. S. P. Sen and A. C. Leopold, *Physiol. Planatarum* **7**, 98 (1954).
34. H. Sprince, *J. Chromatog.* **3**, 97 (1960).
35. K. Schreier and K. Guedke, *Naturwissenschaften* **43**, 130 (1956).
36. L. Reio, *J. Chromatog.* **1**, 338 (1958).
37. L. Reio, *J. Chromatog.* **4**, 458 (1960).
38. L. Reio, *J. Chromatog.* **13**, 475 (1964).
39. D. L. Gumprecht, *Anal. Chem.* **36**, 1154 (1964).
40. D. L. Gumprecht, *J. Chromatog.* **18**, 336 (1965).
41. D. L. Gumprecht and F. Schwartzenburg, Jr. *J. Chromatog.* **23**, 134 (1966).
42. D. L. Gumprecht, *J. Chromatog.* **30**, 528 (1967).
43. S. Marcinkiewicz, J. Green, and D. McHale, *J. Chromatog.* **10**, 42 (1963).
44. J. Green and S. Marcinkiewicz, *J. Chromatog.* **10**, 354 (1963).
45. S. Marcinkiewicz and J. Green, *J. Chromatog.* **10**, 366 (1963).
46. S. Marcinkiewicz and J. Green, *J. Chromatog.* **10**, 372 (1963).
47. J. Green and S. Marcinkiewicz, *J. Chromatog.* **10**, 389 (1963).
48. D. Robinson, J. N. Smith, and R. T. Williams, *Biochem. J.* **50**, 221 (1951).
49. M. D. Armstrong, K. N. F. Shaw, and P. E. Wall, *J. Biol. Chem.* **218**, 293 (1956).
50. H. G. Bray, W. V. Thorpe, and K. White, *Biochem. J.* **46**, 271 (1951).
51. R. J. Boscott, *Chem. & Ind.* (*London*) p. 472 (1952).
52. E. Kelemen, B. Tanos, and D. Halmagyi, *Biochem. J.* **47**, 138 (1950).
53. T. L. Dao, *Endocrinology* **61**, 242 (1957).
54. C. W. Porter, E. L. Clark, and G. E. Block, *J. Lab. Clin. Med.* **54**, 471 (1959).
55. A. Bevenue, J. R. Wilson, E. F. Potter, M. K. Song, H. Beckman, and G. Mallett, *Bull. Environ. Contam. Toxicol.* **1**, 257 (1966).
56. A. Bevenue, M. L. Emerson, L. J. Casarett, and W. L. Yauger, Jr., *J. Chromatog.* **38**, 467 (1968).
57. G. Safina and C. Petronici, *Conserve Deriv. Agrumari* (*Palermo*) **3**, 172 (1954).
58. R. K. Ibrahim and G. H. N. Towers, *Arch. Bioch. Biophys.* **87**, 125 (1960).
59. M. H. Anwar, *Anal. Chem.* **35**, 1974 (1963).
60. R. B. Smyth and G. G. McKeown, *J. Chromatog.* **16**, 454 (1964).
61. C. H. Yang and S. H. Wender, *J. Chromatog.* **8**, 82 (1962).
62. E. D. Barber, E. Sawicki, and S. P. McPherson, *Anal. Chem.* **36**, 2442 (1964).
63. H. Ruckenbrod, *Planta* **46**, 19 (1954).
64. R. L. Hossfeld, *J. Am. Chem. Soc.* **73**, 852 (1951).
65. N. D. Cheronis, "Semimicro and Macro Organic Chemistry: A Laboratory Manual." Crowell, New York, 1942.
66. R. G. K. Murthy, M. A. Siva Samban, and J. S. Aggewal, *J. Chromatog.* **32**, 519 (1968).
67. G. Wagner, *Arch. Pharm.* **286**, 269 (1953)
68. H. E. Wright, Jr. , W. W. Burton, and R. C. Berry, Jr., *Arch. Biochem. Biophys.* **86**, 94 (1960).

69. T. Fuleki and F. J. Francis, *J. Chromatog.* **26**, 404 (1967).
70. J. C. Dawe, F. A. Denz, and T. H. Kennedy, *Biochem. J.* **64**, 777 (1956).
71. E. C. Bate-Smith, *Biochem. Soc. Symp.* (*Cambridge, Engl.*) **3**, 62 (1949).
72. G. W. Gaffney, K. Schreier, N. DiFerrante, and K. I. Altman, *J. Biol. Chem.* **206**, 695 (1954).
72a. J. N. Smith, *Biochem. J.* **60**, 436 (1955).
73. W. Steck and S. H. Wender, *J. Chromatog.* **19**, 564 (1965).
74. P. A. Hedin, J. P. Minyard, Jr., and A. C. Thompson, *J. Chromatog.* **30**, 43 (1967).
75. W. H. Chang, R. L. Hossfeld, and W. M. Sandstrom, *J. Am. Chem. Soc.* **74**, 5766 (1952).
76. M. Lederer, *Australian J. Sci.* **11**, 208 (1949).
77. P. Colombo, D. Corbetta, A. Pirotta, and G. Ruffini, *J. Chromatog.* **6**, 467 (1961).
78. J. H. Freeman, *Anal. Chem.* **24**, 955 and 2001 (1952).
79. C. M. Myers and J. N. Smith, *Biochem. J.* **56**, 498 (1954).
80. R. A. Evans, W. H. Parr, and W. C. Evans, *Nature* **164**, 674 (1949).
81. K. S. Kirby, E. Knowles, and T. White, *J. Soc. Leather Trades' Chemists* **37**, 283 (1953).
82. Z. Vacek, S. Stota, and J. Stanek, *J. Chromatog.* **19**, 572 (1965).
83. S. Husain, *J. Chromatog.* **18**, 197 (1965).
84. H. S. Chogvill and D. E. Bissing, *Anal. Chem.* **32**, 440 (1960).
85. R. W. Matthews, *J. Chromatog.* **20**, 191 (1965).
86. G. B. Crump, *J. Chromatog.* **10**, 21 (1963).
87. D. Locke and J. Sherma, *Anal. Chim. Acta* **25**, 312 (1961).
88. I. T. Clark, *J. Chromatog.* **15**, 65 (1964).
89. R. B. Smyth and G. G. McKeown, *J. Chromatog.* **5**, 395 (1961).
90. A. Bevenue and H. Beckman, *Residue Rev.* **19**, 83 (1967).
91. M. R. Verma, R. Dass, and P. K. Gupta, *J. Chromatog.* **3**, 197 (1960).
92. D. Schleede, *Brenstoff-Chem.* **36**, 78 (1955).
93. M. E. Fewster and D. A. Hall, *Nature* **168**, 78 (1951).
94. C. F. Van Sumere, H. Teuchy, and F. Parmentier, *J. Chromatog.* **6**, 481 (1961).
95. E. Sundt, *J. Chromatog.* **6**, 475 (1961).
96. W. M. Anderson, G. B. Carter, and A. J. Landua, *Anal. Chem.* **31**, 1214 (1959).
97. G. Challa and P. H. Hermans, *Anal. Chem.* **32**, 778 (1960).
98. N. H. Reinking and A. E. Barnabeo, *Anal. Chem.* **37**, 395 (1965).
99. J. Franc, *Collection Czech. Chem. Commun.* **23**, 655 (1958).
100. J. Franc, *J. Chromatog.* **3**, 317 (1960).
101. D. E. Hathway, *Nature* **177**, 747 (1956).
102. D. E. Hathway, *Biochem. J.* **63**, 380 (1956).
103. R. W. Keith, D. LeTourneau, and D. Mahlum, *J. Chromatog.* **1**, 534 (1958).
104. C. Genest and R. A. Chapman, *J. Assoc. Offic. Agr. Chemists* **42**, 436 (1959).
105. E. F. L. J. Anet and T. M. Reynolds, *Nature* **172**, 1188 (1953).
106. A. B. Svendson, *Pharm. Acta Helv.* **26**, 253 (1951).
107. L. C. Mitchell, *J. Assoc. Offic. Agr. Chemists* **40**, 592 (1957).
108. L. V. Golosova, *Zh. Anal. Khim.* **14**, 748 (1959).
109. M. E. Fewster and D. A. Hall, *Nature* **168**, 78 (1951).
110. J. H. Wilkinson, *Biochem. J.* **63**, 601 (1956).
111. A. H. Williams, *Chem. & Ind.* (*London*), p. 120 (1955).
112. J. S. Challice and A. H. Williams, *J. Chromatog.* **21**, 355 (1966).
113. A. G. Newcombe and S. G. Reid, *Nature* **172**, 455 (1953).
114. J. Barnabas, *Naturwissenschaften* **41**, 453 (1954).

115. J. Green and S. Marcinkiewicz, *Nature* **176**, 1172 (1955).
116. J. Green, S. Marcinkiewicz, and P. R. Watt, *J. Sci. Food Agr.* **6**, 274 (1955).
117. L. Jurd, *J. Chromatog.* **4**, 369 (1960).
118. T. K. Sundaram, R. Radhakrishnamurthy, E. R. B. Shanmuggasundaram, and P. S. Sorma, *Proc. Soc. Exptl. Biol. Med.* **84**, 544 (1953).
119. G. J. Kapadia, J. R. Mosby, G. G. Kapadia, and T. B. Zalucky, *J. Chromatog.* **12**, 420 (1963).
120. C. Steelink, *Nature* **184**, 720 (1959).
121. G. L. Haberland, F. Bruns, and K. I. Altman, *Biochim. Biophys. Acta* **15**, 578 (1954).
122. H. G. Bray, V. M. Craddock, and W. V. Thorpe, *Biochem. J.* **60**, 225 (1955).
123. I. A. Pearl and P. F. McCoy, *Anal. Chem.* **32**, 132 (1960).
124. B. N. Ames and H. K. Mitchell, *J. Am. Chem. Soc.* **74**, 252 (1952).
125. E. A. H. Roberts and D. J. Wood, *Biochem. J.* **49**, 414 (1951).
126. W. Keup, *Z. Physiol. Chem.* **291**, 223 (1952).
126a. M. R. Verma, R. Dass, and P. K. Gupta, *J. Chromatog.* **3**, 197 (1960).
127. R. S. Asquith, *Nature* **168**, 738 (1951).
128. W. J. Burke, A. D. Potter, and R. M. Parkhurst, *Anal. Chem.* **32**, 727 (1960).
129. D. J. Sessa and J. J. Rackis, *J. Chromatog.* **23**, 487 (1966).
130. D. W. Grant, *J. Chromatog.* **10**, 511 (1963).
131. H. H. Brownell, *J. Chromatog.* **19**, 556 (1965).
132. D. G. Roux, *Nature* **168**, 1041 (1951).
133. E. Boyland and J. B. Solomon, *Biochem. J.* **63**, 679 (1956).
134. C. Genest and R. A. Chapman, *J. Assoc. Offic. Agr. Chemists* **42**, 436 (1959).
135. I. A. Pearl and P. F. McCoy, *Anal. Chem.* **32**, 1407 (1960).
136. W. M. Azouz, D. V. Parke, and R. T. Williams, *Biochem. J.* **59**, 410 (1955).
137. H. G. Bray, S. P. James, and W. V. Thorpe, *Biochem. J.* **64**, 38 (1956).
138. M. Trojna and J. Hubacek, *Chem. Listy* **52**, 87 (1958).
139. H. A. Schroeder, *J. Chromatog.* **30**, 537 (1967).

11

NATURALLY OCCURRING PIGMENTS

I. Fat-Soluble Chloroplast Pigments

A. Introduction

Adsorption chromatography with the novel development or formation of a chromatogram had its beginning in 1906 when Michael Tswett described the separation and isolation of the green and yellow chloroplast pigments extracted from green plants. In his milestone paper (*1*), which has recently been reevaluated (*2*) and translated into English (*3*), Tswett described his column chromatographic method in detail, the definitive resolution of the mixture of chloroplast pigments and the nomenclature that is still widely used today. In this same paper (*1*), Tswett reported his results with capillary analysis, the technique introduced earlier by Goppelsroeder (*4*) based on the capillary rise of solvents and solutes in filter paper. Tswett realized that this procedure lacked the critical development step and the adsorption phenomenon and cautioned that it should not be confused with adsorption chromatography.

Subsequent studies of the chloroplast pigments led to the discovery of radial paper chromatography in 1939 (*5*). Despite these and a few other early efforts, the earnest application of paper chromatography to the study of plant pigments began only around 1952. Since then, several hundred papers involving many modifications of paper chromatography have appeared. Those published through 1963 have been extensively reviewed by Sesták (*6*).

B. Preparation of Sample

1. *Plant Material*

Paper chromatography has been used to study the pigments in many kinds of plants, including the spore and seed plants as well as algae and colored bacteria. It is important to define clearly the plant material as to plant group, species, part of the plant being used, and conditions for growth. Growth conditions are known to affect greatly the kinds and amounts of pigments. It is desirable to use fresh plant material which was grown under known conditions and collected in its vegetative stage. Preserved, frozen, canned, brined, injured, killed (anesthetized or ground), or nonvegetative materials often yield additional pigments, due perhaps to alterations of some of the natural pigments (*7*). For repetition of earlier chromatographic observations, it is essential to use the same plant material as employed before. Convenient sources of the chloroplast pigments of higher plants are fresh spinach purchased in a market and cocklebur (*Xanthium*) grown under greenhouse conditions.

2. *Extraction of Pigments*

If spurious chromatographic observations are to be avoided, the pigments must be extracted from the plant using extraction solvents and other conditions that avoid alteration of the pigments by oxidation, reaction with acids, and enzymatic hydrolysis. Only the purest grade solvents should be used. Because the solvents must remove the water as well as the pigments from fresh plant material, liquids that are miscible with water are desirable. Acetone, methanol, ethanol, or methanol plus petroleum ether have often been used. Some plants do not yield their pigments to acetone as readily as to methanol plus petroleum ether (*8*). Treatment of the plants initially with boiling water often improves the extraction. Many leaves readily yield their pigments to acetone in a blender (see below). In solutions of pure alcohol, chlorophylls undergo rapid and complete oxidation (allomerization), but this is retarded in alcohols containing water (*9*) or petroleum ether (*10*). Complete extraction of the pigments from dry plant material may require preliminary hydration followed by extraction with the solvents mentioned above (*7*). In addition to the pigments, large quantities of colorless fatty substances (fats, hydrocarbons, fatty acids, resins, waxes, etc.) as well as hydrophilic pigments and colorless material are always removed from the plants during the extraction procedure.

Many workers transfer the pigments to petroleum ether prior to spotting on the paper. This eliminates the greater part of the undesirable hydrophilic material from the extract (*11–14*). Colorless fatty substances extracted from the plant will of course be transferred to the petroleum ether along with the fat-soluble pigments. Some of these substances may accompany the pigments during chromatography and affect the shapes of the chromatographic zones (*6*). They would also be eluted from the paper along with the separated pigments. In some cases, the initial extract is concentrated and spotted directly on the paper (*14*, *15*).

For the extraction of small quantities of fresh, hydrated plant material that is easily disintegrated, the following procedure (*16*) is recommended. Two grams of plant material (remove petioles and large veins from leaves) are placed in a chilled blender with 40 ml cold acetone or methanol and blended for 2 minutes. The mixture is centrifuged, and the clear green supernate is transferred to a 250 ml separatory funnel. (The residue can be reextracted, and the second supernate combined with the first if a more quantitative removal of pigments is required.) The green solution is diluted with cold, high boiling (65–110°C) petroleum ether and 100 ml of 10% aqueous NaCl solution. The mixture is shaken and allowed to separate. The lower layer is discarded unless it is quite green, in which case it can be reextracted with petroleum ether (or in some cases with diethyl ether), which is then combined with the original upper layer. The petroleum ether is then evaporated below 40°C with a rotary evaporator and vacuum pump, and the residue is dried with a high vacuum and stored, if necessary, under vacuum. The quantity of extraction liquid must be kept large relative to the weight of the plant material (at least twenty times) because rapid dehydration and extraction retard many alterations of the pigments (*16*).

This procedure has worked successfully with spinach and cocklebur leaves. It may be desirable to use a different amount of starting material for some other plants in order to obtain an extract with the desired pigment concentration. Some kinds of algae and some other leaf materials must be first placed in boiling water for 1–2 minutes and then quickly cooled before their pigments can be extracted in this way. Although this heating improves the extractability of the pigments by coagulating proteins and extracting certain water-soluble substances, it has the disadvantage that it usually increases the pheophytins and partially isomerizes the chlorophylls (to chlorophylls a′ and b′) and carotenoids. In the case of cocklebur, the heating sometimes produces an additional nonacidic, strongly adsorbed green pigment with an absorption spectrum similar to that of chlorophyll a (*16*). As an alternative to heating, the plant material can be crushed and ground in a mortar with a mixture of sand, a trace of magnesium or calcium carbonate, and a little of the extraction solvent prior to the extraction (*1*, *14*, *17*). It is unlikely, however, that this procedure entirely avoids alterations of the pigments.

Some plants are quite tough and are not easily disintegrated in a blender. It is best to cut these up into small pieces (if necessary), heat and cool them, and leach out the pigments with several fresh portions of excess solvent.

The extract residue prepared above is dissolved in 1 ml of petroleum ether (65–110°C) or 1 ml diethyl ether–petroleum ether (1:1 v/v) to prepare the final solution. Generally, several microliters of this final solution, 1.0 ml of which contains virtually all the pigments in 2.0 gm of plant material, is applied to dry paper in the form of a spot or streak (*18*, *19*). Evaporation of the solvent can be increased in a stream of cold air or heated nitrogen (*20*). Evaporation of the solvent over sulfuric acid (*21*) is undesirable because the pigments may be decomposed. A special applicating device has been described (*22*) for spotting a petroleum ether sample on wet paper for partition chromatography. Because of the lability of the pigments, it is best to prepare the extract immediately before chromatography. Formation of chlorophyll isomers and alterations of the carotenoids often occur in pigment solutions even if they are kept in darkness below 0°C (*14*).

If one wishes to study the yellow carotenoid pigments alone, the green chlorophylls [as well as much colorless material (*16*)] can be removed by saponification of the plant extract. To an acetone extract of 2.0 gm of plant material prepared as above, add 20 ml of a 10% solution of KOH in methanol. After 20 minutes, the reaction mixture is diluted with 40 ml diethyl ether–petroleum ether (65–110°C) (1:1 v/v) and 200 ml 10% NaCl solution. The upper, golden yellow layer is separated, washed twice with 100 ml portions of water, and evaporated below 40°C. The final solution is prepared by dissolving the residue in 1 ml diethyl ether–petroleum ether (1:1 v/v). Alternatively, the chlorophylls can be removed with phosphotungstic acid (*23*), or they can be fixed on top of a column of activated magnesia, which permits the carotenoids to be separated and eluted (*7*, *24*). The principal xanthophylls of diatoms, dinoflagellates, and brown algae are readily decomposed by alkalies; hence, the chlorophylls should not be removed from extracts of these plants by saponification (*8*, *25*).

C. Paper

Many kinds of commercially available filter papers have been utilized for separation of the chloroplast pigments by adsorption chromatography (*6*). A comparative study (*16*) of Eaton–Dikeman No. 301, Munktells and Whatman No. 1, 4, and 3 MM papers indicated that Whatman No. 1 provided the most uniform zones as well as moderate solvent flow and no formation of pheophytins from the chlorophylls. With a given chromatographic solvent, all the papers exhibited very similar adsorption capacity and sequence of the pigments zones. Washing the papers with petroleum ether (20–40°C) previous to the migrations was without detectable effect, and in fact, most workers employ commercial filter papers without preliminary treatment or drying (*6*).

Cellulose Chromagram sheets (Distillation Products Industries, Rochester, N. Y.) have proved to be somewhat superior to paper for separations of leaf extract and saponified leaf extract in some systems (*18*). These sheets contain purified nonfibrous cellulose in a 160 μ layer without binder on a plastic film backing and can be used with paper chromatographic techniques.

Papers impregnated with adsorbents such as alumina (*5*, *26*), calcium carbonate (*26*), sucrose (*26–28*), and calcium hydroxide (*29*) have been employed. These papers are usually activated above 100°C before use. In many cases, the pigments cannot be eluted from impregnated papers in an unaltered state (*30*). Papers impregnated with active adsorbents such as silica gel (e.g., Whatman SG 81) should be avoided because of possible alteration of the pigments (*31*). Glass fiber paper (*32*) and paper containing neutralized ion-exchange resins (*33*) have also provided effective separations.

Filter paper has also been employed for pigment separations by partition and reversed-phase partition chromatography. For the former, the paper is impregnated with water (*22*, *34*) or aqueous alcohols (*34*) and washed with a nonpolar solvent. For the latter, the paper is impregnated with petroleum jelly (*34*), paraffin oil (*35*), or olive oil (*36*) and washed with aqueous alcohols. In reversed-phase systems, the pigments are separated in the reverse sequence. Separations by partition chromatography are more troublesome and no better than those based on adsorption (*7*). Moreover, solvents in reversed-phase systems often dissolve some of the fatty stationary phase which then becomes a contaminant of the separated pigments (*6*).

Separations in paper and in columns of cellulose are quite similar with respect to resolution and time of solvent flow (*37*, *38*). However, because a zone moves through a very much smaller cross-sectional area of sorbent during paper chromatography compared to column chromatography, much smaller amounts can be separated on paper. For the same degree of separation, columns 10 cm in diameter, as used in preparative work (*39*), could be loaded with 1,000–10,000 times more pigment than in a spot on paper.

D. Solvents

Although no universal solvent has been described for the separation of all the pigments in every type of plant material, many different solvents have been tested and found to successfully separate certain chloroplast pigments. These solvents have ranged from completely nonpolar to mixtures of nonpolar solvents containing a little polar solvent. The ratio of polar to nonpolar constituent as well as the nature of each are critical factors in designing a solvent for a particular separation. For example, nonpolar solvents such as petroleum ether serve usually to separate only the nonpolar carotenes from the original spot of leaf extract (*16*, *40*, *41*). As the polarity of the solvent is increased by adding, for example, small amounts of *n*-propanol, acetone, or diethyl ether, the movement and separation of the xanthophylls and chlorophylls is increased (*42*). In quite polar solvents all the pigments move with the solvent front (*43*). As a practical guide, the order of polarity or eluting power of various liquids is available in the literature (*44*) (see also Chapter 3, Section V, A). For mixtures of liquids, the polarity varies primarily with the concentration of the more polar substances (*7*).

Wolf (*6*, *45*) has classified solvents into four general types for the separation of the pigments of spinach on Whatman No. 1 paper based on the sequence of the carotenes, chlorophylls, and unspecified xanthophylls, and Stanishevskaya (*46*) has made a detailed, critical comparison of twenty-seven solvent systems found in the literature. Listed below are examples of solvent systems which have provided good pigment separations. Other solvents will be described in Section H below.

1. *Petroleum Ether (100–140°C)–Benzene–Chloroform–Acetone–Isopropanol (50:35:10:0.5:0.17 v/v)*

This solvent gives sharp separations of numerous carotenoid and chlorophyll pigments by ascending development for 24 hours in vacuum on Schleicher and Schuell No. 2071 paper previously dried at 50°C (*47*). Stanishevskaya (*46*) claims that this solvent is the best of those he has studied for one-dimensional pigment separations. Sesták (*48*) has separated the pigments from wheat leaves with this solvent by ascending development from streaks of mixture for 60–80 minutes at room temperature on Whatman No. 3 paper. The separated zones are in the order carotenes (least sorbed), pheophytin, lutein, violaxanthin, chlorophyll a, and chlorophyll b plus neoxanthin (most sorbed).

2. *Toluene–Isopropanol (400:1 v/v)*

This solvent was used by Sesták (*49*) to separate the pigments of green algae by ascending development from streaks of mixture in the dark for a distance of 20 cm (about 1 hour) on Whatman No. 1 or 3 paper. The pigments are in the order carotenes (least sorbed), lutein, violaxanthin, chlorophyll a, neoxanthin, and chlorophyll b. Attempts to employ this solvent to resolve spinach extract placed on the paper as a spot were unsuccessful in the author's laboratory because of poorly separated, double tailing zones. It has been shown (*16*, *18*, *33*, *37*) that zones of this type are characteristic of separations of leaf extract in paper and cellulose sheets with many solvents when the sample is applied as a spot, especially when the loading is rather heavy. Application of the initial zone as a uniform streak across a narrow sheet of paper obviates this problem.

3. *Toluene*

Toluene separates the pigments of *Chlorella* in the order carotenes (least sorbed), lutein, violaxanthin, chlorophyll a, chlorophyll b, and neoxanthin by centrifugally accelerated radial chromatography at 600 rpm for 12.5 minutes at 25°C under a nitrogen atmosphere on Whatman 3MM paper (*50*).

4. *Petroleum Ether–Ethanol (20:1–20:1.5 v/v)*

This solvent has been used to separate chlorophylls a and b by ascending development prior to their elution and quantitative estimation by colorimetry (*41*).

5. *Benzene–Petroleum Ether (3:1–2:1 v/v)*

This solvent in the first direction followed by solvent 4 in the second direction serves for the two-dimensional separation of all the leaf pigments prior to their quantitative estimation (*51*).

6. *Petroleum Ether–10% Acetone*

This solvent gives the following R_F values for carotenoids by circular development on Schleicher and Schuell No. 287 kieselguhr paper (*52*): neoxanthin, 0.25; violaxanthin, 0.62; lutein, 0.71; zeaxanthin, 0.73; cryptoxanthin, 0.94; β-carotene, 0.98.

E. Chromatographic Techniques

Most solvents used for the chromatography of pigments are highly volatile and flammable. It is therefore necessary to use tightly closed chromatographic chambers and to work in a well-ventilated area away from any flames or exposed electrical contacts. A large glass jar lined with filter paper and fitted with a glass top is used for ascending development. The chamber is pre-equilibrated for 30 minutes prior to chromatography and wrapped in aluminum foil to prevent light decomposition of the pigments (*16*). Pigment decomposition can also be prevented by developing at low temperature (*12, 53, 54*) or by using an inert atmosphere in the chamber (*53, 54*). Rectangular paper is formed into a cylinder, clipped, and stood in the solvent in the bottom of the chamber. Alternatively, the paper can be formed into a cone and stood on its base in the solvent (*55*).

Cellulose Chromagram sheets are developed by the ascending method in rectangular thin-layer chromatography tanks or museum jars lined with filter paper and wrapped in aluminum foil (*18*, *56*).

Descending paper chromatography has been less frequently used (*15*, *57*) and offers no advantages over the ascending technique for pigment separations.

For two-way paper chromatography, the sample is applied in the corner of a square of paper which is formed into a cylinder and developed with the first solvent by the ascending method. The paper is then dried for 5 minutes and formed into a second cylinder with the axis at right angles to that of the first one. This cylinder is then developed with a second solvent in another equilibrated chamber (*58*). Because this technique is slower than one-dimensional chromatography, it is not usually desirable for the labile chloroplast pigments.

Radial chromatographic development is carried out in covered jars as described for the one-way migrations, but the paper, cut in the shape and dimensions shown in Fig. 4, is supported in a horizontal position on a crystallizing dish, with the tab bent downward to dip into the wash liquid in the bottom of the jar (*24*). Although development times are a little longer than for one-way migrations because of the narrow tab, this arrangement of the paper leads to greater flow of solvent at the trailing zone boundaries relative to the leading zone boundaries, usually forming sharper zones (*24*). Radial development can be speeded up by centrifugation (*50*). In another arrangement, the paper is held in a horizontal position between glass plates, and solvent is fed to the paper by means of a cord wick (*59*) or hypodermic syringe (*60*) through a small hole in the top plate.

The effects of changes in temperature (*29*, *61*) and the saturation of the air in the developing tank (*57*, *62*) have been studied in relation to the degree and speed of separation and decomposition of the pigments.

F. Detection and Identification of Separated Pigments

One undesirable feature of paper chromatography is the small quantity of material that can be isolated from the chromatographic zones. Because of this, the properties that may be used as a basis for description and identification of the separated pigments are severely restricted. Pigments separated by paper chromatography have been identified according to their color in visible light, by color reactions with acids, by spectral absorption properties, and by chromatographic comparison (co-chromatography) with authentic preparations of the pigments obtained by large-scale columnar chromatography.

With visual detection, the shapes and colors of the zones must be marked or photographed as soon as the chromatogram is developed, because the colors fade as the pigments decompose upon exposure to air. Carotenes appear yellow-orange, lutein and violaxanthin yellow, neoxanthin lemon yellow, chlorophyll a green, chlorophyll b yellow-green, and pheophytins olive-green to gray. Other carotenoids are yellow to orange to red. In ultraviolet light, chlorophylls may be observed as red-violet fluorescent zones (*14*). R_F values cannot be used for identification purposes because the mobility of each pigment is a function of its concentration (Fig. 1), and the change in mobility is not uniform. That is, the more adsorbed pigments migrate relatively faster than those less adsorbed when the concentration is increased (*16*).

When chromatograms are exposed to the vapors of concentrated HCl, neoxanthin becomes blue-green, violaxanthin becomes blue, and lutein and carotenes remain yellow (*16*, *18*). The exact nature of these color reactions with the adsorbed pigments is not fully understood. This test is most effective when applied to carotenoids separated from saponified extracts because the chlorophylls, which become gray-green when exposed to HCl vapors, often obscure one or more of the xanthophylls on leaf-extract chromatograms.

If sufficient quantities of the separated pigments are present, each zone can be cut out and eluted with ethanol or diethyl ether for determination of spectral absorption curves. For identification, the shapes of the curves and the wavelengths of the maxima are compared to literature results for authentic pigments (*39*, *63*, *64*). Absorption maxima for the most common carotenoids in ethanol are β-carotene, 452, 480 mμ; lutein, 446, 474 mμ; violaxanthin, 442, 470 mμ; and neoxanthin, 437, 466 mμ. Maxima for chlorophylls a and b in ethyl ether are, respectively, 428.5, 660.5 and 452.5, 642.0 mμ. It must be emphasized that these values are for the pure pigments and that a single chromatographic separation does not provide pigments free of all other pigments and colorless substances (*65–67*). Repeated paper chromatography of the eluted pigments on other papers or in other solvents can be employed for further purifications (*68–71*). It is usually more desirable to use a small column or thin layer of an adsorbent other than cellulose so that the residual impurities are strongly adsorbed or decomposed (*25*). During these added purification steps, the likelihood of pigment decomposition is, of course, increased (*70*). For detection of contamination by colorless substances, specific absorption coefficients, infrared absorption, and nuclear magnetic resonance are most useful (*39*, *63*, *72*). For these determinations, the pigments must be isolated in milligram quantities, and this is usually accomplished by column chromatography.

Identification by chromatographic comparison, which can be carried out with very small quantities, is performed by developing the test extract, single authentic pigment spots, and mixtures of the knowns and unknown next to each other in several systems in parallel (*49*, *73*, *74*). Samples with similar pigment contents should be used so that the resultant patterns can be easily compared. The optimum amounts of pigments for comparative developments on various papers have been experimentally determined (*48*). Popova (*61*) has suggested four solvents for comparative development which separate in turn carotenes, oxidized carotenoids, neoxanthin, and chlorophylls a and b. Columnar procedures for the separation and isolation of pure, crystalline chlorophylls (*39*) and carotenoids (*8*) for use as authentic reference compounds have been described.

The separation and detection of the individual pigments after paper chromatography is intimately related to the loading of the initial zone. Obviously, the isolation and identification of minor pigments is more difficult than that of major ones. If the system is overloaded, the zones of the major pigments will be easily detectable but will be poorly separated from each other and may overrun or trail into and obscure the zones of the minor pigments. With light loading, the separation of the zones is high, but their detectability is low. As a rough rule, the paper can be considered to be overloaded when the number of separated zones is less than that observed with lower loading (i.e., two or more zones have run together). The paper is underloaded when the number of detectable zones is less than that obtained with a slightly higher loading (*37*).

G. Quantitative Methods

Paper chromatographic methods for the determination of the separated pigments are at best semiquantitative. The most frequently used procedure involves washing of the separated pigments from the paper with an elutriant like acetone or ethanol followed by photometry (*22*, *41*, *51*, *61*, *75*). Significant amounts of the pigments are retained by the paper during the elution step, amounting to about 10% (*48*) or 20% (*76*) for the chlorophylls and 15–20% for the carotenoids (*76*). Moreover, Bacon (*77*) found that pure chlorophyll a was altered to the extent of 17% when eluted from Whatman No. 1 paper on which it had been spotted and left in the dark for 11 minutes. He suggests that the pigments are readily susceptible to alteration between the paper chromatographic development step and the elution step, when the pigments are spread out on the paper and separated from any protective colorless material in the sample, and when exposure to air allows the evaporation of the solvents. This has also been proved to be true on thin layers of siliceous adsorbents (*31*, *77*).

For the quantitative estimation of the separated pigments while still on the paper, densitometry (*46*, *78*), reflectance (*79*), spot size (*62*, *80–82*), and change in mobility (*46*, *83*, *84*) have been used with only limited success.

H. Detailed Experiments

Chromatographic examination of many plants growing in various parts of the world under different external conditions has indicated that certain combinations of pigments are typical of species in each major taxonomic group (*7*, *8*, *85*). Due to space limitations, only procedures for the separation of the pigments in flowering plants and some algae can be given.

1. *Separation of Pigments of Flowering Plants (Spermatophytes)*

A great variety of organisms, ranging from seed plants to green algae, are known to contain the same pigment mixture (*7*, *8*). For the separation of these pigments, 2 μl of an extract of either 2.0 gm spinach leaves or 25 ml of a suspension of *Chlorella pyrenoidosa* Chick (Emerson Strain, containing e.g. 6×10^7 cells/ml) is developed by ascending chromatography on cellulose Chromagram sheets with a modification of Hager's solvent (*30*), petroleum ether (20–40°C)–benzene–chloroform–acetone–isopropanol (or *n*-propanol) (50:35:10:5:0.17) (*18*, *86*). The chloroform is purified before use by two washings with equal volumes of water followed by drying over Drierite. After about 40 minutes, the separation shown in Fig. 1A is obtained. From top to bottom, the zones are carotenes ($\beta \pm \alpha$), lutein ($\pm$ zeaxanthin), violaxanthin, chlorophyll a, and chlorophyll b plus neoxanthin. Figure 1 demonstrates that with increased loading, the zones are pushed forward and the resolution is decreased. With Hager's solvent (containing 0.5 parts of acetone) on cellulose Chromagram sheets, the pigments are in the same sequence, but the separation of chlorophylls a and b is not complete even at low loadings. Development with diethyl ether–petroleum ether (20–40°C) (1:1 v/v) plus 0.25% *n*-propanol (*18*) yields a separation in the sequence carotenes, lutein, chlorophyll a plus violaxanthin, chlorophyll b, neoxanthin (Fig. 1B). It should be noted that, as is often the case, the sequence of the zones can be changed by a change in solvent.

Figure 2 shows the sequence and separation of pigments in 2 μl of saponified extract on cellulose sheets with three solvents defined in the legend (*18*). In each case, four distinct major yellow zones were formed, and the pigments were identified by spectral absorption properties and

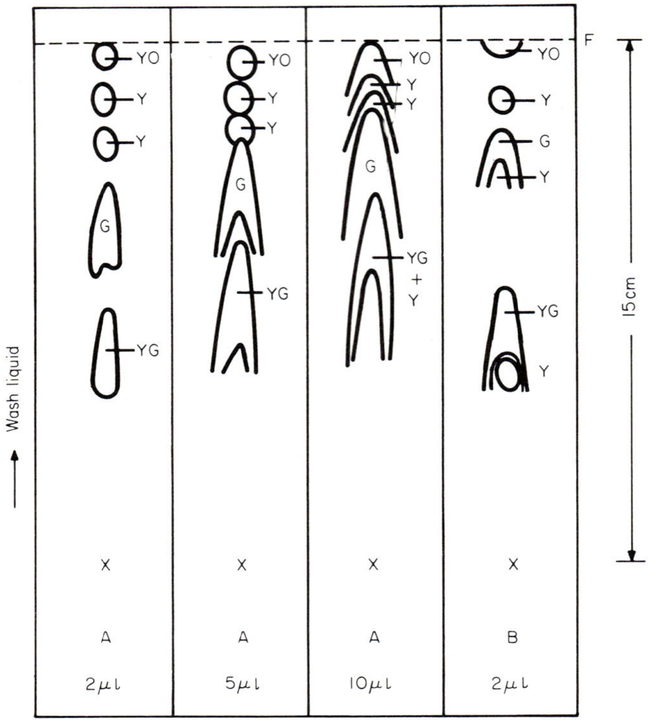

FIG. 1. Pigments in various amounts of leaf extract separated by one-dimensional migration on cellulose sheets with (A) petroleum ether (20–40°C)–benzene–chloroform–acetone–isopropanol (50:35:10:5:0.17) and (B) diethyl ether–petroleum ether (20–40°C) (1:1) plus 0.25% *n*-propanol as the solvent. F = front; f = faint; G = green; O = Orange; × = starting point; Y = yellow (*18*).

reaction with HCl vapors. With solvent A, one or two additional, unidentified minor carotenoids are often observed between violaxanthin and lutein.

Two-way chromatography of 3 μl of extract on Whatman No. 1 paper with petroleum ether (20–40°C) plus 1% *n*-propanol followed by petroleum ether (20–40°C)–chloroform (3:1 v/v) gives the separation shown in Fig. 3 (*58*). Use of solvents A and B (*87*), described above for one-way development, or solvent A followed by petroleum ether (20–40°C) plus 1% *n*-propanol (*88*), provides a two-way separation of the same pigments on cellulose Chromagram sheets.

Radial development of 10 μl of extract for 1–2 hr on Whatman No. 1 paper with petroleum ether (20–40°C) plus 1.5% *n*-propanol gives the

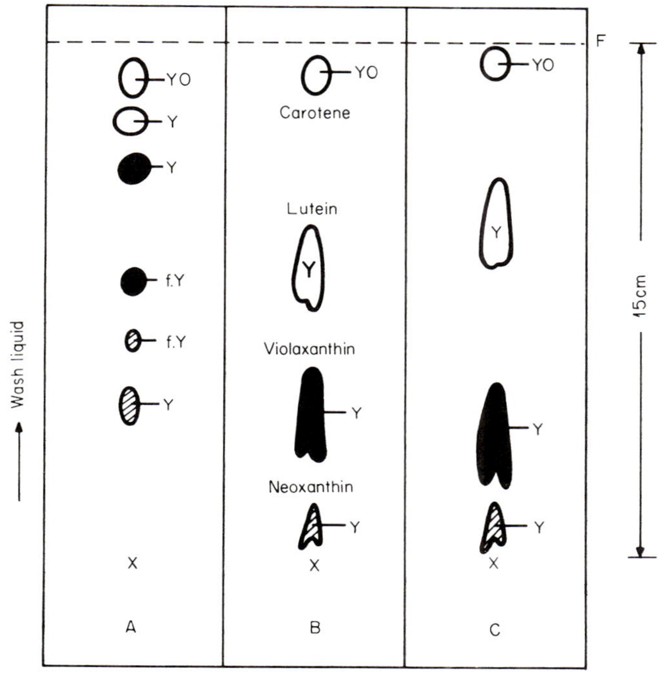

FIG. 2. Pigments in a 2 μl of saponified leaf extract separated by one-dimensional migration on cellulose sheets with (A) petroleum ether (20–40°C)–benzene–chloroform–acetone–isopropanol (50:35:10:5:0.17), (B) carbon tetrachloride plus 0.25% *n*-propanol, and (C) benzene–petroleum ether (20–40°C) (2:1) as the solvent. (See the legend in Fig. 1 for symbols and abbreviations.) Solid areas = blue by HCl vapor; hatched areas = blue-green by HCl vapor (*18*).

separation shown in Fig. 4 (*24*). Zeaxanthin, if present, will be mixed with the lutein in this system. These two pigments can be separated by radial development on paper impregnated with magnesium oxide–hyflosupercel (2:1) with 20% acetone in petroleum ether as the solvent (*52*). Figures 1–4 indicate that leaves and some green algae contain only two green chlorophylls and four yellow carotenoids as major pigments.

2. *Separation of Pigments in Certain Algae*

Jeffrey (*73*, *88a*) has separated the pigments in Siphonales, diatoms (Bacillariophyceae), golden-brown flagellates (Chrysophyceae) and dinoflagellates by two-dimensional ascending chromatography at room temperature in the dark on 22 cm squares of Whatman No. 3 mm paper.

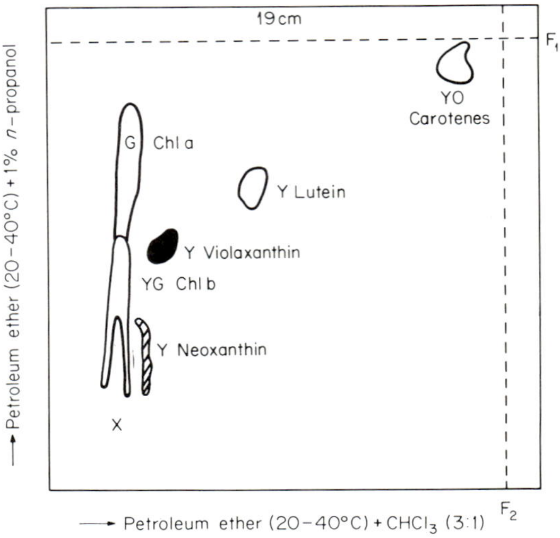

FIG. 3. Pigments in 3 μl of leaf extract separated by two-way chromatography on Whatman No. 1 paper. (See legends to Figs. 1 and 2 for symbols and abbreviations.) (*58*).

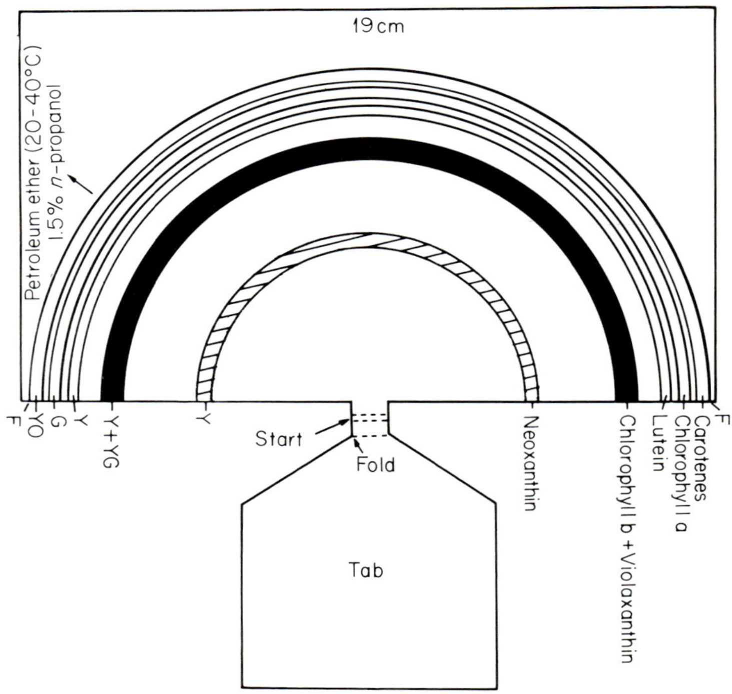

FIG. 4. Pigments in 10 μl of leaf extract separated by radial chromatography on Whatman No. 1 paper with petroleum ether (20–40°C) plus 1.5% *n*-propanol as the solvent. (See legends to Figs. 1 and 2 for symbols and abbreviations.) (*24*).

Development was with petroleum ether (60–80°C) plus 4% *n*-propanol followed by 30% $CHCl_3$ in petroleum ether (60–80°C) (v/v). Table I shows R_F values for the pigments present in these organisms.

TABLE I

APPROXIMATE R_F VALUES OF PIGMENTS FROM MARINE ALGAE SEPARATED BY TWO-DIMENSIONAL CHROMATOGRAPHY (*73*, *88a*)

	R_F [a]	
	First dimension	Second dimension
Pigments	4% *n*-propanol in petroleum ether (60–80°)	30% chloroform in petroleum ether (60–80°)
Carotenes	0.96	0.96 (0.93)
Chlorophyll a	0.84 (0.76)	0.29 (0.21)
Chlorophyll b	0.65	0.10
Chlorophyll c	0.20 (0.10)	0.00
Diatoxanthin	0.57	0.60
Diadinoxanthin	0.54 (0.71)	0.44 (0.54)
Dinoxanthin	0.54 (0.73)	0.44 (0.48)
Fucoxanthin	0.49	0.28
Neofucoxanthin A and B	0.49	0.08
Neodinoxanthin and neodiadinoxanthin	0.52[b] (0.59)	0.30[b] (0.19)
Neoxanthin	0.32	0.05
Siphonein	0.58	0.38
Violaxanthin	0.64	0.50
Lutein	0.80	0.86
Siphonaxanthin	0.30	0.05
Peridinin	0.41[b] (0.49)	0.23 (0.19)
Neoperidinin	0.41[b] (0.43)	0.23 (0.10)

[a] Figures in parentheses show different R_F values determined in separate studies.
[b] Estimates from (*73*).

I. ANOMALOUS SEPARATIONS IN PAPER

The presence of strongly adsorbed colorless substances or overloading of the initial spot can lead to the formation of zones with double tails as well as to multiple zonation (i.e., one pigment yields more than one zone) in both one-way (*16*) and two-way (*58*) paper chromatography (see Chapter 2, Figs. 2 and 3). These anomalous zones should not be interpreted as additional chloroplast pigments not actually present in the plant extract. In this regard, it is often helpful to compare chromatograms of the extract before and after saponification, since this treatment removes much of these colorless impurities (*89*). Or, the chromatogram can be initially developed with petroleum ether, in which case the carotenes plus most of the colorless impurities move with the solvent front. After drying, development with a more polar solvent may yield less distorted zones of the remaining pigments (*61*). As mentioned above, another remedy is to apply the initial zone as a streak rather than a spot.

In order to reproduce the separations herein described, great care must be taken to control many individual variables such as selection, treatment, and extraction of the organism, selection of the paper, the developing solvent and the solvent from which the pigments are spotted, the atmosphere and temperature in the chamber and the development procedure. The composition of the solvent mixture is especially critical if one is attempting to repeat published results. For example, a separation obtained with a solvent containing a certain fraction of petroleum ether may be unsuccessful if another fraction is substituted. It is interesting that although the pigments are more soluble in higher-boiling fractions of petroleum ether, the use of a lower-boiling fraction can increase their R_F values in certain systems (*90*). Because the standardization of paper chromatography is such a difficult problem (*91*), it is likely that some preliminary experiments will be required before satisfactory results are obtained.

II. Various Other Porphyrins

A. Introduction

Many pyrrole pigments other than chlorophylls are found in nature. The porphyrins, many of which contain propionic acid groups rather than ethyl groups as in the chlorophylls, are soluble in water or dilute alkalies and react with methanol plus sulfuric or hydrochloric acid to yield neutral esters which are insoluble in water but soluble in most organic solvents. They are usually much more stable than the chlorophylls, and mixtures of different porphyrins as well as some position isomers can be successfully separated on paper. The chromatography and chemistry of free porphyrins, porphyrin esters, dicarboxylic porphyrins, and iron complexes of porphyrins (hemes) have been reviewed (*92–94*).

B. Preparation of Sample

1. *Extraction of Porphyrins from Natural Materials*

For the extraction of both hemes and porphyrins from body fluids and tissues and plant tissues, organic solvents such as ethyl acetate, diethyl ether, or acetone acidified with glacial acetic or hydrochloric acid are often employed. Protein associated with the porphyrin is precipitated in these solvents and may be removed by filtering or centrifugation. The final extract is dissolved in diluted ammonia before application to the paper or adsorbed on talc and eluted with diluted ammonia or ammonia–acetone for spotting. The grinding of biological samples as part of the extraction procedure was used by Fisher (*95*) and later by Schwartz and Wikoff (*96*) in their widely followed method for the extraction of coproporphyrin and protoporphyrin from blood cells with 4:1 ethyl acetate–glacial acetic acid. The isolation of heme and uro-, copro-, and protoporphyrins after initial extraction with ethyl acetate–acetic acid (*93*, *97*) and the isolation of the porphyrin pigments of Molluscan shells prior to chromatography (*98*) have been described in detail.

The extraction of porphyrins from urine is as follows (*99*, *100*). To 10 ml urine in a centrifuge tube is added 2 ml 10% lead acetate. The precipitate is centrifuged and washed with 5 ml water. The pigments are extracted from the precipitate with 1 ml 15% HCl. The extract is dried *in vacuo* over pellets of NaOH, the residue is taken up in 0.2 ml 10 *M* NH_4OH, and a portion of this solution is spotted on paper for chromatography.

An improvement in the extraction of porphyrins from red cells has been recently published (*101*).

2. *Preparation of Porphyrin Methyl Esters* (*102*)

Two milliliters blood, 10 ml urine, or 5 gm feces are extracted with ethyl acetate–acetic acid (v/v). The extracts are concentrated *in vacuo* under nitrogen which is being passed through pyrogallol to remove any traces of oxygen. Twenty-five milliliters dry methanol and dry HCl gas are added to the residue, and esterification is allowed to proceed for 1 hour with refluxing. Twenty-five milliliters chloroform is then added, and the excess methanol–HCl is washed out with small amounts of water to neutrality. The chloroform extract is taken to dryness, and the methyl esters are redissolved in 30 μl chloroform for chromatography.

The spot of free (acidic) porphyrins which remains at the origin in paper chromatography under some conditions can be esterified on the paper with diazomethane gas or by the slow addition of a solution of diazomethane in cyclohexane from a capillary pipet to the immediate area of the spot (*103*).

C. Paper

Whatman No. 1 paper has been used for the chromatography of free porphyrins (*104–106*), dicarboxylic porphyrins (*107*), and porphyrin esters (*108–111*); S. and S. 2043b for free prophyrins (*112, 113*); and Whatman No. 3 for chlorophyll derivatives (*114*) and for free porphrins, esters, and metal complexes of both (*103*). Whatman No. 3 yields minimum tailing, maximum resolution between adjacent spots, and sharpest definition of the spots when compared with Whatman papers No. 1, 2, and 4 and is recommended for general use (*103*).

Reversed-phase chromatography on Whatman No. 1 coated with silicone has been used for the separation of hemins or their methyl esters (*115*).

D. Solvents (see also Tables II and III)

1. *Lutidine–H_2O (NH_3 Atmosphere)*

Lutidines (dimethylpyridines) mixed with water are widely used for the separation of porphyrin acids in paper. The original method (*104*)

TABLE II

R_F Values of Porphyrins

Compound	R_F [a]				
	1	2	3	4	5
Uroporphyrin	0.26	0.19	—	—	0.90
Coproporphyrin	0.54	0.47	—	—	0.25
Hemin	0.7	0.63	—	—	—
Protoporphyrin	0.84	0.75	0.75	0.96	—
Mesoporphyrin	0.86	0.81	0.82	0.96	—
Hematoporphyrin	0.87	0.77	0.30	0.05	0.07
Deuteroporphyrin	0.88	0.79	—	—	—
Hexacarboxylic porphyrin	—	—	—	—	0.50
Aetioporphyrin	1.0	—	—	—	—
All porphyrin esters	1.0	1.0	—	—	—
Mesochlorin	—	—	0.88	0.96	—
Chlorin a_2 (methyl ester)	—	—	0.55	0.77	—
Pyrrochlorin	—	—	0.96	0.96	—
Rhodochlorin	—	—	0.90	0.96	—
Dioxyprotoporphyrin	—	—	0.34	0	—
Dioxymonovinylhydroxy-ethyldeuteroporphyrin	—	—	0.26	0	—
Monovinylhydroxyethyl-deuteroporphyrin	—	—	0.40	0	—
Pyrroporphyrin	—	—	0.92	0.96	—
Rhodoporphyrin	—	—	0.77	0.96	—

[a] Solvents: (1) 2,6-lutidine–H_2O, NH_3 vapor (*112*); (2) 2,4-lutidine–H_2O, NH_3 vapor (*113*); (3) kerosene–chloroform (4:2.6 v/v) (*108*); (4) trichloroethylene (*129*); (5) 0.1 M LiCl, NH_3 atmosphere (*106*).

TABLE III

R_F VALUES OF PORPHYRIN METHYL ESTERS

Compound	R_F [a]		
	1	2	3
Uroporphyrin I	0.17	0.37	—
Coproporphyrin I	0.47	0.37	—
Coproporphyrin III	0.67	0.34	—
Protoporphyrin IX	0.84	—	0.51
Mesoporphyrin IX	0.89	—	0.59
Deuteroporphyrin IX	0.88	—	0.59
Uroporphyrin III	—	0.40, 0.54	—
Protoporphyrin I	—	0.29	—
Protoporphyrin III	—	0.22	—
Hematoporphyrin IX dimethyl ester	—	—	0.59
Monovinylmonohydroxyethyl-deuteroporphyrin IX	—	—	0.20
Hematoporphyrin IX	—	—	0.03

[a] Solvents: (1) kerosene–chloroform (4.0:2.6 v/v), then same direction kerosene–*n*-propanol (5:1 v/v) (*108*); (2) 80% ethanol v/v (*130*); (3) kerosene–tetrahydropyran–methyl benzoate (5:1.4:0.35 v/v) (*107*).

employing a mixture of 2,4- and 2,5-lutidine saturated with water for development in an ammonia atmosphere separates porphyrins in a sequence inversely related to the number of carboxyl groups in the molecule. Many modifications of this classic procedure have been reported (e.g., *105*, *112*, *113*), but only minor improvements seem to have been achieved. With 2,4- and 2,5-lutidines, two phase systems are obtained with water, but 2,6-lutidine is miscible with water so that a one-phase system is obtained. Development with 2,6-lutidine–H_2O (5:3 or 3.5) in an ammonia atmosphere separates free porphyrins as well as mixtures of coporphyrin isomers (*105*, *116*, *117*). The 2,6-lutidine method was introduced by Kehl and Stich (*112*) and has since been modified (*118*, *119*) and used widely (*94*, *116*, *117*, *120–122*) for the separation of isomers (see Fig. 5).

2. *0.1 M LiCl (NH_3 Atmosphere).*

This solvent separates free porphyrins according to the number of carboxyl groups in a sequence the reverse of that in lutidine solvents (*106*).

3. *Kerosene–Chloroform* (*100:65 v/v*) *and Kerosene–n-Propanol* (*5:1*)

When these solvents are used in succession for one way development, porphyrin methyl esters will be separated in the sequence uroporphyrin

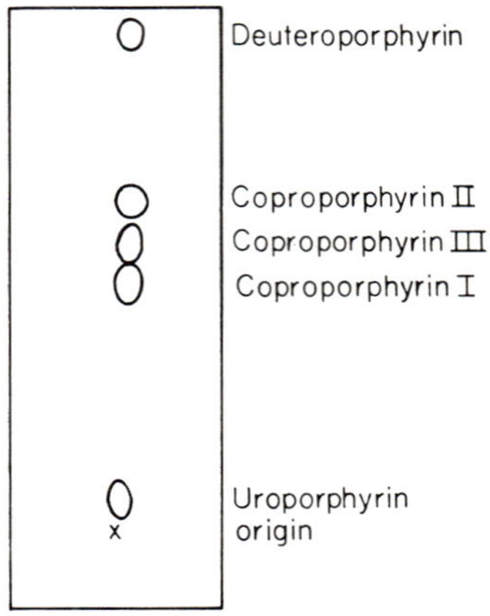

FIG. 5. Separation of porphyrins, including coproporphyrin isomers I, II and III, by 2,6-lutidine. Detection under ultraviolet light. After J. E. Falk (*93*) and J. E. Falk *et al.* (*120*).

I, coproporphyrin I, coproporphyrin III, protoporphyrin IX, and mesoporphyrin IX. The paper is dried at 105°C for about 4 minutes after the first run, and the chamber is saturated with the same solvents as used for development in each case (*108*). Better resolution is obtained if this pair of solvents is used for two-dimensional development (*123*, *124*). The uroporphyrin isomers, which are not separated by this method, are partially separated by washing in the same direction with kerosene–chloroform (4:6) plus 1% (v/v) of ethanol and then with kerosene–dioxane (4:1.5 v/v). The sequence is: uroporphyrins I plus II ($R_F = 0.02$) and III plus IV ($R_F = 0.5$) (*117*). Development with kerosene–dioxane solvents also separates uroporphyrin octamethyl esters I and III (*110*, *111*).

4. *Kerosene–Tetrahydropyran–Methyl Benzoate* (*5:1.4:0.35 v/v*)

Development with this solvent followed by transverse development with water–acetonitrile–*n*-propanol–pyridine (38:10:20:5) after treatment of the paper with silicone separates dimethyl esters of dicarboxylic porphyrins (*107*) (see Fig. 6).

5. *H_2O–n-Propanol–Pyridine* (*5.5 : 0.1 : 0.4*)

This solvent in an atmosphere of water and pyridine vapors on silicone treated paper, or 2,6-lutidine–water (3.3:2.7) in an atmosphere of water vapor, separates various hemins (Fig. 7). Solvents composed of

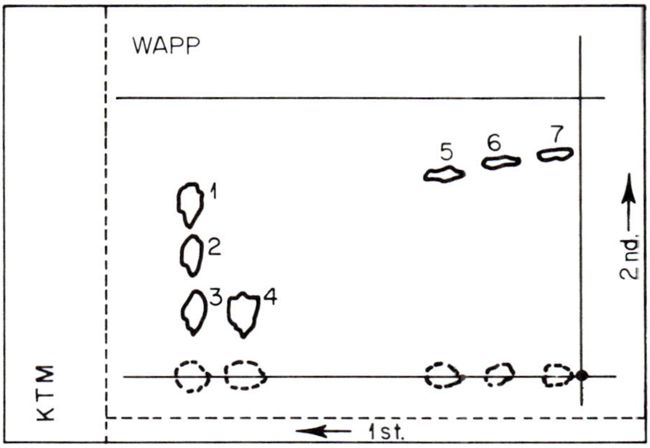

FIG. 6. Paper chromatogram of dimethyl esters of the following dicarboxylic porphyrins: (1) hematoporphyrin IX dimethyl ether; (2) deuteroporphyrin IX; (3) mesoporphyrin IX; (4) protoporphyrin IX; (5) monovinylmonohydroxyethyldeutero-hematoporphyrin IX; (6) a by-product from deuteroporphyrin preparation; (7) hematoporphyrin IX. The dotted spots resulted from the first development. KTM = kerosene–tetrahydropyran–methyl benzoate (5:1.4:0.35 v/v); WAPP = H_2O–acetonitrile–*n*-propanol–pyridine (38:10:20:5 v/v) with Dow-Corning silicone No. 550 as stationary phase. After T. C. Chu and E. J. Chu (*107*).

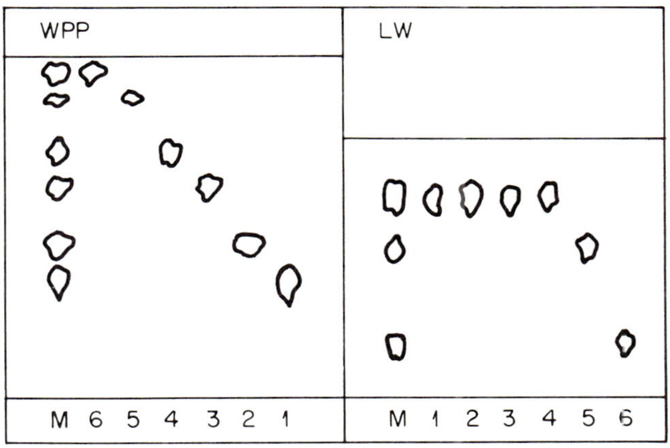

FIG. 7. Paper chromatograms of hemins developed with H_2O–*n*-propanol–pyridine (5.5:0.1:0.4 v/v), with Dow-Corning silicone No. 550 as stationary phase, and 2,6-lutidine–H_2O (3.3:2.7 v/v)—(1) protohemin; (2) mesohemin; (3) deuterohemin; (4) hematohemin; (5) coprohemin I or III; (6) urohemin I; (M) mixture of 1–6. After T. C. Chu and E. J. Chu (*115*).

chloroform–hexane–formic acid, pyridine–H_2O, pyridine–isopropanol-NH_3, and acetic acid–H_2O–toluene can also be used for the separation of certain hemins and protoporphyrin esters in similar reversed-phase systems (*125*, *126*).

6. *Trichloroethylene and 80% Ethanol*

R_F values in these solvents are also shown in Tables II and III (*129*, *130*).

E. Techniques

One dimensional, ascending development in conventional, vapor-saturated tanks is usually employed (*105–108*, *112*, *113*, *115*), although descending (*104*), circular (*114*, *127*), horizontal (*103*), two dimensional (*107*, *123*, *124*), and paper strip (*100*, *128*) methods have also been used.

F. Detection and Quantitative Evaluation of Spots

Once separated by chromatography, porphyrins are easily located by their natural colors or fluorescence or their intense orange to red fluorescence upon radiation with 400 mμ ultraviolet light. Detection of amounts as low as 0.005 μg of porphyrins (*103*) and 3×10^{-4} μg of hemins (*128*) is possible. Qualitative identification of separated compounds is based upon visible or ultraviolet absorption spectra recorded in neutral solvents or aqueous mineral acids (*93*, *112*). Quantitative estimation is achieved by visual or photometric comparison methods on the paper (see Chapter 4) or by elution of the porphyrin spots with chloroform, saponification with HCl, followed by fluorimetry (*102*).

G. Detailed Experiment

This detailed description serves to illustrate one successful resolution of representative porphyrins, porphyrin esters, and metal complexes (*103*).

Whatman No. 3 paper is employed. Sample spots containing as little as 5×10^{-9} gm of free porphyrin or 4×10^{-8} gm of nonfluorescent metal complex are applied. The sheet is supported in a horizontal position in a large desiccator, the walls of which are lined with heavy filter paper wetted with solvent to keep the atmosphere saturated, and developed in the direction of the grain of the paper with carbon tetrachloride containing 30% (v/v) isooctane. After the separation, the spots are located by wetting the sheet with isooctane and illuminating it with longwave ultraviolet light. The following R_F values are obtained in this system: mesoetioporphyrin, 0.63; nickel complex of deoxophyllerythroetioporphyrin (DPEP-Ni), 0.54; DPEP, 0.47; DPEP-V, 0.20; nickel complex of methyldeoxophyllerythrin (DP-Ni-Me), 0.23; DP-Me, 0.15; DP-V-Me, 0.08; nickel complex of dimethylmesoporphyrin ($MP\text{-}Ni\text{-}Me_2$), 0.12; $MP\text{-}Me_2$, 0.10; copper complex of dimethylprotoporphyrin ($PP\text{-}Cu\text{-}Me_2$), 0.13; $PP\text{-}Ni\text{-}Me_2$, 0.10; and $PP\text{-}Me_2$, 0.07.

Mixtures of porphyrincarboxylic acids with aetioporphyrins and esters are spotted in one corner of a square of paper and developed as above. The solvent is removed by evaporation, and the unresolved spot of acids at the origin is reacted with diazomethane in cyclohexane. After a few seconds, the paper is developed in a transverse direction with the same solvent, and the newly formed esters are resolved along one edge of the sheet. Semiquantitative determinations are possible by visual comparison with a series of standards ($\pm 30\%$) or by measurement of spot areas ($\pm 10\%$).

III. Flavonoids

A. Introduction

Paper chromatography has been widely used for the separation and identification of flavonoid pigments. The term flavonoid, as used here, will comprise the phenolic, water soluble plant pigments sometimes known as anthoxanthins (e.g., flavones, flavonols, flavanones, isoflavones, catechins, tannins, biflavonyls, etc.), as well as the closely related anthocyanins. These compounds are easily separated on paper, and most are readily detected without chromogenic reagents because of their natural colors. In addition, the R_F or R_M values of the majority of flavonoids are directly related to their chemical structures (e.g., the numbers of hydroxyl and sugar substituents, the degree of chelation and their molecular weight) (*131*). The characteristic nature of these R_F values together with selective chromogenic spray reagents are of great value in the identification of the flavonoids. Several reviews on the chromatography of these compounds have appeared (*131–134*).

B. Preparation of Sample

Flavones are extracted by immersing the fresh plant material into boiling aqueous ethanol, methanol, or water (*131*). The extracts are filtered, evaporated *in vacuo* to a small volume, and refiltered, if necessary, before application to the paper. Anthocyanins are extracted with methanol or water plus 1% HCl. In both cases, leaf extracts are partitioned with petroleum ether before concentration to remove any chlorophylls or carotenoids which might cause streaked chromatograms.

Pure flavones are dissolved in ethanol and pure anthocyanins in 1% methanolic HCl prior to chromatography (*131*, *133*). Flavonoids in fruit juices are determined after extraction with methanol for 15–20 hours (*135*).

C. Paper

Whatman No. 1 (and to a lesser extent No. 3 or No. 3MM) paper is commonly employed. The paper is impregnated with phosphate or borate (*136*) buffers for the separation of partially methylated flavones, the resultant R_F values varying with the pH of the buffer (*137*).

D. Solvents

A wide variety of solvents for one-dimensional chromatography is listed in Tables IV–VI. Solvents for the cationic anthocyanins generally contain mineral acid to insure their stability. The most important, well-standardized systems are (*132*): the upper layer of *n*-butanol–acetic acid–H_2O (4:1:5 v/v) for the separation of all classes of flavonoids; acetic acid–concentrated HCl–H_2O (30:3:10 v/v) for most flavonoid aglycones (*138*); phenol–H_2O (4:1 v/v), especially for partially methylated derivatives (*139*); water, either alone (*140*, *141*) or containing 2–6% acetic acid (*141*) or 1% HCl, for flavonoid glycosides and flavanones; and the top layer of *n*-butanol–2 *N* HCl (1:1 v/v) for anthocyanins. The relative merits of some of these systems have been discussed (*142*).

The following solvent combinations are suitable for two dimensional separations:

1. *First direction*: the upper layer of *n*-butanol–acetic acid–H_2O (4:1:5 v/v); *second direction*: 2% acetic acid (*143*), ethyl acetate–acetic acid–H_2O (50:2:50 v/v) (*144*), or 15% acetic acid (*133*).
2. *First direction*: *n*-butanol–27% acetic acid (1:1 v/v); *second direction*: *m*-cresol–acetic acid–H_2O (50:2:48 v/v) (*145*).
3. *First direction*: *n*-butanol–2 *N* HCl (1:1); *second direction*: 1% HCl [for anthocyanins (*133*)].

For the separation of some new anthocyanidins by two dimensional chromatography on microcrystalline cellulose thin layers, a solvent was devised (methanol–concentrated HCl–H_2O, 190:1:10 v/v) which would probably be useful for paper chromatography (*146*).

E. Techniques

One-dimensional chromatography, either ascending or descending, is usually employed, and the two methods have been compared (*147*).

TABLE IV

R_F VALUES OF NATURALLY OCCURRING ANTHOCYANIDINS (AGLYCONES OF ANTHOCYANINS) (*132–134, 166*)

Compound	R_F [a]					Colors[b]	
	1	2	3	4	5	Visible light	Ultra-violet
Apigeninidin	0.75	0.74	—	—	0.44	Y	YB
Luteolinidin	0.61	0.56	—	—	0.35	O	RB
Tricetinidin	0.46	0.38	—	—	0.28	OR	—
Pelargonidin	0.68	0.80	0.80	0.82	0.33	R	OR
Cyanidin	0.49	0.68	0.69	0.69	0.22	M	P
Peonidin	0.63	0.71	0.72	0.87	0.30	M	P
Rosinidin	0.76	0.77	—	—	0.39	M	P
Delphinidin	0.32	0.42	0.35	0.52	0.13	Pu	Ma
Petunidin	0.46	0.52	0.45	0.75	0.20	Pu	Ma
Pulchellidin	0.50	0.48	—	—	0.24	Pu	—
Malvidin	0.60	0.58	0.53	0.90	0.27	Pu	Ma
Hirsuitidin	0.78	0.66	—	—	0.36	Pu	Ma
Capensinidin	0.88	0.79	—	—	—	Pu	—

[a] Solvents: (1) acetic acid–concentrated HCl–H_2O (30:3:10 v/v); (2) *n*-butanol–acetic acid–H_2O (4:1:5 v/v, upper layer); (3) *n*-butanol–2 *N* HCl (v/v); (4) *m*-cresol–5.5 *N* HCl–acetic acid (1:1:1 v/v); (5) formic acid–concentrated HCl–H_2O (5:2:3 v/v).

[b] Colors: See key to Table VII.

TABLE V

R_F VALUES OF ANTHOCYANINS (*132–134*)

Compound	R_F [a]			
	1	2	3	4
Pelargonidin glycosides[b]				
Aglycone (0)	0.80	0.80	0.00	—
3-rhamnoside (1)	0.71	0.64	0.22	0.53
5-glucoside (1)	0.51	0.49	0.18	0.57
7-glucoside (1)	0.46	0.51	0.15	—
3-rhamnoside-5-glucoside (2)	0.46	0.24	0.39	0.70
3-glucoside (1)	0.44	0.38	0.14	0.35
3-galactoside (1)	0.39	0.37	0.13	0.33
3-rutinoside (2)	0.37	0.30	0.22	0.44
3-sambubioside (2)	0.37	0.34	0.31	0.60
3-sophoroside (2)	0.36	0.30	0.38	0.65

Table continued

TABLE V (*continued*)

Compound	R_F [a] 1	2	3	4
3-lathyroside (2)	0.35	0.31	0.34	0.60
3-(2^G-glucosylrutinoside) (3)	0.33	0.15	0.63	0.73
3,5-diglucoside (2)	0.31	0.14	0.23	0.45
3-gentiobioside (2)	0.30	0.26	0.21	0.47
3,7-diglucoside (2)	0.30	0.10	0.38	0.70
5-glucoside-3-rutinoside (3)	0.29	0.13	0.40	0.58
5-glucoside-3-sophoroside (3)	0.25	0.10	0.60	0.68
3-gentiotrioside (3)	0.25	0.10	0.35	0.52
5-glucoside-3-sambubioside (3)	0.24	0.18	0.43	—
7-glucoside-3-sophoroside (3)	0.18	0.04	0.73	0.73
Cyanidin glycosides				
3-monoglucoside	0.38	0.25	0.07	0.26
3-monogalactoside	0.37	0.24	0.07	0.26
3-rhamnoglucoside	0.37	0.25	0.19	0.43
3-xyloglucoside	0.36	0.24	0.24	0.51
3-diglucoside	0.33	0.22	0.34	0.61
3,5-diglucoside	0.28	0.06	0.16	0.40
3-rhamnoglucosido-5-glucoside	0.25	0.08	0.36	0.59
Peonidin glycosides				
3-monoglucoside	0.41	0.30	0.09	0.33
3,5-diglucoside	0.31	0.10	0.17	0.44
3-rhamnoglucosido-5-glucoside	0.29	0.12	0.37	0.60
Delphinidin glycosides				
3-monoglucoside	0.26	0.11	0.03	0.18
3-rhamnoglucoside	0.30	0.15	0.11	0.37
3,5-diglucoside	0.15	0.03	0.08	0.32
Petunidin glycosides				
3-monoglucoside	0.35	0.14	0.04	0.22
3-rhamnoglucoside	0.35	0.16	0.13	0.42
3,5-diglucoside	0.24	0.04	0.08	0.32
3-rhamnoglucosido-5-glucoside	0.23	0.06	0.37	0.61
Malvidin glycosides				
3-monoglucoside	0.38	0.15	0.06	0.29
3,5-diglucoside	0.31	0.03	0.13	0.42
3-rhamnoglucosido-5-glucoside	0.30	0.05	0.40	0.63

[a] Solvents: (1) *n*-butanol–acetic acid–H_2O (4:1:5 v/v, upper layer); (2) *n*-butanol–2 *N* acetic acid (1:1 v/v, upper layer); (3) H_2O–12 *N* HCl (97:3 v/v); (4) H_2O–acetic acid–12 *N* HCl (82:15:3 v/v).

[b] Figure in parentheses indicates number of sugar residues.

TABLE VI

R_F VALUES OF FLAVONOID COMPOUNDS (*131*, *134*, *156*, *163*, *167*, *168*)

Compound	R_F [a]										
	1	2	3	4	5	6	7	8	9	10	11
Flavone aglycones											
Acacetin	0.91	0.91	0.88	0.75	0.94	0.80	0.00	0.00	—	—	—
Apigenin	0.89	0.83	0.88	0.33	0.87	0.89	0.00	0.00	—	—	—
Diosmetin	0.85	0.80	0.86	0.47	—	—	—	—	—	—	—
Luteolin	0.78	0.66	0.66	0.17	—	—	—	—	—	—	—
Tricin	0.73	0.72	0.87	—	—	—	—	—	—	—	—
Saponaretin	0.56	0.86	0.80	—	—	—	—	—	—	—	—
Vitexin	0.43	0.82	0.63	—	—	—	—	—	—	—	—
Chrysoeriol	—	—	—	0.47	—	—	—	—	—	—	—
Norwogonin	—	—	—	0.51	0.85	0.83	0.00	0.00	—	—	—
Baicalein	—	—	—	0.61	0.97	0.83	—	—	—	—	—
Chrysin	—	—	—	0.78	0.86	—	0.00	0.00	—	—	—
Tectochrysin	—	—	—	0.93	0.98	0.94	—	0.27 ?	—	—	—
Auranetin	0.91	—	0.93	—	0.92	0.99	0.34	0.63	—	—	—
Flavonol aglycones											
Kampferol	0.83	0.55	0.58	0.30	0.90	0.77	0.04	0.10	—	—	—
Morin	0.79	0.73	0.34	—	0.71	0.58	0.13	0.27	—	—	—
Isorhamnetin	0.74	0.53	0.66	—	—	—	—	—	—	—	—
Fisetin	0.73	0.58	0.32	0.09	—	—	—	—	—	—	—
Rhamnetin	0.72	0.53	0.66	—	0.92	0.73	0.03	0.08	—	—	—
Quercetin	0.64	0.41	0.29	0.10	0.81	0.67	0.04	0.07	—	—	—
Azaleatin	0.48	0.49	0.50	—	—	—	—	—	—	—	—
Myricetin	0.43	0.28	0.13	—	0.78	0.45	—	0.12	—	—	—
Robinetin	0.40	0.36	0.18	—	0.41	0.58	0.03	0.08	—	—	—
Quercetagetin	0.22	—	0.06	0.00	0.17	0.61	0.13	0.19	—	—	—
Gossypetin	0.21	—	0.09	—	0.59	0.51	0.07	0.12	—	—	—
Nortangeretin	0.76	—	0.92	—	0.79	0.60	0.04	0.10	—	—	—
Patuletin	0.78	—	0.13	—	0.81	0.76	0.06	0.10	—	—	—
Flavonol glycosides											
Astralagin	0.70	—	0.74	—	—	—	—	0.43	0.13	—	—
Nicotiflorin	0.54	—	0.64	—	—	—	—	0.54	0.23	—	—
Robinin	0.40	—	0.73	—	—	—	—	0.75	0.54	—	—
Avicularin	0.70	—	0.61	—	—	—	—	0.31	0.07	—	—
Reynoutrin	0.65	—	—	—	—	—	—	0.32	0.06	—	—
Isoquercitrin	0.58	—	0.54	—	0.40	0.79	0.24	0.37	0.08	—	—
Hyperin	0.55	—	0.56	—	—	—	—	0.35	0.09	—	—
Quercetrin	0.72	—	0.58	—	0.50	0.79	0.45	0.49	0.19	—	—
Rutin	0.45	—	0.46	—	0.15	0.83	0.45	0.51	0.23	—	—
Quercimeritrin	0.37	—	0.40	—	0.50	0.80	0.27	0.07	0.00	—	—
Spiraeoside	0.48	—	0.33	—	—	—	—	0.13	0.00	—	—

Table continued

TABLE VI (*continued*)

Compound	R_F [a]										
	1	2	3	4	5	6	7	8	9	10	11
Dactylin	0.38	—	0.63	—	—	—	—	0.62	0.27	—	—
Cannabiscitrin	0.47	—	0.32	—	—	—	—	0.25	0.05	—	—
Myricitrin	0.60	—	0.39	—	0.35	0.87	—	0.44	0.15	—	—
Gossypin	0.87	—	0.81	—	0.02	—	—	—	—	—	—
Gossypitrin	0.49	—	0.39	—	0.11	0.54	0.17	0.14	—	—	—
Flavone glycosides											
Cosmetin	0.65	—	0.78	—	—	—	—	0.25	0.04	—	—
Rhoifolin	0.58	—	0.74	—	—	—	—	0.46	0.09	—	—
Apiin	0.57	—	0.75	—	—	—	—	0.42	0.06	—	—
Saponarin	0.42	—	0.65	—	—	—	—	0.64	0.31	—	—
Galuteolin	0.82	—	0.65	—	—	—	—	0.07	0.00	—	—
Linarin	0.61	—	0.55	—	—	—	—	0.60	0.14	—	—
Naringin	0.70	—	0.75	—	0.51	0.86	0.77	0.80	—	—	—
Neohesperidin	0.67	—	0.75	—	0.38	0.88	0.72	0.81	—	—	—
Flavanone aglycones											
Butin	0.91	—	—	—	0.98	—	—	0.55	0.20	0.64	0.84
Eriodictyol	—	—	—	0.29	—	—	—	—	—	0.56	0.85
Naringenin	0.89	—	—	0.49	—	—	—	—	0.16	0.66	0.69
Sakuranetin	0.90	—	—	—	—	—	—	—	0.09	0.65	—
Homoeriodictyol	0.88	—	—	0.61	—	—	—	—	0.14	0.67	—
Hesperitin	0.89	—	0.96	0.60	0.97	0.94	—	0.50	0.11	0.67	—
Pinocembrin	0.97	—	0.98	0.80	0.99	0.97	—	0.58	—	—	—
Isoflavone aglycones											
Daidzein	0.92	—	—	0.35	—	—	—	—	0.08	0.62	—
Formononetin	0.94	—	—	0.72	—	—	—	—	0.07	0.67	—
Genistein	0.94	—	—	0.41	—	—	—	—	0.04	0.59	—
Prunetin	0.94	—	—	—	—	—	—	—	0.03	0.67	—
Tectirigenin	0.90	—	—	—	—	—	—	—	0.09	0.70	—
Irigenin	0.91	—	—	—	—	—	—	—	0.16	0.79	—
Irigenol	0.54	—	—	—	—	—	—	—	0.02	0.33	—
Orobol	—	—	—	0.20	—	—	—	—	—	—	—
Pratensein	—	—	—	0.54	—	—	—	—	—	—	—
Biochanin A	—	—	—	0.77	—	—	—	—	—	—	—
Podospicatin	—	—	—	0.86	—	—	—	—	—	—	—
Isoflavone glucosides											
Genisten	0.67	—	—	—	—	—	—	—	0.25	0.75	—
Tectiridin	0.67	—	—	—	—	—	—	—	0.38	0.83	—
Iridin	0.66	—	—	—	—	—	—	—	0.48	0.87	—
Chalcone aglycones											
Butein	0.78	0.51	0.66	—	—	—	—	—	0.01	0.19	0.83
Stillopsidin	—	—	—	—	—	—	—	—	0.01	—	0.65
Okanin	0.56	0.38	—	—	—	—	—	—	—	0.08	0.56

Table continued

TABLE VI (*continued*)

Compound	R_F [a]										
	1	2	3	4	5	6	7	8	9	10	11
Glucosides											
Coreopsin	0.56	—	0.64	—	—	—	—	—	0.05	0.43	0.56
Stillopsin	—	—	—	—	—	—	—	—	0.02	—	0.47
Marein	0.45	—	—	—	—	—	—	—	—	0.22	0.38
Aurone aglycones											
Sulphuretin	0.80	—	0.70	—	—	—	—	—	0.01	0.19	0.80
Aureusidin	0.57	—	0.29	—	—	—	—	—	0.01	0.10	0.57
Leptosidin	0.76	—	0.80	—	—	—	—	—	0.01	0.19	0.76
Maritimetin	0.53	—	0.13	—	—	—	—	—	—	0.05	0.53
Glucosides											
Sulphurein	—	—	0.69	—	—	—	—	—	0.03	0.41	0.49
Aureusin	0.28	—	0.35	—	—	—	—	—	0.01	0.16	0.28
Cernuoside	—	—	0.45	—	—	—	—	—	0.02	0.25	0.49
Leptosin	—	—	0.73	—	—	—	—	—	0.03	0.33	0.51
Maritimein	—	—	—	—	—	—	—	—	—	0.21	0.43

[a] Solvents: (1) *n*-butanol–acetic acid–H_2O (4:1:5 v/v, upper phase); (2) acetic acid–concentrated HCl–H_2O (30:3:10 v/v); (3) phenol–H_2O (saturated); (4) benzene–acetic acid–H_2O (125:72:3); (5) ethyl acetate–H_2O (saturated); (6) isopropanol–H_2O (6:4 v/v); (7) *n*-heptane–*n*-butanol–H_2O (29:14:57 v/v); (8) acetic acid–H_2O (15:85 v/v); (9) water; (10) acetic acid–H_2O (30:70 v/v); (11) *n*-butanol–acetic acid (73:27 v/v).

Two-dimensional (*148*) and circular development (*149*, *150*) can also be used, although the former often leads to faded, diffuse spots with anthocyanins (*133*). Special care must be taken with some systems in terms of preparation and aging of the solvent and equilibration of the paper (*131–133*), especially when determining R_F values of new compounds in order to relate chromatographic behavior to chemical structure (*151*).

Unidimensional multiple paper chromatography combined with DEAE-cellulose column chromatography has been used for the identification and determination of flavonoid derivatives in pharmaceutical mixtures (*152*).

F. DETECTION

Some flavonoids and most anthocyanins can be seen on paper chromatograms as colored spots in daylight. Examination under ultraviolet light, with and without ammonia vapor, is also of value because of the resultant characteristic coloration or fluorescence of most of these compounds (*132, 133, 142, 153*). In addition, the original color of these compounds in visible and ultraviolet light may be altered or enhanced with the following color reagents (*132, 133, 154–157*) (See also Tables IV and VII): (1) 5% ethanolic aluminum chloride (*158, 159*); (2) 2% aqueous ferric chloride (*160*); (3) 5% aqueous sodium carbonate; (4) 1% sodium borohydride in isopropanol (*161*) or sodium borohydride–HCl vapor; (5) 1% aqueous ferric chloride–1% aqueous potassium ferricyanide (v/v) (*162*); (6) diazotized sulfanilic acid (*163*). Quinonechloroimides, Erlich's reagent, and *p*-toluenesulfonic acid are of diagnostic value as spray reagents for predicting the structures of various plant flavonoids (*164*).

G. Detailed Experiments

Lists of R_F values in many solvent systems of most of the known naturally-occurring flavonoids and related compounds have been published (*131–134, 154, 155, 163, 165*). A representative selection of these is presented in Tables IV–VI (see also Fig. 8) (*131–134, 156, 163, 166–168*).

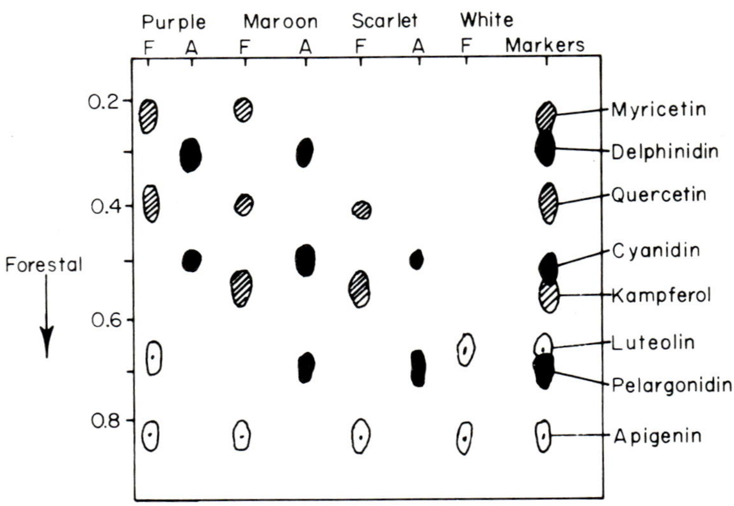

FIG. 8. Separation of flavonoid aglycones of color mutants of *Verbena* garden hybrids with Forestal solvent, acetic acid–concentrated HCl–H_2O (3:3:1 v/v). F = flavone extract (ethyl acetate), A = anthocyanidin extract (amyl alcohol). After J. B. Harbone (*131*). Solid spots = red to mauve in visible light; hatched spots = bright yellow in ultraviolet light; white spots = dull brown in ultraviolet light.

TABLE VII

THE COLORS AND COLOR REACTIONS OF FLAVONOIDS ON PAPER (*131–133*, *163*)

Reagent	Light source	Color reactions[a] in the following flavonoid classes										
		Flavone	Flavonol	Isoflavone	Flavanone	Aurone	Chalcone	Pelargonidin glycosides	Cyanidin glycosides	Peonidin glycosides	Delphinidin and petunidin glycosides	Malvidin glycosides
None	Visible	pY	pY	C	C	bY	Y	O or OR or OY	M	P	Pu	Ma
None	Ultraviolet	dB	bY	fPu	C	bY	dB	dR or FY	dM or bR	dP or FR	dPu	dMa or FCe
NH_3 vapor	Ultraviolet	bG	bY	fPu	C or pY	bR	dR					
$AlCl_3$	Visible	pY	Y	C	C	pY	Y	–	+	–	+	–
$AlCl_3$	Ultraviolet	FG	FY	FY	FG	FG	FO					
Na_2CO_3	Visible	bY	Y	pG	G	O	R					
$NaBH_4$	Visible	C	C	C	M	C	C					
$FeCl_3$–$K_3Fe(CN)_6$	Visible	Bl	Bl	Bl	Bl	Bl	Bl					
Sulfanilic acid	Visible	BO	BO	RB or BO or Pu	O or B or P							

[a] Key: B = brown; b = bright; Bl = blue; C = colorless; Ce = cerise; d = dull; F = fluorescent; f = faint; G = green; M = magenta; Ma = Mauve; O = orange; P = pink; p = pale; Pu = purple; R = red; Y = yellow; + = positive color change; – = no color change.

The general method of separating plant flavonoids by paper chromatography consists of applying spots or streaks of concentrated plant extract on the paper, developing with a suitable solvent, and spraying with a color reagent. If necessary, the separated zones are cut out, eluted, and rechromatographed in other solvents (*131*, *133*). In one specific example (*163*), 10–20 μg of flavonoid aglycones in ethanol is applied as a 0.5 $\times$ 2 cm band on Whatman No. 1 paper which is pre-equilibrated for at least 1 hour and then developed in benzene–acetic acid–H_2O (125:72:3 v/v) by the descending technique at 20$\pm$2°C for 3.5–4 hours (40 cm). Location and identification is by spraying with one or more of the reagents listed above. Separations are excellent, R_F values are reproducible within $\pm$0.02 and vary for compounds having the same substitution pattern in the order flavanone > isoflavone > flavone.

References

1. M. Tswett, *Ber. Deut. Botan. Ges.* **24**, 384 (1906).
2. H. H. Strain and J. Sherma, *J. Chem. Educ.* **44**, 235 (1967).
3. H. H. Strain and J. Sherma, *J. Chem. Educ.* **44**, 238 (1967).
4. F. Goppelsroeder, "Capillaranalyse." Birkhaüser, Basel, 1901.
5. W. G. Brown, *Nature* **143**, 377 (1939).
6. Z. Šesták, *Chromatog. Rev.* **1**, 193 (1959); **7**, 65 (1965).
7. H. H. Strain and W. A. Svec, *Advan. Chromatog.* **8**, 119 (1969).
8. H. H. Strain, *Ann. Priestly Lectures* **32** (1958).
9. R. Willstätter and A. Stoll, "Untersuchungen über Chlorophyll." J. Springer, Berlin, 1913.
10. H. H. Strain, *Agr. Food Chem.* **2**, 1222 (1954).
11. R. Douin, *8th Congr. Intern. Botan., Paris, 1954, Rappt. Commun.* Sect. 11/12, p. 22.
12. Z. Šesták, *Cesk. Biol.* **7**, 153 (1958).
13. R. Harder and W. Koch, *Arch. Mikrobiol.* **21**, 1 (1954).
14. Z. Šesták, *in* "Paper Chromatography" (I. M. Hais and K. Macek, eds.), p. 721. Academic Press, New York, 1963.
15. M.-R. Michel-Wolwertz and C. Sironval, *Biochim. Biophys. Acta* **94**, 330 (1965).
16. H. H. Strain, J. Sherma, F. L. Benton, and J. J. Katz, *Biochim. Biophys. Acta* **109**, 1 (1965).
17. Y. Chiba and I. Noguchi, *Cytologia (Tokyo)* **19**, 41 (1954).
18. J. Sherma and G. Zweig, *J. Chromatog.* **31**, 439 (1967).
19. T. N. Godnev and N. K. Akulovich, *Biokhim. i Fiziol. Rast., Akad. Nauk Beloroussk. SSR, Inst. Biol., Sb. Nauchn. Rabot*, p. 42 (1958).
20. H. Züllig, *Verhandl. Intern. Ver. Limnol.* **14**, 263 (1961).
21. H. I. Virgin, *Physiol. Plantarum* **13**, 155 (1960).
22. R. Monéger, *Rev. Gen. Botan.* **64**, 593 (1957).
23. V. P. Goguadze and N. V. Vitul'skaya, *Soobshch. Akad. Nauk Gruz.* **42**, 609 (1966).
24. H. H. Strain, J. Sherma, F. L. Benton, and J. J. Katz, *Biochim. Biophys. Acta* **109**, 23 (1965).
25. H. H. Strain, W. M. Manning, and G. Hardin, *Biol. Bull.* **86**, 169 (1944).

26. A. Jensen, O. Aasmundrud, and K. E. Eimhjellen, *Biochim. Biophys. Acta* **88**, 466 (1964).
27. J. M. Anderson, *U. S. At. Energy Comm., Rept.* **UCRL-8870** (1959).
28. J. B. Wolff and L. Price, *Arch. Biochem. Biophys.* **72**, 293 (1957).
29. Y. Yamamoto and Y. Tomita, *Mem. Fac. Agr., Kagoshima Univ.* **3**, 16 (1957).
30. A. Hager and T. Bertenrath, *Planta* **58**, 564 (1962).
31. H. H. Strain, J. Sherma, and M. Grandolfo, *Anal. Chem.* **39**, 926 (1967).
32. L. Brauner and A. Hager, *Planta* **51**, 115 (1958).
33. J. Sherma and H. H. Strain, *Anal. Chim. Acta* **40**, 155 (1968).
34. H. H. Strain, *J. Phys. Chem.* **57**, 638 (1953).
35. A. Angapindu, H. Silberman, P. Tantivatana, and J. R. Kaplan, *Arch. Biochem. Biophys.* **75**, 56 (1958).
36. G. Nunez, *Bull. Soc. Chim. Biol.* **36**, 411 (1954).
37. H. H. Strain, J. Sherma, and M. Grandolfo, *Anal. Biochem.* **24**, 54 (1968).
38. H. H. Strain, *in* "Chromatography" (E. Heftmann, ed.), p. 669. Reinhold, New York, 1967.
39. H. H. Strain and W. A. Svec, *in* "The Chlorophylls" (L. P. Vernon and G. R. Seely, eds.), p. 21. Academic Press, New York, 1966.
40. L. Bauer, *Naturwissenschaften* **39**, 88 (1952).
41. D. I. Sapozhnikov, I. A. Bronshteĭn, and T. A. Krasovskaya, *Biokhimiya* **20**, 286 (1955).
42. L. Márkus, *Agrokem. Talajtan* **1**, 291 (1952).
43. M. Asami, *Botan. Mag. (Tokyo)* **65**, 217 (1952).
44. M. J. Sabacky, L. B. Jones, H. D. Frame, and H. H. Strain, *Anal. Chem.* **34**, 306 (1962).
45. F. T. Wolf, *Bull. Torrey Botan. Club* **85**, 128 (1958).
46. E. M. Stanishevskaya, *Vestsi Akad. Navuk Belarusk. SSR, Ser. Biyal. Navuk* No. 4, 52 (1962).
47. A. Hager, *Planta* **48**, 592 (1957).
48. Z. Šesták, *Biol. Plant., Acad. Sci. Bohemoslov.* **6**, 132 (1964).
49. Z. Šesták and M. Baslerová, *Plant Cell Physiol. (Tokyo)*, Spec. Issue, p. 423 (1963).
50. J. M. Anderson, *J. Chromatog.* **4**, 93 (1960).
51. D. I. Sapozhnikov, A. N. Mayevskaya, and I. A. Popova, *Fiziol. Rast.* **6**, 376 (1959).
52. T. Katayama, *Nippon Suisan Gakkaishi* **30**, 490 (1964).
53. A. H. Sporer, S. Freed, and K. M. Sancier, *Science* **119**, 68 (1954).
54. S. Freed, K. M. Sancier, and A. H. Sporer, *J. Am. Chem. Soc.* **76**, 6006 (1954).
55. Y. Osawa, *Nature* **180**, 705 (1957).
56. J. Sherma and G. Zweig, *Sci. Am.* **220**, 124 (1969).
57. M. Holden, *Biochim. Biophys. Acta* **56**, 378 (1962).
58. H. H. Strain, J. Sherma, F. L. Benton, and J. J. Katz, *Biochim. Biophys. Acta* **109**, 16 (1965).
59. M. J. Hendrickson, R. R. Berueffy, and A. R. McIntyre, *Anal. Chem.* **29**, 1810 (1957).
60. D. A. Rappoport, C. R. Calvert, R. K. Loeffler, and J. H. Gast, *Anal. Chem.* **27**, 820 (1955).
61. I. A. Popova, *Tr. Botan. Inst., Akad. Nauk SSSR, Ser. 4: Eksperim. Botan.* **16**, 154 (1963).
62. Y. Katayama and S. Shida, *Cytologia (Tokyo)* **22**, 360 (1957).
63. H. H. Strain, *Carnegie Inst. Wash. Publ.* **490** (1938).
64. P. Karrer and E. Jucker, "Carotinoide." Birkhäuser, Basel, 1948.

65. H. H. Strain, M. R. Thomas, H. L. Crespi, M. I. Blake, and J. J. Katz, *Ann. N. Y. Acad. Sci.* **84**, 617 (1960).
66. H. H. Strain, M. R. Thomas, and J. J. Katz, *Biochim. Biophys. Acta* **75**, 306 (1963).
67. F. C. Pennington, H. H. Strain, W. A. Svec, and J. J. Katz, *J. Am. Chem. Soc.* **86**, 1418 (1964).
68. A. A. Shlyk, V. L. Kaler, and G. M. Podchufarova, *Biokhimiya* **26**, 259 (1961).
69. A. A. Shlyk, V. A. Mashankov, G. N. Nikalayeva, I. V. Prudnikava, and T. V. Kukhtsenka, *Vestsi Akad. Navuk Belarusk. SSR, Ser. Biyal. Navuk* No. 3, 37 (1961).
70. A. A. Shlyk, V. L. Kaler, and G. M. Podchufarova, *Byul. Inst. Biol., Akad. Nauk Belorussk. SSR* **5**, 141 (1959).
71. A. P. Losev, *Fiziol. Rast.* **11**, 1098 (1964).
72. H. H. Strain, M. R. Thomas, H. L. Crespi, and J. J. Katz, *Biochim. Biophys. Acta* **52**, 517 (1961).
73. S. W. Jeffrey, *Biochem. J.* **80**, 336 (1961).
74. Z. Šesták, *Preslia* **35**, 123 (1963).
75. H. Wittmann, *Arch. Mikrobiol.* **25**, 272 (1957).
76. S. W. Jeffrey and M. B. Allen, *Limnol. Oceanog.* **12**, 533 (1967).
77. M. F. Bacon, *Biochem. J.* **101**, 34C (1966).
78. I. G. Mokhnachev and L. G. Serdyuk, *Izvest. Vysshikh Uchebn. Zavedenii, Pishchevaya Tekhnol.* No. 5, 147 (1962).
79. K. Yamaguchi, S. Fukushima, and M. Itō, *J. Pharm. Soc. Japan* **76**, 339 (1956).
80. Y. Katayama and S. Shida, *Mem. Fac. Agr., Univ. Miyazaki* **3**, No. 1, 11 (1961).
81. K. Toyoda, *Botan. Mag. (Tokyo)* **73**, 398 (1960).
82. S. Knodo and I. Mori, *J. Pharm. Soc. Japan* **75**, 519 (1955).
83. R. W. Schramm and S. Organiściak-Matuszak, *Acta Soc. Botan. Polon.* **27**, 299 (1958).
84. H. Fukasawa and K. Mito, *Seiken Ziho* **8**, 4 (1957).
85. H. H. Strain, *in* "The Biochemistry of Chloroplasts" (T. W. Goodwin, ed.), Vol. 1, p. 387. Academic Press, New York, 1966.
86. J. Sherma and G. Zweig, *J. Chromatog.* **31**, 589 (1967).
87. J. Sherma, *Anal. Letters* **1**, 189 (1967).
88. J. Sherma and H. H. Strain, unpublished results (1968).
88a. S. W. Jeffrey, *Biol. Bull.* **135**, 141 (1968); S. W. Jeffrey and F. T. Haxo, *ibid.* **135**, 149 (1968).
89. C. Audigié, *Rev. Gen. Botan.* **65**, 581 (1958).
90. S. W. Jeffrey, personal communication (1968).
91. I. M. Hais, *J. Chromatog.* **33**, 25 (1968).
92. J. E. Falk, "Porphyrins and Metalloporphyrins: Their General, Physical and Coordination Chemistry and Laboratory Methods." Elsevier, Amsterdam, 1964.
93. J. E. Falk, *J. Chromatog.* **5**, 277 (1961).
94. J. E. Falk, *Brit. Med. Bull.* **10**, 211 (1954).
95. H. Fisher and H. Orth, "Die Chemie des Pyrrols," Vol. 2, Part 1. Akad. Verlagsges., Leipzig, 1937.
96. S. Schwartz and H. M. Wikoff, *J. Biol. Chem.* **194**, 563 (1952).
97. E. I. B. Dresel and J. E. Falk, *Biochem. J.* **63**, 72 (1956).
98. R. E. H. Nicholas and A. Comfort, *Biochem. J.* **45**, 208 (1949).
99. S. Nishikawa, *J. Japan. Biochem. Soc.* **24**, 52 (1952).
100. L. M. Corwin and J. M. Orten, *Anal. Chem.* **26**, 608 (1954).
101. R. J. Beard, *Clin. Chim. Asta* **18**, 151 (1967).

102. R. Kehl and B. Günter, *Z. Physiol. Chem.* **297**, 254 (1954); *Naturwissenschaften* **41**, 118 (1954).
103. M. Blumer, *Anal. Chem.* **28**, 1640 (1956).
104. R. E. H. Nicholas and C. Rimington, *Scand. J. Clin. & Lab. Invest.* **1**, 12 (1949).
105. L. Ericksen, *Scand. J. Clin & Lab. Invest.* **5**, 155 (1953).
106. T. K. With, *Scand. J. Clin. & Lab. Invest.* **9**, 395 (1957).
107. T. C. Chu and E. J. Chu, *J. Biol. Chem.* **208**, 537 (1954).
108. T. C. Chu, A. A. Green, and E. J. Chu, *J. Biol. Chem.* **190**, 643 (1951).
109. C. Rimington and A. Benson, *J. Chromatog.* **6**, 350 (1961).
110. P. A. D. Cornford and A. Benson, *J. Chromatog.* **10**, 141 (1963).
111. A. M. D. C. Battle and A. Benson, *J. Chromatog.* **25**, 117 (1966).
112. R. Kehl and W. Stich, *Z. Physiol. Chem.* **289**, 6 (1951).
113. R. Kehl and W. Stich, *Z. Physiol. Chem.* **290**, 151 (1952).
114. M. J. Hendrickson, R. R. Berueffy, and A. R. McIntyre, *Anal. Chem.* **29**, 1810 (1957).
115. T. C. Chu and E. J. Chu, *J. Biol. Chem.* **212**, 1 (1955).
116. E. I. B. Dresel, A. Benson, and B. C. Knight, *Biochem. J.* **63**, 87 (1956).
117. J. E. Falk and A. Benson, *Biochem. J.* **55**, 101 (1953).
118. L. Ericksen, *Scand. J. Clin. & Lab. Invest.* **10**, 319 (1958).
119. V. Holeček, *Chem. Listy* **51**, 315 (1957).
120. J. E. Falk, E. J. B. Dresel, A. Benson, and B. C. Knight, *Biochem. J.* **63**, 87 (1956).
121. J. E. Falk, *Biochem. Soc. Symp. (Cambridge, Engl.)* **12**, 17 (1954).
122. J. E. Falk, *Ciba Found. Symp., Biosyn. Porphyrins Porphyrin Metab.* p. 63 (1955).
123. L. Bogorad and S. Granick, *J. Biol. Chem.* **202**, 793. (1953).
124. R. L. Veech, G. A. Rogeness, and H. Weil-Malherbe, *Biochem. J.* **105**, 1209 (1967).
125. M. Morrison and E. Stotz, *J. Biol. Chem.* **228**, 123 (1957).
126. J. L. Connelly, M. Morrison, and E. Stotz, *J. Biol. Chem.* **233**, 743 (1958).
127. D. A. Rappoport, C. R. Calvert, R. K. Loeffler, and J. H. Gast, *Anal. Chem.* **27**, 820 (1955).
128. G. Y. Kennedy, *Scand. J. Clin. & Lab. Invest.* **5**, 281 (1953).
129. J. Barrett, *Biochem. J.* **64**, 626 (1956).
130. G. Serchi, *Ann. Chim. (Rome)* **43**, 253 (1953).
131. J. B. Harbone, *J. Chromatog.* **2**, 581 (1959).
132. J. B. Harbone, *in* "Chromatography" (E. Heftmann, ed.), p. 677. Reinhold, New York, 1967.
133. J. B. Harbone, *J. Chromatog.* **1**, 473 (1958).
134. J. B. Harbone, "Comparative Biochemistry of Bioflavonoids." Academic Press, New York, 1967.
135. K. S. Szotyori and E. W. Jurics, *Z. Lebensm. Untersuch.-Forsch.* **135**, 192 (1967).
136. C. A. Wachtmeister, *Acta Chem. Scand.* **5**, 976 (1951).
137. M. K. Seikel, *J. Am. Chem. Soc.* **77**, 5685 (1955).
138. E. C. Bate-Smith, *Sci. Proc. Roy. Dublin Soc.* **27**, 165 (1956).
139. E. C. Bate-Smith, *Biochem. Soc. Symp. (Cambridge, Engl.)* No. 3, 62 (1949).
140. E. A. H. Roberts and D. J. Wood, *Biochem. J.* **53**, 332 (1953).
141. E. A. H. Roberts, R. A. Cartwright, and D. J. Wood, *J. Sci. Food Agr.* **7**, 637 (1956).
142. Y. Abe and K. Hayashi, *Botan. Mag. (Tokyo)* **69**, 577 (1956).
143. E. A. H. Roberts, *in* "Chemistry of the Vegetable Tannins," p. 87. Soc. Leather Trades' Chemists, London, 1956.
144. Y. Oshima, T. Nakabayashi, and H. Imagawa, *Nippon Nogeikagaku Kaishi* **25**, 478 (1952).

145. T. A. Geissman, *in* "Moderne Methoden der Pflanzenanalyse" (K. Paech and M. V. Tracey, eds.), Vol. III, p. 450. Springer, Berlin, 1955.
146. D. B. Mullick, *J. Chromatog.* **39**, 291 (1969).
147. R. M. Acheson, J. L. Harper, and I. H. McNaughton, *Nature* **178**, 1283 (1956).
148. K. Egger, *J. Chromatog.* **5**, 74 (1961).
149. K. S. Pankajamini and T. R. Seshadri, *Proc. Indian Acad. Sci.* **A36**, 157 (1952).
150. L. Ponniah and T. R. Seshadri, *J. Sci. Ind. Res. (India)* **12B**, 605 (1953).
151. E. C. Bate-Smith, *in* "Methods in Polyphenol Chemistry" (J. B. Pridham, ed.), p. 73. Pergamon Press, Oxford, 1964.
152. V. Amormino, G. Bellomonte, and E. Cingolani, *Ann. Ist. Super Sanita* **4**, 105 (1968).
153. J. B. Harbone and H. S. A. Sherratt, *Experientia* **13**, 486 (1957).
154. L. Reio, *J. Chromatog.* **1**, 338 (1958); **4**, 458 (1960).
155. R. Haensel, *in* "Papierchromatographie in der Botanik" (F. Linskens, ed.), 2nd ed., p. 228. Springer, Berlin, 1959.
156. H. W. Casteel and S. H. Wender, *Anal. Chem.* **25**, 508 (1953).
157. B. H. Koeppen, *J. Chromatog.* **18**, 604 (1965).
158. P. Dupuis and J. Puisais, *Compt. Rend.* **240**, 1802 (1955).
159. T. B. Gage and S. H. Wender, *Anal. Chem.* **22**, 708 (1950).
160. F. G. C. Forsyth and N. W. Simmonds, *Proc. Roy. Soc.* **B142**, 549 (1954).
161. E. Eigen, M. Blitz, and E. Gunsberg, *Arch. Biochem. Biophys.* **68**, 501 (1957).
162. G. M. Barton, R. S. Evans, and J. A. F. Gardner, *Nature* **170**, 249 (1952).
163. E. Wong and A. O. Taylor, *J. Chromatog.* **9**, 449 (1962).
164. D. G. Roux, *J. Chromatog.* **10**, 473 (1963).
165. E. Lederer and M. Lederer, "Chromatography." Elsevier, Amsterdam, 1957.
166. E. C. Bate-Smith, *Biochem. J.* **58**, 123 (1954).
167. T. B. Gage, C. D. Douglass, and S. H. Wender, *Anal. Chem.* **23**, 1582 (1951).
168. R. Paris, *Bull. Soc. Chim. Biol.* **34**, 767 (1952).

12

PESTICIDES

I. Introduction

Ever since Paul Müller discovered the insecticidal properties of DDT in 1939, it has become imperative to develop sensitive analytical methods for micro quantities of pesticide residues and their metabolites in foods and biological materials. Paper chromatography seemed to be the ideal technique, and its application to problems in pesticide analyses has been described in detail by Mitchell (*1*), Block *et al.* (*2*), Getz (*2a*), and McKinley (*3*).

In 1961, the first application of gas chromatography to pesticide analyses was reported, and more recently thin-layer chromatography has become very popular due to its rapidity and high resolving power. These three techniques, gas–liquid, thin-layer, and paper chromatography, as they pertain to pesticide analyses, have been compared and evaluated (*4*, *5*).

Although gas–liquid chromatography has the advantage of greater sensitivity for pesticide analyses, paper and thin-layer chromatography are extremely useful tools for the confirmation of results obtained from relatively nonspecific gas chromatographic detectors. From an examination of R_F values and characteristic *in situ* color reactions on the respective media, identification of many pesticides and metabolites becomes feasible (*6*).

The favored technique for the chromatography of water-insoluble pesticides is reversed-phase paper chromatography. Detection techniques are by visual inspection with ultraviolet light or a modified silver nitrate spray followed by exposure to ultraviolet light.

For its great simplicity, ascending chromatography is recommended, using 8 × 8 inch squares of filter paper and a metal tank already described in Chapter 3 (see Fig. 11 of that chapter), designed by Mitchell (*1*). Also, Mitchell's monograph "Ascending Paper Chromatography: A Way To Do It" (*1*) is a model of brevity and clarity and should be on the laboratory bench of every pesticide chromatographer. A complete guide to the pesticide literature has been compiled and the suggested reading list may serve to start a library in the pesticide field (*7*).

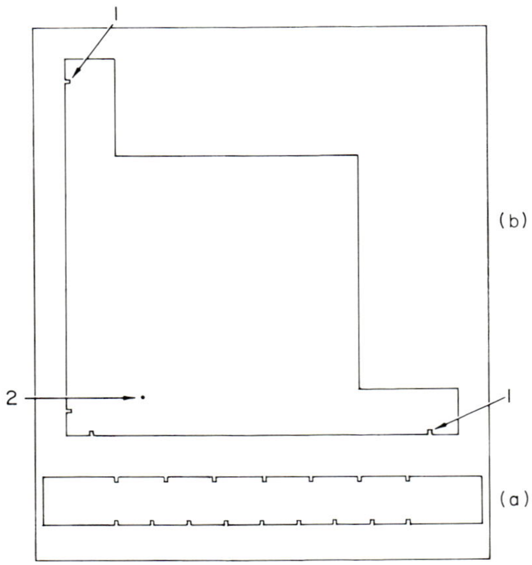

FIG. 1. Templates for marking papers; (a) one-dimensional; (b) two-dimensional; (1) are notches for reference compounds for each direction in two-dimensional chromatography; (2) origin for two-dimensional chromatography (*8*); adapted with permission of *J. Assoc. Offic. Anal. Chemists.*

An accessory for ascending chromatography on 8 $\times$ 8 inch papers are two simple templates, depicted in Fig. 1 (*8*). Template (a) on the bottom of the picture is a 9 $\times$ 1 inch strip of plastic or metal, notched 1.5 inches from each edge and equidistant on the lateral edges, accommodating seven to nine spots for one-dimensional development. The upper template (b) is cut from an 8 inch square of stainless steel and in practice is lined up along the sides of the filter paper. Notches (1) are for reference compounds in each direction and a small hole (2) marks the origin for two-dimensional chromatography.

The use of paper chromatography for clean-up of pesticide residues and their metabolites from fatty tissue, especially from insects, involves the ascending or descending development of the crude extract with acetonitrile as irrigating solvent (*9*). A miniature chromatographic tank fashioned from a short length of 1 inch glass tubing, serves for descending development. A paper strip technique is similar to the one depicted for elution in Chapter 4, Fig. 1. The insecticide residue and a marker dye [1 μl of 0.5% (w/v) *N*, *N'*-dimethyl-*p*-1-naphthylazoaniline] move with the solvent front, while the interfering lipid material remains near the origin.

II. Chlorinated Insecticides

A. Sample Preparation

1. *Fatty Samples*

Probably the most important clean-up procedure for the extraction of chlorinated pesticides is the one involving animal fat, including dairy products. The principle is the preferential extraction of the nonpolar pesticides into acetonitrile, followed by a two-step column chromatographic clean-up. Detailed instructions are as follows (*10*, *11*).

Dissolve sample of fat (about 5 gm) into 25 ml petroleum ether (30–60°C) and extract the pesticides with several portions of acetonitrile. Pour acetonitrile layer into excess water and reextract pesticides with petroleum ether. Dry the petroleum ether fraction over anhydrous Na_2SO_4 and evaporate to a small volume on a steam bath, but preferentially under reduced pressure in order to minimize possible loss of pesticides (aldrin*, dieldrin, endrin). The concentrate is transfered to a column of separate layers of anhydrous Na_2SO_4 and activated Florisil. (Florisil as obtained from the manufacturer is heated at 650°C for 2 hours, then after 4 days heated again at 130°C for 5 hours and stored at that temperature until ready for use.) The pesticides are eluted with 200 ml each of ethyl ether–petroleum ether (6:94) and (15:85). The latter fraction contains, among other pesticides, endrin and dieldrin and may require an additional column clean-up—magnesia–Celite column with 150 ml of petroleum ether as eluant (*12*).

When eggs were analyzed for chlorinated pesticides, it was found that the (15:85) eluted fraction had to be saponified with dilute KOH prior to the second column cleanup (*11*).

2. *Nonfatty Foods* (*13*)

The sample of fresh fruits or vegetables (100 gm) and 10 gm of Celite are blended at high speed with 200 ml acetonitrile. Filter mixture on a

* For formulas of pesticides, the reader is referred to Spencer, "Guide to the Chemicals Used in Crop Protection" (*11a*).

Büchner funnel, add 100 ml petroleum ether, shake vigorously, add 10 ml NaCl saturated solution, and about 600 ml water. Shake gently and let phases separate; discard aqueous phase. Wash with an additional two portions of 100 ml water, and dry petroleum ether fraction over Na_2SO_4. The final extract may be concentrated and is chromatographed on activated Florisil as described in Section A, 1.

B. Chromatographic Systems

Whatman No. 1 filter paper for chromatography is being used almost exclusively for the separation of chlorinated pesticides. As a matter of caution, the filter paper should not be handled with bare hands until the chromatography is completed, and the use of clean cotton gloves is recommended (*14*). Since the paper has to be impregnated with a stationary phase for successful resolution of chlorinated pesticides, this section will treat the subjects of paper and solvents together rather than separately (*14*, *15*).

1. *Aqueous Solvent System*

Immobile solvents: (5%, v/v) refined soybean oil or mineral oil (USP heavy) in diethyl ether; *mobile solvents*: acetone–water (75:25); methanol–water (75:25); 2-methoxyethanol (methylcellosolve)–water (75:25). Over 114 pesticides have been examined with this solvent system and some of the data are summarized in Table I (*14*, *16*).

It is recommended that the chromatographic paper be washed with water by ascending chromatography prior to spotting of the samples and air-dried overnight or at 70°C for 30 minutes. The washing procedure minimizes the appearance of a dark ghost front near the solvent front after reaction with the silver nitrate spray reagent (see below) A modification of the washing procedure consists of impregnating the paper with the immobile phase and "washing" the paper with the moving phase, e.g., 75% aqueous acetone (*17*). Considerable improvement can be demonstrated by minimizing the curtain effect due to paper impurities reacting with silver nitrate. This improvement results in better resolution and identification of suspected chlorinated pesticide residues in butter fat.

2. *Nonaqueous Systems*

Immobile solvents: 10% (v/v) 2-phenoxyethanol in diethyl ether; 35% *N*, *N*′-dimethylformamide in diethyl ether; *mobile solvent*: 2, 2, 4-trimethylpentane; see Table I for R_F values of selected chlorinated pesticides.

Dieldrin and endrin, not normally separated by the above solvents can be resolved by ascending chromatography in about 1.25 hours with

TABLE I

R_F VALUES AND APPROXIMATE SENSITIVITY (TO SILVER NITRATE REAGENT[a]) OF CHLORINATED PESTICIDES (*16*)

Common or trade name	R_F Values[b] Aqueous solvent system	Nonaqueous solvent system	Approximate minimum detectable quantity (μg)
Aldrin	0.37	1.0	0.2
BHC	0.79; 0.83; 0.88; 0.93	0.16; 0.28; 0.68; 0.77	0.2–2.0
Mitox	0.62	0.78	1.0
Chlordane	Streak	0.96	0.2
Chlorobenzilate	1.0	0.50	0.2–0.5
DDE	0.48	0.98	0.1
2,4-D	0.95	0	0.05–0.2
DDT	0.60	0.87	0.1
Dieldrin	0.60	0.89	0.1
Endrin	0.62	0.83	0.1
Heptachlor	0.48	0.97	0.1
Heptachlor epoxide	0.72	0.87	0.1–0.5
Kelthane	0.76	0.37	0.1
Lindane	0.83	0.65	0.1
Methoxychlor	0.89	0.51	0.2
PCNB	0.48	0.80	0.2–1.0
Perthane	0.59	Not determined	0.5
Strobane	Streak	Streak	0.2
Thiodan	0.55; 0.78	0.48; 0.95	0.5
Rhothane	0.75	Streak	0.2
Toxaphene	Streak	Streak	0.2

[a] Color reagent: $AgNO_3$–phenoxyethanol; ultraviolet irradiation (see Section I,C).

[b] Solvent systems: aqueous—mineral oil, 75% acetone (see Section I,B); nonaqueous—phenoxyethanol, 2,2,4-trimethylpentane (see Section I,B).

the following solvent system (*18*)—*immobile solvent*: 25 ml of Ucon Lubricant 50 HB 5100 (Union Carbide Chemicals, New York, N. Y. 10017) diluted with ethyl ether to 500 ml; *mobile solvent*: 2, 2, 4-trimethylpentane. The R_F values are approximately 0.7 for dieldrin and 0.8 for endrin.

C. DETECTION

Silver Nitrate–2-Phenoxyethanol Reagent (*14*)

This is the most universal and sensitive reagent for organochlorine pesticides. It is made from 1.7 gm $AgNO_3$ dissolved in 5 ml distilled water; 10 ml of 2-phenoxyethanol (for chromatograms developed with the nonaqueous system) or 20 ml (for aqueous system) and a drop of 30% H_2O_2 are added and acetone to a final volume of 200 ml. This solution may be used as a spraying or dipping reagent. After treatment of the developed chromatogram, the paper is held under an ultraviolet light source (e.g., 30 watt germicidal lamp) at a distance of 5 inches above the surface of the paper. One minute on each side and 5 additional minutes for each side is the correct exposure for full development without undue darkening of the background. Black spots on a light gray background result (see Fig. 2). To lighten the background even further, we have washed the developed chromatograms successively in acetone, 20% (w/v) sodium thiosulfate in water, finally running water, each wash being interspersed with drying in air or oven. The finished chromatograms may be kept indefinitely in a laboratory notebook or in subdued light.

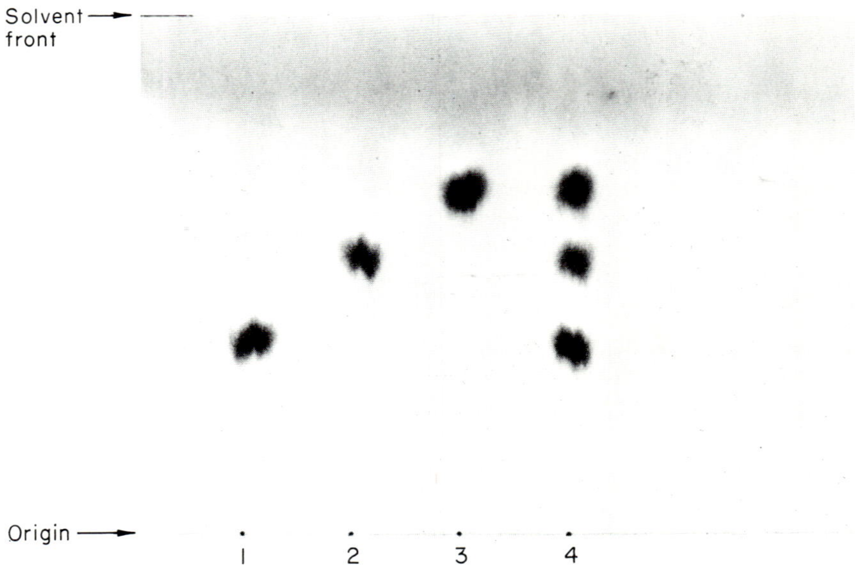

FIG. 2. One-dimensional chromatogram of mixture of (1) DDT, (2) DDD, and (3) Lindane; reversed phase chromatography; mobile solvent, 75% aqueous acetone (*4*).

The sensitivity of the silver nitrate color reagent ranges from 0.05 to 2 μg chlorinated pesticide per spot; sensitivity for a series of pesticides has been summarized in Table I (*16*).

The unsprayed chromatograms may also be viewed directly after solvent development and drying under ultraviolet light, and many pesticides, even those not containing chlorine, show up as black (quenched) or fluorescent spots (*14*); no concise information on the sensitivity or selectivity of this technique has been reported.

D. Quantitative Estimation

A mixture of DDD and DDT was separated on Whatman No. 1 paper using the aqueous system of 70% (v/v) aqueous acetone as mobile solvent described in Section B, 1 (*19*). Guide strips were developed with the silver nitrate reagent (cf. above) and parallel spots corresponding to the R_F values of the guide spots were eluted with acetone and analyzed by the well known Schechter–Haller method (*20*).

For the quantitative estimation of chlorinated insecticides in soil, the immobile phase incorporates the silver nitrate color reagent and is prepared as follows (*21*): 3 gm $AgNO_3$ is dissolved in 250–600 ml *N*, *N'*-dimethylformamide, 50 ml 2-phenoxyethanol is added and sufficient amounts of diethyl ether to a final volume of 1 liter; *n*-hexane is used as mobile solvent. After ascending development, the chromatogram is first air dried overnight, then humidified, and finally irradiated with ultraviolet light. The unknown and standard spots are outlined with a soft pencil, cut out of the paper, and weighed to the nearest milligram. Calibration curves are drawn by plotting weights in milligrams against the logarithm of the concentration.

Using Michell's aqueous system and 75% (v/v) aqueous acetone as mobile solvent, mixtures of DDT–DDD–lindane and DDT–DDE–DDD–Kelthane (Figs. 2 and 3) may be resolved and quantified by the maximum spot density method described in Chapter 4 using a photoelectric densitometer. Standard curves are drawn by plotting the product of area and density versus micrograms of DDT (*4*), for example; see Fig. 4.

E. Detailed Experiment

The experimental procedure described below has been taken from a laboratory course given at the University of California, Davis, to a group of dairy laboratory technicians for the analyses of chlorinated presticides in milk and daiy products.

1. Chromatographic papers (Whatman No. 1, 8 × 8 inch) are prewashed by suspending them in a solvent trough containing water and allowing the water to reach the top of the papers by ascending migration; the papers may be left in this manner overnight.
2. The papers are dried in a forced-air oven at 70°C.

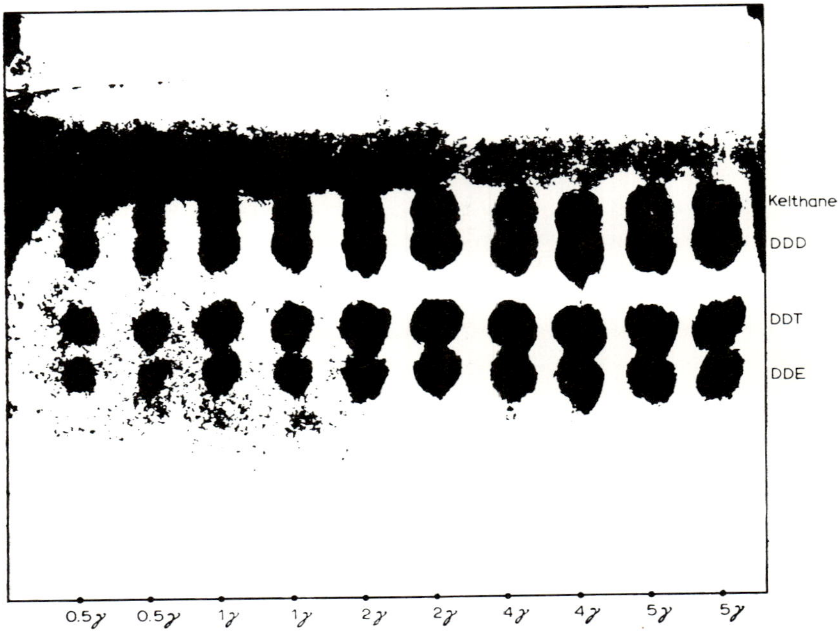

FIG. 3. Paper chromatography of chlorinated pesticides for quantitative determination (*4*).

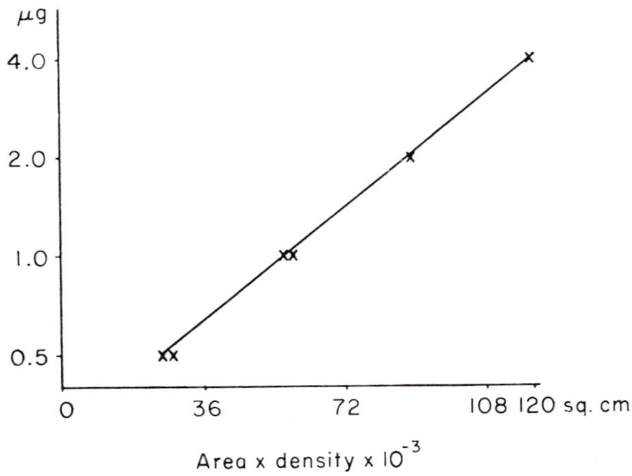

FIG. 4. DDT quantitative analysis by paper chromatography; area × density method (*4*).

3. The origins for spots are ruled with a hard pencil 1 inch from the lower edge with a ruler or template as depicted in Fig. 1.

4. Below each spot a code number is penciled in.

5. The paper is taped to a glass plate so that the origins extend over the edge.

6. Place 60 watt electric bulb close to the paper surface; the heat from bulb will keep the spots small.

7. Apply spots in the following manner: *For standards*: spot 1 μl increments using a lamda micropipette, ranging from 1 to 5 μg. *For samples*: (see Section A, 1 for clean-up) to the dry cleaned up sample in conical centrifuge tube, 1–2 drops of ethyl ether are added to effect solution of the sample. The samples are spotted in 1 μl increments until all of the sample has been applied. For best results, it is recommended to keep the spot size at the origin as compact as possible.

8. The paper is removed from the glass plate to a dish containing the immobile solvent in solution. The whole sheet of paper up to the line of origin is immersed until completely wet and is removed immediately.

9. The paper is suspended on a glass rod in a fume hood by means of chromatographic clips (see Chapter 3, Fig. 7) and allowed to air dry; this takes 1 minute or less.

10. To the solvent trough inside the chromatography chamber (Chapter 3, Fig. 11), 50 ml of the mobile solvent is added, and the papers are suspended, right side up, until the lower edge just touches the surface of the solvent. The tank is sealed and development allowed to proceed for about 2 hours.

11. After solvent development (the solvent front should be about 2 inches from the upper edge of the sheet), the papers are removed from the tank and air dried for several minutes.

12. The developed and dried chromatogram is dipped into the silver nitrate color reagent solution (see Section C).

13. The paper sheet is oven dried at 70°C for 5 minutes.

14. The paper is placed under the ultraviolet light source as described above.

15. After the spots are fully developed (black), the papers are washed with acetone, thiosulfate, and water, as described in Section C, the spots are marked by encircling them with a soft pencil, and maximum spot density is read by a photoelectric densitometer (see Chapter 4).

III. Organophosphate Insecticides

A. Preparation of Samples—Fruits and Leafy Vegetables

Based on recommended usage and known modes of metabolism, organophosphate insecticide residues and their possible metabolites are more usually encountered in green leafy vegetables and fruits than in milk and dairy products. The general extraction and clean-up procedure is as follows.

The sample is homogenized and extracted with excess of acetonitrile. The acetonitrile is evaporated in a flash evaporator at a water bath temperature of 50°C, and the residual aqueous phase extracted into ethyl acetate. The crude extract is cleaned up chromatographically on acid-treated charcoal (Norit A, treated with 10% HCl in 50% ethanol, filtered, and activated at 130°C for 48 hours). The organophosphates are eluted from the charcoal–Celite column (3:2, w/w) first with 100 ml 25% chloroform in ethyl acetate, followed by 100 ml chloroform. The final eluate is evaporated just to dryness and taken up in a suitable volume of ethyl acetate. For Guthion, benzene must be substituted for chloroform. This procedure has been found to be suitable for malathion, parathion, Diazinon, Guthion, Systox, and Trithion (*22*).

A procedure suitable for parathion, methyl parathion, EPN, and their respective oxons is as follows (*23*).

The original acetonitrile extract is evaporated until only water remains, and sufficient amounts of acetonitrile are added to a concentration of 40%. One milliliter 1 *N* HCl is added, and the extract chromatographed on a column of polyethylene-coated alumina (Kensington Scientific, Berkeley, Calif.). The organophosphates are eluted with 100 ml 40% acetonitrile in water containing 1 ml 1 *N* HCl. A second chromatographic clean-up follows on a column of 5 gm magnesol (Canada Colors and Chemicals, Toronto, Ontario) and 30 gm anhydrous Na_2SO_4, eluted with chloroform. The final eluate is evaporated to a suitable volume.

For the analysis of Guthion in fruits and vegetables, an initial acetone extract was reextracted into chloroform, washed with water and evaporated to a small volume for direct spotting on paper chromatograms (*24*).

B. Paper

Whatman No. 1 chromatographic paper treated with an immobile phase (see Section C below) and prewashed with water for the removal of interfering impurities, has been usually employed for one- and two-dimensional paper chromatography of organophosphate insecticides and their metabolites (*25*). Whatman paper 3 MM (*26*) and Whatman No. 20 paper (*27*) have also been used. Acetylated paper, 100%, (Schleicher and Schuell) has been used for a variety of organophosphate insecticides (*28*).

C. Chromatographic Systems

Table II lists a variety of chromatographic systems in which the paper is impregnated with a stationary phase (polar or nonpolar) to be used with a suitable mobile solvent. Most of the systems found in this table are suitable for one- and two-dimensional paper chromatography.

For the separation of Dursban and possible metabolites, the paper is not impregnated but must be equilibrated with the vapors of the solvents very thoroughly prior to solvent development (*29*) (see Table III). R_F values of organophosphate insecticides employing the solvent systems described in Table II are listed in Table IV. Figure 5 shows the two-dimensional separation of eleven organophosphate insecticides.

A novel technique permits the separation of organophosphate insecticides of widely different R_F values on a single sheet of paper by impregnating the same paper with two stationary phases of different polar characteristics, e.g., 10% light mineral oil and 10% Epon resin 828 (*26*). The lower one-third of Whatman 3 MM paper is impregnated with the oil while the upper two-thirds with the resin. The developing solvent is 35% aqueous acetonitrile. Separation is effected of MTST (tetraethyl monothiopyrophosphate), Diazoxon, Sulfotepp, and Diazinon.

D. Detection Methods

A detailed review of many commonly used chromogenic reagents for organophosphate insecticides has been published (*30*) to which the reader may want to refer. Below are described the most useful reagents for the detection of organophosphates and their metabolites.

1. *N-Bromosuccinimide–Fluorescein* (*31*)

This reagent is made up of two solutions: (a) 0.002 *M* recrystallized *N*-bromosuccinimide in 1,1,1-trichloroethane and (b) 0.0003 *M* fluorescein in ethanol adjusted to pH 7. The developed chromatogram is lightly sprayed with (a), air dried, and sprayed with (b). Thiophosphates show up as yellow-green fluorescent spots on a light pink background. Sensitivity ranges from 2 to 10 μg.

Alternately, the thoroughly dried chromatograms, especially those treated with dimethylformamide and formamide, are exposed to bromine vapors for 30 seconds and sprayed with the dilute fluorescein solution (*25*). Under short wave ultraviolet light, spots appear as dark blue to black spots.

2. *Ultraviolet light* (*2537 Å*)

Some organophosphate insecticides show up as quenched areas (*25*). The chromatograms may also be blueprinted by exposing them for

TABLE II

CHROMATOGRAPHIC SYSTEMS FOR PAPER CHROMATOGRAPHY OF ORGANOPHOSPHORUS INSECTICIDES

No.	Stationary phase	Mobile solvent	Paper	Application	Reference
1	Dimethylformamide (20%)	2,2,4-Trimethylpentane	Whatman 1	Solvent 1, Fig. 5	*22, 25*
2	Heavy mineral oil (10%)	50% Dimethylformamide in water	Whatman 1	Solvent 2, Fig. 5	*25*
3	10% Formamide in acetone	2,2,4-Trimethylpentane	Whatman 1	Guthion, Phosdrin	*25*
4	30% Formamide in acetone	2,2,4-Trimethylpentane	Whatman 1	Guthion, Malathion, methylparathion, Phosdrin, Systox	*25*
5	Mineral oil (8%) (BDH)	Methanol–H_2O–0.5% NH_4OH	Whatman 1	See Table IV	*28*
6	Mineral oil (2%) (BDH)	70% Aqueous acetone	Scheicher and Schuell	See Table IV	*28*
7	Silicone 550 (5%)	Acetone–ethanol–H_2O (1:1:2)	Whatman 1[a]	Parathion, methylparathion, EPN	*23*
8	Silicone 550 (5%)	10% Aqueous acetone	Whatman 1[a]	Oxons of parathion, methylparathion, EPN	*23*
9	Heavy mineral oil (10%)	40% Aqueous acetonitrile	Whatman 1	Second dimension after solvent 1	*22*
10	20% Formamide in acetone	Benzene–chloroform (6:4)	Whatman 20	Two-dimensional with solvent 11	*27*
11	20% Formamide in acetone	Benzene–chloroform (9:1)	Whatman 20	See above and Table IV	*27*
12	30% Dimethylformamide in acetone	*n*-Hexane	Whatman 20	Two-dimensional with solvent 13	*24*
13	Liquid paraffin (5%)	50% Diethylformamide	Whatman 20	See above	*24*

[a] Paper grade not given in original paper but assumed to be Whatman No. 1.

TABLE III

R_F VALUES OF DURSBAN AND RELATED COMPOUNDS WITH VARIOUS CHROMATOGRAPHIC SYSTEMS (*29*)

Name	R_F [a]							
	1	2	3	4	5	6	7	8
3,5,6-Trichloro-2-pyridinol	0.80	0.90	0.76	0.78	0.68	0.94	0.78	0.62
O,*O*-Diethyl-*O*-3,5,6-trichloro-2-pyridylphosphorothioate (Dursban)	0.92	0.96	0.94	0.94	0.98	0.94	0.00	0.94
O-Ethyl-*O*-3,5,6-trichloro-2-pyridyl phosphorothioate	0.86	0.92	0.76	0.78	0.78	0.84	0.86	0.70
Diethyl-3,5,6-trichloro-2-pyridyl phosphate	0.80	0.88	0.90	0.90	0.95	0.90	0.00	0.92
Ethyl-3,5,6-trichloro-2-pyridyl phosphate	0.88	0.90	0.75	0.79	0.60	0.95	0.87	0.67
3,5,6-Trichloro-2-pyridyl phosphate	0.82	0.85	0.59	0.59	0.48	0.07	0.96	0.00
Sodium chloride	0.78	0.69	0.38	0.35	0.29	0.00	0.90	0.08
Sodium bicarbonate	0.00	—	0.00	0.00	0.00	0.00	0.00	0.00
5-Chloro-2,3,6-(1*H*)-pyridinetrione	0.72	0.00	0.00	0.00	0.00	0.64	0.84	0.15

[a] Solvent systems: (1) *tert*-butanol–acetic acid–H_2O (22:9:69 v/v); (2) ethanol–*n*-propanol–NH_4OH–H_2O (30:30:5:35); (3) ethanol–*n*-propanol–NH_4OH–H_2O (40:40:5:15); (4) isopropanol–NH_4OH–H_2O (75:2:23); (5) acetonitrile–NH_4OH–H_2O (80:2:18); (6) acetic acid: benzene (v/v); (7) *n*-propanol–NH_4OH–H_2O (10:2:88); (8) *n*-propanol–NH_4OH–H_2O (90:2:8).

TABLE IV

R_F VALUES OF ORGANOPHOSPHORUS INSECTICIDES (*25, 27, 28*)

Compound	R_F [a]							
	1	2	3	4	5	6	10	11
Co-Ral	—	—	—	—	0.81	0.23	—	—
Delnav	0.80, 0.95	0.18, 0.39[b]	—	—	—	—	—	—
Diazinon	0.48, 0.98	0.24, 0.96	—	—	0.77	0.38	—	—
Dibrom	—	—	—	—	0.78	—	—	—
Dipterex	—	—	—	—	0.78	—	—	—
Di-Syston	—	—	—	—	0.61	0.23	—	—
EPN	0.69	0.34	—	—	0.68	0.25	—	—
Ethion	—	—	—	—	0.63	0.22	—	—
Guthion	0.10	0.89	0.51	0.30	0.77	0.30	—	—
Korlan (Trolene)	—	—	—	—	0.51	0.21	—	—
Malathion	0.62	0.81	—	0.90	0.81	0.34	—	—
Menazon	—	—	—	—	—	—	0.09	0.04
Methylparathion	0.35	0.81	—	0.72	—	—	—	—
Nemacide (VC-13)	—	—	—	—	0.53	0.22	—	—
Parathion	0.75	0.51	—	—	0.75	0.24	—	—
Phosdrin	0.11, 0.04	0.98	0.15, 0.04	0.07, 0.01	0.84	0.56, 0.63	—	—
Rogor (Dimethoate)	—	—	—	—	0.79	0.54	0.78	0.65
Rogor-oxon	—	—	—	—	—	—	0.18	0.06
Ronnel	0.96	0.12	—	—	—	—	—	—
Systox (Demeton)	0.96, 0.59	0.34, 0.90	—	0.62	0.73, 0.84	0.32, 0.46	—	—
Systox-*S*-methylsulfoxide	—	—	—	—	—	—	0.41	0.14
Systox-*S*-methylsulfone	—	—	—	—	—	—	0.61	0.36
TEPP	—	—	—	—	0.76	0.71	—	—
Thimet	—	—	—	—	0.61	0.25	—	—
Trithion	0.96	0.06	—	—	0.51	0.20	—	—
Vamidothion	—	—	—	—	—	—	0.70	0.40
Vamidothionsulfoxide	—	—	—	—	—	—	0.09	0.04

[a] Two R_F values indicate two isomers. Numbers refer to solvent system in Table II.

[b] Indicates R_F range.

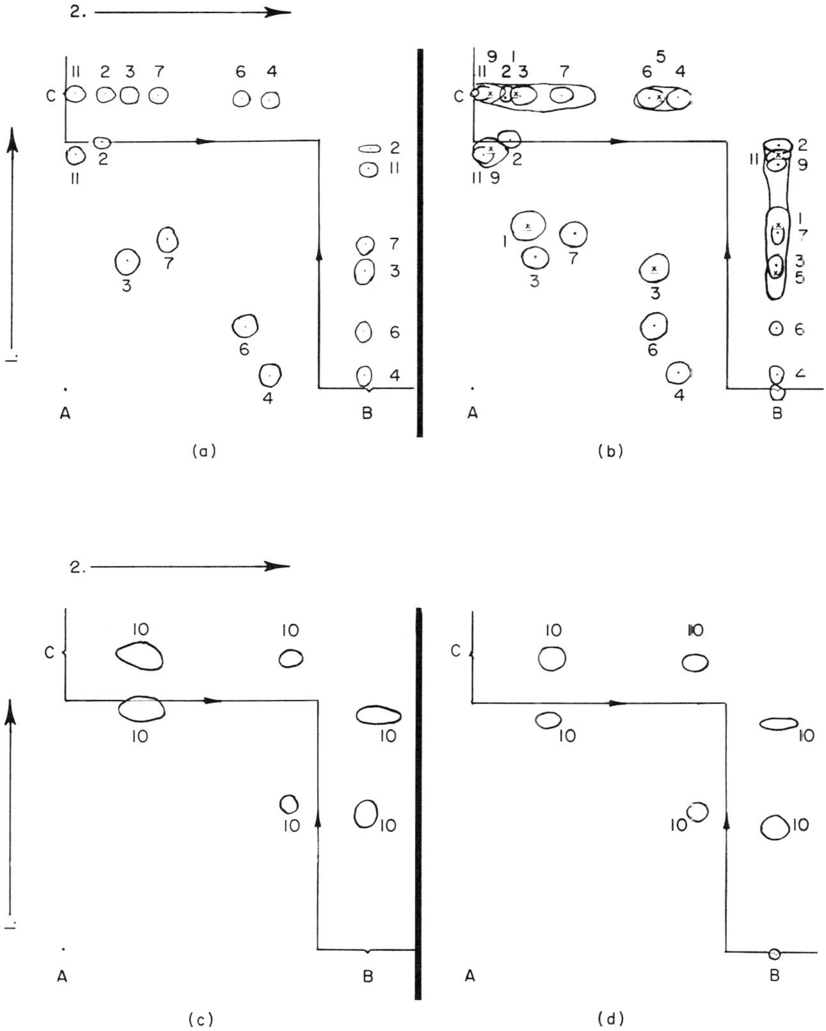

FIG. 5. Two-dimensional chromatography of eleven organophosphorus insecticides; solvent systems (1) and (2) described in Table II; (A) indicates the origin for two-dimensional chromatography; (B) and (C) are markers for first and second direction, respectively; legends for spots: (1) Delnav; (2) Diazinon; (3) EPN; (4) Guthion; (5) malathion; (6) methyl parathion; (7) parathion; (8) Phosdrin; (9) Ronnel; (10) Systox; (11) Trithion. Detection: (5a) Ultraviolet light (2537 Å), followed by ammoniacal silver nitrate and germicidal lamp. (5b) Chromogenic agent, ammoniacal silver nitrate, and germicidal light; * = quenched areas under short ultraviolet; × = not visible under ultraviolet until after color development. (5c) Two-dimensional chromatography of isomers of Systox; chromogenic agent, bromine and fluorescein. (5d) Isomers of Systox; chromogenic agent, ammoniacal silver nitrate and germicidal lamp. Reproduced by courtesy of *J. Assoc. Offic. Agric. Chemists* (*25*).

30 minutes to ultraviolet light from a germicidal lamp and placing them on top of blueprint paper; the blueprint paper is developed in running cold water; white spots on a blue background indicate the position of the phosphates.

3. *Ammoniacal Silver Nitrate* (*25*)

One hundred seventy milligrams silver nitrate is dissolved in 1 ml water, and 5 ml concentrated NH_4OH is added and final volume made up to 200 ml with 95% ethanol. This reagent and subsequent exposure to ultraviolet light visualizes most organophosphorus insecticides with the exception of Phosdrin.

4. *Silver Nitrate–Bromphenol Blue* (*32*)

Ten milliliters 2% aqueous silver nitrate solution is mixed with 90 ml 0.4% bromphenol blue in acetone. The dried chromatogram is sprayed and placed in an oven at 50°C for 10 minutes. The paper is then dipped into 0.01% citric acid in water. Thiophosphates are visualized as blue spots on a white background; the colors are stable for several hours and the sensitivity ranges from 1 to 5 μg per spot.

5. *2,6-Dibromo-N-chloro-p-quinononeimine* (*DCQ reagent*) (*33*)

The chromatogram is sprayed with a 0.5% solution of DCQ in cyclohexane and heated at 110°C for 7 minutes; thiophosphate insecticides show up as colored spots with a sensitivity ranging from 1 to 20 μg per spot as shown in the tabulation below.

Compound	Quantity (μg)	Color
Systox	1	Red-brown
Isosystox	2	Yellow
Parathion	2	Red-brown
Trithion	1	Orange-brown
Thimet	1	Orange-brown
Malathion	2	Orange-brown
Guthion	2	Orange-brown
Dipterex	20	None
Phosdrin	20	None
Ronnel	5	Red-brown

6. *Ferric Chloride–Salicylsulfonic Acid* (*28*)

This reagent consists of (a) 0.1% (w/v) ferric chloride solution in 80% ethanol prepared fresh daily and (b) 1% (w/v) salicylsulfonic acid solution in 80% ethanol. The dried chromatograms are sprayed with (a), air dried, and then sprayed with (b). Thiophosphates must first be converted to the corresponding oxons by exposure to bromine vapors; white spots on a mauve background result, provided the residual pH of the paper is between 1.5 and 2.5. TEPP and Phosdrin appear without bromine exposure. On acetylated paper, phosphate esters disappear on drying, whereas thiophosphates appear as intense yellow spots. On exposure to air, the phosphates reappear as white spots on a mauve background.

For Dipterex, the paper must first be exposed to NH_3 vapors, presumably converting it to DDVP. Dibrom and Schradan give only slight reactions with these reagents on Whatman No. 1 or Schleicher and Schuell (S. and S.) acetylated papers.

7. *Bromine–Congo Red* (*34*)

The dried chromatogram is exposed to bromine vapors generated from a 10% (v/v) solution of bromine in CCl_4 for 20 seconds; the paper is then aerated to remove all traces of bromine and sprayed with 0.4% (v/v) solution of Congo red in 50% aqueous ethanol; blue spots on a red background for thiophosphates at a sensitivity of 0.5 μg result. Similar results are obtained with Metanil yellow [sodium salt of 4-(3-sulfophenylazo)diphenylamine], yellow RFS [disodium salt of 4-sulfo-4′(sulfomethylamino)azobenzene] or methyl orange, yielding the colors in the tabulation (*34a*).

Spray reagent	Color of spot	Background color
Metanil yellow	Purple	Orange-yellow
Yellow RFS	Red	Yellow
Methyl orange	Pink	Yellow

8. *Alkaline Hydrolysis* (*23*)

For nitrophenyl derivatives like parathion, EPN, and paraoxon, the dried chromatograms after exposure to bromine vapors are sprayed with 5% aqueous KOH and yellow spots on a white background result.

9. *Enzymatic Detection* (*30*)

This technique is based on the inhibition of the enzymatic hydrolysis of acetylcholine to acetic acid and choline by organophosphorus insecticides and some of their metabolites.

a. "*Sandwich Method*" (*35*). An enzyme–indicator solution is composed of 20 ml pooled human blood serum, 60 ml water, 2 ml 0.1 *N* NaOH, and 3.25 ml bromthymol blue indicator solution (0.15 gm bromthymol blue in 25 ml 0.1 *N* NaOH). The substrate solution is 2% (w/v) acetylcholine bromide in water. A blank strip of chromatographic paper is sprayed with the enzyme–indicator solution, and the developed, air-dried chromatogram is laid on top of the moist strip; both strips are sandwiched between two glass plates and incubated at room temperature or at 37°C, where an incubator oven is available, for 15 minutes. The strips are separated and the blank strip, originally sprayed with the enzyme–indicator solution, is sprayed with the substrate solution. Spots containing inhibitor (insecticides or certain metabolites) remain blue on a yellow background after about 10 minutes. The sensitivity of this detection technique is in the submicrogram range.

b. Direct method 1 (*35*). The dried chromatogram is sprayed directly with the enzyme–indicator solution and incubated for 30 minutes in an atmosphere of high humidity. The chromatogram is then sprayed with the substrate solution, and blue spots on a yellow background result within 10 minutes, indicating the position of cholinesterase inhibitors.

Thinono insecticides (P $\rightarrow$ S) are converted to their corresponding thiol (oxons) derivatives, which are more potent anti-cholinesterase compounds. This is accomplished directly on the developed chromatogram by exposing it to bromine vapors for 30 seconds, aerating at 50–60°C, and proceeding as above.

c. Direct method 2 (*36*). Cholinesterase was obtained by homogenizing 3 gm beef liver with 27 ml water, centrifuging the homogenate, and using the supernate immediately or storing by freezing. The dried chromatogram is treated with bromine as described above and sprayed with liver homogenate supernate. After 20 minutes, the chromatogram is sprayed with detection solution (25 mg 1-naphthyl acetate in 10 ml ethanol, 50 mg Azoene fast blue RR salt in 20 ml water plus 20 ml pH 7.0 buffer). The immediate appearance of yellow spots on a brown background indicate the location of as little as 0.5 μg of Trithion, EPN, Ethion, parathion, Co-Ral, methyl parathion, or Guthion. Schradan could not be detected by this method.

E. Quantitative Estimation

None of the color reagents described in Section III,D above produce well defined and lasting colors on chromatograms to make an *in situ* densitometric method for quantitative estimation practicable. Elution techniques, followed by total organophosphorus analysis or specific colorimetric reactions for specific pesticides may be recommended. However, in most cases the quantities separated by paper chromatography are not sufficiently large to permit spectrophotometric determinations. The best use for paper chromatography of organophosphorus insecticides and metabolites is the confirmation of suspected compounds analyzed by conventional quantitative methods. Also, the studies on the metabolism of pesticides have been greatly advanced by the paper chromatography of radio-tagged insecticides.

F. Detailed Experiment

1. *One-Dimensional Procedure* (*25*)

One microliter portions of standards containing 10 μg of organophosphorus insecticides are spotted along a premarked pencil line (see Fig. 1) on prewashed sheets of 8 × 8 inch Whatman No.1 papers. The dipping tank (*37*) is filled with a solution of the immobile solvent (Table II); the dipping tank depicted in Fig. 6 has a capacity of about

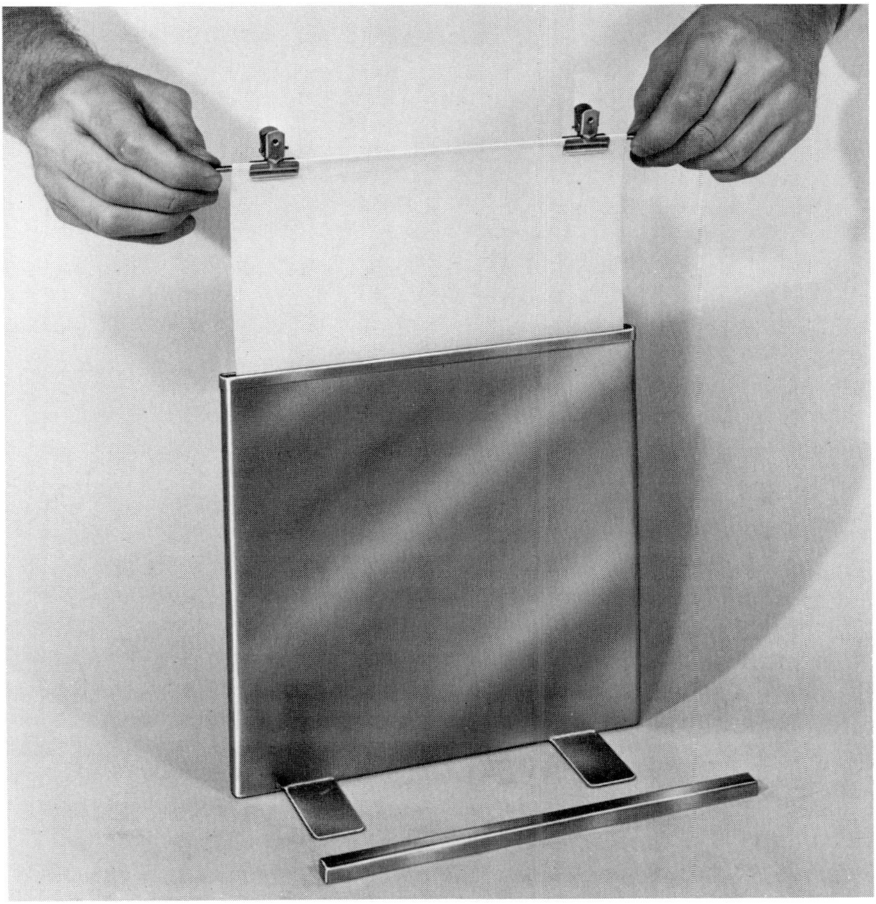

FIG. 6. Dipping tank for impregnating papers for reversed-phase chromatography (*37*). Reproduced by courtesy of A. H. Thomas Co.

370 ml. A 2 inch paper binder clip is secured to the lower edge nearer the applied spots, and the sheet is inverted and dipped into the tank up to the starting line. The paper is immediately removed, the upper edge is clipped to a glass rod, and placed on a drying rack (Fig. 7) inside a fume hood. As soon as practicable, the treated sheet is transfered to the chromatographic tank (Chapter 3, Fig. 11), which is sealed with the glass cover and masking tape. Ascending development time is about 75 minutes with the nonaqueous solvent system.

FIG. 7. Drying rack for supporting 8 × 8 inch chromatographic papers attached to support rods used in chromatographic tanks (Fig. 11, Chapter 3). Reproduced by courtesy of A. H. Thomas Co.

At the end of the run, the papers are removed on the glass support rod to the drying rack and dried in the hood overnight. Sometimes it is necessary to drive off the last traces of immobile solvent in an oven at 50°C; otherwise interference with certain color reagents results (e.g., Section III,D,4). The dried chromatograms are first viewed under ultraviolet light, and the visible spots marked with a pencil; the sheets may then be blueprinted (see Section III,D,2). They are then treated with bromine vapors, followed by treatment with a suitable color reagents (see Section III,D).

2. *Identification of Imidan Metabolites* (*38*)

A cleaned up acetone extract from a cotton plant treated with Imidan-^{14}C, along with Imidan standards and suspected metabolites are spotted on Whatman No. 4 paper, impregnated with 12% (v/v) glutaronitrile in acetone as the stationary phase and developed with isopropyl ether saturated with glutaronitrile. After solvent development, the papers are dried and sprayed with DCQ reagent (see Section III,D,5) to visualize Imidan and its oxygen analog. Other metabolites are detected with silver nitrate reagent (see Section I,C) followed by exposure to ultraviolet light at 2537 Å. Radioactive regions were detected by means of a 2π scanner and autoradiography (see Chapter 4).

An alternative solvent system that will characterize the more polar metabolites of Imidan, is ethanol–H_2O–NH_4OH (80:15:5 v/v). The R_F values of Imidan and possible metabolites are shown in Table V.

IV. Carbamate Insecticides

Although the development of carbamate insecticides in the early 1960's looked promising, to date there are still only a few commercial carbamate insecticides of any significance. Thus, the literature on the paper chromatography of carbamates is limited to a few publications. Yet it seems that paper and thin-layer chromatography are the choice methods of analysis rather than gas chromatography due to the high instability of carbamates on a gas chromatographic column at elevated temperatures.

A. SAMPLE PREPARATION (*39*)

For the analysis of carbaryl (Sevin) residues in wine, 100 ml of the sample are extracted with excess of chloroform, and the chloroform–ethanol layer is separated in a separatory funnel. The extract is evaporated under partial vacuum, and the residue dissolved in 50% aqueous methanol. Any remaining waxes are extracted into petroleum ether (30–60°C), and the methanol phase is extracted with an excess of reagent grade benzene. The benzene layer is evaporated to dryness and redissolved into 0.1 ml methanol and is ready for spotting on paper.

TABLE V

R_F VALUES OF IMIDAN AND SEVERAL POSSIBLE METABOLITES IN TWO SOLVENT SYSTEMS (*38*)

Compound	E/G system[a]				Aqueous system $AgNO_3$–ultraviolet
	DCQ[b]		$AgNO_3$–ultraviolet[b]		
	Authentic compound	Extracted compound	Authentic compound	Extracted compound	Authentic compound
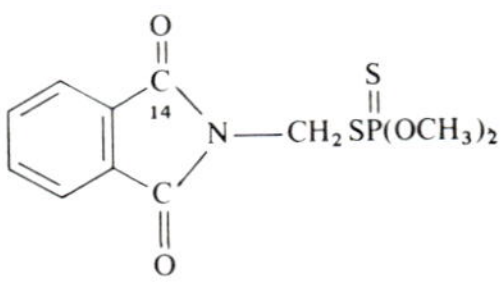 I	0.46 ± 0.05	0.46 ± 0.05	0.46 ± 0.05	0.46 ± 0.05	0.91
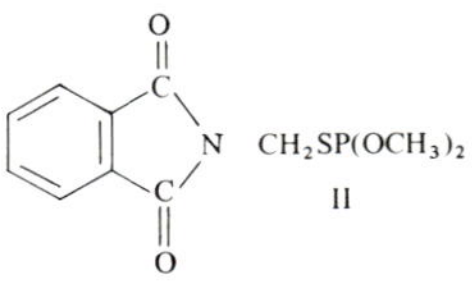 II	0.08	0.08	0.08	0.08	0.84
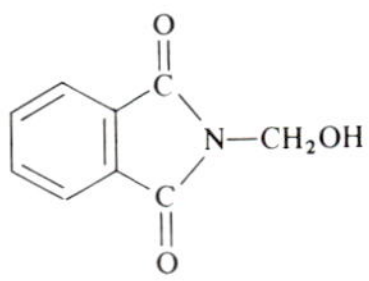 III	None	None	0.25	0.25	0.86
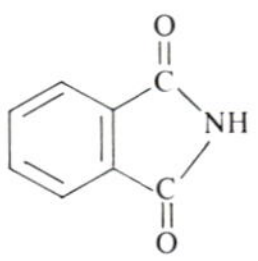 IV	None	None	0.58	0.58	0.86
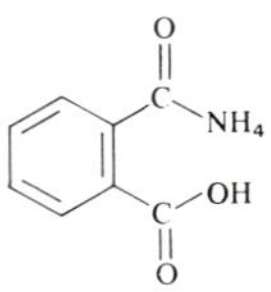 V	None	None	0–0.17	0–0.17	0.24

Table continued

TABLE V (*continued*)

Compound	E/G system[a]				Aqueous system $AgNO_3$–ultraviolet
	DCQ[b]		$AgNO_3$–ultraviolet[b]		
	Authentic compound	Extracted compound	Authentic compound	Extracted compound	Authentic compound
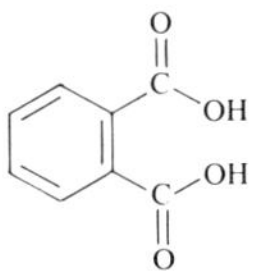 VI	None	None	0–0.17	0–0.17	0.24
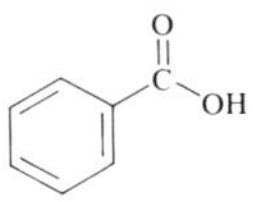 VII	None	None	0	None	0.71
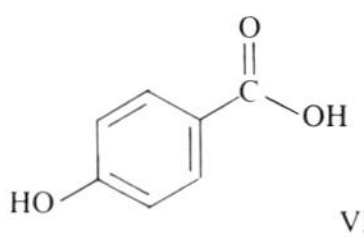 VIII	None	None	0	None	0.50
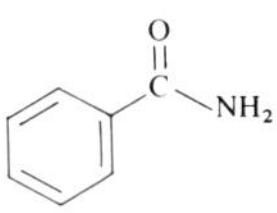 IX	None	None	0.12	None	0.76

[a] E/G system: glutaronitrile–isopropyl ether; aqueous system: ethanol–H_2O–NH_4OH (80:15:5 v/v).
[b] DCQ reagent D,5; $AgNO_3$ reagent D,3.

B. Paper Chromatographic Systems

1. *Aqueous System* (*39*)

Whatman No. 20 paper and the ascending development using 20% aqueous methanol will separate carbaryl and the metabolic breakdown product 1-naphthol (see Fig. 8), yielding an R_F of 0.37 for naphthol and 0.65 for Sevin.

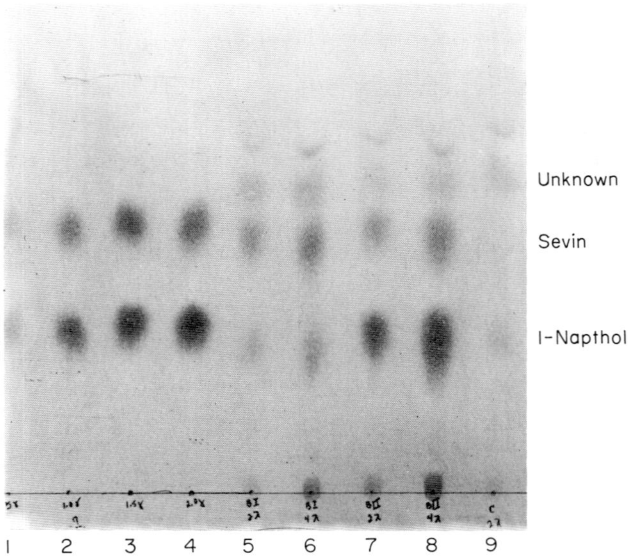

FIG. 8. Separation of Sevin and 1-naphthol by one-dimensional chromatography (*39*).

2. *Diphase System* (*40*)

Whatman No. 4 paper is coated with 10% (v/v) glutaronitrile in acetone and developed by descending technique with diisopropyl ether saturated with glutaronitrile. A 30 minute development time is sufficient for the chromatography of single carbamate insecticides—Dimetan, Pyrolan, Dimetilan, and Isolan.

C. DETECTION—*p*-NITROBENZENEDIAZONIUM FLUOBORATE

1. *Dipping Reagent* (*39*)

The dried chromatogram is first sprayed evenly with 1.5 *N* methanolic NaOH–1-butanol (v/v) to hydrolyze carbaryl to 1-naphthol. After air drying, the paper is dipped into 0.01 % reagent dissolved in methanol–ethyl ether (v/v). Blue spots result which can be evaluated quantitatively by the maximum spot density method (see below). The limit of detection by this method is 0.5 μg of carbaryl.

2. *Spray Reagent* (*40*)

The dried chromatogram is first sprayed with 15% (w/v) aqueous KOH, held over steam for 1 minute and then sprayed with 1 *N* acetic acid in methanol. After drying, the chromatogram is finally sprayed with 0.1% (w/v) methanolic reagent solution. The orange-red spots may be further intensified with a 15% (w/v) aqueous NaOH spray.

D. Quantitative Estimation (*39*)

Calibration curves for carbaryl and 1-naphthol are obtained by developing chromatograms using the aqueous system (see Section IV,B,1) for amounts ranging from 0.5 to 2.0 μg. The colored spots are read directly with a photoelectric densitometer using a 570 mμ filter, and a plot of percent transmission against the logarithm of the concentration yields a straight line, in accord with the findings of Block *et al.* (*2*).

V. Herbicides

A. Organochlorine Herbicides

1. *Preparation of Samples*

a. Water Samples (*41*). MCPA (4-chloro-2-methylphenoxyacetic acid), MCPB (butyric acid analog), 2,4-D, 2.4-DB, and 2,4,5-T are extracted from a 100 ml sample, acidified with 5 ml 6 *N* H_2SO_4, with diethyl ether. The ether solution is extracted with 150 ml 8% (w/v) sodium sulfate and 5 ml 2.5 *N* NaOH. Ten milliliters 5 *N* H_2SO_4 are added to the alkaline solution which is reextracted with an excess of ether. The ether solution is passed through a column containing anhydrous $NaSO_4$. The solvent is removed at moderate temperatures and the residue is dissolved in a small amount of ethyl acetate, ready for spotting on paper for chromatography.

When dalapon is suspected in the water sample, the extraction is begun by adding 1 ml 2.5 *N* NaOH to the original sample and evaporating to 30 ml on a water bath. The solution is cooled and 40 ml of concentrated H_2SO_4 is carefully added. The solution is extracted with ether, dried over anhydrous Na_2SO_4, evaporated to dryness, and taken up in small volume of ethyl acetate.

b. Soil Samples (*41*). A 100 gm sample is mixed with a small amount of water in a mortar to a paste, and a further 100 ml water is added. Twenty milliliters 6 *N* H_2SO_4 and 200 ml ethyl ether are added to a well stoppered glass container and agitated for 1 hour. The slurry is centrifuged, the aqueous layer decanted, and the water phase discarded. The remaining procedure is as above for the water sample.

c. Wheat Samples (*42*). The reader is referred to the original reference for the separation and clean-up of phenoxy acids and esters from wheat. The procedure is laborious and involves solvent extraction and column chromatography. In the final step, the acids are esterified to the corresponding methyl esters with diazomethane.

2. *Paper*

Whatman No. 3 and Whatman No. 1 chromatographic papers, the latter washed three times with water, have been used for the separation of phenoxy acids, esters, and dalapon (*41*, *42*). Whatman No.1 paper previously treated with ammonia vapor has also been recommended (*43*).

3. *Solvent Systems* (see Tables VI and VII for R_F values)

a. *Immobile phase*, 10% mineral oil in ether; *mobile phase*, 50% acetonitrile in water—good separation for methyl esters and compact spots; also lighter background with silver nitrate chromogenic reagent than with solvent (b).

b. *Immobile phase*, 35% dimethylformamide in ethyl ether; *mobile phase*, 2,2,4-trimethylpentane—for methyl esters of phenoxy acids.

c. *Immobile phase*, 10% mineral oil in ether; *mobile phase*, 65% acetonitrile in water—this system is suitable for esters of phenoxy acids (*42*).

d. *t*-Butanol–NH_4OH–H_2O (18:2:40), diluted to 200 ml with ethyl methyl ketone (*41*)—suitable for quantitative estimation by reflectance densitometry (see below).

e. Liquid paraffin–benzene–acetic acid (10:30:20 v/v), diluted to 200 ml with cyclohexane (see Fig. 9)—suitable for semiquantitative estimation of phenoxy acids by visual comparison or spot area method (*41*).

f. 1-Butanol saturated with water.

g. Butanol saturated with 1.5 N NH_4OH.

h. 1-Butanol saturated with 3.0 N NH_4OH (*43*).

4. *Detection*

a. Silver Nitrate Reagent (*41*). 1.7 gm silver nitrate is dissolved in 5 ml water, 20 ml 2-phenoxyethanol is added and the mixture is diluted to 200 ml with acetone; one drop of 30% H_2O_2 is added and the mixture is set aside overnight; decant the solution before use. The developed and dried chromatograms (80°C for 10 minutes) are dipped into this reagent by one continuous motion, the solvent is dried off in a current of air, and the two sides of the paper are irradiated with a germicidal lamp for 10 minutes each. The paper is washed successively in water, 0.5% sodium thiosulfate, and water, and dried in an oven at 80°C. These chromatograms are ready for densitometric evaluation (see below).

b. Bromcresol Purple (0.04% w/v) in Formaldehyde–Ethanol (1:5 *v/v*). The chromatograms are sprayed with this reagent and exposed to NH_3 vapors; yellow spots on a purple background result but fade rapidly. The spots may be regenerated periodically by an exposure to NH_3 (*43*).

TABLE VI

R_F VALUES OF PHENOXY ACID HERBICIDES (*41*, *42*)

Compound	R_F [a]		
	a[b]	b[b]	e
2,4-D	0.64	0.39	0.58
MCPA	0.50	0.58	0.68
MCPB	—	—	0.81
2,4,5-T	0.48	0.50	0.66
2,4-DB	0.34	0.73	0.79
2,4,5-TP	0.31	0.82	—
Dalapon	—	—	0.45

[a] Solvent systems described in text.
[b] Resolved as the methyl ester.

TABLE VII

R_F VALUES OF 2,4-DICHLOROPHENOXYACETIC ACID ESTERS (*42*)

2,4-D ester	R_F [a]		
	1	2	3
Isooctyl	0.32	—	—
Isopropyl	0.54	0.43	—
Butoxyethanol	0.53	0.28	—
n-Butyl	0.44	—	0.60
Propyleneglycol butyl ether	0.42	—	0.76

[a] Solvent system	Immobile phase	Mobile phase
1	10% mineral oil in ether	65% acetonitrile in H_2O
2	10% phenoxyethanol in ether	Isooctane
3	50 dimethylformamide in ether	Isooctane

5. *Quantitative Estimation* (*41*)

One microliter portions of the phenoxyalkoxy acids containing 2–12 μg of each acid are spotted on Whatman No. 3 paper. The chromatogram is developed for about 100 minutes with the *t*-butanol–NH_3 solvent (Section V,A,3,d). At the end of the run, the papers are first air dried and then heated in an oven at 80°C. The papers are dipped

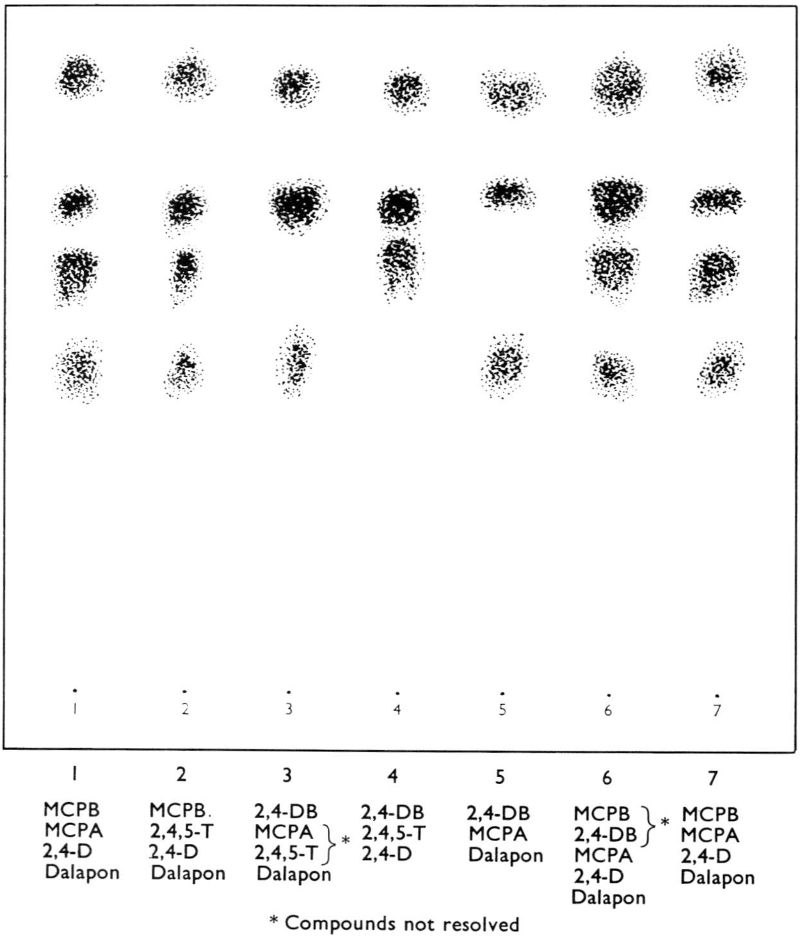

FIG. 9. One-dimensional chromatography of chlorinated herbicides on Whatman No. 1 paper developed with solvent (e), described in text (*41*). Reproduced by courtesy of *The Analyst*.

into the chromogenic reagent [silver nitrate (Section V,A,4)] and exposed to ultraviolet light from a germicidal lamp and washed in sodium thiosulfate as described in the previous section. The final spots are orange-brown of nearly elliptical shape on an off-white background, whose quality depends on the thoroughness of the washings. The sensitivity of the reflectance densitometer (Model X.110 fitted with

Mark III head; Evans Electroselenium, Ltd.) is adjusted to maximum sensitivity. The spots are scanned and minimum deflection readings are recorded; the logarithm of the galvanometer reading (reflectance) versus micrograms of herbicide yield a straight line. Unknown samples are read and the concentration calculated from the standard curve. The method is claimed to yield a deviation of ±5%.

B. Symmetrical Triazines

Atrazine and simazine residues in corn may be determined by paper chromatography semiquantitatively with a sensitivity of 0.1 ppm (*44*). The finely ground up samples of corn (shucks, kernel, and cob) are extracted with chloroform–NH_4OH (200:10 v/v) in a high speed blender, the slurry is centrifuged, washed with several portions of chloroform, and dried over anhydrous Na_2SO_4.

A portion representing 10–20 gm of sample is evaporated to dryness and transfered with a minimum of petroleum ether (30–60°C) saturated with water to a Celite 545–10% aqueous $HClO_4$ (4:4 w/v) column. The herbicides are eluted with 100 ml chloroform saturated with water. The chloroform solution is evaporated and further cleaned up on a Florisil column using as elutrient 10% acetone in petroleum ether. The final eluate is evaporated to dryness and made up to a small volume with acetone or ethyl ether.

Samples and standards (containing 1, 2, and 3 μg) are spotted on 8 × 8 inches Whatman No. 1 chromatographic paper. The paper is impregnated with 15% methanol in petroleum ether (stationary phase) and immediately transfered to the development tank (Chapter 3, Fig. 11), where it is developed with the mobile phase, redistilled trichloroethane–methanol–2,2,4-trimethylpentane (20:5:75 v/v). The developed chromatogram is air dried and sprayed with the $AgNO_3$–2-phenoxy-ethanol reagent (see Section I,C) and exposed to ultraviolet light. Atrazine (R_F approximately 0.7) and simazine (approximately 0.5) are well resolved and may be estimated semiquantitatively by visual comparison.

C. 3-Aminotriazole (ATA)

1. *Sample Preparation*

Plant tissue homogenates prepared with water are heated to 100°C for 2 minutes to precipitate protein and filtered (*45*). The filtrate is passed through an ion-exchange column of Amberlite IR-120 (H^+) and aminotriazole eluted with 3 *M* NH_4OH. An alternate procedure specifies the use of 80% ethanol as extractant (*46*). A more elaborate clean-up procedure has been described for the separation of aminotriazole and metabolites from vegetable crops, and the reader may want to refer to the original reference (*45*).

2. *Chromatographic Systems*

Whatman No. 1 has been recommended throughout for the one- and two-dimensional paper chromatography of aminotriazole and metabolites. The following tabulation gives solvents and observed R_F values reported for aminotriazole (*45–48*):

Solvent	R_F
(1) Methanol–formic acid–H_2O (80:15:5 v/v)	0.60
(2) Isopropanol–NH_4OH–H_2O (6:2:2 v/v)	0.75
(3) 80% Phenol	0.78
(4) *n*-Butanol–acetic acid–H_2O (27:7:17)	0.55
(5) 80% Pyridine	0.72
(6) Ethanol–H_2O–*n*-butanol (1:1:4 v/v)	0.47
(7) Dioxane–H_2O (8:2)	Not given

For two-dimensional chromatography, solvent (2) in one direction and for the second direction, *N*,*N*-dimethylformamide–H_2O (8:2 v/v) was employed.

3. *Color Reagents*

a. Azo Dye with Phenol. The dried chromatogram is first sprayed with 5% purified KNO_2 in water, followed quickly with a saturated solution of chromatographic grade phenol in 20% (v/v) HCl (*46*). Both solutions should be prepared fresh before each use. A yellow spot for ATA results.

b. Azo Dye with "H-Acid" (*45*). The dry chromatogram is first sprayed with 1% $NaNO_2$ in 80% aqueous acetone, followed by 20% (v/v) concentrated HCl in acetone (prepare fresh). After 10 minutes, the paper is sprayed with 0.25% disodium salt of "H-acid" (8-amino-1-naphthol-3,6-disulfonic acid). Aminotriazole at a concentration of 2 $\mu g/cm^2$ appeared immediately as a bright red spot against a white background. This color reaction was suitable for quantitative estimation by scanning with a stepwise photometer using a green filter (Chromaphor, Gamma Instruments, Great Neck, N. Y.). Concentration of ATA is proportional to the area under the curve, resulting in an accuracy of 5–10%.

D. Monuron (*49, 50*)

A sample of plant tissue is hydrolyzed with 20% NaOH for 12 hours and steam distilled. The resultant *p*-chloroaniline is extracted into hexane, reextracted with 1 *N* HCl, and finally the pH is adjusted to greater than 11 with NaOH. The amine is reextracted into a small amount of diethyl ether.

A portion of the ether solution containing 1–10 μg of *p*-chloroaniline is spotted on Whatman No. 1 filter paper and developed by the ascending technique with methanol–isoamyl alcohol–benzene–2 *N* HCl (35:17.5: 35:12.5 v/v). It is recommended that the paper be prewashed with the solvent mixture and dried over a steam plate. The amine is detected with one of the following spray reagents: (1) *Ehrlich reagent*—1 gm of *p*-dimethylaminobenzaldehyde mixed with 95 ml ethanol and 5 ml of concentrated HCl; brilliant yellow which fluoresces brilliantly under ultraviolet light, (2) *diazotization and coupling*—the dried chromatogram is first sprayed with 1% (w/v) $NaNO_2$ in 1 *N* HCl (freshly prepared), followed immediately by 0.2% *N*-(1-naphthyl)-ethylenediamine HCl in ethanol; magenta colored dye is formed.

Caution should be exercised in the interpretation of the results from hydrolyzed plant tissue, since an interfering amine has been identified as *o*-aminoacetophenone resulting from the hydrolysis of tryptophan (*49*). This interference ($R_F = 0.50$) may be eliminated by column chromatography of the two azo dyes on packed cellulose powder, eluting first the *o*-aminoacetophenone dye with 1 *N* HCl and *p*-chloroaniline with 1 *N* HCl–glacial acetic acid (1:1). Quantitative determination is made spectrophotometrically at 560 mμ (*51*).

The above methods should be amenable to the analysis of Diuron, Neburon, and other related substituted urea and anilide herbicides.

E. Maleic Hydrazide (*52*)

Whatman No. 1 and the ascending development with the following solvents have been employed for the separation of maleic hydrazide (MH) and indoleacetic acid: (1) 1-butanol–acetic acid–H_2O (5:1:4 v/v) (R_F for MH = 0.59); (2) isopropanol–concentrated NH_4OH–H_2O (7:1:2) ($R_F = 0.38$).

The color reagent 1% ferric chloride–1% $K_3Fe(CN)_6$ (aqueous) (v/v) produces a blue color for MH within several minutes after spraying, but with time the background of the chromatogram turns blue, and the contrast of the spots is lost. The reagent detects as little as 0.2 μg MH.

For quantitative estimation, the spots due to MH are outlined with a pencil and traced on a sheet of transparent Celluloid. The spots on the Celluloid are excised and weighed. With solutions of 0.21–4.64 μg of MH, the spot area increases directly with the logarithm of the concentration. By means of a calibration curve it is possible to estimate the content of MH in plant tissues with an accuracy of about 35%.

VI. Fungicides

A. Paper

Fiberglass paper (Reeve Angel & Co.) impregnated with formamide (*53*), Schleicher and Schuell 2043 b, and Whatman No. 1 chromatographic papers (*54, 55*) have been used to chromatograph a variety of fungicides and oxidation products. Specific applications will be found in Section B and Tables VIII and IX.

TABLE VIII

R_X VALUES[a] OF DITHIOCARBAMATE FUNGICIDES (*53*)

Fungicide	R_X
Ferbam	1.0
Thiram	1.0
Ziram	1.0
Maneb	0.37
Nabam	0.37
Zineb	0.37

[a] R_X = position relative to that of Ferbam. Solvent system is that of Section VI,B,1,a.

TABLE IX

R_F VALUES OF FUNGICIDES AND SENSITIVITY OF BIOAUTOGRAPHIC DETECTION (*54*)

Compound	Trade or common name	Sensitivity (μg)	R_F[a]
Tetramethylthiuram disulfide	Thiram	0.1	0.82
Polyethylenethiuram disulfide	Polyram	2.0	0.83
Zinc dimethyldithiocarbamate	Ziram	0.1	0.80
Ferric dimethyldithiocarbamate	Ferbam	0.1	0.81
Zinc ethylenebisdithiocarbamate	Zineb	5	Streaks
Methylarsinebisdimethyldithiocarbamate	Urbazit	0.1	0.83
Triphenyltin acetate	Brestan	2.5	0.91
N-Trichloromethylthiotetrahydrophthalimide	Captan	2.0	0.89
Copper oxychloride	—	5	0.22
Methoxyethylmercuric chloride	Ceresan	25	0.80
Quinoneoximebenzoyl hydrazone	Cerenox	10	0.78

[a] Solvent: *n*-butanol–acetic acid–H_2O (4:1:1 v/v) on S. and S. 2043b paper; ascending development.

B. SOLVENT SYSTEMS (see also Tables VIII and IX)

1. *For Glass Fiber Paper* (*53*)—these solvent systems are useful for the separation of dithiocarbamate fungicides.

	Stationary phase	Mobile phase
(a)	5% Formamide in acetone (v/v)	Chloroform
(b)	0.5% Formamide in acetone	Petroleum ether (30–60°C)
(c)	20% Formamide in acetone	*n*-Hexane–chloroform (75:25 v/v)

2. *n-Butanol–acetic acid–H_2O* (4:1:1 v/v) (*54*) for a variety of fungicides, including inorganic compounds such as copper oxychloride.
3. *n-Butanol–ethanol–1.5 N ammonium carbonate buffer* (120:33:57 v/v) will separate nabam ($R_F = 0.09$), ethylenethiourea (0.64), and ethylenethiuram monosulfide (0.82) (*55*).

C. DETECTION

1. *Sodium Azide–Iodine Reagent* (*53*, *55*)

The dried chromatogram is first sprayed with 3% (w/v) sodium azide in water, then exposed to I_2 vapors for 1 minute, and finally sprayed with a 1% starch solution (3 gm soluble starch dissolved in 200 ml hot water to which is added 100 ml saturated aqueous NaCl). Dithiocarbamate fungicides show up as white spots on a blue background. The color reaction depends on the ability of a number of sulfur-containing compounds to catalyze the following reaction (*55*):

$$I_2 + 2\ NaN_3 \rightarrow 2\ NaI + 3\ N_2\ .$$

2. *Grote Reagent* (See Block *et al.*, *2*, p. 337)

One-half gram of sodium nitroprusside is dissolved in 10 ml water; 0.5 gm of hydroxylamine HCl is added followed by 1 gm of $NaHCO_3$. After evolution of the gas has ceased, 2 drops of bromine are added, the excess bromine is removed by aeration, and the solution is filtered and made up to a final volume of 25 ml. Sulfur compounds show up as characteristically colored spots after the chromatogram is sprayed with Grote reagent; e.g., ethylenethiuram monosulfide is yellow-green, ethylene thiourea is blue, and nabam is orange (*55*).

3. *Bioautographic detection* (*54*) (see also Chapter 13)

A chromatogram is developed with the butanol solvent (*2*) and dried thoroughly at room temperature for at least 12 hours. The chromatogram is then sprayed with a suspension of *Stemphylium consortiale* (Thüm.) in nutrient solution*. The chromatogram is laid on a glass plate and incubated in a moist atmosphere for 72 hours at 27°C. Fungicides appear as distinct areas on a light gray background. The bioautograms may be fixed in ethanol and kept indefinitely. Sensitivity of this technique ranges from 0.1 to 5 μg, depending on the specific fungicide (see Table IX).

VII. Miscellaneous Pesticides and Related Products

A. Pyrethrins

1. *Preparation of Samples* (*56*)

Samples of pyrethrum flowers are ground to a fine powder and extracted by percolation at room temperature with petroleum ether (40–60°C), followed by further purification as described by McClellan (*57*). The final residue containing the pyrethrin II fraction is dissolved in ethanol and spotted on Whatman No. 1 paper.

A solution of the pyrethrin II acids in chloroform may be brominated, presumably to the dibromochrysanthemum dicarboxylic acid, by a reaction with bromine in chloroform for 3 hours.

2. *Solvents*

a. *95% Ethanol–concentrated* NH_3 (19:1 v/v), descending development on Whatman No. 1 paper—chrysanthemumdicroboxylic acid has an R_F value of 0.27 and the brominated product 0.47 (*56*).

b. *Petroleum ether* (*80–100°C*) *saturated with methanol*—using Whatman No. 1 paper strips and a specially constructed glass chamber allowing rapid equilibration in an evacuated system, yields the following R_F values (*58*): pyrethrin I, 0.85; pyrethrin II, 0.70; (±)-*trans*-chrysanthemic acid, 0.54; (±)-*cis*-chrysanthemic acid, 0.56; chrysanthemum dicarboxylic acid, streak from origin; methylpyrethrate, 0.73; α-DL-*trans*-allethrin, 0.82; allethrolone, 0.09;

3. *Detection*

a. Potassium permanganate–benzidine (*58*). The chromatogram is first dipped into a 0.1% aqueous $KMnO_4$ solution, washed with water, and partially dried. The paper is then treated with 0.5% aqueous

* Nutrient solution: 50 gm sucrose, 5 gm $NaNO_3$, 1.25 gm KH_2PO_4, 1.25 gm $MgSO_4 \cdot 7H_2O$, 500 ml water; pH adjusted to 4.6.

benzidine hydrochloride. Blue spots for pyrethrins and related compounds result.

b. Ultraviolet Photography (*58*). Due to the absorption of ultraviolet light in the 230 mμ region by pyrethrins, the chromatograms are laid on top of light-sensitive document paper and exposed to a Hanovia Chromatolite ultraviolet lamp. The developed prints show the pyrethrins as pale spots on a dark background. This method will only detect a minimum of 50 μg of material and does not reveal spots due to chrysanthemic and pyrethric acids.

c. Acid–Base Indicator (*56*). The thoroughly dried chromatogram (95°C for 5 minutes) is sprayed with 0.04% aqueous solution of the sodium salt of bromcresol purple. Pyrethrin acids show up as yellow spots on a blue background.

4. *Quantitative Estimation* (*56*)

Whatman 3 MM paper was washed with NH_4OH, water, and ethanol. About 100 mg of the pyrethrin II acid fraction is applied as a streak onto the washed paper and developed with ethanol–ammonia (Section VII,A,2,a). The regions containing chrysanthemum dicarboxylic acid were eluted with 1 *N* NH_4OH. The eluate is treated with caustic soda solution, acidified with HCl, saturated with NaCl, and extracted with freshly purified ethyl ether. The ether solution is evaporated to dryness and dissolved in 2 ml of ethanol and diluted with 20 ml of CO_2-free distilled water. This solution is titrated with 0.02 *N* NaOH to a thymol blue indicator endpoint (first appearance of purple tint).

B. Rotenone (*59*)

Rotenone, a naturally occuring insecticide, is extracted from leaves of *Tephrosia vogelii* Hook f. and roots of *Derris elliptica* (Wall.) Benth., dried at 70–80°C and ground with acetone for 16 hours. The filtered extracts are evaporated to a small volume and portions are applied to Whatman No. 1 filter paper strips.

Commercial rotenone is recrystallized twice from CCl_4 and twice from 95% ethanol. The recrystallized materials are dried 2 hours at 70°C.

Ascending development is accomplished with methanol–benzene–acetic acid (5:1:1 v/v). The developed and dried chromatograms are sprayed with 5 *N* KI–85% H_3PO_4 (1:15 v/v) (mix just before use). The colors produced by rotenone and related compounds are listed below; the R_F values for rotenone is 0.88 and for elliptone, 0.80. The colors were tested at 50 μg/cm² on Whatman No. 1 paper.

Compound	Color produced	
	Within 0.5 hours	After 6 hours
Rotenone	Blue	Blue
Elliptone	Pink-violet	Violet-purple
Isorotenol	None	Light tan
Deguelin	None	Light tan
Dihydrodeguelin	None	Light tan
Dehydrodeguelin	None	Light tan
Tephrosin	None	Light tan
Toxicarol	None	Light tan
Sumatrol	None	Faint blue

C. Insecticide Synergists (*60*)

3,4-Methylenedioxyphenyl insecticide synergists may be identified by reversed-phase paper chromatography. Whatman No. 1 filter paper is impregnated with 3% purified paraffin oil in petroleum ether (60–70°C) and chromatographic development proceeds with aqueous solutions of acetic acid, methanol, acetonitrile, and dimethylformamide (Table X). Qualitative identification and quantitative estimation are accomplished by direct ultraviolet spectrophotometry with a scanning device adapted to a Beckman DU spectrophotometer (*2*) at a set wavelength of 289 $m\mu$. This method is applicable to the analyses of insecticide formulations containing pyrethrins or DDT.

TABLE X

R_F Values of 3,4-Methylenedeoxyphenyl Synergists by Reversed-Phase Paper Chromatography (*60*)

Compound	R_F [a]			
	1	2	3	4
Piperonyl butoxide	0.54	0.21	0.25	0.24
Piperonyl sulfoxide	0.81	—	0.61	0.75
Piperonyl cyclonene	0.42	0.12	0.22	0.28
Sesamin	0.84	0.64	Tailing	0.77
Sesamolin	0.78	0.47	0.55	0.76

[a] Solvents: (1) 40% (v/v) aqueous acetic acid; (2) 30% (v/v) aqueous acetic acid; (3) 35% aqueous acetonitrile; (4) 50% aqueous dimethylformamide.

D. Insect Repellant (*61*)

MGK Repellant 11, [2,3,4,5-bis (Δ^2-butenylene) tetrahydrofurfural], may be separated from its corresponding acid and alcohol by biphase paper chromatography. Whatman No. 1 filter paper is impregnated with 20% dimethylformamide in ethyl ether, and the chromatogram is developed by ascending technique with mixed octanes as mobile solvent. The finished chromatogram is sprayed with (1) four parts 1% (w/v) aqueous KIO_4 and (2) one part 1 gm $KMnO_4$ and 2 gm Na_2CO_3 in 100 ml water; (1) and (2) are mixed just prior to spraying. The paper is impregnated with the immobile phase by spraying with a glass atomizer (see Chapter 3). The following R_F values are observed: aldehyde, 0.83; alcohol, 0.44; acid, 0.07.

E. Rodenticides (*62*)

The rodenticides ANTU [1-(1-naphthyl)2-thiourea] and warfarin are separated on Whatman No. 1 paper which is impregnated with 4% dimethylacetamide in ethyl ether (the immobile phase is sprayed on the paper). The mobile solvent for ascending chromatography is mixed octanes. The spots for ANTU ($R_F = 0.35$) and warfarin (0.14) are visible under ultraviolet light as blue and purple spots, respectively.

F. Thiourea (*63*)

For the analysis of thiourea in oranges, the sample of whole fresh fruit is first extracted by the stripping technique (*64*) with excess water; the volume is concentrated, taken up in acetone, and evaporated to dryness in a nitrogen atmosphere. The solution must be filtered before evaporation. The residue is again dissolved in acetone and portions are spotted on Whatman No. 1 filter paper. The chromatograms are developed by the ascending technique with 1-butanol saturated with water. The chromatogram, after drying, is treated with the sodium azide–iodine reagent (see Section V,C,1), revealing thiourea as a white spot on a blue background ($R_F = 0.3$). To confirm the presence of thiourea, another chromatogram is sprayed with Grote reagent (Section V,C,2), producing a blue color for thiourea spots.

VIII. Food and Feed Additives

A. Antioxidants

1. *Method 1—Reversed-Phase Chromatography* (*65, 66*)

Whatman No. 1 paper is impregnated with the stationary phase, 10% soybean oil in ethyl ether or 7% liquid paraffin in petroleum ether as described in Section II,E of this chapter. The following four antioxidants are separated by one- and two-dimensional chromatography: BHA (butylated hydroxyanisole, BHT (butylated hydroxytoluene), PG (*n*-propylgallate), and NDGA (nordihydroguaiaretic acid). For two-dimensional chromatography, the first solvent is 80% (v/v) aqueous methanol and the second solvent is water. As seen in Table XI, NDGA and PG are not resolved with the methanol solvent, but separation takes place with water as solvent.

TABLE XI

R_F AND R_{PG}[a] VALUES OF VARIOUS ANTIOXIDANTS (*65, 67*)

	R_F		R_{PG}[b]
Compound	1	2	3
BHT	0.09	0.00	0.38
BHA	0.56	0.00	0.65
NDGA	0.75	0.05	0.81
PG	0.75	0.36	1.00
DPPD	—	—	0.34
Santoquin (6-ethoxy-2,2,4-trimethyl-1,2-dihydroquinoline)	—	—	0.65
DTBH (2,5-di-*t*-butylhydroquinone)	—	—	0.60
α-Tocopherol	—	—	0.19

[a] R_{PG} = distance moved relative to PG.

[b] Solvent systems: (1) immobile phase, soybean oil; mobile phase, 80% methanol; (2) immobile phase, soybean oil; mobile phase, water; (3) acetylated paper; solvent, 60% acetone.

Another novel technique (*65*) employs the use of two solvents in the same direction; the first solvent, 80% methanol is permitted to migrate two-thirds of the normal distance of the 8 × 8 inch paper sheets, after which the chromatogram is dried. Water is then used as the second solvent in the same direction traveling the normal distance of the paper. Excellent resolution of the four antioxidants is effected as shown in Fig. 10.

The chromogenic agent consists of 0.5% (w/v) phosphomolybdic acid in 95% ethanol; the dried chromatogram is sprayed with this reagent and immediately, without further drying, exposed to ammonia vapors. The sensitivity of this reagent is 1–2 μg for BHA, NDGA, and PG and 20 μg for BHT; the spots of these antioxidants are blue.

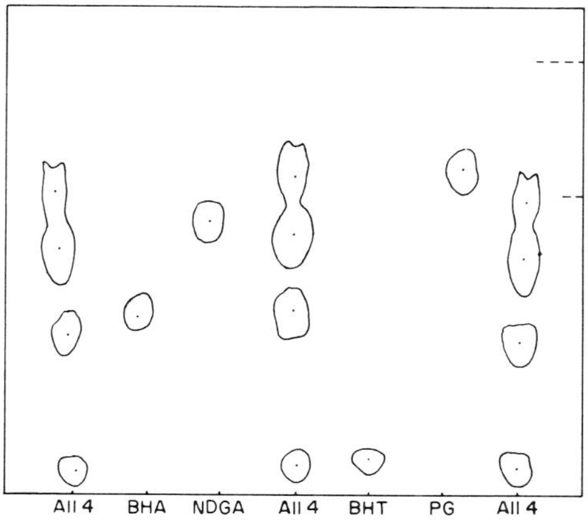

FIG. 10. One-dimensional chromatography of antioxidants developed with two solvents in the same direction; first solvent, 80% aqueous methanol; second solvent, water (*65*). Redrawn by courtesy of *J. Assoc. Offic. Agric. Chemists.*

2. *Method 2—Chromatography with Acetylated Paper* (*67*)

Whatman No. 1 chromatographic paper is acetylated according to the methods described in Chapter 3. Eight different antioxidants are separated by ascending development with 60% (v/v) aqueous acetone (see Table XI). With the phosphomolybdate reagent (see above) all spots appear blue with the exception of DPPD (*N*,*N*′-diphenyl-*p*-phenylenediamine), which gives a brown spot.

3. *Quantitative Determination of DPPD* (*68*)

Benzene solutions of purified DPPD were chromatographed by the ascending method on Whatman No. 1 paper with 80% aqueous methanol as developing solvent. The dry chromatogram was then sprayed with concentrated HNO_3, and the zone for DPPD was visible as a red band at an R_F between 0.7 and 0.85. The zones corresponding to DPPD were cut out of the paper, cut into small segments and placed into 12 ml concentrated HNO_3 in a glass-stoppered tube. The tube was shaken for 75 seconds and the solution filtered through glass wool directly into colorimeter tubes. The color is read exactly 2 minutes after the addition of the HNO_3 at 490 mμ. The range for the standard curve is 5–25 μg DPPD per 12 ml.

B. Arseno Organic Compounds (*69*)

Two-dimensional paper chromatography on Whatman No. 1 filter paper will separate six arseno organic compounds (Fig. 11)—arsanilic acid (I), arsenosobenzene (II), arsphenamine (III), 3-nitro-4-hydroxyphenylarsonic acid (IV), 4-nitrophenylarsonic acid (V), and *p*-ureidobenzenearsonic acid (VI). The first solvent is acetonitrile–HNO_3–H_2O (78:2:20 v/v); for the second direction the solvent is acetonitrile–NH_4OH–H_2O (65:5:30 v/v). Detection is accomplished by one of three methods.

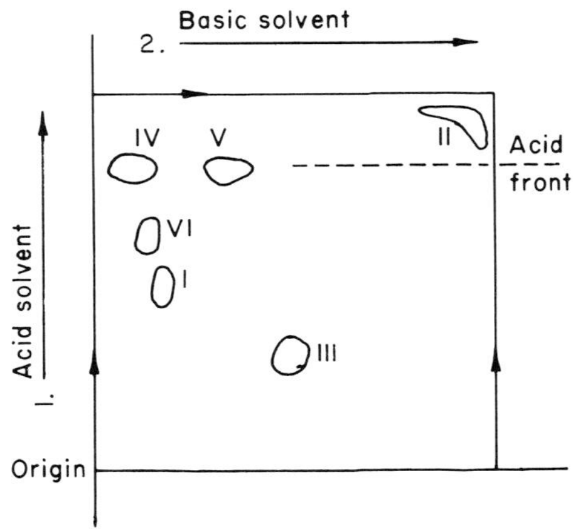

Fig. 11. Two-dimensional chromatography of arseno organic compounds; solvents and legends for compounds described in text (*69*). Reproduced by courtesy of *J. Assoc. Offic. Agric. Chemists.*

1. *Blueprinting*

The developed and dried chromatogram is placed on the sensitized surface of blueprint paper and exposed to the light from a germicidal lamp for 25 minutes. After the paper is washed in cold water, the arsenicals show up as white spots on a blue background.

2. *Ammoniacal Silver Nitrate.*

The spray reagent is prepared by dissolving 170 mg $AgNO_3$ in 1 ml water; 5 ml NH_4OH is added, and the solution is made up to 200 ml with 95% ethanol. The sprayed chromatogram is viewed in daylight and ultraviolet light.

3. *Pyrogallol*

Thirteen milligrams of pyrogallol are dissolved in 200 ml 95% ethanol. The chromatogram which is first sprayed with reagent (*2*) is then treated with the pyrogallol solution and viewed in daylight and ultraviolet light.

References

1. L. C. Mitchell, *J. Assoc. Offic. Agr. Chemists* **40**, No. 4, 999 (1957).
2. R. J. Block, E. L. Durrum, and G. Zweig, "A Manual of Paper Chromatography and Paper Electrophoresis," 2nd rev. ed. Academic Press, New York, 1958.

2a. M. E. Getz, *Residue Rev.* **2**, 9 (1963).

3. W. P. McKinley, *in* "Analytical Methods for Pesticides, Plant Growth Regulators, and Food Additives" (G. Zweig, ed.), Vol. I, Chapter 10, pp. 227–252. Academic Press, New York, 1963.
4. G. Zweig, *Chromatog. Rev.* **6**, 110 (1964).
5. G. Zweig, *in* "Essays of Toxicology" (F. R. Blood, ed.), Vol. 2, pp. 155–198. Academic Press, New York, 1970.
6. D. E. Coffin, *J. Assoc. Offic. Anal. Chemists* **49**, No. 3, 638 (1966).
7. W. R. Benson and H. A. Jones, *J. Assoc. Offic. Anal. Chemists* **50**, 22 (1967).
8. L. C. Mitchell, P. A. Mills, and F. E. Yarnall, *J. Assoc. Offic. Agr. Chemists* **43**, No. 3, 748 (1960).
9. J. J. Menn, M. E. Eldefrawi, and H. T. Gordon, *Agr. Food Chem.* **8**, No. 1, 41 (1960).
10. P. A. Mills, *J. Assoc. Offic. Agr. Chemists* **44**, No. 2, 171 (1961).
11. J. H. Onley and P. A. Mills, *J. Assoc. Offic. Agr. Chemists* **45**, No. 4, 983 (1962).

11a. E. Y. Spencer, *Can., Dept. Agr., Publ.* **1093**, 483 (1968).

12. M. Eidelman, *J. Assoc. Offic. Agr. Chemists* **45**, No. 3, 672 (1962).
13. P. A. Mills, J. H. Onley, and R. A. Gaither, *J. Assoc. Offic. Agr. Chemists* **46**, No. 2,, 186 (1963).
14. L. C. Mitchell, *J. Assoc. Offic. Agr. Chemists* **41**, No. 4, 781 (1958).
15. P. A. Mills, *J. Assoc. Offic. Agr. Chemists* **42**, No. 4, 734 (1959).
16. L. C. Mitchell, *J. Assoc. Offic. Agr. Chemists* **46**, No. 6, 988 (1963).
17. R. A. Moffitt, B. Burger, and W. Meckler, *J. Assoc. Offic. Agr. Chemists* **44**, No. 4, 795 (1961).
18. L. C. Mitchell, *J. Assoc. Offic. Agr. Chemists* **45**, No. 3, 682 (1962).
19. W. P. McKinley, G. Savary, and C. Webster, *J. Assoc. Offic. Agr. Chemists* **44**, No. 2, 193 (1961).
20. R. Miskus, *in* "Analytical Methods for Pesticides, Plant Growth Regulators, and Food Additives" (G. Zweig, ed.), Vol. 2, pp. 97–107. Academic Press, New York, 1964.
21. J. P. San Antonio, *J. Assoc. Offic. Agr. Chemists* **43**, No. 3, 721 (1960).
22. M. E. Getz, *J. Assoc. Offic. Agr. Chemists* **45**, No. 2, 393 (1962).
23. D. E. Coffin and W. P. McKinley, *J. Assoc. Offic. Agr. Chemists* **46**, No. 2, 223 (1963).
24. N. A. Smart, *Analyst* **92**, 779 (1967).
25. L. C. Mitchell, *J. Assoc. Office. Agr. Chemists* **43**, 810 (1960).
26. A. Cortes and D. R. Gilmore, *J. Chromatog.* **19**, 450 (1965).
27. N. A. Smart and A. R. C. Hill, *J. Chromatog.* **30**, 626 (1967).
28. H. F. MacRae and W. P. McKinley, *J. Assoc. Offic. Agr. Chemists* **44**, No. 2, 207, (1961).

29. G. N. Smith and F. S. Fischer, *Agr. Food Chem.* **15**, 182 (1967).
30. R. R. Watts, *Residue Rev.* **18**, 105 (1967).
31. J. W. Cook, *J. Assoc. Offic. Agr. Chemists* **37**, 984 (1954).
32. M. E. Getz, *J. Assoc. Offic. Agr. Chemists* **45**, 393 (1962).
33. J. J. Menn, W. R. Erwin, and H. T. Gordon, *Agr. Food Chem.* **5**, No. 8, 601 (1957).
34. A. Irudayasamy and A. R. Natarajan, *Analyst* **90**, No. 1073, 503 (1965).
34a. M. C. Dutt and P. H. Seow, *Agr. Food Chem.* **11**, No. 6, 467 (1963).
35. M. E. Getz and S. J. Friedman, *J. Assoc. Offic. Agr. Chemists* **46**, 707 (1963).
36. W. P. McKinley and P. S. Johal, *J. Assoc. Offic. Agr. Chemists* **46**, 840 (1963).
37. L. C. Mitchell and P. A. Mills, *J. Assoc. Offic. Agr. Chemists* **41**, 481 (1958).
38. J. J. Menn and J. B. McBain, *Agr. Food Chem.* **12**, No. 2, 162 (1964).
39. G. Zweig and T. E. Archer, *Agr. Food Chem.* **6**, No. 12, 910 (1958).
40. R. Miskus, M. E. Eldefrawi, D. B. Menzel, and W. A. Svoboda, *Agr. Food Chem.* **9**, No. 3, 190 (1961).
41. D. C. Abbott, H. Egan, E. W. Hammond, and J. Thomson, *Analyst* **89**, 480 (1964).
42. G. Yip, *J. Assoc. Offic. Agr. Chemists* **47**, No. 2, 343 (1964).
43. L. S. Bark and R. J. T. Graham, *Analyst* **84**, 454 (1959).
44. A. Major, Jr., *J. Assoc. Offic. Agr. Chemists* **45**, No. 3, 679 (1962).
45. D. Racusen, *Arch. Biochem. Biophys.* **74**, 106 (1958).
46. F. D. Aldrich and S. R. McLane, Jr., *Plant Physiol.* **32**, 153 (1957).
47. R. W. Storherr and J. Onley, *J. Assoc. Offic. Agr. Chemists* **45**, No. 2, 382 (1962).
48. L. C. Mitchell, *J. Assoc. Offic. Agr. Chemists* **43**, No. 1, 87 (1960).
49. W. E. Bleidner, *Agr. Food Chem.* **2**, No. 13, 682 (1954).
50. H. Y. Young and W. A. Gortner, *Anal. Chem.* **25**, 800 (1953).
51. W. K. Lowen, W. E. Bleidner, J. J. Kirkland, and H. L. Pease, *in* "Analytical Methods for Pesticides, Plant Growth Regulators, and Food Additives" (G. Zweig, ed.), Vol. 4, pp. 157–170. Academic Press, New York, 1964.
52. W. A. Andreae, *Can. J. Biochem. Physiol.* **36**, 71 (1958).
53. W. P. McKinley and S. A. Magarvey, *J. Assoc. Offic. Agr. Chemists* **43**, No. 3, 717 (1960).
54. H. C. Weltzien, *Naturwissenschaften* **45**, 228 (1958).
55. G. D. Thorn and R. A. Ludwig, *Rec. Trav. Chim.* **79**, 160 (1960).
56. H. C. Hodgson, *J. Sci. Food Agr.* **8**, 347 (1957).
57. D. B. McClellan, *in* "Analytical Methods for Pesticides, Plant Growth Regulators, and Food Additives" (G. Zweig, ed.), Vol. 2, pp. 399–413. Academic Press, New York, 1964.
58. J. R. Quayle, *Nature* **178**, 375 (1956).
59. N. E. Delfel, *Agr. Food Chem.* **13**, 56 (1965).
60. M. Beroza, *Anal. Chem.* **28**, 1550 (1956).
61. L. C. Mitchell, *Agr. Food Chem.* **5**, No. 10, 748 (1957).
62. L. C. Mitchell, *J. Assoc. Offic. Agr. Chemists* **40**, No. 4, 1034 (1957).
63. W. P. McKinley and R. Yasin, *J. Assoc. Offic. Agr. Chemists* **43**, No. 4, 829 (1960).
64. W. W. Thornburg, *in* "Analytical Methods for Pesticides, Plant Growth Regulators and Food Additives" (G. Zweig, ed.), Vol. 1, Chapter 5, p. 87. Academic Press, New York, 1963.
65. L. C. Mitchell, *J. Assoc. Offic. Agr. Chemists* **40**, No. 3, 909 (1957).
66. B. R. Roy, S. N. Mitra, and P. N. Sen Gupta, *Current Sci. (India)* **29**, 132 (1960).
67. B. A. Dehority, *J. Chromatog.* **2**, 384 (1959).
68. B. A. Dehority, *J. Chromatog.* **2**, 81 (1959).
69. L. C. Mitchell, *J. Assoc. Offic. Agr. Chemists* **42**, 684 (1959).

13

ANTIBIOTICS AND VITAMINS

I. Antibiotics

A. Introduction (*1*)

Antibiotics are a group of substances, produced by living organisms (bacteria, molds, and fungi) or now, in some cases, synthetically, which have a common biological property—the power to destroy other disease-causing organisms. These compounds have diverse chemical properties and structures (many of which are still unknown) and are difficult to classify in this manner. Paper chromatography is used to compare, identify, and classify antibiotics (*2, 3*) to check their purity during manufacture, to separate mixtures of antibiotics and their derivatives and degradation products, and to follow their fate after application to an organism. The paper chromatography of antibiotics has been reviewed earlier in greater detail than is deemed necessary here (*4, 5*).

B. General Directions

1. *Preparation of Sample*

The crude, fermented cultivation broth of the test microorganism may be chromatographed directly, in which case bioautography (see below) is usually employed for detection. Crude broths may contain one antibiotic or a mixture of chemically similar or dissimilar antibiotics (*1*). Alternatively, the fermentation broth or solutions of antibiotics from cultures can be purified by sorption on silica columns and elution with diethyl ether and ethyl acetate (*6*) or on cation-exchange resins with elution by 1 N NH_4OH (*7*) prior to paper chromatography.

2. *Detection*

a. *Bioautography* (*1*, *5*). This method is specific for active antibiotics and does not detect artifacts, decomposition products, or impurities which are biologically inactive. Biologically active artifacts are, of course, detected (*8*). Bioautography can therefore be used for impure or unknown antibiotics while chemical or physical detection methods are usually limited to pure substances.

In practice, the chromatogram is put on the surface of an agar plate which has been inoculated with a microorganism sensitive to the antibiotic being studied. After incubation for 15–20 hours at an appropriate temperature (about 37°C), clear zones become visible in the agar layer into which the antiobiotic diffused and consequently inhibited the growth of the microorganism. The rest of the surface, where the microorganism grows, appears opaque.

Most workers employ *Staphylococcus aureus* or *Bacillus subtilis* as the test organism for bioautography, although many others have been used. If several different antibiotics are present or the antibiotics are unknown, detection with several or many microorganisms must be employed (*9*, *10*).

The time of incubation can be shortened and the growth inhibition zones made more conspicuous by spraying the agar plate with a solution of 2, 3, 5-triphenyltetrazolium chloride and a solution of 2,6-dichlorophenolindophenol after 4 hours of incubation. After an additional 30 minutes of incubation, inhibition zones are visible as blue spots on a colorless background (*11*) (Fig. 1).

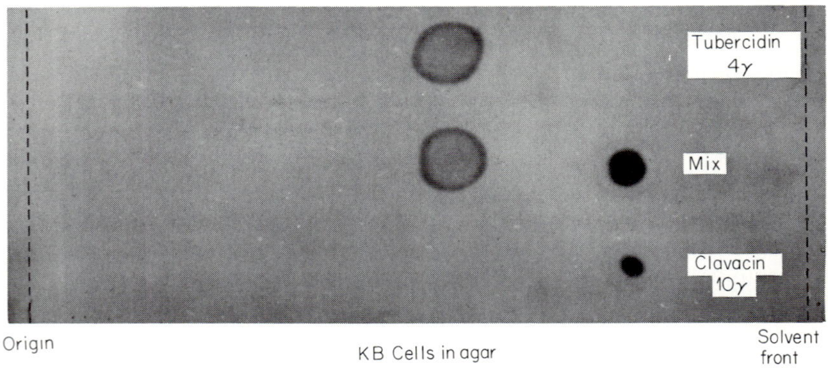

FIG. 1. Bioautograph of antitumor substances. Detection with 2,6-dichlorophenolindophenol after incubation with agar innoculated with KB cancer cells. Courtesy of G. B. Whitfield, Upjohn Co.

The agar test plates can be photographed under polarized light (*12*) or by direct contact with sensitive photographic paper. Color reprints of the zones are made by applying a 1% solution of methylene blue containing 1% of phenol to the bioautogram, removing the excess with filter paper, and then printing on common filter paper (*4*).

Bioautographic detection is used for antibacterial, antifungal, cytostatic [plates inoculated with cultures of animal cells (*13*)], and antiviral antibiotics. Chromatograms of agents which induce λ-bacteriophage in *Escherichia coli* can be developed using bioautographs with a lysogenic test system (*14*). Bioautography is also applied to quantitative analysis (*6*). Inhibitory zone diameters are measured conveniently with a modified microfilm reader projection device and are related to the concentration of the zones (*14a*).

In another, more sensitive, bioautographic procedure (*15*), the developed chromatogram is air dried, carefully immersed in a soluble agar medium inoculated with appropriate microorganism, and incubated in a moist atmosphere. The growing microbial culture on the paper is then colored by an available method.

b. *Other Methods.* Some antibiotics fluoresce in or absorb ultraviolet light and can be detected in this manner. Labeled antibiotics are detected by their radioactivity. Chemical detection tests are useful if the antibiotics form new compounds on the paper which are colored, decolorize the detection reagent, fluoresce under ultraviolet light, or absorb ultraviolet light (*5*). Chemical detection methods for antibiotics have been reviewed (*16, 17*).

For quantitative determination, ultraviolet densitomery (*18*) or extraction followed by colorimetry is used. For example (*7*), the eluate from an ion-exchange column containing 4–12 μg of neomycin (an amino sugar antibiotic complex) is spotted on Whatman No. 4 paper along with standard solutions of neomycins B and C and neamine free bases. The paper is rolled and soaked in methanol–acetic anhydride–CCl_4 (3:2:95 v/v) for at least 12 hours to form *N*-acetyl derivatives, and is then air dried for 1 hour. The paper is then developed for 24–36 hours by descending flow of *n*-butanol–H_2O–piperidine (84:16:2 v/v), air dried, and then humidified at 37°C for 1–2 hours in a chamber with a water-saturated atmosphere. The *N*-acetyl neomycins are well separated in the sequence neomycin C, B, and neamine (least sorbed) (Fig. 2). The paper is next rolled inside the chamber and quickly transferred to a cylinder, and filtered Cl_2–CCl_4 reagent is added (Cl_2 produced from 100 ml concentrated HCl + 50 gm $KMnO_4$ is passed through H_2O and H_2SO_4 and absorbed in 1 liter CCl_4). The cylinder is kept in the dark for 20 minutes at room temperature, the paper is removed and air dried

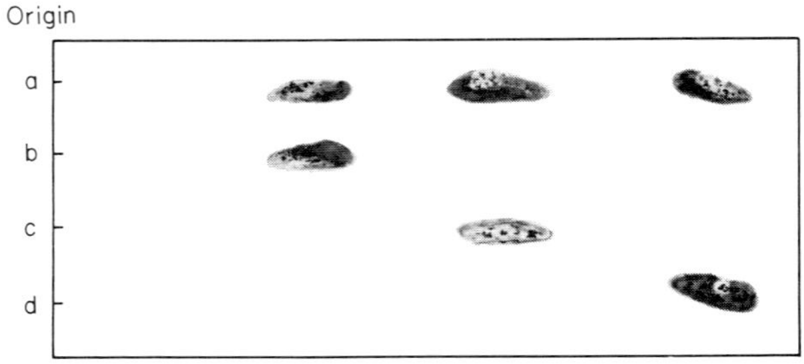

FIG. 2. Chromatogram of neomycin solutions which are *N*-acetylated after application to the paper. Solvent: *n*-butanol–H_2O–piperidine (84:16:2 v/v), 24-hour development at 34°C. (a) Eluate from Amberlite IRC-50 ion-exchange column after adsorption of neomycins B and C and neamine salts; (b) neomycin C free base; (c) neomycin B free base; (d) neamine free base. After Majumdar and Majumdar (*7*).

at 4–6°C for 30 minutes to remove excess Cl_2. The paper is sprayed with the reagent solution (0.14 gm starch + 0.036 gm KI dissolved by heating in 3.5 ml H_2O and added quickly to 50 ml pyridine) and transferred to a water-saturated atmosphere to produce bluish pink spots on a white background. Using these spots as a guide, areas corresponding to the spots are removed from unsprayed chromatograms and cut into small pieces. These are extracted for 30 minutes with 4 ml H_2O and 1 ml starch (1 gm)–KI (0.25 gm)–HCl (1 ml, 5 *N*)–H_2O (100 ml) reagent. The absorbances of the colored solutions are measured at 570 mμ against solutions prepared from areas of blank paper, and the amounts of neomycins are determined by reference to standard curves.

3. *Techniques*

All the usual techniques (see Chapter 3) are used for the chromatography of antibiotics on unmodified paper. Radial development with various solvents has been used for the rapid detection of chlortetracycline, sulfathiazole, chloramphenicol, nitrofurantoin, and cryogenine in some pharmaceutical preparations (*18a*). Ion-exchange paper has also been employed for antibiotic separations (*19*).

Solvent systems are composed of one or several components (*3*), for example butanol–acetic acid–KCN (100:1:0.05 v/v/w), which separates chloramphenicol, tetracycline, neomycin, and penicillin (*20*). Salting out chromatography employs increasing concentrations (0.5%–saturated) of NH_4Cl as the mobile phase (*1, 21*). pH chromatography employs papers adjusted to various pH values by impregnation with buffers and a single organic solvent saturated with H_2O as the mobile phase (*1, 6, 22*). This procedure allows separations (Fig. 3), as well as the study of the ionic nature of an antibiotic (*23*) and its classification (see below).

FIG. 3. Bioautograph illustrating the separation of natural penicillins at different pH values of the stationary phase, optimum being at pH 6.0. Development with water-saturated $CHCl_3$. After V. Betina (*5*).

Since many antibiotics are quite unstable, conditions must be chosen so that inactivation or isomerization of the compound does not occur during chromatography (*1*).

C. Systematic Analysis, Classification, and Identification

1. *Chromatographic Spectra* (*Summarized Chromatograms*)

Various authors have proposed standard sets of solvent systems in which the R_F values of a new or unknown antibiotic can be determined and compared with those of reference antibiotics. Using these R_F values, together with additional data (physical, chemical, and biological properties, ultraviolet and infrared spectra), a reference antibiotic can usually be found which is identical with or at least similar to the unknown, so that identification or classification can be made. The chromatographic results are usually presented as chromatographic spectra (plots of R_F in relation to each system used) (*3*) and can be collected on punched cards for use with sorting machines. As mentioned above, fermentation broths can be chromatographed directly, or the antibiotics can be first purified by solvent extraction or column chromatography (*24*).

Although many schemes for the classification and systematic analysis of antibiotics have been suggested [employing, for example, six (*25*), eight (*24*, *26*), ten (*27*), thirteen (*28*), fourteen (*29*), and twenty (*3*, *30*) systems], space permits the description of only one, which combines sequential and simultaneous analysis (*5*, *31*). This scheme illustrates a typical series of solvents recommended for the paper chromatography of antibiotics.

In the first step, antibiotics are analyzed simultaneously in four principal solvent systems with decreasing polarity and ability to form hydrogen bonds—(1) distilled H_2O; (2) *n*-butanol saturated with H_2O; (3) ethyl acetate saturated with H_2O; (4) benzene saturated with H_2O— and are classified into five classes and fourteen subclasses based on the resultant R_F values in these numbered solvent systems. The classification is further refined by a second step using additional solvent systems (Table I).

A chromatographis spectrum (plots of R_F values in eight solvents) is prepared for each antibiotic on the basis of which it is characterized and classified in greater detail (Fig. 4). Structurally related antibiotics (tetracyclines, streptomycins, etc.) have, of course, quite similar chromatographic spectra. According to this classification system, the hydrophilic character of the antibiotics generally decreases from class Ia to Vc. Detailed results for sixty-two known antibiotics from actinomycetes, fungi, and lichens have been reported (*31*), and more than 300 antibacterial preparations were divided into five groups according to such a paper chromatographic scheme (*31a*).

Recently numerous solvents proposed for summarized chromatograms have been evaluated relative to each other to determine which supplies the most information concerning the identification of antibiotic substances. Water, butanol, 3% aqueous NH_4Cl, benzene–methanol (4:1 v/v), and butanol–methanol–H_2O (4:1:2 v/v) proved most informative, and R_F values of 35 antibiotics are listed in these and seven other solvent systems (*31b*).

2. *Salting Out Chromatography*

Some authors classify antibiotics using a salting out paper chromatography technique in which nine solvents with different concentrations of NH_4Cl in H_2O (0, 0.5%, 1%, 2%, 3%, 5%, 10%, 20%, and saturated) are used (*5*, *21*). Depending upon the shapes (straight line, parabola, etc.) and positions of plots of R_F versus percentage of NH_4Cl for each antibiotic, classification into six groups is made. Fifty-two antibiotics from actinomycetes and fungi have been so classified (*5*).

TABLE I

CLASSIFICATION OF ANTIBIOTICS INTO CLASSES (PART A) AND INTO SUBCLASSES (PARTS B, C, D, E, AND F) (*5, 31*)

PART A

R_F values in principal systems	Classes I	II	III	IV	V
R_{F_1}	>0.00	>0.60	0.31–0.60	0.00–0.30	0.00–1.00
R_{F_3}	0.00–0.30	>0.30	$>R_{F_1}$	$>R_{F_1}$	>0.60
R_{F_2}	0.00	>0.00	$>R_{F_4}$	$\geqslant R_{F_4}$	>0.60
R_{F_4}	0.00	0.00	0.00–0.60	0.00–0.60	>0.60

PART B—Class I

R_F values	Subclass Ia	Ib	Ic
R_{F_1}	>0.60	0.31–0.60	⩽0.30

PART C—Class II

Subclass IIa	IIb	IIc
$R_{F_1} > R_{F_2} > R_{F_3}$	$R_{F_1} > R_{F_2} < R_{F_3}$	$R_{F_1} < R_{F_2} < R_{F_3}$

PART D—Class III

R_F values	Subclass IIIa	IIIb
R_{F_4}	0.00	0.05–0.60

PART E—Class IV

R_F values	Subclass IVa	IVb	IVc
R_{F_2}	$>R_{F_1}$	$>R_{F_1}$	$>R_{F_1}$
R_{F_3}	$>R_{F_1}$	0.00	$>R_{F_4}$
R_{F_4}	0.00	0.00	0.05–0.60

Table continued

TABLE I (*continued*)

PART F—Class V

R_F values	Subclass Va	Vb	Vc
R_{F_1}	>0.60	0.31–0.60	0.00–0.30

Additional Solvents :

For class I alkaloids

a. Methanol–H_2O (4:6 v/v)
b. *n*-Propanol–H_2O (4:6 v/v)
c. Methanol–aqueous 3% NH_4Cl (7:3 v/v)
d. Methyl ethyl ketone–*n*-butanol–H_2O (30:5:65 v/v)

For classes II and III

e. 3% NH_4Cl in H_2O
f. Isoamyl acetate–methanol–99% formic acid–H_2O (65:20:5:10 v/v, upper layer)
g. *n*-Butyl acetate–methyl ethyl ketone–0.15 *M* phosphate buffer, pH 7.4 (50:25:5 v/v, upper layer)
h. Ethyl acetate–*n*-hexane–0.15 *M* phosphate buffer, pH 6.0 (65:15:20 v/v, upper layer)

For classes IV and V

i. Isoamyl acetate–methanol–99% formic acid–H_2O (4:2:1:3 v/v, bottom layer)
j. *n*-Butanol–methanol–H_2O (4:1:5 v/v, bottom layer)
k. Methanol–*n*-hexane (6:4 v/v, bottom layer)
l. Benzene–cyclohexane–0.15 *M* phosphate buffer, pH 7.4 (5:35:60 v/v, bottom layer)

3. *pH Chromatograms*

The mobility of antibiotics in water-saturated ether, ethanol, $CHCl_3$, or a similar solvent (Fig. 5) is determined on paper buffered in the pH range 2–10. Plots of R_F versus pH are of the following forms: (1) S-shaped with the maximum in the acidic range (for acidic antibiotics, e.g., benzylpencillin); (2) S-shaped with the maximum in the alkaline range (basic antibiotics, e.g., erythromycin); (3) increasing to a maximum (the neutral point) and then decreasing (amphoteric antibiotics, e.g., chlortetracycline); (4) straight line (neutral antibiotics, e.g., chloramphenicol) (see Fig. 5). For those compounds forming S-shaped curves, the R_F values depend upon the pH of the paper, the pK value of the antibiotic and its distribution coefficient between the stationary and mobile phases. This last fact indicates that the optimal pH value for the extraction of an antibiotic from an aqueous solution into an inmiscible organic solvent

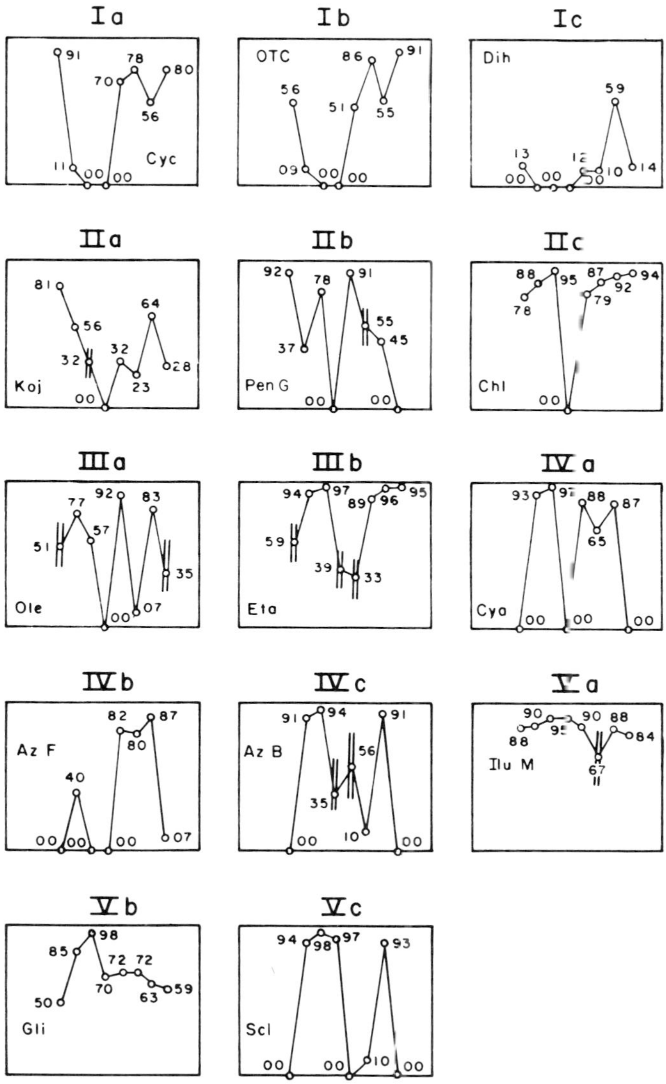

FIG. 4. Representative chromatographic spectra of antibiotics belonging to subclasses Ia through Vc.Cyc = Cycloserine, OTC = oxytetracycline, Dih = dihydrostreptomycin, Koj = kojic acid, Pen G = penicillin G, Chl = chloramphenicol, Ole = oleandomycin, Eta = etamycin, Cya = cyclopaldic acid, AzF = azalomycin F, AzB = azalomycin B, Ilu M = illudin M, Gli = gliotoxin, Scl = sclerotiorin. Values given are $R_F \times 100$. Solvent systems for subclasses Ia, Ib, Ic: 1,2,3,4 a,b,c,d (see text); for IIa, IIb, IIc, IIIa, IIIb: 1,2,3,4,e,f,g,h; for IVa, IVb, IVc, Va, Vb, Vc: 1,2,3,4,i,j,k,l. After V. Betina (*5*).

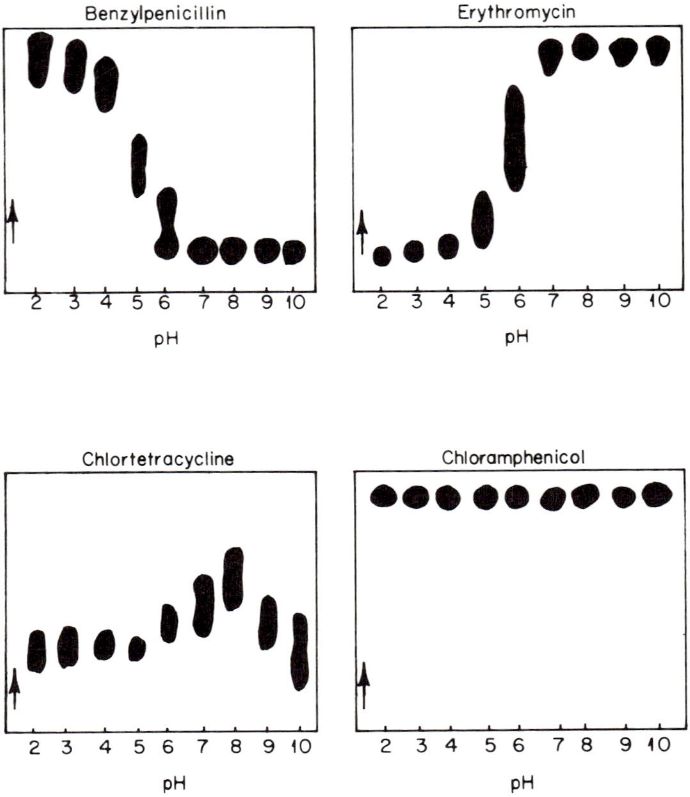

FIG. 5. pH Chromatogram of four antibiotics. Solvent systems (all H_2O saturated): amyl or butyl acetate (for benzylpenicillin), ethyl acetate (for chloramphenicol and erythromycin), and *n*-butanol (for chlortetracycline). Bioautographic detection with *Bacillus subtilis*. After V. Betina (*5*).

would be that pH where the R_F on the pH chromatogram is greatest; conversely, the best pH value for the reextraction into water is the minimum of the R_F–pH curve. This procedure has been used, therefore, to determine the ionic character of antibiotics, devise extraction techniques, and to compare, characterize, and identify new and unknown antibiotics (*5*, *22*, *32*).

D. Classes of Antibiotics

1. *Penicillins*

a. *Preparation of Sample.* Penicillins in fermentation cultures can be extracted from the culture filtrate with diethyl ether, converted to hydroxamic acids by the action of hydroxylamine in basic aqueous solution, concentrated, and applied to the paper (*33*). Tissue samples are homogenized with water and a portion of filtered extract is applied as the initial zone (*34*).

d. *Detection.* Bioautography is done with *B. subtilis*, *S. aureus*, or hay bacillus No. 6633 (*34*). For chemical detection, the paper is suspended over aqueous NH_3 for 1 hour and then passed through freshly prepared $5\text{–}15 \times 10^{-4}$ *N* iodine solution in acetone, evaporated, and sprayed with 1% starch–1% KI to develop white spots on a violet background. Penicillins or their penicilloic acids are detected at a sensitivity of about 10 γ (*35*). Iodine–sodium azide reagent (1 gm sodium azide in 10% aqueous 0.1 *M* iodine solution) applied after spraying with 0.5% soluble starch solution and drying is sensitive to as little as 0.2 γ penicillin (*35a*). Hydroxamic acids are detected with $FeCl_3$ and evaluated photometrically after extraction of the colored spots into *n*-butanol (*33*).

c. *Technique.* The chromatography chamber must be well sealed and lined with wet paper to insure good equilibration (*4*).

b. *Chromatographic Systems.* 1. For the chromatography of hydroxamic acids, develop with diisopropyl ether–isopropyl alcohol–0.1 *M* aqueous potassium hydrogen phthalate (85:15:2.4 v/v) on Whatman No. 4 paper impregnated with 0.1 *M* potassium hydrogen phthalate and the lower third with pH 6.2 phthalate buffer (*33*) (Fig. 6).

2. Natural penicillins are separated by development with water-saturated diethyl ether on paper impregnated with pH 6.2 buffer; quantitation is based on zone diameters (*36*).

3. Penicillin metabolites are separated on paper treated with 2.5% phosphate buffer (pH 6.8–7.0) using butanol–ethanol–H_2O (4:1:5 v/v), containing a little Sudan III to indicate the solvent front. 6-Aminopenicillanic acid and penicillin are thus separated (*35*).

4. Ten different pharmaceutical penicillins have been compared by chromatography on paper impregnated with pH 4.5 phosphate buffer and developed with butyl acetate–*n*-butanol–acetic acid–H_2O (80:15:40:24 v/v) and *n*-butanol–*sec*-butanol–H_2O–acetone (12:12:18:5 v/v) (*37*).

5. Mixtures of penicillin G and M and adipylpenicillin or *p*-aminobenzylpenicillin are separated by development with methanol on paper treated with Na_2CO_3 (*38*).

6. Development with butanol–acetic acid–H_2O (4:1:5 v/v) on Whatman No. 4 paper separates cephalosporin C (most sorbed), cephalosporin N, 6-aminopenicillanic acid, and dimethoxyphenylpenicillin (least sorbed). Detection is with a starch–iodine reagent similar to that described above (*39*). The same system has been used to study the biosynthesis of cephalosporin C (*40*).

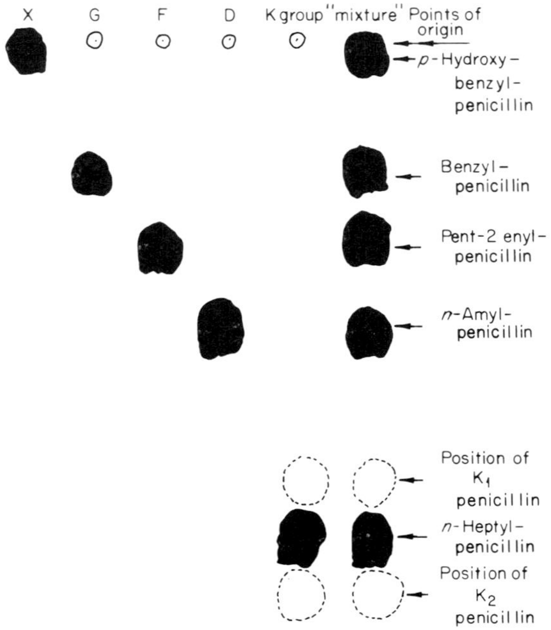

FIG. 6. Chromatogram of various penicillin types; 100 μg per spot. After Baker *et al.*, (*33*).

The R_F values of penicillins on paper chromatograms generally increase in the order: X, G, F, V, dihydro-F, K (*1*, *41*).

2. *Cephalosporins*

Cephalosporins C (aminoadipoyl-7-aminocephalosporanic acid) and N (penicillin N) are related to penicillins; cephalosporins P are steroids. They are found together along with other antibiotics in fermentation broths (*4*).

a. *Detection.* All cephalosporins are detected by bioautography with *B. subtilis*; cephalosporins C and N are detected chemically with ninhydrin (*5*) and starch–iodine reagents. Ultraviolet light causes rapid inactivation of cephalosporin C (*42*).

b. *Chromatographic Systems.* The separation of cephalosporins C and N is described above. Cephalosporins N and P are separated using *n*-propanol–H_2O (7:3 v/v) (*43*); in this system cephalosporin C is mixed with N, but penicillins G and V are resolved. Cephalosporins C, N, and P are mutually separated in methanol–*n*-propanol–H_2O (6:2:1 v/v) (paper impregnated with 0.75 *M* phosphate buffer, pH 4.0) (*44*), propanol–0.013 *M* sodium citrate, pH 5.5 (7:3 v/v) (*45*), and *n*-butanol–acetic acid–H_2O (4:1:4, top layer, or 60:15:25 v/v) (*9, 40*). Cephalosporins P_1–P_5 are chromatographed in ethanol–0.01 *N* acetic acid–H_2O (2.5:2:2.5 v/v) (*46*).

7-Aminocephalosporanic acid derivatives have been chromatographed, and R_F values have been reported in the following systems: *n*-butanol–ethanol–H_2O (4:1:5 v/v, top layer), paper untreated or buffered to pH 6; ethyl acetate saturated with aqueous 0.1 *M* sodium acetate buffer, pH 5.2, paper treated with the same buffer or a pH 7 buffer (*47, 48*).

c. *Applications.* The metabolism of oral cephalothin and related cephalosporins has been studied in the rat using compounds labeled with ^{14}C and radiochromatography with methyl ethyl ketone saturated with water and *n*-butanol saturated with 5 *N* NH_3 as developers (*49*).

Quantitative chromatography of the desacetyl derivative and lactone of cephalosporin C in water-saturated methyl ethyl ketone has been reported (*50*).

3. *Streptomycins, Neomycins, and Kanamycins*

These are a group of structurally similar, basic, water-soluble antibiotics.

a. *Detection.* Bioautography with *S. aureus* or *B. subtilis.* Chemically by spraying with 0.1% diacetyl–2.5% alcoholic α-napthol–20% KOH (1:1:1 v/v) to yield red zones with a sensitivity of 5 γ for streptomycin and related compounds (*51*).

b. *Systems* (*see also Table II*). 1. *n*-Butanol (water saturated)–2% piperidine–2% *p*-toluenesulfonic acid has been used for the separation and characterization of the four known natural streptomycins (streptomycin, dihydrostreptomycin, hydroxystreptomycin, and mannosidostreptomycin) (*52*), bis-α-hydroxystreptomycylamine (*53*) and *N'*-alkylstreptomycylamines (*54*) (Fig. 7) (also see *4, 55-57*).

2. *n*-Butanol (water saturated)–2% *p*-toluenesulfonic acid separates various antibiotics as shown in Table II and Fig. 8.

3. Butanol–pyridine–H_2O (6:4:3 v/v), descending development on Whatman No. 1 paper overnight separates *N*-acetylneomycins in the following sequence: neomycin B + framycetin (least sorbed), neamine, catenulin, neomycin C + kanamycin (*58, 59*).

c. *Applications.* Paper chromatography has been employed for the study of streptomycin biosynthesis (*60*), and in the discovery (*52*), isolation, comparison (*61*), and identification (*62*) of antibiotics of this type.

TABLE II

R_F VALUES OF STREPTOMYCINS, NEOMYCINS, AND KANAMYCINS

Compound	R_F [a] 1	2	3	4
Streptomycin	0.49	1.0	—	0.38
Mannosidostreptomycin	0.10	0.4	—	—
Dihydrostreptomycin	0.40	0.7	—	0.34
Hydroxystreptomycin	0.41	—	—	—
Streptothricin I	0.27	—	—	—
Streptothricin II	0.16	—	—	—
Streptothricin III	0.22	—	—	—
Neomycin B	0.41	0.75–0.85	0.26	—
Neomycin C	0.53	0.26	0.30	—
Neomycin complex	—	0.3–0.45	—	0.05
Kanamycin A	—	0.35	—	—
Kanamycin B	—	0.6	—	—
Mannosidodihydrostreptomycin	—	0.25	—	—
Neomycin A (Neamine)	—	0.85	0.43	—
Catenulin	—	0.29	0.52	—
Kanamycin	—	—	0.57	0.15
Paromomycin	—	—	—	0.23
Viomycin	—	—	—	0.16
Streptothricin	—	—	—	0.27
Vancomycin	—	—	—	0.40

[a] Solvents: (1) *n*-butanol–methanol–H_2O–*p*-toluenesulfonic acid (10:40:20:1 v/v/v/w) (*4*); (2) *n*-butanol (H_2O saturated)–2% *p*-toluenesulfonic acid (*4, 55–57*); (3) *n*-propanol–acetic acid–H_2O (9:1:10 v/v) (compounds as sulfate salts) (*58*); (4) methanol–2% aqueous NaCl (2:1 v/v), paper impregnated with 5% Na_2SO_4 (*1*).

d. *Quantitative Analysis* (*51*). Streptomycin sulfate and its accompanying compounds, bis-α-hydroxystreptomycylamine and streptomycin monoaldehyde ammonium, are separated by development of a linear initial zone containing 1500 units on paper buffered to pH 7 (0.013 *M* phosphate buffer) with the filtered upper layer of a solution containing 15 gm NaOH–150 ml H_2O–30 gm pentachlorophenol–850 ml butanol and shaken for 1 hour. The paper is not equilibrated before the run, but the chamber contains water-saturated butanol. After 18–20 hours, the paper is dried, washed two to three times with diethyl ether and cut in half. One half is sprayed with α-napthol reagent (above) to locate the zones. The corresponding zones in the other half are cut out, eluted with 5–50 ml pH 4.5 acetate buffer, and determined by colorimetry at 525 mμ after adding 0.3 ml 20% KOH, 0.4 ml 5% α-napthol, and

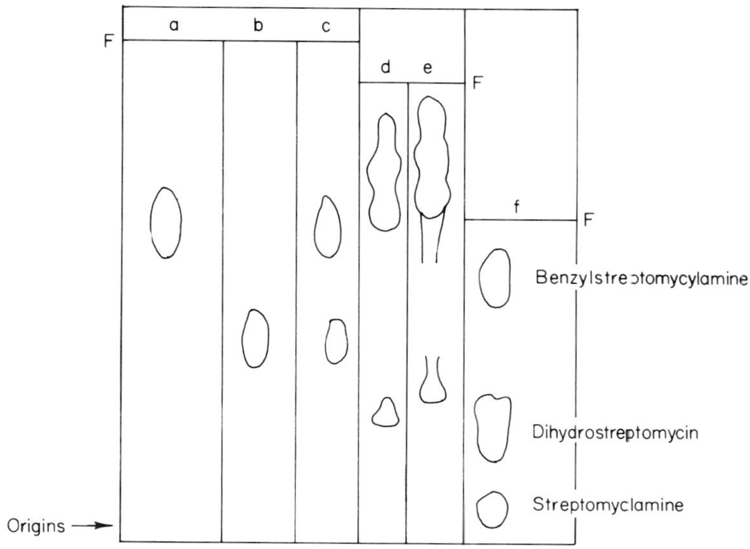

FIG. 7. Chromatogram of (a) dihydrostreptomycin (10 μl sample containing 600 units per milliliter), (b) streptomycin (10 μl sample containing 200 units per milliliter), (c) mixture of streptomycin and dihydrostreptomycin, (d) *N'*-*n*-octylstreptomycylamine preparation, (e) *N'*-*n*-decylstreptomycyclamine preparation, (f) hydrogenation products of bis-α-hydroxystreptomycylamine sulfate. Solvent: *n*-butanol–2% piperidine–2% *p*-toluenesulfonic acid. Detection: (a–e) by bioautography against *Micrococcus pyrogenes* or *B. subtilis*; (f) by streptidine spray. F = solvent front. After I. A. Solomons and P. P. Regna (*53*) and W. A. Winsten *et al.* (*54*).

0.3 ml 0.05% diacetyl to a 5 ml portion of each eluate. The quantitative determination of neomycin by colorimetry (Section A, 2, b) and bioautography (*6*) has been described.

e. R_F *values.* R_F values for twenty basic water-soluble antibiotics in six systems have been published (*63*).

4. *Gentamicin Complex Components*

The components are separated in the sequence C_{1a} (most sorbed), C_2, and C_1 on paper strips by development for 5 hours with the lower phase of $CHCl_3$–methanol–17% NH_4OH (2:1:1 v/v) in a tank equilibrated for several hours with the upper phase. After chromatography, the components are detected and quantified by bioautography against *S. aureus* or by treating the paper with 0.25% ninhydrin in pyridine–acetone followed by heating at 105°C for several minutes and reading the color intensities on an integrating scanner (*64–66*) (Fig. 9).

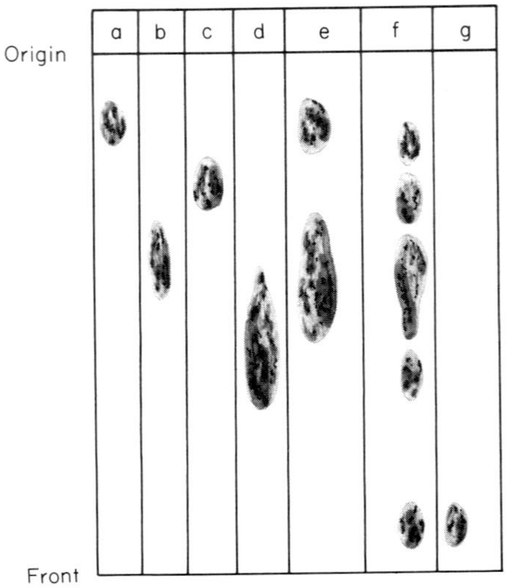

FIG. 8. Bioautogram of preparations developed with *n*-butanol (water saturated)–2% *p*-toluenesulfonic acid· H_2O. (a) 0.4 unit mannosidostreptomycin; (b) 0.2 unit streptomycin A; (c) 0.4 unit dihydrostreptomycin A; (d) 0.4 unit neomycin A; (e) 0.4 unit each of streptomycin A and mannosidostreptomycin; (f) 1.0 unit of antibiotic 136; (g) 0.8 unit of known streptothricin sulfate. After D. H. Peterson and L. M. Reineke (*57*).

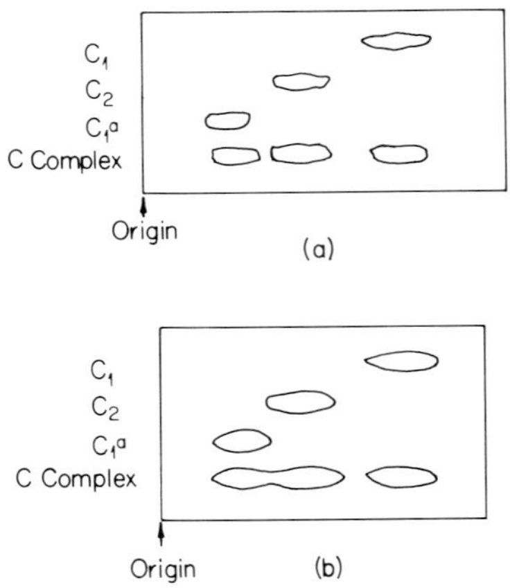

FIG. 9. Ninhydrin-sprayed chromatogram (a) and bioautograph (b) of gentamicins run on Whatman No. 1 paper using lower phase of $CHCl_3$–methanol–17% NH_3 (2:1:1 v/v). After G. H. Wagman *et al.* (*65*).

A similar separation (*67*) in the same sequence but in only 45 munutes is obtained on silicic acid–glass fiber sheets (ChromAR sheet 500, Mallinckrodt Chemical Works, St. Louis, Mo.), which are handled much like a sheet of paper (*68*). Bioautography against *Sarcina lutea* was used with these sheets.

5. *Tetracyclines*

a. *Detection.* Bioautography against *B. subtilis*, *S. aureus*, and *Klebsiella pneumoniae*, by fluorescence in ultraviolet light after exposure to NH_3 vapor or by color reaction with $FeCl_3$ reagent (0.05% in 0.01 *N* HCl) (*4, 5*).

b. *Chromatographic Systems* (see also Fig. 10). Table III shows R_F values of the tetracycline group of antibiotics for various systems, including the widely used system of Selzer and Wright (*69*) which is used by the British government to provide legal identity tests for tetracyclines (*70*). Results in their system are greatly dependent upon the way in which the paper is impregnated (*4*). Tetracyclines are generally chromatographed in acid media in order to suppress the formation of epimers during chromatography. The separation and quantitative determination of the epimer of tetracycline has been accomplished by a

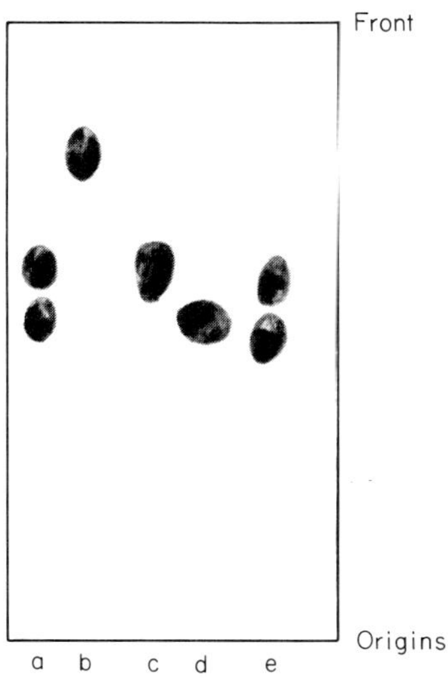

FIG. 10. Chromatogram resulting from the development of tetracycline antibiotics with $CHCl_3$–*n*-butanol (4:1 v/v) saturated with pH 4.5 buffer on paper impregnated with this buffer. Detection by fluorescence under ultraviolet light. (b) Chlortetracycline; (c) demethylchlortetracycline; (d) tetracycline; (a, e) mixtures of tetracyclines. After M. Urx *et al.* (*73*).

TABLE III

R_F VALUES OF TETRACYCLINE ANTIBIOTICS

Compound	R_F [a]				
	1	2	3	4	5
Epitetracycline	0.05	—	—	0.15	0.14
Epichlortetracycline	0.08	—	—	—	0.33
Oxytetracycline	0.13	0.36	—	0.27	0.24
Tetracycline	0.28	0.29	0.47–0.50	0.39	0.20
Chlortetracycline	0.50	0.58	0.70–0.73	0.48	0.62
Demethylchlortetracycline	—	0.51	0.56–0.60	0.25	0.39
Demethyltetracycline	—	—	0.38–0.41	—	0.22
Epidemethylchlortetracycline	—	—	—	0.17	0.20
Isotetracycline	—	—	—	0.22	—
Epianhydrotetracycline	—	—	—	0.40	—
Epianhydrodemethylchlortetracycline	—	—	—	0.44	—
Isochlortetracycline	—	—	—	0.48	—
Anhydrotetracycline	—	—	—	0.62	—
Anhydrodemethylchlortetracycline	—	—	—	0.71	—
Epidemethyltetracycline	—	—	—	—	0.10
Epioxytetracycline	—	—	—	—	0.12
Dehydrochlortetracycline	—	—	—	—	0.39

[a] Conditions: (1) $CHCl_3$–nitromethane–pyridine (10:20:3 v/v), paper wetted with pH 3.5 McElvaine buffer and blotted dry (*69*); (2) methyl butyl ketone–ethyl acetate–butanol (80:80:35 v/v), Whatman No. 1 paper impregnated with a pH 4.6 buffer, 0.2 *M* Na_2HPO_4–0.1 *M* citric acid (9.35:10.65 v/v) (*72*); (3) $CHCl_3$–*n*-butanol (4:1 v/v) saturated with McIlvain's buffer, pH 4.5, paper impregnated with the same buffer saturated with the organic phase, circular development (*73*); (4) *n*-butanol–acetic acid–H_2O (4:1:5 v/v, top layer) (*4, 74*); (5) $CHCl_3$–*n*-butanol (9:1 v/v), paper impregnated with 0.3 *N* H_3PO_4 and 0.1 % trichloroacetic acid (*2, 75*).

modification of the method of Selzer and Wright (*71*). Circular development with methanol–acetic acid–H_2O (3:0.25:7) is also useful (*71a*, also see *72–75*), for the separation of tetracyclines.

Tetracyclinemethylenelysine, which is detected with ninhydrin, is separated from tetracycline hydrochloride by ascending chromatography in butanol–acetic acid–H_2O (75:15:10 v/v, $R_F = 0.05$) or butanol–pyridine–H_2O (1:1:1 v/v, $R_F = 0.02$) (*76*).

c. *Applications and Quantitative Analysis.* Chlortetracycline (*77*) and tetracycline (*78*) in feed mixtures have been separated and determined using paper chromatography.

After detection of separated tetracyclines and their epimers and analogs in ultraviolet light, the compounds may be eluted from the paper with

0.1 *N* HCl, converted into anhydro derivatives, and determined by spectrophotometry (*79*). Oxytetracycline acidic decomposition products have been separated by circular chromatography (butanol–acetic acid–H_2O, 4:1:5 v/v, on paper impregnated with pH 3 phosphate buffer). The fluorescent spots are eluted with methanol–*N* HCl (5:1 v/v) and the extinction of the eluate at 360 mμ is referred to a standard curve (*80*).

d. *Separation of Tetraphenylborate–Tetracycline Complexes.* A method has been reported for the separation of low concentrations (< 200 μg/l) of tetracycline compounds from nutrient liquids. A 2 ml portion of mixture solution is adjusted to pH 2.0 and 1 ml of saturated tetraphenylborate solution is added. The solution is centrifuged, and the precipitate of complexes is dissolved in ethyl acetate or butyl acetate to a final total concentration of 1000 μg/ml. Ten microliters of this solution are spotted on dry Whatman No. 4 paper impregnated with 3.7% Chelaton III. The paper is moistened with the aqueous phase of the solvent (butanol–$CHCl_3$, 9:1 v/v, saturated with 0.15 *M* phosphate buffer, pH 3) and developed downward for 30 cm (6 hours) with the organic phase. The R_F values of the complexes are: anhydrotetracycline, 0.79–0.88; chlortetracycline, 0.57–0.68; and tetracycline, 0.34–0.42 (*81*).

6. *Saturated and Unsaturated* (*Polyene*) *Macrolide Antibiotics*

a. *Saturated Macrolides.* These compounds are antibiotics which contain a macrocyclic lactone ring and an amino sugar residue. Their separation is therefore based partly on differences in the structure of the nonbasic moiety and partly on the dissociation constants of the basic group. They are detected bioautographically against *B. subtitis* (*82*).

A detailed study (*82*) of systems for the separation of macrolides indicated the following generalizations : (a) the sequence of separation of those antibiotics studied is usually magnamycin (least sorbed), methymycin, picromycin, oleandomycin, and erythromycin; (b) for any given solvent (composed of a nonpolar–polar pair), the R_F is higher the greater the percentage of polar component (alcohol); (c) R_F values increase with the number of carbon atoms of the alcohol but decrease with an increase of the number of carbon atoms of the nonpolar component; (d) the pH value of the organic phase also has a great effect on R_F values (it was found that buffering the solvent instead of the paper leads to more compact zones).

Based on this study, the following solvents are recommended (*72*) for separations on Schleicher and Schuell (S. and S.) 2043b paper strips: methanol–benzene (2:7 v/v) buffered to various pH values (1–5) with diisopropylamine–acetic acid, pyridine–acetic acid, or pyridine–oxalic acid; and methanol–dichloroethane (1:8 v/v) buffered with pyridine–

acetic acid. Prior to buffering, all organic solvents are saturated with water by shaking with one-tenth volume of a 30% aqueous NaCl solution.

Some earlier systems used for the separation of macrolide antibiotics are shown in Table IV (*83-85*) and Fig. 11. Six systems for the chromatography of nonactine (*86*) and thirteen for relomycins (*87*) have also been reported.

TABLE IV

R_F VALUES OF MACROLIDE ANTIBIOTICS

Compound	R_F [a]			
	1	2	3	4
Erythromycin	0.27	—	0.04	0.11
Methymycin	0.56	—	—	—
Carbomycin	0.71	—	0.68	0.79
Oleandomycin	—	0.05	—	—
3-Monoacetyloleandomycin (MAO)	—	0.10	—	—
2-MAO	—	0.25	—	—
2,3-Diacetyloleandomycin (DAO)	—	0.35	—	—
1-MAO	—	0.50	—	—
1,3-DAO	—	0.65	—	—
1,2-DAO	—	0.90	—	—
Triacetyloleandomycin	—	0.95	—	—
Acumycin	—	—	0.69	0.80
Angolamycin	—	—	0.16	0.20
Tylosin	—	—	0.07	0.16
Foromacidin A	—	—	0.22	0.43
Foromacidin B	—	—	0.66	0.78
Foromacidin C	—	—	0.74	0.80
Narbomycin	—	—	0.85	0.91
Picromycin	—	—	0.47	0.62

[a] Solvents: (1) nonyl alcohol–CCl_4–*n*-propanol (75:75:2 v/v) (*83*); (2) benzene–cyclohexane (2:1 v/v, formamide saturated), paper pretreated with 50% formamide in methanol (*84*); (3) benzene, paper pretreated with formamide (*85*); (4) benzene–$CHCl_3$ (3:1 v/v), paper pretreated with formamide (*85*).

b. *Polyenes*. The lactone ring of these compounds contains from four to seven conjugated double bonds. They are acidic, basic, or amphoteric, and in most cases possess antifungal activity and are produced by actinomycetes (*1*, *5*). Examples include the tetraene (e.g., pimaricin), pentaene (e.g., kapacidin), and heptaene [e.g., the candinin–candicidin group, hamycin, and trichomycin (*88*)] antibiotics.

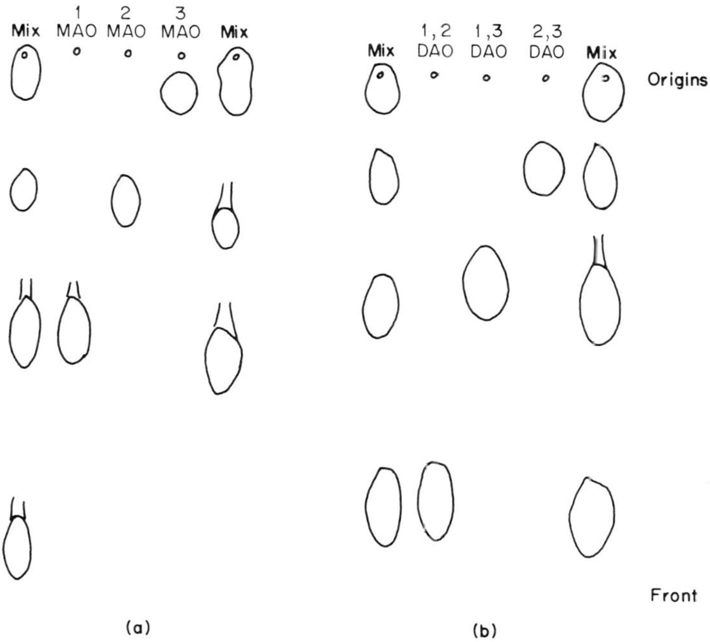

FIG. 11. *B. subtilis* bioautographic plates showing separation of (a) the 3-mono-acetyloleandomycins (MAO) plus oleandomycin base and triacetyloleandomycin by benzene–cyclohexane (2:1 v/v) and (b) the diacetyloleandomycins (DAO) plus oleandomycin base by benzene–cyclohexane (1:1 v/v). After T. M. Lees *et al.* (*84*).

Solvent systems for the polyenes (*5*) are usually based on *n*-butanol: *n*-butanol–ethanol–H_2O (5:1:4 v/v), *n*-butanol–acetic acid–H_2O (20:1:25 v/v), *n*-butanol–acetone or methanol–H_2O (1:1:1 or 1.5; 4:1:2 v/v), *n*-butanol–25% NH_4OH–H_2O (4:5:20 v/v). Other solvents include: *n*-propanol–H_2O (7:3 or 8:2 v/v), acetone–H_2O (1:1 v/v), 60% aqueous isopropanol, and pyridine–ethyl acetate–H_2O (2.5:6:7 v/v).

7. *Polypeptide Antibiotics*

This group contains acidic, basic, and amphoteric antibiotics with macrocyclic structures which contain amino acid residues. Examples are antibiotics of the following families: quinoxalines, bacitracins, polymyxins, and sporidesmolides.

a. *Systems.* A systematic procedure has been described for the separation and identification of seventeen polypeptide antibiotics employing fourteen solvent systems (*29*). Three of these were *n*-butanol–acetic acid–H_2O in different proportions (74:3:25, 79:6:15, and 4:1:5 v/v) and three were *tert*-butanol–acetic acid–H_2O (65:3:32, 55:6:4, and 70:6:24 v/v). Three were acetone–acetic acid–H_2O (20:6:74, 80:3:17, and 60:13:17 v/v). Two of the others were basic systems: acetone–H_2O (7:3 v/v) and *tert*-butanol–H_2O (8:2 v/v), with NH_3 in the atmosphere.

Impregnated paper (with glycine at pH 2.5) developed with butanol has been used for the separation of polymyxins into components A, B, D, and E (*89*).

b. *Detection.* Detection of the polypeptide antibiotics is by bioautography (*90*) or color reactions (*91*), e.g., with ninhydrin.

c. *Recent Applications.* The basic polypeptide antibiotic edeine is separated into four components—component A, $R_F = 0.05$; B, 0.25; C, 0.12; D, 0.50—on paper buffered with phosphate–citrate buffer, pH 5.0 (48.5 ml 0.1 *M* citric acid + 51.5 ml 0.2 *M* Na_2HPO_4) and developed with phenol–pH 5 buffer (1:1 v/v) (*90*).

Fifteen solvents were used for the chromatography of antibiotics of the ostreogrycin complex and related compounds on Whatman No. 1 paper during the course of structural studies (*92*). Paper chromatography is very often used in such studies of polypeptide antibiotics (*5*). A general discussion of peptide and polypeptide antibiotics has very recently been published (*93*).

8. *Actinomycins*

These are polypeptide antibiotics which have been studied in great detail by Brockmann and co-workers (*94*). They used circular chromatography with *n*-dibutyl ether saturated with 10% sodium naphthalene-1,6-disulfonate, and *n*-dibutyl ether–*n*-butanol (3:2 v/v) or butyl acetate (1:3 v/v) saturated with 10% sodium *m*-cresotinate on paper pretreated with the same solution, and bioautographic detection. A review of early work in this field is available (*95*).

Later workers have employed 10% aqueous sodium *o*-cresotinate–dibutyl ether–*sym*-tetrachloroethane (4:3:1 or 3:2:1 v/v) and 10% aqueous sodium salicylate–dibutyl ether–butyl acetate (4:1:3 or 2:1:1 v/v) (*96–98*).

Paper chromatography is successfully applied in studies of the biosynthesis and biogenesis of actinomycins (*96*, *98*, *99*).

9. *Rifomycins*

Rifomycins A, B, C, D, and E are separated by development with water containing 1% ascorbic acid and 3% NH_4Cl (*100*). Rifomycins B, O, S, and SV are separated by reversed-phase chromatography on paper impregnated with 2-octanol by development with 0.07 *M* phosphate buffer, pH 8.6. A quantitative determination is possible by eluting the antibiotics with 0.07 *M* phosphate buffer, pH 7.3, containing 0.05%

sodium ascorbate and measuring the absorbtion of the eluates at 425 mμ (for B and O) and 445 mμ (S and SV) (*101*).

10. *Novobiocin* (*102*)

a. *Paper.* After spotting the samples, impregnate the paper by spraying with 0.2 *N* phosphate buffer, pH 7.7–7.8, and air dry 3–5 minutes before development.

b. *Solvent.* Benzene–hexane–methyl ethyl ketone–ethanol (45:39:13:3 v/v).

c. *Technique.* Circular development for 2 hours on paper of 40 mm diameter. Detection is under ultraviolet light.

d. *R_F Values.* Novobiocin, 0.19; isonovobiocin, 0.29; decarbamoyl-novobiocin, 0.42, four unknown degradation products are also separated.

Quantitative Analysis. Cut out zones, elute for 1 hour with 10–20 ml 0.1 *N* NaOH, filter the eluates, and measure spectrophotometrically at 307 mμ against a 0.1 *N* NaOH blank. Compare to standard reference compounds. Accuracy is ±3%.

II. Vitamins

The vitamins, like the antibiotics, are defined and classified biologically rather than chemically. Chromatography has proved useful for the isolation of these substances from test materials, for preliminary purification, and for the ultimate identification and determination of the vitamins.

A. General Techniques

1. *Detection* (*103*)

The following vitamins are detected after separation by inspection in ordinary light: Vitamin A aldehyde, A acid, K_3, B_2, B_{12}, and folic acid; in short wave (254 mμ) ultraviolet light: A, D, B_1, B_2, B_6, niacin, niacinamide, B_{12}, and ascorbic acid; in long wave (365 mμ) ultraviolet light: A, B_2, B_6, B_{12}, folic acid, and ascorbic acid.

The following color reagents are generally useful for the detection of vitamins: phosphomolybdic acid (10% in ethanol, freshly prepared) detects vitamins A, D, E, K, and coenzyme Q; $SbCl_5$ (20% in $CHCl_3$) detects vitamins A, D, E, and K.

2. *Systems*

Water-soluble vitamins (e.g., B_1, B_2, B_6, B_{12}, niacin and niacinamide, pantothenic acid and pantothenol, biotin, folic acid) are separated on ordinary chromatography paper, for example by multiple circular development with butanol–acetone–ethanol–H_2O (5:1:1:5 v/v) (*104*). Ion-exchange paper can also be employed (*105*).

Fat-soluble vitamins (e.g., A, D, E, K) are chromatographed in reversed-phase and adsorption systems. As an example, listed below are some R_F values on silicic acid impregnated glass fiber paper developed with (a) cyclohexane, (b) 2% ether in isooctane, and (c) 1% methanol in cyclohexane (*106*).

Compound	R_F		
	a	b	c
Vitamin A palmitate	0.0	0.46	0.87
α-Tocopheryl acetate	0.32	0.83	0.91
α-Tocopherol	0.42	0.79	0.91
Vitamin D_3	0.07	0.55	0.74
Vitamin A alcohol	0.0	0.0	0.0

Paper treated with paraffin and developed with acetic acid–H_2O (84:16 v/v) can also be used (*107*), and the R_F values are: vitamin D_2, 0.38; vitamin D_3, 0.36; dihydrovitamin D_2, 0.29; vitamin A, 0.30; vitamin A acetate, 0.01; DL-α-tocopherol, 0.13.

An attempt (*108*) to simultaneously separate fat- and water-soluble vitamins by two-dimensional chromatography was successful only with the water-soluble vitamins and α-tocopherol (Fig. 12). Table V shows R_F values in the solvents employed and the detection tests and quantities used. Centrifugal chromatography has been used to separate both fat- and water-soluble vitamins (*109*).

B. Water-Soluble Vitamins

The preliminary fractionation of vitamins (and other water-soluble substances) in crude biological extracts is carried out in isopropanol–pyridine–H_2O–glacial acetic acid (8:8:4:1 v/v); high solute concentrations and the presence of salts are usually no problem (*110*).

Multivitamin preparations can be analyzed by means of paper chromatography using amyl alcohol–acetic acid–H_2O (6:3:3 v/v) as the solvent. The vitamins as well as degradation products of thiamine and lactoflavin are separated (*111*).

1. *Vitamin B_1 (Thiamine)*

a. *Paper*. Whatman No. 1 or 3MM.

b. *Solvents*.

1. pH 5 buffer–H_2O–propanol (1:1:3 v/v), paper equilibrated for 6 hours in vapors and developed for 40 hours (*112*).
2. Water-saturated butanol, paper previously extracted with flowing wet butanol and dried (*113*).

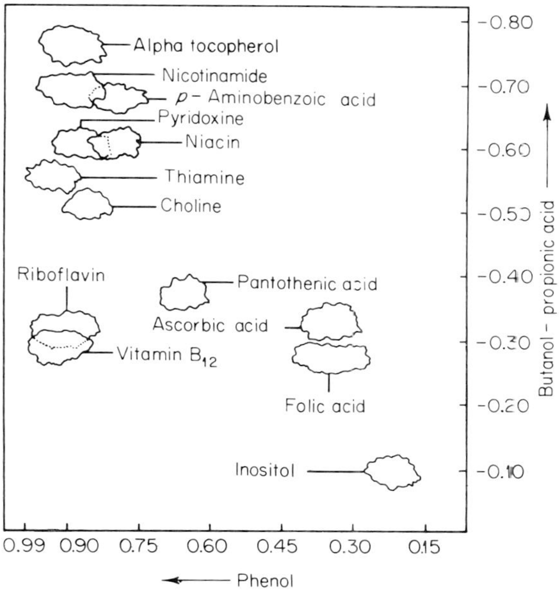

FIG. 12. Chromatomap of vitamins and related compounds showing migration in buffered water-saturated phenol and butanol–propionic acid. After E. L. Gadsden, *et al.* (*108*).

3. Isopropanol–H_2O (5:1 v/v)—separates thiamine and hexamethyldiamino-2-propanol (*114*).
4. *n*-Butanol–acetic acid–H_2O (4:1:5 v/v).
5. *sec*-Butanol–pyridine–H_2O–acetic acid (30:30:30:1 v/v).
6. Propanol-1 *M* sodium acetate, pH 5–H_2O (7:2:1 v/v).
7. Isobutyric acid–NH_4OH (concentrated)–0.1 *M* NaH_2PO_4–H_2O (198:3:40:59 v/v).
8. Propanol–1 *M* acetate buffer, pH 5–H_2O (7:1:2 v/v).
9. Butanol saturated with water, paper (Schleicher and Schuell 2043b Mgl) impregnated with pH 3.5 phosphate–citrate buffer; radial development (*115*).

c. *R_F Values* (*116, 117*).

Compound	R_F in the following solvents				
	4	5	6	7	8
Thiamine pyrophosphate	0.04	0.22	0.09	0.54	0.06
Thiamine monophosphate	0.09	0.30	0.20	0.70	0.27
Thiamine	0.25	0.55	0.44	—	—
Thiamine HCl	—	—	—	0.84	0.60

TABLE V

R_F VALUES OF SOME VITAMINS AND RELATED COMPOUNDS (*108*)

Compound	R_F [a] 1	R_F [a] 2	Quantity used (μg)	Color[b] D_1	D_2	D_3	D_4	D_5	D_6	D_7	D_8	D_9
Vitamin A	0.90	—	250	—	—	—	—	—	—	—	—	y[d]
α-Tocopherol	0.89	0.77	250[c]	b	—	—	—	—	—	—	—	—
Menadione	0.94	—	250	—	—	—	—	—	—	—	—	y[d]
Thiamine	0.93	0.55	250	—	f	y	—	—	—	—	—	—
Riboflavin	0.91	0.32	20	—	—	—	—	—	—	—	—	y, f
Niacin	0.83	0.68	250	—	—	—	bn	p	o	—	—	—
Nicotinamide	0.85	0.69	250	—	y	—	bn	—	y	—	—	—
Pyridoxine	0.87	0.60	250	—	—	—	—	—	—	bn	—	f
Pantothenic acid	0.66	0.38	250	—	—	pu	—	—	—	—	—	—
Biotin	0.79	0.78	250	—	—	—	bn	—	—	—	—	—
Inositol	0.21	0.12	50	bn	—	—	—	—	—	—	—	—
Choline	0.87	0.52	250	—	—	—	bn	—	—	—	—	—
p-Aminobenzoic acid	0.80	0.69	250	—	—	—	—	—	—	bn	bn	—
Folic acid	0.34	0.30	12.5	—	—	—	—	—	—	—	—	f
Vitamin B_{12}	0.92	0.31	50	—	—	—	—	—	—	—	—	p
Vitamin C	0.34	0.34	250	bn	—	—	—	—	—	—	—	—

[a] Solvents: 1 = phenol–buffer (100:25 v/v; aqueous buffer: 6.3% sodium citrate, 3.7% KH_2PO_4); 2 = *n*-butanol–propionic acid–H_2O [freshly prepared from equal volumes of solution A (1246 ml *n*-butanol + 84 ml water) and solution B (620 ml propionic acid + 790 ml water)]. Paper: Whatman No. 1 (descending). Time of run: 18–22 hours (solvent 1); 14–16 hours (solvent 2). Temperature of run: 24° ± 0.5°C. Detection: D_1 = ammoniacal silver nitrate; D_2 = ferricyanide–nitroprusside; D_3 = ninhydrin; D_4 = iodine vapor; D_5 = 2,6-dichlorophenolindophenol; D_6 = cyanogen bromide; D_7 = ferric chloride; D_8 = phenol–hypochlorite reagent; D_9 = light; order for multiple detection: D_9 (U.V.), D_3, D_4, D_7 or D_1.

[b] y = Yellow; b = black; bn = brown; p = pink; o = orange; pu = purple; f = fluorescent in ultraviolet light.

[c] In milligrams.

[d] After phenol run only.

d. *Detection.* (a) Spray with 55% ethanol–2.5% $K_3[Fe(CN)_6]$–10% NaOH (5:0.1:5 v/v) and view under ultraviolet light (*112, 116, 117*). This reagent is more sensitive than CNBr or phosphomolybdic acid for detecting thiamine (*118*). One percent $K_3[Fe(CN)_6]$–15% NaOH–H_2O (3:20:40) is also used (*115*). (b) Treat with I_2 vapor at 15–18°C for 4 hours (*114*).

e. *Quantitative determination.* Photodensitometry (*114*) on the paper after detection by method (b), spectrophotometry at 324 mμ(*115*), radioactive scanning or radioautography of ^{14}C-labeled thiamine (*116*), and conversion to thiochromes followed by elution from the paper and fluorimetry (*117*) have been used.

f. *Applications*. Thiamine has been determined in fruit musts and juices (*112*), breakfast cereal (*113*), extracts of microorganisms (*116*), and with pyridoxine in the presence of ascorbic acid, rutin, and procaine (115). The biosynthesis and metabolism of thiamine and its phosphate esters have been studied (*117*).

2. *Vitamin B_2 (Riboflavin)*

The decomposition of riboflavin caused by $KMnO_4$ (*119*) and microorganisms found in the soil (*120*) have been studied with butanol-acetic acid–H_2O (4:1:5 v/v) as the solvent.

For the separation of riboflavin, flavin mononucleotide, and flavin-adenine dinucleotide, development with *n*-butanol–pyridine–H_2O–glacial acetic acid (3:3:3:1 v/v) on wedge-shaped paper is recommended (*121*). These substances, extracted from yeast, have R_F values of 0.67, 0.37, and 0.30, respectively. The zones are marked in ultraviolet light, washed with $CHCl_3$, cut out, eluted with buffers of pH 5.5, 3.5, and 2.8, respectively, and determined fluorimetrically against standard riboflavin (0.08 μg/ml). Recovery of the flavins is about 80%.

Flavines and ribolumazines in flavinogenic and nonflavinogenic strains of *Ashbya gossypii* have been separated in *tert*-butanol–33% (w/v) trimethylamine–H_2O (60:0.95:39.05 v/v) and the upper alcoholic phase of *n*-butanol–1 *N* HCl–H_2O (100:2:148 v/v) to which was added 50 ml *tert*-butanol, 2.5 ml 0.01 N HCl and 40 ml H_2O. The R_F of riboflavin in the latter solvent is 0.24. Quantitation of the zones was by fluorimetry (*122*). A similar solvent (*n*-butanol–*tert*-butanol–1 *N* HCl–H_2O, 100:40:0.15:0.148 v/v), in which riboflavin has R_F 0.295, was used earlier to separate lumazine and isoalloxazine derivatives (*123*). Table VI shows R_F values of some flavins (*124*) on Whatman No. 1 paper after detection by ultraviolet light.

TABLE VI

R_F VALUES OF FLAVINS (*124*)

Compound	R_F [a]			
	1	2	3	4
Riboflavin	0.25	0.30	0.16	0.80
Riboflavin phosphate	0.04	0.56	—	0.15
Lumichrome	0.65	0.05	0.28	—
Flavinadenine dinucleotide	0.03	0.41	0.00	0.17
Lumiflavin	0.41	0.15	—	—

[a] Solvents: (1) *n*-butanol–acetic acid–H_2O (4:1:5 v/v, upper phase); (2) 5% Na_2HPO_4 (w/v); (3) methanol; (4) phenol–*n*-butanol–H_2O (16:3:10 w/v/v).

3. *Pteridines*

Identification of the brightly colored pteridines can be performed by paper chromatography utilizing the following solvent systems: water-saturated butanol, isopropanol–2% ammonium acetate (1:1 v/v), *n*-propanol–1% NH_3 (2:1 v/v), *n*-butanol–acetic acid–H_2O (4:1:1 v/v), *n*-propanol–ethyl acetate–H_2O (7:1:2 v/v), 5% acetic acid, 3% NH_4Cl, and 3% urea (*125*). Riboflavin and sepiapterin, both of which are present in amphibian integument, have the same R_F values in these solvents but are separated in *n*-butanol–1% sodium borate with respective R_F values of 0.37 and 0.47.

The variation of R_F on Whatman No. 1 paper as a function of the pH of the solvent (ammonium acetate, 0.1 *N*) was determined for a series of pterins. The curve for each pterine is characteristic and can be used for identification and to choose the solvent pH giving the best separation (*126*). R_F values for some pteridines, isoalloxazines, and pyrimidines on Whatman No. 1 paper with *n*-butanol–5 *N* acetic acid (7:3 v/v) and 3% aqueous NH_4Cl as solvents have been listed (*127*).

4. *Folic Acid Derivatives*

Folic acid and related compounds (e.g., pteroic acid, teropterin, rhizopterin) are chromatographed in the following solvents (*128*) for which R_F values have been listed (*102*): 15% $Na_2HPO_4 \cdot 12H_2O$ solution, 0.1 *M* phosphate buffer (pH 7.0), 5% citric acid–NH_4OH buffer (pH 9.0)–isoamyl alcohol, and water-saturated *sec*-butanol containing 3% acetic acid.

Folic acid (and its conjugated derivatives) exhibits no fluorescence. It is detected with high sensitivity and estimated by bioautography with *Streptococcus faecalis*. The chromatogram is dried, spread on a glass plate with an agar support innoculated with *S. faecalis*, and incubated 18 hours at 37°C. Planimetric estimation of the growth areas gives, after comparing with standards, the amount of folic acid (*129*).

5. *Vitamin* B_6 (*Pyridoxines*)

The pyridoxine vitamins are light sensitive (*129a*) and chromatography in the dark is recommended. A useful solvent is H_2O–acetone–*tert*-butanol–diethylamine (20:35:40:5 v/v) (*130*) in which the following R_F values obtain: pyridoxol, 0.31; pyridoxaloxime, 0.55; pyridoxal, 0.64; pyridoxamine, 0.70; and 4-pyridoxic acid, 0.80.

Pyridoxal phosphates have been prepared and chromatographed, and it was found that the R_F values of pyridoxol, pyridoxamine, pyridoxal, and their corresponding 5-phosphates are a function of the pH of the developing solvent—pyridine–butanol–H_2O (1:2:saturated) adjusted with HCl or NH_3 (*131*). Spots are located under ultraviolet light, the fluorescence of pyridoxal being greatly enhanced by exposure of the strip to concentrated NH_3 vapors. The areas of the chromatographed spots are found to decrease with increasing pH. Four of the six vitamin B_6 congeners are speparated by development with either pH 6 or pH 9 solvents; the R_F values ± 1 standard deviation are as given in the tabulation below.

Compound	R_F	
	pH 6	pH 9
Pyridoxamine 5-phosphate	0.07 ± 0.03	0.06 ± 0.01
Pyridoxal 5-phosphate	0.16 ± 0.01	0.11 ± 0.02
Pyridoxamine 2HCl	0.33 ± 0.02	0.49 ± 0.03
Pyridoxal HCl	0.75 ± 0.01	0.51 ± 0.01
Pyridoxol HCl	0.79 ± 0.01	0.57 ± 0.03
Pyridoxol 5-phosphate	0.80 ± 0.02	0.64 ± 0.03

After separation of pyridoxol, pyridoxal, and pyridoxamine (which together act as vitamin B_6) or their phosphate derivatives as above or by two-dimensional development with isopropanol–H_2O (4:1 v/v, adjusted to pH 3.5 with acetic acid) followed by amyl alcohol–acetone–H_2O (2:1:2 v/v, upper phase), quantitative determination by fluorimetry and photography is possible. Conversion to pyridoxic acid, which is fifty times more fluorescent, increases the sensitivity of the method (*132*).

Pyridoxyl derivatives of amino acids are separated by two-dimensional chromatography with water-saturated collidine–lutidine (1:1 v/v) followed by *n*-butanol–acetic acid–H_2O (4:1:5 v/v). Detection is under ultraviolet light and by spraying with ninhydrin to detect free amino groups (see Chapter 5) or with 2,4-dichloroquinonechlorimide to detect the phenolic groups of the pyridoxyl derivatives. R_F values of eighteen acid derivatives in these two solvents have been listed (*133*).

6. *Vitamin B_{12} (Cyanocobalamin)*

a. *Paper*. Whatman No. 1 (*134*) and 2 (*135*).

b. *Solvents*. (a) $CHCl_3$–phenol–butanol–H_2O (5:1:2:1 and 4:1:1.5:0.5 v/v; $R_F = 0.50$ and 0.55, respectively, at 27–29°C) (*136*). (b) Butanol saturated with H_2O (*135*).

c. *Quantitation*. Densitometry: 10–200 γ of vitamin B_{12} is determined with a 2–10% error (*134*).

Bioautography : B_{12} in liver extract is determined by placing the chromatogram on an agar plate innoculated with *Lactobacillus leichmannii*

313. After incubation, the median growth rate under the strip compared with a logarithm curve for standards (1.25–5 γ/ml vitamin B_{12}) is used to calculate the concentration of B_{12} (*135*).

7. *Vitamin B_{15} (Pangamic Acid)* (*136a*).

a. *Solvent.* Multiple development with phenol–isoamyl alcohol–H_2O (1:1:1) for 24–30 hours each.

b. *Quantitation.* Dried papers are treated with CH_3I vapors at 45°C for 2 hours, the excess CH_3I is removed at 100°C and the papers are dipped into 0.1 *N* $AgNO_3$. The excess Ag(I) solution is removed with water, and the AgI is eluted from the paper with 15 *N* HNO_3 and determined by colorimetry with dithizone at 620 mμ. The reported sensitivity is 10–15 γ for vitamin B_{15} in blood, tissue extracts, and urine.

C. Fat-Soluble Vitamins

A rapid method for the separation and identification of naturally occurring quinones and fat-soluble vitamins involves development with various solvents on S. and S. No. 288 alumina paper. Table VII and Fig. 13 show results for pure compounds and plant extracts (*137*). See also Section 1 above and Tables II, IV, and V in Chapter 8.

1. *Vitamin A*

Vitamin A esters, vitamin A alcohol, vitamin A aldehyde, vitamin A acid, and related compounds have been separated on paper impregnated with either silicone (*138*) or with saturated dibasic calcium phosphate followed by 2 *N* NH_4OH, washing with water, and drying at 120°C (*103*). Solvents include ethanol–0.1 *M* phosphate buffer (1:1 v/v, pH 8), in which R_F values are vitamin A, 0.52, 0.85; A aldehyde, 0.20; and A acid, 0.57, in the former case (*138*), and petroleum ether or $CHCl_3$ in the latter.

Vitamins A_1 and A_2 and allied substances are separated by reversed-phase circular paper chromatography on Whatman No. 1 paper impregnated with 3% Vaseline and developed for 4–5 hours with 90% methanol (*139*). Results are shown in Table VIII.

2. *Vitamin D_2* (*140*)

a. *Paper.* Chromatography paper is impregnated with 2% formamide for 20 minutes, and washed with H_2O, $A1_2(SO_4)_3$ solution, and dilute NH_4OH.

b. *Solvent.* Petroleum ether plus 6% acetone.

c. *Detection.* Hot $SbCl_3$ solution in $CHCl_3$.

d. *Quantitation.* Elute unknown and standard spots with 1 ml $CHCl_3$ and measure color formed with $SbCl_3$ in $CHCl_3$; recovery about 80%.

TABLE VII

R_F VALUES FOR NATURALLY OCCURRING QUINONES[a], FAT-SOLUBLE VITAMINS, AND PLANT PIGMENTS ON SCHLEICHER AND SCHUELL NO. 288 PAPER (*137*)

Compounds	Cyclo-hexane	Cyclo-hexane–benzene (7:3)[b]		Cyclo-hexane–benzene (3:7)[b]	Benzene	Benzene–$CHCl_3$ (1:1)[b]		$CHCl_3$	Iso-octane	Petroleum ether (b.p. 30–60°C)	CCl_4	Toluene[b]	
		1	2	1		1	2					1	2
β-Carotene (C)	0.51	0.81	0.98	0.84	0.92	0.87	0.95	0.94	0.13	0.32	0.79	0.86	0.90
Vitamin K_1 (K_1)	0.20	0.50	0.90	0.78	0.87	0.85	0.94	0.91	0.08	0.10	0.53	0.79	0.87
Plastoquinone A (PQ_9)	0.08	0.18	0.23	0.77	0.82	0.85	0.94	0.91	—	0.05	0.18	0.78	0.87
Coenzyme Q_{10} (Q_{10})	0.05	0.09	0.16	0.47	0.70	0.83	0.90	0.91	—	—	0.13	0.55	0.85
α-Tocopherol (T)	—	0.05	0.13	0.26	0.46	0.48	0.64	0.82	—	—	0.10	0.32	0.78
α-Tocopherylquinone (TQ)	—	—	0.05	0.05	0.14	0.41	0.50	0.73	—	—	0.05	0.09	0.65
Vitamin D_2 (D_2)	—	—	0.05	0.08	0.13	0.40	0.41	0.59	—	—	—	0.09	0.52
Vitamin A (A)	—	—	—	0.04	0.11	0.29	0.31	0.51	—	—	—	0.09	0.38
Chlorophylls (Chl) a + b	—	—	—	—	—	0.04	0.04	0.08	—	—	—	—	—
Xanthophylls (X)	—	—	—	—	—	—	—	0.08	—	—	—	—	—
Pheophytin (Ph)	—	—	—		0.17	0.47	0.58	0.77	—	—	—	0.09	0.74

[a] Detection of quinones: Chromatograms are first immersed in neotetrazolium chloride solution and then left in borohydride solution as long as necessary to yield distinct red to violet spots (usually 20–90 sec.).

[b] (1) Walls of tank lined with solvent-wetted filter paper; (2) Walls of tank not lined.

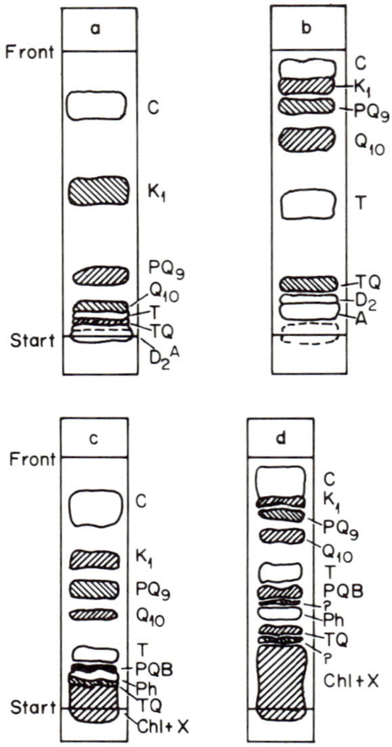

FIG. 13. Paper chromatograms of pure lipid-soluble vitamins and quinones (a and b), and of acetone extracts from spinach leaves (c and d). Solvent for a and c is cyclohexane–benzene (7:3 v/v) and for b and d, benzene. See Table VII for abbreviations. PQB = plastoquinone B. Dark spots indicate quinones. After H. G. Lichtenthaler (*137*).

3. *Vitamins E* (*Tocopherols*) *and K*

These vitamins have been separated on cellulose paper treated with Vaseline, liquid paraffin, olive oil, silicone fluid, and zinc carbonate, and on glass fiber paper coated with silica gel and fluorescein (*141*). Table IX lists some results in such systems (*141–144*). Spots are usually detected under ultraviolet light or by the Emmerie–Engel reagent: spray with a 1:1 (v/v) mixture of 0.5% (w/v) ethanolic 2,2′-dipyridyl solution and 0.2% (w/v) ethanolic $FeCl_3$ solution to yield red spots on a white background. The tocopherols reduce the iron to Fe^{+2}, which combines with dipyridyl to form the red color.

Nitrosotocopherols (*145*) are chromatographed on zinc carbonate paper with 30% benzene in cyclohexane (v/v) or the same paper impregnated with liquid paraffin and developed with 93% ethanol in H_2O (v/v). R_F values are shown in the tabulation below (*146*).

Nitroso derivative	$ZnCO_3$ paper	$ZnCO_3$–paraffin paper
β-Tocopherol	0.70	0.42
γ-Tocopherol	0.77	0.19
ε-Tocopherol	0.70	0.60
η-Tocopherol	0.70	0.31
σ-Tocopherol (7-nitroso)	0.25	0.79
σ-Tocopherol (5-nitroso)	0.46	0.48

TABLE VIII

REVERSE-PHASE CIRCULAR PAPER CHROMATOGRAPHIC RESOLUTION OF VITAMINS A_1, A_2, AND THEIR DERIVATIVES (*139*)

Substance	Band as observed in ultraviolet light	R_F values[a]	λ_{max} in petroleum ether (mμ)
Vitamin A_1 palmitate	Blue fluorescence	0.00	325
Anhydrovitamin A_1	Bright yellow fluorescence	0.20 ± 0.01	350, 370, 390
Isoanhydrovitamin A_1	Blue fluorescence	0.29 ± 0.03	330, 350, 370
Vitamin A_1 acetate	Blue fluorescence	0.35 ± 0.03	325
Vitamin A_1 aldehyde	Dull red absorbance	0.55 ± 0.01	370
Vitamin A_1 acid	Dull red absorbance	0.65 ± 0.02	350
Vitamin A_1 alcohol	Blue fluorescence	0.82 ± 0.03	325
Anhydrovitamin A_2	Bright yellow fluorescence	0.33 ± 0.01	350, 370, 390
Vitamin A_2 acetate	Green fluorescence	0.39 ± 0.02	350
Vitamin A_2 aldehyde	Dull red absorbance	0.67 ± 0.03	385
Vitamin A_2 alcohol	Green fluorescence	0.93 ± 0.01	350
β-Carotene (see Chapter XI)	Orange colored band	0.00	—

[a] Mean values of ten observations.

Two-dimensional reversed-phase chromatography utilizing these two solvents separates all six derivatives (Fig. 14). Quantitative analysis is based on colorimetry at 520 mμ after application of the Emmerie–Engle reagent and elution from the paper with ethanol. Alternatively, elution is with isooctane–ethanol (1:1 v/v) followed by reaction with 0.2 ml bathophenanthrone, 0.1 ml $FeCl_3$, 0.3 ml and H_3PO_4 and colorimetry at 534 mμ (*141*).

TABLE IX

R_F VALUES OF TOCOPHEROLS, VITAMINS K, AND RELATED LIPIDS

Compound	R_F[a]						
	1	2	3[b]	4	5	6	7
α-Tocopherol	0.50	—	0.33 (*d*)	0.10	0.31	0.4–0.5	—
β-Tocopherol	0.72	—	—	0.22	0.37	—	—
γ-Tocopherol	0.72	—	0.49 (*d*)	0.22	0.37	—	—
δ-Tocopherol	0.84	—	0.62 (*d*)	0.42	0.45	—	—
α-Tocopherylquinone	—	0.84	0.58	—	—	—	—
Vitamin K_1	—	0.59	—	—	—	—	—
Vitamin K_2	—	0.29	—	—	—	—	—
Coenzyme Q_{10}	—	0.27	—	—	—	ca. 0.3	0.06
Coenzyme Q_9	—	0.36	—	—	—	—	0.11
Coenzyme Q_8	—	0.42	—	—	—	—	0.18
Coenzyme Q_7	—	0.49	—	—	—	—	0.23
Coenzyme Q_6	—	0.54	—	—	—	—	0.31
Coenzyme Q_{10} hydroquinone	—	—	—	—	—	—	0.26
Coenzyme Q_9 hydroquinone	—	—	—	—	—	—	0.41
Coenzyme Q_8 hydroquinone	—	—	—	—	—	—	0.55
Coenzyme Q_7 hydroquinone	—	—	—	—	—	—	0.60
Coenzyme Q_6 hydroquinone	—	—	—	—	—	—	0.66
ε-Tocopherol	—	—	—	0.53	—	—	—
Hydrogenated ε-tocopherol	—	—	—	0.22	—	—	—
Tocol	—	—	—	0.65	0.50	—	—

[a] Solvents: (1) 75% ethanol, paper coated with Vaseline (*142*); (2) *n*-propanol–H_2O (4:1 v/v), paper impregnated with silicone (*143*); (3) ethanol–H_2O (3:1 v/v), paper impregnated with silicone (*143*); (4) 65% (v/v) ethanol in H_2O, paper impregnated with a 5% (v/v) solution of liquid paraffin in petroleum ether (*144*); (5) 90% (v/v) ethanol in H_2O, Whatman No. 4 paper impregnated with a 5% (v/v) solution of olive oil in petroleum ether (*144*); (6) 2,2,4-trimethylpentane–isopropyl acetate (200:0.5 v/v), silica-coated glass fiber paper (*141*); (7) *n*-propanol–H_2O (7:3 v/v), paper impregnated with silicone (*143*).

[b] *d* = *d*-tocopherols.

A detailed description has been published for the separation and analysis of seven tocopherols in oils, foods, and feeding stuffs employing two-way chromatography on zinc carbonate-impregnated paper (first solvent: 30% v/v benzene in cyclohexane; second solvent: 75% aqueous ethanol after dipping the paper in liquid paraffin), removal of the areas of paper known to contain the tocopherols and extraction and colorimetry by the Emmerie–Engel method (*147*). Methods for the separation and determination of tocopherols in animal tissues (*148*), feces (*149*), and human and animal serum and plasma (*150*) utilizing $ZnCO_3$ paper have also been published.

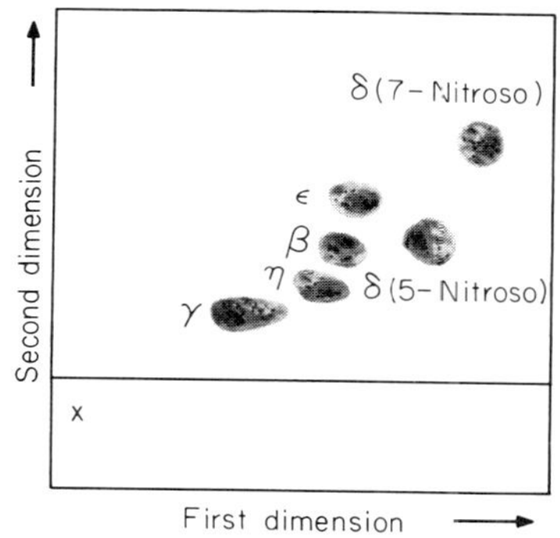

FIG. 14. Two-dimensional chromatogram of six nitrosotocopherols on paraffined zinc carbonate-impregnated Whatman No. 1 paper. Development is with 30% benzene in cyclohexane (v/v) followed by 93% ethanol in H_2O (v/v). Developed zones are orange. The mixture is spotted at the origin (×) on unparaffined paper; the paraffin impregnation is made only above the horizontal line. After S. Marcinkiewicz and J. Green (*146*).

D. VITAMIN C (ASCORBIC ACID)

Chromatographic and extraction techniques for vitamin C assay have been reviewed (*151*), and the chromatographic method has been compared with titrimetric and photometric methods (*152*).

1. *Preparation of Sample*

For the determination of ascorbic acid in urine, a measured volume of urine containing 1–5 μg ascorbic acid is mixed with 1–2 ml 4% oxalic acid and centrifuged for 5 minutes. Portions of 10-100 μl of supernate are spotted on the paper with drying between applications (*153*). Determinations in food (*154, 155*), skin (*156*), blood, and lemon juice (*157*) have also been made.

a. *Paper.* Untreated chromatography paper or paper pretreated with 3% HPO_3 (*155, 158*), trilon B (*156*), or 5% oxalic acid (*157*).

b. *Systems.* (1) Upper layer of a mixture of 300 ml 2% aqueous oxalic acid and 200 ml butanol (*153*); (2) butanol–acetic acid–H_2O (4:1:5 v/v) (*156*); (3) methyl ethyl ketone saturated with H_2O, or acetonitrile–acetone–10% metaphosphoric acid (80:5:15 v/v with H_3PO_3 paper) (*155*); (4) butanol plus a few crystals of KCN; (5) benzyl alcohol–acetic acid–H_2O (4:1:5 v/v, lower phase; the paper is impregnated with the upper phase); R_F of ascorbic acid–0.22 (*155a*); (6) phenol–acetic acid–H_2O (99:1:100 v/v); Whatman No. 1 paper impregnated with 3% metaphosphoric acid; descending development for 45 hr at 20°C with KCN crystals added to the tank; *L*-ascorbic (R_F = 0.43) and *D*-iso-ascorbic acid (0.47) are separated (*155b*).

c. *Technique*. Develop in a CO_2 or other inert or reducing atmosphere because of the susceptibility of ascorbic acid to oxidation (*156*, *157*). Oxidation while drying the paper after the run is reduced by using a vacuum high-frequency induction heater (*155*) or drying in a CO_2 atmosphere.

d. *Color Reagents*. (1) 1% ammoniacal $AgNO_3$—Detects 0.45 γ of ascorbic acid (*156*); (2) 15% ammonium molybdate in 1% NH_4OH adjusted to pH 3.8 with H_2SO_4 (*153*); (3) 2,6-dichloroindophenol—Spray with a 0.1% ethanolic solution; ascorbic acid shows up as white spots; (4) iodine–starch—Spray with a 0.001–0.005% iodine solution containing a trace of KI in a 0.4% starch solution; ascorbic acid becomes white on a blue background; (5) Nessler reagent—Mix a solution of 11.5 gm HgI_2 and 8 gm KI in 50 ml H_2O with 50 ml of 6N NaOH; set solution aside, then decant the liquid from the precipitate and spray the the dry chromatogram. Ascorbic acid appears as a stable black spot (*155a*).

2. *Quantitative Determination* (*153*).

The spot [$R_F = 0.39$ in solvent (1)] is eluted with 5 ml 1% sodium acetate (pH 3.5) in 0.67% oxalic acid, mixed with 2 ml dichloroindophenol (prepared by dissolving 15.2 mg in boiling H_2O and filtering after standing overnight in the cold). Sixty milligrams $NaHCO_3$ is then added and the mixture diluted to 500 ml with H_2O. The eluted color is read at 510 mμ versus a reagent blank. Recoveries average 99% for 4–20 μg ascorbic acid.

Alternatively, the ascorbic acid spot [$R_F = 0.41$ in solvent (4)] is cut out, extracted with 5 ml of 3% acetic acid under CO_2 and titrated with 0.01% *N*-bromosuccinimide in the presence of 5 ml 4% KI and starch. Errors range from -3% to $+2.6\%$ for 0.5–50 γ ascorbic acid (*157*). Extraction with 2 ml of 1% oxalic acid followed by adjustment to pH 3.5 and titration with 0.001 *N* 2,6-dichloroindophenol to a deep rose end point gives good results for amounts up to 10–12 μg (*158a*).

3. *Determination of Ascorbic Acid as its Osazone*

Ascorbic acid osazones (salmon colored) are formed with 2,4-diphenylhydrazine, dissolved in ethyl acetate, and separated by repeated ascending chromatography with dichloromethane–chlorobenzene (6:4 v/v) and formic acid–H_3PO_4–H_2O (2:1:2 v/v) as solvents. Quantitative evaluation ($\pm 10\%$) is performed by densitometry or by eluting the spots with dichloromethane and measuring the absorbance of the eluate (*159*).

References

1. M. Vondráček, *in* "Chromatography" (E. Heftmann, ed.), p. 717. Reinhold, New York, 1967.
2. P. D. Reshetov, N. O. Blinov, and A. S. Khokhlov, *Antibiotiki* **8**, 104 (1963).
3. N. O. Blinov, E. F. Oparysheva, G. Z. Yakubov, A. V. Puchnina, Yu. M. Khokhlova, E. M. Kleiner, N. G. Fed'kina, K. M. Khryashcheva, and A. S. Khokhlov, *Molekul. Khromatogr., Akad. Nauk SSSR, Inst. Fiz. Khim.* p. 137 (1964).
4. M. Vondráček and I. M. Hais, *in* "Paper Chromatography" (I. M. Hais and K. Macek, eds.), p. 686. Academic Press, New York, 1963.
5. V. Betina, *Chromatog. Rev.* **7**, 119 (1965).
6. R. A. Radzhapov, A. B. Silaev, and N. S. Agre, *Antibiotiki* **11**, 909 (1966).
7. M. K. Majumdar and S. K. Majumdar, *Anal. Chem.* **39**, 215 (1967).
8. I. Szilágyi, T. Vályi-Nagy, and T. Keresztes, *Nature* **205**, 1225 (1965).
9. C. A. Claridge and D. L. Johnson, *Antimicrobial Agents Ann.* p. 682 (1962).
10. Z. Baráth, V. Betina, and P. Nemec, *J. Antibiotics (Tokyo)* **A17**, 144 (1964).
11. V. Betina and L. Pilátová, *Cesk. Mikrobiol.* **3**, 202 (1958).
12. N. A. Drake, *J. Am. Chem. Soc.* **72**, 3803 (1950).
13. E. G. Toporova, *Antibiotiki* **7**, 90 (1962).
14. B. Heinemann, A. J. Howard, and Z. J. Hollister, *Appl. Microbiol.* **15**, 723 (1967).

14a. E. Stutter, H. Steiniger, and G. Bradler, *Jena Rev. (English Transl.)* **13**, 20 (1968).

15. Gy. Csobán and G. Szabó, *Kiserl. Orvostud.* p. 1 (1962).
16. S. N. Litvinenko, *Lab. Delo* p. 39 (1962).
17. N. O. Blinov and A. S. Chochlov, *Antibiotiki* **8**, 751 (1963).
18. W. T. Sokolski, N. J. Eilers, and T. E. Eble, *Antibiot. Chemotherapy* **9**, 436 (1959).

18a. M. Bauer, *et al.*, *Farmacia (Bucharest)* **16**, 293 (1968).

19. E. Addison and R. G. Clark, *J. Pharm. Pharmacol.* **15**, 286 (1963)
20. F. P. de Franca and G. de Oliveira Dias, *Rev. Brasil. Farm.* **47**, 67 (1966).
21. P. Nemec, V. Betina, and L. Kovacicova, *Folia Microbiol. (Prague)* **6**, 277 (1961).
22. V. Betina and P. Nemec, *Chem. Zvesti* **15**, 853 (1961).
23. D. E. Dykhovichnaya, *Antibiotiki* **8**, 939 (1963).
24. D. Doskocilova and M. Vondráček, *Antibiotiki* **6**, 649 (1961).
25. P. Nemec, V. Betina, and L. Kovacicova, *Biologia* **16**, 375 (1961).
26. N. Ishida, *J. Antibiotics (Tokyo)* **A5**, 481 (1952).
27. V. R. Sevcik, M. Podojil, and A. Vrtiskova, *Cesk. Mikrobiol.* **2**, 175 (1957).
28. N. O. Blinov, N. G. Fed'kina, E. F. Oparysheva, and A. S. Khokhlov, *Izv. Akad. Nauk SSSR, Ser. Biol.* **29**, 533 (1964).
29. N. Snell, K. Ijichi, and J. C. Lewis, *Appl. Microbiol.* **4**, 13 (1956).
30. N. O. Blinov, G. Z. Yakubov, L. A. Vetlugina, and N. Khokhlova, *Mikrobiologiya* **30**, 642 (1961).
31. V. Betina, *J. Chromatog.* **15**, 379 (1964).

31a. N. O. Blinov, *et al.*, *Antibiotiki* **14**, 275 (1969).

31b. J. Souto and A. Gonzalez De Valesi, *J. Chromatog.* **46**, 274 (1970).

32. V. Betina, *Chem. Zvesti* **15**, 661, 848, and 859 (1961).
33. P. B. Baker, F. Dobson, and A. J. P. Martin, *Analyst* **75**, 651 (1950).
34. L. N. Astanina and L. M. Yakobson, *Lab. Delo* p. 666 (1967).
35. M. Goldman and R. A. Day, *Ohio J. Sci.* **67**, 190 (1967).

35a. H. Hellberg, *J. Assoc. Offic. Anal. Chemists* **51**, 552 (1968).

36. R. R. Goodall and A. A. Levi, *Analyst* **72**, 277 (1947).
37. R. Fischer and H. Lautner, *Arch. Pharm.* **294**, 1 (1961).

38. H. Freytag, *Z. Anal. Chem.* **188**, 217 (1962).
39. R. Thomas, *Nature* **191**, 1161 (1961).
40. A. L. Demain and J. F. Newkirk, *Appl. Microbiol.* **10**, 321 (1962).
41. S. Yamatodani, *J. Antibiotics (Tokyo)* A**3**, 400 (1950).
42. A. L. Demain, *Nature* **210**, 426 (1966).
43. F. Kawanagh, D. Tunin, and G. Wild, *Arch. Biochem. Biophys.* **77**, 268 (1958).
44. J. L. Ott, C. W. Godzeski, D. Pavey, J. D. Farran, and D. R. Horton, *Appl. Microbiol.* **10**, 515 (1962).
45. R. P. Elander, J. F. Stauffer, and M. P. Backus, *Antimicrobial. Agents Ann.* p. 91 (1961).
46. H. S. Burton and E. P. Abraham, *Biochem. J.* **50**, 168 (1952).
47. B. Loder, G. C. F. Newton, and E. P. Abraham, *Biochem. J.* **79**, 408 (1961).
48. Glaxo Group Ltd., Neth. Patent Appl. 6, 606, 820 (1966); *Chem. Abstr.* **67**, 82219y (1967).
49. H. R. Sullivan and R. E. McMahon, *Biochem. J.* **102**, 976 (1967).
50. R. P. Miller, *Antibiot. Chemotherapy* **12**, 689 (1962).
51. M. Albu-Budai and H. W. Unterman, *Rev. Chim. (Bucharest)* **18**, 173 (1967).
52. W. Grundy, A. L. Whitman, M. E. Hanes, and J. C. Sylvester, *Antibiot. Chemotherapy* **1**, 309 (1959).
53. I. A. Solomons and P. P. Regna, *J. Am. Chem. Soc.* **72**, 2974 (1950).
54. W. A. Winsten, C. I. Jarowski, F. X. Murphy, and W. A. Lazier, *J. Am. Chem. Soc.* **72**, 3969 (1953).
55. B. E. Leach, W. H. DeVries, H. A. Nelson, W. G. Jackson, and J. S. Evans, *J. Am. Chem. Soc.* **73**, 2797 (1951).
56. M. J. Cron, D. L. Johnson, F. M. Palermiti, Y. Perron, H. D. Taylor, D. F. Whitehead, and I. R. Hooper, *J. Am. Chem. Soc.* **80**, 752 (1958).
57. D. H. Peterson and L. M. Reineke, *J. Am. Chem. Soc.* **72**, 3598 (1950).
58. K. L. Rinehart, A. D. Argoudelis, W. A. Gross, A. Sohler, and D. P. Schaffner, *J. Am. Chem. Soc.* **82**, 3938 (1960).
59. S. C. Pan and J. D. Dutcher, *Anal. Chem.* **28**, 836 (1956).
60. M. S. Walker and J. B. Walker, *Biochim. Biophys. Acta* **136**, 272 (1967).
61. W. F. Phillips and H. S. Ragheb, *J. Chromatog.* **19**, 147 (1965).
62. T. Miyaki, H. Tsukiura, M. Wakae, and H. Kawaguchi, *J. Antibiotics (Tokyo)* **A15**, 15 (1962).
63. S. Kondo, M. Sezaki, and M. Shimura, *J. Antibiotics (Tokyo)* **B17**, 1 (1964).
64. G. H. Wagman, E. M. Oden, and M. J. Weinstein, *Appl. Microbiol.* **16**, 624 (1968).
65. G. H. Wagman, J. A. Marquez, and M. J. Weinstein, *J. Chromatog.* **34**, 210 (1968).
66. G. H. Wagman, J. V. Bailey, and M. M. Miller, *J. Pharm. Sci.* **57**, 1319 (1968).
67. G. H. Wagman and J. V. Bailey, *J. Chromatog.* **41**, 263 (1969).
68. J. Sherma and G. S. Lippstone, *J. Chromatog.* **41**, 220 (1969).
69. G. B. Selzer and W. W. Wright, *Antibiot. Chemotherapy* **7**, 292 (1957).
70. Anonymous, *Federal Register* **32**, No. 11, 522–523 (1967).
71. C.-S. Huang and C.-I. Chang, *Yao Hsueh Hsueh Pao* **11**, 837 (1964).
71a. E. Grigorescu, M. Gafencu, and E. Bilbie, *Farmacia (Bucharest)* **16**, 431 (1968).
72. G. Cantarelli, *Farmaco (Pavia), Ed. Prat.* **17**, 728 (1963).
73. M. Urx, J. Vondráčková, L. Kovařík, O. Horský, and M. Herold, *J. Chromatog.* **11**, 62 (1963).
74. R. G. Kelly and D. A. Buyske, *Antibiot. Chemotherapy* **10**, 604 (1960).
75. J. R. D. McCormick, N. O. Sjoalander, U. Hirsch, E. R. Jensen, and A. P. Doerschuk, *J. Am. Chem. Soc.* **79**, 4561 (1957).

76. C. Garber, J. Dobrecky, and R. C. D. DeCarnevale Bonino, *Rev. Soc. Quim. Mex.* **11**, 121 (1967).
77. I. Ciznar and V. Krcmery, *Vet. Med. (Prague)* **9**, 253 (1964).
78. A. Miyake and T. Konoshita, *J. Pharm. Soc. Japan* **75**, 1358 (1955).
79. G. Coppi, *Farmaco (Pavia), Ed. Prat.* **15**, 407 and 411 (1960).
80. L. Mazor and M. K. Papay, *Acta Pharm. Hung.* **32**, 59 (1962).
81. J. Vondráčková and O. Štrauchová, *J. Chromatog.* **32**, 780 (1968).
82. M. Igloy, A. Mizsei, and I. Horvath, *J. Chromatog.* **20**, 295 (1965)
83. W. T. Sokolski, S. Ullmans, H. Koffler, and P. A. Tetrault, *Antibiot. Chemotherapy* **4**, 1057 (1954).
84. T. M. Lees, J. de Muria, and W. H. Boegmann, *J. Chromatog.* **5**, 126 (1961).
85. H. Bickel, E. Gäumann, R. Hütter, W. Sackmann, E. Vischer, W. Voser, A. Wettstein, and H. Zähner, *Helv. Chim. Acta* **45**, 1396 (1962).
86. K. H. Wallhaeusser, G. Huber, G. Nesemann, P. Praeve, and K. Kepf, *Arzneimittel-Forsch.* **14**, 356 (1964).
87. H. A. Whaley, E. L. Patterson, A. C. Dornbuss, E. J. Backus, and N. Bohonos, *Antimicrobial Agents Chemotherapy* p. 45 (1963).
88. P. V. Divekar, V. C. Vora, and A. W. Khan, *J. Antibiotics (Tokyo)* **A19**, 63 (1966).
89. H. A. Nash and A. R. Smashey, *Arch. Biochem.* **30**, 237 (1951).
90. E. Borowski, H. Chmara, and E. Jareczek-Morawska, *Biochim. Biophys. Acta* **130**, 560 (1966).
91. E. Roets and H. Vanderhaeghe, *Pharm. Tijdschr. Belg.* **44**, 57 (1967).
92. G. R. Delpierre, F. W. Eastwood, G. E. Gream, D. G. I. Kingston, P. S. Sarin, Lord Todd, and D. H. Williams, *J. Chem. Soc., C* p. 1653 (1966).
93. M. Bodanszky and D. Perlman, *Science* **63**, 352 (1969).
94. H. Brockmann and H. Gröne, *Naturwissenschaften* **40**, 222 (1953); *Chem. Ber.* **87**, 1036 (1954).
95. E. Katz and I. M. Pugh, *Appl. Microbiol.* **9**, 263 (1961).
96. E. Katz, *Ann. N. Y. Acad. Sci.* **89**, 304 (1960).
97. I. B. Mironova, V. S. Kuznecova, and T. I. Orlova, *Antibiotiki* **8**, 273 (1963).
98. T. Yoshida, A. Mauger, B. Witkop, and E. Katz, *Biochem. Biophys. Res. Commun.* **25**, 66 (1966).
99. O. Ciferri, A. Albertini, and G. Cassani, *Biochem. J.* **90**, 82 (1964).
100. P. Sensi, A. M. Greco, and R. Ballotta, *Antibiot. Ann.* p. 262 (1959–1960).
101. G. Sferruzza and R. Rangone, *Farmaco (Pavia), Ed. Prat.* **19**, 486 (1964).
102. V. B. Korchagin, V. V. Stepushkina, and Z. E. Voinova, *Antibiotiki* **11**, 107 (1966).
103. G. Katsui, *in* "Chromatography" (E. Heftmann, ed.), p. 699. Reinhold, New York, 1967.
104. G. Barbiroli, *Minerva Dietol.* **7**, 257 (1967).
105. V. Prelog, A. M. Gold, G. Talbot, and A. Zamojski, *Helv. Chim. Acta* **45**, 4 (1962).
106. J. J. Wren, *J. Chromatog.* **4**, 173 (1960).
107. J. W. Copius Peereboom, J. B. Roos, and H. W. Beekes, *J. Chromatog.* **5**, 500 (1961).
108. E. L. Gadsden, C. H. Edwards, and G. A. Edwards, *Anal. Chem.* **32**, 1415 (1960); *J. Chromatog.* **6**, D32 (1961).
109. Z. Deyl, J. Rosmus, J. Blattna, and O. Manousek, *Tech. Publ., Stredisko Tech. Inform. Potravinar. Prymyslu* **139**, 359 (1963).
110. H. T. Gordon, W. W. Thornburg, and L. N. Werum, *J. Chromatog.* **9**, 44 (1962).
111. M. Danilovic, B. Lavica, and Z. Savic, *Arch. Farm. (Belgrade)* **16**, 13 (1966).
112. D. Kusewicz, *Zeszyty Nauk. Politech. Lodz., Chem. Spozywcza* **10**, 27 (1966).
113. H. N. Ridyard, *Analyst* **91**, 328 (1966).

114. O. Schettino, *Boll. Chim. Farm.* **103**, 561 (1964).
115. V. Koen and N. Ruseva, *Farmatsiya (Sofia)* **18**, 4 (1968).
116. H. Y. Neujahr, *Acta Chem. Scand.* **20**, 786 (1966).
117. L. M. Lewin and R. Wei, *Anal. Biochem.* **16**, 29 (1966).
118. K. Matsui, K. Tazoe, and F. Murakami, *Bitamin* **20**, 91 (1960).
119. S. Svobodová-Leblová, J. V. Koštíř, and I. M. Hais, *J. Chromatog.* **14**, 451 (1964).
120. Y. Kato and H. Makida, *Gifu Yakka Daigaku Kiyo* **14**, 38 (1964).
121. K. Maslowski, *J. Chromatog.* **18**, 609 (1965).
122. P. Cerletti, R. Strom, S. Giovenco, D. Barra, and M. A. Giovenco, *J. Chromatog.* **29**, 182 (1967).
123. R. Strom, S. Giovenco, M. G. Giordano, M. A. Giovenco, and A. Marzoli, *Giorn. Biochim.* **14**, 414 (1965).
124. I. G. White and G. J. Lincoln, *Biochem. J.* **76**, 301 (1960).
125. J. D. Taylor and G. J. Proksch, *J. Chromatog.* **27**, 509 (1967).
126. M. Barial, *J. Chromatog.* **28**, 492 (1967).
127. R. M. Creswell and H. C. S. Wood, *J. Chem. Soc.* p. 4768 (1960).
128. K. Iwai, *Bitamin* **27**, 97 (1963).
129. H. Gajcy, *Acta Physiol. Polon.* **13**, 447 (1962).
129a. S. Ikeda, T. Oka, N. Ohishi and S. Fakui, *Bitamin* **38**, 109 (1968).
130. V. W. Rodwell, B. E. Volcani, M. Ikawa, and E. E. Snell, *J. Biol. Chem.* **233**, 1548 (1958).
131. W. R. Gustavson, G. Ledin, Jr., and A. Furst, *J. Chromatog.* **24**, 288 (1966).
132. H. Kraut and U. Imhoff, *Nahrung* **12**, 29 (1968).
133. C. Taruno, *J. Chromatog.* **14**, 201 (1964).
134. M. Covello and O. Schettino, *Farmaco (Pavia), Ed. Prat.* **20**, 581 (1965).
135. H. Kakol, J. Bogucka, and H. Gajcy, *Med. Doswiadczalna Mikrobiol.* **16**, 69 (1964).
136. A. Sauciuc, L. Ionescu, and M. Albu-Budai, *Rev. Chim. (Bucharest)* **18**, 237 (1967).
136a. A. L. Kechkhuashvili, *Soobshch. Akad. Nauk Gruz.* **49**, 357 (1968).
137. H. J. Lichtenthaler, *J. Chromatog.* **13**, 166 (1964).
138. S. Futterman, *J. Biol. Chem.* **237**, 677 (1962).
139. F. B. Jungalwala and H. R. Cama, *J. Chromatog.* **8**, 535 (1962).
140. V. A. Devyatnin and I. A. Solunina, *Molekul. Khromatogr., Akad. Nauk SSSR, Inst. Fiz. Khim.* p. 112 (1964).
141. S. J. Atlas and K. G. Pinter, *Anal. Biochem.* **17**, 258 (1966).
142. F. Brown, *Biochem. J.* **51**, 237 (1952); **52**, 523 (1952).
143. R. L. Lester and T. Ramasarma, *J. Biol. Chem.* **234**, 672 (1959).
144. J. Green, S. Marcinkiewicz, and D. McHale, *J. Chromatog.* **10**, 158 (1963).
145. S. Marcinkiewicz, J. Green, and D. McHale, *J. Chem. Soc., C* p. 1431 (1966).
146. S. Marcinkiewicz and J. Green, *Analyst* **84**, 304 (1959).
147. Report of the Vitamin E Panel, *Analyst* **84**, 356 (1959).
148. A. T. Diplock, J. Green, E. E. Edwin, and J. Bunyan, *Biochem. J.* **76**, 563 (1960).
149. R. F. Bayfield, J. D. Barrett, and R. H. Falk, *J. Chromatog.* **28**, 363 (1967).
150. R. F. Bayfield, R. H. Falk, and J. D. Barrett, *J. Chromatog.* **36**, 54 (1968).
151. E. DeRitter, *J. Assoc. Offic. Agr. Chemists* **48**, 985 (1965).
152. E. Kevei, O. Petro, and J. Szarfoldi, *Elelmiszertudomany* p. 45 (1967).
153. M. Zobel and A. Teutloff, *Aerztl. Lab.* **13**, 293 (1967).
154. H. Sawai, K. Mine, K. Ito, K. Sato, M. Yamazaki, and T. Fujisawa, *Bitamin* **27**, 21 (1963).
155. T. Fujisawa, M. Yamazaki, and H. Sawai, *J. Vitaminol. (Kyoto)* **9**, 236 (1963).
155a. G. I. Kazarinova, *Ref. Zh., Khim.* **19GD**, 1969, Abstract No. 12G223.

155b. J. Pazarincevic and A. F. Damanski, *Qual. Plant. Mater. Vegetabiles* **16** (1–4), 210–214 (1968).
156. B. N. Rapoport and S. B. Grube, *Nauchn. Zap. Gor'kovsk. Nauchn.-Issled. Inst. Dermatol. i Venerol.* p. 31 (1962).
157. M. A. Kassem, A. Sina, and H. A. El-Sourady, *Farmaco (Pavia), Ed. Prat.* **22**, 152 (1967).
158. T. Miki and Y. Sahashi, *Bitamin* **25**, 171 (1962).
158a. G. Gorbach, I. Hekal, and L. Schmid, *Deut. Lebensm.-Rundschau* **64**, 203 (1968).
159. K. S. Szotyori, *Nahrung* **11**, 129, 139, and 421 (1967); *Elelmiszervizsgalati Kozlemen.* **13**, 148 and 209 (1967).

14

MISCELLANEOUS ORGANIC COMPOUNDS

Although this volume is not intended to cover a complete review of the literature of paper chromatography, it seems desirable to describe briefly some work which has been done in fields to which separate chapters have not been devoted. Some overlap in the coverage of compounds types is unavoidable; for example, organic sulfur and nitrogen compounds will be mentioned here, and nitroamines and sulfur amino acids are covered in Chapter 5.

I. Hydrocarbons

A. Olefins

Olefins can be separated by means of addition compounds formed by reacting with mercuric acetate and methanol. A gaseous or liquid mixture is dissolved in methanol to a concentration of 0.5 gm of each compound in 100 ml solvent. Excess solid mercuric acetate is added, and the solution is allowed to stand for 30 minutes (2 hours for higher olefins) at room temperature with frequent stirring. Ten to twenty microliter portions are then spotted on the paper. Lower olefins and diolefins are separated by development with *n*-propanol saturated with a mixture of 1.5 *N* ammonia and 1.5 *N* ammonium carbonate (1:1 v/v). Higher olefins and diolefins are separated on paper impregnated with the ammonia–ammonium carbonate solution and developed with *n*-butanol; some approximate R_F values in this system are: ethylene, 0.1; propylene, 0.15; isobutylene, 0.2; butene-1, 0.3; cyclohexene, 0.45; *n*-pentene-1, 0.5; styrene, 0.55; *n*-hexene-1, 0.7; and isooctene, 0.8. Location of the spots is by drying the paper, liberating the mercuric ions in an HCl atmosphere, and detecting these with $(NH_4)_2S$. Densitometry is useful for quantification (*1*). Mercury adducts of styrols have also been chromatographed in similar systems (*2*).

B. Aromatic Hydrocarbons

1. *Reversed-Phase Chromatography*

Reversed-phase chromatography is suitable for separation of aromatic hydrocarbons (*3*). The paper is impregnated by passing it through a mixture of di-*n*-butyl ether or decahydronaphthalene with benzene (2:1 v/v) and removing the excess between other sheets of filter paper. After the benzene evaporates, spot the hydrocarbons on the impregnated paper in 0.5–2.0 μg amounts from benzene solution in a nitrogen atmosphere. The paper is developed by the ascending technique for 16–18 hours with 0.50 *M* 1-(2,3-dihydroxypropyl)-theobromine (DHPTBr) or 0.05 *M* 1, 3, 7, 9-tetramethyluric acid (TMUA). These compounds act as solubilizing agents for the hydrocarbons through the formation of π-electron complexes. The zones are detected by observing their fluorescence under ultraviolet light. R_F values for a number of compounds may be found in Table I.

TABLE I

R_F Values of Some Aromatic Hydrocarbons (*3*)

Hydrocarbon	R_F [a]			
	1	2	3	4
Anthracene	0.30	0.21	0.22	0.16
Chrysene	Start	Start	Start	Start
1:2;5:6-Dibenzanthracene	0.12	0.05	0.12	0.15
Anthanthrene	0.31	0.16	0.30	0.17
2:3-Benzofluorene	0.17	0.06	0.05	Start
Fluoranthene	0.38	0.26	0.36	0.24
20-Methylcholanthrene	0.11	0.08	0.10	0.06
Pyrene	0.45	0.31	0.37	0.28
3:4-Benzopyrene	0.26	0.14	0.23	0.15
1:2;3:4-Dibenzopyrene	0.19	0.09	0.17	0.11
1:2;4:5-Dibenzopyrene	0.20	0.11	0.20	0.12
1:2;6:7-Dibenzopyrene	Start	Start	Start	Start
3:4;9:10-Dibenzopyrene	0.17	0.07	0.14	0.09

[a] The stationary phase for columns 1 and 2 is di-*n*-butyl ether; for columns 3 and 4 it is decahydronaphthalene. The solvent for columns 1 and 3 is 0.50 *M* DHPTBr; for columns 2 and 4 it is 0.05 *M* TMUA.

Stilbene derivatives have been separated on Whatman No. 1 paper impregnated with 30% formamide in ethanol, 30% kerosene in gasoline, or 30% dimethylformamide in ethanol with solvents composed of cyclohexane–pyridine (5:3 v/v), butanol–pyridine–H_2O (5:3:3 v/v), butanol–10% NH_3 (85:15 v/v), or 80% ethanol. Spraying with 1% *p*-dimethylaminobenzaldehyde causes the formation of colored spots (mostly orange, brown, or yellow), which often fluoresce when viewed under ultraviolet light. The R_F values and structures of thirty-six compounds have been listed (*4*).

The amounts of polycyclic hydrocarbons in basic foodstuffs (*5*) and milk (*6*) have been quantitatively estimated with the aid of partition chromatography on paper.

2. *Chromatography on Acetylated Paper.*

Paper with a 28.0% acetyl content is prepared as follows: sixteen strips of Whatman No. 1 paper (24 × 6 inch) are rolled together, immersed in an acetylating mixture composed of 700 ml redistilled acetic anhydride, 1500 ml thiophene-free benzene, and 3.4 gm 92% H_2SO_4 and allowed to stand for 24 hours in ethanol with occasional stirring. The papers are then washed with more ethanol and then in running water for 6 hours and are finally allowed to stand in distilled H_2O for 2 hours and then air dried. R_F values (*7*) on this paper with two different solvents are given in the tabulation below.

Compound	R_F values[a]	
	A	B
3:4-Benzopyrene	0.06	0.18
Chrysene	0.16	0.40
1:2-Benzanthracene	0.26	0.50
Perylene	0.25	0.53
3:4-Benzophenanthrene	0.35	0.57
Pyrene	0.40	0.64

[a] Solvents (A) methanol–ether–H_2O (4:4:1 v/v); (B) ethanol–toluene–H_2O (17:4:1 v/v).

Compounds are located on these papers by viewing under ultraviolet light; nonfluorescent zones are detected by spraying with tetrachlorophthalic anhydride (a 2% solution in acetone–chlorobenzene, 10:1 v/v) followed by viewing under ultraviolet light. Table II lists R_F values, relative to 3:4-benzopyrene, for other aromatic hydrocarbons on acetylated paper. A recent study indicates that for 3:4-benzopyrene, the relation between R_F and the percentage of acetyl content (x) of the paper is $R_F = ax + b$, where $a = -0.028$ and $b = -0.835$. For other polynuclear hydrocarbons, the values of a and b are different (*7a*).

TABLE II

R_F VALUES ON ACETYLATED PAPER (*7*)

Compound	R_F (relative to 3:4-benzopyrene)[a]	
	A	B
Benzo-(*c*)-phenanthrene	2.96	5.86
1-Methylbenzo-(*c*)-phenanthrene	2.64	5.20
2-Methylbenzo-(*c*)-phenanthrene	3.38	6.95
3-Methylbenzo-(*c*)-phenanthrene	2.90	5.76
4-Methylbenzo-(*c*)-phenanthrene	2.40	3.47
5-Methylbenzo-(*c*)-phenanthrene	3.25	5.84
6-Methylbenzo-(*c*)-phenanthrene	3.66	7.44
1:2-Benzanthracene	2.65	4.44
3-Methyl-1:2-benzanthracene	3.30	5.90
4-Methyl-1:2-benzanthracene	2.82	5.00
5-Methyl-1:2-benzanthracene	2.74	3.80
6-Methyl-1:2-benzanthracene	3.00	4.52
7-Methyl-1:2-benzanthracene	3.45	5.95
9-Methyl-1:2-benzanthracene	1.93	2.94
10-Methyl-1:2-benzanthracene	2.80	3.64
1′-Methyl-1:2-benzanthracene	2.04	2.50
2′-Methyl-1:2-benzanthracene	2.98	5.25
3′-Methyl-1:2-benzanthracene	3.20	5.70
4′-Methyl-1:2-benzanthracene	2.50	3.34

[a] Solvent A: ethanol–toluene–H_2O (17:4:1 v/v); solvent B: methanol–ether–H_2O (4:4:1 v/v).

3. *Applications*

A study has been made of the paper chromatographic behavior of thirty-two aza-heterocyclic hydrocarbons on Whatman 3 MM paper. R_F values are reported for forty-five developing solvents (*8*). 3:4-Benzopyrene has been quantitatively determined by measurement of the total fluorescence intensity of a separated spot with a recording double-beam fluorimeter and reference to a straight line plot of the logarithm of the peak area versus the logarithm of the concentration for standard solutions (*8a*). Styrene in amounts as low as 0.5γ is determined by development with butanol–diethylamine–H_2O (5:4:1 v/v) on a slow paper for 15 hours followed by detection with alcoholic diphenylcarbazide (*8b*).

II. Alcohols

A. Free Alcohols

Monohydric alcohols are usually separated after conversion to derivatives [e.g., 3, 5-dinitrobenzoates, xanthates, nitrophthalates, diphenates (*9*)] because the lower alcohols are volatile and the higher ones are quite difficult to detect on paper with high sensitivity. The series of normal aliphatic alcohols from *n*-heptyl through *n*-dodecyl has been conveniently separated, however, by solubilization chromatography on Amberlite SA-2 cation-echange paper in the sodium form (*10*). An initial zone of mixture containing 5 μl of each undiluted alcohol is developed downward for 36 cm with aqueous acetic acid or methanol in an equilibrated tank. Detection of alcohols after drying the paper is by spraying with vanadium oxinate reagent to yield light red zones on a blue-black background. The reagent is prepared by dissolving 0.40 gm 8-hydroxyquinoline in 25 ml each of xylene and glacial acetic acid and heating on a steam bath to 55°C; 0.20 gm ammonium metavanadate is added with stirring, and the resultant blue-black solution is filtered through paper after cooling. The solution is stable for 3 days. Development of a mixture with 75% methanol resolves all the alcohols except C_7 and C_8. The R_F values are: C_7 plus C_8, 0.93; C_9, 0.75; C_{10}, 0.45; C_{11}, 0.25; and C_{12}, 0.17. The two lower alcohols are separated by separate development with 6.5 *M* acetic acid or 70% methanol.

Compounds containing free hydroxyl groups can be detected after chromatography by selective conversion to dinitrobenzoate esters on the paper (*11*). The air-dried chromatogram, which is first steamed for 2 minutes if a butanol solvent was used, is sprayed uniformly with a 2% solution of 3,5-dinitrobenzoyl chloride in toluene and hung in the air to dry overnight. After this, the paper is steamed for 2 minutes and then drawn repeatedly and quickly back and forth through an aqueous 0.5% solution of NaOH in a crystallizing dish for 30 seconds. The paper is then immediately rinsed in a pan of tap water until the wash water dripping from the paper does not feel slippery. The wet paper is then drawn repeatedly through an aqueous 0.002% solution of rhodamine in a crystallizing dish for 15 seconds and allowed to air dry, after which it is examined under ultraviolet light in the dark. Table III shows that this method provides a sensitive means of detecting many aliphatic and aromatic hydroxy compounds. Also detected, but not shown in the table, are phenols, monosaccharides, carbohydrate derivatives, hydroxy acids, esters, ketones, and amino compounds. The technique does not detect compounds which are either too volatile or too soluble in the dilute NaOH wash solution.

TABLE III

DETECTION[a] OF 1-HYDROXY COMPOUNDS AS 3,5-DINITROBENZOATE ESTERS (*11*)

Compound	R_F	Solvent[b]	Smallest amount detected (μg)
Primary Aliphatic Alcohols			
1-Octanol	0.64	MH	17
3-Phenyl-1-propanol	0.31	MH	1.9
2-Phenoxyethanol	0.86	BW	5.6
o-Hydroxyethylguaiacol	0.11	LW	0.2
Cinnamyl alcohol	0.25	MH	0.6
p-Methoxybenzyl alcohol	0.13	MH	0.6
Secondary Aliphatic Alcohols			
Benzhydrol	0.34	MH	0.6
Benzoin	0.23	MH	1.9
Cholesterol	0.90	MH	50
Tertiary Aliphatic Alcohol			
Triphenylmethanol	0.57	MH	1.9
Glycols			
Ethylene glycol	0.52	BW	5.6
Diethylene glycol	0.62	BW	1.9
Triethylene glycol	0.63	BW	1.9
Polyhydroxy Compounds			
Glycerol	0.35	BW	1.9
Pentaerythritol	0.36	BW	0.6
Sugar Alcohols			
Dulcitol	0.09	BW	0.6
Mannitol	0.08	BW	0.6
Sorbitol	0.08	BW	0.6
Dulcitol	0.23	BAW	0.6
Mannitol	0.21	BAW	0.6
Sorbitol	0.19	BAW	0.6

[a] Not detected: *n*-butanol, 1- and 2-phenylethanol, 2-methoxyetnanol, benzyl alcohol, α-terpineol, polyethylene glycol 400.

[b] MH = heptane saturated with methanol; LW = petroleum ether saturated with water; BW = *n*-butanol saturated with water; BAW = *n*-butanol–acetic acid–H_2O (4:1:5 v/v).

B. ALCOHOL DERIVATIVES

1. *3,5-Dinitrobenzoates*

Methods for preparing the 3,5–dinitrobenzoates of alcohols have already been published (*9, 12*). These derivatives are spotted from benzene solution on paper moistened with 50% dimethylformamide (DMF) in acetone. The paper is air dried until just damp and developed with decalin saturated with DMF in a tank saturated with DMF vapor.

After development, the dry paper is sprayed with 0.2% aqueous $SnCl_2$ in concentrated HCl, dried again for 1 hour, and sprayed with 0.3% ethanolic *p*-dimethylaminobenzaldehyde (DMAB) in concentrated HCl. The alcohol benzoates appear as yellow spots which are cut out, extracted with 2 ml of 0.1% ethanolic DMAB in concentrated HCl, and assayed by colorimetry (*13*). This and a similar procedure (*14*) have been used to determine alcohols in brandy and other liquors; methyl, ethyl, propyl, isobutyl, isopentyl, hexyl, and heptyl alcohols were found.

2. *Xanthates*

The alcohols are converted into their respective xanthates by dissolving 0.10 mg of KOH in 0.10 ml of the sample and adding 0.15 ml of CS_2. The precipitate is dissolved in 5 drops of acetone and the solution applied to Whatman No. 3 MM or S. and S. 2043 paper for descending development for 6–8 hours with butanol–pentanol–H_2O–5% aqueous NH_3 (5:5:5:9 v/v) or butanol–H_2O–5% aqueous NH_3 (10:9:1 v/v). The spots are revealed with ammonium molybdate solution. Spot areas are proportional to the logarithm of the alcohol content. Separation and estimation within $\pm 10\%$ is obtained for C_1 to C_5 alcohols (*15*).

Alkyl xanthates must be chromatographed in solvent systems that are sufficiently alkaline to avoid their destruction. *n*-Butanol saturated with 2% KOH, 5% $NaHCO_3$, concentrated NH_3, or 30% K_2CO_3, or *n*-butanol saturated with water on paper impregnated with pH 12 phosphate buffer are often used (*16*) (Fig. 1).

The separation of both xanthates and 3,5-dinitrobenzoates of higher aliphatic alcohols has been studied in great detail and the effects of many variables on the determination of these compounds have been enumerated (*17*).

3. *Mercury Adducts*

Adducts formed by monohydric alcohols with Hg(II) acetate plus allyl methyl ether may be separated by ordinary (for C_1–C_8 alcohols) or reversed-phase paper chromatography (C_1–C_{18}). Detection is as described above for olefin adducts. The method has a sensitivity of 0.01–0.5 γ of alcohol. Procedures for the formation of the adducts and the selection of suitable chromatographic systems have been presented in detail (*18*).

4. *Alkyl 2,4-Dinitrobenzyl Ethers* (*18a*)

Aliphatic alcohols are separated in the form of these derivatives on Whatman No. 2 paper in the following systems: (1) paper impregnated with a 50% solution of dimethylformamide (DMF) in methanol and developed with cyclohexane; (2) paper impregnated with DMF and

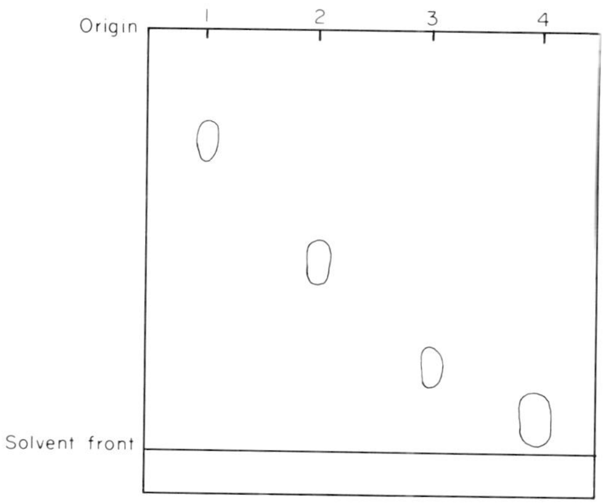

FIG. 1. Migration of alkylxanthates in *n*-butanol saturated with 30% K_2CO_3 solution, paper impregnated with 10% K_2CO_3 solution. (1) Methylxanthate; (2) ethylxanthate; (3) *n*-propylxanthate; (4) dodecylxanthate. After J. Gasparič and J. Borecký (*16*).

developed with cyclohexane–benzene (25:1 v/v); (3) paper impregnated with a 3% solution of paraffin oil in hexane and developed with DMF–methanol–H_2O (3:2:1 v/v). Detection is by the Janovsky reaction (4 gm sodium are dissolved in 50 ml ethanol and diluted with an equal volume of acetone before spraying) or Ehrlich reagent after previous reduction of the nitro groups by reaction with $SnCl_2$. The R_F values in each system are tabulated below.

Alkyl 2,4-dinitrobenzyl ethers	R_F		
	1	2	3
Methyl	0.20	0.31	—
Ethyl	0.35	0.51	—
n-Propyl	0.50	0.67	—
Isopropyl	0.49	—	—
n-Butyl	0.62	0.78	—
Isobutyl	0.61	—	—
sec-Butyl	0.61	—	—
n-Amyl	0.72	0.85	—
Isoamyl	0.71	—	—
n-Hexyl	0.80	—	—
n-Heptyl	0.86	—	—
n-Octyl	—	—	0.67
n-Nonyl	—	—	0.57
n-Decyl	—	—	0.44
n-Dodecyl	—	—	0.23

C. Glycols and Polyhydroxy Compounds (see Table III)

1. *Solvents*

Noncondensed glycols are separated by ascending development with H_2O–ethyl acetate–ethanol (1:12:2 v/v), in which the following R_F values are obtained: glycerol, 0.22; ethylene glycol, 0.55; diethylene glycol, 0.63; propylene glycol, 0.72; sorbitol, 0.1–0.19; inositol, 0.0 (Fig. 2). Polyglycols are resolved by two-dimensional development with butanol–acetic acid–H_2O (4:1:5 v/v) followed by ammoniacal butanol (*19*).

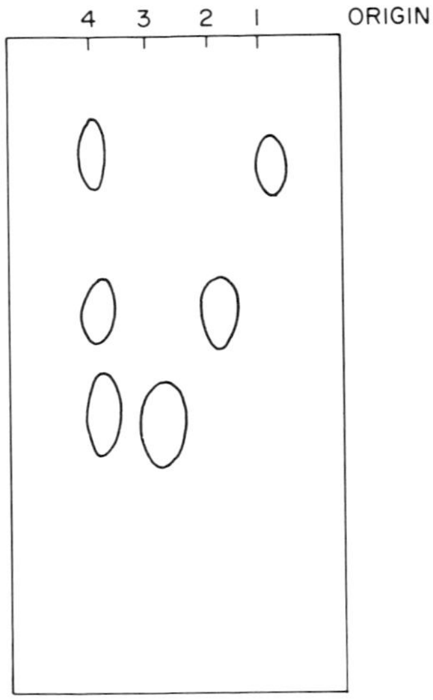

Fig. 2. Separation of polyvalent alcohols in ethyl acetate–ethanol–H_2O (12:2:1 v/v). (1) Glycerol, (2) ethylene glycol, (3) 1,2-propylene glycol, (4) mixture of 1–3. After K. Macek (*12*).

Other recommended solvents (*20*) are the organic layers of the following mixtures, which are shaken and settled overnight: (1) acetone–benzene–H_2O (10:10:1 v/v); (2) butyl acetate–H_2O (1:1 v/v); (3) butanol–H_2O (1:1 v/v).

2. *Detection*

The following spray reagents have been used to detect polyols on paper chromatograms (*19*, *20*):

1. Ammoniacal $AgNO_3$ plus heating at 100°C for 2–5 minutes —black spots on a white background.

2. Benzidine periodate—blue spots.
3. Potassium iodobismuthate (sensitivity 1–3 γ of polyglycol) —orange-rose spots.
4. Diazotized *p*-nitroaniline hydrochloride followed by 15 % NaOH.
5. Isovanillin–*p*-toluenesulfonic acid.
6. *p*-Dimethylaminobenzaldehyde–*p*-toluenesulfonic acid.

3. *Quantitative Determination*

Twenty to two hundred micrograms of polyglycerols are applied to Whatman No. 1 paper and developed downward with butanol–acetic acid–H_2O (12:3:4 v/v) for 12–15 hours at 20°C. After drying at 105°C, the paper is sprayed with $AgNO_3$ solution in acetone followed by 2 *N* ethanolic KOH and $Na_2S_2O_3$ solution. Spot areas are used for quantitative estimation of the series as far as heptaglycerol; 2 μg of glycerol and 15–25 μg of the higher glycerols can be detected (*21*).

Glycerol in culture fluid has been estimated by chromatography and planimetry of spots detected with $AgNO_3$ (*22*) and plasma glycerol by chromatography, elution, and spectrophotometry (*23*).

D. Miscellaneous Alcohols

Methods have been described for the separation and identification of aliphatic 3-chloro-2-hydroxypropyl ethers. The ethers are separated on impregnated paper (for example with formamide) with mixed solvents (e.g., $CHCl_3$–cyclohexane, 22:10 or 1:1 v/v) and then converted on the paper to glycidyl derivatives which are detected by treatment with heated vapors of tertiary heterocyclic amines such as pyridine (*24*).

Two-phase systems, based on the salt-induced separation of miscible solvents, are proposed for the separation of amino acids, alkaloids, azo dyes, etc. A series of sugar alcohols were separated with acetone–sodium acetate–H_2O (*25*).

1-Methoxy-2(3)-hydroxy-3(2)-bromocyclohexanes and their derivatives, including bromodiols and diols, have also been separated by paper chromatography (*26*).

III. Oxo Compounds

A. Uncombined Ketones

Aliphatic aldehydes and ketones are preferably chromatographed as derivatives, e.g., 2,4-dinitrophenylhydrazones. Aromatic oxo compounds can be directly chromatographed, for example by solubilization chromatography on Amberlite SB-2 quaternary ammonium anion-exchange paper in the chloride form (*27*). Table IV shows R_F values

TABLE IV

VALUES OF R_F FOR UNCOMBINED KETONES (*27*)[a]

Ketone[b]		R_F[c] 8.0	10.0	12.0	14.0	16.0	18.0
1. Phenyl-2-propanone	a	0.52	0.61	0.64	0.80	0.86	0.88
	b	0.41	0.48	0.54	0.69	0.73	0.80
2. 4′-Methylacetophenone	a	0.39	0.40	0.60	0.69	0.74	0.83
	b	0.30	0.33	0.48	0.62	0.67	0.76
3. *trans*-4-Phenyl-3-buten-2-one	a	0.28	0.36	0.49	0.63	0.76	0.81
	b	0.17	0.21	0.34	0.51	0.64	0.69
4. 1-Phenyl-1,3-butanedione	a	0.10	0.19	0.28	0.53	0.58	0.67
	b	0.04	0.08	0.16	0.35	0.46	0.54
5. Hexanopenhone	a	0.05	0.09	0.40	0.50	0.61	0.74
	b	0.00	0.00	0.08	0.37	0.44	0.64
6. Penyl 2-thienyl ketone	a	0.06	0.11	0.25	0.38	0.46	0.55
	b	0.00	0.00	0.14	0.28	0.38	0.47
7. 4′-Phenylacetophenone	a	0.05	0.06	0.12	0.26	0.38	0.52
	b	0.00	0.00	0.00	0.00	0.28	0.40
8. 2-Tridecanone	a	0.01	0.01	0.05	0.19	0.36	0.64
	b	0.00	0.00	0.00	0.00	0.00	0.23

[a] Amberlite SB-2 anion-exchange paper (Reeve Angel Co.).

[b] a = R_F front; b = R_F rear.

[c] Figures at heads of columns indicate molarity of aqueous methanol solvent.

obtained when 5 μl of 15% (w/v) solutions of each ketone in methanol are developed with aqueous methanol solutions by downward development in saturated tanks for 1.5–3.0 hours (about 34 cm). Detection is by spraying the wet papers with freshly prepared 3,5-dinitrobenzoic acid solution (1 gm of acid is dissolved in 20 ml methanol and then 10 ml of 15% NaOH is added) to yield blue-purple zones on a yellow background. Every pair of ketones, except numbers 7 and 8, may be separated by development with an appropriate concentration of methanol.

B. 4-NITROPHENYLHYDRAZONES

Table V shows R_F values for 4-nitrophenylhydrazine derivatives of aromatic carbonyl compounds on paper impregnated with formamide or dimethylformamide and developed upward with various solvents (*28*). The derivatives are prepared as described elsewhere (*29*). For quantitative determination, the spots are cut out of the paper and extracted in micro soxhlet extractors with 95% ethanol for 4–6 hours. The extracts are evaporated to an appropriate volume for measurement by ultraviolet absorption spectroscopy (*28*). A series of hydroxybenzophenone derivatives have been separated in similar chromatographic systems (*30*).

TABLE V

R_F VALUES OF 4-NITROPHENYLHYDRAZONES OF AROMATIC CARBONYL COMPOUNDS[a] *(28)*

Compound	R_F				Color
	I_1S_1	I_2S_2	I_3S_2	I_4S_3	
Benzaldehyde	0.43	0.44	0.41	0.37	Rust
Salicylaldehyde	0.07	0.34	0.27	0.06	Rust
m-Hydroxybenzaldehyde	0.04	0.24	0.12	0.02	—
p-Hydroxybenzaldehyde	0.03	0.23	—	0.02	Deep purple
o-Methoxybenzaldehyde	0.29	0.30	—	—	Light purple
Anisaldehyde	0.27	0.26	—	—	Pink
2-Hydroxy-3-methoxy-benzaldehyde	0.04	0.20	—	—	Purple
Vanillin	0.03	0.03	—	—	—
Veratraldehyde	0.08	0.11	—	—	Purple
Piperonal	0.17	0.24	—	—	Orange
Acetophenone	0.56	0.45	0.45	0.43	Purple-red-purple
2-Hydroxyacetophenone	0.05	0.35	0.38	0.07	Purple
3-Hydroxyacetophenone	0.05	0.24	0.15	0.02	—
4-Hydroxyacetophenone	0.05	0.03	0.13	0.02	Lavender
3,4-Dimethylacetophenone	0.72	0.53	0.50	0.29[b]	Deep pink
2,4-Dimethylacetophenone	0.86	0.50	0.20	0.41[b]	—
2,5-Dimethylacetophenone	0.81	0.50	—	0.37[b]	—
Benzalacetophenone	—	0.54	—	—	—
Propiophenone	0.76	0.44	0.46	0.26[b]	Rust
Butyrophenone	0.84	0.60	0.50	0.46	Rust
Benzophenone	0.94	0.58	—	0.94	Brown
o-Hydroxybenzophenone	0.89	0.14	—	0.15	—
	0.22	0.47	—	0.72	—
m-Hydroxybenzophenone	0.93	0.54	—	0.83	—
p-Hydroxybenzophenone	0.06	0.14	0.27	0.03	Deep pink
2-Hydroxy-5-methyl-benzophenone	0.09	0.34	—	—	—
2-Naphthaldehyde	0.02	0.08	—	—	Brown
Cinnamaldehyde	0.60	0.43	—	0.07	—
Furfural	0.08	0.33	—	—	—
5-Methyl-2-furfural	0.14	0.40	—	—	—

[a] Paper: Schleicher and Schuell 2043 B gl. Impregnation: I_1 = 35% solution of formamide in ethanol; I_2 = 50% solution of dimethylformamide; I_3 = 25% solution of dimethylformamide; I_4 = 20% solution of formamide. Solvents: S_1 = cyclohexane–benzene–dipropylene glycol (70:30:3 v/v); S_2 = dibutyl ether–*N*,*N*-dimethylformamide–tetrahydrofuran (85:15:4 v/v); S_3 = cyclohexane–benzene–formamide (15:12:7 v/v). Detection: potassium hydroxide.

[b] Smears.

C. Quinones

Anthraquinones can be readily detected by their fluorescence in ultraviolet light with sensitivities as low as 0.1 μg. Reversed-phase chromatographic systems can often be used (*31*) (see Table VI), as can paper impregnated with polar organic solvents such as dimethylformamide (*32*). Anthraquinone derivatives may be determined spectrophotometrically after elution from the paper.

TABLE VI

R_F Values of Some Anthraquinone Derivatives[a] (*31*)

Compound	R_F
Anthraquinone	0.57
1-Fluoroanthraquinone	0.81
2-Fluoroanthraquinone	0.75
1-Chloroanthraquinone	0.64
2-Chloroanthraquinone	0.50
1-Bromoanthraquinone	0.64
2-Bromoanthraquinone	0.55
1-Iodoanthraquinone	0.59
2-Iodoanthraquinone	0.50
1-Chloro-2-fluoroanthraquinone	0.65
1-Chloro-2-bromoanthraquinone	0.31
1-Chloro-2-iodoanthraquinone	0.28
1-Fluoro-2-chloroanthraquinone	0.71
1-Bromo-2-chloroanthraquinone	0.33
1-Iodo-2-chloroanthraquinone	0.30
1-Chloro-5-fluoroanthraquinone	0.75
1-Chloro-5-bromoanthraquinone	0.70
1-Chloro-5-iodoanthraquinone	0.68
2-Chloro-3-fluoroanthraquinone	0.44
2-Chloro-3-bromoanthraquinone	0.19
2-Chloro-3-iodoanthraquinone	0.17

[a] Paper: Soviet paper M. Impregnation: 5% paraffin oil in light petroleum. Solvent: 60% acetic acid saturated with paraffin oil. Detection: ultraviolet light.

tert-Alkyl benzoquinones and hydroxybenzoquinones are separated in Whatman No. 1 paper impregnated with 40% dimethylformamide in benzene and developed with heptane or butyl acetate–heptane (4:1 v/v). Detection is by reduction with H_2S and then spraying with ferric ferricyanide (*32a*).

D. AZINES (*33*)

Table VII summarizes mean R_F values (12–15 runs) of various aromatic carbonyl azines which were produced by reacting the carbonyls with hydrazine or hydrazine dihydrochloride. Ascending development at 20°C was employed, and detection of the spots was by means of ultraviolet light or by spraying with one of the following reagents: (a) 10% $FeCl_3$–1% potassium ferricyanide (1:1 v/v), freshly mixed; (b) 5% aqueous tetraethylammonium hydroxide; (c) 0.1% *p*-dimethylaminocinnamaldehyde in 95% ethanol plus 1% HCl by volume.

TABLE VII

PAPER CHROMATOGRAPHIC PROPERTIES OF AZINES OF AROMATIC CARBONYLS[a] (*33*)

Substance	$R_F \times 100$						
	A	B	C	D	E	F	G
Benzalazine	54	65	56	75	—	83	42
2-Hydroxybenzalazine	2	2	3	53	21	2	—
Salicylaldazine	31	23	24	45	—	56	—
2-Methoxybenzalazine	31	36	41	67	55	69	28
Anisaldazine	22	21	22	37	—	49	17
Vanillin azine	0	—	—	0	11	2	—
4-Carboxybenzalazine	0	—	—	—	—	—	—
2-Hydroxy-3-methoxybenzalazine	3	—	4	41	32	12	—
2,3-Dimethoxybenzalazine	20	—	—	64	61	68	—
4-Acetoaminobenzalazine	0	—	—	1	11	1	—
4-Dimethylaminobenzalazine	8	—	—	22	44	79	—
Veratraldazine	4	—	—	4	13	13	—
3,4,5-Trimethoxybenzalazine	4	—	—	30	17	16	—
Piperonal azine	9	9	8	41	37	32	—
Acetophenone azine	62	—	—	80	—	92	52
2-Hydroxyacetophenone azine	33	44	38	64	59	68	—
4-Hydroxyacetophenone azine	0	—	—	0	17	2	—
2,5-Dihydroxyacetophenone azine	2	—	—	5	31	1	—
2,4-Dihydroxyacetophenone azine	—	—	—	3	8	—	—
2,6-Dihydroxyacetophenone azine	—	—	—	5	5	—	—
4-Aminoacetophenone	—	—	—	—	4	—	—
4-Aminoacetophenone azine hydrochloride	3	—	—	2	—	—	—
2,4-Dimethylacetophenone azine	84	—	4	97	—	—	58
2,5-Dimethylacetophenone azine	86	—	—	96	—	95	67
3,4-Dimethylacetophenone azine	79	—	—	92	—	95	57
Propiophenone azine	78	84	75	91	—	—	60

Table continued

TABLE VII (*continued*)

Substance	$R_F \times 100$						
	A	B	C	D	E	F	G
Butyrophenone azine	86	—	—	92	—	98	73
Benzophenone azine	57	70	55	85	71	95	42
4-Hydroxybenzophenone azine	0	—	—	43	27	7	—
2-Hydroxy-5-methylbenzophenone azine	41	—	—	94	70	93	—
2-Hydroxy-4-methoxybenzophenone hydrazone	19	—	—	—	—	20 49	—
2-Hydroxy-4-methoxybenzophenone azine	12	—	—	92	42	71	—
2,4-Dihydroxybenzophenone azine	—	—	—	—	5	—	—
Terephthaladazine	30	—	—	—	14 74	—	—

[a] Paper: Schleicher and Schuell a gl. Conditions: A = paper impregnated with 75% *N,N*-dimethylformamide–25% acetone, *N,N*-dimethylformamide–cyclohexane (20:80 v/v) solvent; B = paper impregnated with 50% *N,N*-dimethylformamide–50% acetone, *N,N*-dimethylformamide–cyclohexane (20:80 v/v) solvent; C = paper impregnated with 50% *N,N*-dimethylformamide–50% acetone, Decalin–*N,N*-dimethylformamide (80:20 v/v) solvent; D = paper impregnated with 75% *N,N*-dimethylformamide–25% acetone, cyclohexane–*N,N*-dimethylformamide–ethyl acetate–H_2O (120:120:20:40 v/v) solvent; E = paper impregnated with 75% *N,N*-dimethylformamide–25% acetone, dibutyl ether–*N,N*-dimethylformamide–tetrahydrofuran (85:15:4 v/v) solvent; F = paper impregnated with 25% *N,N*-dimethylformamide–75% ethyl alcohol (95%), cyclohexane–*N,N*-dimethylformamide–carbon tetrachloride (20:1:4 v/v) solvent; G = paper impregnated with 10% methylformanilide–90% ethyl alcohol (95%), cyclohexane–methylformanilide (4:1 v/v) solvent.

E. 2, 4-Dinitrophenylhydrazones

These derivatives are most widely employed. They are separated on unmodified paper using petroleum ether (100–115°C)–diethyl ether (95:5 v/v), water saturated (*34*); *n*-butanol–3% aqueous NH_3 (1:1 v/v) and Veronal buffer (pH 8.6) (for keto acids, *35*); or butanol–propanol–NH_3 (28%)–H_2O (6:1:2:1 v/v) (*36*) as the mobile phase. Aldehyde hydrazones have been separated on paper impregnated with dimethylformamide (DMF) and developed with cyclohexane saturated with DMF (*37*) (Fig. 3) or petroleum ether–hexane (3:1 v/v) saturated with DMF (*38*) (Fig. 4). The separation of C_1–C_8 aldehyde and ketone 2,4-dinitrophenylhydrazones has been effected using a 10 cm development with nitroethane–acetonitrile–H_2O–methanol (1:2:3:3 v/v) on paper coated with light mineral oil, or development with light mineral oil–

isooctane (1:2 v/v) on untreated paper in a methanol-saturated chamber. Two-dimensional development using the latter system first is especially effective (*38a*).

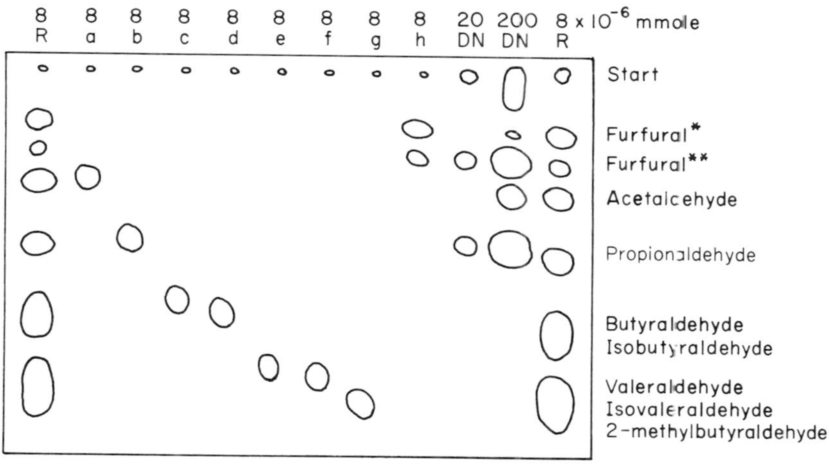

FIG. 3. Paper chromatogram of pure 2,4-dinitrophenylhydrazones of (a) acetaldehyde, (b) propionaldehyde, (c) butyraldehyde, (d) isobutyraldehyde, (e) valeraldehyde, (f) isovaleraldehyde, (g) 2-methylbutyraldehyde, (h) furfural, (R) their mixture, and (DN) 2,4-dinitrophenylhydrazine. The solvent was cyclohexane saturated with dimethylformamide on Whatman No. 3 paper pretreated with dimethylformamide. The single asterisk indicates the trans form of furfural hydrazone, and the double asterisk indicates the cis form. After P. Ronkainen (*37*).

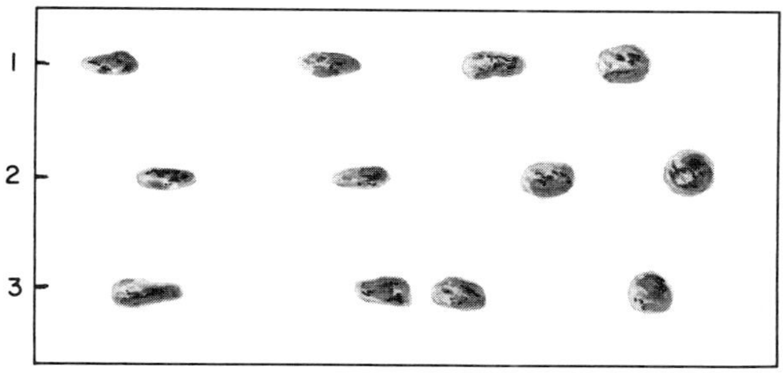

FIG. 4. Chromatogram of 2,4-dinitrophenylhydrazones of substituted acetaldehyde derivatives developed with hexane–petroleum ether (1:3 v/v) saturated with dimethylformamide on Whatman No. 4 paper impregnated with dimethylformamide. (1) phenyloxy-, propoxy-, amyloxy-, and heptyloxyacetaldehyde; (2) tolyloxy-, *p-tert*-butylphenyloxy-, hexyloxy-, and nonyloxyacetaldehyde; (3) benzyloxy-, butoxy-, cyclohexyloxy-, and octyloxyacetaldehyde. After V. Ulbrich and J. Makeš (*38*).

2,4-Dinitrophenylhydrazones are naturally yellow in color and appear as dark spots under ultraviolet light. Increased sensitivity (to 1–3 μg) is often attained by spraying with 10% KOH or 1% alcoholic NaOH, which turns the spots to brown or violet.

Mobility data for aliphatic and cyclic oxo-2,4-dinitrophenylhydrazones has already been published (*39*, *40*), as has an elaborate scheme which affords the separation of a complete mixture of four homologous series of oxo compounds as well as the identification of an unknown member of one of these series (*41*).

The light-absorption maxima of the 2,4-dinitrophenylhydrazones of seven classes of aliphatic carbonyl compounds separated in four chromatographic systems have been directly measured on paper and are useful for differentiation between these classes (*42*).

Quantitative analysis usually involves elution of the separated hydrazones from the paper with ethanol or methanol followed by spectrophotometry at 360–390 mμ (*39*, *43*).

F. Applications

Determinations of oxo compounds via paper chromatography have been made on the following materials: fruits (*43*), bark (*44*), tea (*45*), hops (*46*), polluted air (*47*), sherry (*48*), and essential oils (*49*). In most cases, the carbonyls are converted to 2,4-dinitrophenylhydrazones before chromatography.

IV. Organic Peroxides

A. Aliphatic Peroxides (*50*)

Because of the volatility of many of these compounds, development of the paper between two glass plates clamped together by large spring clips in a tank saturated with vapors of the mobile phase is employed.

The paper (Whatman No. 3 MM) is either (a) untreated, (b) treated with a 5% (v/v) solution of silicone oil in petroleum ether (80–100°C) and oven dried at 110°C for 1 hour, or (c) treated with a solution of ethylene glycol in acetone (5 or 20% v/v) and dried in a stream of air. With the treated paper, it is necessary to cover the glass plates with silicone solution and allow the petroleum ether to evaporate before use in order to get proper solvent flow. Samples are transferred to the paper by means of a heated platinum loop.

After the run, the paper is removed from between the plates, the solvent front is marked, and the paper is dried in a stream of air for 1–2 minutes and immediately sprayed with one of the following reagents:

a. *Ferrous thiocyanate.* Dissolve 0.7 gm ferrous ammonium sulfate in 10 ml of a solution prepared to contain 5 gm ammonium thiocyanate and 1 ml concentrated H_2SO_4 in 100 ml H_2O—Hydrogen peroxide and organic peroxides yield red spots.

b. *p-Phenylenediamine* and acetaldehyde in aqueous acetic acid—spots are pale red.

c. *o-Tolidine.* A dilute solution containing ferrous ammonium sulphate in aqueous acetic acid; enclose paper between plates immediately after spraying;—blue-green color with compounds containing the —OOH group.

d. *HI.* A 10% (v/v) solution in glacial acetic acid—black or brown spots on a light brown background.

A solvent for use with untreated paper is H_2O–ether–*n*-butanol (1:10:10 v/v), in which R_F values are H_2O_2, 0.50; hydroxyperoxides, 0.95; and hydroperoxides, 1.0. Solvents for the development of chromatograms utilizing treated papers are shown in Table VIII (*50, 51*).

Dialkyl peroxides are separated in the same way (*52*) on silicone-treated paper using H_2O–methanol solvents; spraying with ferrous thiocyanate gives red spots with as little as 100 μg of these compounds. R_F values are tabulated below.

	R_F	
	10% H_2O in methanol	15% H_2O in methanol
Di-*n*-hexyl peroxide	0.43	0.21
Dicyclohexyl peroxide	—	0.44
Di-*n*-pentyl peroxide	0.58	0.35
Di-*n*-butyl peroxide	0.70	0.55
Alkyl hydroperoxides	1.0	1.0

B. Peroxybenzoic Acids (*53*)

Isomeric monosubstituted peroxybenzoic acids are separated on acetylated paper (25% acetylation, from Binzer, Germany) or on paper impregnated by standing overnight in a 20% solution of *N*-methylformamide in acetone and dried in air (Table IX). The papers are allowed to equilibrate for 30 minutes in the tank before immersion in the mobile phase for ascending development. For spot location, the chromatogram is sprayed with *p*-aminodimethylaniline hydrochloride solution (1 gm reagent, 1 ml glacial acetic acid, 10 ml ethanol, 10 ml H_2O) to yield red

TABLE VIII

R_F VALUES OF ORGANIC PEROXIDES (*50, 51*)

Compound	R_F [a]				
	I	II	III	IV	V
Hydrogen peroxide	0.05	0	0	0.68	0.83
Methyl hydroperoxide	0.36	0	0	—	—
Ethyl hydroperoxide	0.38	0	0	—	—
tert-Butyl hydroperoxide	Spreads across front	0.35	0.22	—	—
n-Butyl, *sec*-butyl, isobutyl hydroperoxide	—	—	0.31	—	—
tert-Amyl hydroperoxide	—	—	0.43	—	—
n-Heptyl hydroperoxide, 3-methylhexyl-3-hydroperoxide	—	—	0.80	—	—
Cyclohexane hydroperoxide	—	—	0.30	—	—
Cyclohexene hydroperoxide	—	—	0.14	—	0.46
n-Heptane 3,3′-dihydroperoxide	—	—	0.07	—	—
2,5-Dimethylhexane–2,5-dihydroperoxide	—	—	0.20	—	—
Cumene hydroperoxide	—	—	0.53	0.82	—
Peracetic acid	0.26	0.05	0	—	—
Perpropionic acid	0.54	0.16	0.11	—	—
Per-*n*-, per-isobutyric acid	—	0.37	0.30	—	—
Per-*n*-valeric acid	—	0.57	0.50	—	—
Per-*n*-hexoic acid	—	0.71	—	—	—
Percrotonic acid	—	0.23	—	—	—
Diacetyl peroxide	0.37	0.18	—	—	—
Hydroxymethyl hydroperoxide	0.09	0	0	—	—
2-Hydroxyethyl hydroperoxide	0.22	0	0	—	—
Dihydroxymethyl peroxide	0.25	—	—	—	—
tert-Butyl, 2-hydroxyethyl peroxide	—	—	0.55	0.80	—
n-Heptyl, 2-hydroxyethyl peroxide	—	—	0.92	—	—
Cumyl, 2-hydroxy-2-phenylethyl peroxide	—	—	0.76	—	—
tert-Butyl perbenzoate	—	—	1.0	—	—
Di-*n*-heptyl peroxide	—	—	0.95	0.018	—
$(Me_3COO)_2CHMe$	—	—	0.95	0.49	—
Di-2-hydroxyethyl peroxide	—	—	—	0.88	—
Tetrahydrofuran hydroperoxide	—	—	—	—	0.76
Benzoyl peroxide	—	—	—	—	0.19
Isochroman peroxide	—	—	—	—	0.22
Succinyl peroxide	—	—	—	—	0.82
Isochroman hydroperoxide	—	—	—	—	0.50
Tetralin hydroperoxide	—	—	—	—	0.36

[a] I = stationary phase, glycol from a 5% solution; mobile phase, *n*-butanol in petroleum ether (80–100°C) (1:9 v/v). II = stationary phase, glycol from a 20% solution; mobile phase, chloroform in petroleum ether (80–100°C) (1:1 v/v). III = stationary phase, same as II; mobile phase, diethyl ether in petroleum ether (80–100°C) (5:95 v/v). IV = stationary phase, silicone from 5% solution; mobile phase, H_2O–ethanol–chloroform (20:17:2 v/v). V = stationary phase, acetylated paper; mobile phase, ethyl acetate–dioxane–H_2O (20:45:46 v/v).

zones, or with KI–starch (5 ml glacial acetic acid, 5 ml saturated KI solution and 5 ml 5% starch solution) to yield permanent black zones. Unsaturated fatty acid peroxides (*54*) and synthetic hydroxyhydroperoxides of thymine (*55*) have also been separated by paper chromatography.

TABLE IX

R_b VALUES[a] OF ISOMERIC MONOSUBSTITUTED PEROXYBENZOIC ACIDS (*53*)

Peroxybenzoic acid	R_b values[b] I	II
o-Chloro	0.51	1.17
m-Chloro	0.91	1.30
p-Chloro	0.80	1.42
o-Bromo	0.66	1.06
m-Bromo	1.23	1.20
p-Bromo	1.14	1.42
o-Nitro	0.11	0.33
o-Fluoro	0.57	0.89
p-Fluoro	0.10	1.06
m-Methyl	1.57	1.53
p-Methyl	1.31	1.36

[a] $R_b = \frac{\text{migration distance of substituted peroxybenzoic acid}}{\text{migration distance of peroxybenzoic acid}}$; the R_F of peroxybenzoic acid is usually 0.35.

[b] I = 25% acetylated paper; solvent, hexane saturated with dioxane and dimethylformamide. II = *N*-methylformamide-impregnated paper; solvent, benzene.

V. Nitro Compounds

A. NITROFURAN DERIVATIVES (*56*)

These substances are dissolved in alcohol by heating at a concentration of 1 mg/ml, and 0.02–0.04 ml samples are applied to S. and S. 2043b Mgl paper which is either untreated, or impregnated with a 30% methanolic solution of dimethylformamide (DMF), blotted between two filter papers and air-dried for 15 minutes. The papers are equilibrated with the solvent in the chamber (overnight for untreated paper, 30 minutes for DMF paper) and then developed by the descending technique.

Spots on the dried chromatograms are visible in daylight, in ultraviolet light, or after spraying with a 1% solution of *p*-dimethylaminobenzaldehyde in 8% HCl to yield yellow, red, or orange colored zones.

Solvents with a wide range of pH and polarity characteristics are appropriate for use with untreated paper: *n*-butanol–acetic acid–H_2O (4:1:5 v/v); *n*-butanol–2.5% NH_3 (1:1 v/v); butyl acetate–H_2O (1:1 v/v); carbon tetrachloride–acetic acid (50:1 v/v); benzene–methanol–H_2O (2:1:1 v/v); and petroleum ether (60–70°C)–methanol–H_2O (2:1:1 v/v). With DMF-impregnated paper, chloroform, benzene, or cyclohexane–benzene (1:4 v/v) are recommended.

B. Nitrophenols

1. *Separations on Untreated Paper* (*57*)

Isomeric mononitrophenols are separated on Whatman No. 1 paper equilibrated for 1 hour with the aqueous phase of benzene–diethylamine–H_2O (3:2:5 v/v) and then developed (ascending) with the organic phase. The nitrocompounds are detected visually in daylight or in ultraviolet light or are sprayed with diazotized sulphanilic acid. R_F values in the system are as follows: *o*-nitrophenol, 0.44; *m*-nitrophenol, 0.65; *p*-nitrophenol, 0.10; 2, 4, 6-trinitrophenol, 0.89.

2. *Separations on Impregnated Paper* (*58*)

Nitro- and nitrosophenols are chromatographed on papers impregnated with α-bromonaphthalene and developed with α-bromo–naphthalene–saturated 80% acetic acid by the ascending method. Nitro compounds give lemon yellow to red-orange spots by spraying with 1% KOH in alcohol–H_2O (1:1 v/v); nitroso compounds give blue spots with aqueous 0.5–1.0% $Na_2[Fe(NH_3)(CN)_5]$. Other suitable systems are formamide-treated paper, with hexane, benzene, or chloroform, alone or mixed (19:1) with acetic acid, as the mobile phase (Fig. 5). For nitroderivatives of dihydroxybenzenes, the solvent system *n*-propanol–NH_3 (2:1 v/v) is most suitable. R_F values of forty-nine compounds have been tabulated (*58*). A theoretical discussion of the separation of isomers of aromatic nitro compounds by reversed-phase partition chromatography in terms of differences in the polarity of the separated compounds has been published (*59*).

3. *Separations on Alumina-Impregnated Paper* (*60*)

A series of substituted nitrophenols have been chromatographed on various commercial papers impregnated with 2–25% alumina in order to study the relationship between R_F values and molecular structure. The solvent employed was cyclohexane, and detection was in ultraviolet light. R_F values for thirty-eight compounds on five different alumina papers have been listed; although a great number of these values are 1.0 or 0.0, many separation possibilities are evident.

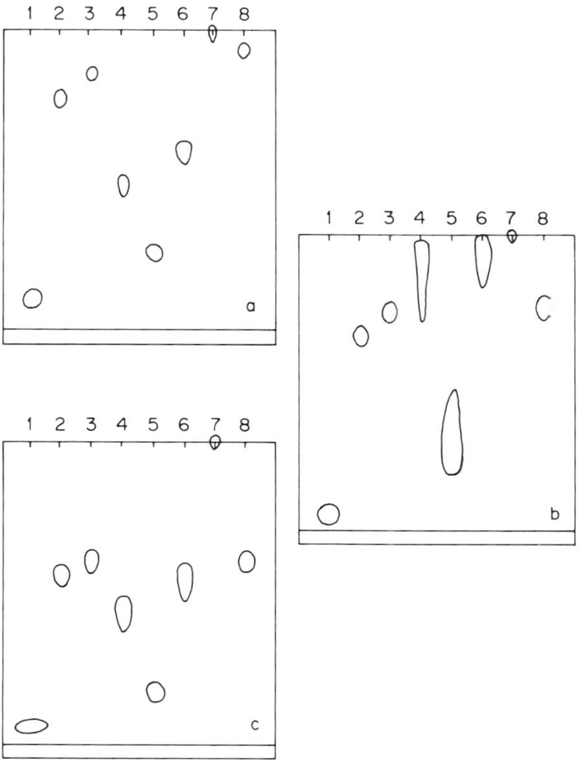

FIG. 5. Chromatograms of nitrophenols on formamide-treated papers developed with (a) benzene–acetic acid (19:1 v/v), (b) $CHCl_3$, and (c) $CHCl_3$–acetic acid (19:1 v/v) (all formamide saturated). (1) 2-nitrophenol, (2) 3-nitrophenol, (3) 4-nitrophenol, (4) 2,4-dinitrophenol, (5) 2,5-dinitrophenol, (6) 2,6-dinitrophenol, (7) 2,4,6-trinitrophenol, (8) 4-nitrosophenol. After J. Gasparič (*58*).

C. Isomers of Nitrobenzoic Acid (*61*)

The ortho, meta, and para isomers of nitrobenzoic acids are separated by descending chromatography on Whatman No. 1 paper with methanol–benzene–*n*-amyl alcohol–H_2O (35:35:17.5:12.5 v/v) containing 0.25 gm sodium EDTA per 100 ml for 6–8 hours. To the mixture of acids (0.01 gm in 10 ml ethanol) is added 8 ml H_2O, 1 ml HCl, and 4 ml 15% $TlCl_4$ solution. This mixture is centrifuged for 1 hour at room temperature and diluted to 25 ml, and 70–80 μl is applied to the paper. After development, the chromatogram is air dried for several hours and spots of the respective isomers appear at R_F 0.75, 0.63, and 0.50. The zones can be eluted and quantitatively determined by spectrophotometry.

Another solvent for the chromatography of nitrocarboxylic acids is *n*-butanol–pyridine–H_2O (5:3:3 v/v) (*62*).

D. Nitrated Aromatic Hydrocarbons

Nitrated dye intermediates are separated on paper impregnated with kerosene and developed with ethanol–acetic acid–H_2O (20:1:14 v/v) (*63*). The R_F values of some substituted toluenes in this system are: 2,3-dinitro, 0.73; 2,6-dinitro, 0.55; 2,5-dinitro, 0.44; 3,4-dinitro, 0.72; 3,5-dinitro, 0.45; 2,4-dinitro, 0.52; and 2,4,6-trinitro, 0.36.

Nitrate explosives have been separated on acetylated paper using butanol–acetic acid–H_2O (4:1:1 v/v) as the mobile phase. Detection was in ultraviolet light or by treatment with diphenylamine–H_2SO_4 solution (*64*).

Table X lists some R_F values in other systems which have been recommended (*65*) for the chromatography of nitro compounds.

TABLE X

R_F Values of Some Nitro Compounds (*65*)

	R_F [a]				
System	A	B	C	D	E
1-(*p*-Nitrophenyl)ethanol	0.20	0.92	0.93	0.92	0.30
2-(*p*-Nitrophenyl)ethanol	0.15	0.91	0.90	0.90	0.25
p-Nitrobenzoic acid	0.01	0.71	0.43	0.81	0.32
p-Nitrophenylacetic acid	0.02	0.60	0.42	0.81	0.34
p-Nitrotoluene	0.92	0.93	0.90	0.72	0.91
p-Nitroethylbenzene	0.91	0.92	0.91	0.70	0.90

[a] Systems: A = *n*-hexane–diisopropyl ether (5:1 v/v) run on paper treated with formamide-saturated ether and dried; B = butan-2-one–2 *N* NH_3 (1:1 v/v); C = *n*-butanol–NH_3 (specific gravity 0.88)–H_2O (4:1:5 v/v); D = ethanol–H_2O (7:3 v/v) on paper treated with 5% (v/v) olive oil in ether and dried; E = *n*-hexane–diisopropyl ether (5:1 v/v) on paper treated with a saturated solution of Carbowax 4000 in ether–ethanol (50:1) and dried. Chromatograms were run on Whatman No. 4 paper until the solvent fronts had moved 12 inches.

VI. Alkaloids*

A. Preparation of Samples

Alkaloids are extracted from vegetable matter with aqueous acids; the extract is made alkaline, and the alkaloids are reextracted with ether or

* Reviews covering this topic have been published earlier (*66–68*).

chloroform. Alternatively, dried and ground material is made alkaline and extracted directly with chloroform. Extractions of stable alkaloids can be made with hot ethanol in a Soxhlet extractor. Heat labile alkaloids (e.g., pyrrolizidine alkaloids) are extracted in an apparatus (*69*) which enables continuous flow of room temperature solvent through the plant-material, absorption of dissolved alkaloids from the solvent onto a cation-exchange resin bed, and washing and recirculation of used solvent. The bases are eluted from the column with 0.8 N NH_3.

For extraction from animal material, the procedure is similar. For example, *Datura* alkaloids are extracted for forensic toxicological examination by saturating a slurry of 50 gm tissue and 100 ml 5% acetic acid with solid $(NH_4)_2SO_4$. The mixture is kept at 100°C for 30 minutes and then filtered. The residue is washed three times with 50 ml 5% acetic acid, and the combined filtrates are extracted with diethyl ether. The ether phase is extracted with HCl, and this solution is adjusted to pH 7.5 with NH_3, saturated with $(NH_4)_2SO_4$, and extracted with ether–chloroform (3:1 v/v). This extract is evaporated to dryness, dissolved in chloroform, and spotted for chromatography (*70*).

Extraction of alkaloids from edible oils is achieved by adding 25 ml hexane, 20 ml ethanol, and 5 ml HCl to about 50 ml oil and warming for 10 minutes. The ethanol layer is separated, washed with three 10 ml portions of hexane, made alkaline with 10% aqueous NH_3, and extracted three times with chloroform. The combined chloroform extracts are washed with water, and the alkaloids are extracted three times with 20 ml 33% acetic acid. This extract is made alkaline, the alkaloids are reextracted into chloroform, and the chloroform is removed by evaporation. The residue is dissolved in 2 ml chloroform, and a portion containing 50–100 μg alkaloids is applied to the paper for chromatography (*70a*).

B. Detection

The following are general spray reagents for alkaloids possessing a tertiary or quaternary nitrogen atom. (a) *Dragendorff reagent.* Mix a solution containing 0.6 gm bismuth subnitrate dissolved in 2 ml concentrated HCl and 10 ml H_2O with a solution containing 6 gm KI in 10 ml H_2O. Add 22 ml dilute HCl (7 ml concentrated HCl + 15 ml H_2O) and dilute the mixture with 400 ml H_2O (*71*). (b) *Potassium iodoplatinate.* Spray with a (1:1 v/v) mixture of 1.1% aqueous KI solution and 0.135% hexachloroplatinic acid (*72*). Viewing the chromatograms under ultraviolet light can augment these reactions. Some individual groups of alkaloids are most suitable detected with more selective reagents (*66*).

1. *Tobacco Alkaloids*

In the König reaction, the paper is placed in an atmosphere of cyanogen bromide gas for 1 hour and then sprayed with an acidic ethanolic solution of *p*-aminobenzoic acid.

2. *Ergot Alkaloids*

Many of these compounds produce an intense blue fluorescence in ultraviolet light. Chromatograms may be sprayed with a solution of 1 gm *p*-dimethylaminobenzaldehyde in a mixture of 95 ml of ethanol and 5 ml HCl followed by careful heating at 80°C for a few minutes to yield blue zones.

3. *Rauwolfia Alkaloids*

Many of these compounds have a green fluorescence in ultraviolet light which can be enhanced by spraying with sodium nitroprusside. Most also produce characteristic colors with cinnamic aldehyde.

C. Paper, Solvents, and Techniques (see also Table XI)

One way linear development is usually used to separate alkaloids by paper chromatography. Multiple development with solvents of increasing polarity has been used (*66*), as has elevated temperature to speed development (*73*). Volatile alkaloids are chromatographed as salts or at low temperatures (*66*). Paper is impregnated in an aqueous solution of $NiSO_4$ and $NaNO_2$, dried at 100°C for 1–5 minutes and developed with a xylene–methyl ethyl ketone mixture for alkaloids such as strychnine, brucine, atropine, and hyoscyamine (*73a*).

Because of the presence of an amino group, alkaloids are basic compounds. They are chromatographed either as ionized salts in acid systems (with aqueous solvents containing an organic or inorganic acid on papers impregnated with a buffer solution) or as undissociated bases in alkaline systems (usually on paper impregnated with formamide). Below are specific systems are recommended for the chromatography of individual groups of alkaloids (*66*).

1. *Tobacco Alkaloids*

Tobacco alkaloids are volatile and are usually chromatographed as salts with *tert*-amyl alcohol–0.2 *M* acetate buffer, pH 5.7 (1:1 v/v) on paper impregnated with the same buffer, or with methanol–isoamyl alcohol–benzene–0.2 *M* acetate buffer (31:15:50:8 v/v) on paper impregnated with 0.2 *M* ammonium tartrate solution. Low temperature chromatography in a refrigerator has also been recommended. Anabasis alkaloids may also be separated through the respective *N*-oxides (*74*).

2. *Tropine Alkaloids*

These compounds are developed with ethyl acetate–25% formic acid (4:3 v/v) on paper equilibrated with the aqueous phase for 14 hours. The alkaloid esters are separated on paper impregnated with formamide using benzene–chloroform (3:2 v/v) as the mobile phase.

3. *Opium Alkaloids*

Opium alkaloids are developed with isobutanol–toluene (1:1 v/v, water saturated) on paper impregnated with pH 3.5 Kolthoff buffer, with chloroform on formamide–impregnated paper, or with isobutanol–acetic acid–H_2O (100:20:36–40 v/v) on paper impregnated with 2% $(NH_4)_2SO_4$ (*75*)(see Fig. 6).

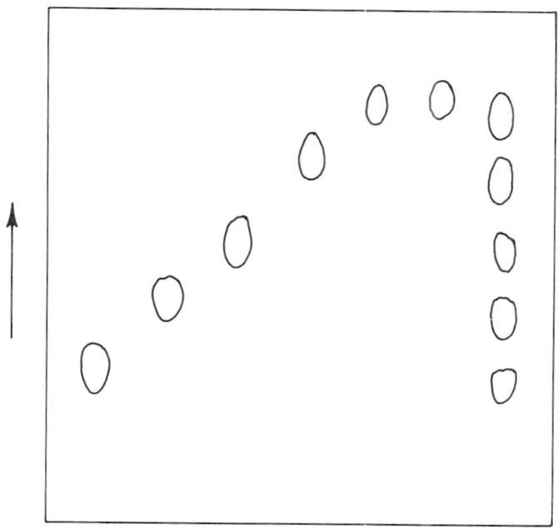

FIG. 6. Chromatogram of (left to right) morphine, codeine, cotarnine, thebaine, narcotine, papaverine and a mixture of the alkaloids. After Nadeau *et al.* (*75*).

4. *Ergot Alkaloids*

These compounds are developed with benzene–cyclohexane (1:1 v/v) on paper impregnated with formamide plus ammonium formate, or with formamide–water (2:3 v/v, adjusted to an appropriate pH with formic acid) on paper impregnated with dimethylphthalate (reversed-phase chromatography).

5. *Rauwolfia Alkaloids*

These compounds are developed with heptane–methyl ethyl ketone (1:1 v/v) on formamide-impregnated paper; use (4:1 v/v) for less polar alkaloids.

TABLE XI R_F VALUES OF ALKALOIDS

Compound	R_F [a]										
	1	2	3	4	5	6	7	8	9	10	11
Genista alkaloids											
Cytisine	0.17	0.12	—	—	—	—	—	—	—	—	—
Methylcytisine	0.23	0.10	—	—	—	—	—	—	—	—	—
Anagyrine	0.31	0.24	—	—	—	—	—	—	—	—	—
Lupanine	0.66	0.37	—	—	—	—	—	—	—	—	—
Hydroxylupanine	0.50	0.22	—	—	—	—	—	—	—	—	—
Lupinine	0.72	0.33	—	—	—	—	—	—	—	—	—
Epilupinine	0.76	0.36	—	—	—	—	—	—	—	—	—
Sparteine	0.59	0.45	—	—	—	—	—	—	—	—	—
Hydroxysparteine	0.43	0.35	—	—	—	—	—	—	—	—	—
17-Oxosparteine	0.82	0.37	—	—	—	—	—	—	—	—	—
Retamine	0.45	0.14	—	—	—	—	—	—	—	—	—
l-α-Isosparteine	0.52	0.56	—	—	—	—	—	—	—	—	—
Calycotomine	0.36	0.26	—	—	—	—	—	—	—	—	—
Solonaceous alkaloids											
Oscine	—	—	0.045	0.07	—	—	—	—	—	—	—
Tropine	—	—	0.08	0.028	—	—	—	—	—	—	0.55
Belladonnine	—	—	0.20	0.38	—	—	—	—	—	—	—
Scopolamine	—	—	0.29	0.85	—	—	—	—	—	—	—
Meteloidine	—	—	0.39	0.66	—	—	—	—	—	—	—
Hyoscyamine	—	—	0.57	0.65	—	—	—	—	—	—	0.42
Norhyoscyamine	—	—	0.60	0.56	—	—	—	—	—	—	—
Tigloidine	—	—	0.71	0.83	—	—	—	—	—	—	—
Valeroidine	—	—	0.62	0.71	—	—	—	—	—	—	—
Apoatropine	—	—	0.81	0.82	—	—	—	—	—	—	—
Papaveraceae alkaloids											
Berberine	—	—	—	—	0.63	0.20	—	—	—	—	—

Coptisine											
Corysamine	—	—	—	—	0.77	0.51	—	—	—	—	—
Palmatine	—	—	—	—	0.59	0.32	—	—	—	—	—
Chelerythrine	—	—	—	—	0.57	—	—	—	—	—	—
Chelilutine	—	—	—	—	0.67	—	—	—	—	—	—
Chelirubine	—	—	—	—	0.57	—	—	—	—	—	—
Macarpine	—	—	—	—	0.51	—	—	—	—	—	—
Sanguilutine	—	—	—	—	0.62	—	—	—	—	—	—
Sanguinarine	—	—	—	—	0.46	—	—	—	—	—	—
Sanguirubine	—	—	—	—	0.42	—	—	—	—	—	—
Aporheine	—	—	—	—	0.63	—	0.30	—	—	—	—
Armepavine	—	—	—	—	0.65	—	0.40	—	—	—	—
Glaucamine	—	—	—	—	0.60	—	0.62	—	—	—	—
Mecambrine	—	—	—	—	0.58	—	0.55	—	—	—	—
Morphine	—	—	—	—	0.36	—	0.13	—	—	—	0.36
Narcotine	—	—	—	—	0.69	—	0.82	—	—	—	—
Papaverine	—	—	—	—	0.71	—	0.81	—	—	—	0.27
Rhoeagenine	—	—	—	—	0.64	—	0.44	—	—	—	—
Rhoeadine	—	—	—	—	0.73	—	0.52	—	—	—	—
Roemeridine	—	—	—	—	0.33	—	0.08	—	—	—	—
Thebaine	—	—	—	—	0.63	—	0.34	—	—	—	—
Allocryptopine	—	—	—	—	0.72	—	0.42	—	—	—	—
(−)-β-Hydrastine	—	—	—	—	0.76	—	0.64		—	—	—
Hydrastinine	—	—	—	—	0.51	—	0.18	—	—	—	0.15
Chelidonine	—	—	—	—	0.75	—	0.61	—	—	—	—
Cryptopine	—	—	—	—	0.73	—	0.34	—	—	—	—
Muramine	—	—	—	—	0.77	—	0.48	—	—	—	—
Protopine	—	—	—	—	0.69	—	0.40	—	—	—	—
Stylophylline	—	—	—	—	0.70		0.83	—	—	—	—
Stylopine	—	—	—	—	0.75	—	0.76	—	—	—	—
Tetrahydroberberine	—	—	—	—	0.80	—	0.76	—	—	—	—
Miscellaneous alkaloids											
Theobromine	—	—	—	—	—	—	—	0.43	—	—	—
Theophylline	—	—	—	—	—	—	—	0.65	—	—	—
Caffeine	—	—	—	—	—	—	—	0.74	—	—	0.73

Table continued

TABLE XI (*continued*)

Compound	R_F [a]										
	1	2	3	4	5	6	7	8	9	10	11
Emetine	—	—	—	—	—	—	—	0.30	0.21	0.37	—
Veratridine	—	—	—	—	—	—	—	0.82	0.24	—	—
Cevadine	—	—	—	—	—	—	—	0.87	—	0.84	—
Quinine	—	—	—	—	—	—	—	0.70	0.51	0.40	0.03
Quinidine	—	—	—	—	—	—	—	0.72	—	0.47	—
Cinchonine	—	—	—	—	—	—	—	—	—	0.51	0.03
Eserine	—	—	—	—	—	—	—	0.57	—	—	0.41
Sparteine	—	—	—	—	—	—	—	0.16	—	>0.90	—
Pilocarpine	—	—	—	—	—	—	—	0.16	0.29	—	—
Brucine	—	—	—	—	—	—	—	0.27	0.73	—	—
Strychnine	—	—	—	—	—	—	—	0.40	0.48	0.09	—
Colchicine	—	—	—	—	—	—	—	—	>0.80	—	0.83
Demecolcine	—	—	—	—	—	—	—	0.49	>0.80	—	—
Lobeline	—	—	—	—	—	—	—	0.87	0.26	0.66	—
Aconitine	—	—	—	—	—	—	—	0.84	0.70	0.54	—
Cocaine	—	—	—	—	—	—	—	—	0.52	0.71	0.58
Ephedrine	—	—	—	—	—	—	—	—	0.00	0.47	0.63
Vincamine	—	—	—	—	—	—	—	—	0.72	—	—
Heroine	—	—	—	—	—	—	—	—	0.76	—	0.33
Hordenine	—	—	—	—	—	—	—	—	—	0.27	—
Hydrastinine	—	—	—	—	—	—	—	—	0.00	—	—
Physostigmine	—	—	—	—	—	—	—	—	—	0.08	—

[a] (1) Isopropanol–HCl–H_2O (8:1:1 v/v), S. and S. 2045 bM paper (*97*). (2) *n*-Butanol–acetic acid–H_2O (50:2:17 v/v), S. and S. 2045 bM paper (*97*). (3) Butanol saturated with McIlvaine buffer (pH 3.8), Whatman No. 1 paper saturated with the buffer (*98*). (4) Same as (3) with pH 6.8 buffer. (5) *n*-Butanol–acetic acid–H_2O (10:1:3 v/v), Whatman No. 1 paper (*99*, *100*). (6) Ethanol–H_2O (3:2 v/v), Whatman No. 1 paper (*99*). (7) *n*-Butanol saturated with water, Whatman No. 1 paper impregnated with 0.15 *M* KH_2PO_4 (*99*, *100*). (8) *n*-Butanol–H_2O–citric acid (50:50:1 v/v/w), paper impregnated with citrate buffer (pH 3.8) (*66*). (9) Chloroform, paper impregnated with formamide plus 1% acetic acid (*66*). (10) Cyclohexane–diethylamine (9:1 v/v), paper impregnated with formamide (*66*). (11) 5% Acetic acid, Whatman

6. *Purine Alkaloids*

Purine alkaloids are developed with *n*-butanol–H_2O–acetic acid (4:1:1 v/v) on Whatman No. 1 paper (*76*).

7. *Cannabis Alkaloids*

These alkaloids are developed with cyclohexane saturated with dimethylformamide on paper impregnated with dimethylformamide saturated with cyclohexane (*77*). These alkaloids give yellow to brown zones on a white background when sprayed with diazotized *p*-nitroaniline. R_F values: cannabidiol acid, 0.05; cannabidiol, 0.15; cannabinol, 0.43; tetrahydrocannabinol, 0.65; pyrahexyl, 0.84.

8. *Curare Alkaloids*

Conjugate curares are separated by development with *n*-butanol–acetic acid–H_2O (4:1:5 v/v) on Whatman No. 1 paper. R_F values: decamethonium iodide, 0.42; gallamine iodide, 0.30; *d*-tubocurarine chloride, 0.22; succinylcholine chloride, 0.12 (*78*). Two-dimensional chromatography is performed in the first direction with ethyl accetate–pyridine–H_2O (7.5:2.3:1.65 v/v) and in the second with methyl ethyl ketone saturated with H_2O and 1–3% methanol (*75*).

R_F values for the first five groups in the systems mentioned have been listed elsewhere (*66*). R_F values for other groups are given in Table XI.

Besides those employing papers impregnated with buffers or formamide, other systems employing modified paper are suitable for alkaloid separations. R_F–pH relationships have been determined for some alkaloids using (a) paper strips impregnated with the liquid ion-exchanger di-(2-ethylhexyl) orthophosphoric acid with aqueous citric acid solutions as the mobile phase (*79*), and (b) liquid ion-exchangers (e.g., oleic acid and a diluting solvent) as the developing solvent in combination with paper impregnated with aqueous buffer solutions (*80*). Papers impregnated with solid organic (*81–84*) and inorganic (*85*)(see Table XI) ion-exchangers and alumina (*86*) have also been employed. Reversed-phase systems employing papers impregnated with solubilizing agents (e.g., tetramethyluric acid) and developed with aqueous buffers are suitable for the separation of isoquinoline alkaloids (*86a*).

D. Quantitative Analysis

Spot areas measured with a planimeter have been related to concentration for the quantitative estimation of quinine and papaverine in suppositories. It is claimed that the mean error is −0.6 to +1.3% (*84*).

Densitometry is widely used after detection with the Dragendorff reagent, potassium iodoplatinate, or phosphomolybdic acid plus $SnCl_2$ (*87*). For example, *Belladonna* alkaloids, after separation by paper chromatography on Whatman No. 2 paper by development with butanol–acetic acid–H_2O (100:3:65 v/v) and detection with Dragendorff reagent, were determined by densitometry at 530 mμ. Beer's Law was followed for 5–25 γ of atropine and hyoscyamine (*88*). The main opium alkaloids were determined after development with butyl acetate–butanol–acetic acid–H_2O (100:8:52:40 v/v) and spraying with a modified Dragendorff reagent (*88a*).

Methods applied to eluted spots such as the three below are usually more accurate (*66*). (a) Spectrophotometry in the visible range—For example, a test solution of tropane alkaloids (0.02 ml containing 0.2–0.5% total alkaloids) is subjected to chromatography on wedge-shaped paper strips with butanol–acetic acid–H_2O (4:1:1 v/v) for 20 hours. The paper is dried in air, drawn through Dragendorff reagent, pressed between filter papers, and dried overnight in the dark. The orange alkaloid zones are cut out, extracted with acetone, and the extinction of each extract is measured at 395–400 mμ (*89*). (b) Spectrophotometry in the ultraviolet range e.g., for opium alkaloids (*90*). (c) Fluorimetry—For example, a separated serpentine zone is cut out, eluted with ethanol, and the extract is measured fluorimetrically for comparison with calibration curves (*91*). Fluorimetry has also been applied to the microdetermination of hydrastine and hydrastinine after separation by radial development (*91a*).

Three modifications of the elution method are of interest: The color produced by the Dragendorff reagent is bleached by NH_3, the bleached spot is cut out, and the alkaloid is extracted from the paper with a buffer. An indicator is added and the alkaloid–indicator complex is extracted from the aqueous solution with chloroform and determined by spectrophotometry (*92, 93*). Alternatively, the separated spots may be extracted with 2% H_2SO_4 and the concentration determined by spectrophotometry at 485 mμ with Tropaeolin 000 as indicator (*93a*). Finally, the chromatogram may be treated with ^{32}P-labeled phosphomolybdic acid, the excess reagent is washed out, and the spots are measured radiometrically (*66*).

E. Applications

A number of schemes for the systematic analysis of a great number of alkaloids belonging to different groups have been reported in the literature. Because of space limitations, interested readers are referred to Macek (*94*) for details. Paper chromatography has also been applied to structural studies of alkaloids (*66*). Structural information is obtained by determining acidic and basic hydrolysis products (*95*), functional groups, and less successfully, the basic skeleton (*96*) of the alkaloids.

Table XI lists some recently published R_F values for various alkaloids (*66, 85, 97-100*). Other collections of mobility data are also available in the literature (e.g., *9*, *101*, *102*).

VII. Synthetic Dyes

A. Introduction (*103*)

Paper chromatography is used extensively to check the purity of dye intermediates and the final products (*103a*) and to follow the progress of reactions for control during manufacture. A generally recommended solvent for dye intermediates is *n*-propanol–H_2O (7:3 v/v). Intermediates are often fluorescent, e.g., those of the naphthalene series, and also give a variety of intense colors when sprayed with certain reagents, such as diazotized *p*-nitraniline, or buffered chromotropic acid after treatment with gaseous nitrous acid. Intermediates containing sulfonic acid groups are sprayed with pinacryptol yellow to yield a variety of different colors when viewed under ultraviolet light.

Dyes and brighteners are removed from fibers by extraction with a suitable solvent. It is important that the fiber itself not be extensively dissolved during the extraction of the dye. Extractions from polyacrylonitrile, polyamide, or cellulose fibers are often performed with a mixture of dimethylformamide–H_2O (2 or 3:1 v/v) at 100°C on a water bath. The extract is then reduced to a suitable volume in a steam-jacketed glass evaporator.

Unknown dyes are classified and identified by chromatographing the mixture and then (a) comparing the results to a library of standard chromatograms, (b) subjecting the spots on the paper to comparative chemical tests with strong acids, alkalies, etc., and noting the color changes, or (c) eluting the spots and determining visible and ultraviolet absorption spectrophotometric curves.

Azo and anthraquinone disperse dyes are separated on S. and S. 589 paper with ascending eluents of various pyridine–H_2O mixtures (e.g., 1:3 v/v) or by reversed-phase chromatography on silicone-treated paper equilibrated overnight with the bottom layer of a chloroform–acetic acid–H_2O mixture and then developed with the upper layer.

Stilbene-based fluorescent brightening agents are separated using *n*-butyl acetate–H_2O–pyridine (2:3:5 v/v) as the mobile phase. The extraction and chromatography steps are carried out in the dark to preserve the brighteners in their trans forms.

Relationships between dye structures and chromatographic behavior, developing solvents for use with different types of dyes, and the quantitative paper chromatography of dyes have been described (*104*). Simple experiments illustrating the separation of dyes and indicators by paper chromatography in test tubes have been devised for use by students (*105*).

B. Direct Dyes

Direct (substantive) dyes are separated by thin-layer chromatography on silica gel or paper chromatography with benzyl alcohol–dimethylformamide–H_2O (3:2:2 v/v) (*103*). Plain and acetylated paper (Ederol type 202), impregnated with formamide (7.5 ml in 142.5 ml ethanol), with 2% *n*-hexadecyl alcohol in ethanol, with paraffin oil (15 ml in 135 ml *n*-hexane), and with silicone oil (10 gm in 30 ml $CHCl_3$) have also been used for the ascending chromatography of dyes with butanol–pyridine–H_2O (3:4:3 v/v) and amyl alcohol–pyridine–NH_4OH (1:1:1 v/v) (*106*).

C. Food Dyes

Isolation of dyes from foodstuffs often calls for considerable ingenuity, such as absorption onto wool fibers and subsequent removal with dilute NH_3 (*103*). Circular chromatography is often applied to the identification and assay of food colors in foods, for example, with 2% trisodium citrate in 5% NH_4OH as the solvent (*107*). The following solvents have been recommended for lineal development of food dyes on Whatman No. 1 paper, and R_F values for many dyes in these solvents have been reported (*108*):

a. *n*-Butanol–acetic acid–H_2O (20:5:12 and 4:1:5 v/v).
b. Isobutanol–ethanol–H_2O (3:2:2 v/v); to 99 ml of this mixture add 1 ml concentrated NH_3.
c. Phenol–H_2O (80:20 w/w).
d. Methyl ethyl ketone–acetone–H_2O–NH_3 (70:30:30 v/v); or the same mixture containing also 0.2 parts NH_3.
e. Ethyl acetate–pyridine–H_2O (11:5:4 v/v).
f. Trisodium citrate (2 gm) in H_2O (95 ml) and NH_3 (5 ml).
g. Propanol–5 *N* NH_3 (4:1 v/v).

Two-dimensional separations of food dyes have been carried out with butanol–acetic acid–H_2O (4:1:5 v/v) followed by a solution containing 2% disodium EDTA and 5% NH_3 (*108a*) R_F values for twenty-nine permitted water-soluble food colors in twelve solvent mixtures proposed by the British Standards Institute have also been published (*109*).

D. Pharmaceutical Dyes

Synthetic colorants in drugs have been separated by ascending paper chromatography in darkness using butanol–ethanol–NH_3–H_2O

(50:25:10:25 v/v) (*110*). Neutral, acidic, and alkaline developing agents with impregnated paper have also been employed (*111*). Preliminary separation of the dyes in gelatin capsules prior to their identification by chromatography on thin layers of cellulose has been obtained by passage of an aqueous solution of the extracted dyes through a stack of Whatman AE 81 aminoethylcellulose paper disks and elution with dilute NH_3 (*112*).

Methyl violet and crystal violet in various commercial pharmaceutical preparations may be differentiated by paper chromatography. The dye (about 0.1 g) is boiled for 10 minutes with 10 ml acetic anhydride, whereby methyl violet is acetylated but crystal violet is not. After cooling, a spot is submitted to ascending development on S. and S.2043b paper for 2 hours with 1% aqueous NaCl. A light green spot of the acetyl derivative of methyl violet ($R_F = 0.41$–0.50) is separated from a violet spot of crystal violet (0.11–0.17) (*112a*).

E. Cosmetic Dyes (*113*)

Fluorescein dyes from cosmetics have been separated by column chromatography on Celite eluted with 1:1 ethanol–H_2O. The eluates were evaporated to dryness, dissolved in 1% aqueous NH_3, and separated by two-dimensional ascending chromatography using 1% aqueous NH_3 saturated with isoamyl alcohol followed by butanol–ethanol–H_2O–NH_3 (200:40:88:2 v/v).

Cosmetic color lakes are decomposed by heating with aqueous, alcoholic sodium versenate and excess NaOH. Acidification liberates the dye, which is extracted with pentanol and then reextracted into 2% NH_3. The dye is compared with standards by chromatography on Whatman No. 1 paper with butanol–ethanol–H_2O (2:1:1 v/v) or the two solvents described above.

Dyes in a wax–fat base (e.g., lipsticks and eye makeup) can be analyzed without preparing extracts by drawing a line with the sample to be tested along the origin of a strip of paper and developing with methyl ethyl ketone–acetone–H_2O–NH_3 (70:20:20:0.2 v/v). The chromatogram is examined in daylight and under ultraviolet light.

F. Photosensitizing Dyes

Various thiazine and other photosensitizing dyes have been chromatographically analyzed (*114*) in three solvent systems: (a) *tert*-butanol–acetone–0.2 *M* HCl (35:35:30 v/v, pH 1.9); (b) *tert*-butanol–acetone–0.2 *M* NH_3–0.2 *M* HCl (35:35:21:9 v/v, pH 9.0); (c) *tert*-butanol–*n*-butanol–0.2 *M* NH_3 (4:3:3 v/v, pH 10.6). Development was carried out in the dark on wedge-shaped strips of Whatman 3 MM paper

(4 × 45 cm) in a Chromatobox (see Chapter 3) sealed with Parafilm in order to minimize solvent evaporation. Development times were 18 hours for solvents (a) and (b), and 40 hours for (c). Commercial preparations were found to contain up to 60% of isomeric contaminants.

Thiazine dyes have also been analyzed by circular chromatography with cyclohexane-based solvents which were allowed to continually evaporate from the edge of the paper (*115*).

G. Azo Dyes (see also Fig. 3, Chapter 3)

Acetylation of disperse hydroxyazo dyes results in characteristic changes in R_F and R_M values which indicate the number and type of hydroxy groups present as an aid in structure elucidation of the dyes. Whatman No. 3 MM paper impregnated with a 10% solution of 1-bromonaphthalene in chloroform is used with a pyridine–H_2O (2:1 v/v) mixture saturated with 1-bromonaphthalene as the mobile phase (*116*). R_F values for eighty-eight water-insoluble azo dyes in this chromatographic system have been tabulated (*117*).

Other studies have been made of the correlation between the R_F values of azo dyes and their structure (*118*) and their affinity to nylon 6 polyamide fiber (*119*). Various bisazodyes have been chromatographed on Whatman No. 1 paper with pyridine–33% NH_3–H_2O (1:1:3 v/v) as the mobile phase (*120*), direct azo dyes with butanol–pyridine–H_2O (2:3:3 v/v) (*121*), and mono- and bisazo derivitives of chromotropic acid with 0.1 *N* HCl as the mobile phase (*121a*).

H. Fast Dye Bases and Basic Dyes

For the separation, identification, and purity control of fast dye bases (the active coupling components of azo dyes), chromatography on Whatman No. 3 paper impregnated with a 20% solution of formamide in ethanol is used, mixtures of heptane–benzene (2:1 or 1:2 v/v) being the mobile phase. Detection is performed by *p*-dimethylaminobenzaldehyde (1% solution in alcohol–HCl, 95:5 v/v). Results of the analyses of eighty-seven commercial bases have been reported (*122*). Alternatively, tetrahydrofuran–H_2O (80:54 v/v) and acetylated paper with detection by diazotization and spraying with Naphthol AS is suitable (*121*).

Basic dyes for polyacrylonitrile fibers have been analyzed for purity by development with methanol–amyl alcohol–benzene–H_2O (31:65:46:8 v/v) or pyridine–H_2O (1:4 v/v) on untreated paper; R_F values for thirty dyes have been listed (*123*). Basic dyes are also chromatographed with methanol–5 *N* NH_3 (8:2 v/v), in which solvent the R_F values for forty-five dyes have been given (*121*).

I. Dichloroquinoxaline and Vinylsulfone Reactive Dyes

These dyes are chromatographed on Whatman No. 3 MM paper with *n*-butanol–dimethylformamide–H_2O (11:3:10 or 11 v/v), *n*-propanol–H_2O (2:1 v/v), or *n*-propanol–acetic acid–H_2O (5:2:3 v/v) as solvents. R_F values for fifty-four dyes have been reported (*124*, *125*).

J. Oil-Soluble Dyes

Azo, anthraquinone, and fluorescent oil-soluble dyes, which are used in many petroleum products, can be separated and identified by reversed-phase partition chromatography on paper impregnated with liquid paraffin or *n*-cetane with 80% ethanol as the mobile phase (*126*).

K. Metal–Dye Complexes

1:1 Cr complex dyes have been separated on untreated paper by development with *n*-butanol–acetic acid–H_2O (8:1:5 v/v) (*127*); development with 60% aqueous acetic acid on untreated paper also separates 1:1 Cr complex dyes, and on partially acetylated paper separates 2:1 Cr complex dyes (*121*).

L. Membrane Chromatography of Dyes

Various dyes have been separated by ascending chromatography on "Synpor" type 8 or 9 nitrocellulose membranes (Chemapol, Prague). Development of a 0.2 mm wide streak of dye mixture (0.1 μl of an ethanolic solution containing 0.1–0.5% of each dye) for 15 minutes in a corked Erlenmeyer flask with 96% ethanol–chloroform–acetic acid (20:20:1 v/v) yields separated zones in the sequence shown in Fig. 7. Two-dimensional development of a 0.6 mm initial spot of mixture with this solvent followed by 96% ethanol–25% NH_3 (10:1 v/v), each for 15 minutes, also separates these dyes (Fig. 7). Because of their uniform microporous structure, these membranes allow the application of very minute samples so that high quality separations are achieved in a short time (*128*).

M. Identification of Dye Groups

A full scheme of qualitative analysis of dye classes by means of mixed techniques, including paper chromatography, has been described (*129*).

VIII. Sulfur Compounds

A. Detection

The following reagents are useful for the detection of organic sulfur compounds:

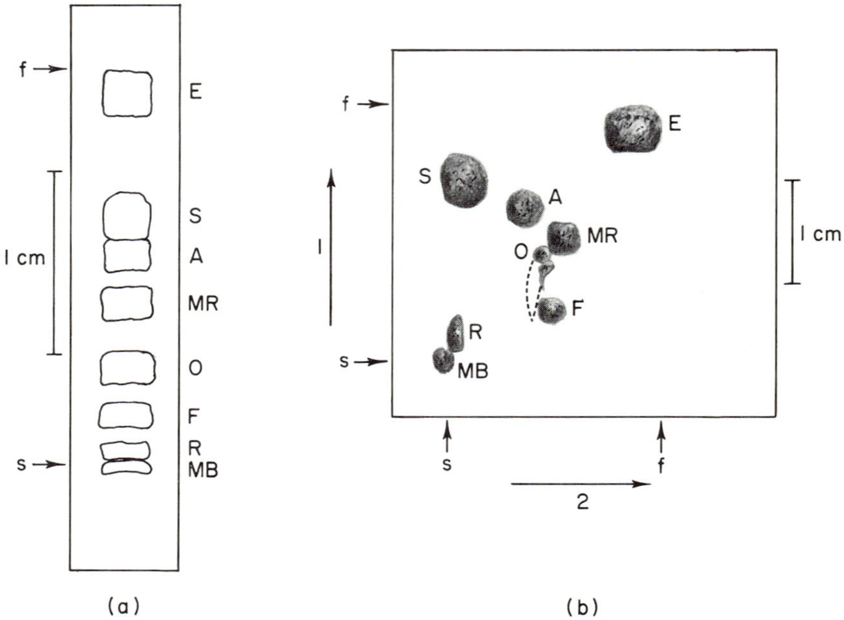

FIG. 7. (a) Microchromatography of dyes on Synpor 9 membranes by development with 96% ethanol–$CHCl_3$–acetic acid (20:20:1 v/v) for 15 minutes. E = eosin, O = orange G, A = *p*-aminoazobenzene, MR = methyl red, F = fuchsin, S = sudan yellow, R = rhodamine B, MB = methylene blue, s = start, f = solvent front. (b) Two-dimensional chromatography on Synpor 8. Direction 1: solvent in (a); direction 2: 96% ethanol–25% NH_3 (10:1 v/v), 15 minutes in both directions. After J. I. Přistoupil and M. Kromlová (*128*).

a. *Fluorescein-1,3,6,8-tetramercuritetraacetate*—detects thio compounds such as mercaptopyrimidines after separation on Whatman No. 3 paper with butanol–ethanol–H_2O (*130*).

b. *Hydrogen selenide* readily detects sulfur, sulfite, thiosulfate, disulfides, and disulfide bonds in proteins; there is no reaction with sulfate and thiocyanate. Dry strips of paper are put into a vertical glass tube and air is removed by a stream of inert gas. The tube is filled with H_2Se, prepared from Al_2Se_3 and HCl, and saturated with H_2O vapor from a wash bottle at room temperature. Reaction times vary from 20 to 120 minutes. After treatment, the H_2Se is again replaced by an inert gas and absorbed in NaOH. Hydrogen selenide is poisonous, so care should be exercised (*131*).

c. *Bromine vapor*. The dried chromatogram is drawn through a solution of aniline (3% v/v) in petroleum ether (40–60°C), and the solvent is allowed to evaporate. On exposing to the vapors of bromine contained in an open dish, a positive test (generally a blue or mauve color) is obtained within a few minutes. The method has a sensitivity of the order 2–5 $\mu g/cm^2$ for classes of organic sulfur compounds in which the sulfur atom is at its lowest oxidation level; thiols, sulfides, disulfides, oxygenated derivatives of disulfides, thiocarbonyls, sulfenamides, and sulfinamides give a positive test. The test is also selective for sulfur amino acids with detection limits of 10–750 $\mu g/cm^2$ (*132*).

d. *2,2′-Dithiobis-(5-nitropyridine)*. A spray reagent composed of 0.03% of the reagent in acetone is quite selective for thiols, the only interference being produced by strong reducing agents such as sodium bisulfite. The yellow color of the spots is stable for more than 24 hours. Cysteine is detected with a sensitivity of 0.2 μg (*133*).

e. *Periodate* (*134*). The paper on a glass plate is sprayed with a 0.1% solution of sodium (meta) periodate in water (for aliphatics) or in 8 *N* acetic acid (for aromatics). After 4 minutes, the color is developed by spraying with a 0.5% solution of benzidine in *n*-butanol–acetic acid (4:1 v/v). The final chromatogram shows white spots (where oxidizable substances consumed periodate) on a blue background (oxidized benzidine). The test is positive for compounds containing divalent sulfur (e.g., aliphatic and aromatic thiols, sulfides, disulfides, and aliphatic thioketones, thiamides and thioureas). These compounds give no reaction with the hydriodic acid test (*135*).

B. Aniline Derivatives

A series of halogen-substituted anilines, aminothiophenols, and diaminodiphenyl disulfides have been separated on S. and S. 2043b paper impregnated with a 25% solution of dimethylformamide in benzene and developed with cyclohexane–benzene–ethanol (6:2:1 v/v). Spots were detected with Ehrlich's Reagent (1 gm *p*-dimethylaminobenzaldehyde in a mixture of 95 ml ethanol and 5 ml HCl). R_F values for sixty compounds have been reported (*136*), some of which are as follows: 4-thiocyanatoaniline, 0.30; 4,4′-diaminodiphenyl disulfide, 0.04; 2-fluoro-4-thiocyanatoaniline, 0.20; 2-iodo-4-mercaptoaniline, 0.27; 2,5-dibromo-4-thiocyanatoaniline, 0.40; and 2, 3, 5-trichloro-4-mercaptoaniline, 0.45.

C. Thioureas

About seventy ureas and thioureas were chromatographed on Whatman No. 1 paper by ascending development with (a) *n*-butanol–

acetic acid–H_2O (4:1:2 v/v), (b) *n*-butanol–95% ethanol–H_2O (4:1:1 v/v), and (c) $CHCl_3$–methanol–H_2O (7:5:1 v/v). Tabulated below are some

Compound	R_F a	R_F b	R_F c
Thiourea	0.56	0.41	0.43
N-Methylthiourea	0.54	0.43	0.46
N-Ethylthiourea	0.59	0.48	0.51
N-Isopropylthiourea	0.65	0.53	0.55
N-Allylthiourea	0.61	0.51	0.52
N,*N'*-Diethylthiourea	0.88	0.66	0.62

of the R_F values obtained (*137*). *N*-substituted thioureas are separated with water saturated $CHCl_3$ as the mobile phase (Fig. 8).

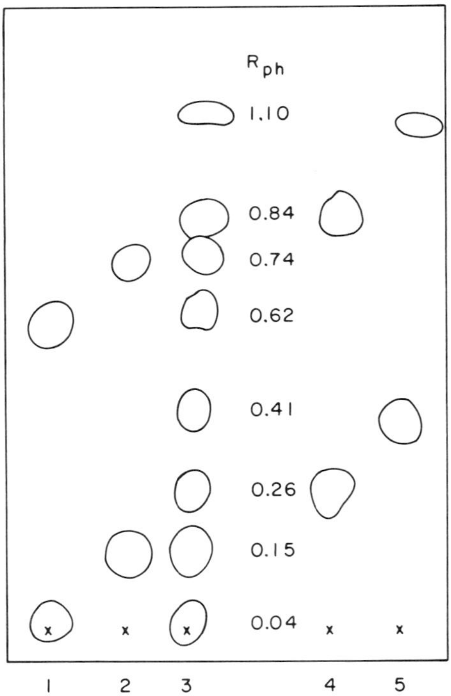

FIG. 8. Separation of *N*-substituted thioureas with $CHCl_3$–H_2O; detection with Grote's reagent. (1) Methyl- and β-methallyl-, (2) ethyl- and α-methallyl-, (3) mixture, (4) allyl- and *n*-butyl-, (5) isopropyl- and β-phenylethyl-. R_{ph} values indicate migration relative to *N*-phenylthiourea. After A. Kjaer and K. Rubstein (*137a*).

Thioureas are detected with one of the following color reagents (*137, 138*):

a. Mixture of equal volumes of 10% NaOH, 10% sodium nitroprusside, 10% potassium ferricyanide diluted with 3 volumes of water.
b. Ninhydrin (see Chapter 5).
c. Grote's reagent—0.5 gm sodium nitroprusside is dissolved in 10 ml H_2O, and 0.5 gm hydroxylamine hydrochloride and 1.0 gm $NaHCO_3$ are added. Two drops bromine are added, and the excess is removed by heating. The solution is filtered, diluted to 25 ml, and diluted 1:1 with saturated Na_2CO_3 before spraying.
d. Ehrlich's reagent (see Section VIII, B above).
e. Feigl's reagent—Spray the dried chromatogram with 1% starch solution and while still damp with a solution of 3 gm sodium azide in 100 ml 0.1 *N* iodine; colorless spots on a blue background.
f. Ammoniacal $AgNO_3$—gray spots.
g. Diazotized sulfanilic acid—orange-red spots.
h. Acidified $KMnO_4$—colorless spots on a purple background.

The R_F values of fifty thioureas in $CHCl_3$ saturated with H_2O have been published (*9*).

D. Aromatic Sulfenamides and Sulfinamides

A series of these compounds has been examined by paper chromatography (*139*). Sulfenamides are best separated on paraffined paper (untreated or zinc carbonate–fluorescein-impregnated (*140*) paper is drawn through a 3% v/v solution of liquid paraffin in petroleum ether, 40–60°C, suspended vertically, and dried) using 65% methyl alcohol as the mobile phase. Sulfinamides are better separated on phenoxyethanol paper (the paper is treated as above using a 10% v/v solution of phenoxyethanol in acetone and 16 hours drying) developed with phenoxyethanol saturated *n*-hexane. Spot location is with *p*-dimethylaminobenzaldehyde solution, 0.1% v/v in ethanol (sensitivity, 2–5 μg). When viewed under ultraviolet light, sulfenamides on zinc carbonate–fluorescein paraffined paper appear mauve on a light yellow fluorescent background; the spots can be cut out, leached with solvent, and identified and estimated by spectroscopy.

E. Sulfonamides

Various sulfonamide drugs and sulfanilamide preparations have been separated and identified on untreated paper developed with *n*-butanol–

acetic acid–H_2O (4:1:2 v/v), *n*-butanol–NH_3–H_2O (10:1:1 v/v/v), and *n*-propanol–NH_3 (7:3 v/v). The compounds are detected under ultraviolet light or in daylight after spraying with Ehrlich's reagent (see Section VIII, B) or *p*-dimethylaminocinnamaldehyde (0.25% solution in 1:1 2 *N* HCl–ethanol) (*141*). Alternatively, 2,4-dinitrobenzene–sulfenyl derivatives of the sulfanilamides are prepared and separated on untreated paper with toluene–acetone–methanol–10% NH_3 (15:5:5:0.2 v/v) as solvent or on paper impregnated with 1% aqueous NaF with xylene–$CHCl_3$–isopropanol–methanol–concentrated NH_3 (8:5:3:5:0.25 v/v) (*142*).

Sulfonamides are also separated on paper impregnated with a pH 8.29 borate buffer solution, dried 4–5 hours at 20°C, spotted with 0.05–0.1 mg of sample, and developed twice with butanol–methanol–$CHCl_3$–2% borax solution (25:25:50:5 v/v) (*143*). "Filtrak" FN 3 paper, either unimpregnated or impregnated with various buffers (e.g., 0.55 ml 0.1 *M* citric acid–19.45 ml 0.2 *M* Na_2HPO_4, or 4 gm KH_2PO_4–4 gm $Na_2CO_3 \cdot 10H_2O$ in 200 ml H_2O) and developed for 24 hours with isopropanol–H_2O (9:1 v/v), butanol–acetone–NH_3 (15:2.5:5 v/v), or butanol saturated with a buffer also gives good separations (*143a*). Detection is by spraying with 1% NaOCl–5% phenol or fluorescein dissolved in acetone (followed by viewing under ultraviolet light), or by diazotization and coupling with *N*-naphthylethylenediamine. In the latter case, the colored spots may be eluted with a mixture of 50% acetic acid and formamide (1:1 v/v) and determined photometrically (*144*) or by measurement of spot areas (*145*).

F. Mercaptotriazines (*146*)

Dithioammelide, thioammeline, and trithiocyanuric acid have been separated by ascending chromatography with the solvents methanol–5% NH_3 (1:4 v/v) (respective R_F values on Filtrak–Niederslag paper No. 16 and 4: 0.56, 0.45, 0.67) and methanol–dimethylformamide (2:1 v/v) (R_F = 0.67, 0.42, 0.86, on FN paper No. 2/L). The separated substances are detected by spraying with 0.05% $KMnO_4$ (yellow spots) and 0.1% *o*-toluidine solution in ethanol (blue spots).

G. Dithiocarbamates and Thiurams (*147*)

These compounds are separated by reversed-phase chromatography on Whatman No. 2 paper impregnated with 2% acrylonitrile–butadiene copolymer in 2:1 benzene–acetone as the stationary phase, with mixtures of organic solvents and H_2O as the mobile phase [e.g., benzene–ethanol (3:7 v/v), methanol–ethylenedichloride or chlorobenzene–H_2O (6:4:1 v/v), and ethanol–benzene–H_2O (7:3:1 v/v)]. Two percent $CuSO_4$ in

1% H_2SO_4 mixed with an equal volume of methanol is the coloring agent, or the chromatogram is exposed to iodine or IBr vapor to form brown spots.

H. Aliphatic sulfonates (*148*)

Alkane, alkene, 2- and 3-hydroxyalkane, and 2- and 3-oxoalkane sulfonates (C_8–C_{18}) and some alkyl sulfates may be separated and identified on S. and S. 2043b paper impregnated with 5% dodecanol in benzene (w/w) using methanol–25% NH_3 (1:1 or 3:2 w/w) solutions saturated with dodecanol as the solvent. This method is useful for the analyses of mixtures of sulfonation and sulfation products (Fig. 9).

I. Thiocyanates

The migration behavior of eleven organic thiocyanate compounds has been studied on paper impregnated with dimethylformamide (DMF) by development with xylene or cyclohexane–benzene (3:1 v/v). For detection, the chromatogram was sprayed with $Na_2S \cdot 9H_2O$ reagent (5 gm dissolved in 20 ml H_2O–80 ml ethanol), dried, and sprayed with 5% aqueous $FeCl_3$ (40 ml) made up to 100 ml with boiled 20% HNO_3; the sensitivity is 5 μg (*149*).

Thiocyanato derivatives of a series of phenols, phenyl ethers, and phenyl ketones have been chromatographed on DMF paper with nine different developing solvents. R_F values and color reactions of these compounds are tabulated (*150*).

A mixture of thiourea, ammonium thiocyanate, and sodium sulfide is separated by ascending development with 1% aqueous $Pb(NO_3)_2$ ($R_F = 0.70$, 0.81, and 0.36, respectively) or pyridine–methanol–H_2O (1:2:1 v/v) ($R_F = 0.71$, 0.87, and 0.41) on strips of S. and S. 2043a paper. Thiourea is detected with 0.1 *M* $HgNO_3$, thiocyanate with 0.12 *M* $Fe(NO_3)_3$ and sulfide with 0.1 *M* $AgNO_3$ followed by 4 *M* HNO_3 (*151*).

J. Sulfonic Acids

The sulfonic acids in fresh water diatoms, labeled with ^{35}S by culturing for 4 days, were extracted from the harvested and washed cells with boiling 80% and 100% ethanol; the extract was evaporated under nitrogen and redissolved in 80% ethanol–$CHCl_3$ to prepare the sample for two-dimensional chromatography on Whatman No. 4 paper in phenol–H_2O (100:38 w/w) and *n*-butanol–propionic acid–H_2O (142:71:100 v/v). Cysteinolic acid and sulfopropanediol were separated and identified (*152*).

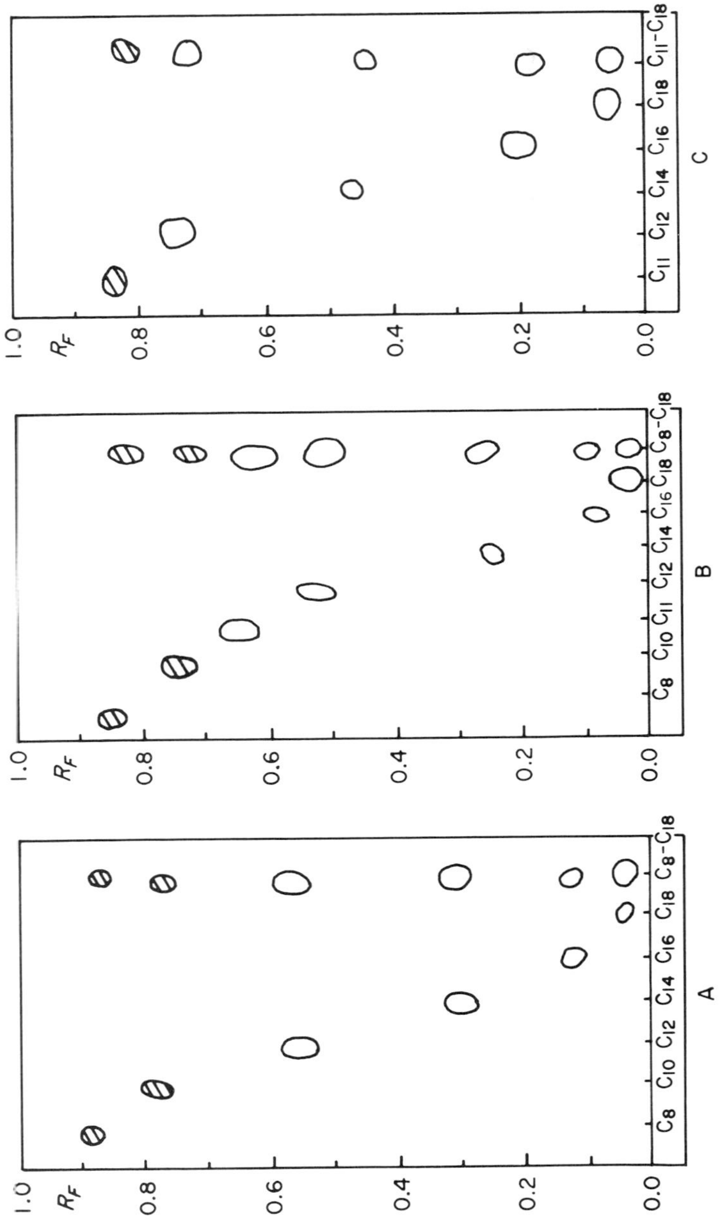

1.0
R_F
0.8
0.6
0.4
0.2
0.0
C_8 C_{10} C_{12} C_{14} C_{16} C_{18} C_8–C_{18}
A
C_8 C_{10} C_{11} C_{12} C_{14} C_{16} C_{18} C_8–C_{18}
B
C_{11} C_{12} C_{14} C_{16} C_{18} C_{11}–C_{18}
C

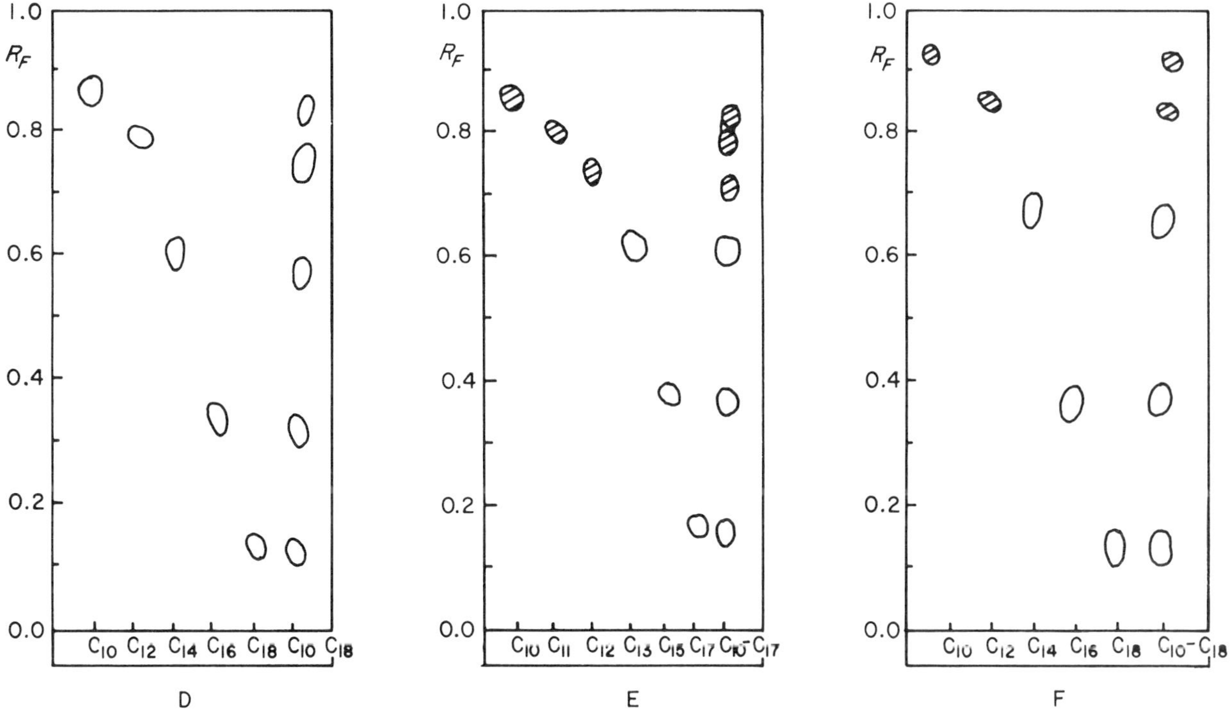

FIG. 9. Separation of aliphatic sulfonates on S. and S. 2043b paper impregnated with 5% dodecanol solution and developed with methanol–25% NH_3 (1:1 w/w). (A) Sodium *n*-alkyl sulfates (C_8–C_{18}), (B) sodium *n*-alkane sulfonates (C_8–C_{18}), (C) sodium *n*-alkene-2 sulfonates (C_{11}–C_{18}), (D) sodium 3-hydroxy-*n*-alkane sulfonates (C_{10}–C_{18}), (E) sodium 2-oxo-*n*-alkane sulfonates (C_{10}–C_{17}), (F) sodium 3-oxo-*n*-alkane sulfonates (C_{10}–C_{18}). Clear spots are yellow, cross-hatched spots orange. After F. Püschel and D. Prescher (*148*).

Amino, hydroxy, and aminohydroxy sulfonic acids of the benzene and napthalene series are chromatographed on Whatman No. 1 paper with butanol–acetic acid–H_2O (4:1:5 v/v) in a chamber saturated for 14 hours. Detection is with $NaNO_2$ and H_2SO_4 (*153*).

1- and 2-naphthylamine sulfonic acids are sucessfully separated on Whatman No. 1 or 4 paper impregnated with 5% $NaHCO_3$ and developed with *n*-propanol–H_2O (2:1 v/v) saturated with $NaHCO_3$. Detection is by diazotization with nitrous acid fumes and coupling with 2-naphthol-3,6-disulfonic acid or resorcinol (*154*). R_F values are tabulated below.

Compound	R_F
1-Naphthylamine	1.00
1-Naphthylamine-*N*-sulfonic acid	0.96
1-Naphthylamine-2-sulfonic acid	0.95
1-Naphthylamine-4-sulfonic acid	0.77
1-Naphthylamine-7,*N*-disulfonic acid	0.37
1-Amino-2-naphthol-*N*,4-disulfonic acid	0.31
1-Naphthylamine-4,*N*-disulfonic acid	0.25
1-Naphthylamine-2,4-disulfonic acid	0.21
1-Naphthylamine-4,7-disulfonic acid	0.11
1-Naphthylamine-2,4,7-trisulfonic acid	0.03
2-Naphthylamine	1.00
2-Amino-1-naphthol-*O*-sulfonic acid	0.93
2-Naphthylamine-*N*-sulfonic acid	0.90
2-Naphthylamine-1-sulfonic acid	0.83
2-Amino-1-naphthol-*N*-sulfonic acid	0.64
2-Naphthylamine-1,*N*-disulfonic acid	0.52
2-Naphthylamine-8,*N*-disulfonic acid	0.35
2-Naphthylamine-6,*N*-disulfonic acid	0.25
2-Naphthylamine-1,6-disulfonic acid	0.18

Sodium salts of 1-aminoanthraquinonemonosulfonic acids and their bromination products are best separated by descending or circular development of samples (0.005 ml of 0.05–0.1% aqueous solutions) with 5–23% aqueous HCl or ethanol–HCl (3:1 v/v). Detection is by coloration of the spots or their fluorescence in ultraviolet light (365 mμ) after 5–10 minutes irradiation or treatment with ammonia vapor (*155*).

R_F values for various other classes of sulfonic acids in systems recommended for their separation and methods for their detection are shown in Table XII (*156*, *157*). Spectroscopic studies of the separated spots are often useful for the identification and characterization of unknown acids (*158–159*).

TABLE XII

R_F VALUES OF SULFONIC ACIDS

Compound	R_F [a]					
	1	2	3	4	5	6
o-Aminonaphtholsulfonic acid isomer						
1,2,0	0.93	—	—	—	—	—
1,2,4	0.61	—	—	—	—	—
1,2,5	0.39	—	—	—	—	—
1,2,6	0.44	—	—	—	—	—
1,2,7	0.47	—	—	—	—	—
1,2,8	0.27	—	—	—	—	—
2,1,0	0.93	—	—	—	—	—
2,1,4	0.70	—	—	—	—	—
2,1,5	0.41	—	—	—	—	—
1,2 Naphthoquinone derivative						
1-diazide	—	0.90	0.98	—	—	—
4-sulfonic acid 1-diazide	—	0.51	0.65	—	—	—
5-sulfonic acid 1-diazide	—	0.41	0.52	—	—	—
6-sulfonic acid 1-diazide	—	0.36	0.47	—	—	—
7-sulfonic acid 1-diazide	—	0.41	0.56	—	—	—
8-sulfonic acid 1-diazide	—	0.55	0.73	—	—	—
2-diazide	—	0.92	0.97	—	—	—
4-sulfonic acid 2-diazide	—	0.45	0.71	—	—	—
5-sulfonic acid 2-diazide	—	0.40	0.62	—	—	—
Naphthalene derivative						
-1-sulfonic acid	—	—	—	0.71	0.57	0.78
-2-sulfonic acid	—	—	—	0.64	0.49	0.78
-1,5-disulfonic acid	—	—	—	0.86	0.70	0.46
-1,6-disulfonic acid	—	—	—	0.81	0.65	0.47
-1,7-disulfonic acid	—	—	—	0.86	0.70	0.51
-2,6-disulfonic acid	—	—	—	0.76	0.58	0.46
-2,7-disulfonic acid	—	—	—	0.76	0.60	0.46
-1,3,5-trisulfonic acid	—	—	—	0.96	0.76	0.27
-1,3,6-trisulfonic acid	—	—	—	0.88	0.72	0.24
-1,3,7-trisulfonic acid	—	—	—	0.96	0.76	0.27
-1,3,5,7-tetrasulfonic acid	—	—	—	0.98	0.86	0.15

[a] (1) *n*-Propanol–5% $NaHCO_3$ (2:1 v/v); Whatman No. 1 paper; detection by ultraviolet light (*156*). (2) Butanol–pyridine–H_2O (3:1:1 v/v); Whatman No. 4 paper; detection by spraying chromatogram with 10% resorcinol and exposing to NH_3 vapor (*156*). (3) Solvent and detection same as (2); Whatman No. 4 paper impregnated with 5% solution of $NaHCO_3$ (*156*). (4) HCl–H_2O (1:3 v/v); Whatman No. 4 paper; detection with 0.05% aqueous pinacryptol yellow (*157*). (5) Saturated aqueous K_2SO_4 ; paper and detection same as (4) (*157*). (6) *n*-Propanol–NH_3 (2:1 v/v); paper and detection same as (4) (*157*).

IX. Methylphenylpolysiloxanes

Table XIII shows R_F values for these compounds in two reversed-phase systems (*160*). The compounds are detected by drawing the chromatograms through 98% fuming HNO_3 and immediately immersing them in distilled water. After drying they are ironed with a gently warmed flat iron.

TABLE XIII

R_F VALUES OF METHYLPHENYL POLYSILOXANES[a] (*160*)

Siloxane	R_F I_1	R_F I_2
$[(CH_3)_2(C_6H_5)Si]_2O$	0.71	0.77
$[(CH_3)_2SiO]_2OSi(C_6H_5)_2$	0.80	0.84
$[(CH_3)_2SiO]_3OSi(CH_3)(C_6H_5)$	0.17	0.14
$[(CH_3)_3SiO]_2Si(C_5H_5)_2$	0.49	0.52
$[(CH_3)_3SiO]_2Si(CH_3)(C_6H_5)OSi(CH_3)_2$	0.17	0.13
$(C_6H_5)(CH_3)_2SiOSi(C_6H_5)_2Cl$	0.82	0.88
cis-$[(CH_3)(C_6H_5)SiO]_3$	0.88	0.91
trans-$[(CH_3)(C_6H_5)SiO]_3$	0.97	0.92
$[(CH_3)(C_6H_5)_2Si]_2O$	0.93	0.94
$[(CH_3)_2SiO]_3[(C_6H_5)_2SiO]$	0.43	0.29
$[(CH_3)_2SiO]_2[(CH_3)(C_6H_5)SiO]_2$	0.63	0.57
$(CH_3)_3SiOSi(C_6H_5)_2OSi(CH_3)_2OSi(CH_3)_3$	0.36	0.23
$[(CH_3)_3SiOSi(CH_3)(C_6H_5)]_2O$	0.39	0.33
$[(C_6H_5)_2(OC_2H_5)Si]_2O$	0.76	0.92
$[(CH_3)_3SiOSi(C_6H_5)(CH_3)]_2OSi(CH_3)_2$	0.25	0.17
$[(CH_3)(C_6H_5)SiO]_4$	0.80	0.85
$[(CH_3)_3SiOSi(C_6H_5)_2]_2O$	0.75	0.70
$(CH_3)_3SiO[Si(CH_3)(C_6H_5)O]_3Si(CH_3)_3$	0.36	0.28
$[(CH_3)_3SiOSi(CH_3)(C_6H_5)OSi(CH_3)(C_6H_5)]_2O$	0.41	0.29
$[(CH_3)_3SiOSi(CH_3)(C_6H_5)OSi(CH_3)(C_6H_5)]_2OSi(CH_3)(C_6H_5)$	0.49	0.33

[a] Paper: Whatman No. 1. Impregnation: I_1 = 5% solution of Apiezon L (Carlo Erba) in toluene; I_2 = 5% solution of silicone elastomer (VChZ-Kolin) in toluene. Solvent: H_2O–dimethylformamide (1:8 v/v).

X. Phosphorus Compounds

A. PHOSPHORIC ESTERS

1. *Isolation from Blood* (*161*)

After centrifuging a blood sample, the precipitated erythrocytes are washed at 0°C with 0.9% NaCl and mixed with an equal volume of distilled water. The hemolyzate is mixed with an equal volume of 20%

trichloroacetic acid for 30 minutes at 4°C and centrifuged to separate protein. The clear supernate can be used directly or the high concentration of acid can be removed by double extraction with equal volumes of diethyl ether, which is in turn removed *in vacuo*. The alkaline earths are complexed by adding a small amount of disodium EDTA.

2. *Systems*

Sugar phosphates are generally chromatographed on prewashed, equilibrated paper with acidic solvents to avoid their hydrolysis. The solvents below have proved useful (*162–164*). To the first two, 0.05 gm of the tetrasodium salt of EDTA is added for each 100 ml of solvent.

1. *n*-Butanol–*n*-propanol–acetone–80% (w/v) formic acid–30% (w/v) trichloroacetic acid (40:20:25:25:15 v/v, run twice in the same direction).
2. Methanol–90% formic acid–H_2O (80:15:5 v/v).
3. Isobutyric acid–*N* NH_3 (10:6 v/v).
4. Isopropyl ether–*n*-butanol–90% (w/w) formic acid (3:3:2 v/v, run twice).
5. *n*-Propanol–concentrated NH_3–H_2O (6:3:1 v/v).
6. *n*-Butanol–acetic acid–H_2O (60:15:25 v/v).
7. *tert*-Butanol–50% formic acid–H_2O (16:1:4 v/v).
8. *n*-Propanol–ethyl acetate–H_2O (7:1:2 v/v).

Table XIV shows *R* values of sugar phosphates and related compounds in some of these solvents, and Figs. 10 and 11 illustrate phosphate separations by one- and two-dimensional development.

DEAE ion-exchange paper has been used to separate anionic galactose 1-phosphate (R_F = 0.0 with H_2O as the solvent) from galactose (1.0) (*165*).

3. *Detection* (*161–163*)

Phosphoric esters are detected via the phosphoric group or the sugar or alcohol component in the following ways:

a. *The Hanes and Isherwood Acid–Molybdate Reagent.* The chromatogram is sprayed with a solution containing 5 ml 60% $HClO_4$, 10 ml 1 *N* HCl, 25 ml 4% ammonium molybdate and is adjusted to 100 ml with water. It is then dried for 7 minutes at 85°C and viewed under ultraviolet light. To use as a dip reagent, part of the H_2O is replaced with acetone. This test hydrolyzes the ester and detects the liberated inorganic phosphate. Certain polyols (inositols, lactones, monosaccharides, etc) give a positive reaction and can be confused with organophosphate compounds (*166*).

TABLE XIV

R VALUES OF PHOSPHATES AND RELATED COMPOUNDS (*162, 162a*)

Compound (abbreviation)	*R* value[a]					
	R_a (1)	R_F (2)	R_b (3)	R_a (4)	R_F (5)	R_F (6)
Adenosine 5′-phosphate (AMP-5′)	0.70	—	1.00	0.00	0.26	0.20
Adenosine 3′-phosphate (AMP-3′)	0.80	—	1.22	0.00	0.30	0.22
Adenosine diphosphate (ADP)	0.24	—	0.82	0.00	0.18	0.05
Adenosine triphosphate (ATP)	0.12	—	0.75	0.00	0.15	0.04
Inosine monophosphate (IMP)	0.49	—	0.58	0.00	0.17	0.17
Inosine diphosphate (IDP)	0.19	—	0.43	0.00	0.13	0.09
Inosine triphosphate (ITP)	0.07	—	0.35	0.00	0.10	0.05
Ribulose 5-phosphate	0.55	0.61	—	—	—	—
Xylulose 5-phosphate	0.55	0.52	—	—	—	—
Erythrose 4-phosphate	0.57	0.64	—	—	—	—
Phosphoribulose pyrophosphates	0.40	0.36[b]	—	—	—	—
Guanosine 3′-phosphate (GMP-3′)	0.56	—	0.70	0.00	0.12	0.18
Guanosine 5′-phosphate (GMP-5′)	0.51	—	0.56	0.00	0.10	0.18
Guanosine diphosphate (GDP)	0.05	—	0.39	0.00	0.07	0.05
Guanosine triphosphate (GTP)	0.02	—	0.26	0.00	0.06	0.02
Cytidine triphosphate (CTP)	0.07	—	0.52	0.00	0.11	0.03
Uridine diphosphate glucose (UDPG)	0.13	—	0.25	0.00	0.60	0.05
Diphosphopyridine nucleotide (DPN)	0.10	—	1.02	0.00	Decomposes	0.04
Triphosphopyridine nucleotide (TPN)	0.05	—	0.71	0.00	Decomposes	0.01
Phosphate (PO_4^{-3})	1.00	0.72	0.68	1.00	0.19	0.28
Pyrophosphate ($P_2O_7^{-4}$)	0.52	—	0.52	0.02	0.11	0.11
Creatine phosphate (Cr-P)	0.98[c]	—	0.86[c]	0.62[c]	0.17	Decomposes
Glyceraldehyde 3-phosphate (GAP)	0.73	0.60	0.40	0.23	Decomposes	0.15
2,3-Diphosphoglyceric acid (2,3-PPGA)	0.54	—	0.38	0.01	0.04	0.12
Dihydroxyacetone phosphate (DHAP)	0.71	0.65	0.68	0.23	Decomposes	0.05
Phosphoenol pyruvate (PEP)	1.06	—	0.62	0.73	0.26	0.27
3-Phosphoglycerate (3-PGA)	0.79	—	0.45	0.32	0.23	0.18

2-Phosphoglycerate (2-PGA)	0.74	—	0.45	0.20	0.26	0.16
Glycerol 1-phosphate (G1-1-P)	0.76	—	0.65	0.52	0.35	0.14
Glycerol 2-phosphate (G1-2-P)	0.83	—	0.65	0.52	0.35	0.14
Ribose 5-phosphate (R-5-P)	0.56	0.57	0.53	0.04	0.25	0.18
Fructose 1-phosphate (F-1-P)	0.52	—	0.40	0.01	0.27	0.12
Fructose 6-phosphate (F-6-P)	0.50	—	0.48	0.03	0.31	0.19
Fructose 1,6-diphosphate (F-1,6-PP)	0.35	—	0.37	0.01	0.09	0.08
Glucose 1-phosphate (G-1-P)	0.41	—	0.48	0.02	0.30	0.20
Glucose 6-phosphate (G-6-P)	0.41	0.40	0.46	0.02	0.26	0.12
6-Phosphogluconic acid (6-PG)	0.49	0.53	0.42	0.02	0.22	0.12
Sedoheptulose 7-phosphate (S-7-P)	0.41	0.43	0.46	0.00	0.28	0.12
Ribulose 1,5-diphosphate (Ru-1,5-PP)	0.46	0.54	0.35	0.00	0.06	0.08
Glucose (G)	0.59	—	0.69	0.16	0.77	0.22
Fructose (F)	0.68	—	0.77	0.21	0.98	0.26
Ribose (R)	0.86	—	0.84	0.40	0.78	0.30
Adenine (Ad)	1.40	—	1.60	0.30	0.72	0.60
Hypoxanthine (Hy)	1.15	—	1.10	0.31	0.66	0.48
Adenosine (Ao)	1.00	—	1.57	0.16	0.84	0.49
Inosine (I)	0.75	—	0.96	0.16	0.65	0.36
Adenosine tetraphosphate	0.00	—	0.48	—	—	—
3′-5′-cyclic AMP	0.53	—	1.00	—	—	—
Phosphoglycolic acid	0.90	—	0.51	—	—	—
Phospholactic acid	1.10	—	0.64	—	—	—
Phosphoglycolaldehyde	1.08	—	0.41	—	—	—
Propanediol phosphate	1.22	—	0.85	—	—	—
Phosphohydroxypyruric acid	0.68	—	0.68	—	—	—
1,3-Diphosphoglyceric acid	0.75[c]	—	0.34	—	—	—
Glucose 1,6-diphosphate	0.25	—	0.25	—	—	—
Acetyl-CoA	0.50	—	1.13	—	—	—
2-Deoxyglucose 6-phosphate	0.64	—	0.54	—	—	—
Phosphoribonic acid	0.50	—	0.31	—	—	—

[a] $R_a = R_F$ relative to phosphate; $R_b = R_F$ relative to AMP-5′. Solvents corresponding to column numbers are given in the text.
[b] Streaked zone.
[c] Partial decomposition.

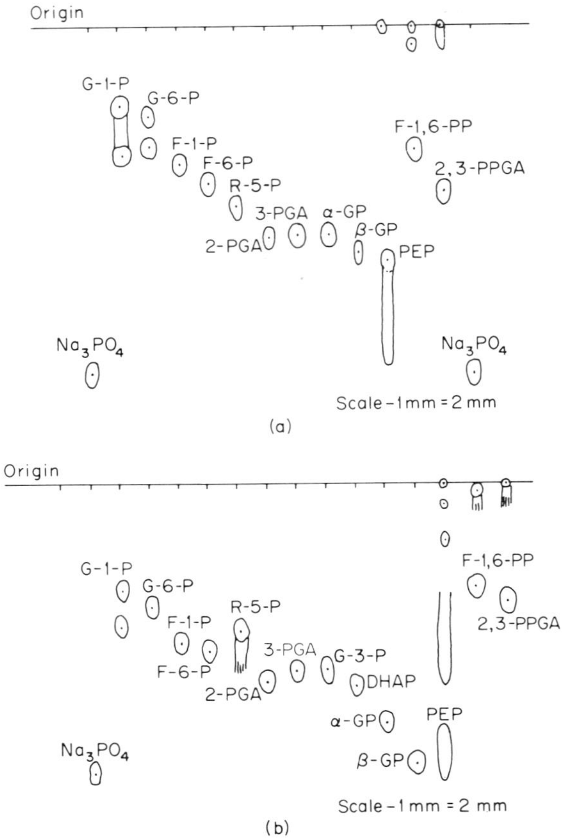

FIG. 10. Paper chromatogram of sugar phosphates and three-carbon phosphates on formic acid-washed paper by descending development with (a) solvent 7 (see text) for 39.5 hours, and (b) solvent 8 for 15.5 hours, and then solvent 7 for 20 hours. Most abbreviations used are in Table XIV; α-GP and β-GP = DL-α and β-disodium glycerophosphate, respectively; G-3-P = DL-glyceraldehyde 3-phosphate, diethyl acetate, barium salt. After M. H. Subers *et al.* (*164*).

b. *$FeCl_3$–Sulfosalicyclic Acid.* The chromatogram is sprayed with 0.1 % $FeCl_3$ in 80 % ethanol, dried at room temperature, and sprayed with 1 % sulfosalicyclic acid in 80 % ethanol. The color of the spots depends upon the developing solvent used. Detects 1–2 γ of phosphorus per square centimeter.

c. *$HClO_4$*. The chromatogram is dipped into a solution of 1.5 ml 63 % $HClO_4$ in 20 ml acetone and dried for 5 minutes at 20–22°C and

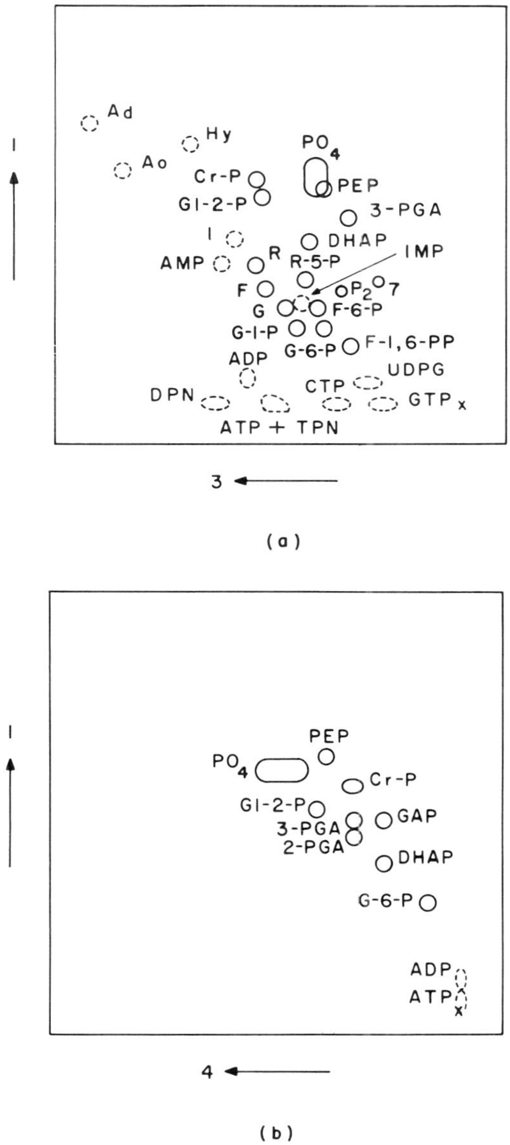

FIG. 11. Map of phosphorus compounds in (a) solvent 1 (see text) twice followed by solvent 3, and (b) in solvent 1 twice followed by solvent 4 twice. × = origin; spots visible under ultraviolet light are shown in broken outline; abbreviations used are in Table XIV. After T. Wood (*162*).

2.5 minutes at 90°C. The resultant spots can be directly scanned using reflection densitometry, a linear relationship existing between the amount of sugar phosphate (0–0.20 μmoles) and the peak area corresponding to the spot density. Accuracy is $\pm 10\%$.

d. *3,4-Dinitrobenzoic Acid.* The paper is dipped through a 1% solution in acetone, the solvent is evaporated and dipped through a solution of aqueous 10% tetraethylammonium hydroxide–10% ethanolic glycerol (1:2 v/v). Yields violet spots which blacken with time for ketopentose phosphates.

e. *Enzymatic Hydrolysis.* Enzymatic hydrolysis using *Candida utilis* detects stable phosphate compounds including nucliotides (see Chapter 9) (*167*).

4. *Quantitation*

Besides direct densitometry as described above, esters can be determined by measurement of the ultraviolet absorption of eluted zones (*162*) or by wet-ashing the spots, decomposing the pyrophosphate, and determining the inorganic phosphorus content (*161*, *162*). Phosphoric esters are often labeled with radioactive carbon or phosphorus so that autoradiography or counting methods (*165*) may be used for detection and quantitative analysis.

5. *Alkyl Phosphates and Phosphites*

Phosphites and dialkyl phosphates are detected on paper by the acid–molybdate method (*168*) (above) or with isatin reagent, and trialkyl phosphates with a reagent containing ferrous ammonium sulfate (0.7 gm) and ammonium thiocyanate (5 gm) in concentrated H_2SO_4 (1 ml) and H_2O (100 ml) (*169*).

Tables XV and XVI (*170*) show R_F values and conditions for the chromatography of assorted alkyl phosphates and ethyl esters of condensed phosphoric acids, respectively.

A scheme for the analysis of all alkyl phosphates and alkyl phosphites has been suggested employing propanol–NH_3 (4:1 v/v), isoamyl alcohol–pyridine–NH_3 (3:2:1 v/v) and *n*-amyl alcohol–70% formic acid (1:1 v/v) as developing solvents (*169*).

6. *Inositol Phosphates* (*171*)

Inositol hexaphosphate is hydrolyzed enzymically to yield inositol mono- to pentaphosphates. The compounds are purified by ion exchange and precipitation methods and converted to their ammonium salts, which are separted by ascending chromatography with butanol–propanol–2 *N* NH_4OH (1:1:1 v/v) R_F values compared to ammonium phosphate: monophosphate, 0.86; diphosphate, 0.46; triphosphate,

TABLE XV

R_F VALUES OF ALKYL PHOSPHATES[a] (*168*)

Compound	R_F
Ammonium dibenzyl hydrogen phosphate	0.86
Diammonium P^1,P^2-di-*n*-butyl pyrophosphate	0.82
Cyclohexylammonium benzyl *n*-butyl phosphate	0.90
tert-Butyl ethyl hydrogen phosphate	0.88
Diammonium P^1,P^2-diethyl pyrophosphate	0.48
tert-Butyl phosphate	0.28
Ethyl *n*-propyl hydrogen phosphate	0.84
n-Propyl dihydrogen phosphate	0.28
P^1,P^2-Di-*n*-propyl pyrophosphate	0.77
Ammonium diethyl phosphate	0.84
Ammonium ethyl *n*-propyl phosphate	0.80
Cyclohexylammonium isopropyl *n*-propyl phosphate	0.82
Cyclohexylammonium isopropyl *n*-propyl pyrophosphate	0.63
Ammonium *n*-butyl *n*-propyl phosphate	0.82

[a] Paper: Whatman No. 30. Solvent: 2-propanol–NH_3–H_2O (7:2:1 v/v).

TABLE XVI

R_F VALUES OF ETHYL ESTERS OF CONDENSED PHOSPHORIC ACIDS[a] (*170*)

Compound abbreviation[b]	R_F
$P_2(H_4P_2O_7)$	0.13
$P_1(H_3PO_4)$	0.26
E_1P_2	0.29
E_1P_1	0.39
E_2P_3	0.48
E_2P_2	0.59
E_3P_3	0.71
E_2P_1	0.77
E_3P_2	0.86
E_4P_2	0.93
E_4P_3	0.91

[a] Solvent: Isobutanol–isopropanol–ethanol–H_2O–25% NH_4OH (25:20:20:34:0.7 v/v). Paper: S. and S. No. 2043b Mgl.

[b] Example of abbreviations:

```
         O O              O             O
E3P2 = HOPOPOR;   E1P1 = HOPOH;   E2P1 = HOPOR.
         O O              O             O
         R R              R             R
```

0.32; tetraphosphate, 0.25; pentaphosphate, 0.20; and hexaphosphate, 0.16. Spot detection is by method b, Section 3, above.

7. *Aromatic Phosphate Esters*

R_F values for these compounds, including pyridyl phosphates, are given in Table XVII (*172*, *173*). Detection is by the molybdate–acid spray reagent (above) or by inspection in ultraviolet light.

TABLE XVII

R_F VALUES OF AROMATIC AND PYRIDYL PHOSPHATE ESTERS AND RELATED COMPOUNDS (*172*, *173*)

Compound	R_F [a] 1	R_F [a] 2
Diphenyl pyrophosphate	0.70–0.73	—
Ditolyl pyrophosphate	0.78	—
Di(*p*-chlorophenyl) pyrophosphate	0.84	—
Monophenyl phosphate	0.39–0.41	0.08
Monotolyl phosphate	0.46–0.48	—
Mono(*p*-chlorophenyl) phosphate	0.54–0.55	—
Monoethyl phosphate	0.26	—
Monophenyl pyrophosphate	0.30–0.32	—
Inorganic trimetaphosphate	0.25–0.26	—
Inorganic phosphate	0.15–0.16	—
Inorganic pyrophosphate	0.10–0.11	—
4-Chlorophenyl phosphate	—	0.16
β-Naphthylphosphate	—	0.11
Benzyl phosphate	—	0.11
Thymyl phosphate	—	0.32
Phenyl-α-pyridyl phosphate	—	0.62
β-Naphthyl-α-pyridyl phosphate	—	0.64
Benzyl-α-pyridyl phosphate	—	0.64
Thymyl-α-pyridyl phosphate	—	0.76
Di-α-pyridyl phosphate	—	0.53
Methylphenyl phosphate	—	0.57
Phenylbenzyl phosphate	—	0.72
Propylphenyl phosphate	—	0.70

[a] (1) *n*-Propanol–NH_3–H_2O (6:3:1 v/v); Machery-Nagel paper No. 2261 (washed). (2) Isopropanol–NH_3–H_2O (8:1:1 v/v); S. and S. paper No. 2043b (washed).

8. *Miscellaneous Applications*

Metabolites of mevalonate have been purified by ion-exchange and paper chromatography (*174*). Phosphate esters of human erythrocytes have been separated and assayed (based on radioactivity and ultraviolet absorption) on acid-washed Whatman No. 1 paper with isobutyric acid–1 *N* NH_3–0.1 *M* sodium EDTA (50:30:0.8 v/v) as the mobile phase (*161*, *175*).

B. Phosphorus Acids

1. *Alkyl phosphoric Acids*

Mono- and dialkyl phosphoric acids (C_1–C_8 in the ester alkyl group) are separated by the following mobile phases: ethanol–25% NH_4OH–H_2O (6:3:1 v/v), isopropanol–25% NH_4OH–H_2O (7:1:2 v/v), and butanol–25% NH_4OH–H_2O (6:2:2 v/v). The acids are detected with aniline–xylose reagent (5 gm aniline and 5 gm xylose dissolved in several hundred ml of 50% aqueous ethanol; the sprayed paper is heated at 125°C to reveal acids as brown spots). In general, R_F values of monoalkyl-phosphoric acids are smaller than for the dialkyls and increase with the number of carbon atoms in the ester alkyl group (*176*).

2. *Phosphinic, Phosphonous, and phosphonic acids*

a. *Paper*. Whatman No. 1 or No. 3 MM.

b. *Solvents*. (1) Propanol–NH_4OH (2:1 v/v); R_F values by ascending development on paper pretreated with 2% EDTA: Ph_3PO–Ph_3P, 1.00; phenylphosphinic acid, 0.91; phenylphosphonous acid, 0.85; phenyl-phosphonic acid, 0.43 (*177*). (2) Isopropanol–formic acid–H_2O (8:1:1 v/v); R_F values: 2-aminoethylphosphonic acid, 0.23; 1-aminoethyl-phosphonic acid, 0.26; 2-aminoethylphosphate, 0.17 (*178*). (3) $CHCl_3$–pyridine–25% NH_4OH (8:1:5 v/v); for the separation of chloromethyl-phosphinic acids, methylphosphinic acid, and inorganic phosphates and phosphites (*179*).

c. *Detection*. Phosphinic acids yield brown spots upon heating the chromatogram at 180°C and yellow spots with molybdate reagent in ultraviolet light (*179*). Aminophosphonic acids are detected with ninhydrin (see Chapter 5) (*178*). Phosphonous and phosphonic acids may be detected with the molybdate reagent or radiometrically after neutron activation (*177*).

C. Phospholipids

In one recent method (*180*) for the determination of phospholipids, commercial silica gel paper (Whatman SG 81) is equilibrated for 30 minutes with the mobile solvent and then developed for 4.5 hours with diisobutyl ketone–acetic acid–H_2O (40:30:7 v/v). Resultant R_F values are: sphingomyelin, 0.39; lecithin, 0.44; phosphatidylserine, 0.53; and phosphatidylethanolamine, 0.58. Detection is by light spraying with rhodamine G. The spots are viewed under ultraviolet light, marked, cut out, and the phospholipids eluted by refluxing 1 ml of methanol through the paper for 1 minute. After evaporation of the methanol, the lipid is determined by hydrolysis with $HClO_4$, treatment with molybdate and $SnCl_2$ reagents, and color measurement of phosphate at 740 mμ.

XI. Epoxy Resins

Commercial epoxy resins (molecular weight up to 1000) containing various epoxy derivatives of phenols or dihydroxybenzenes as impurities can be analyzed by reversed-phase paper chromatography. The systems used are Whatman No. 3 paper impregnated with a 30% methanolic solution of octane-1,8-diol–dimethylsulfoxide (1:2 v/v) with cyclohexane–dioxane (4:1 or 2:1 v/v, water saturated) as solvent. The developed chromatograms are treated with pyridine, and after removal of excess pyridine, are treated with a chloroplatinate–KI or a $FeCl_3$–KI reagent (*181*).

References

1. W. Huber, *Mikrochim. Acta* p. 44 (1960).
2. G. Schwachula and F. Wolf, *J. Chromatog.* **24**, 297 (1966).
3. M. Stuchlík, L. Krasnec, and L. Csiba, *J. Chromatog.* **30**, 543 (1967).
4. J. Franc, V. Kovář, and F. Mikeš, *Mikrochim. Acta* p. 133 (1966); *J. Chromatog.* **32**, D2 (1968).
5. G. Grimmer and A. Hildebrandt, *J. Chromatog.* **20**, 89 (1965).
6. J. W. Howard and R. T. Teague, Jr., *J. Assoc. Offic. Agr. Chemists* **48**, 315 (1965).
7. T. M. Spotswood, *J. Chromatog.* **3**, 101 (1960).
7a. J. H. Dhont, *Chem. & Tech. (Amsterdam)* **23**, 259 (1968).
8. M. Lederer and G. Roch, *J. Chromatog.* **31**, 618 (1967); S. Caroli and M. Lederer, *J. Chromatog.* **37**, 333 (1968).
8a. P. Engel and R. Tomingas, *Beckman Rept.* p. 18 (1967).
8b. N. I. Kaznina, *Gigiena i Sanit.* **33**, 65 (1968).
9. R. J. Block, E. L. Durrum, and G. Zweig, "A Manual of Paper Chromatography and Paper Electrophoresis," 2nd rev. ed. Academic Press, New York, 1958.
10. J. Sherma and D. E. Thompson, Jr., *Anal. Chim. Acta* **32**, 181 (1965).
11. H. H. Brownell, *J. Chromatog.* **19**, 556 (1965).
12. K. Macek, *in* "Paper Chromatography" (I. M. Hais and K. Macek, eds.), p. 238. Academic Press, New York, 1963.
13. I. A. Egorov and A. K. Rodopulo, *Vinodelie i Vinogradarstvo SSSR* p. 7 (1962)
14. E. L. Mndzhoyan, *Vinodelie i Vinogradarstvo SSSR* p. 9 (1965).
15. B. Růžička, *Chem. Anal. (Warsaw)* **10**, 1165 (1965).
16. J. Gasparič and J. Borecký, *J. Chromatog.* **4**, 138 (1960).
17. V. B. Luk'yanov and V. S. Gavrichev, *Vestn. Mosk. Gos. Univ., Ser. Khim.* p. 25 (1965).
18. L. Syper, *Dissertationes Pharm.* **17**, 229 (1965).
18a. J. Churáček, K. Komárek, V. Vaňásek, and M. Jureček, *Collection Czech. Chem. Commun.* **33**, 3877 (1968); *J. Chromatog.* **43**, D35 (1969).

19. J. Cotte and B. Guillot, *Gattefosse SFPA, Bull. Tech.* p. 59 (1963).
20. P. Fijolka and W. Radowitz, *Plaste Kautschuk* **12**, 207 (1965).
21. J. Zajić, *Sb. Vysoke Skoly Chem.-Technol. v Praze, Oddil Fak. Potravinareske Technol.* **6**, 179 (1962).
22. T. P. Udalova, *Vestn. Mosk. Univ., Ser. VI: Biol., Pochvoved.* **21**, 88 (1966).
23. A. S. W. de Freitas, *Can. J. Biochem.* **45**, 1041 (1967).
24. J. Novak, *J. Chromatog.* **31**, 137 (1967).
25. F. Scheffer and R. Kickuth, *Z. Anal. Chem.* **230**, 254 (1967).
26. A. A. Casselman and R. A. B. Bannard, *J. Chromatog.* **20**, 424 (1965).
27. J. Sherma and L. H. Pignolet, *Anal. Chim. Acta* **34**, 185 (1966).
28. E. D. Barber and E. Sawicki, *Anal. Chem.* **40**, 984 (1968); *J. Chromatog.* **39**, D10 (1969).
29. S. McElvain, "The Characterization of Organic Compounds," p. 199. Macmillan, New York, 1949.
30. V. Kapišinská and H. Mokrošová, *Chem. Prumysl* **42**, 448 (1967); *J. Chromatog.* **34**, D45 (1968).
31. N. S. Dokunikhin, B. N. Kolokolov, and O. A. Egorova, *Zh. Analit. Khim.* **21**, 889 (1966).
32. J. Gasparič and M. Večeřa, *Mikrochim. Acta* pp. 68 and 681 (1958).
32a. I. Buben and J. Pospísil, *J. Collection Czech. Chem. Commun.* **34**, 2000 (1969).
33. E. D. Barber, *J. Chromatog.* **27**, 398 (1967).
34. K. T. Finley, *J. Chromatog.* **19**, 443 (1965).
35. E. Papke and R. Pohloudek-Fabini, *Pharmazie* **19**, 27 (1964).
36. T. Shinohara, *Kagaku Keisatsu Kenkyusho Hokoku* **18**, 10 (1965).
37. P. Ronkainen, *J. Chromatog.* **27**, 384 (1967).
38. V. Ulbrich and J. Makeš, *J. Chromatog.* **15**, 371 (1964).
38a. M. C. Burnett, *J. Chem. Educ.* **43**, 385 (1966).
39. K. Macek, *in* "Paper Chromatography" (I. M. Hais and K. Macek, eds.), p. 279. Academic Press, New York, 1963.
40. D. A. Forss and W. Stark, *Anal. Chem.* **36**, 941 (1964).
41. A. M. Gaddis and R. Ellis, *Anal. Chem.* **31**, 870 (1959).
42. D. A. Forss, P. R. Edwards, B. J. Sutherland, and R. Birtwistle, *J. Chromatog.* **16**, 460 (1964).
43. A. Pribela and J. Vasatko, *Biologia* **22**, 96 (1967).
44. R. Kinget, *Planta Med.* **14**, 460 (1966); *Lloydia* **31**, 17 (1968).
45. I. G. Mokhnachev, V. Kamenshchikova, and V. S. Kovtun, *Izv. Vysshikh Uchebn. Zavedenii, Pishchevaya Tekknol.* p. 58 (1966).
46. L. O. Spetsig, *J. Inst. Brewing* **72**, 266 (1966).
47. C. R. Engle and E. Sawicki, *J. Chromatog.* **37**, 508 (1968).
48. A. K. Rodopulo and I. A. Egorov, *Vinodelie i Vinogradarstvo SSSR* **25**, 6 (1965).
49. H. Fuerst and G. Feustel, *Reichstoffe, Aromen, Koerperpflegemittel* **14**, 317 (1964).
50. J. Cartlidge and C. F. H. Tipper, *Anal. Chim. Acta* **22**, 106 (1960).
51. A. Rieche and M. Schulz, *Angew Chem.* **70**, 694 (1958).
52. G. Dobson and G. Hughes, *J. Chromatog.* **16**, 416 (1964).
53. R. Kavčič, B. Plesničar, and B. Hristovski, *J. Chromatog.* **22**, 381 (1966).
54. I. Filipek, *Tech. Publ. Stredisko Tech. Inform. Potravinar. Prumyslu* **161**, 84 (1959–1961).
55. M. Daniels and A. Grimson, *Biochim. Biophys. Acta* **142**, 292 (1967).
56. V. Cohen, *J. Chromatog.* **23**, 446 (1966).
57. R. W. Matthews, *J. Chromatog.* **20**, 190 (1965).

58. J. Gasparič, *J. Chromatog.* **13**, 459 (1964); **15**, 83 (1964).
59. L. S. Bark and R. J. T. Graham, *Talanta* **11**, 839 (1964).
60. W. Kemula and J. Geisler, *Roczniki Chem.* **38**, 1691 (1964).
61. C. Germano, *Boll. Chim. Farm.* **101**, 225 (1962).
62. J. Franc, *Collection Czech. Chem. Commun.* **23**, 655 and 2018 (1958).
63. J. Franc and J. Knížek, *Collection Czech. Chem. Commun.* **24**, 2299 (1959).
64. G. Krien, *Explosivstoffe* **11**, 207 (1963).
65. J. Chakraborty and J. N. Smith, *Biochem. J.* **93**, 389 (1964).
66. K. Macek, *in* "Chromatography" (E. Heftmann, ed.), p. 606. Reinhold, New York, 1967.
67. E. Graf, *Mitt. Deut. Pharm. Ges.* **36**, 237 (1966).
68. V. E. Papendick, J. W. Sutherland, and D. E. Williamson, *Anal. Chem.* **41**, 190R (1969).
69. A. R. Mattocks, *Nature* **191**, 1281 (1961); *J. Chem. Soc., C* p. 329 (1967).
70. S. N. Tewari, *Arch. Kriminol.* **140**, 61 (1967).
70a. P. S. Nataraja Sarma and V. V. Nithyanandan, *Res. Ind., New Delhi* **12**, 167 (1967).
71. H. Kaiser and H. Jori, *Arch. Pharm.* **287**, 222 (1954).
72. D. Waldi, "Chromatographie," 2nd ed. Merck, Darmstadt, 1959.
73. H. V. Street, *Acta Pharmacol. Toxicol.* **19**, 325 (1962).
73a. I. Turkovic, *J. Pharm. Belg.* [N. S.] **23**, 283 (1968).
74. Ya. I. Pakanaev and A. S. Sadykov, *Vopr. Estestv. Nauk (Bukhara: Tashkent)* p. 5 (1966).
75. G. Nadeau, G. Sobolewski, L. Fiset, and C. G. Farmilo, *J. Chromatog.* **1**, 327 (1958); G. B. Marini-Bettolo and G. C. Casinovi, *J. Chromatog.* **1**, 411 (1958).
76. H. H. Cornish and A. A. Christman, *J. Biol. Chem.* **228**, 315 (1957).
77. T. W. M. Davis, C. G. Farmilo, and M. Osadchuk, *Anal. Chem.* **35**, 751 (1963).
78. A. Fiori and M. Marigo, *J. Chromatog.* **31**, 171 (1967).
79. E. Soczewiński and M. Rojowska, *J. Chromatog.* **27**, 206 (1967).
80. E. Soczewiński, G. Matysik, and H. Szumilo, *Separation Sci.* **2**, 25 (1967).
81. J. E. Jaffe and D. Kirkpatrick, *Psychopharmacol. Bull.* **3**, 49 (1966).
82. H. Witkowski, *Bull. Soc. Amis Sci. Lettres Poznan* **B18**, 29 (1964–1965).
83. J. Dadlez and Z. Wojciak, *Bull. Soc. Amis Sci. Lettres Poznan* **C15**, 129 (1965).
84. J. Pasich and E. Pilawska, *Farm. Polska* **23**, 105 (1967).
85. I. D. Coussio, G. B. Marini-Bettolo, and V. Moscatelli, *J. Chromatog.* **11**, 238 (1963).
86. V. Ivanov, N. Nikolov, and A. Yaneva, *Farmatsiya (Sofia)* **17**, 20 (1967).
86a. M. Stuchlík and L. Krasnec, *J. Chromatog.* **36**, 522 (1968).
87. P. H. List, *Naturwissenschaften* **41**, 454 (1954).
88. M. S. Shipalov, Yu. M. Shilov, and S. D. Mekhtikhanov, *Prikl. Biokhim. i Mikrobiol.* **2**, 707 (1966).
88a. M. S. Shipalov and S. D. Mekhtikhanov, *Prikl. Biokhim. i Mikrobiol.* **4**, 444 (1968).
89. K. Pfordte, *J. Chromatog.* **21**, 495 (1966).
90. V. A. Danel'yants and S. Kh. Mushins'ka, *Farm. Zh. (Kiev)* **19**, 34 (1964).
91. N. Ya. Tsarenko and M. S. Schraiber, *Farm. Zh. (Kiev)* **21**, 16 (1966).
91a. W. Wiśniewski and T. Gorta, *Acta Polon. Pharm.* **25**, 427 (1968).
92. W. Deckers and A. Mueller, *J. Chromatog.* **17**, 618 (1965).
93. H. Gertig, Z. Kowalewski, and K. Nowaczyk, *Herba Polon.* **12**, 111 (1966).
93a. B. Zsadon, L. Decsei, and E. Dezseri, *Herba Hung.* **6**, 17 (1967).
94. K. Macek, *in* "Paper Chromatography" (I. M. Hais and K. Macek, eds.), p. 595. Academic Press, New York, 1963.

95. K. Macek, M. Semonsky, S. Vanecek, and A. Cerny, *Naturwissenschaften* **42**, 647 (1964).
96. A. Resplandy and C. Sannie, *Compt. Rend.* **241**, 65 (1955).
97. R. Barnasconi, S. Gill, and E. Steinegger, *Pharm. Acta Helv.* **40**, 253 (1965).
98. M. S. Karawya and S. I. Balbaa, *J. Pharm. Sci. U. Arab Rep.* 4, 57 (1963); *J. Chromatog.* **22**, D11 (1966).
99. J. Slavík, *Collection Czech. Chem. Commun.* **32**, 4434 and 4435 (1967).
100. J. Slavík and J. Appelt, *Collection Czech. Chem. Commun.* **30**, 3690 (1965).
101. S. Pfeifer, *J. Chromatog.* **24**, 364 (1966).
102. E. Soczewiński, *Bull. Acad. Polon. Sci., Ser. Sci. Chim.* **13**, 213 (1965).
103. W. Voeglein, W. A. Vetterli, and S. Napier, *Ciba Rev.* p. 24 (1966).
103a. J. Tušl, *J. Chromatog.* **37**, 546 (1968).
104. H. Zukriegel and O. Jily, *Textil-Praxis* **22**, 332 (1967).
105. W. J. Owen and G. Agar, *School Sci. Rev.* **48**, 231 (1966).
106. S. Logar, N. Mesicek, J. Perkavac, and M. Pepar, *Kem. Ind. (Zagreb)* **17**, 473 (1968).
107. M. Hansens and Y. DeRudder-Tack, *Pharm. Tijdschr. Belg.* **44**, 125 (1967).
108. M. H. E. Griffiths, *J. Food Technol.* p. 64 (1966); *J. Chromatog.* **30**, D6 (1967); J. Dobrecky and R. C. D. DeCarnevale Bonino, *SAF-YBI* **7**, 165 (1967).
108a. J. Dobrecky and R. C. D. DeCarnevale Bonino, *Rev. Asoc. Bioquim. Arg.* **32**, 139 (1967).
109. D. Pearson and R. Walker, *J. Assoc. Public Analysts* **3**, 45 (1965).
110. J. Alary, C. Luu Duc, and A. Coeur, *Bull. Trav. Soc. Pharm. Lyon.* **10**, 78 (1966).
111. V. Popovici, A. Schweiger, and A. Spitzer, *Farmacia (Bucharest)* **13**, 569 (1965).
112. R. Berret, A. Gavaudan, and J. Hirtz, *Ann. Pharm. Franc.* **25**, 365 (1967).
112a. A. Fischer, *Chromatographia*, **1**, 481 (1968).
113. B. Legatowa, *Roczniki Panstwowego Zakladu Hig.* **16**, 453 and 541 (1965); **17**, 21 (1966).
114. J. S. Bellin and M. E. Ronayne, *J. Chromatog.* **24**, 131 (1966).
115. K. B. Taylor, *J. Histochem. Cytochem.* **8**, 248 (1960); *Stain Technol.* **36**, 73 (1961).
116. J. Gasparič and I. Gemzová, *J. Chromatog.* **35**, 362 (1968).
117. J. Gasparič, I. Gemzová, and D. Šnobl, *Collection Czech. Chem. Commun.* **31**, 1712 (1966); *J. Chromatog.* **24**, D5 (1966).
118. A. Lorinc, F. Peter, and O. Dobozy, *Acta Chim. Acad. Sci. Hung.* **47**, 419 (1966).
119. T. Shimizu and K. Oi, *Sen-i Gakkaishi* **21**, 491 (1965).
120. A. Arcoria and G. Scarlata, *Ann. Chim. (Rome)* **58**, 34 (1968); *J. Chromatog.* **39**, D3 (1969).
121. E. G. Kiel and G. H. A. Kuypers, *Tex* **22**, 468, 533, 607, and 779 (1963).
121a. J. Siemroth and I. Hennig, *Talanta* **15**, 765 (1968).
122. J. Gasparič, *Z. Anal. Chem.* **218**, 113 (1966).
123. S. Logar, J. Perkavec, and M. Perpar, *Mikrochim. Acta* p. 496 (1967); *J. Chromatog.* **30**, D14 (1967).
124. A. Cee and J. Gasparič, *Mikrochim. Acta* p. 453 (1968); *J. Chromatog.* **39**, D13 (1969).
125. A. Cee and J. Gasparič, *Collection Czech. Chem. Commun.* **33**, 1094 (1968); *J. Chromatog.* **39**, D14 (1969).
126. I. Shimizu, *Yukagaku* **17**, 35 (1968).
127. S. Logar, N. Mesicek, J. Perkavac, and M.Perpar, *Kem. Ind. (Zagreb)* **16**, 342 (1967).
128. T. I. Přistoupil and M. Kramlová, *J. Chromatog.* **34**, 21 (1968).
129. G. H. A. Kuypers and E. G. Kiel, *Tex* **23**, 365 (1964).
130. J. Havir, A. Fidler, and R. Husak, *Acta Chim. Acad. Sci. Hung.* **50**, 39 (1966).

131. M. Wroński, *J. Chromatog.* **24**, 480 (1966).
132. R. F. Bayfield, V. Clarke, and E. R. Cole, *J. Chromatog.* **19**, 370 (1965); R. F. Bayfield and E. R. Cole, *ibid.* **40**, 470 (1969).
133. D. R. Grassetti and J. F. Murray, Jr., *J. Chromatog.* **41**, 121 (1969).
134. R. Stephan and J. G. Erdman, *Nature* **203**, 749 (1964).
135. C. DeMarco, *Nature* **198**, 683 (1963).
136. K. Papke and R. Pohloudek-Fabini, *Pharmazie* **22**, 486 (1967); *J. Chromatog.* **36**, D69 (1968).
137. L. Fishbein, *Rec. Trav. Chim.* **84**, 470 (1965); *J. Chromatog.* **22**, D2 (1966).
137a. A. Kjaer and K. Rubstein, *Acta Chem. Scand.* **7**, 528 (1953); *Nature* **17**, 840 (1953).
138. B. K. Patnaik, B. C. Dash, D. P. Das, and G. N. Mahapatra, *Current Sci. (India)* **35**, 595 (1966).
139. R. F. Bayfield, V. Clarke, and E. R. Cole, *J. Chromatog.* **26**, 132 (1967).
140. Report of the Analytical Methods Committee, Vitamin E Panel, *Analyst* **84**, 356 (1959).
141. N. I. Trofimova, N. G. Molchanova, L. V. Kashina, and A. I. Shafiev, *Tr. Estestvennonauchn. Inst. Perm. Univ.* **11**, 119 (1965); P. Millburn, R. L. Smith, and R. T. Williams, *Biochem. J.* **105**, 1283 (1967).
142. R. Tulus and A. Guran, *Istanbul Univ. Fen Fak. Mecmuasi* **C28**, 114 (1963).
143. G. P. Sementovs'ka, *Zb. Aspirants'k. Prats. Odes'k. Tekhnol. Inst. Kharch. ta Kholodil'n. Prom.* **1**, 56 (1963).
143a. J. Vasic, R. Popovic, and S. Ramic, *Glasnik Drustva Hemicara Technol. SR Bosne Hercegorine* **15**, 99 (1967).
144. M. T. van der Venne, and J. B T'Siobbel, *J. Pharm. Belg.* [N,S], **18**, 557 (1963).
145. I. M. Roushdi, H. Abdine, and W. S. Abdel Sayed, *J. Pharm. Sci. U. Arab. Rep.* **5**, 91 (1974).
146. H. Bieling, W. Alms, and A. Pogadl, *Z. Chem.* **5**, 376 (1965).
147. K.-C. Lu and H.-S. P'o, *Hua Hsueh Tung Pao* **11**, 55 (1963).
148. F. Püschel and D. Prescher, *J. Chromatog.* **32**, 337 (1968).
149. H. Brückner, G. Göckeritz, and R. Pohloudek-Fabini, *J. Chromatog.* **22**, 490 (1966).
150. R. Pohloudek-Fabini and K.-D. Lüss, *J. Chromatog.* **24**, 291 (1966).
151. M. Qureshi and N. M. Kishore, *Anal. Chem.* **37**, 1267 (1965).
152. W. F. Busby, *Biochim. Biophys. Acta* **121**, 160 (1966).
153. D. Dimitrov, B. N. Panaiotova, and A. Draganov, *Godishnik Khim.-Tekhnol. Inst.* **10**, 145 (1963).
154. Z. Vrba and Z. J. Allan, *Collection Czech. Chem. Commun.* **33**, 4427 (1968); *J. Chromatog.* **42**, D31 (1969).
155. E. V. Sokolova and A. A. Cherkasskii, *Zavodsk. Lab.* **34**, 523 (1968); *J. Chromatog.* **43**, D46 (1969).
156. I. L. Volodarskii and V. V. Kozlov, *Zh. Analit. Khim.* **23**, 763 and 764 (1968); *J. Chromatog.* **38**, D97 and D98 (1968).
157. J. Latinák, *Mikrochim. Acta* p. 350 (1966).
158. A. Demian and J. Reichel, *Rev. Roumaine Chim.* **11**, 263 (1966).
159. J. Reichel and D. Bader, *Rev. Roumaine Chim.* **11**, 1323 (1966).
160. J. Franc and J. Ceeová, *Collection Czech. Chem. Commun.* **33**, 1572 (1968); *J. Chromatog.* **40**, D20 (1969).
161. A. Bukureshchliev, *Vutreshni Bolesti, Uatrechni Bolesti, Nauchno Druzhestvo Internistite* **3**, 573 (1964).
162. T. Wood, *J. Chromatog.* **35**, 352 (1968); **6**, 142 (1961).
162a. T. Wood, Unpublished results, personal communication, 1970.

163. T. Wood and D. E. Abrahams, *Anal. Biochem.* **22**, 117 (1968).
164. M. H. Subers, A. I. Schepartz, and R. P. Koob, *J. Chromatog.* **23**, 319 (1966).
165. S. Hayman, M. F. Lou, L. O. Merola, and J. H. Kinoshita, *Biochim. Biophys. Acta* **128**, 474 (1966).
166. A. Szente and T. Posternak, *J. Chromatog.* **31**, 614 (1967).
167. F. Jungnickel, *J. Chromatog.* **34**, 560 (1968).
168. G. M. Blackburn, J. S. Cohen, and Lord Todd, *J. Chem. Soc., C* p. 239 (1966).
169. N. I. Gabov and A. I. Shafiev, *Zh. Analit. Khim.* **21**, 1107 (1966).
170. W. Wieker, H. P. Walter, and E. Thilo, *Chem. Ber.* **97**, 2385 (1964).
171. P. Sequi, A. Marchesini, and E. Galante, *Ric. Sci. Suppl.* **36**, 183 (1966).
172. W. Kampe, *Chem. Ber.* **98**, 1038 (1965).
173. N. Hennrich and F. Cramer, *J. Am. Chem. Soc.* **87**, 1121 (1965).
174. D. N. Skilleter and R. G. O. Kekwick, *Anal. Biochem.* **20**, 171 (1967).
175. B. S. Vanderheiden, *Bioch. Biophys. Res. Commun.* **21**, 265 (1965).
176. H.-Y. Lu, K.-H. Yeh, and C.-Y. Yuan, *Hua Hsueh Hsueh Pao* **30**, 471 (1964).
177. A. Siuda, *Nukleonika* **10**, 459 (1965).
178. H. Shimizu, Y. Kakimoto, T. Nakajima, A. Kanazawa, and I. Sano, *Nature* **207**, 1197 (1965).
179. N. I. Gabov, *Zh. Analit. Khim.* **22**, 814 (1967).
180. T. A. Kilroe-Smith, *J. Chromatog.* **41**, 116 (1969).
181. I. Wiesner, *Collection Czech. Chem. Commun.* **32**, 4216 (1967).

15

INORGANIC ANALYSIS

This chapter describes the separation and estimation of inorganic ions by paper chromatography. It deals mainly with cations but it also considers anions, metal complexes, radioactive elements, and inorganic phosphates briefly. Schemes of qualitative analyses have been included which should be of special interest to students. Recent advances in the area of inorganic chromatography involve the use of paper impregnated with adsorbents, ion exchangers, or selective partitioning agents and developers containing organic solvents, liquid ion exchangers, complex-forming organic reagents, and organic and mineral acids in various combinations. Reviews have been published recently which discuss the use of paper chromatography in inorganic analysis (*1–7*).

I. Cations

A. Preparation of Sample

Standard solutions of the common metals are prepared by dissolving reagent grade nitrates or chlorides in water, 0.1 *M* HNO_3 , or 0.1 *M* HCl to a final metal concentration of 0.05–0.2 *M*. Some ions require the presence of larger amounts of acid or other reagents to stabilize the solutions. For example, Sn(II), Sn(IV), Sb(III), and As(III) are prepared by dissolving chlorides in 4–6 *M* HCl; oxides of the rare earths and related compounds are dissolved in 0.5–6 *M* HNO_3 ; Mo(VI) is prepared as a solution of molybdic acid made slightly basic with NaOH; and Ti(IV) is often used as a solution of $TiCl_4$ in 0.2 *M* H_2SO_4 and 0.3% hydrogen peroxide (*8–11*). Five to about 50 μl of test solution is spotted at the origin, the exact amount depending upon the ultimate sensitivity of the detection method. Overloading the paper may cause tailing, especially with cationic complexes such as the amines [$Ni(NH_3)_x^{2+}$ etc.] which are strongly adsorbed on the negatively charged filter paper (*12*).

For the systematic analysis of unknowns, samples are prepared by procedures generally employed in classical inorganic qualitative analysis. In one method, a small amount (about 0.2 gm) of an unknown sample is boiled with 2 ml of 2 *N* HNO_3, adding a few drops of H_2O_2 if dichromate, chromate, or permanganate ions are present. The solution is centrifuged after cooling, if necessary, and the clear supernate is diluted with water until the acid concentration is about 0.1 *N* (*13*).

Anions such as tartrate, oxalate, formate, etc., may interfere with the chromatographic analysis of cations and can be removed by carbonate fusion (NH_4^+ may be lost as well), exchange for acetate on a column of resin (*14*, *15*), or precipitation (*16*).

Organic or biological samples are generally ashed before chromatographic analysis. For example, in the determination of Fe, Mn, Zn, and Cu in plant material by paper chromatography (on acid-washed Whatman No. 1 paper developed with butanol–H_2O–HCl, 100:17:23 v/v) and direct reflectance densitometry, the sample is dry-ashed to remove organic matter, and the elements of interest are separated from phosphates and other cations on an anion-exchange column (*16a*).

B. Paper

1. *Untreated Chromatographic Paper*

Whatman No. 1 paper has been most widely used for the separation of inorganic cations. Other papers that have been recommended include Whatman No. 2, 3, 3 MM, 4, 40, and 540; Schleicher and Schuell (S. and S.) No. 2040a, 2040b, 2043a, 2043c, 2045a, and 2045c; Binzer No. 202 (German); Toyo 5B (Japanese) and Munktells OB (*17*).

2. *Acid-Washed Paper*

Some workers believe a preliminary washing with 2–3 *N* HCl (or the solvent to be used for the chromatography) for several days by descending flow is essential, especially for quantitative work, in order to remove traces of metal ions in the manufactured paper (*18–20*). As an alternative, the paper can be soaked in dilute HCl for several days (*21*) or the eluate from a blank paper developed in parallel can be analyzed (*22*). Acid-washed grades of paper are now commercially available from distributors such as H. Reeve Angel and Co. (Whatman papers).

3. *Modified Papers*

The selectivity of chromatography paper such as Whatman 1 can be enhanced by modifying it in various ways. For example, Ag(I) or Ni(II)

(R_F = 1.0) and Cu(II) (0.41) are separated from twenty-seven other metals which precipitate at the origin by development with 10% NH_4OH solution on paper impregnated with strontium chromate (*23*). The separation of Ti(IV) and V(V), which are strongly complexed and remain at the origin, from about thirty other elements which move with the solvent front is obtained by development with 3–5 *M* $HClO_4$ on paper impregnated with phenylbenzohydroxamic acid and 2-octanone (*9*). Paper pretreated with 1% cellulose acetate solution, impregnated with 0.2 *M* 2-thenoyltrifluoroacetone and developed with 0.1 *M* acetate buffer at various pH values separates Pb–Cu, Co–Cu, Zn–Cu, and Fe–Co–Ni (*23a*).

The small natural ion-exchange capacity of paper can be increased by impregnating the paper with synthetic inorganic ion exchangers. The separation of Th(IV) or Ti(IV), Fe(III), and U(VI) is achieved by development with 2 *M* HCl–4 *M* NH_4Cl (1:1 v/v) on paper impregnated with zirconium phosphate (*24*) (see Fig. 1). Cs(I) or Tl(I) (R_F = 0.025),

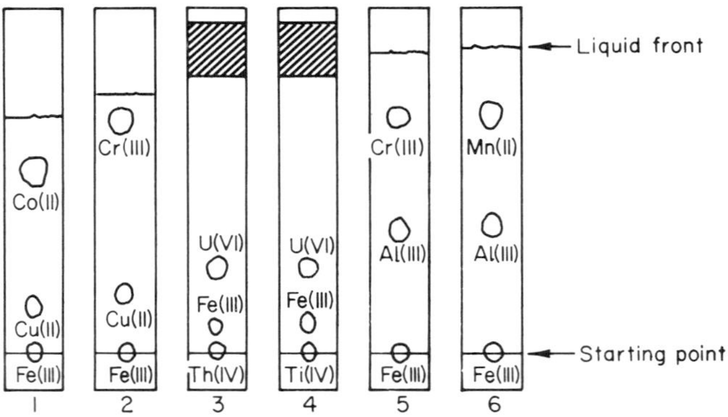

FIG. 1. Separations on zirconium phosphate paper developed with (1) and (2) 0.035 *M* HCl–0.045 *M* NaCl (1:1 v/v); (3) and (4) 2 *M* HCl–4 *M* NH_4Cl (1:1 v/v); (5) and (6) 0.05 *M* HCl. The shaded area represents the localization of several cations, including La, Ce, Ni, Co, Cr, and Al. After G. Alberti and G. Grassini (*24*).

Hg(I) (0.42), Ag(I) (0.62), and Au(III) (0.68) are separated by development with 0.5 *M* NH_4NO_3 in 1 *M* HNO_3 on ammonium molybdophosphate paper; nineteen other ions tested in this system move with R_F 0.75–1.0 (*25*). Numerous separations, such as Au(III) from thirty-six metal ions and Sb(III) from thirty-five metal ions result from development with *n*-butanol–HCl–10% NH_4Cl (7:2:1 v/v) on stannic phosphate paper (*26*).

Stannic phosphate paper is prepared as follows. Make 0.50 *M* $SnCl_4 \cdot 5H_2O$ and 1.00 *M* $(NH_4)_2HPO_4$ solutions in water. Dip paper strips in the stannic chloride solution for 3 seconds. Remove excess reagent by placing the strips on a paper sheet. Then pass the strips through the phosphate solution for 5 seconds. Drain the strips and dry at room temperature. After 1–2 hours, wash twice with water, air dry, and use for chromatography.

Other inorganic exchangers which have been used for paper chromatography include zirconium selenite (*27*), ammonium tungstate (*28*), titanium (IV) tungstate (*28a*), stannic tungstate and selenite (*29*), cerium (IV) phosphate (see Chapter 3, Fig. 4), and zirconium arsenate.

Filter paper impregnated with synthetic organic ion-exchange resins was first produced by Lederer in 1955 (*30*). Since then, such papers have become commercially available (H. Reeve Angel Co.) and have been extensively evaluated for inorganic separations (*31–43a*). Various types of solvents have been used with these papers: 0.50 *M* HCl for the separation of As(III) ($R_F = 0.91$) and Sn(IV) (0.40) from more than twenty other ions on hydrogen-form sulfonic acid cation-exchange paper (SA-2)(*34*); 1 *N* $HClO_4$ containing 5% HF for the separation of Ta(V) (0.0) from Hf, Zr, Ti, Nb, Mo, W, and U on quaternary ammonium anion exchange paper (SB-2) (*41*); 10% lactic acid solution (w/w) at pH 3 for the separation of La(III), Eu(III), Lu(III), and UO_2(II), and at pH 1.5 for La(III), Sm(III), Eu(III), Tb(III), Dy(III), and UO_2(II) on hydrogen-form SA-2 paper; 2 *N* sodium acetate–2 *N* acetic acid (1:1 v/v) for the separation of Pb(II) (0.06), Ag(I) (0.13), and Tl(I) (0.24) from each other on hydrogen-form carboxylic paper (WA-2) (*42*); 0.125 *M* NH_3 triacetic acid in 3.0 *M* NH_3, pH 11.0, for the separation of Ag(I) (0.77) and Tl(I) (0.21) from twenty other metals on chloride-form SB-2 paper (*32*); 90% tetrahydrofuran–10% 6 *N* HNO_3 (v/v) for the separation of Au(III) (0.91) and UO_2(II) (0.13) from ten other metals on hydrogen-form SA-2 paper (*39*); 95% tetrahydrofuran–5% 6 *M* HCl–0.5 *M* trioctylphosphine oxide for the separation of Sc(III) (0.6) from other rare earths onhydrogen-form SA-2 paper (*43*); 1.0 *M* NH_4Cl for the separation of Hg(II) (0.04), Bi(III) (0.11), Cd (0.52), Cu(II) (0.27), and Pb (0.47) on sodium-form SA-2 paper (*43a*); and HCNS (*37*) (see Fig. 2). Complexing, partitioning, and ion-exchange effects in these solvents, combined with the ion-exchange and adsorption effects of the paper lead to very high selectivities for inorganic ions (*39*).

Cellulose ion-exchangers (cellulose paper impregnated with ion-exchange cellulose fibers) in combination with aqueous or mixed developers are also useful for inorganic separations. Hg(II) is separated from sixteen other metals by development with 0.010 *M* NH_4SCN–

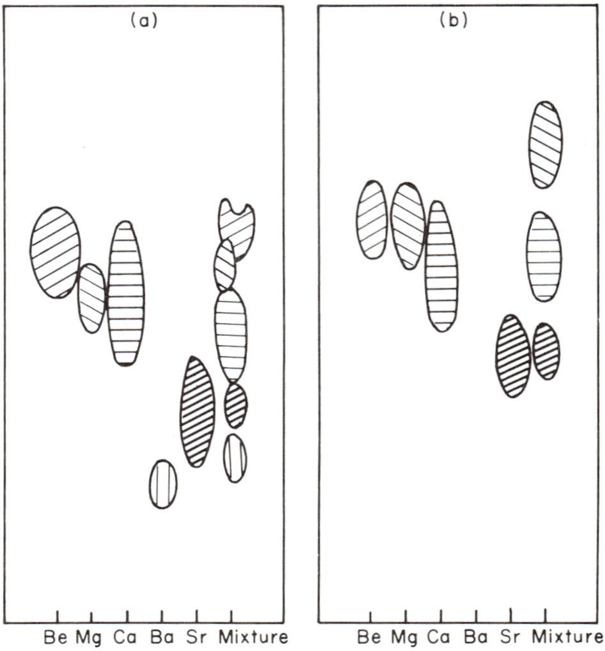

FIG. 2. Migration of alkaline earth ions on Machery Nagel sulfonic cellulose paper with (a) 0.5 *N* HCNS and (b) 1.0 *N* HCNS. Note that the R_F values change somewhat with the loading and presence of other alkaline earths. After G. Bagliono, *et al.* (*37*).

0.10 *M* HCl on diethylaminoethylcellulose paper (Whatman No. DE 81) (*10*), and UO_2(II), Cu(II), Zn(II), Co(II), Cd(II), and Fe(III) are separated by development with acetone–2 *N* HCl and ethanol–2 *N* HCl mixtures on Whatman cellulose phosphate P-81, diethylamino paper DE-20, and aminoethylcellulose AE-30 (*44*, *45*). Development with 80% methanol–2*M* HCl on P-81 paper separates sodium (R_F = 0.63) and potassium (0.40) (*45a*).

Paper impregnated with liquid ion-exchangers has also been employed for metal ion separations. Such paper is similar to the solid resin impregnated papers mentioned above except that the use of partly organic solvent mixtures is impossible (*46*). On paper impregnated with di-(2-ethylhexyl) orthophosphoric acid and developed with HCl or acetic acid, the separations shown in Fig. 3 are achieved (*47*). Various anion exchangers such as tri-*n*-octylamine and triisooctylamine are also useful (*46*, *47*). The liquid ion exchanger can instead be put into the solvent: for example, benzene solutions of tri-*n*-octylamine hydrochloride

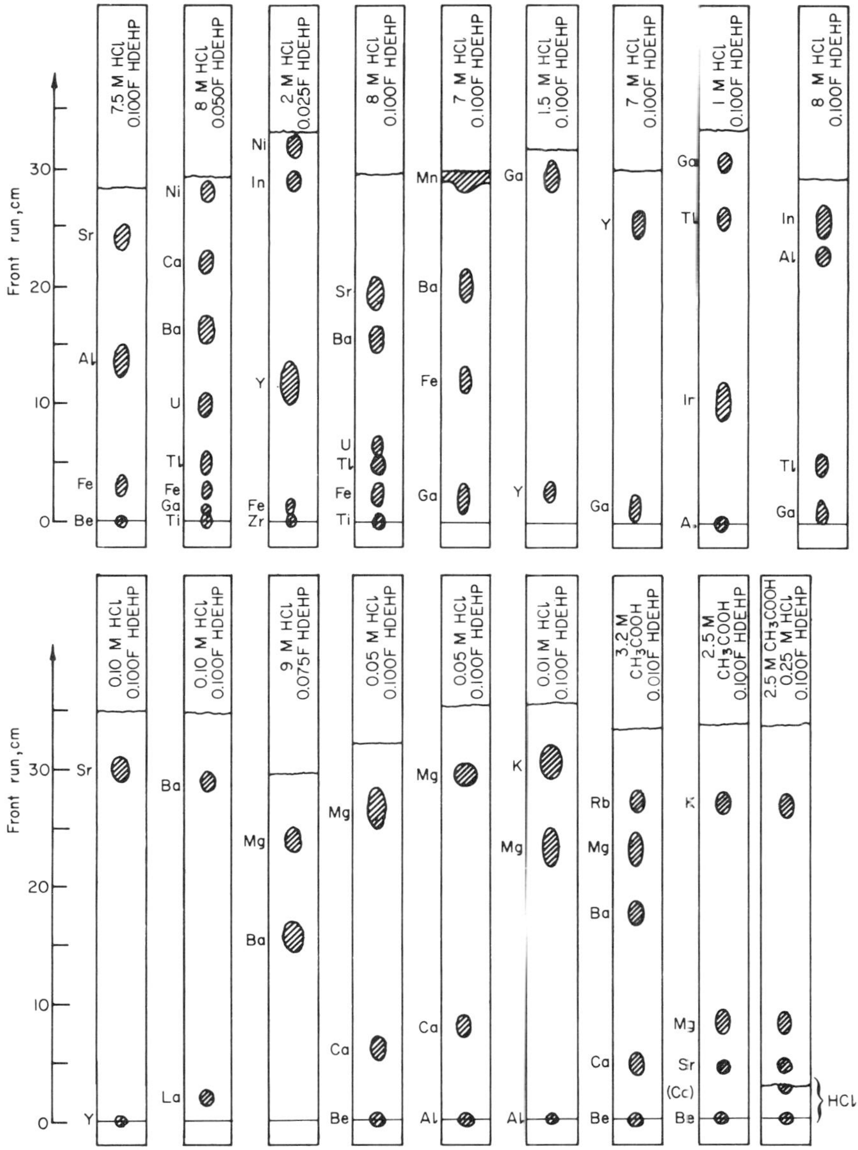

FIG. 3. Chromatographic separations obtained by ascending development on Whatman No. 1 paper treated with di(2-ethylhexyl) orthophosphoric acid (HDEHP) in cyclohexane. The conditions are reported in the figure. The last diagram refers to double development, the former with acetic acid, the latter with HCl. After E. Cerrai, (*47*).

(0.1 *M*) are used to develop moist papers impregnated with HCl (*47a*).

Ion-exchange papers impregnated with various extractive and gravimetric reagents and developed with complex-forming solvents exhibit very great selectivities for metal ions. Thus, Cd(II) ($R_F = 0.35$) was separated from twenty-five other ions when developed with 0.5 *M* HCl on SA-2 cation exchange paper impregnated with phenylbenzohydroxamic acid (*48*).

R_F values for inorganic ions on styrenated paper (Whatman No. 41 with 18% styrene graft) developed with *n*-butanol saturated with 1 *N* HCl are tabulated below (*49*).

Ion	R_F	Ion	R_F
As(III)	0.70	Ni(II)	0.19
Sb(III)	0.72	Cd(II)	0.57
Sn(II)	0.82	Bi(III)	0.72
Fe(III)	0.26	Cu(II)	0.20
Al(III)	0.18	MoO_4^{-2}	0.42
Zn(II)	0.66	Be(II)	0.27
Au(III)	0.86	Co(II)	0.17
Hg(II)	0.81	Mn(II)	0.20

C. Solvents

Most solvents for the separation of the common inorganic cations by partition chromatography are similar to those mentioned above for use with modified papers. Usually, the solvents consist of a polar constituent, such as an alcohol, ester, ketone, or ether, combined with water and a mineral acid or other complexing agent such as benzoylacetone or antipyrine. However, aqueous solvents such as NH_3–NH_4Cl and acetate buffers are also useful (*50, 51*). The selection of the solvent is generally empirical, although some attempts have been made to theoretically predict the best solvent for a particular separation (*52, 53*). In this section is listed a few of the recommended solvents for successful separations of mixtures of common metals on unmodified paper. For a comprehensive list of solvents for specific separations, the reader is referred to Tables I, II, and VI, and to other sources (*17, 54, 55*).

1. *Solvents for the Systematic Analysis of Cations*

(a) *n*-Butanol–benzoylacetone–HNO_3 (5 gm of benzoylacetone is dissolved in 50 ml of *n*-butanol and shaken with 50 ml of 0.1 *M* HNO_3; the upper organic layer is used for development), (b) *S*-collidine–0.4 *M*

HNO_3 (v/v), and (c) dioxane–antipyrine (1 gm antipyrine is dissolved in a solution of 100 ml dioxane, 1 ml concentrated HNO_3, and 2.5 ml water) (*13*), or ethanol–HCl–H_2O (15:4:1 v/v) followed by tetrahydrofuran–HNO_3–H_2O (17:1:2 v/v) for two-dimensional development (*56*).

2. *Methanol–2 to 10 M Nitric Acid (2:1)*

This solvent is used to develop sixty-seven inorganic ions (*57*). As the amount of acid is increased, the migration rates of many of the cations decreases allowing separations not previously accomplished. Correlations between the dielectric constant of mixtures of various alcohols with formic acid in ratios 1:9 to 9:1 (v/v) and R_F values for many metal ions have also been made (*58*).

3. *n-Butanol–HCl*

Butanol is shaken in a separatory funnel with 3 *N* HCl, and the organic phase is used for the separation of Pb, Hg(II), Bi, Cu, and Cd (*59*). Using different concentrations of HCl, a larger variety of mixtures can be separated. Presaturation of the paper with this solvent has a marked effect on the movement of Bi and Cd (*60*).

A mixture of 80 ml *n*-butanol and 20 ml concentrated HCl has been used to separate cation–oxine complexes (*61*). A novel modification of the *n*-butanol–3 *N* HCl solvent has been to saturate the organic phase with H_2S gas, thus causing Cd^{2+} to move to an R_F of 0.76, while Cu, Pb, Bi, and Hg remain at the origin (*62*).

4. *Acetone–H_2O–concentrated HCl (7:2:1 v/v)*

This solvent resolves the mixtures Ba–Sr–Ca–Mg–Be and Li–Na–K. In the ratio 40:10:5 v/v, it is a good universal solvent for many cations (*63*).

5. *n-Butanol–2.5 N HCNS (140:40 v/v)*

This single-phase solvent is used for rapid horizontal chromatography of heavy metal ions. The movement of Bi(III) in this solvent seems to be due to adsorption rather than partition (*64*).

6. *Two-solvent One-Dimensional System for Co, Ni, Cu, and Zn (65)*

(a) *n*-Butanol–H_2O-concentrated HCl (60:15:5 v/v); descending development overnight; Zn moves ahead of the other ions. (b) Acetone–3-methyl-2-butanone–6 *N* HCl (25:21:3 v/v); Co–Ni–Cu will separate. (See Section F1 below.)

7. *Two-Dimensional System for Heavy Metal Ions (66)*

(a) Acetone–8 *N* acetic acid (9:1 v/v). (b) 90% Ethanol–2 *N* acetic acid (8:2 v/v).

TABLE I

R_F VALUES OF METAL IONS ON UNTREATED PAPER SUCH AS WHATMAN NO. 1 (*8*, *17*, *77–79*)

Metal	R_F [a]												
	1	2	3	4	5	6	7	8	9	10	11	12	13
Ag(I)	0.02	0.0	0.0	0.26–0.57	0.0	0.18	0.12	—	—	0.10	0.0 –0.68	—[b]	0.0 –1.0
Al(III)	0.37	0.04	0.21	0.0 –0.03	0.42	0.10	0.15	—	—	—	0.21–0.45	0.53	0.88–1.0
As(III)	0.50	0.46	0.83	0.83–0.88	0.65	0.45	0.15	0.13	0.27	—	0.79–0.90	0.33	—
Ba(II)	0.04	0.0	0.09	0.0	0.08	0.06	0.17	—	—	—	0.0 –0.31	0.51	0.74–0.87
Bi(III)	0.94	0.41	0.76	0.41–0.60	0.81	0.20	0.34	0.30	—[b]	—	0.44–0.62	0.30	0.88–1.0
Ca(II)	—	—	0.18	0.01–0.04	0.27	0.12	0.20	—	—	—	0.23–0.38	0.50	0.86–1.0
Cd(II)	1.0	0.37	0.97	0.82–0.93	0.82	0.15	0.28	0.22	0.32	1.0	0.64–0.76	0.45	0.86–1.0
Co(II)	0.32	0.03	0.22	0.65–0.82	0.38	0.12	0.24	—	—	—	0.58–0.76	0.51	0.87–0.96
Cr(III)	0.47	0.04	0.22	0.0 –0.03	0.42	0.10	0.65	—	—	—	0.21–0.45	0.50	0.93–1.0
Cu(II)	0.47	0.05	0.31	0.65–0.76	0.46	0.15	0.63	0.06	0.20	0.74	0.56–0.68	0.40	0.82–1.0
Fe(III)	0.56	0.05	0.42	0.89–0.98[c]	0.73	0.13	0.73	0.38	0.80	0.92	0.80–1.0	0.62	0.81–1.0
Hg_2(II)	0.08	—	—	0.84–0.90[c]	0.0	0.50	0.09	—	—	—	0.0 –0.72	0.50[c]	0.82–0.94
Hg(II)	1.0	0.06	1.0	—	0.90	0.50	0.83	0.74	0.36	—	0.61–0.76	0.60	—
K(I)	0.08	0.06	0.19	0.0 –0.03	0.14	0.11	0.23	—	—	—	—	—	—
Mg(II)	0.33	0.07	0.24	0.01–0.05	0.40	0.13	0.20	—	—	—	0.0 –0.21	0.50	0.86–0.96
Mn(II)	0.36	0.06	0.24	0.15–0.22	0.48	0.13	0.25	—	—	—	0.46–0.66	0.42	0.71–1.0
Na(I)	—	—	0.24	0.0 –0.06	0.24	0.11	0.25	—	—	—	—	—	—
Ni(II)	0.34	0.03	0.22	0.01–0.06	0.36	0.12	0.24	—	—	—	0.21–0.39	0.46	0.81–1.0

Pb(II)	0.16	0.0	0.39	0.12–0.23	0.33	0.09	0.18	—	—	—	0.21–0.62	0.38	0.83–0.94
Sb(III)	0.85	—[b]	0.99	0.88–0.95[c]	0.81	0.0	0.13	0.21	0.68	—	0.58–1.0[c]	0.63	0.81–0.98
Sn(II)	0.97	0.55	1.0	0.86–0.99	0.84	0.70	0.13	—[b]	0.83	—	0.68–0.75	0.60	0.68–0.87
Sn(IV)	—	—	1.0	0.0	0.84	0.65	0.15	—	—	—	0.61–0.78	—	0.71–0.87
Sr(II)	0.11	0.0	0.09	0.0	0.15	0.08	0.18	—	—	—	0.23–0.32	0.42	0.75–0.87
Zn(II)	0.93	0.51	0.98	0.80–0.93	0.84	0.12	0.30	—	—	—	0.64–0.75	0.57	0.81–0.96
UO_2(II)	—	—	—	—	—	—	—	0.10	0.30	0.28	0.54–0.72	0.53	0.81–0.93
WO_4^{-2}	—	—	—	—	—	—	—	—	—	0.0	—	—	0.0 –1.0
MoO_4^{-2}	—	—	—	—	—	—	—	—	—	0.69	—	—	0.64–0.91
Zr(IV)	—	—	—	—	—	—	—	—	—	0.0	0.0 –0.68	0.60[b]	0.0
Th(IV)	—	—	—	—	—	—	—	—	—	0.11	0.16–0.44	—[b]	0.87–1.0
VO(II)	—	—	—	—	—	—	—	—	—	0.13	—	—[b]	0.87–0.96
Au(III)	—	—	—	—	—	—	—	—	—	1.0	0.38–0.55	—	—
Pt(IV)	—	—	—	—	—	—	—	—	—	0.77	—	—	—
Pd(II)	—	—	—	—	—	—	—	—	—	0.57	0.43–0.60	—	0.75–0.90
Tl(I)	—	—	—	—	—	—	—	—	—	—	0.0 –0.59	0.71	0.31–1.0
Fe(II)	—	—	—	—	—	—	—	—	—	—	0.75–1.0	—	—

[a] Solvents: (1) ethanol–5 *M* HCl (9:1 v/v); (2) 100 ml *n*-butanol saturated by shaking with 100 ml 1 *M* HCl; (3) 100 ml *n*-butanol saturated by shaking with 100 ml 4 *M* HCl; (4) di-*n*-propyl ketone–10 *M* HCl (85:15 v/v); (5) diethyl ether–methanol–H_2O-concentrated HCl (50:30:15:4 v/v); (6) 95 ml *n*-butanol + 5 ml acetoacetic ester saturated by shaking with 100 ml 0.1 *M* HNO_3 ; (7) *n*-butanol–H_2O–glacial acetic acid–acetoacetic ester (50:35:10:5 v/v); (8) methyl ethyl ketone–10% aqueous HCl (9:1 v/v); (9) cyclohexanone–30% aqueous HCl (9:1 v/v); (10) chloroform–acetone–amyl alcohol–concentrated HCl (47.6:23.8:23.8:4.5) (radial development); (11) formic acid–concentrated HCl–acetone (3:3:4 v/v); (12) cyclohexanol–formic acid (3:7 v/v); (13) HCl–H_3PO_4–H_2O (10:1:9 v/v).

[b] Trailing zone.

[c] Trace at origin.

TABLE II

R_F VALUES OF METAL IONS ON MODIFIED PAPERS

Metal	R_F [a]												
	1	2	3	4	5	6	7	8	9	10	11	12	13
Be(II)	0.05–0.90	0.00–0.18	0.00–0.90	—	0.98	0.98	0.37	—	—	0.26–0.52	0.18, 0.46	—	—
Sc(III)	0.0	0.0	0.10	0.01	0.99	0.93	—	0.69	0.05	—	0.04	—	—
Y(III)	0.98	0.98	0.98	0.98	—	—	—	0.90	0.79	0.0 –1.0	0.23	—	—
La(III)	0.97	0.98	0.98	0.98	0.94	0.95	—	1.0	—	0.0 –0.07	0.14	0.06	—
Ti(IV)	0.0	0.0	0.0	0.0	0.95	0.95	0.0	—	0.11	0.0 –0.37	—	—	—
Zr(IV)	0.01	0.0	0.02	0.0	0.94	0.96	—	0.19	0.02	0.0 –0.10	0.03	—	—
Th(IV)	0.96	0.30	0.98	0.45	0.95	0.93	0.59	0.06	0.0	0.0 –0.07	—	—	—
V(IV)	—	—	—	—	—	—	—	—	0.0	0.22–0.40	—	—	—
V(V)	0.03	0.03	—	—	—	—	0.76	—	—	—	—	—	0.05
Cr(III)	0.98	0.98	0.99	0.99	0.98	0.98	0.66	1.0	1.0	0.0 –0.20	—	—	—
Mo(VI)	0.0	0.0	0.0	0.0	—	—	0.61	—	0.0	0.83–0.97	—	—	—
W(VI)	0.0	0.0	0.0	0.0	0.06	0.0	—	—	—	—	—	—	—
U(VI)	0.0	0.0	0.0	0.01	0.95	0.13	0.94	0.05	0.0	0.68–0.90	0.7[b]	—	—
Mn(II)	0.08	0.12	0.10	0.30	0.98	0.95	0.23	—	0.90	0.21–0.36	0.64	—	—
Fe(III)	0.0	0.0	0.0	0.01	0.50–0.99	0.01	0.88	1.0	0.08	0.87–1.0	0.0	0.51	0.88
Co(II)	0.0	0.0	0.0	0.01	0.94	0.55	0.64	0.95	0.92	0.54–0.75	0.58	0.39	0.55
Ni(II)	0.79	0.99	0.80	0.99	0.98	0.95	0.58	1.0	0.96	0.00–0.20	0.57	0.39	0.0
Pd(II)	0.0	0.09	0.0	0.10	0.0	0.0	—	1.0	—	0.50–0.85	0.51[b]	—	—
Pt(IV)	0.01	0.01	—	—	0.01	0.01	—	—	0.02	0.77–0.94	0.28[b]	—	—
Cu(II)	0.0	0.63	0.03	0.84	0.82	0.03	0.74	1.0	0.76	0.60–0.76	0.21	0.61	0.55
Ag(I)	0.07	0.97	0.13	0.98	0.0	0.31	0.40	—[c]	—	0.0 –0.09	0.17	—	—
Au(III)	0.0	0.08	—	—	0.0	0.0	—	—	0.0	0.94–1.0	0.12[b]	0.10	—
Zn(II)	0.02	0.0	0.02	0.01	0.10	0.02	0.23	1.0	0.02	0.80–0.92	0.35	0.04–0.51	—
Cd(II)	0.0	0.23	0.02	0.56	0.01	0.01	0.63	1.0	0.02	0.85–1.0	0.21	—	0.70
Hg(II)	0.0	0.12	0.01	0.22	0.0	0.0	0.83	0.10	0.0	0.0 –0.90	0.05	—	—
Al(III)	0.98	0.23–0.87	0.98	0.52–0.91	0.97	0.98	0.27	1.0	1.0	0.0 –0.20	0.0 –0.52	—	—

Ga(III)	0.03	0.01	0.02	0.02	0.95	0.0 –0.09	—	—	—	0.90–1.0	0.5[b]	—	—
In(III)	0.0	0.0	0.0	0.0	0.0	0.0	—	—	0.0	0.68–0.92	—	—	—
Tl(I)	0.64	0.98	0.66	0.98	0.0	0.0	0.26	—	—	0.0 –0.22	0.33	—	—
Ge(IV)	0.96	0.98	0.96	0.98	0.96	0.99	—	—	—	—	—	—	—
Sn(II)	0.0	0.0	0.0	0.0	0.01	0.01	0.90	—	0.0	—	—	—	—
Pb(II)	0.21	0.98	0.45	0.99	0.0	0.30	0.48	1.0	—	0.04–0.47	0.15	—	—[c]
As(III)	0.90	0.99	0.90	0.99	0.92	0.20	0.30	—	—	0.0 –0.66	—	—	—
Sb(III)	0.01–0.28	0.08	0.0 –0.30	0.16–0.40	0.03	0.0	0.0	—	—	0.88–1.0	0.48[b]	—	—
Bi(III)	0.0	0.09	0.01	0.23	0.01	0.0	0.10	0.95	—	0.45–0.80	0.02	—	—[c]
Se(IV)	0.91	0.91	0.91	0.93	0.88	0.12	—	—	—	0.0 –0.16	—	—	—
Te(IV)	0.0 –0.82	0.02–0.82	0.0 –0.80	0.30–0.80	0.20	0.05	—	—	—	0.85–1.0	—	—	—
Sn(IV)	—	—	—	—	—	—	0.0	—	—	—	—	—	—
K(I)	—	—	—	—	—	—	0.05	—	—	0.0 –0.07	—	—	—
Na(I)	—	—	—	—	—	—	0.08	—	—	—	—	—	—
Ca(II)	—	—	—	—	—	—	0.13	—	—	0.0 –0.08	0.49	—	—
Sr(II)	—	—	—	—	—	—	0.13	—	—	0.0	0.48	0.16	—
Ba(II)	—	—	—	—	—	—	0.14	—	—	0.0	0.32	0.06	—
Mg(II)	—	—	—	—	—	—	0.19	—	—	0.0 –0.20	0.79	—	—
Li(I)	—	—	—	—	—	—	0.19	—	—	—	—	—	—
Ce(III)	—	—	—	—	—	—	0.64	—	—	0.0 –0.08	0.16	0.03	—
Hg_2(II)	—	—	—	—	—	—	0.70	—	—	0.95–1.0	—	—	—
Eu(III)	—	—	—	—	—	—	—	0.92	—	—	—	—	—

[a] Solvent: (1) 1 *M* aqueous NH_4SCN, Whatman No. 1 paper impregnated with 0.2 *M* tri-*n*-butylamine in benzene; (2) 7 *M* aqueous NH_4SCN, stationary phase same as 1; (3) 1 *M* aqueous NH_4SCN, Whatman No. 1 paper impregnated with 0.2 *M* tri-*n*-octylamine in benzene; (4) 7 *M* aqueous NH_4SCN, stationary phase same as 3 (*80*); (5) 0.5 *M* HBr, Whatman No. 1 paper impregnated with 0.2 *M* tri-*n*-octylamine hydrobromide in benzene; (6) 5 *M* HBr; stationary phase same as 5 (*81*); (7) *n*-butanol–15% aqueous trichloroacetic acid–1-phenyl-1,3-butanedione (100:100:1 v/v/w), carboxymethylcellulose paper (*82*); (8) 3 *M* HNO_3, paper impregnated with 20% by volume tri-*n*-butyl phosphate in CCl_4 (*72*); (9) 0.5 *M* HCl, paper impregnated with 0.1 *M* bis (di-*n*-hexyl-phosphinyl) methane in CCl_4 (*72*); (10) *n*-butanol–dioxane–10 *M* HCl (6:2:2 v/v), Whatman No. 1 paper impregnated with stannic tungstate (*29*); (11) 3 *N* sodium acetate–3 *N* acetic acid, Amberlite WA-2 hydrogen-form carboxylic acid paper (*42*); (12) acetic acid–2.4 *N* HCl (30:70 v/v), Amberlite sulphonic resin paper, hydrogen form (*31*); (13) acetone–6 *N* HCl (9:1 v/v), Amberlite sulphonic resin paper, hydrogen form (*40*).

[b] Solvent is 6 *N* HCl.

[c] Trailing zone.

8. *Solvents Containing Benzyl Alcohol or HF*

Studies of sixty inorganic ions (*66a*) indicate that solvents containing benzyl alcohol (e.g., benzyl alcohol–ethanol–6 *N* HCl and benzyl alcohol–dioxane–8 *N* HCl, both 2:2:1 v/v) or HF plus HCl (especially butanol–40% HF–6 *N* HCl, 15:1:4 or 15:1:3) are effective in separating multicomponent mixtures on Whatman No. 1 paper.

D. Techniques

Conventional, simple techniques of paper chromatography are recommended for the separation of inorganic cations. Since many solvents contain corrosive mineral acids, it is recommended to use all-glass apparatuses. The ascending technique with test tubes or all-glass cylinders is useful for survey work or qualitative analysis, and horizontal development of paper disks with Petri dishes (*67*) is recommended for solvents requiring good vapor saturation (see Chapter 3, Figure 15). Some special techniques for the chromatography of inorganic cations are described below.

1. *Impregnated Paper.*

For reversed-phase paper chromatography (often termed precipitation or extraction chromatography), mentioned briefly above, the paper is impregnated with a cation complexing or precipitating agent prior to spotting of the sample and solvent development. The order of migration of the ions is found to be inversely proportional to the stability and solubility of the complexes which are formed on the paper (*68*).

A filter paper disk is immersed for 1 hour in 80 ml 95% ethanol containing 1.2 gm 8-hydroxyquinoline and is dried by heating. The developing solvent is propionic acid–2% acetic acid–*n*-butanol–H_2O (1:20:4:40 v/v). Development is horizontal, and definite zones for the following mixture are visible without further color development: Cu–Fe(III)–Ni–Co–Cd–Mn(II)–Mg (*69*).

An analogous technique is the impregnation of filter paper with dithizone (diphenylthiocarbazone) and 0.1 *N* HNO_3–acetone (10:1 v/v) as the developing solvent. A mixture of Hg–Cu–Bi–Cd is resolved into colored zones by horizontal development (*70*).

Paper treated with a commercial mixture of substituted oximes (Lix-64, General Mills Co.) and developed with a solution 0.5 *M* in $(NH_4)_2SO_4$ and 0.18 *M* in NH_4HF_2 separates Cu ($R_F = 0.0$) from Fe(II), Fe(III), Co, Ni, Mn(II), Cr(III), Mo, W, and V(IV) (*70a*).

By a novel technique in which the paper is only partially treated with a precipitating agent (CdS), it is possible to separate Pb from Hg(II), Hg(I), Ag, Cu(II), and Bi(III) (*71*). A zone of CdS is painted horizontally across the filter paper above the point of application. The mixture of cations is spotted and developed with isopropanol–H_2O–acetic acid (50:45:1 v/v) to which is added 5 gm ammonium acetate. The dried chromatogram is sprayed with a solution of $(NH_4)_2S$, and a brown-black spot for PbS appears 2–4 cm above the CdS zone.

Paper impregnated with bis-(di-*n*-hexylphosphinyl) methane (HDPM) or other neutral organophosphorous compounds and developed with different molarities of HCl, HNO_3, and $HClO_4$ resolves mixtures of many metal cations including Na–Li, alkaline earths, and lanthanides (see Table II). Sheets or disks of paper are soaked in 0.1 *M* HDPM in CCl_4. After the solvent evaporates, the paper is spotted and developed by the ascending or radial technique. The zones are located by spraying with 8-hydroxyquinoline and then exposing the paper to fumes of ammonia (*72*).

Paper impregnated with a mixture of trioctyl phosphate and 2-thenoyltrifluoroacetone (e.g., 2:3 v/v) and developed with 0.05 or 0.075 *M* HCl serves to separate Am(III), Ce (III) and La (*72a*). Paper treated with triallyl phosphate serves to separate UO_2(II), Cu(II), and Fe(III) when 0.1–4 *N* HCl is used as the developing solvent (*72b*).

2. *Two-Dimensional Method Combining Ion-Exchange and Partition Techniques* (*73*)

On cellulose ion-exchanger papers of the type mentioned above, it is possible to combine the ion exchange and partition techniques in the form of a two-dimensional chromatogram. In the first direction, the pH of the developer is such that the ion-exchange characteristics of the paper are suppressed. The second development is with a neutral salt so that the ion-exchange mechanism is operative. In this way, a mixture of eight ions (see Fig. 4) is separated by development on cellulose phosphate paper (Whatman P-81) with *n*-butanol–HCl followed by 1 *N* $MgCl_2$. After the first run, the spots are in the positions expected on unmodified cellulose, except for iron which is complexed by the phosphate groups and remains at the origin. In the second direction, the ions migrate as expected on cation-exchange paper. The flat shape of the cadmium zone in the $MgCl_2$ direction is attributed to the occurrence of a magnesium–ammonium boundary at this point. The separated spots are detected by spraying with a pentacyanoamminoferrate–rubeanic acid complex, which produces characteristic colors with many cations (see Section E,1,e).

3. *Wet Start Technique*

Papers loaded with synthetic organic ion exchange resins and "strong" cellulose ion exchange papers (e.g., cellulose phosphate) are prepared by

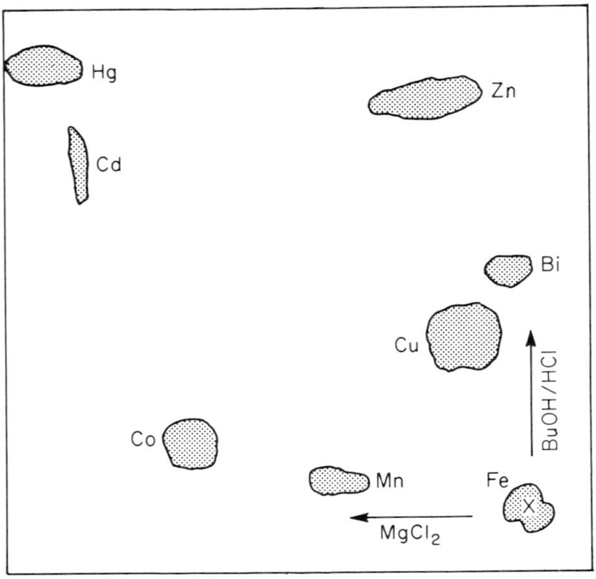

FIG. 4. Two-dimensional separation of metal ions on a 14 inch square of phosphorylated cellulose paper containing dihydrogen phosphate groups in the monoammonium salt form. Development is by the downward flow of the organic phase of a mixture of *n*-butanol saturated with 2 *N* HCl (partition chromatography) followed by 1 *N* magnesium chloride (ion-exchange chromatography). After the first run, the sheet is dried over ammonia to neutralize excess acid and to convert the paper to the original form (*73*).

washing with (or soaking in) a suitable salt for conversion to the desired ionic form, washing out the excess salt with water, and drying prior to chromatography. "Weak" cellulose exchangers cannot be washed free of excess electrolyte without disturbing the established equilibrium so that the wet start development technique must be employed (*74*). The paper is converted by the downward flow of solvent, removed from the tank while wet, lightly blotted, spotted, replaced in the tank, and developed before any appreciable drying can occur (*10*). A disadvantage here is the absence of a physical solvent front, and mobilities are best recorded relative to some standard marker.

4. *Frontal Development of the Solvent*

In many systems involving unmodified and modified papers, solvents, especially those containing an organic solvent and a mineral acid, can be altered as they pass over the dry paper, leading to the formation of two solvent fronts. The area between the origin and the first front contains all the components of the solvent, while the area between the fronts usually contains dehydrated organic solvent. The first front is often termed the acid front because the acid component of the solvent does not reach the top part of the paper (*52, 75*). Many workers are unaware of these fronts and do not realize that many successful separations of inorganic ions are the result of solvent gradients of this type produced on the paper. In some cases, however, effective separations are impossible because frontal development of the solvent precludes the attainment of equilibrium conditions by the usual development techniques. Ossicini and Lederer (*76*) have measured the heights of such fronts in various ion-exchange paper systems and suggest that in some cases the front effect can be overcome by allowing the solvent to run 5 or 10 cm past the origin and then placing the spot at a measured distance behind the front (*42*). Migration distances are then corrected by this distance before the calculation of R_F values (*39*). It is claimed that stirring the gas phase equilibrated with the solvent inside the tank decreases the steepness of the elution gradient caused by chromatographic demixing of the solvent (*76a*).

E. Detection of spots

A description of some useful general color reagents and confirmatory sprays for specific ions will be given. The reader is cautioned not to interpret the indicated colors too literally, since higher concentrations of the ions may change a color, for example, from red to pink. The majority of reagents are chosen for their greatest sensitivity but may react, as well, with other ions which are not listed. By examination of the color produced with a specific reagent and the R_F value obtained in a particular solvent, identifications of separated ions can often be made. Likewise, even though two or more ions might overlap after development, the use of selective spray reagents often allows each of the ions to be identified.

1. *General Spray Reagents.*

a. *8-Hydroxyquinoline* (*Oxine, 8-Quinolinol*). One-half gram reagent in 100 ml 60% ethanol. The chromatogram is sprayed and exposed to NH_3 vapors; it is observed in daylight and under ultraviolet. A bright fluorescence is exhibited by Al, Be, Ca, Sr, Ge, Mg, Zn, Sn, Cd, and Ba, but most other cations show up as dark spots (*83, 84*). A modification of this reagent consists of the addition of 0.1 gm kojic acid (see Table III). As and Sb are not detected.

TABLE III

APPEARANCE OF METAL IONS AFTER OXINE-KOJIC ACID SPRAY (*13*)

Metal	Color	
	Daylight	Ultraviolet
Ag	Light brown	Yellow
Hg(I)	Yellow-green	Black
Pb	Pale yellow	Dark gray
Cu	Green	Black
Bi	Pale yellow	Dark brown
Cd	Pale yellow	Bright yellow
Hg(II)	Bright yellow	Dark brown
As	Colorless	—
Sb	Colorless	—
Sn(II)	Pale yellow	Dirty yellow
Sn(IV)	Colorless	Blue
Co	Yellow-brown	Black
Ni	Yellow-green	Black
Fe(III)	Black	Black
Al	Pale yellow	Whitish yellow
Cr	Yellow-green	Violet
Mn	Yellow-brown	Black
Zn	Pale yellow	Bright yellow
Ba	Colorless	Whitish blue
Sr	Colorless	Whitish blue
Ca	Colorless	Whitish blue
Mg	Colorless	White
Na	Colorless	Violet
K	Colorless	Violet

b. *Hydrogen Sulfide*. Spraying with 2% aqueous yellow ammonium sulfide, followed by HCl, reveals colored sulfides of Ag, Tl, Pb, Cu, Cd, Co, Ni, Hg, As, Fe, Sb, V, Bi, Sn, Pd, Au, and Pt (*33*).

c. *Quercetin*. A 0.2% alcoholic solution followed by concentrated NH_4OH causes colors to form with Ag, Al, Bi, Ca, Cd, Co, Cr, Cu, Hg, Fe, Mg, Mn, Ni, Pb, Sb, Sn, U, and Zn. Many of these spots also fluoresce under ultraviolet light (*85*).

d. *Alizarin*. A saturated ethanolic solution reacts with many cations to give violet to red colors (*55*). After spraying, the chromatogram is exposed to ammonia and then heated or sprayed with glacial acetic acid to eliminate the background color. Positive color reaction is given with Ag, Hg, Pb, Cu, Cd, Bi, Fe, Cr, Zn, Mn, Co, Ni, Ca, Ga, In, Be, Zr, Ti, rare earths, Ce, Sc, Th, Pd, Pt, and U.

e. *Rubeanic Acid.* A 0.5% solution in 96% ethanol followed by exposure to ammonia is a good general detection reagent for most of the heavy metals (*55*). Pentacyanoamminoferrate–rubeanic acid complex is prepared by dissolving 0.70 gm trisodium pentacyanoamminoferrate in 20 ml water and pouring the resulting solution into a solution of 0.25 gm rubeanic acid in 10 ml ethanol. This mixture is shaken for 15 minutes, filtered and sprayed on the paper followed by washing with 0.2 *N* acetic acid. The reagent must be made fresh daily. Colors are observed with Cu, Ag, Au, Zn, Cd, Hg, Tl, rare earths, Sn, Pb, Ti, Zr, Hf, Th, Bi, Mo, U, Mn, Fe, Co, Ni, and Pd (*86*).

f. *Glyoxal-bis(2-hydroxyanil).* A 1% solution in methanol containing 3% KOH forms colors with thirty-four metal ions including Ni, Co, Cu, Cd, Zn, Be, Mo, Ti, U, V, W, and Zr (*87*).

2. *Confirmatory Sprays for the Identification of Individual Cations* (*17, 55*).

a. *Tannic Acid.* 10% in water; Ag gives a brown spot, Ti gives a red-orange spot after spraying with 2 *N* acetic acid.

b. *8-Hydroxyquinoline.* (See Section 1a above). Fluorescence in ultraviolet after exposure to NH_3 : Al, yellow-green; Ca, yellow-green to blue-green; Cd, yellow; Ge, yellow; La, green; Ba, Sr, Mg, bluish-green; Yb, dark (yellow in visible light); Sn(II), Sn(IV), Zn, yellow. The bright fluorescence of Al, Zn, Sn (II), and Sn (IV) is unquenched with acetic acid spray.

c. *Sodium Dithionite, $Na_2S_2O_4$.* 2% aqueous solution. As (III) gives an orange-brown spot; As (V), yellow; Bi, dark brown; Sb (III) and (V), orange-brown.

d. *Stannous Chloride–Potassium Iodide.* 5.6 gm $SnCl_2 \cdot 6H_2O$ dissolved in 10 ml concentrated HCl is diluted to 100 ml with water and 0.2 gm KI is added. Te, Au gives a gray to purple black spot (sensitivity, 0.1 μg for gold); Rh, brown; Se, Pt, orange.

e. *Sodium Rhodizonate.* 0.1% aqueous solution. Sr gives a red-orange spot which is decolorized by spraying with a mixture of 2 *N* HCl–2 *N* acetic acid; Ba gives a red-orange spot which is not decolorized; Pb gives a violet spot which is colorless on pink background with NH_3.

f. *Quinalizarin.* 0.05% solution in 0.1 *N* NaOH. Be gives a blue spot (sensitivity, 0.7 μg).

g. *Alizarin.* The paper is placed in an NH_3 atmosphere for 10 minutes and then sprayed with saturated alizarin in alcohol followed by 1 N acetic acid. Rare earths, Zr, Th, Hf, Mo, and Cr give blue-purple spots; Sc gives a red-purple spot and Ti orange-pink.

h. *Dipicrylamine.* With 1% aqueous solution, Cs and Rb give red spots. With 1% solution in 2 *N* Na_2CO_3, sprayed with 2 *N* HCl after drying, K gives an orange-red spot.

i. *Rubeanic acid.* (See Section 1e above). Co gives a brown spot; Cu, olive green; Ni, blue; Pd and Pt, red-purple.

j. *Potassium Ferrocyanide.* 5% (w/v) in water. Fe(II) gives a blue spot; Fe(III), green; Cu, U, brown.

k. *Diphenylcarbazone.* Saturated solution in ethanol. Hg_2(II) and Hg(II) give a violet blue spot.

l. *Zinc Uranyl Acetate.* Saturated solution in 2 *N* acetic acid; filtered before use. Li and Na yield blue-green fluorescence under ultraviolet light.

m. *Quercetin.* 0.2% solution in ethanol. Cr gives a green spot (sensitivity, 0.06 μg); U(IV) and (VI), brown; V(IV) and (V), yellow-brown.

n. *6-Nitro-2-naphthylamine-8-sulfonic acid.* A 2% (w/v) solution in water causes a blue fluorescent spot specific for Sn(II).

o. *Sodium Tetraphenylboron* (*87a*). The chromatogram is dipped into a 2% solution of the reagent in water, the paper is washed with about 50 ml water to remove excess reagent, dried with low heat, and brushed or sprayed lightly with a solution of bromophenol blue (0.1%) and mercuric chloride (2.5%) in 80% ethanol adjusted to the highest acidity which will still produce a blue color on dry paper. Amounts of potassium as low as 0.025 μmole yield yellow spots on a blue background. Cs, Rb, Tl, Ag, and NH_4^+ also react; the interference of NH_4^+ is avoided by using a basic reagent solution (2% sodium tetraphenylboron–6 *M* NaOH, 3:1 v/v).

p. *2-Mercapto-5-anilino-1, 3, 4-thiadiazole* (*87b*). The dried paper is sprayed with a 1% solution of the reagent in ethanol to produce the indicated colors down to be stated detection limit: Cu, green, 1.3 μg; Hg, yellow, 1.8; Co, brown, 0.8; Ni, brown, 1.2; Zn, white, 2.6.

Additional color reagents may be found in references *13*, *17*, *54*, *55*, and *88*.

F. Quantitative Methods

Paper chromatography is uniquely suited for the separation of fairly complex mixture of cations and their subsequent quantitative estimation after elution from the paper or directly on the paper. The theory and practice of quantitative paper chromatography have been discussed in detail in Chapter 4, and similar principles are applicable to inorganic quantitative paper chromatography.

In this section, several procedures are described in detail, while others are mentioned only briefly. With some practice the analyst should be able to devise individual quantitative procedures to meet his own particular problems and requirements.

1. *Elution methods—The Determination of Co, Ni, Cu, and Zn.*

In a procedure developed by Frierson (*65*), 10–40 μl of a solution containing 2.5–10 μg Co and Ni and 25–100 μg Cu and Zn are applied as a narrow band on acid-washed Whatman No. 3MM paper strips. After a 1 hour equilibration, the papers are developed for 17–22 hours by descending development with *n*-butanol–HCl (see Section C6a). A control strip is sprayed with dithizone to locate the exact position of Zn, which is the fastest moving ion in this solvent. Zn is eluted by descending migration (see Chapter 3) with 0.1N HCl until 1 to 2 ml has been collected.

The remainder of the chromatogram (above Zn) is redeveloped for 3–5 hours with the ketone–HCl solvent (see Section C6b). Cu, the migration of which can be followed since it forms a bright yellow zone in this solvent, is permitted to reach almost the bottom of the strip. A control strip is sprayed with ammoniacal rubeanic acid. Sections corresponding to Ni (near origin) are cut out and eluted with 0.1 *N* HCl, while Co (near center) and Cu (near bottom) are eluted with 3 *N* HCl.

a. *Determination of Co.* To the Co eluate is added 0.5 ml 20% citric acid; it is neutralized to litmus paper with NH_4OH, and then made just acid with a drop of 6 *N* HCl. One milliliter of color reagent (50 mg 1-nitroso-2-naphthol, 10 ml water, 0.5 ml 1 *M* NaOH, heat to dissolve; dilute to 100 ml) is added. The developed color is extracted into 5 ml isoamyl alcohol in a separatory funnel; 5 drops of concentrated NH_4OH are added, and the mixture is shaken. The absorbance of the alcohol layer is read at 550 mμ (mean deviation 1.71%).

b. *Determination of Ni.* To the eluate, 0.5 ml citric acid solution is added and the pH adjusted to about 9.5 with NH_4OH. Two milliliters of 0.2% aqueous diethyldithiocarbamate · $3H_2O$ is added, diluted to 10 ml with water, and the Ni complex extracted in a separatory funnel with 6 ml isoamyl alcohol. After allowing the phases to separate for 30 minutes, the absorbance of the alcohol layer is read at 325 mμ against an appropriate paper blank (mean deviation 0.86%).

c. *Determination of Cu and Zn.* The eluates are transferred to 50-ml volumetric flasks, and 20 ml borate buffer (pH 9) and 3 ml Zincon solution (0.130 gm Zincon dissolved in 2 ml 0.1 *M* NaOH, diluted to 100 ml with water) are added. The pH of the solutions is adjusted to 8.5–9.5 with NH_4OH, and after dilution with water, the absorbance is read at 620 mμ against appropriate paper blanks (mean deviation: Cu, 1.43%; Zn, 0.95%).

d. *Other Methods and Ions.* A similar colorimetric determination of Cu, Co, and Ni, following separation and elution, is accomplished using dimethylglyoxime for Ni, Nitroso-R salt for Co, and dithizone for Cu (*89*). Se is separated from many other ions on tin tungstate ion-exchange paper, eluted with water, and determined colorimetrically by the starch–iodide method (*90*). Ti is isolated by chromatography on Whatman No. 1 paper, eluted with 5% H_2SO_4, and determined by colorimetry with sulfosalicylic acid (*91*). Fe(II) and Fe(III) are separated from each other on Whatman No. 3 MM paper, eluted with 1% HCl, and determined with 1, 10-phenanthroline (*92*). Pb, Cu, and Bi are separated with ethanol containing 10% 5 *N* HCl, sprayed with 0.5% oxine reagent, exposed to NH_3 fumes, eluted with constant-boiling HCl, and determined in the ultraviolet region at 270, 275, and 325 mμ, respectively (*92a*).

Alkaline earth metals are titrated after elution with disodium EDTA and Eriochrome Black T as the indicator (*93*); eluted rare earth metals are back titrated with $LaCl_3$ after adding excess EDTA (*94*). Direct titration on paper, without prior elution, in a vessel containing the excised paper segment and a sufficient amount of water to immerse the paper completely has been studied by Lacourt *et al.* (*95-97*). The reported accuracy for V(V) is ±4–5% and for Mo, ±3.5%.

2. *Determination on the Paper*

These methods involve the measurement of the color (*16a*) or fluorescence of the separated zones by visual or spectrophotometric methods. Results are at best semiquantitative and details will not be offered here. A general discussion of *in situ* methods is found in Chapter 4, and procedures for the reflectance method for the estimation of Ni and the total color density method for Co are given in reference (*54*).

Besides colorimetry, other instrumental methods, such as atomic absorption, X-ray fluorescence, polarography, and flame photometry, are useful in combination with paper chromatography for micro determinations (*98, 99*). Such methods, along with solvent extraction (*100*), are used for the estimation of traces of metals in ancient objects (*101*), biological materials (*102*), ores (*100*), and geochemical samples (*103*).

G. Systematic Analysis of Common Cations

A complete systematic analysis of the inorganic cations may be performed by paper chromatography. A proposed scheme after the usual separation into groups is given below (*88*).

1. *Group I—Ag(I), Hg_2(II), Pb(II)*

A spot of the mixed chlorides is placed in the center of a circular filter paper, followed by two drops of water and two drops of 5*N* NH_4OH. The paper is exposed to H_2S. Hg is a central black spot; then AgS, followed by Pb. This is an example of capillary separation rather than paper chromatography.

2. *Group IIA—Bi(III), Cd(II), Cu(II), Pb(II), Hg(II)*

Chlorides of these cations are separated in *n*-butanol saturated with 3 *N* HCl. Strips are sprayed with dithizone. Pb is detected with rhodizonic acid. Results are given in the tabulation.

Ion	R_F	Color (dithizone)
Cu	0.20	Brown
Pb	0.27	Blue
Bi	0.60	Purple
Cd	0.77	Purple
Hg	0.81	Pink

3. *Group IIB—As(III), Sb(III), Sn(II)*

Separated as chlorides using acetylacetone (saturated with H_2O)–HCl–acetone (75:0.5:25 v/v); 100% humidity must be maintained in chromatographic chamber. The damp chromatogram is sprayed with dithizone. R_F values are as follows: As, 0.20; Sb, 0.50; Sn, 1.0.

4. *Group IIIA—Fe(III), Al(III), Cr(III)*

Metals are dissolved in 5 *N* HCl. Chromatography is with glacial acetic acid containing 25% methanol. After chromatography, paper is cut in half longitudinally and treated as follows: (*a*) exposed to NH_3, then alizarin; Al is red, Fe is purple; (*b*) sprayed with Na_2O_2, then benzidine (0.05% in 10% acetic acid); Cr is blue. The difficulty of this method involves the separation of Al and Cr.

5. *Group IIIB—Ni(II), Mn(II), Co(II), Zn(II)*

Salts are dissolved in diluted HCl and chromatographed with acetone–8% HCl–5% H_2O. The chromatogram is sprayed with an alcoholic solution of a mixture of 0.05% alizarin, 0.05% rubeanic acid, and 0.4% salicylaldoxime. An alternate color reagent is diphenylcarbazide. The results are tabulated below.

Ion	R_F	Color
Ni	0.07	Blue
Mn	0.30	Brown
Co	0.60	Orange-brown
Zn	0.90	Purple

6. *Group IV—Ca(II), Sr(II), Ba(II)*

Salts are separated as chlorides using pyridine–1% (w/v) KCNS–20% H_2O. Chromatgram is sprayed with sodium rhodizonate for Sr and Ba, alizarin for Ca.

7. *Group V—Li(I), Na(I), K(I), NH_4^+*

The solvent for this group is phenol saturated with 20% HCl, and the color reagent is 10% sodium cobaltinitrite in 5% acetic acid followed by water wash. Yellow spots of the following R_F values are the result: NH_4^+, 0.11; K, 0.19; Rb, 0.27; and Cs, 0.43.

8. *Analysis without Prior Separation*

Pollard (*83*) and Elbeih (*13*) have proposed a scheme for the systematic analysis of cations without prior separation by precipitation. Three solvent systems are utilized—benzoylacetone, collidine, and dioxane (see Section C1a). Two sets of chromatograms are run in each solvent system; one chromatogram from each tank is first sprayed with oxine–kojic acid and examined in daylight and under ultraviolet light. The resultant characteristic colors are listed in Table III, and the positions of the cations are indicated in Table IV. The same papers are next sprayed with ammoniacal H_2S, and different colored sulfides are produced (Table V). Since there is some overlap of ions in the three solvent systems, it may be necessary to apply specific confirmatory spray tests for suspected cations. These reagents are found in Section E. Other schemes for the systematic analysis of cations have been suggested (*87*, *104–107*).

H. Miscellaneous Metal Ions

Table VI contains solvent systems for the separation of groups of metals and metals in different valence states (*19*, *59*, *92*, *108–126b*). Most of these separations are performed by one-way ascending or descending development on Whatman No. 1 paper or on equivalent grade.

Separations of the platinum metals and gold are obtained by one-dimensional development on unmodified paper with methyl amyl ketone–HCl (7:3 v/v) or acetophenone–acetone–HCl (5:2:3 v/v) (*127*),

TABLE IV

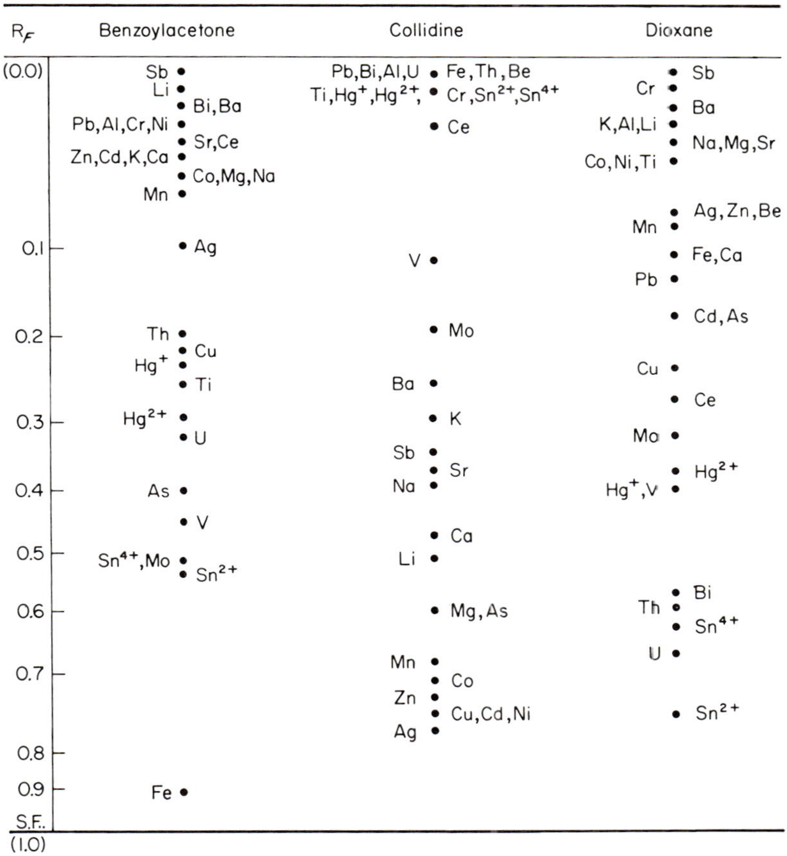

TABLE V

COLOR DETECTION OF METAL IONS[a]

Color	Cations
Yellow	Cd, As, Th, Ti, Sn(II)
Light brown	Mn, Ce
Brown	Ag, Cu, Hg(II), V
Chocolate brown	Pb
Orange	Sb, U
Pale green	Cr
Black	Hg_2(II), Bi, Fe, Co, Ni

[a] Appearance of color in daylight after spraying with kojic acid–oxine mixture then with ammoniacal hydrogen sulfide water (*13*).

TABLE VI

PAPER CHROMATOGRAPHY OF MISCELLANEOUS METAL IONS

Solvent	R_F	Reference
tert-Butanol–acetone–H_2O–concentrated HNO_3 (40:40:12:8 v/v)	Hg(I)(0.0), Ag(0.25), Pb(0.50), Hg(II)(0.81)[a]	*108*
Methyl *n*-propyl ketone–10 *N* HCl (85:15)	Pb(0.19), Ag(0.59), Hg(I)(0.84)	*109*
	Pb(0.19), Bi(0.40), Cu(0.60), Cd(0.75), Hg(II)(0.80)	*109*
n-Butanol–isopropanol–5 *N* HCl (45:45:10 v/v)	Pb(0.03), Cu(0.12), Bi(0.65), Cd(0.73), Hg(II)(0.85)	*110*
tert-Butanol–$CHCl_3$–8 *N* HCl–acetylacetone (45:45:8:2 v/v)	As(III)(0.20), Sb(III)(0.50), Sn(II)(1.0)[a]	*108*
Glacial acetic acid–acetone–H_2O (8:1.5:2.5)	As(III)(0.59), Sb(III)(0.76), Sn(II)(0.92)	*111*
Acetone–isobutanol–12 *M* HCl (60:38:2)	Th = Zr(0.0), Pd(II)(0.29), Co(0.45), U(VI)(0.53)	*112*
Methyl isobutyl ketone–10 *N* HCl (85:15 v/v)	Ni(0.04), Mn(II)(0.13), Co(0.52), Zn(0.58), MoO_4^{2-}(0.90)	*109*
Acetic acid–concentrated HCl–pyridine (80:20:6 v/v)	Cr(III)(0.26), Al(0.30), Ni(0.39), Mn(II)(0.57), Co(0.68), Zn(0.82), Fe(III)(0.92)	*113*
Methanol–ethanol (1:1 v/v)	Ba(0.28), Sr(0.49), Ca(0.62), Mg(0.80)[a]	*114*
	K(0.08), Na(0.23), NH_4^+(0.32), Li(0.65)	*115*
Phenol saturated with 20% HCl	NH_4^+(0.11), K(0.19), Rb(0.27), Cs(0.41)	*116*
Methyl *n*-propyl ketone–concentrated HCl (7:3 v/v)	Ru(0.07), Rh = Ir(0.11), Pd(0.56), Pt(0.81), Os(0.89), Au(0.96)	*59*
Isobutanol–40% HF (20:1 v/v)	Nb(0.60), Ta(0.90)	*117*
n-Butanol–concentrated HCl–30% H_2O_2–H_2O (10:1:1:8 v/v)	Ti(0.09), V(0.27), Mo(VI)(0.79)	*118*
n-Butanol–3.5 *N* HCl (7:3 v/v)	Au > Pt > Pd > Ru, Rh	*119*
30 ml acetone–ether (1:1 v/v) to which is added 0.25 gm NH_4CNS in 0.4 ml H_2O and 0.3 ml concentrated HCl; filtered before use	La(0.08), Ce(0.11), Pr(0.16), Nd(0.20), Sm(0.31), Gd(0.44), Y(0.48), Dy(0.50), Er(0.56), Yb(0.59), Th(0.90)	*120*
n-Butanol–acetylacetone–acetic acid–H_2O (100:30:5:65 v/v)	Ac(0.08), Nd(0.22), La(0.31), Ce = Pr(0.38), Gd(0.43), Sm(0.47), Eu(0.49), Yb = Y(0.59), Er(0.60), Dy(0.62), Sc(0.97)	*121*
n-Butanol saturated with 7 *M* $LiNO_3$ and 2 *N* HNO_3	La(0.40), Ce(0.46), Pr(0.51), Y(0.58), Sc(1.0)	*122*
Ether–methanol–concentrated HCl–H_2O (50:30:4:15)	As(III)(0.55), As(V)(0.75),	*19*
	Cr(III)(0.3), Cr(VI)(0.8),	*19*
	Cu(II)(0.35), Cu(I)(0.9)	*19*
(50:30:2:20)	Co(III)(0.0), Co(II)(0.3)	
4 *M* HCl–*n*-butanol–acetic acid–acetone (1:1:1:1 v/v)	Fe(II)(0.17), Fe(III)(0.68)	*92*

Table continued

TABLE VI (*continued*)

Solvent	R_F	Reference
Acetic acid–H_2O–ethyl acetate (1:1:1 v/v)	Sb(III)(0.65–0.71), Sb(V)(1.0)	*123*
Ethanol	Mo(VI)(0.0), Mo(V)(0.9–1.0)	*124*
Acetic anhydride–glacial acetic acid (3:1)	Mn(II)(0.16), Mn(VII)(0.48), Mn(III)(0.94)	*125*
Ether–methanol–concentrated HCl–H_2O (50:30:4:15)	U(IV)(0.05), U(VI)(0.55)	*19*
Ethanol–acetic acid–20% (w/v) aqueous sodium acetate (5:2:3)	V(V)(0.50), V(IV)(0.80)	*126*
Ethanol–acetic acid–40% (w/v) aqueous sodium acetate (10:1:2)	V(IV)(0.68), V(III)(0.97)	*126*
Ethyl acetate–methyl salicylate (1:1 v/v)	Sn(II)(0.31), Sn(IV)(0.97)[a]	*126a*
Propanol or butanol + 1% concentrated HNO_3	Tl(III)(1.0), Tl(I)(0.0)[a]	*126a*
Benzene + trichloroacetic acid	Hg(I)(0.0), Hg(II)(1.0)	*126b*

[a] Circular development.

and on tri-*n*-butyl phosphate-impregnated paper with aqueous mineral acids (*128*, *129*). Two-dimensional chromatography (*130*) and chromatography combined with precipitation (*131*) is also useful. Ethyl acetate–HCl–acetic acid (6:3:1 v/v) separates all the platinum metal chlorides except Rh and Ir, and these are resolved in methyl nitrate–HCl–ethanol (1:1:8 v/v) (*130*).

II. Common Anions

A. Preparation of Sample

Samples are prepared in a manner similar to that for cations. Soluble samples are dissolved in a small volume of water or dilute acid or alkali and spotted directly. Samples are concentrated by precipitating the anions and redissolving in a smaller volume, or the precipitates can be applied directly to the paper (*132*). Interfering cations are removed by ion exchange (*133*). Biological materials are usually fused with carbonate or ashed (*134*).

B. Paper

Anions are usually separated on Whatman No. 1 paper or an equivalent grade, although ion exchange paper (*135–137*) and other modified papers have been used. Preliminary washing of the paper with the mobile phase or acid is often employed, especially when separating phosphates (*138–140*).

C. Solvents

For the systematic analysis of the common anions, a general solvent is ethanol–pyridine–H_2O–concentrated. NH_4OH (60:20:16:4 v/v) (see Table VII). Although a number of anions congregate near the point of application, specific spray reagents (see below) may be used for confirmatory tests. Besides this general anion solvent, several other solvents for specific mixtures are recommended.

1. *Methanol*–1 *N ammonia* (*8:2 v/v*) on acid-washed Whatman No. 1 paper separates AsO_4^{-3} and AsO_3^{-3}. The R_F values are :Na_3AsO_3, 0.51; $(NH_4)_3AsO_3$, 0.57; Na_2HAsO_4, 0.22; Na_3AsO_4, 0.21; $(NH_4)_3AsO_4$, 0.25. This solvent yields a wet front at R_F 0.65 (*141*) (Fig. 5).

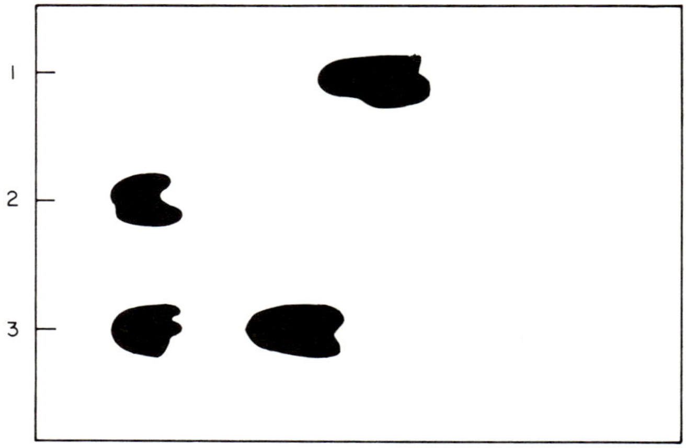

FIG. 5. Separation of arsenite and arsenate ions by ascending development with methanol-1 *N* NH_4OH (8:2 v/v) for 18 cm (140 minutes). Detection: 2 *N* $AgNO_3$ in 10% NH_4OH [As(III) is yellow, As(V) is brown; under UV light, blue absorbancy]. 1 = Na_3AsO_3 ; 2 = Na_3AsO_4 ; 3 = mixture of 1 + 2. After V. Mikeťuková, *et al.* (*141*).

2. *Acetone*–H_2O–*concentrated* NH_3 (*6:1:1 v/v*). This solvent is used for the separation of iodide (R_F = 0.95), iodate (0.35) and tellurite (0.05) (*142*).

3. *n-Butanol*–1.5 *N* NH_3 (*1:1 v/v*). This solvent is used for the separation of F^- (R_F = 0.0), Cl^- (0.10), Br^- (0.16), I^- (0.30), and CNS^- (0.45) by ascending development (*143*).

4. *n-Butanol*–1.5 *N* NH_3–*pyridine* (*2:2:1 v/v*). This solvent gives the following R_F values by descending chromatography on Whatman No. 1 paper: CrO_4^{-2} (0.0), IO_3^- (0.09), Cl^- (0.24), BrO_3^- = NO_2^- (0.25), Br^- (0.36), NO_3^- (0.40), ClO_3^- (0.42), I^- (0.47), SCN^- (0.56) (*103*).

5. *Isopropanol*–H_2O (*3:1 v/v*). This solvent separates IO_3^- ($R_F = 0.20$), BrO_3^- (0.47), ClO_2^- (0.52), and ClO_3^- (0.68) by ascending development on Arches No. 320 paper (French) (*144*).

6. *Isopropanol*–H_2O–*pyridine–concentrated* NH_3 (*15:2:2:2 v/v*). Ten microliters of a mixture 1% in each anion is spotted on strips of Whatman No. 1 paper and developed with the solvent for 16 hours in a chromatobox. The bottom part (0–11 cm) of a strip is sprayed with 0.2 *N* aqueous $AgNO_3$, and the zone of Cl^- appears as a dark spot under ultraviolet light at R_F 0.25. The middle region (11–24 cm) of the strip is sprayed with a fresh mixture of equal parts of 3 *M* aqueous HCl–acetone (1:1 v/v) and 5% ethanolic diphenylamine, and chlorite is revealed immediately as a blue spot at 0.36 and chlorate after drying at 0.54. The rest of the strip is sprayed with saturated aqueous sodium acetate, dried, and oversprayed with 0.2% aqueous methylene blue to locate perchlorate as a violet spot at 0.71 (*145*).

7. *Isopropanol–acetone*–H_2O–*potassium acetate* (*5 ml:2 ml:3 ml:0.2 gm*). This solvent is used to separate sulfur oxyacids by the descending rear phase technique: $K_2S_2O_3$ ($R_F = 0.1$), $K_2S_3O_6$ (0.32), $K_4S_4O_6$ (0.45), $H_2S_4O_6$ (0.65), $H_2S_5O_6$ (0.74), $H_2S_6O_6$ (0.82) (*146*).

8. *Three-solvent system for systematic analysis* (*147*). (a) Acetone–H_2O (10:1 v/v). For Br^-, Cl^-, I^-, CNS^-, ClO_3^-, BrO_3^-, and NO_3^-. (b) *n*-Butanol–methanol–H_2O (1:3:1 v/v). For F^-, NO_3^-, $S_2O_3^{2-}$, SO_4^{2-}, CrO_4^{2-}, IO_3^-, AsO_3^{3-}, $Fe(CN)_6^{4-}$, and BO_2^-. (c) *n*-Butanol saturated with 2 *N* HNO_3. For $C_2O_4^{2-}$, AsO_4^{3-}, and PO_4^{3-}.

D. Techniques

Conventional ascending and descending techniques are recommended for the separation of the common anions, as well as the usual precautions for prior equilibration of chromatograms with the aqueous solvent phase. All-glass apparatuses are mandatory for solvents containing strong acids.

E. Detection of Spots

A general detection reagent for most common anions consists of 50 ml ammoniacal $AgNO_3$, 50 ml saturated aqueous solution of 1-naphthylamine-5-sulfonic acid (Laurent's acid), and 0.05 gm fluorescein; characteristic colors are produced when treated chromatograms are viewed in daylight and in ultraviolet light (see Table VIII). For the systematic analysis of common anions, the technique of Elbeih (*148*) is recommended in which twelve chromatograms are developed with the general solvent described above. After drying, the strips are sprayed with the following reagents:

TABLE VII

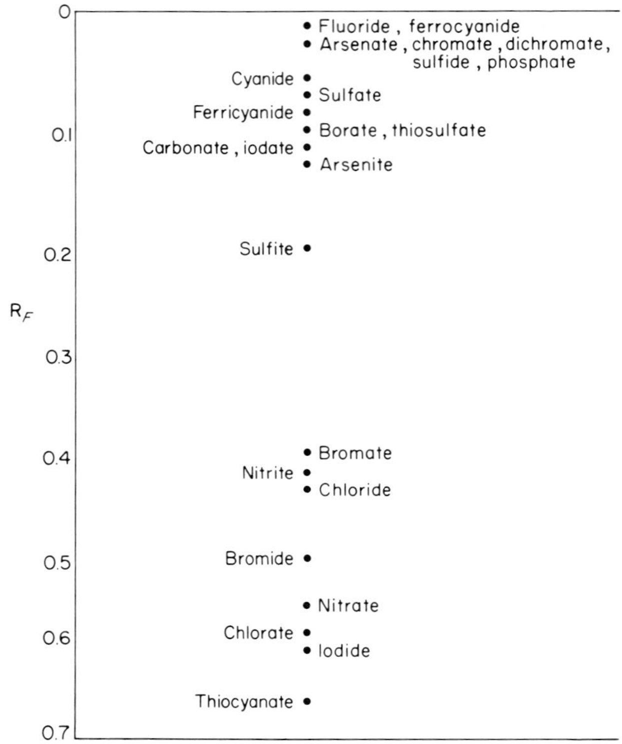

1. *Strip 1—General Reagent.* A mixture of ammoniacal $AgNO_3$ solution (50 ml), saturated aqueous Laurent's acid (50 ml) and fluorescein (50 mg) is used as a spray. The sprayed strips are allowed to air-dry for 2 hours and examined in visible and ultraviolet light (see Table VIII).

2. *Strip 2—Fluoride, Chlorate, Bromate, and Nitrate.* The strip is first sprayed with aqueous 1% zirconium nitrate–0.1% alizarin red S (dilute HNO_3) (v/v). A yellow spot on purple background shows the presence of F^-. The dried strip is next sprayed with 0.02% brucine in 2 *N* H_2SO_4 and warmed gently. The colors are noted, and the strip is then sprayed with 2 *N* NaOH. The tabulated ions are thus confirmed.

Ion	Acid spray	Base spray
BrO_3^-	Deep red	Blood red
NO_3^-	Red, then yellow	Orange-yellow
ClO_3^-	Reddish-brown	Blood red

TABLE VIII

COLOR DETECTION OF ANIONS WITH SILVER NITRATE[a]

Anions	Appearance in daylight	Appearance under ultraviolet light: Bright spots	Dark spots	Color
Fluoride	Pale gray		+	Gray
Ferrocyanide	Bright yellow	++		Bright yellow
Arsenate	Green-yellow		+	Green-gray
Chromate	Brown		+++	Dark gray
Dichromate	Brown		+	Green-gray
Sulfide	Dirty green		+++	Dark green-gray
Phosphate	Yellow-green	+++		Bright green-yellow
Cyanide	Bright green-yellow	+++		Bright green-yellow
Sulfate	Bright yellow	+++		Bright yellow
Ferricyanide	Pink		+++	Dark red-violet
Borate	Pale gray			Blue
Thiosulfate	Brown		+++	Dark violet-gray
Carbonate	Pale green-yellow	+		Green-yellow
Iodate	Yellow	++		Bright yellow
Arsenite	Dirty yellow		+++	Dark green-gray
Sulfite	Pale brown		+++	Dark gray
Bromate	Yellow	+		Yellow
Nitrite	Pale yellow	+		Yellow
Chloride	Gray	++		Bright yellow
Bromide	Dark gray		+++	Dark gray
Nitrate	Yellow	+		Yellow
Chlorate	Pale yellow	+		Yellow
Iodide	Red		+++	Dark violet
Thiocyanate	Pink	++		Bright yellow

[a] Silver nitrate–Laurent's acid–fluorescein (*148*).

3. *Strip 3—Arsenate and Nitrite.* A solution of 10% KI in 2 *N* HCl followed by gentle heating produces a brown spot for AsO_4^{3-}. The strip is next sprayed with 0.1% *o*-coumaric acid in ethanol producing a dark green spot under ultraviolet light for nitrite.

4. *Strip 4—Sulfate and Chloride.* A solution of Laurent's acid in 50% ethanol produces a violet spot for SO_4^{2-} when viewed under ultraviolet light. After the Laurent's acid treatment, the paper is sprayed with 0.1 *N* $AgNO_3$, washed with water, dried, and sprayed with a fresh solution of $(NH_4)_2S_x$ to produce a black spot for Cl^-.

5. *Strip 5—Cyanide, Bromide, and Thiocyanate.* A spray consisting of a 2% guaiacol in ethanol–0.1% aqueous $CuSO_4$ (10:1 v/v), produces a permanent blue spot for CN^-. The strip is next sprayed with 5%

$Fe(NO_3)_3$ in 0.5 *N* HNO_3 and warmed, producing a yellow spot for Br^- and blood -red spot for CNS^-.

6. *Strip 6—Borate and Iodide.* The strip is first sprayed with 2 *N* HCl, allowed to dry, and sprayed with a saturated solution of turmeric powder. After gentle warming, a pale red spot for borate results. The strip is next sprayed with 5% KNO_2, dried, and sprayed with a slightly acid starch solution; a deep blue spot for I^- appears.

7. *Strip 7—Ferrocyanide and Ferricyanide.* A solution of 5% hydrated $FeCl_3$ in 2% (v/v) HCl produces a dark blue spot for $Fe(CN)_6^{4-}$ and bright tan spot for $Fe(CN)_6^{3-}$.

8. *Strip 8—Chromate and Dichromate.* A spray of 0.2% diphenylcarbazide in 2 *N* acetic acid produces a red-violet color for both ions.

9. *Strip 9—Arsenite.* An ammoniacal $AgNO_3$ solution produces a bright yellow spot for arsenite ions.

10. *Strip 10—Iodate.* A 1% phenylhydrazine hydrochloride in 50% ethanol spray, followed by gentle warming, produces a brown-red spot for iodate.

11. *Strip 11—Sulfide, Thiosulfate, and Phosphate.* A spray containing 5% ammonium molybdate in 2 *N* H_2SO_4 is used. The papers are dried and then sprayed with a 5% aqueous sodium sulfite solution. The colors that are produced before and after sulfite treatment are tabulated below.

Ion	Before SO_3^{2-}	After SO_3^{2-}
Sulfide	Violet-blue	Deep blue
Thiosulfate	Violet-blue	Deep blue
Phosphate	Yellow	Green turning to blue-gray on warming

12. *Strip 12—Sulfite and Carbonate.* A solution of 0.1% each of kojic acid and *o*-coumaric acid in ethanol, followed by a spray of ammoniacal silver nitrate, is used. A dark brown spot appears for carbonate. The spot due to sulfite fluoresces while the strip is still wet and turns to brown on drying.

Another complete scheme of anion analysis has been suggested (*147*, *149*).

III. Miscellaneous Inorganic Substances

A. Inorganic Phosphates

Hettler has reviewed the paper chromatography of condensed phospates through 1957 (*150*). Recent work on the complex metaphosphates of Ba, Co, and Cu (*151*), condensed phosphates (*152, 153, 153a*) and phosphorus halides (*154*) has appeared in the literature.

Chromatography is usually by ascending or two-dimensional development on washed Whatman No. 1 paper or an equivalent. Detection is usually by spraying the dried chromatograms with ammonium molybdate (1 gm dissolved in 20 ml concentrated HCl and diluted to 200 ml with ethanol while cooling the solution) to form phosphomolybdic acid, followed by reduction to molybdenum blue by spraying with stannous chloride (40% $SnCl_2$ in concentrated HCl, diluted 1:2000 with 1 *N* H_2SO_4). Quantitative estimation is obtained by direct densitometry on the paper after spraying as above or colorimetry after elution (*155*). Table IX shows R_F values for condensed phosphates in three recommended solvents. Cyclic phosphates are also successfully separated by development with isopropanol–isobutanol–H_2O–25% aqueous NH_3 (40:20:39:1 v/v) (*152*).

TABLE IX

R_F VALUES OF INORGANIC PHOSPHATES (*150*)

Phosphate		R_F [a] 1	2	3
Monophosphate	$(PO_4)^{-3}$	0.69	0.73	0.79
Diphosphate	$(P_2O_7)^{-4}$	0.44	0.53	0.68
Triphosphate	$(P_3O_{10})^{-5}$	0.29	0.39	0.58
Tetraphosphate	$(P_4O_{13})^{-6}$	0.17	0.29	0.47
Pentaphosphate	$(P_5O_{16})^{-7}$	0.11	0.22	0.36
Hexaphosphate	$(P_6O_{19})^{-8}$	0.07	0.16	0.25
Heptaphosphate	$(P_7O_{22})^{-9}$	0.04	0.11	0.15
Octaphosphate	$(P_8O_{25})^{-10}$	—	0.08	—
Trimetaphosphate	$(P_3O_9)^{-3}$	0.20	0.32	0.39
Tetrametaphosphate	$(P_4O_{12})^{-4}$	0.08	0.18	0.22

[a] Solvents: (1) isopropanol–H_2O–trichloracetic acid–20% NH_3 (75:25:5:0.3 v/v/w/v); (2) isopropanol–H_2O–20% trichloracetic acid–25% NH_3 (70:10:20:0.3 v/v); (3) methanol–trichloracetic acid (to 100 ml trichloroacetic acid, diluted to 500 ml, add 22.7 ml NH_3)–diluted acetic acid (20 ml 96% acetic acid plus 80 ml water) (60:10.3:5 v/v).

B. METAL COMPLEXES

Many of the solvents recommended above for the paper chromatography of metals are effective because of their ability to selectively complex cations in the mixture during development. (e.g., the separation of

Ti, U, Th, and Zr as thiocyanate complexes by development with the upper layer of butanol–1% KSCN–HNO_3, 14:14:1 v/v (*155a*).

Metal complexes can also be formed prior to chromatography and then spotted and separated, although it is often questionable whether the migrating metals are in the form of chelates or simple cations resulting from the dissociation of the complex in the mobile phase (*61*). In any case, metal oxinates, diphenylcarbazides (*155b*), dithizonates, picrates, acetylacetonates, etc., can be separated by common paper chromatographic techniques.

One example is the separation of cationic ethylenediamine and ammonia complexes of Co, Ni, and Cr (*156*). The complexes are dissolved in water, spotted on paper strips, and developed downward with *n*-butanol–acetic acid–pyridine–H_2O (30:6:24:20 v/v). Separated spots are visualized by staining the strips with 0.2% ninhydrin in acetone followed by heating at 85°C for 10–15 minutes. Table X shows the results in this system.

TABLE X

CHROMATOGRAPHIC SEPARATION AND DETECTION OF Co, Ni AND Cr AND THEIR CATIONIC COMPLEXES (*156*)

Compound[a]	Color[b]	R_F value[c]	Amount taken (μg)
$[Co(en)_2Cl_2]Cl$	Violet-brown (intense and deep shade)	0.33	8
$[Co(en)_3]Cl_3$	Pink-violet (deep shade)	0.21	8
$[Cr(en)_3]_2(SO_4)_3$	Violet (deep shade)	0.168	8
$[Co(NH_3)_6]Cl_3$	Distinct yellow (weaker and lighter shade)	0.165	15
$[Ni(en)_3Cl_2$	Violet (very deep and intense)	0.21	5
$[Co(NH_3)_3(H_2O)]Br_3$	Yellow pinkish (very light shade)	0.184	15
$[Co_4(en)_6(OH)_6]Cl_6$[d]	Pink-brown	0.20	8
$[Cr_4(en)_6(OH)_6]Cl_6$[d]	Violet-pink (intense and deep)	0.23	8
$CoCl_2$[e]	Pink-yellow	0.77	40
$NiCl_2$[e]	Yellow (very light shade)	0.78	40
$Cr_2(SO_4)_3 \cdot K_2SO_4$	Green-gray (very light and dull)	0.20	60

[a] (en) = ethylenediamine.
[b] 0.2% ninhydrin in acetone plus heating.
[c] *n*-Butanol–acetic acid–pyridine–H_2O (30:6:24:20 v/v) solvent.
[d] Streaked zones.
[e] Detected spots became more prominent on storage.

The R_F values (1 *N* aqueous ethylamine as solvent) for twenty-three inert Co (III) complexes with ionic charges of +3, +2, +1, 0, −1, and −3 increase with decreasing charge, suggesting a chromatographic means of determining the charge on such a complex (*157*). Ion pair formation between metal complexes and various anions is conveniently studied by partition and ion-exchange paper chromatography. On ion-exchange paper, the slope of a plot of R_M versus log [eluent] indicates the charge of the ion under study (*158*).

C. Radioactive elements

Solvents and chromatographic techniques for the separation of radionuclides are usually analogous to those described in previous sections of this chapter. The chromatography, however, differs by the unique detection technique, measurement of radioactivity with Geiger–Müller or scintillation counters or autoradiography with X-ray film (see Chapter 4 for details). Additional specificity is imparted by counting the same chromatogram immediately after development and several days later. Due to the rapid decay of short half life isotopes, a different scanning pattern is obtained. Furthermore, by placing absorbers of varying thickness (e.g., aluminum foil or lead sheeting) between film and chromatogram, one may be able to distinguish between soft and hard β emitters (^{14}C and ^{32}P). For further details on this subject, the reader is referred to a review of the techniques, apparatus, and applications of quantitative paper chromatography of radioactive elements (*159*).

Separation of Tc(IV), Tc(V), and Tc(VII) is obtained by ascending development on Whatman 3MM paper with 0.3–1.5 *N* HCl. The chromatograms are scanned for radioactivity by counting every 5 mm section of the paper under a thin-window Geiger–Müller counter (*160*).

Radioisotopes of Cs, Rb, Sr, and Y are separated by descending development for 70 minutes with a solution 1.5 *M* in NH_4NO_3 and 2 *M* in HNO_3 on strips of Whatman No. 4 paper impregnated with ammonium phosphotungstate. Detection is obtained by passing the strip under a Geiger–Müller counter with a mica window. R_F values are: ^{134}Cs, 0.04; ^{91}Y, 0.25; ^{86}Rb, 0.53; ^{89}Sr, 0.94 (*161*).

Nickel is separated from other metals on Whatman No. 42 paper with acetone–HCl–H_2O (87:8:5 v/v) and located as a blue complex with alizarin red S and aqueous NH_3. The strip is immersed for approximately 3 hours in aqueous $NH_4{}^{35}S$ to precipitate $Ni^{35}S$, which is located by autoradiography, removed, and counted by liquid scintillation methods (*162*). Sulfur anions labeled with ^{35}S have also been analyzed (*163*).

REFERENCES

1. G. Nickless, *Advan. Chromatog.* **7**, 121 (1968).
2. G. Nickless and F. H. Pollard, *in* "Chromatography" (E. Heftmann, ed.), p. 735. Reinhold, New York, 1967.
3. R. Neidlein, R. Gren, J. Hoff, and P. Krause, *Deut. Apotheker-Ztg.* **105**, 1287 (1965).
4. G. Nickless, *Lab. Pract.* **16**, 1238 (1967).
5. A. A. Lur'e, *Usp. Khim.* **37**, 104 (1968).
6. G. Zweig, *Anal. Chem.* **40**, 490R (1968); G. Zweig, R. M. Moore, and J. Sherma, *ibid.* **42**, 349R (1970).
7. E. Heftmann, *Anal. Chem.* **38**, 31R (1966); **36**, 14R (1964); **34**, 13R (1962).
8. M. Qureshi and F. Khan, *J. Chromatog.* **34**, 222 (1968).
9. J. S. Fritz and J. Sherma, *J. Chromatog.* **25**, 153 (1966).
10. J. Sherma and A. D. Finck, *Anal. Chim. Acta* **43**, 503 (1968).
11. J. S. Firtz and L. H. Dahmer, *Anal. Chem.* **37**, 1272 (1965).
12. W. F. Pickering, *Anal. Chim. Acta* **15**, 337 (1956).
13. I. I. M. Elbeih and M. A. Abou-Elnaga, *Anal. Chim. Acta* **17**, 397 (1957).
14. F. H. Pollard and J. F. W. McOmie, "Chromatographic Methods of Inorganic Analysis." Butterworth, London and Washington, D.C., 1953.
15. H. Seiler, E. Sorkin, and H. Erlenmayer, *Helv. Chim. Acta* **35**, 120 (1952).
16. H. T. Gordon and C. A. Hewel, *Anal. Chem.* **27**, 1471 (1955).

16a. R. A. Webb, D. G. Hallas, and H. M. Stevens, *Analyst* **94**, 794 (1969).

17. F. H. Pollard and G. Nickless, "Handbook of Analytical Chemistry," pp. 10–48. McGraw-Hill, New York, 1963.
18. F. H. Pollard, G. Nickless, and A. J. Banister, *Analyst* **81**, 577 (1956).
19. F. H. Pollard, J. F. W. McOmie, and A. J. Banister, *Chem. & Ind. (London)* p. 1598 (1955).
20. F. H. Pollard, J. F. W. McOmie, A. J. Banister, and G. Nickless, *Analyst* **82**, 780 (1957).
21. F. H. Pollard, J. F. W. McOmie, and J. V. Martin, *Analyst* **81**, 353 (1956).
22. A. Lacourt, G. Sommereyns, and G. Wantier, *Analyst* **77**, 943 (1952).
23. M. Qureshi and M. A. Khan, *Z. Anal. Chem.* **232**, 194 (1967).

23a. S. Kawamura, T. Fujimoto, and M. Izawa, *J. Chromatog.* **34**, 72 (1968).

24. G. Alberti and G. Grassini, *J. Chromatog.* **4**, 83 (1960).
25. G. Alberti, *J. Chromatog.* **31**, 177 (1968).
26. M. Qureshi, I. Akhtar, and K. N. Mathur, *Anal. Chem.* **39**, 1766 (1967).
27. M. J. Nunes Da Costa and M. A. S. Jeronimo, *J. Chromatog.* **5**, 546 (1961).
28. Y. B. Hai, *Acta Chim. Sinica* **31**, 266 (1965).

28a. M. Qureshi and W. Husain, *Separation Sci.* **4**, 197 (1969).

29. M. Qureshi, K. N. Mathur, and A. H. Israili, *Talanta* **16**, 503 (1969).
30. M. Lederer, *Anal. Chim. Acta* **12**, 142 (1955).
31. L. Ossicini and M. Lederer, *J. Chromatog.* **34**, 292 (1968), and previous papers in the series.
32. J. Sherma, *Talanta* **9**, 775 (1962); **10**, 787 (1963).
33. F. Hilgeman, K. Shimomura, and H. F. Walton, *Separation Sci.* **4**, 111 (1969).
34. J. Sherma, *Anal. Chem.* **36**, 690 (1964).
35. J. Sherma, *Talanta* **11**, 1373 (1964).
36. J. Sherma and C. W. Cline, *Anal. Chim. Acta* **30**, 139 (1964).
37. G. Bagliano, L. Ossicini, and M. Lederer, *J. Chromatog.* **21**, 471 (1966).
38. J. Sherma, D. A. Goldstein, and J. P. Gutai, *Separation Sci.* **2**, 691 (1967).

39. J. Sherma, *Separation Sci.* **2**, 177 (1967).
40. J. Sherma, *Anal. Chem.* **39**, 1497 (1967); *J. Chromatog.* **35**, D53 (1968).
41. G. Bagliano and L. Ossicini, *J. Chromatog.* **21**, 499 (1966).
42. M. Lederer, *J. Chromatog.* **29**, 306 (1967); M. Lederer and R. Marini-Bettólo, *ibid.* **43**, 149 (1969).
43. J. Sherma, unpublished results (1969).
43a. N. Sugawara, *Kagaku Keisatsu Kenkyusho Hokoku* **20**, 176 (1967).
44. M. Lederer and V. Moscatelli, *J. Chromatog.* **13**, 194 (1964).
45. M. Lederer and L. Ossicini, *J. Chromatog.* **13**, 188 (1964).
45a. J. G. van Raaphorst and H. H. Haremaker, *Talanta* **17**, 345 (1970).
46. U. A. T. Brinkman, G. DeVries, and E. Van Dalen, *J. Chromatog.* **22**, 407 (1966).
47. E. Cerrai, "Reversed-Phase Partition Chromatography in Inorganic Chemistry," CISE Doc. Serv., CISE Rept. No. 103. Segrate, Milano, Italy, 1966.
47a. S. Przeslakowski, *Roczniki Chem.* **42**, 975 (1968).
48. J. Sherma, *Anal. Chim. Acta* **36**, 138 (1966).
49. J. R. A. Anderson, S. Dilli, J. L. Garnett, and E. C. Martin, *Nature* **201**, 772 (1964).
50. M. Qureshi and F. Khan, *Chemist-Analyst* **56**, 44 (1967).
51. E. Pluchet and M. Lederer, *J. Chromatog.* **3**, 290 (1960).
52. R. A. Guedes de Carvalho, *J. Chromatog.* **4**, 353 (1960).
53. H. Hartkamp and H. Specker, *Naturwissenschaften* **42**, 534 (1955); *Z. Anal. Chem.* **158**, 92 and 161 (1957).
54. R. J. Block, E. L. Durrum, and G. Zweig, "A Manual of Paper Chromatography and Paper Electrophoresis," 2nd rev. ed., p. 410. Academic Press, New York, 1958.
55. B. Kakáč, *in* "Paper Chromatography" (I. M. Hais and K. Macek, eds.), p. 733. Academic Press, New York, 1963.
56. A. Schneer-Erdey and T. Toth, *Talanta* **11**, 907 (1964).
57. T. Nascutiu, *Rev. Roumaine Chim.* **9**, 273 (1964).
58. M. Quershi and M. A. Khan, *Talanta* **13**, 117 (1966).
59. F. H. Burstall, G. R. Davies, G. P. Linstead, and R. A. Wells, *J. Chem. Soc.* p. 516 (1950).
60. J. A. Coch-Frugoni, *J. Chromatog.* **2**, 69 (1959).
61. W. A. Reeves and T. B. Crumpter, *Anal. Chem.* **23**, 1576 (1951).
62. M. Ziegler and H. Winkler, *Z. Anal. Chem.* **166**, 241 (1959).
63. A. K. Majumdar and B. K. Pal, *Z. Anal. Chem.* **174**, 429 (1960).
64. E. C. Martin, *Anal. Chim. Acta* **22**, 228 (1960).
65. W. J. Frierson, D. A. Rearick, and J. H. Yoe, *Anal. Chem.* **30**, 468 (1958).
66. H. S. R. Barreto, R. C. R. Barreto, and I. R. Pinto, *J. Chromatog.* **5**, 5 (1961).
66a. G. Popa, T. Nascutiu, and V. Illescu, *Rev. Roumaine Chim.* **13**, 315 and 447 (1968).
67. G. DeVries, M. J. Hardonk, and K. H. J. Buschow, *Anal. Chim. Acta* **21**, 568 (1959).
68. H. Nagai, *Nippon Kagaku Zasshi* **78**, 285 (1957).
69. H. Nagai, *Kumamoto J. Sci.* **A3**, 167 (1957).
70. H. Nagai, *Nippon Kagaku Zasshi* **80**, 617 (1959).
70a. E. Cerrai and G. Ghersini, *Analyst* **94**, 599 (1969).
71. M. Ziegler, *Z. Anal. Chem.* **174**, 323 (1960).
72. J. W. O'Laughlin and C. V. Banks, *Anal. Chem.* **36**, 1223 (1964).
72a. N. Cvjetićanin, *J. Chromatog.* **34**, 520 (1968).
72b. C. Liteanu, *Rev. Roumaine Chim.* **13**, 899 (1968).
73. C. S. Knight, *Nature* **183**, 165 (1959).
74. C. S. Knight, *Advan. Chromatog.* **4**, 61 (1967).
75. A. Gruene, *Chimica (Milan)* **11**, 173 and 213 (1957).

76. L. Ossicini and M. Lederer, *J. Chromatog.* **17**, 387 (1965).
76a. H. G. Bungenberg de Jong and J. T. Hoogeveen, *Koninkl. Ned. Akad. Wetenschap., Proc.* **B64**, 1, 18, 167, and 183 (1961); **B63**, 228, 243, and 383 (1960).
77. Chromatographic Data Section, Tables 1 and 2, *J. Chromatog.* **15**, D1, 2 (1964).
78. M. Qureshi and F. Khan, *Anal. Chem.* **39**, 1329 (1967).
79. M. Qureshi and M. A. Khan, *J. Indian Chem. Soc.* **41**, 673 (1964).
80. S. Przeszlakowski, *Chem. Anal. (Warsaw)* **12**, 60 (1967).
81. S. Przeszlakowski and E. Soczewiński, *Chem. Anal. (Warsaw)* **11**, 898 (1966).
82. I. I. M. Elbeih and M. A. Abou-Elnaga, *Chemist-Analyst* **56**, 99 (1967).
83. F. H. Pollard, J. F. W. McOmie, and I. I. M. Elbeih, *J. Chem. Soc.* pp. 466 and 470 (1951).
84. F. H. Pollard and J. V. Martin, *Analyst* **81**, 348 (1956).
85. A. Weiss and S. Fallab, *Helv. Chim. Acta* **37**, 1253 (1954).
86. D. Harrop and E. F. G. Herington, *Analyst* **81**, 499 (1956).
87. H. G. Möller and N. Zeller, *J. Chromatog.* **14**, 560 (1964).
87a. G. E. Knudson, *J. Chromatog.* **38**, 164 (1968).
87b. E. Popper, E. Florean, and P. Marcu, *Rev. Roumaine Chim.* **13**, 909 (1968).
88. E. Blasius and W. Gottling, *Z. Anal. Chem.* **162**, 423 (1958).
89. N. M. Chistyakov and Z. I. Bagoveshchskaya, *Gigiena i Sanit.* **28**, 58 (1963).
90. M. Qureshi and K. N. Mathur, *Anal. Chim. Acta* **41**, 560 (1968).
91. M. Qureshi, J. P. Rawat, and F. Khan, *J. Chromatog.* **34**, 237 (1968).
92. M. Qureshi, I. Akhtar, and K. G. Varshney, *Anal. Chem.* **38**, 1415 (1966).
92a. R. K. Barua and N. K. Baishya, *Current Sci. (India)* **37**, 434 (1968).
93. J. B. Headridge and R. J. Magee, *Talanta* **1**, 117 (1958).
94. G. D. Eliseeva and O. A. Kul'skaya, *Ukr. Khim. Zh.* **32**, 220 (1966).
95. A. Lacourt, *Ind. Chim. Belge* **20**, 399 (1955).
96. A. Lacourt, G. Sommereyns, and J. Soerte, *Mikrochem. Ver. Mikrochim. Acta* **38**, 348 (1951).
97. A. Lacourt, G. Sommereyns, A. Stadler-Denis, and G. Wantier, *Mikrochem. Ver. Mikrochim. Acta* **40**, 268 (1953).
98. H. Taylor and F. E. Beamish, *Talanta* **15**, 497 (1968).
99. M. Covello and G. Ciampa, *J. Chromatog.* **20**, 201 (1965).
100. R. R. Lever, *Australasian Inst. Mining & Met. Proc.* **210**, 139 (1964).
101. A. Lewandowski and M. Tominska, *Ochrona Zabytkow (Warsaw)* **16**, 17 (1963).
102. D. Tocksteinova, *Vet. Casopis* **11**, 537 (1962).
103. J. Plamondon, *Econ. Geol.* **63**, 76 (1968).
104. E. Pfeil, *Oesterr. Chemiker-Ztg.* **65**, 177 (1964).
105. E. Pfeil, *Proc. 2nd Chromatog. Symp., Brussels*, 1962 p. 203.
106. S. Harasawa, *J. Chem. Soc. Japan* **72**, 236, 295, and 352 (1951); **73**, 59, 207, 210, and 240 (1952).
107. Z. V. Zirakishvili, *Sb. Tr., Nauchn.-Issled. Inst. Gigieny Truda i Profzabolevanii, Gruz. SSR* **9**, 171 (1963).
108. J. G. Surak and R. J. Martinovich, *J. Chem. Educ.* **32**, 95 (1955).
109. G. Carleson, *Acta Chem. Scand.* **8**, 1673 (1954).
110. W. R. Walker and M. Lederer, *Anal. Chim. Acta* **5**, 191 (1951).
111. T. Barnabas and J. Barnabas, *Naturwissenschaften* **44**, 61 (1957).
112. S. K. Data and P. Ghose, *Z. Anal. Chem.* **158**, 347 (1957).
113. E. Pfeil, *Chem. Labor.* **5**, 177 (1957).
114. T. Barnabas, M. G. Badve, and J. Barnabas, *Anal. Chim. Acta* **12**, 542 (1955).
115. T. Barnabas, M. G. Badve, and J. Barnabas, *Naturwissenschaften* **41**, 478 (1954).

116. A. E. Steel, *Nature* **173**, 315 (1954).
117. I. A. P. Scott and R. J. Magee, *Talanta* **1**, 329 (1958).
118. M. Lederer, *Anal. Chim. Acta* **8**, 491 (1953).
119. E. Blasius and M. Fisher, *Z. Anal. Chem.* **177**, 412 (1960).
120. R. S. Lauer and N. S. Poluektov, *Zavodsk. Lab.* **25**, 391 (1959).
121. M. Lederer, *Compt. Rend.* **236**, 1557 (1953).
122. J. Danon and M. Levi, *J. Chromatog.* **3**, 193 (1960).
123. M. Qureshi and M. A. Khan, *Anal. Chem.* **35**, 2050 (1963).
124. H. M. Stevens, *Anal. Chim. Acta* **14**, 126 (1956).
125. H. M. Stevens, *Anal. Chim. Acta* **16**, 435 (1957).
126. H. M. Stevens, *Anal. Chim. Acta* **15**, 51 (1956).
126a. P. B. Janardhan and A. Paul, *Separation Sci.* **2**, 597 (1967).
126b. P. B. Janardhan and A. Paul, *Indian J. Chem.* **7**, 66 (1969).
127. C.-L. Kao and H. H. Chang, *Hua Hsueh Hsueh Pao* **29**, 421 (1963).
128. C. T. Hu and H. C. Hsieh, *K'o Hsueh T'ung Pao* p. 54 (1963).
129. C. T. Hu and H. C. Shi, *Hua Hsueh Hsueh Pao* **30**, 352 (1964).
130. W. Noddack and S. Sanir, *Chemiker Ztg.* **88**, 738 (1964).
131. C.-L. Kao and M. Sun, *K'o Hsueh T'ung Pao* p. 434 (1964).
132. K. Yamaguchi, *J. Pharm. Soc. Japan* **74**, 1276 (1954).
133. H. G. Linskens, "Papierchromatographie in der Botanik." Springer, Berlin, 1955.
134. E. Maly, *J. Chromatog.* **19**, 206 (1965).
135. J. Sherma, *J. Chem. Educ.* **39**, A839 (1962).
136. L. Ossicini, F. Saracino, and M. Lederer, *J. Chromatog.* **16**, 524 (1964).
137. T. Oshima and T. Yoshino, *Kagaku To Kogyo (Osaka)* **40**, 440 (1966).
138. C. S. Hanes and F. A. Isherwood, *Nature* **164**, 1107 (1948).
139. J. P. Ebel, *Mikrochim. Acta* p. 679 (1954).
140. A. E. R. Westman, A. E. Scott, and J. T. Pedley, *Chem. Can.* **4**, 189 (1952).
141. V. Miketuková, J. Kohlíček, and K. Kácl, *J. Chromatog.* **34**, 284 (1968).
142. A. Moghissi, *J. Chromatog.* **15**, D2 (1964).
143. M. Lederer, *Australian J. Sci.* **11**, 174 (1949); *Science* **110**, 115 (1949).
144. Y. Servigne, *Compt. Rend.* **239**, 272 (1954).
145. B. L. Harrison and D. H. Rosenblatt, *J. Chromatog.* **13**, 271 (1964).
146. F. H. Pollard, *Brit. Med. Bull.* **10**, 187 (1954).
147. Z. Tamura and M. Miyazaki, *Bunseki Kagaku* **5**, 566 (1956).
148. I. I. M. Elbeih and M. A. Abou-Elnaga, *Anal. Chim. Acta* **23**, 30 (1960).
149. Z. Tamura and M. Miyazaki, *Bunseki Kagaku* **12**, 181 (1963).
150. H. Hettler, *J. Chromatog.* **1**, 389 (1958).
151. R. C. Mehrotra and V. S. Gupta, *J. Polymer Sci.* **A2**, 3959 (1964).
152. L. A. Ionova and N. N. Postnikov, *Khim. Prom.* p. 198 (1969).
153. M. A. Rommel and R. A. Keller, *J. Chromatog.* **18**, 349 (1965); *Talanta* **14**, 1205 (1967).
153a. D. O. De Shazer and J. A. Ursick, *Arch. Oral Biol.* **13**, 1163 (1968); W. Feldmann and I. Grunze, *Z. Anorg. Allgem. Chem.* **360**, 225 (1968); J. B. Gill and S. A. Riaz, *J. Chem. Soc., A* p. 183 (1969).
154. F. H. Pollard, G. Nickless, and M. T. Rothwell, *J. Chromatog.* **11**, 383 (1963).
155. M. J. Smith, *Anal. Chem.* **31**, 102 (1959).
155a. M. R. Verma and P. K. Gupta, *Current Sci. (India)* **37**, 194 (1968).
155b. F. E. Clanet, *Comm. Energie At. (France), Rappt.* **3411**, 1–133 (1968).
156. C. Singh, *J. Chromatog.* **18**, 194 (1965).
157. H. Yoneda, *Bull. Chem. Soc. Japan* **40**, 2442 (1967).

158. M. Mazzei and M. Lederer, *J. Chromatog.* **40**, 197 (1969).
159. F. Pochian and C. Rossi, *Chromatog. Rev.* **4**, 1 (1962).
160. S. K. Shukla, *J. Chromatog.* **21**, 92 (1966).
161. J. Prášilová and F. Šebesta, *J. Chromatog.* **14**, 555 (1964).
162. E. L. Chiotis, G. A. Welford, and R. S. Morse, *Mikrochim. Acta* p. 297 (1968).
163. J. Zamfir and M. Szabo, *Rev. Roumaine Chim.* **13**, 219 (1968).

AUTHOR INDEX

Numbers in parentheses are reference numbers and indicate that an author's work is referred to although his name is not cited in the text. Numbers in italics show the page on which the complete reference is listed.

Aasmundrud, O., 323(26), *351*
Abbott, D. C., 379(41), 380(41), 381(41), *396*
Abdel Sayed, W. S., 478(145), *498*
Abdel-Tawab, G. A., 176(29), 196(29), *197*
Abdine, H., 478(145), *498*
Abdullah, M., 71(144), *87*, 154(29), *170*
Abe, Y., 342(142), 348(142), *353*
Abelson, D. M., 96(13), *108*, 227(46), 231(46), *245*
Abish, L., 185(84), 187(84), *198*
Abou-Elnaga, M. A., 501(13), 507(13), 511(82), 516(13), 518(13), 522(13), 523(13), 527(148), 529(148), *534*, *536*, *537*
Abraham, E. P., 409(46, 47), *434*
Abrahams, D. E., 485(163), *499*
Acheson, R. M., 342(147), *354*
Ackermann, G., 12(32), *28*, 98(28), 104(56), 105(65), *108*, *109*
Adams, W. S., 250(8), 257(8), 270(8), *281*
Addison, E., 45(45), *85*, 400(19), *433*
Agar, G., 470(105), *497*
Aggewal, J. S., 300(66), *316*
Agnew, N. H., 118(19), *146*
Agrawal, H. C., 120(41), *147*
Agre, N. S., 397(6), 399(6), 400(6), 411(6), 415(6), *433*
Akhtar, I., 502(26), 520(92), 522(92), 524(92), *534*, *536*
Akulovich, N. K., 322(19), *350*
Akune, S., 256(31a), 257(31a), *281*
Alam, M., 164(68), 165(68), *171*
Alary, J., 471(110), *497*
Alberti, G., 47(75), 48(76), *85*, 502(24, 25), *534*
Albertini, A., 418(99), *435*
Albu-Budai, M., 409(51), 410(51), 425(136), *434*, *436*
Aldrich, F. D., 383(46), 384(46), *396*
Alexos, M., 45(58), *85*
Allan, P. W., 51(94), *85*
Allan, Z. J., 482(154), *498*
Allen, F. W., 249(5), *281*
Allen, M. B., 328(76), *352*
Allen, W. M., 224(61), 229(61), *246*
Allgeier, H., 155(34), *170*
Alms, W., 478(146), *498*
Altman, K. I., 300(72), 312(72, 121), *317*, *318*
Altmann, S. M., 176(28), 185(28), *197*
Amato, H., 51(93), *86*, 152(11), *169*
Ambler, R. J., 141(150), *150*
Amerine, M. A., 154(14), 158(14), *169*
Ames, B. N., 312(124), *318*
Amormino, V., 347(152), *354*

Anastassiadis, P. A., 216(36), *245*
Anderson, J. M., 323(27), 325(50), 326(50), *351*
Anderson, J. R. A., 41(24), *84*, 506(49), *535*
Anderson, M. E., 71(146), *87*
Anderson, P. J., 168(85), *171*
Anderson, W. M., 301(96), 312(96), *317*
Andreae, W. A., 385(52), *396*
Androulidakis, N., 117(11), 120(11), *146*, 284(7), 290(7), 291(7), 293(7), *315*
Anet, E. F. L. J., 308(105), *317*
Anfinsen, C. B., 140(141), *150*
Angapindu, A., 323(35), *351*
Ansorge, S., 135(122), *149*
Antaniades, E. P., 48(81), *86*
Antonopoulos, C. A., 48(77), *85*
Antoszewski, R., 92(6), *108*
Anwar, M. H., 298(59), 313(59), *316*
Aoyagi, S., 259(59), *282*
Appelt, J., 466(100), 469(100), *497*
Appert, H. E., 13(36), *28*
Arbuckle, R., 141(158), *150*
Archer, A. A. P. G., 290(31), 292(31), *316*
Archer, T. E., 375(39), 377(39), 378(39), 379(39), *396*
Arcoria, A., 472(120), *497*
Arfin, S. M., 177(31), *197*
Argoudelis, A. D., 409(58), 410(58), *434*
Arkhipovich, N. A., 162(63), *171*
Armstrong, M. D., 284(9), 296(49), *315*, *316*
Arnold, A., 167(83), *171*
Arnstein, H. R. V., 102(41), *109*
Aronoff, S., 99(32), *109*
Asami, M., 324(43), *351*
Asatoor, A. M., 41(26), *84*, 111(5), *146*
Ashley, B. D., 177, 185(82), 187(82), *197*, *198*
Ashton, F. M., 99(31), 100(31), *108*
Ashwell, G., 164(76), 166(76), *171*
Asimov, I., 251(15), *281*
Aspinall, G. O., 164(70, 71), 165(70, 71), *171*
Asquith, R. S., 313(127), *318*
Astanina, L. N., 407(34), *433*
Atlas, S. J., 428(141), 429(141), 430(141), *436*
Audigié, C., 334(89), *352*
Awapara, J., 118(21), *146*
Axelrod, L. R., 201(8), 207(17), 209(17), 210(17), *244*, *245*
Ayres, P. J., 216(30), *245*
Azouz, W. M., 307(136), 312(136), 314(136), *318*

Backus, E. J., 416(87), *435*
Backus, M. P., 409(45), *434*
Bacon, M. F., 35, *84*, 328, 329(77), *352*
Bader, D., 482(159), *498*
Badve, M. G., 522(114, 115), 524(114, 115), *536*
Bagliano, G., 45(47), *85*, 503(37, 41), 504(37), *534*, *535*
Bagoveshchskaya, Z. I., 520(89), *536*
Bailey, G. F., 96(19), *108*
Bailey, J. L., 141(152), *150*
Bailey, J. V., 411, 413(66, 67), *434*
Bailey, R. E., 34(5), 74(166), *84*, *88*
Bailey, R. W., 163(65), *171*
Baishya, N. K., 520(92a), *536*
Bajaj, V. R., 130(112a), *149*
Baker, C. G., 142, *150*
Baker, P. B., 467(33), *433*
Baker, R. G., 185(83), 187(83), *198*
Baker, W., 191(97), 193(97), *199*
Balbaa, S. I., 466(98), 469(98), *497*
Ballance, P. E., 172(1), 175(1), 177(1), 185(1), 187(1), *196*
Ballotta, R., 418(100), *435*
Bambalov, G. K., 154(19), *169*
Banaszak, J., 71(149, 152), 72(149), *87*
Bandklajder Sznelwar, R., 132(117), *149*
Banister, A. J., 501(18, 19, 20), 522(19), 524(19), 525(19), *534*
Banks, C. V., 511(72), 513(72), *535*
Bannard, R. A. B., 120(28), *147*, 447(26), *495*
Baráth, Z., 398(10), *433*
Barber, E. D., 299(62), *316*, 448(28), 449(28), 451(33), *495*
Barber, M., 141(159), *150*
Barbier, M., 127(108), *149*
Barbiroli, G., 127(96), 129(96), *148*, 156(38), 159(38), *170*, 419(104), *435*
Barial, M., 424(126), *436*
Bark, L. S., 107(70), *109*, 380(43), *396*, 458(59), *496*

Barker, H. A., 177(39), 179(47), 181(47), 183(39, 47), 191(39), 192(39), *197*
Barker, N. T., 257(39), 272(39), *281*
Barnabas, J., 311(114), *317*, 522(111, 114, 115), 524(111, 114, 115), *536*
Barnabas, T., 522(111, 114, 115), 524(111, 114, 115), *536*
Barnabeo, A. E., 301(98), 311(98), 312(98), *317*
Barnasconi, R., 466(97), 469(97), *497*
Barra, D., 423(122), *436*
Barrett, A. J., 154(23), 158(23), *170*
Barrett, G. C., 78, 79, *88*
Barrett, J., 340(129), *353*
Barrett, J. D., 430(149, 150), *436*
Barreto, H. S. R., 507(66), *535*
Barreto, R. C. R., 507(66), *535*
Barrollier, J., 61(112), *86*
Barton, G. M., 348(162), *354*
Barua, R. K., 520(92a), *536*
Basch, A., 231(70), *246*
Baslerová, M., 324(49), 328(49), *351*
Bass, S. T., 257(34), 258(34), *281*
Bassl, A., 113(34), *147*
Bate-Smith, E. C., 18(46), *28*, 300(71), *317*, 342(138, 139), 343(166), 347(151), 348(166), *353*, *354*
Battle, A. M. D. C., 336(111), 338(111), *353*
Bauer, K., 146(182), *151*, 256(32), *281*
Bauer, L., 324(40), *351*
Bauer, M., 400(18a), *433*
Baumann, E., 113(34), 122(75), 127(75), 132(75), 139(75), *147*, *148*
Baumgartner, W. E., 104(54), *109*
Bayer, E., 1(2), *4*, 11, *28*
Bayfield, R. F., 430(149, 150), *436* 475(132), 477(139), *498*
Bayne, W., 71(146), *87*
Beale, D., 139(136), *150*
Beamish, F. E., 520(98), *536*
Beard, R. J., 335(101), *352*
Beaven, M. A., 26(57), *29*
Bechtler, G., 164(72), *171*
Beck, J. H., 34(5), *84*
Becker, A., 72, *87*
Becker, E., 173(13), *197*
Becker, J., 7(7a), *27*
Beckett, A. H., 26, *29*
Beckman, H., 301(90), *317*
Beckman, H. F., 297(55), *316*
Beekes, H. W., 420(107), *435*
Beer, J. Z., 105(61), *109*, 158(43), *170*
Belen'Kaya, T. V., 161(54), *170*
Bell, C. E., 118(17), *146*
Bell, K., 141(147), *150*
Bellin, J. S., 471(114), *497*
Bellis, D. M., 240(99), *247*
Bellomonte, G., 347(152), *354*
Benassi, C. A., 120(25), *147*, 284(11), 285(11), 286(11, 21), 291(1), *315*
Benigni, J., 120(29), *147*
Bennet-Clark, T. A., 283(1), 286(14), *315*
Bennett, W. D., 168(52), *170*
Benraad, T. J., 202(9), *244*
Benson, A., 336(109, 110, 111), 337(116, 117, 120), 338(110, 111, 117, 120), *353*
Benson, A. A., 45(49), *85*, 105(60), *109*
Benson, W. R., 356(7), *395*
Benton, F. L., 22(55), 23(55), 24(55), 25(55), *29*, 321(16), 322(16, 24), 324(16), 325(16), 326(24, 58), 327(16), 330(58), 331(24), 332(24, 58), 333(16, 58), *350*, *351*
Berenson, G. S., 123(79, 80), *148*
Bereznikov, V. M., 57(107), *86*
Berg, A., 41(23), *84*
Berg, E. W., 49(87), *86*
Bergel, F., 142(165), 144(165), *150*
Berger, B. A., 257(36), 260, 272(36), *281*
Bergmann, F., 173(9), 195(9), *197*
Bergmann, H., 257(45), 277(45), *282*
Bergren, W. R., 45(56), *85*
Berliner, D. L., 216(34), *245*
Berlingozzi, S., 63(121), *87*
Bermes, E. W., Jr., 71(150, 151), *87*
Bernasconi, R., 216(29), 244(29), *245*
Berner, P., 238(90), 240(90), *246*
Bernhard, K., 185(84), 187(84), *198*
Beroza, M., 390(60), *396*
Berret, R., 471(112), *497*
Berridge, N. J., 177(42), 183(42), *197*
Berry, H. K., 138(131), *149*, 174(18), *197*, 284(8), *315*
Berry, R. C., Jr., 300(68), 310(68), *316*
Bertenrath, T., 323(30), 329(30), *351*
Berueffy, R. R., 326(59, 114), 340(114), *351*, *353*
Bessman, S. P., 127(102), 129(102), *149*
Betina, V., 81(193), *88*, 397(5), 398(5, 10), 399(5), 400(21, 22), 401, 402(5, 21, 25,

31), 403(5, 31), 405, 406(5, 22, 32), 408(8), 413(5), 415(5), 416(5), 417(5), 418(5), *433*
Bevenue, A., 297(55, 56), 301(90), *316*, *317*
Beyreder, J., 232(76), *246*
Beyrich, T., 52(98), *86*
Bharucha, F. R., 173(12), *197*
Bhavanandan, V. P., 164(70, 71), 165(70, 71), *171*
Bickel, H., 416(85), *435*
Bickoff, E. M., 231(68), *246*
Bieganowska, J., 7(10), *27*
Bieling, H., 478(146), *498*
Biernat, J. F., 143(172), *151*
Bilbie, E., 414(71a), *434*
Biondi, L., 127(103), *149*
Birkofer, L., 167(81), *171*
Biró, A., 73(161), *88*
Biro, J., 80(187), *88*
Birtwistle, R., 454(42), *495*
Biserte, G., 122(72), 139(72), *148*
Bissing, D. E., 301(84), 309(84), *317*
Bitancourt, A. A., 26(65), 27(65), *29*
Black, D. K., 127(107), *149*
Blackburn, G. M., 490(168), 491(168), *499*
Blackburn, S., 120(47, 54), *147*
Blake, G. G., 105(64), *109*
Blake, M. I., 327(65), *352*
Blasius, E., 518(88), 520(88), 522(119), 524(119), *536*, *537*
Blattna, J., 420(109), *435*
Bleidner, W. E., 384(49), 385(49, 51), *396*
Blinov, N. O., 397(2, 3), 399(17), 400(3), 401(3), 402(3, 28, 30, 31a), 414(2), 415(2, 3), *433*
Blitz, M., 348(161), *354*
Black, G. E., 297(54), *316*
Block, R. J., 8, 9(17), *27*, 33(4), 53(4), 58(110), 60(110), 64, 73(4), 83, *84*, *86*, 96, 98, 102, 107, *108*, 120, 121(44), 122(44), 123(44), 126(44), 127(44), 132(44), 133(44), 134(44), 138, 139(44), *147*, 152(1), 153(1), 158(1), 163(1), *169*, 178, 179, 184(104), 186(104), *199*, 200(6), 201, 202(6), 216(6), 229, 236, 237(6), 242(6), 243(6), 244(6), *244*, 249(3), 250, 251(3), 252(3), 253(3), 255(3), 257(3), 258, *281*, 355, 379, 390(2), 395(2), *395*, 442(9), 443(9), 477(9), *494*, 506(54), 518(54), 520(54), *535*
Blombäck, M., 141(155), *150*
Blumer, M., 335(103), 336(103), 340(103), *353*
Bodanszky, M., 142, *150*, 418(93), *435*
Boegmann, W. H., 416(84), 417(84), *435*
Bogorad, L., 338(123), 340(123) *353*
Bogucka, J., 425(135), 426(135), *436*
Bohley, P., 135(122), *149*
Bohonos, N., 416(87), *435*
Bojesen, E., 216(35), *245*
Boldingh, J., 185(80), 187(80), 191(80), 193(80), *198*
Bol Raap, A., 231(67), *246*
Bondy, P. K., 227(46), 231(46), *245*
Bongiovanni, A. M., 226(45), 227(45), *245*
Bonner, D. M., 286(20), *315*
Bonner, T. G., 155(31), 158(31), *170*
Borecký, J., 113(10), 114(10), 116(38), 120(38), *146*, *147*, 444(16), 445, *494*
Borelius, E., 48(77), *85*
Born, H. J., 105(58), *109*
Borowiecki, M. T., 133(119), *149*
Borowski, E., 418(90), *435*
Boscott, R. J., 297(51), 300(51), *316*
Bottomley, S., 126(87), *148*
Boulton, A. A., 96(14, 20), *108*, 139(138, 139), *150*
Boulton, G., *108*
Bourne, E. J., 155(31), 158(31), *170*
Bowman, G., 121(60), *148*
Boyce, K., 231(65), *246*
Boyd, G. A., 99(29), *108*
Boyland, E., 118(18), 120(18), *146*, 314(133), *318*
Boylen, J. B., 93(10), *108*
Brad, I., 120(53), *147*
Bradbeer, J. W., 257(38), 261, *281*
Bradler, G., 399(14a), *433*
Bradlow, H. L., 56(104), *86*
Brauner, L., 323(32), *351*
Bray, H. G., 297, 300(50), 307(137), 312(50, 122), 313(50), *316*, *318*
Bremner, J. M., 120(27), *147*
Brenner, A., 122(71), *148*
Bricas, E., 141(159), *150*
Bridges, P. M., 104(49), 105(49), *109*
Bridges, R. G., 105(57), *109*
Briner, G. P., 257(48), *282*
Brink, V. C., 35(12), *84*
Brinkman, U. A. T., 42(30), *84*, 504(46), *535*

Broadley, K. J., 117(13, 14), 120(14), *146*
Brochmann-Hanssen, E., 125(82), *148*
Brockmann, H., 418(94), *435*
Broda, E., 176(29), 196(29), *197*
Brodasky, T. F., 57(107), *86*
Broich, J. R., 219(56), 229(56), *245*
Bronshteĭn, I. A., 324(41), 325(41), 328(41), *351*
Brown, B., 71(145), *87*
Brown, F., 177(36), 183(36), 191(36, 92), 192(36, 92), *197*, *198*, 428(142), 430(142), *436*
Brown, R. D., 257(33), *281*
Brown, W. G., 2, *5*, 319(5), 323(5), *350*
Brownell, H. H., 83(199), *89*, 314(131), *318*, 442(11), 443(11), *494*
Brückner, H., 479(149), *498*
Bruns, F., 312(121), *318*
Bryant, F., 183(53), *198*
Brzuszkiewicz, H., 51(92), *86*
Buben, I., 450(32a), *495*
Buch, M. L., 183(51), 191(51), 192(51), *198*
Buchanan, J. G., 251(16), *281*
Buchanan, M. A., 175(25), 176(25), 185(25), 187(25), 196(25), *197*
Buckingham, D. A., 142(162), *150*
Budwig, J., 194(98), 195(98), *199*
Budzynski, A. Z., 104(53), 105(61), *109*
Bukes, H. W., 203(16), 207(16), 218(16), 229(16), *245*
Bukureshchliev, A., 484(161), 490(161), 493(161), *498*
Bungenberg de Jong, H. G., 515(76a), *536*
Bunyan, J., 430(148), *436*
Buras, E., 41(21), *84*
Burger, B., 358(17), *395*
Burke, W. J., 313(128), *318*
Burma, D. P., 67(137), *87*, 126(88), *148*
Burness, A. T. H., 191(91), 192(91), *198*
Burnett, M. C., 453(38a), *495*
Burrows, I. E., 127(98), 129(98), *148*
Burstall, F. H., 507(59), 522(59), 524(59), *535*
Burstein, S., 57(105), *86*, 227(47), *245*
Burton, H. S., 409(46), *434*
Burton, R. B., 52(96), *86*, 227(49), *245*
Burton, W. W., 300(68), 310(68), *316*
Burzynski, S., 138(129a), *149*
Busby, W. F., 479(152), *498*
Buschow, K. H. J., 64(122), *87*, 512(67), *535*
Bush, I. E., 7(9), *27*, 96(15, 16, 17), 98(26), 107(26), *108*, 139(138), *150*, 152(9), *169*, 200(2, 5), 202(2, 13, 14), 217(2), 226(2), 231(64), 232(75), 241(102), 244(102), *244*, *246*, *247*
Butler, G. W., 45(52), *85*
Butterworth, K. R., 45(40), *85*
Buyske, D. A., 414(74), *434*
Buzina, T. P., 142(169), *151*

Cabot, H. M., 143(171), *151*
Cadet, J., 250(10a), *281*
Cahnmann, H. J., 142(163), *150*
Cain, J. C., 159(44), 162(61), *170*, *171*
Caldwell, P. C., 77(173), *88*
Calleri, A., 154(21), 158(21), *170*
Calvert, C. R., 326(60), 340(127), *351*, *353*
Cama, H. R., 426(139), 429(139), *436*
Campbell, I. G., 104(53), *109*
Cantarelli, G., 414(72), 415(72), *434*
Capp, G., 48(81), *86*
Cardinal, E. V., 104(54), *109*
Carles, J., 26, *29*
Carleson, G., 522(109), 524(109), *536*
Carnegie, P. R., 141(160), *150*
Caroli, S., 441(8), *494*
Carter, G. B., 301(96), 312(96), *317*
Cartlidge, J., 454(50), 455(50), 456(50), *495*
Cartwright, R. A., 342(141), *353*
Casarett, L. J., 297(56), *316*
Casinovi, G. C., 463(75), 467(75), *496*
Cassani, G., 418(99), *435*
Casselman, A. A., 83(199), *89*, 120(28), *147*, 447(26), *495*
Cassidy, H. G., 16(41), 21(41), *28*
Casteel, H. W., 345(156), 348(156), *354*
Caster, C., 164(78), *171*
Catelli, P., 124(86), *148*
Cavallini, D., 185(69, 74), *198*
Cayen, M. N., 216(36), *245*
Cecil, R., 80(189), *88*
Cee, A., 120(31), *147*, 473(124, 125), *497*
Ceeová, J., 484(160), *498*
Cerbulis, J., 194(100), *199*
Cerletti, P., 423(122), *436*
Cerny, A., 468(95), *497*

Cerrai, E., 47(74a), *85*, 504(47), 505(47), 512(70a), *535*
Chaikoff, I. L., 237(87), *246*
Chain, E. B., 102(44), *109*
Chakrabortty, H. C., 67(137), *87*
Chakraborty, J., 460(65), *496*
Challa, G., 301(97), *317*
Challice, J. S., 308(112), 313(112), 314(112), *317*
Chan, B. G., 159(44), 162(61), *170*, *171*
Chan, S. K., 141(146), *150*
Chandlier, E. L., 160(51), *170*
Chang, C.-I., 414, *434*
Chang, H. H., 522(127), *537*
Chang, W. H., 300(75), 306(75), *317*
Chapman, R. A., 306(104), 314(134), *317*, *318*
Chard, N. E., 96(20), *108*, 139(139), *150*
Charezinski, M., 51(92), *86*
Chargaff, E., 251(12), 258, *281*, *282*
Chayen, R., 177(30), 195(30), *197*
Cheffel, R. I., 183(58, 59, 60), *198*
Chen, H., 71, *87*
Cherkasskii, A. A., 482(155), *498*
Chernov, V. N., 57(107), *86*
Cheronis, N. D., 299, *316*
Chiang, E., 48(81), *86*
Chiba, Y., 321(17), *350*
Chiotis, E. L., 104(55), *109*, 533(162), *538*
Chipault, J. R., 41(28), *84*, 176(27), 195(27), *197*
Chistyakov, N. M., 520(89), *536*
Chittum, J. W., 177(38), 183(38), *197*
Chmara, H., 418(90), *435*
Chochlov, A. S., 399(17), *433*
Chogvill, H. S., 301(84), 309(84), *317*
Chortyk, O. T., 78(183), 79(183), *88*
Chou, T.-J., 161(56), *170*
Christensen, G. M., 290(32), 293(32), *316*
Christensen, T. B., 164(70), 165(70), *171*
Christman, A. A., 467(76), *496*
Chu, E. J., 335(107), 336(108, 115), 337(107, 108), 338(107, 108), 339, 340(107, 108, 115), *353*
Chu, T. C., 335(107), 336(108, 115), 337(107, 108), 338(107, 108), 339, 340 (107, 108, 115), *353*
Chulkova, Z. S., 121(59), *148*
Chunekamrai, N., 138(131), *149*, 174(18), *197*, 284(8), *315*
Churáček, J., 52(97), *86*, 444(18a), *494*
Churý, J., 232(74), *246*
Ciampa, C., 520(99), *536*
Cico, S., 118(16), *146*
Ciferri, O., 418(99), *435*
Cingolani, E., 347(152), *354*
Ciznar, I., 414(77), *435*
Clanet, F. E., 532(155b), *537*
Claridge, C. A., 398(9), 409(9), *433*
Clark, E. L., 297(54), *316*
Clark, I. T., 301(88), 310(88), *317*
Clark, R. G., 45(45), *85*, 400(19), *433*
Clarke, V., 475(132), 477(139), *498*
Cline, C. W., 503(36), *534*
Coch, J. A., 143(171), *151*
Coch-Frugoni, J. A., 507(60), *535*
Cociumian, L., 64(123), *87*
Coda, H., 143(171), *151*
Coeur, A., 471(110), *497*
Coffin, D. E., 355(6), 364(23), 366(23), 371(23), *395*
Cohen, J. S., 490(168), 491(168), *499*
Cohen, V., 457(56), *495*
Cohn, W. E., 248(1), 257(1), *281*
Cole, E. R., 475(132), 477(139), *498*
Collins, P. M., 164(74), 166(74), *171*
Collman, J. P., 142(162), *150*
Colombo, P., 300(77), 311(77), *317*
Comberg, R., 73(162), *88*
Comfort, A., 335(98), *352*
Connelly, J. L., 340(126), *353*
Connors, K. A., 51(89), *86*
Consden, R., 2, *5*, 10(24), 11(24), 21(24), *28*, 55(101), *86*
Consonni, V., 154(21), 158(21), *170*
Cook, E. S., 174(19), 194(19), *197*
Cook, J. W., 365(31), *396*
Coppi, G., 415(79), *435*
Coppini, D., 284(11), 285(11), 286(11), 291(11), *315*
Corbetta, D., 300(77), 311(77), *317*
Cormick, J., 141(149), *150*
Cornford, P. A. D., 336(110), 338(110), *353*
Cornish, H. H., 467(76), *496*
Cortes, A., 364(26), 365(26), *395*
Corwin, L. M., 335(100), 340(100), *352*
Cotte, J., 446(19), *495*
Coulomb, R., 105(59), *109*
Coussia, I. D., 466(85), 467(85), 469(85), *496*

Covello, M., 425(134), *436*, 520(99), *536*
Coward, R. F., 120(40), *147*
Cowles, E. J., 290(26, 32), 293(26, 32), 296(26), *316*
Craddock, V. M., 312(122), *318*
Cramer, F., 492(173), *499*
Crespi, H. L., 327(65, 66, 72), *352*
Creswell, R. M., 424(127), *436*
Crombie, W. M., 172(1), 175(1), 177(1), 185(1), 187(1), *196*
Cron, M. J., 410(56), *434*
Crook, E. M., 176(28), 185(28), *197*
Crossman, J. K., 82(197), *89*
Crump, G. B., 301(86), *317*
Crumpter, T. B., 507(61), 532(61), *535*
Csiba, L., 439(3), *494*
Csobán, Gy., 399(15), *433*
Cullis, C. F., 177(40), 183(40), 191(40), 192(40), *197*
Curtain, C. C., 45(53), *85*
Custer, G., III., 141(152a), *150*
Cvjetićanin, N., 513(72a), *535*
Czerniak, Z., 138(129a), *149*

Dadlez, J., 467(83), *496*
Dahmer, L. H., 500(11), 511(11), *534*
Dalferes, E. R., Jr., 123(80), *148*
Dalgliesh, C. E., 284(10), 286(10, 16), *315*
Dallas, M. S. J., 81(191), *88*
Daly, J. W., 120(29), *147*
Dalziel, A. M., 104(54), *109*
Damanski, A. F., 431(155b), *437*
Danel'yants, V. A., 468(90), *496*
Daniels, M., 457(55), *495*
Danilovic, M., 420(111), *435*
Danon, J., 522(122), 524(122), *537*
Dao, T. L., 297(53), *316*
Das, B. C., 141(159), *150*
Das, D. P., 477(138), *498*
Das, J. M., 122(66), *148*
Dash, B. C., 477(138), *498*
Dass, R., 301(91), 313(126a), *317*, *318*
Data, S. K., 522(112), 524(112), *536*
Datta, S. P., 176(28), 185(28), *197*
Dauben, W. G., 237(87), *246*
David, S., 45(44), *85*, 172(8), *197*
Davidek, J., 156(37), *170*
Davies, G. R., 507(59), 522(59), 524(59), *535*
Davis, C. N., 13(36), *28*
Davis, T. W. M., 467(77), *496*
Davis, V. E., 118(21), *146*
Dawe, J. C., 300(70), *317*
Dawson, J. T., Jr., 284(12), 290(12), 292(12), *315*
Day, R. A., 407(35), *433*
DeAntoni, A., 120(25), *147*
Debro, J. R., 81(196), *89*
de Camargo Fonesca Moraes, E., 132(117), *149*
DeCarnevale Borino, R. C. D., 414(76), *435*, 470(108, 108a), *497*
Decker, P., 70(142), 71(143), *87*, 286(24), 290(24), 292(24), *316*
Deckers, W., 468(92), *496*
Decsei, L., 468(93a), *496*
Dedic, G., 121(61), *148*
de Franca, F. P., 400(20), *433*
de Freitas, A. S. W., 447(23), *495*
Dehority, B. A., 393(67, 68), *396*
DeJong, E., 164(78), *171*
Dekker, C. A., 251(16), *281*
deKlonia, H., 35(10), *84*
Delfel, N. E., 389(59), *396*
deLigny, C. L., 13, *28*, 120(55), 123, *147*, *148*
Dellacha, J. M., 139(135), *150*
de los Angeles Cervantes, M., 217(42), *245*
Delpierre, G. R., 418(92), *435*
Demain, A. L., 407(40), 408(42), 409(40), *434*
DeMarco, C., 475(135), *498*
Demian, A., 482(158), *498*
de Moor, P., 202(10), *244*
de Muria, J., 416(84), 417(84), *435*
Denef, C., 202(11), *244*
Denison, F. W., 183(56), *198*
Dent, C. E., 121(62), *148*
Denz, F. A., 300(70), *317*
de Oliveira Dias, G., 400(20), *433*
De Ritter, E., 431(151), *436*
DeRudder-Tack, Y., 470(107), *497*
Descimon, H., 250(11), *281*
De Shazer, D. O., 531(153a), *537*
DeThomas, A. V., 51(90), 66(131), 75(168), *86*, *87*, *88*
Deuticke, B., 250(7), 257(7), 270(7), *281*
DeVay, J. E., 26, *29*

DeVries, G., 64(122), *87*, 105(66), *109*, 504(46), 512(67), *535*
DeVries, W. H., 410(55), *434*
Devyatnin, V. A., 426(140), *436*
DeWachter, R., 57(106), 73(159a), *86*, *88*
Deyl, Z., 139(137), 140(137, 139a), 145(175), *150*, *151*, 420(109), *435*
Dezseri, E., 468(93a), *496*
Dhont, J. H., 441(7a), *494*
Dieckert, J. W., 48(80), *85*, 155(36), *170*, 229(53), 230, 235(53), 236(53), *245*
DiFerrante, N., 300(72), 312(72), *317*
Diggs, J. W., 45(57), *85*
Dilli, S., 41(24), *84*, 506(49), *535*
Dimitrov, D., 482(153), *498*
Dimler, R. J., 67(135), *87*
Diplock, A. T., 430(148), *436*
Dittrich, S., *88*
Divekar, P. V., 416(88), *435*
Dixon, G. H., 127(101), 129(101), *149*
Dobozy, O., 472(118), *497*
Dobrecky, J., 414(76), *435*, 470(108, 108a), *497*
Dobson, F., 407(33), *433*
Dobson, G., 455(52), *495*
Döring, H. J., 252(19), *281*
Dogramadzhyan, A. D., 161(55), *170*
Doerschuk, A. P., 414(75), *434*
Dohan, F. C., 284(12), 290(12), 292(12), *315*
Dokunikhin, N. S., 450(31), *495*
Dominguez, O. V., 216(34), *245*
Donike, M., 167(81), *171*
Donnell, G. N., 45(56), *85*
Dorfman, R. I., 213(19), *245*
Dornbuss, A. C., 416(87), *435*
Doskocilova, D., 402(24), *433*
Douglass, C. D., 345(167), 348(167), *354*
Douin, R., 320(11), *350*
Dowling, E. J., 172(7), 177(41), 183(41), *197*
Draganov, A., 482(153), *498*
Drake, N. A., 216(27), *245*, 399(12), *433*
Dreisbach, R. H., 250(7), 257(7), 270(7), *281*
Dresel, E. I. B., 335(97), 337(116), *352*, *353*
Dreyer, W. J., 140(141), *150*
Duee, E. D., 257(34a), *281*
Dumont, E., 35(13), *84*
Duncan, R. E. B., 191(88), 192(88), *198*
Dunn, D. B., 257(41), 274(41), *282*
Dunn, M. S., 79(185), *88*
Dupuis, P., 348(158), *354*
Durkee, A. B., 290(25), 292(25), *316*
Durrum, E. L., 8, 9(17), *27*, 33(4), 53(4), 64(4), 73(4), 83(4), *84*, 96(18), 102(18), 104(18), 107(18), *108*, 120(44), 121(44), 122(44), 123(44), 126(44), 127(44), 132(44), 133(44), 134(44), 138(44), 139(44), *147*, 152(1), 153(1), 158(1), 163(1), *169*, 178(104), 179(104), 184(104), 186(104), *199*, 200(6), 201(6), 216(6), 229(6), 236(6), 237(6), 242(6), 243(6), 244(6), *244*, 249(3), 250(3), 251(3), 252(3), 253(3), 255(3), 257(3), 258(3), *281*, 355(2), 379(2), 390(2), 395(2), *395*, 442(9), 443(9), 477(9), *494*, 506(54), 518(54), 520(54), *535*
Durso, D. F., 154(25), *170*
Dutcher, J. D., 409(59), *434*
Dutt, M. C., 371(34a), *396*
Dutta, N. L., 238(95), 244(95), *246*
D'yachkov, G. S., 155(30), 158(30), 161(30), *170*
Dykhovichnaya, D. E., 401(23), *433*

Eastoe, J. E., 120(50), *147*
Eastwood, F. W., 418(92), *435*
Ebel, J. P., 525(139), *537*
Eberlein, W. R., 226(45), 227(45), *245*
Eble, T. E., 399(18), *433*
Edman, P., 141(155), *150*
Edström, J. E., 257(48), *282*
Edstrom, R. D., 93(11a), *108*
Edwards, C. H., 229(52), *245*, 420(108), 421(108), 422(108), *435*
Edwards, G. A., 420(108), 421(108), 422(108), *435*
Edwards, P. R., 454(42), *495*
Edwards, R. W. H., 213(18), 217(39, 43), 222(59), 229(59), *245*, *246*
Edwin, E. E., 430(148), *436*
Egan, H., 379(41), 380(41), 381(41), *396*
Egge, H., 164(72), *171*
Egger, K., 347(148), *354*
Eglite, G., 133(119a), *149*
Egorou, I. A., 444(13), 454(48), *494*, *495*
Egorova, O. A., 450(31), *495*

Eidelman, M., 357(12), *395*
Eigen, E., 253(22), *281*, 348(161), *354*
Eigner, E. A., 103(46), *109*
Eilers, N. J., 399(18), *433*
Eimhjellen, K. E., 323(26), *351*
Ekert, B., 257(40), 274(40), *282*
Elander, R. P., 409(45), *434*
Elbeih, I. I. M., 501(13), 507(13), 511(82), 515(83), 516(13), 518(13), 522(13, 83), 523(13), 527(148), 529(148), *534*, *536*, *537*
Eldefrawi, M. E., 357(9), 378(40), *395*, *396*
El Dein, M. S. N., 93(53), 161(53), *170*
El Hawary, M. S. F., 185(71), *198*
El-Hehyawi, M. G. E., 66(132), *87*
Eliseeva, G. D., 520(94), *536*
Elliott, W. H., 232(80), *246*
Ellis, R., 454(41), *495*
Ellis, S., 143(170), *151*
Elmore, D. T., 249(6), *281*
El-Sourady, H. A., 431(157), 432(157), *437*
Emerson, M. L., 297(56), *316*
Engel, P., 441(8a), *494*
Engle, C. R., 454(47), *495*
Epstein, E., 96(21), *108*
Erdman, J. G., 475(134), *498*
Ericksen, L., 335(105), 337(105, 118), 340(105), *353*
Ericksson, S., 237(85), *246*
Erlenmayer, H., 501(15), *534*
Erwin, W. R., 370(33), *396*
Esau, P., 154(14), 158(14), *169*
Evans, J. S., 410(55), *434*
Evans, R. A., 300(80), *317*
Evans, R. S., 348(162), *354*
Evans, W. C., 300(80), *317*

Fairbairn, J. W., 94(12), 103(47), 105(12, 67), 106(12, 47), 107(12, 47), *108*, *109*
Fakui, S., 424(129a), *436*
Falk, J. E., 334(92, 93, 94), 335(93, 97), 337(94, 117, 120, 121, 122), 338(93, 117, 120), 340(93), *352*, *353*
Falk, R. H., 430(149, 150), *436*
Fallab, S., 516(85), *536*
Farkas, W. R., 259(50), *282*
Farmer, A., 45(62), *85*
Farmilo, C. G., 463(75), 467(75, 77), *496*
Farran, J. D., 408(44), *434*
Faulkner, W. R., 138(134), *150*
Fazekas, A. T., 202(12), *244*
Fazekas, I. Gy., 202(12), *244*
Feast, A. A. J., 155(33), *170*
Fed'kina, N. G., 397(3), 400(3), 401(3), 402(3, 28), 415(3), *433*
Fedoroňko, M., 153(12), *169*
Feinberg, J. G., 50, *86*
Feldmann, W., 531(153a), *537*
Fernandez, M., 37(17), *84*
Ferrier, R. J., 164(75), 166(75), *171*
Feustel, G., 454(49), *495*
Fewster, M. E., 301(93), 308(109), 314(93), *317*
Fidler, A., 474(130), *497*
Fiebig, E. C., 75(170), *88*
Fielitz, C., 143(171), *151*
Fijolka, P., 446(20), *495*
Filipek, I., 457(54), *495*
Finck, A. D., 500(10), 514(10), *534*
Fink, K., 178(46), 181(46), 183(46), *197*, 250(8), 257(8), 270(8), *281*
Fink, R. M., 178(46), 181(46), 183(46), *197*
Finley, K. T., 452(34), *495*
Fiori, A., 467(78), *496*
Fischer, A., 286(23), *315*, 471(112a), *497*
Fischer, D., 35(8), *84*
Fischer, E., 191(65), 193(65), *198*
Fischer, F. S., 365(29), 367(29), *396*
Fischer, R., 407(37), *433*
Fiset, L., 463(75), 467(75), *496*
Fishbach, H., 67(136), *87*
Fishbein, L., 76(172), *88*, 120(33), *147*, 476(137), 477(137), *498*
Fisher, H., 335, *352*
Fisher, M., 522(119), 524(119), *537*
Fitelson, J., 183(64), 191(64), *198*
Fletcher, J. C., 127(104), 130(104), *149*
Flikke, M., 249(4), 259(4), *281*
Floch, H., 115(43), *147*
Flood, H., 2, *5*
Floodgate, D. G., 174(20), 185(77), *197*, *198*
Florean, E., 518(87b), *536*
Foner, H. A., 3(11), *5*
Fontaine, A. V., 139(135), *150*
Forchielli, E., 213(19), *245*
Forss, D. A., 454(40, 42), *495*
Forsyth, F. G. C., 348(160), *354*
Fowden, L., 140(141a), *150*

Fraga, A., 143(171), *151*
Frame, H. D., 324(44), *351*
Franc, J., 301(99, 100), *317*, 440(4), 460(62, 63), 484(160), *494*, *496*, *498*
Francis, F. J., 300(69), *317*
Franck, A., 81(194), *88*
Franglen, G., 90, 91(3), 92(3), 94(3), 95(3), 97(3), 98(3, 23), *108*
Frank, M., 102(44), *109*
Franklin, M. J., 111(3), *146*
Franks, F., 73(159a), *88*
Franzen, E., 136(127), *149*
Freed, S., 325(53, 54), *351*
Freeman, J. H., 300(78), 312(78), *317*
Freese, E., 259(53), *282*
French, D., 71(144), *87*, 154(29), *170*
Frew, R. G., 12(34b), *28*
Frey, H. P., 12(32), *28*
Freytag, H., 407(38), *434*
Fricke, H., 78(180a), *88*
Friedman, S. J., 371(35), 372(35), *396*
Frierson, W. J., 101, *109*, 507(65), 519, *535*
Fries, N., 253(20), *281*
Fritz, J. S., 500(9, 11), 502(9), 511(11), *534*
Frontali, N., 185(69, 74), *198*
Fuerst, H., 454(49), *495*
Fujimoto, T., 502(23a), *534*
Fujioka, T., 232(81), 233(81), 234(81), *246*
Fujisawa, T., 431(154, 155), 432(155), *436*
Fujito, Y., 141(153), *150*
Fukasawa, H., 329(84), *352*
Fukushima, D. K., 56(104), *86*
Fukushima, S., 329(79), *352*
Fuleki, T., 300(69), *317*
Fuller, W. H., 118(15), *146*
Funahashi, S., 111(6), *146*
Furlong, N. B., 45(54), *85*
Furst, A., 18(44), *28*, 127(105), *149*, 425(131), *436*
Fuse, K., 254(23), 257(23), *281*
Futterman, S., 426(138), *436*

Gabbay, K. H., 216(38), 229(38), *245*
Gabov, N. I., 490(169), 493(179), *499*
Gabriel, T. F., 257(42), 274(42), *282*
Gaddis, A. M., 454(41), *495*
Gadsden, E. L., 420(108), 421, 422(108), *435*
Gaedke, K., 191(94), 193(94), *198*
Gäumann, E., 416(85), *435*
Gafencu, M., 414(71a), *434*
Gaffney, G. W., 300(72), 312(72), *317*
Gage, T. B., 345(167), 348(159, 167), *354*
Gaither, R. A., 357(13), *395*
Gaitonde, M. K., 127(95, 99), 129(99), 133(99), *148*, *149*
Gajcy, H., 424(129), 425(135), 426(135), *436*
Gal, A. E., 162(64), *171*
Galanos, D., 15(39), 16(39), *28*
Galante, E., 490(171), *499*
Galindo, J., 138(133), *150*
Gallagher, T. F., 56(104), *86*
Gallimore, J. C., 99(29), *108*
Gallo-Torres, H. E., 235(84), 236(84), *246*
Ganis, F. M., 200(4), *244*
Gapanhuk, E., 141(154), *150*
Garber, C., 414(76), *435*
Garcia-Hernandez, M., 174(17), *197*
Gardell, S., 48(77), *85*
Gardner, J. A. F., 348(162), *354*
Garnett, J. L., 41(24), *84*, 506(49), *535*
Gasparič, J., 8(13), *27*, 53(99a), *86*, 110(1), 111(1, 7, 8), 113(1, 35, 36, 37, 38, 39), 115(1), 116(1), 118(1, 8), 120(30, 31, 32), *146*, *147*, 444(16), 445, 450(32), 458(58), 459, 472(116, 117, 122), 473(124, 125), *494*, *495*, *496*, *497*
Gast, J. H., 326(60), 340(127), *351*, *353*
Gaull, G. E., 127(95, 99), 129(99), 133(99), *148*, *149*
Gavaudan, A., 471(112), *497*
Geiger, U. P., 122(75), 127(75), 132(75), 139(75), *148*
Gavrichov, V. S., 444(17), *494*
Geisler, J., 458(60), *496*
Geissman, T. A., 342(145), *354*
Gellerman, J. L., 41(27), *84*
Gemzová, I., 120(30, 32), 113(36), *147*, 472(116, 117), *497*
Genest, C., 306(104), 314(134), *317*, *318*
George, H., 124(83), *148*
Gerlach, E., 250(7), 252(19), 257(7), 270(7), *281*
Germano, C., 459(61), *496*
Gertig, H., 468(93), *496*
Gestetner, B., 217(41), 231(41), *245*
Getz, M. E., 355, 364(22), 366(22), 370(32), 371(35), 372(35), *395*, *396*

Gheorghiu, V., 120(53), *147*
Ghersini, G., 47(74a), *85*, 512(70a), *535*
Ghosal, S., 111(9), *146*
Ghose, P., 522(112), 524(112), *536*
Gibree, N. B., 232(75), *246*
Giddings, J. C., 8(14), 9(19), 10(14), 11(19), 12(14), 13(14), 15(19), 16(19), 18(19), 27(67), *27*, *28*, *29*
Gierke, T. D., 17(43a), *28*
Gil, F. P., 126(91), *148*
Gill, J. B., 531(153a), *537*
Gill, S., 466(97), 469(97), *497*
Gilmore, D. R., 364(26), 365(26), *395*
Gioia de Coch, M. N., 143(171), *151*
Giordano, M. G., 423(123), *436*
Giovannini, M., 154(21), 158(21), *170*
Giovenco, M. A., 423(122, 123), *436*
Giovenco, S., 423(122, 123), *436*
Giri, K. V., 63(119), *87*, 132(113), *149*, 254(25), *281*
Glauser, E. M., 141(158), *150*
Glauser, S. C., 141(158), *150*
Glazer, A. N., 141(148), *150*
Gocan, S., 73(158, 159), *88*, 132(116), *149*
Godnev, T. N., 322(19), *350*
Godoy, C. H., 154(17), 159(17), *169*
Godzeski, C. W., 408(44), *434*
Göckeritz, G., 479(149), *498*
Goggins, J. F., 35(9), *84*
Goguadze, V. P., 322(23), *350*
Gold, A. M., 419(105), *435*
Goldman, M., 407(35), *433*
Goldstein, D. A., 503(38), *534*
Golosova, L. V., 308(108), *317*
Gonzalez De Valesi, A., 402(31b), *433*
Good, T. A., 164(69), *171*
Goodall, R. R., 407(36), *433*
Goodban, A. E., 183(50), *198*
Goodman, P. J., 154(20), *169*
Goodwin, L., 111(4), *146*
Goppelsroeder, F., 319, *350*
Gorbach, G., 121(61), *148*, 432(158a), *437*
Gordon, A. H., 2, *5*, 55(101), *86*
Gordon, H. T., 77(174), *88*, 357(9), 370(33), *395*, *396*, 420(110), *435*, 501(16), *534*
Gordon, M., 138(133), *150*
Gordon, M. P., 77(177), *88*
Gorodetskii, V. K., 154(15), 158(15), *169*
Gorta, T., 468(91a), *496*
Gortner, W. A., 384(50), *396*
Gottling, W., 518(88), 520(88), *536*
Gower, D. B., 224(60), 229(60), *246*
Graf, E., 460(67), *496*
Graham, O., 118(21), *146*
Graham, R. J. T., 380(43), *396*, 458(59), *496*
Grandolfo, M., 323(31, 37), 324(37), 328(37), 329(31), *351*
Granger, M., 138(131), *149*, 174(18), *197*, 284(8), *315*
Granick, S., 338(123), 340(123), *353*
Grant, D. W., 314(130), *318*
Grant, L., 96(20), *108*, 139(139), *150*
Grassetti, D. R., 475(133), *498*
Grassini, G., 45(47), *85*, 502, *534*
Gream, G. E., 418(92), *435*
Greco, A. M., 418(100), *435*
Green, A. A., 336(108), 337(108), 338(108), 340(108), *353*
Green, J., 7(6, 8), 17(6), 19, *27*, 61(114), *86*, 296(43, 44, 45, 46, 47), 311(115, 116), *316*, *318*, 428(144, 145), 429(146), 430(144, 148), 431, *436*
Green, J. H., 257(39), 272(39), *281*
Green, J. W., 82(197), *89*
Greenspan, F. S., 45(64), *85*
Gren, R., 500(3), *534*
Griffith, D., 57(107), *86*
Griffiths, M. H. E., 470(108), *497*
Grigorescu, E., 414(71a), *434*
Grimmer, G., 440(5), *494*
Grimson, A., 457(55), *495*
Grisebach, H., 259(56), *282*
Gritsch, J., 120(53), *147*
Gröne, H., 418(94), *435*
Grogory, G. F., 93(9), 95(9), *108*
Gross, W. A., 409(58), 410(58), *434*
Grube, M., 238(98), *247*
Grube, S. B., 431(156), 432(156), *437*
Gruene, A., 515(75), *535*
Grumet, G., 172(4), 194(4), *197*
Grundy, W., 409(52), *434*
Grunze, I., 531(153a), *537*
Guedes de Carvalho, R. A., 506(52), 515(52), *535*
Guedke, K., 291(35), *316*
Günter, B., 333(102), 340(102), *353*
Günzel, C., 241(101), 243(101), 245(101), *247*

Guggolz, J., 231(68), *246*
Guillot, B., 446(19), *495*
Guimaraes, C. V., 238(89), 240(89), *246*
Guinn, G., 161(50), *171*
Gulbaran, E., 154(26), *170*
Gumina, J. I., 121(59), *148*
Gumprecht, D. L., 296(39, 40, 41, 42), 313(39), *316*
Gundlach, D., 100(36), *109*
Gunsberg, E., 348(161), *354*
Gupta, P. K., 301(91), 313(126a), *317*, *318*, 532(155a), *537*
Gupta, V. S., 531(151), *537*
Guran, A., 478(142), *498*
Gustafsson, C., 191(93), 193(93), *198*
Gustavson, W. R., 18(44), *28*, 425(131), *436*
Gustin, T. A., 177(38), 183(83), *197*
Gutai, J. P., 503(38), *534*
Gutand, J., 259(55), *282*
Guyer, M., Jr., 133(119b), *149*

Haack, E., 238(98), *247*
Haavaldsen, L., 251(17), *281*
Haberland, G. L., 312(121), *318*
Habermann, E., 238(93), *246*
Haensel, R., 348(155), *354*
Haer, F. C., 48(82), *86*
Hager, A., 324(47), 323(30, 32), 324(47), 329, *351*
Hahn, G., 48(84), *86*
Hai, Y. B., 503(28), *534*
Haidu, V., 120(53), *147*
Haines, W. J., 216(27), *245*
Hais, I. M., 2(5), *5*, 11(28), 12(28), 16(40), 17(28, 40), 21(28, 54), 23(28, 54), 26(54), *28*, *29*, 38(19), 39(19), 52(19), 53(99), 54(19), 81(191), *84*, *86*, *88*, 120(45), 121(45), 123(45), 127(45), 132(45), 133(45), 141, *147*, 334(91), *352*, 397(4), 399(4), 407(4), 408(4), 410(4), 413(4), 414(4), 423(119), *433*, *436*
Hall, D. A., 301(93), 308(109), 314(93), *317*
Hallas, D. G., 501(16a), 520(16a), *534*
Halmagyi, D., 297(52), *316*
Hamilton, J. G., 48(80), 83(199), *85*, *89*, 154(24), 155(24, 36), 156(24), *170*, 229(53), 230, 235(53, 84), 236(53, 84), *245*, *246*
Hammond, E. W., 379(41), 380(41), 381(41), *396*
Hamnstrom, B., 48(77), *85*
Hanes, C. S., 121(63), 127(63, 109, 110, 111), *148*, *149*, 172(6), 177(6), *197*, 525(138), *537*
Hanes, M. E., 409(52), *434*
Hansen, H. J. M., 231(69), *246*
Hansen, R. G., 257(34), 258(34), *281*
Hansen, S., 127(101), 129(101), *149*
Hansens, M., 470(107), *497*
Happer, D. A., 142(162), *150*
Harasawa, S., 522(106), *536*
Harbone, J. B., 191(97), 193(97), *199*, 341(131, 132, 133, 134), 342(131, 132, 133), 343(132, 133, 134), 345(133, 134), 347(131, 132, 133), 348(131, 132, 133, 134, 135), 349(131, 132, 133), 350(131, 133), *353*, *354*
Harder, R., 320(13), *350*
Hardin, G., 322(25), 327(25), *350*
Harding, N., 61(113), *86*
Hardonk, M. J., 64(122), *87*, 512(67), *535*
Hardy, T. L., 127(93), *148*
Haremaker, H. H., 504(45a), *535*
Harley-Mason, J., 290(31), 292(31), *316*
Harold, F. M., 237(87), *246*
Harper, J. L., 342(147), *354*
Harris, C., 219(55), 220(57), 229(55, 57), *245*
Harris, C. K., 121(63), 127(63), *148*
Harrison, A., 105(57), *109*
Harrison, B. L., 527(145), *537*
Harrop, D., 517(86), *536*
Hartel, J., 41(22), *84*, 221(58), 229(58), 231(67), *246*
Hartkamp, H., 506(53), *535*
Hartley, R. D., 172(2, 3), 191(2, 3, 99), *196*, *199*
Hartmann, T., 120(26), *147*
Harwood, S. E., 155(31), 158(31), *170*
Hashimoto, Y., 42(32), *84*
Hashmi, M. H., 177(40), 183(40), 191(40), 192(40), *197*
Haslewood, G. A. D., 232(82), *246*
Hassall, C. H., 238(94), *246*
Hatchadourian, A., 45(58), *85*
Hathway, D. E., 301(101), 306(102), *317*
Havir, J., 474(130), *497*
Haxo, F. T., 331(88a), 333(88a), *352*

Hayashi, K., 342(142), 348(142), *353*
Hayatsu, R., 254(23), 257(23), *281*
Hayden, A. L., 238(90, 96), 240(90, 96), 243(96), 244(96), *246*
Hayman, S., 485(165), 490(165), *499*
Hayward, S. J., 224(61), 229(61), *246*
Hayes, H., 41(27), *84*
Heacock, R. A., 286(22), 290(30), 292(30), 293(30), 311(30), *315*, *316*
Headridge, J. B., 520(93), *536*
Heathcote, J. G., 136(124), *149*
Hecht, E., 74(163), *88*
Heckemann, H.-J., 113(34), *147*
Hedin, P. A., 300(74), *317*
Hedrick, C. E., 257(36), 260, 272(36), *281*
Heftmann, E., 216(25), 227(50), 228(25), 229(25), 238(90, 96), 240(90, 96), 243(96), 244(96), *245*, *246*, 500(7), *534*
Heidelberger, C., 255(30), *281*
Heinemann, B., 399(14), *433*
Heines, S. V., 1(4), *5*
Hekal, I., 432(158a), *437*
Hellberg, H., 407(35a), *433*
Hellman, L., 56(104), *86*, 174(16), *197*
Hellyer, G. C., 104(49), 105(49), *109*
Hendrickson, M. J., 326(59), 336(114), 340(114), *351*, *353*
Hennig, I., 472(121a), *497*
Hennrich, N., 492(173), *499*
Henriques, O. B., 141(154), *150*
Herbert, M., 185(73, 79), *198*
Herington, E. F. G., 517(86), *536*
Hermans, P. H., 301(97), *317*
Herndon, J. G., 13(36), *28*
Herold, M., 413(73), 414(73), *434*
Herranen, A., 121(60), *148*
Hesse, G., 15(37a), 16(37a), *28*
Hessel, B., 141(155), *150*
Hettler, H., 530(150), 531(150), *537*
Hewel, C. A., 501(16), *534*
Hickman, J., 164(76), 166(76), *171*
Hildebrandt, A., 440(5), *494*
Hilding, A. C., 290(26, 32), 293(26, 32), 296(26), *316*
Hilgeman, F., 503(33), *534*
Hill, A. R. C., 154(20), *169*, 364(27), 366(27), 367(27), *395*
Hirayama, O., 175(23, 24), 185(24, 86), 187(24, 86), 191(24), 193(24), *197*, *198*
Hirsch, U., 414(75), *434*
Hirshfield, H., 45(44), *85*
Hirtz, J., 45(58), *85*, 471(112), *497*
Hiscox, E. R., 177(42), 183(42), *197*
Hiyama, H., 64(125), 66(125), *87*
Hobart, S. R., 41(21), *84*
Hockenhull, D. J. D., 174(20), 185(73, 77, 78, 79), *197*, *198*
Hodgson, H. C., 388(56), 389(56), 392(56), *396*
Högenauer, G., 254(27), *281*
Hoerman, K. C., 141(160), *150*
Hoff, J., 500(3), *534*
Hoffman, C. P., 96(17), *108*
Hoffman, I., 100(34), *109*
Hoffmann, B., 216(23a), *245*
Hoffmann, J., 105(62), *109*
Holden, M., 326(57), *351*
Holder, R. L., 96(14), *108*
Holeček, V., 337(119), *353*
Holland, D. O., 127(93), *148*
Hollemann, J. W., 122(72), 139(72), *148*
Holleman-Dehove, J., 122(72), 139(72), *148*
Hollister, Z. J., 399(14), *433*
Holman, R. T., 41(27), *84*
Holmsen, H., 259(51), *282*
Hoogeveen, J. T., 515(76a), *536*
Hooper, I. R., 410(56), *434*
Hoppe, W., 141(156), *150*
Hordis, C. K., 152(7), *169*
Horitsu, K., 160(50), *170*
Horský, O., 413(73), 414(73), *434*
Horton, D. R., 408(44), *434*
Horvath, L., 17(41a), *28*
Horvath, I., 415(82), *435*
Holt, C. E., 257(33), *281*
Holý, A., 257(46a), 280(46a), *282*
Hossfeld, R. L., 299, 300(64, 75), 306(75), 311(64), *316*, *317*
Hough, L., 158(41), *170*
Howard, A. J., 399(14), *433*
Howard, J. W., 440(6), *494*
Hrdlička, J., 118(22), *146*, 156(37), *170*
Hristovski, B., 455(53), 457(53), *495*
Hruška, K., 35(7), *84*
Hsia, D. Y., 133(118), *149*
Hsieh, H. C., 525(128), *537*
Hu, C. T., 525(128, 129), *537*
Huang, C.-S., 414, *434*
Hubacek, J., *318*
Huber, G., 416(86), *435*

Huber, W., 78(180a), *88*, 439(1), *494*
Hütter, R., 416(85), *435*
Hug, D. H., 27(66), *29*
Huggins, A. K., 51(91), *86*
Hughes, G., 455(52), *495*
Hunger, A., 238(92), 242(92), *246*
Hunter, G. D., 185(79), *198*
Husain, S., 301(83), 309(83), *317*
Husain, W., 503(28a), *534*
Husak, R., 474(130), *497*

Ibrahim, R. K., 298(58), *316*
Idu, S. M., 64(123), *87*
Igloy, M., 415(82), *435*
Ijichi, K., 402(29), 417(29), *433*
Ikawa, M., 424(130), *436*
Ikeda, S., 424(129a), *436*
Illescu, V., 512(66a), *535*
Imagawa, H., 342(144), *353*
Iman, G. M., 93(53), 161(53), *170*
Imhoff, U., 425(132), *436*
Ingle, R. B., 97(22), 98(22), *108*
Ingram, V. M., 140(140), *150*
Ionescu, L., 425(136), *436*
Ionova, L. A., 531(152), *537*
Inouye, T., 133(118), *149*
Inoue, Y., 173(11), 175(24), 185(24), 187(24), 191(11, 24), 193(24), *197*, 259(59), *282*
Irimescu, I., 64(123), *87*
Irudayasamy, A., 371(34), *396*
Irvine, D. G., 71(146), *87*
Irzykiewicz, H., 163(66), 169(66), *171*
Isaksson, L. A., 259(52), *282*
Isherwood, F. A., 172(6), 177(6), 185(72, 76), *197*, *198*, 525(138), *537*
Ishida, N., 402(26), *433*
Ishihara, H., 250(9), *281*
Israili, A. H., 503(29), 511(29), *534*
István, K., 121(58), *148*
Ito, K., 431(154), *436*
Itō, M., 329(79), *352*
Ivanov, V., 467(86), *496*
Iwai, K., 424(128), *436*
Izawa, M., 502(23a), *534*

Jaarma, M., 104(51), *109*
Jack, M., 80(189), *88*
Jackson, H. R., 100(34), *109*
Jackson, W. G., 410(55), *434*
Jacob, T. M., 257(44, 47a), 275(44), *282*
Jacoby, W. B., 286(20), *315*
Jaffe, J. E., 467(81), *496*
Jako, N., 156(39), *170*
James, S. P., 307(137), *318*
Jameson, D., 284(12), 290(12), 292(12), *315*
Janardhan, P. B., 10(21), *28*, 522(126a, 126b), 525(126a, 126b), *537*
Janczarski, I., 120(52), *147*
Janíček, G., 118(22), *146*
Jankulov, J., 65(127), *87*
Jareczek-Morawska, E., 418(90), *435*
Jaroszewicz, K., 255(28), *281*
Jarowski, C. I., 409(54), 411(54), *434*
Jarvis, B. C., 257(38), 261, *281*
Jayle, M. F., 229(51), *245*
Jayme, G., 48(78, 84), *85*, *86*
Jeanes, A., 67, *87*
Jeffrey, J. E., 161(57), *170*
Jeffrey, S. W., 328(73, 76), 331(73, 88a), 333(73, 88a), 324(90), *352*
Jenkins, P., 122(75), 127(75), 132(75), 139(75), *148*
Jenner, F. A., 231(65), *246*
Jensen, A., 323(26), *351*
Jensen, E. R., 414(75), *434*
Jepson, J. B., 286(19), *315*
Jerchel, D., 286(15), *315*
Jeronimo, M. A. S., 503(27), *534*
Jevons, F. R., 159(45), *170*
Jily, O., 469(104), *497*
Jiracek, V., 135(123), *149*
Johal, P. S., 372(36), *396*
Johnson, B. T., 99(30), *108*
Johnson, D. L., 398(9), 409(9), 410(56), *433*, *434*
Johnstone, B. M., 257(48), *282*
Jones, A. R., 172(7), 177(41), 183(41), *197*
Jones, H. A., 356(7), *395*
Jones, J. K. N., 158(41), *170*
Jones, J. W., 101(40), *109*
Jones, L. B., 324(44), *351*
Jones, R. T., 141(149), *150*
Jones, W. J., 12(33), *28*
Jonsen, J., 251(17), *281*
Jori, H., 461(71), *496*
Jork, H., 90, *108*
Joshi, G. V., 173(12), *197*

Jucker, E., 327(64), *351*
Jungalwala, F. B., 426(139), 429(139), *436*
Jungnickel, F., 490(167), *499*
Jurd, L., 311, *318*
Jureček, M., 444(18a), *494*
Jurics, E. W., 342(135), *353*

Kabasakalian, P., 231(70), *246*
Kácl, K., 526(141), *537*
Kadis, B., 219(55), 229(55), *245*
Kaiser, A., 96(21a), *108*
Kaiser, C., 167(81), *171*
Kaiser, F., 238(98), *247*
Kaiser, H., 461(71), *496*
Kakáč, B., 506(55), 516(55), 517(55), 518(55), *535*
Kakimoto, Y., 493(178), *499*
Kakol, H., 425(135), 426(135), *436*
Kalbe, H., 183(63), *198*
Kaldor, G., 255(30), *281*
Kaler, V. L., 327(68, 70), *352*
Kalyankar, G. D., 183(54), *198*, 254(25), *281*
Kamel, K., 141(160), *150*
Kamenshchikova, V., 454(45), *495*
Kamienski, E. S. v., 113(10a), *146*
Kamm, L., 161(52a, 52c), *170*
Kampe, W., 492(172), *499*
Kanaoka, Y., 120(29), *147*
Kanazawa, A., 493(178), *499*
Kao, C.-L., 522(127), 525(131), *537*
Kapadia, G. G., 312(119), *318*
Kapadia, G. J., 312(119), *318*
Kapišinská, V., 118(16), *146*, 448(30), *495*
Kaplan, J. R., 323(35), *351*
Kapoulas, V. M., 15(39), 16(39), *28*
Kapri, A., 64(126), *87*
Karawya, M. S., 466(98), 469(98), *497*
Karger, B. L., 8, *27*
Karrer, P., 327(64), *351*
Karrer, R., 253(21), *281*
Kashina, L. V., 478(141), *498*
Kassem, M. A., 431(157), 432(157), *437*
Katayama, M., 111(6), *146*
Katayama, T., 325(52), 331(52), *351*
Katayama, Y., 326(62), 329(62, 80), *351*, *352*
Katkevich, R. G., 162(62), *171*
Katkevich, Yu. Yu., 162(62), *171*
Kato, Y., 423(120), *436*
Katrukha, G. S. 142, (168), *151*
Katsui, G., 419(103), *435*
Katz, A. M., 140(141), *150*
Katz, E., 418(95, 96, 98), *435*
Katz, J. J., 22(55), 23(55), 24(55), 25(55), *29*, 321(16), 322(16, 24), 324(16), 325(16), 326(24, 58), 327(16, 65, 66, 67, 72), 330(58), 331(24), 332(24, 58), 333(16, 58), *350*, *351*, *352*
Katz, S., 203(15), 204(15), 219(56), 229(56), *245*
Kaufmann, H., 168(86), *171*
Kaufmann, H. P., 194(98), 195(98), *199*
Kavčič, R., 455(53), 457(53), *495*
Kawaguchi, H., 409(62), *434*
Kawamura, S., 502(23a) *534*
Kawanagh, F., 408(43), *434*
Kawerau, E., 63(120), *87*
Kazarinova, G. I., 431(155a), 432(155a), *436*
Kaznina, N. I., 441(8b) *494*
Kechkhuashvili, A. L., 426(136a), *436*
Keerberg, O., 122(65), *148*
Kefford, N. P., 283(1, 5), 286(14), *315*
Kehl, R., 335(102), 336(112, 113), 337(112, 113), 340(102, 112, 113), *353*
Keith, R. W., 306(103), *317*
Kekwick, R. G. O., 492(174), *429*
Kelemen, E., 297, *316*
Keller, M., 122(73), *148*
Keller, R. A., 9(19), 11(19), 12(33, 34, 34a), 15(19), 16(19), 18(19), 27(67), *28*, *29*, 531(153), *537*
Kellner, G., 176(29), 196(29), *197*
Kelly, R. G., 414(74), *434*
Kember, N. F., 45(62) *85*
Kemula, W., 458(60), *496*
Kennedy, E. P., 177(39), 183(39), 191(39), 192(39), *197*
Kennedy, G. Y., 340(128), *353*
Kennedy, T. H., 300(70), *317*
Kenten, R. H., 120(27), *147*
Kepf, K., 416(86), *435*
Keresztes, T., 398(8), *433*
Keston, A. S., 103(48), *109*
Keup, W., 313(126), *318*
Keutmann, E. H., 52(96), *86*, 227(49), *245*
Kevei, E., 431(152), *436*
Khan, A. W., 416(88), *435*

Khan, F., 500(8), 502(23), 506(50), 507(58), 508(8, 78), 520(91), *534*, *535*, *536*
Khan, F. M., 154(18), *169*
Khan, M. A., 508(79), 522(123), 525(123), *536*, *537*
Khattak, M. N., 257(39), 272(39), *281*
Khokhlov, A. S., 397(2, 3), 400(3), 401(3), 402(3, 28), 414(2), 415(2, 3), *433*
Khokhlova, N., 402(30), *433*
Khokhlova, Yu. M., 397(3), 400(3), 401(3), 402(3), 415(3), *433*
Khorana, H. G., 257(43, 47a), 275, *282*
Khryashcheva, K. M., 397(3), 400(3), 401(3), 402(3), 415(3), *433*
Khulbe, K. C., 118(23), *146*
Kibrick, A. C., 78(181a), *88*
Kickuth, R., 168(86a), *171*, 447(25), *495*
Kiel, E. G., 472(121), 473(121, 129), *497*
Kikodse, K., 71(143), *87*
Kilroe-Smith, T. A., 493(180), *499*
Kimball, H. L., 57(105), *86*, 227(47), *245*
Kimura, M., 141(153), *150*
King, H. K., 191(91), 192(91), *198*
Kinget, R., 454(44), *495*
Kingston, D. G. I., 418(92), *435*
Kinoshita, J. H., 485(165), 490(165), *499*
Kirby, K. S., 300(81), *317*
Kirk, M. R., 232(77), *246*
Kirk, P., 81(192), *88*
Kirkland, J. J., 385(51), *396*
Kirkpatrick, D., 467(81), *496*
Kiroku, S., 161(52b), *170*
Kishore, N. M., 479(151), *498*
Kissane, J. Q., 71(152), *87*
Kitahara, S., 64(125), 66(125), *87*
Kitai, R., 141(144), *150*
Kito, K., 141(153a), *150*
Kjaer, A., 476, *498*
Klans, R., 98(27), *108*
Klausner, Y. S., 141(157), *150*
Klein, P. D., 15(38), 20(52), *28*
Kleiner, E. M., 397(3), 400(3), 401(3), 402(3), 415(3), *433*
Kleinschmidt, W. J., 249(2), 257(2), *281*
Klemm, L. H., 48(81), *86*
Kley, H. K., 227(48), *245*
Kliman, B., 202(14a), *244*
Kloppenborg, P. W. C., 202(9), *244*
Kloucek, B., 113(37), *147*
Knauff, R. E., 216(27), *245*
Knight, B. C., 337(116, 120), 338(120), *353*
Knight, C. S., 10(23), *28*, 44(38), 45(43, 48), *84*, *85*, 123, 124(81, 84, 85), 125(81, 84, 85), *148*, 513(73), 514(73, 74), *535*
Knížek, J., 460(63), *496*
Knodo, S., 329(82), *352*
Knolle, H., 48(78), *85*
Knowles, E., 300(81), *317*
Knudson, G. E., 12(33, 34a), *28*, 518(87a), *536*
Knypl, J. S., 92(6), *108*
Kobrde, V., 185(81), 187(81), *198*
Koch, W., 167(81), *171*, 320(13), *350*
Kocher, V., 253(21), *281*
Kocourek, J., 160(49), *170*
Kodicek, E., 290(28), 292(28), *316*
Koehler, K., 143(172), *151*
Koen, V., 421(115), 422(115), 423(115), *436*
Koeppen, B. H., 348(157), *354*
Körös, Z., 251(14), *281*
Koffler, H., 416(83), *435*
Kohlíček, J., 526(141), *537*
Kok, E. C. M., 123(76), *148*
Kolb, J. J., 26(56), *29*
Kolman, Z., 42(34), *84*, 217(54), 229(54), *245*
Kolokolov, B. N., 450(31), *495*
Kolor, M. G., 46(73), 62(116), *85*, *87*
Komárek, K., 444(18a), *494*
Kometani, K., 238(97), *246*
Kompisova, A., 118(16), *146*
Kondo, S., 411(63), *434*
Konoshita, T., 414(78), *435*
Kooh, R. P., 485(164), 488(164), *499*
Koppanyi, Z. H., 127(102), 129(102), *149*
Korchagin, V. B., 419(102), 424(102), *435*
Kornel, L., 231(66), *246*
Korte, F., 80(190), *88*
Koštíř, J. V., 160(49), *170*, 423(119), *436*
Kovacicova, L., 400(21), 402(21, 25), *433*
Kovacs, G., 120(56), *147*
Kovář, V., 440(4), *494*
Kovařík, L., 413(73), 414(73), *434*
Kovtun, V. S., 454(45), *495*
Kowalewski, Z., 468(93), *496*
Kowkabany, G. N., 126(90), 129(90), 131(90), *148*, 152(7), *169*
Kozlov, V. V., 482(156), 483(156), *498*

Kraeger, S. J., 154(24), 155(24), 156(24), *170*
Kramlová, M., 143(174), 145(174), 146(180, 181), *151*, 473(128), 474, *497*
Krasnec, L., 439(3), 467(86a), *494*, *496*
Krasovskaya, T. A., 324(41), 325(41), 328(41), *351*
Krause, P., 500(3), *534*
Kraut, H., 425(132), *436*
Krauze, D., 120(52), *147*
Kraybill, M., 107(69), *109*
Krcmery, V., 414(77), *435*
Krien, G., 460(64), *496*
Krishnamurthy, K., 63(119), 64(124), 65(124), *87*, 132(113), *149*
Krishnaswamy, P. R., 183(54), *198*, 254(25), *281*
Kritchevsky, D., 232(77), *246*
Kritchevsky, T. H., 177(34), *197*
Kruger, S., 49(85), *86*
Krzeczkowska, I., 65(129), *87*, 132(115), 138(129a), *149*, 159(46), *170*
Krzyzanowska, M., 67(136), *87*, 254(26), 257(26), 258(26), *281*
Kučera, Z., 57(108), *86*
Kuczyński, J., 20(49), *28*
Kuhn, R., 164(72), *171*
Kukhtsenka, T. V., 327(69), *352*
Kulhanek, M., 155(35), *170*
Kul'skaya, O. A., 520(94), *536*
Kume, Y., 122(67), *148*
Kun, E., 174(17), *197*
Kuntto, H., 185(75), *198*
Kupfer, D., 213(19), *245*
Kupper, U., 122(75), 127(75), 132(75), 139(75), *148*
Kupressova, V. B., 154(22), 159(22), *170*
Kurup, P. A., 169(89), *171*
Kusewicz, D., 420(112), 422(112), 423(112), *435*
Kushnir, I., 78(183), 79(183), *88*
Kuypers, G. H. A., 472(121), 473(121, 129), *497*
Kuznecova, V. S., 418(97), *435*
Kvamme, E., 174(16), *197*
Kwasik, T., 66(133), *87*

Labarrere, J. A., 41(28), *84*, 176, 195(27), *197*
Lacourt, A., 105(62), *109*, 501(22), 520(95, 96, 97), *534*, *536*
Ladzinska-Kulinska, H., 66(133), *87*
Laland, S., 251(17), *281*
Landor, S. R., 127(107), *149*
Landua, A. J., 301(96), 312(96), *317*
Langer, A., 105(63), *109*
L'Annunziata, M. F., 118(15), *146*
Lanzarini, G., 127(100), 129(100), *149*
Larner, J., 259(60), *282*
Lascombes, S., 26(63), *29*
Laskin, A. I., 233(83), 234(83), *246*
Latinák, J., 482(157), 483(157), *498*
Lauer, R. S., 522(120), 524(120), *537*
Lautner, H., 407(37), *433*
Lavica, B., 420(111), *435*
Lawson, G. J., 172(2, 3), 191(2, 3), *196*
Lazer, L. S., 104(54), *109*
Lazier, W. A., 409(54), 411(54), *434*
Leaback, D. H., 26, *29*
Leach, B. E., 410(55), *434*
Lederer, E., 141(159), *150*, 348(165), *354*
Lederer, M., 18(45), 20(47), 26(64), *28*, *29*, 45(47, 50), 46(72), 47(74a), 73(157, 159a), *85*, *87*, *88*, 152(10), *169*, 177(37), 183(37), 191(87), 192(87), *197*, *198*, 300(76), *317*, 348(165) *354*, 441(8), *494*, 503(30, 31, 37, 42), 504(37, 44, 55), 506(51), 511(31, 42), 515(42, 76), 522(110, 118, 121), 524(110, 118, 121), 525(136), 526(143), 533(158), *534*, *535*, *536*, *537*, *538*
Ledin, G., Jr., 18(44), *28*, 425(131), *436*
Lee, L. A., 183(52), *198*
Lee, W. W., 111(4), *146*
Lees, T. M., 416(84), 417, *435*
Legallais, V., 107(69), *109*
Legatowa, B., 471(113), *497*
Leggio, T., 45(65), *85*, 257(33a), *281*
Leng, C. O., 26(59), *29*
Leonard, C., 138(131), *149*, 174(18), *197*, 284(8), *315*
Leopold, A. C., 290(33), 293(33), 294(33), *316*
Lepeshkina, V. A., 155(30), 158(30), 161(30), *170*
Lerner, S. R., 80(188), *88*
Leshchenko, V. P., 154(16), 158(16), *169*
Lester, R. L., 428(143), 430(143), *436*
Letham, D. S., 251(13), *281*

LeTourneau, D., 306(103), *317*
Letters, A., 71(145), *87*
Levant, A. J., 238(96), 240(96), 243(96), 244(96), *246*
Lever, R. R., 520(100), *536*
Levi, A. A., 407(36), *433*
Levi, M., 522(122), 524(122), *537*
Levine, J., 67(136), *87*
Levy, M., 103(48), *109*
Lewandowski, A., 520(101), *536*
Lewbart, M. L., 232(71, 72), *246*
Lewin, L. M., 421(117), 422(117), 423(117), *436*
Lewis, D., 155(31), 158(31), *170*
Lewis, J. C., 402(29), 417(29), *433*
Liberman, L. A., 191(68), 193(68), *198*
Lichtenthaler, H. J., 426(137), 427(137), 428, *436*
Lieldiens, R., 136(126), *149*
Lincoln, G. J., 423(124), *436*
Linden, J. C., 71(144), *87*, 154(29), *170*
Lindner, K., 122(68), *148*
Linek, K., 153(12), *169*
Lingren, W. E., 40(18), 73(159b), *84*, *88*
Link, W. E., 41(27), *84*
Linskens, H. G., 525(133), *537*
Linstead, G. P., 507(59), 522(59), 524(59), *535*
Lippstone, G. S., 413(68), *434*
Lipton, R., 71(147), *87*, 133(120), *149*
List, P. H., 467(87), *496*
Liteanu, C., 73(158, 159), *88*, 132(116), *149*, 513(72b), *535*
Litvinenko, S. N., 399(16), *433*
Livinston, A. L., 231(68), *246*
Locke, D., 301(87), 310(87), 313(87), *317*
Lockhart, I. M., 111(5), *146*
Loder, B., 409(47), *434*
Loeffler, R. K., 326(60), 340(127), *351*, *353*
Loewenstein, J. M., 45(64), *85*
Loffler, J. E., 183(57), *198*
Loftfield, R. B., 103(46), *109*
Logan, J. E., 83(198), *89*
Logar, S., 470(106), 472(123), 473(127), *497*
Long, A. G., 251(16), *281*
Longland, R. C., 42(33), 43(33), *84*
Lonsdale, D., 138(134), *150*
Lorinc, A., 472(118), *497*
Losev, A. P., 327(71), *352*
Losse, G., 143(172), *151*
Lotz, W. E., 99(29), *108*
Lou, M. F., 485(165), 490(165), *499*
Louwerse, W., 91(4), *108*
Lowe, A. A., 103(45), *109*
Lowen, W. K., 385(51), *396*
Lu, H.-Y., 493(176), *499*
Lu, K.-C., 478(147), *498*
Ludwig, D., 256(31), 258(31), *281*
Ludwig, R. A., 387(55), *396*
Lüss, K.-D., 479(150), *498*
Lugg, J. W. H., 136(127), *149*, 179(48), 183(48, 49), 191(48), 192(48), *197*, *198*
Luk'yanov, V. B., 135(121), *149*, 444(17), *494*
Lundberg, W. O., 41(28), *84*, 176(27), 195(27), *197*
Lur'e, A. A., 500(5), *534*
Luu Duc, C., 471(110), *497*

McBain, J. B., 375(38), 376(38), *396*
McCalla, K., 141(149), *150*
McClellan, D. B., 388(57), 392(57), *396*
McComb, E. A., 117(11), 120(11), *146*, 284(6, 7), 290(7), 291(7), 293(7), *315*
McCormick, J. R. D., 414(75), *434*
McCoy, P. F., 312(123), 315, *318*
McCullough, T., 164(78), *171*
McDonald, H. J., 71, 72(154), *87*
Macek, K., 11(28), 12(28), 17(28), 21(28), 23(28), *28*, 32(3), 38(3, 19), 39(3, 19), 40(3), 41(20), 50(3), 51(3), 52(3, 19), 54(19), 62(3), 67(140), 81(3, 195), 83(3), *84*, *87*, *88*, 152(3), 154(3), 156(3), 157(3), 158(3), 163(3), 165(3), 169(3), *169*, 257(47), *282*, 443(12), 446, 454(39), 460(66), 461(66), 462(66), 466(66), 467(66), 468(66, 94, 95), 469(66), *494*, *495*, *496*, *497*
Macek, V., 136(125), *149*
McElvain, S., 448(29), *495*
McFarren, E. F., 123(78), 126(78), *148*
McGuire, R. L., 177(38), 183(38), *197*
McHale, D., 7(6), 17(6), 19, *27*, 296(43), *316*, 428(144, 145), 430(144), *436*
Machebœuf, M., 183(58, 59, 60), *198*
Maciejewicz, W., 20(50), *28*
McIlroy, R. J., 164(68), 165(68), *171*
McIntyre, A. R., 326(59), 336(114), 340(114), *351*, *353*

Mack, J. H., 80(189), *88*
McKendell, L. V., 71(150), 72(154), *87*
McKenzie, H. A., 141(147), *150*
McKeown, G. G., 298(60), 301(89), 313(60), *316*, *317*
McKinley, W. P., 355, 359(18), 361(19), 364(23, 28), 366(23, 28), 367(28), 370(28), 371(23), 372(36), 386(53), 387(53), 391(63), *395*, *396*
McLane, S. R., Jr., 383(46), 384(46), *396*
Maclaurin, D. J., 82(197), *89*
McMahon, R. E., 409(49), *434*
McNally, S., 168(84), *171*
McNaughton, 342(147), *354*
McNiven, N. L., 57(105), *86*, 227(47), *245*
McOmie, J. F. W., 501(14, 19, 20, 21), 515(83), 522(19, 83), 524(19), 525(19), *534*, *536*
McPherson, S. P., 299(62), *316*
MacRae, H. F., 364(28), 366(28), 367(28), 370(28), *395*
Madigan, P. M., 77(172a), *88*
Magar, N. G., 41(29), *84*
Magarvey, S. A., 386(53), 387(53), *396*
Magasanik, B., 173(14), 191(66), 193(66), 196(14), *197*, *198*
Magee, R. J., 520(93), 522(117), 524(17), *536*, *537*
Mahapatra, G. N., 477(138), *498*
Mahesh, V. B., 202(14), *244*
Mahlum, D., 306(102), *317*
Mahon, M. E., 286(22), 290(30), 292(30), 293(30), 311(30), *315*, *316*
Majer, J., 104(52), *109*
Majer, J. R., 71(146), *87*
Major, A., Jr., 383(44), *396*
Majumdar, A. K., 507(63), *535*
Majumdar, M. K., 397(7), 399(7), *433*
Majumdar, S. K., 397(7), 399(7), *433*
Mak, E. Y. K., 48(81), *86*
Makeš, J., 452(38), 453, *495*
Makida, H., 423(120), *436*
Mallett, G., 297(55), *316*
Maly, E., 525(134), *537*
Mangold, H. K., 41(27), *84*
Mankinen, C. B., 35(8), *84*
Mańko, R., 17(43), *28*
Manley, G., 155(32), *170*
Manning, W. M., 322(25), 327(25), *350*
Manousek, O., 420(109), *435*
Manson, D., 118(18), 120(18), *146*
Manthey, J. A., 249(2), 257(2), *281*
Marchesini, A., 490(171), *499*
Marcinkiewicz, S., 7(8), *27*, 61(114), *86*, 296(43, 44, 45, 46, 47), 311(115, 116), *316*, *318*, 428(144, 145), 429(146), 430(144), 431, *436*
Marcu, P., 518(87b), *536*
Margery Linday, E., 177(30), 195(30), *197*
Margna, U., 17(42), *28*
Margoliash, E., 141(146), *150*
Marigo, M., 467(78), *496*
Mariner, R., 81(192), *88*
Marinetti, G. V., 45(39), *84*, 172(5), *197*
Marini-Bettolo, G. B., 463(75), 466(85), 467(75, 85), 469(85), *496*
Marini-Bettólo, R., 26(64), *29*, 503(42), 511(42), 515(42), *535*
Marjanovic, V., 17(41a), *28*
Márkus, L., 324(42), *351*
Marquez, J. A., 411(65), 412(65), *434*
Marson, R. J., 40(18), 73(159b), *84*, *88*
Martin, A. J. P., 2(8b, 8c), *5*, 7, 8, 19, *27*, 55(101), *86*, 152(8), *169*, 407(33), *433*
Martin, C., 37(17), *84*
Martin, E. C., 41(24), *84*, 506(49), 507(64), *535*
Martin, J. V., 501(21), 515(84), *534*, *536*
Martin, S. L., 238(94), *246*
Martinez, A. P., 111(4), *146*
Martinovich, R. J., 522(108), 524(108), *536*
Martinsson, A., 159(48), *170*
Marutyan, S. A., 161(55), *170*
Marzilli, L. G., 142(162), *150*
Marzoli, A., 423(123), *436*
Mashankov, V. A., 327(69), *352*
Maslowski, K., 423(121), *436*
Mason, M., 229(53a), *245*
Massucci, M. A., 48(76), *85*
Matheson, A. T., 127(110), *149*
Mathur, K. N., 502(26), 503(29), 511(29), 520(90), *534*, *536*
Matsui, K., 422(118), *436*
Matthews, J. S., 217(42), *245*
Matthews, R. W., 301(85), 309(85), *317*, 458(57), *495*
Matthias, W., 120(51), *147*, 156(39a), 157(39a), *170*
Mattocks, A. R., 461(69), *496*
Matysik, G., 7(10a), *27* 467(80), *496*

Mauger, A., 418(98), *435*
Mayer, D., 107(69), *109*
Mayer, J., 111(3), *146*
Mayevskaya, A. N., 325(51), 328(51), *351*
Mazor, L., 415(80), *435*
Mazzei, M., 47(74a), *85*, 533(158), *538*
Meckler, W., 358(17), *395*
Meera Bai, B., 177(33), 185(33), 190(33), 196(33, 103), *197*, *199*
Mefferd, R. B., Jr., 79(186), *88*
Mehrotra, R. C., 531(151), *537*
Meister, A., 124(83), *148*
Mekhtikhanov, S. D., 468(88, 88a), *496*
Melicow, M. M., 71(147), *87*, 133(120), *149*
Meloun, B., 121(57), *148*
Melzer, M. S., 259(53), *282*
Meniga, A., 73(159a), *88*, 145(177), *151*
Menn, J. J., 357(9), 370(33), 375(38), 376(38), *395*, *396*
Mennis, A., 172(4), 194(4), *197*
Menzel, D. B., 378(40), *396*
Merkus, F. W. H. M., 20(48), *28*
Meroa, L. O., 485(165), 490(165), *499*
Mes, J., 161(52a, 52c), *170*
Mesicek, N., 470(106), 473(127), *497*
Mestan, J., 42(34), *84*
Meštan, J., 217(54), 229(54), *245*
Meudt, W., 283(2), 287(2), 290(2), 292(2), *315*
Michael, F., 175(22), 185(22), 187(22), *197*
Michalec, Č., 42(34), 43(36), 48(79), *84*, *85*, 195(102), *199*, 217, 229(54), *245*
Michel-Wolwertz, M.-R., 321(15), 325(15), 326(15), *350*
Michl, H., 252(18), *281*
Miettinen, J. K., 185(75), *198*
Mikeš, F., 440(4), *494*
Miketuková, V., 526(141), *537*
Miki, T., 431(158), *437*
Millburn, P., 478(141), *498*
Miller, H. K., 238(90), 240(90), *246*
Miller, M. M., 411(66), *434*
Miller, O. N., 155(36), *170*
Miller, R. P., 409(50), *434*
Mills, P. A., 356(8), 357(10, 11, 13), 358(15), 373(37), *395*, *396*
Mine, K., 431(154), *436*
Minnis, R., 120(29), *147*
Minshall, E., 97(22), 98(22), *108*
Minyard, J. P., Jr., 300(74), *317*
Mironova, I. B., 418(97), *435*
Miskus, R., 361(20), 378(40), *395*, *396*
Mitchell, H. K., 312(124), *318*
Mitchell, L. C., 59(111), 60, *86*, 308(107), 313(107), *317*, 355(1), 356(1, 8), 358(14, 16), 359(14, 16, 18), 360(16), 361(14), 364(25), 365(25), 366(25), 367(25), 369(25), 370(25), 373(25, 37), 384(48), 391(61, 62, 65), 392(65), 393(65), 394(69), *395*, *396*
Mito, K., 329(84), *352*
Mitoma, C., 286(17), *315*
Mitra, S. N., 391(66), *396*
Mix, H., 127(106), *149*
Miyake, A., 414(78), *435*
Miyaki, T., 409(62), *434*
Miyashita, C., 168(87), *171*
Miyazaki, M., 527(147, 149), 530(147), *537*
Mizsei, A., 415(82), *435*
Mndzhoyan, E. L., 444(14), *494*
Möller, H. G., 517(87), 522(87), *536*
Moffitt, R. A., 358(17), *395*
Moghissi, A., 526(142), *537*
Mokhnachev, I. G., 154(16), 158(16), *169*, 329(78), *352*, 454(45), *495*
Mokrošová, H., 498(30), *495*
Molchanova, N. G., 478(141), *498*
Monéger, R., 322(22), 323(22), 328(22), *350*
Monier, R., 257(40), 274(40), *282*
Monnier, J., 172(8), *197*
Montgomery, R., 183(51), 191(51), 192(51), *198*
Montorsi, M., 284(11), 285(11), 286(11), 291(11), *315*
Moon, M. W., 257(43), 275(43), *282*
Mooney, J. B., 45(63), *85*
Moore, A. M., 93(10), *108*
Moore, R. M., 500(6), *534*
Moore, T. B., 142, *150*
Mori, I., 329(82), *352*
Morris, C. J. O. R., 141(151), *150*
Morrissette, R. A., 41(27), *84*
Morrison, J. F., 257(33a), *281*
Morrison, M., 340(125, 126), *353*
Morse, R. S., 104(55), *109*, 533(162), *538*
Moscarello, M. A., 121(63), 127(63), *148*
Moscatelli, V., 466(85), 467(85), 469(85), *496*, 504(44), *535*
Moses, V., 51(91), *86*
Mosettig, E., 238(90), 240(90), *246*

Mosby, J. R., 312(119), *318*
Motonaga, K., 145(176), *151*
Muehlradt, P., 168(86), *171*
Mukai, J.-I., 256(31a), 257(31a, 35), 259, *281*
Müller, G. C., 121(60), *148*
Mueller, A., 468(92), *496*
Müller, H. R., 253(21), *281*
Müeller, W., 238(93), *246*
Muić, N., 73(159a), *88*, 145(177), *151*
Mukerjee, H., 169(90), *171*
Mukherjee, B., 111(9), *146*
Muller, R., 286(15, 18), *315*
Mullick, D. B., 35(12), *84*, 342(146), *354*
Mumma, R. O., 45(49), *85*
Munier, R., 183(58, 59, 60), *198*
Munk, V., 191(95), 193(95), *198*
Murachi, T., 141(153a), *150*
Murakami, F., 422(118), *436*
Murayama, K., 141(153), *150*
Murphy, F. X., 409(54), 411(54), *434*
Murray, J. F., Jr., 475(133), *498*
Murthy, R. G. K., 300(66), *316*
Mushins'ka, S. Kh., 468(90), *496*
Myers, C. M., 300(79), 312(79), *317*

Nadeau, G., 463(75), 467(75), *496*
Nagai, H., 512(68, 69, 70), *535*
Nair, J. H., 177(43), 179(43), 183(43), *197*
Nair, R. M., 169(89), *171*
Nakabayashi, T., 342(144), *353*
Nakajima, T., 493(178), *499*
Nakanishi, K., 145(176), *151*
Napier, S., 120(48), *147*, 469(103), 470(103), *497*
Narang, S. A., 257, *282*
Narath, A., 100(36), *109*
Nascutiu, T., 507(57), 512(66a), *535*
Nash, H. A., 418(89), *435*
Nataraja Sarma, P. S., 461(70a), *496*
Natarajan, A. R., 371(34), *396*
Nayler, J. H. C., 127(93), *148*
Nazir, D. J., 41(29), *84*
Neher, R., 31(2), *84*, 200, 202(1), 214, 216(1, 26), 226(1, 44), 227(1), 228(1), 229(1), 231, 233(1), 239(1), 244(1), *244*, *245*
Neidlein, R., 500(3), *534*
Nelson, H. A., 410(55), *434*
Nemec, P., 398(10, 11), 400(21, 22), 402(21, 25), 406(22), *433*
Nesemann, G., 416(86), *435*
Nettleton, R. M., Jr., 79(186), *88*
Neujahr, H. Y., 421(116), 422(116), 423(116), *436*
Newcombe, A. G., 310(113), *317*
Newkirk, J. F., 407(40), 409(40), *434*
Newton, G. C. F., 409(47), *434*
Ng, W. G., 45(56), *85*
Niavis, C. A., 185(76), *198*
Niccolai, L., 158(42), *170*
Nicholas, R. E. H., 335(98, 104), 336(104), 340(104), *352*, *353*
Nickless, G., 3(13), *5*, 500(1, 2, 4), 501(17, 18, 20), 506(17), 508(17), 518(17), 531(154), *534*
Niederwieser, A., 122(71), *148*
Nielson, E. D., 216(27), *245*
Niemierko, W., 67(136), *87*, 254(26), 257(26), 258(26), *281*
Nikalayeva, G. N., 327(69), *352*
Nikolov, N., 467(86), *496*
Nishikawa, S., 335(99), *352*
Niskasaari, E., 191(93), 193(93), *198*
Nithyanandan, V. V., 461(70a), *496*
Niyogi, S. K., 45(60), *85*
Noda, M., 173(11), 175(23, 24), 185(24, 86) 187(24, 86), 191(11, 24), 192(11), 193(24), *197*, *198*
Noddack, W., 525(130), *537*
Noguchi, I., 321(17), *350*
Nomoto, M., 141(153), *150*
Northcote, D. H., 43(83), *86*, 152(6), 154(23), 158(6, 23), 164(6), 168(6), *169*, *170*
Northcote, J. P., 161(59), *171*
Nosek, J., 113(35), *147*
Noujaim, A. A., 77(177a), *88*
Novak, J., 447(24), *495*
Nowaczyk, K., 468(93), *496*
Nowicka, E., 132(115), *149*
Nuenke, J. M., 143(170), *151*
Nunes Da Costa, M. J., 503(27), *534*
Nunez, G., 323(36), *351*

Oakey, R. E., 200(3), 231, *244*
Oberneder, G., 70(142), *87*
Occolowitz, J. L., 177(35), 183(35), *197*

Oden, E. M., 411(64), *434*
O'Donnell, L., 56(103), *86*
Ogura, S., 118(20), *146*
Ohara, M., 183(61), *198*
Ohishi, N., 424(129a), *436*
Ohkuma, S., 156(40), *170*
Ohtsuka, E., 257(43), 275(43), *282*
Oi, K., 472(119), *497*
Oka, T., 424(129a), *436*
Okada, M., 238(97), *246*
Okada, S., 154(28), *170*
O'Laughlin, J. W., 511(72), 513(72), *535*
Ollis, W. D., 191(97), 193(97), *199*
Onley, J. H., 357(11, 13), 384(47), *395*, *396*
Oparysheva, E. F., 397(3), 400(3), 401(3), 402(3, 28), 415(3), *433*
Opienska-Blauth, J., 51(92), *86*
Organiściak-Matuszak, S., 329(83), *352*
Orlova, T. I., 418(97), *435*
Orten, J. M., 335(100), 340(100), *352*
Orth, H., 335(95), *352*
Osadchuk, M., 467(77), *496*
Osawa, Y., 65(128), *87*, 325(55), *351*
Oscik, J., 11(29), *28*
Oshima, T., 525(137), *537*
Oshima, Y., 342(144), *353*
Oskaja, V., 133(119a), *149*
Ossicini, L., 21(53), *28*, 45(47, 50), 46(72), 73(157), *85*, *87*, 503(31, 37, 41), 504(37, 45), 511(31), 515(76), 525(136), *534*, *535*, *536*, *537*
Ott, J. L., 408(44), *434*
Overell, B. T., 179(48), 183(48, 49, 53), 191(48), 192(48), *197*, *198*
Overend, W. G., 155(33), 164(74, 75), 166(74, 75), 168(84), *170*, *171*
Owen, W. J., 470(105), *497*
Owens, H. S., 183(50), *198*

Padayatty, J. D., 142(164), 143(164), *150*
Pajetta, P., 127(103), *149*
Pakanaev, Ya. I., 462(74), *496*
Pal, B. K., 507(63), *535*
Palacios, P. V., 126(91), *148*
Palermiti, F. M., 410(56), *434*
Palmer, J. K., 183(62), *198*
Pan, S. C., 214(22, 23), 233(83), 234(83), *245*, *246*, 409(59), *434*
Panaiotova, B. N., 482(153), *498*
Pankajamini, K. S., 347(149), *354*
Pannamperuma, C., 81(192), *88*
Pant, R., 120(41), *147*
Paolini, C., 45(65), *85*, 257(33a), *281*
Papay, M. K., 415(80), *435*
Papendick, V. E., 460(68), *496*
Papke, E., 452(35), *495*
Papke, K., 475(136), *498*
Paris, R., 345(168), 348(168), *354*
Parke, D. V., 307(136), 312(136), 314(136), *318*
Parker, C., 284(12), 290(12), 292(12), *315*
Parkhurst, R. M., 313(128), *318*
Parmentier, F., 301(94), *317*
Parr, W. H., 300(80), *317*
Parrish, J. R., 118(19), *146*
Partlow, E. V., 161(57), *170*
Partridge, S. M., 191(90), 192(90), *198*
Parups, E. V., 100(34), *109*
Pascual, F., 51(90), 66(131), 75(168), *86*, *87*, *88*
Pasich, J., 467(84), *496*
Pasieka, A. E., 83(198), *89*, 127(112), 133(119), *149*
Paskova, J., 191(95), 193(95), *198*
Pasqualini, J. R., 229(51), *245*
Pataki, G., 120(49), 122(71, 73, 74, 75), 127(75), 132(75), 139(74, 75), *147*, *148*
Patnaik, B. K., 477(138), *498*
Patterson, E. L., 416(87), *435*
Paul, A., 10(21), *28*, 522(126a, 126b), 525(126a, 126b), *537*
Paulson, J. C., 154(25), *170*
Pavey, D., 408(44), *434*
Pawlaczyk, J., 45(61), *85*
Payne, A. H., 229(53a), *245*
Pazarincevic, J., 431(155b), *437*
Pazur, J. H., 154(28), *170*
Péaud-Lenoel, C., 45(58), *85*
Pearl, I. A., 312(123), 315, *318*
Pearson, D., 470(109), *497*
Pease, H. L., 385(51), *396*
Pedley, J. T., 525(140), *537*
Peereboom, J. W. C., 203(16), 207(16), 229(16), *245*, 420(107), *435*
Pennington, F. C., 327(67), *352*
Pepar, M., 470(106), *497*
Percheron, F., 152(2), 153(2), 154(2), 155(2), 157(2), 158(2), 159(2), 163(2), 169(2), *169*

Perila, O., 178(45), 183(45), 185(85), 187(85), *197*, *198*
Perisho, C. R., 7(3), *27*
Perkavac, J., 470(106), 472(123), 473(127), *497*
Perlman, D., 418(93), *435*
Perpar, M., 472(123), 473(127), *497*
Perrin, D. D., 216(28), *245*
Perron, Y., 410(56), *434*
Perry, T. L., 127(101), 129(101), *149*
Petek, F., 164(73), *171*
Peter, F., 472(118), *497*
Peters, H., 138(131), *149*, 174(18), *197*, 284(8), *315*
Peterson, D. H., 410(57), 412, *434*
Peterson, P. J., 45(52), *85*, 140(141a), *150*
Peterson, R. E., 202(14a), *244*
Petránek, J., 111(7), *146*
Petro, O., 431(152), *436*
Petronici, C., 298(57), *316*
Peyron, L., 62(117), *87*
Pfaff, J. D., 77(179, 180), *88*
Pfeifer, S., 469(101), *497*
Pfeiffer, C. J., 81(196), *89*
Pfeil, E., 522(104, 105, 113), 524(113), *536*
Pfordte, K., 468(89), *496*
Phares, E. F., 183(56), *198*
Philippi, E. H., 121(61), *148*
Phillips, J. H., 259(52), *282*
Phillips, R. F., 101(38), *109*
Phillips, W. F., 409(61), *434*
Philippu, A. J., 65(130), *87*
Picard, J., 254(24), *281*
Pickering, W. F., 12(346), *28*, 501(12), 511(12), *534*
Pieters, H. P., 34(5), *84*
Pifferi, P. G., 127(100), 129(100), *149*
Pignolet, L. H., 45(70), *85*, 447(27), 448(27), *495*
Pilátová, L., 398(11), *433*
Pilawska, E., 467(84), *496*
Pinter, K. G., 428(141), 429(141), 430(141), *436*
Pinto, A., 224(61), 229(61), *246*
Pinto, A. A., 154(17), 159(17), *169*
Pinto, I. R., 507(66), *535*
Piper, E. A., 102(41), *109*
Pirelli, A., 20(47), *28*, 152(10), *169*
Pirotta, A., 300(77), 311(77), *317*
Piśkala, A., 259(57), *282*
Piskorski, J., 227(46), 231(46), *245*
Plamondon, J., 520(103), *536*
Plesničar, B., 455(53), 457(53), *495*
Pluchet, E., 506(51), *535*
P'o, H.-S., 478(147), *498*
Pocchiari, F., 102(44), *109*
Pochian, F., 533(159), *538*
Podchufarova, G. M., 327(68, 70), *352*
Podojil, M., 402(27), *433*
Poduska, K., 142, *151*
Pogadl, A., 478(146), *498*
Pohl, F., 93(11), *108*
Pohloudek-Fabini, R., 452(35), 475(136), 479(149, 150), *495*, *498*
Pokidova, N. V., 142(169), *151*
Polglase, W. J., 161(57), *170*
Pollard, F. H., 500(2), 501(14, 17, 18, 19, 20, 21), 506(17), 508(17), 515(83, 84), 518(17), 522(19, 83), 524(19), 525(19), 527(146), 531(154), *534*, *536*, *537*
Poluektov, N. S., 522(120), 524(120), *537*
Ponniah, L., 347(150), *354*
Popa, G., 512(66a), *535*
Popova, I. A., 325(51), 326(61), 328(51, 61), 334(61), *351*
Popovic, R., 478(143a), *498*
Popovici, V., 471(111), *497*
Popper, E., 518(87b), *536*
Porteous, J. W., 191(88), 192(88), *198*
Porter, C. W., 297(54), *316*
Porter, W. L., 183(51), 191(51), 192(51), *198*
Poser, H., 52(98), *86*
Pospísil, J., 450(32a), *495*
Posternak, T., 485(166), *499*
Postnikov, N. N., 531(152), *537*
Potter, A. D., 313(128), *318*
Potter, E. F., 297(55), *316*
Powell, R. H., 57(109), *86*
Powell, W. A., 183(52), *198*
Powning, R. F., 163(66), 169(66), *171*
Price, T. D., 71(147), *87*, 133(120), *149*
Praeve, P., 416(86), *435*
Prášilová, J., 533(161), *538*
Prelog, V., 419(105), *435*
Prescher, D., 479(148), 481, *498*
Pribela, A., 454(43), *495*
Price, L., 323(28), 330(28), *351*
Pridham, J. B., 163(65), *171*
Principe, P., 233(83), 234(83), *246*

Priori, O., 191(89), 192(89), *198*
Přistoupil, T. I., 41(25), *84*, 143(174), 145(174, 179), 146(180, 181), *151*, 473(128), 474, *497*
Procházka, Z., 52(95), *86*, 216(24), 227(24), 228(24), *245*
Proksch, G. J., 424(125), *436*
Prudnikava, I. V., 327(69), *352*
Prusíková, M., 225(62), 229(62), *246*
Przeslakowski, S., 506(47a), 511(80, 81), *535*, *536*
Przyborowska, M., 7(5), 11(30), *27*, *28*
Puchnina, A. V., 397(3), 400(3), 401(3), 402(3), 415(3), *433*
Püschel, F., 479(148), 481, *498*
Pugh, I. M., 418(95), *435*
Puisais, J., 348(158), *354*
Pullan, B. R., 101(37), *109*
Pulley, A. P., 71(144), *87*, 154(29), *170*
Pulliam, J. E., 207(17), 209(17), 210(17), *245*
Puro, H., 96(21), *108*

Quayle, J. R., 388(58), 389(58), *396*
Quesnel, V. C., 117(12), *146*
Qureshi, M., 479(151), *498*, 500(8), 502(23, 26), 503(28a, 29), 506(50), 507(58), 508(8, 78, 79), 511(29), 520(90, 91, 92), 522(92, 123), 524(92), 525(123), *534*, *535*, *536*, *537*

Rackis, J. J., 313(129), *318*
Racusen, D., 383(45), 384(45), *396*
Radhakrishnamurthy, B., 123(79), *148*, 164(77), *171*
Radhakrishnamurthy, R., 311(118), *318*
Radowitz, W., 446(20), *495*
Radzhapov, R. A., 397(6), 399(6), 400(6), 411(6), 415(6), *433*
Rafferty, G. A., 164(75), 166(75), *171*
Ragheb, H. S., 409(61), *434*
Ramaley, L., 12(33, 34a), *28*
Ramasarma, T., 428(143), 430(143), *436*
Ramic, S., 478(143a), *498*
Rammelt, R., 257(45), 277(45), *282*
Randerath, K., 4(14), *5*
Rangone, R., 419(101), *435*
Rao, D. V., 242(89a), *246*
Rao, E. V., 242(89a), *246*
Rao, P. L. N., 254(25), *281*
Rapoport, B. N., 431(156), 432(156), *437*
Rappoport, D. A., 326(60), 340(127), *351*, *353*
Rapport, M. M., 177(31), *197*
Ratajewicz, D., 11(25), *28*
Raue, H., 143(172), *151*
Rawat, J. P., 520(91), *536*
Rayo, C. L., 154(17), 159(17), *169*
Rearick, D. A., 507(65), 519(65), *535*
Reddi, K. K., 290(28), 292(28), *316*
Reeck, G., 40(18), *84*
Reeck, W. E., 73(159b), *88*
Rees, D. A., 167(80), *171*
Reeves, W. A., 507(61), 532(61), *535*
Regna, P. P., 409(53), 411, *434*
Regura, R. M., 251(15), *281*
Reichl, E. H., 183(57), *198*
Reichel, J., 482(158, 159), *498*
Reichstein, T., 168(86), *171*, 216(29), 238(92), 242(92), 243(100), 244(29), *245*, *246*, *247*
Reid, R. L., 177(37), 183(37), 191(87), 192(87), *197*, *198*
Reid, S. G., 310(113), *317*
Reindel, F., 141(156), *150*
Reineke, L. M., 56(104), *86*, 410(57), 412, *434*
Reinking, N. H., 301(98), 311(98), 312(98), *317*
Reio, L., 81(195a), *88*, 296(36, 37, 38), 314(36), *314*, 348(154), *354*
Relph, S. J., 103(47), 105(67), 106(47), 107(47), *109*
Remijnse, A. G., 13, *28*, 120(55), *147*
Rendig, V. V., 117(11), 120(11), *146*, 284(6, 7), 290(7), 291(7), 293(7), *315*
Reshetov, P. D., 397(2), 414(2), 415(2), *433*
Resnik, F. E., 183(52), *198*
Resplandy, A., 468(96), *497*
Rettenbacher-Daübner, H., 232(76), *246*
Reynolds, T. M., 308(105), *317*
Rhaese, H.-J., 259(53), *282*
Riaz, S. A., 531(153a), *537*
Ribeiro, L. P., 71(149), 72(149), *87*
Rice, C. B. F., 54(100), 66(100), *86*
Rich, K. M., 45(66), 46(66), *85*
Richards, L. P., 77(177a), *88*
Richter, M., 164(67), *171*

Ridyard, H. N., 420(113), 423(113), *435*
Rieche, A., 455(51), 456(51), *495*
Riffart, W., 70(142), *87*
Rigby, G. J., 240(99), *247*
Rimington, C., 335(104), 336(104, 109), 340(104), *353*
Rinehart, K. L., 409(58), 410(58), *434*
Ritter, F. J., 41(22), *84*, 221(58), 229(58), 231(67), *246*
Riva, A., 127(100), 129(100), *149*
Rive, D. J., 127(97), 129(97), *148*
Roberts, D. J., 117(13, 14), 120(14), *146*
Roberts, E. A. H., 313(125), *318*, 342(140, 141, 143), *353*
Roberts, H. R., 46(73), 61(115), 62(115, 116), *85*, *86*, *87*, 217(40), *245*
Roberts, J. B., 45(51), *85*, 232(75), *246*
Roberts, M., 45(41), *85*
Roberts, R. M., 257(47b), *282*
Robinson, A. E., 26(57), *29*
Robinson, D., 296(48), 314(48), *316*
Rockland, L. B., 79(185), 83(200), *88*, *89*
Robinson, K. S., 284(9), *315*
Robinson, N., 138(132), *150*
Robison, R. S., 217(40), *245*
Roch, G., 26(64), *29*, 441(8), *494*
Rodnight, B., 290(29), 292(29), *316*
Rodopulo, A. K., 444(13), 454(48), *494*, *495*
Rodwell, V. W., 424(130), *436*
Roets, E., 418(91), *435*
Rogeness, G. A., 338(124), 340(124), *353*
Rojowska, M., 126(92), *148*, 467(79), *496*
Rommel, M. A., 12(34), *28*, 531(153), *537*
Ronayne, M. E., 471(114), *497*
Ronkainen, P., 452(37), 453, *495*
Rony, P. R., 20(51), *28*
Roos, J. B., 203(16), 207(16), 218(16), 229(16), *245*, 420(107), *435*
Rosenblatt, D. H., 527(145), *537*
Rosmus, J., 139(137), 140(137), 145(175), *150*, *151*, 420(109), *435*
Ross, H. F., 96(14), *108*
Rossi, C., 102(44), *109*, 257(37), 260, 272(37), *281*, 533(159), *538*
Rossi, L., 257(37), 260, 276(37), *281*
Rostimeshin, V. A., 155(30), 158(30), 161(30), *170*
Roth, D., 27(66), *29*
Roth, M., 77(178), *88*
Rothwell, M. T., 531(154), *537*
Roushdi, I. M., 478(145), *498*
Roux, D. G., 314(132), *318*, 348(164), *354*
Roy, B. R., 391(66), *396*
Rozenberg, M. C., 259(51), *282*
Rubstein, K., 476, *498*
Ruckenbrod, H., 299(63), 314(63), *316*
Rudinger, J., 142(168), *151*
Rüdiger, H., 67(138), *87*
Rüdiger, R., 67(138), *87*
Ruffini, G., 300(77), 311(77), *317*
Ruseva, N., 421(115), 422(115), 423(115), *436*
Rutter, L., 63(118), *87*
Růžička, B., 444(15), *494*
Ryle, A. P., 141(144), *150*

Saarnio, J., 191(93), 193(93), *198*
Sabacky, M. J., 324(44), *351*
Sackmann, W., 416(85), *435*
Sadykov, A. S., 462(74), *496*
Safina, G., 298(57), *316*
Sahashi, Y., 431(158), *437*
Saini, A. S., 130(112a), *149*, 153(13), 160(13), *169*
Salganicoff, L., 107(69), *109*
Salhanick, H. A., 219(55), 220(57), 229(55, 57), *245*
San Antonio, J. P., 361(21), *395*
Sancier, K. M., 325(53, 54), *351*
Sandermann, H., Jr., 259(56), *282*
Sandstrom, W. M., 300(75), 306(75), *317*
Sanecka, M., 51(92), *86*
Sanger, F., 141(142, 143, 144), *150*
Sanir, S., 525(130), *537*
Sankar, D. V. S., 111(9a), *146*
Sannie, C., 468(96), *497*
Sano, I., 286(24), 290(24), 292(24), *316*, 493(178), *499*
Santilli, V., 259(55), *282*
Sapozhnikov, D. I., 324(41), 325(41, 51), 328(41, 51), *351*
Saracino, F., 525(136), *537*
Sarin, P. S., 418(92), *435*
Sarsunova, M., 104(52), *109*
Sato, K., 431(154), *436*
Sautiere, P., 122(72), 139(72), *148*
Sauciuc, A., 425(136), *436*

Savary, G., 361(19), *395*
Savic, Z., 420(111), *435*
Savory, E., 194(101), *199*
Sawai, H., 431(154, 155), 432(155), *436*
Sawicki, E., 77(179, 180), *88*, 133(119b), *149*, 299(62), *316*, 448(28), 449(28), 454(47), *495*
Saxena, S., 127(105), *149*
Scarlata, G., 472(120), *497*
Schaffner, D. P., 409(58), 410(58), *434*
Scharnow, H.-G., 167(82, 83), *171*
Scheffer, D., 168(86a), *171*
Scheffer, F., 447(25), *495*
Scheltino, O., 98(25), *108*
Schenker, E., 238(92), 242(92), *246*
Schepartz, A. I., 485(164), 488(164), *499*
Schepky, G., 73(160), *88*
Schettino, O., 421(114), 422(114), 425(134), *436*
Schilling, J. A., 121(62), *148*
Schindler, O., 243(100), *247*
Schinkmannová, L., 139(137), 140(137), *150*
Schleede, D., 301(92), *317*
Schlender, K. K., 259(60), *282*
Schlenk, J., 41(27), *84*
Schmid, L., 432(158a), *437*
Schmitt, J. H., 127(108a), *149*
Schneer-Erdey, A., 507(56), *535*
Schneider, J. J., 229(63a), 232(71, 72), *246*
Schoental, R., 127(97), 129(97), *148*
Scholtissek, C., 238(91), 243(91), *246*
Schraiber, M. S., 468(91), *496*
Schramm, R. W., 329(83), *352*
Schreglmann, A., 238(93), *246*
Schreier, K., 191(94), 193(94), *198*, 291(35), 300(72), 312(72), *316*, *317*
Schriefers, H., 93(11), *108*
Schroeder, H. A., *318*
Schroeder, W. A., 141(149), *150*
Schülke, B., 232(73), *246*
Schulz, M., 455(51), 456(51), *495*
Schwachula, G., 439(2), *494*
Schwartz, S., 335, *352*
Schwartzenburg, F., Jr., 296(41), *316*
Schwarz, K., 26, 27(65), *29*
Schweiger, A., 471(111), *497*
Schweppe, H., 175(22), 185(22), 187(22), *197*
Scott, I. A. P., 522(117), 524(117), *537*
Scott, A. E., 525(140), *537*
Scott, J. E., 48(77), *85*
Šebesta, F., 533(161), *538*
Šebesta, K., 255(29), 256(29), *281*
Segal, R., 173(9), 195(9), *197*
Seikel, M. K., 342(137), *353*
Seiler, H., 501(15), *534*
Seligson, D., 174(15), 176(15), 185(15), 196(15), *197*
Sell, H. M., 283(3), *315*
Selzer, G. B., 413, 414(69), *434*
Sementovs'ka, G. P., 478(143), *498*
Semonsky, M., 468(95), *497*
Sen, S. P., 290(33), 293(33), 294(33), *316*
Sen Gupta, P. N., 391(66), *396*
Sensi, P., 418(100), *435*
Seow, P. H., 371(34a), *396*
Sequi, P., 490(171), *499*
Serchi, G., 63(121), *87*, 337(130), 340(130), *353*
Serdyuk, L. G., 329(78), *352*
Serlupi-Crescenzi, G., 45(65), *85*, 257(33a), *281*
Servigne, Y., 527(144), *537*
Seshadri, T. R., 347(149, 150), *354*
Sessa, D. J., 313(129), *318*
Šesták, Z., 319, 320(6, 12, 14), 321(14), 322(6, 14), 323(6), 324(6, 48, 49), 325(12), 327(14), 328(48, 49, 74), *350*, *351*, *352*
Sevcik, V. R., 402(27), *433*
Sevcikova, Z., 155(35), *170*
Sezaki, M., 411(63), *434*
Sferruzza, G., 419(101), *435*
Shafieu, A. I., 478(141), 490(169), *498*, *499*
Shainskii, I. I., 121(64), 138(130), *148*, *149*
Shanmuggasundaram, E. R. B., 311(118), *318*
Shapiro, B., 174(15), 176(15), 185(15), 196(15), *197*
Shaw, D. C., 141(147), *150*
Shaw, G., 127(98), 129(98), *148*
Shaw, K. N. F., 290(27), 296(49), *316*
Shepherd, H. G., Jr., 71(151), *87*
Sheppard, G., 169(88), *171*
Sherma, J., 2(6a), 3(10, 12), *5*, 8(12), 11(31), 22(55), 23(55), 24(55), 25(55), *27*, *28*, *29*, 35(11), 45(66, 68, 70, 71), 46(66), 78(81), *84*, *85*, *88*, 301(87), 310(87), 313(87), *317*, 319(2, 3), 321(16), 322(16, 18, 24), 323(18, 31, 33, 37), 324(16, 18, 33, 37),

325(16), 326(18, 24, 56, 58), 327(16, 18), 328(37), 329(18, 31, 86), 330(58, 87, 88), 331(18, 24), 332(24, 58), 333(16, 58), *350*, *351*, *352*, 413(68), *434*, 442(10), 447(27), 448(27), *494*, *495*, 500(6, 9, 10), 502(9), 503(32, 34, 35, 36, 38, 39, 40, 43), 506(48), 511(40), 514(10), 515(39), 525(135), *534*, *535*, *537*
Sherman, J. R., 45(55), *85*
Sherratt, H. S. A., 348(153), *354*
Shiba, T., 142(163), *150*
Shi, H. C., 525(129), *537*
Shida, S., 326(62), 329(62, 80), *351*, *352*
Shifrine, M., 45(42), *85*
Shih, V. E., 77(172a), *88*
Shilov, Yu. M., 468(88), *496*
Shimi, I. R., 93(53), 161(53), *170*
Shimizu, H., 493(178), *499*
Shimizu, I., 473(126), *497*
Shimizu, K., 232(81), 233(81), 234(81), *246*
Shimizu, T., 472(119), *497*
Shimura, M., 411(63), *434*
Shin, Y. S., 136(129), 137(129), *149*
Shinohara, T., 156(40), *170*, 452(36), *495*
Shimomura, K., 503(33), *534*
Shipalov, M. S., 468(88, 88a), *496*
Shkolnik, R. Y., 183(55), *198*
Shlyagin, K. N., 72(153), *87*
Shlyk, A. A., 327(68, 69, 70), *352*
Shozaburo, K., 161(52b), *170*
Shukla, S. K., 533(160), *538*
Shultz, O.-E., 73(162), *88*
Siblíková, O., 201(7), 216(7), 222(7), 231(7), 244(7), *244*
Siegel, H., 75(170), *88*
Siemroth, J., 472(121a), *497*
Sigg, H. P., 216(29), 244(29), *245*
Signor, A., 127(103), *149*
Silaev, A. B., 142(168), *151*, 397(6), 399(6), 400(6), 411(6), 415(6), *433*
Silberman, H., 216(31), *245*, 323(35), *351*
Simmonds, N. W., 348(160), *354*
Simonov, E. F., 135(121), *149*
Simpson, S. A., 216(30, 33), *245*
Sina, A., 431(157), 432(157), *437*
Singh, C., 532(156), *537*
Singh, P. I., 130(112a), *149*
Siperstein, M. D., 237(87), *246*
Sipman, C. B. J., 115(10b), *146*
Sirois, J. C., 290(25), 292(25), *316*
Sironval, C., 321(15), 325(15), 326(15), *350*
Siuda, A., 493(177), *499*
Siva Samban, M. A., 300(66), *316*
Sjoalander, N. O., 414(75), *434*
Sjövall, J., 232(78), 237(85, 86), *246*
Skilleter, D. N., 492(174), *499*
Skinner, C. G., 7(7a), *27*
Skipski, V. P., 177, *197*
Škoda, J., 259(55), *282*
Skolik, J., 93, *108*
Skraba, W. J., 172(7), 177(41), 183(41), *197*
Skrygan, A. I., 161(54), *170*
Slavík, J., 466(99, 100), 469(99, 100), *497*
Sleeman, H. K., 45(57), *85*
Smart, N. A., 364(24, 27), 366(24, 27), 367(27), *395*
Smashey, A. R., 418(89), *435*
Smillie, M., 46(74), *85*
Smith, E. L., 141(148), *150*
Smith, G. H., 145(178), *151*
Smith, G. N., 365(29), 367(29), *396*
Smith, I., 49(86), 50, 74, *86*, 120(46), 127, 129(46), 133(46), *147*, 174(21), *197*, 286(19), *315*
Smith, J. D., 257(41), 274(41), *282*
Smith, J. N., 296(48), 300(72a, 79), 305(72a), 312(79), 314(48), *316*, *317*, 460(65), *496*
Smith, K. C., 249(5), *281*
Smith, L. F., 141(144), *150*
Smith, M. J., 531(155), *537*
Smith, P., 120(40), *147*
Smith, R. L., 478(141), *498*
Smrt, J., 257(46a), 280(46a), *282*
Smyth, R. B., 298(60), 301(89), 313(60), *316*, *317*
Snell, E. E., 424(130), *436*
Snell, N., 402(29), 417(29), *433*
Šnobl, D., 472(117), *497*
Snyder, L. R., 11(27), *28*
Sobolewski, G., 463(75), 467(75), *496*
Soczewiński, E., 6(2), 7(5, 10, 10a), 11(30), 17(43), 20(49, 50), *27*, *28*, 67(134), 73(162a), *87*, *88*, 126(92), *148*, 467(79, 80), 469(102), *496*, *497*, 511(81), *536*
Soerte, J., 520(96), *536*
Sohler, A., 409(58), 410(58), *434*
Sokolova, E. V., 482(155), *498*
Sokolowska, T., 143(172), *151*
Sokolski, W. T., 399(18), 416(83), *433*, *435*

Solberg, Y., 127(94), 131(94), 134(94), *148*
Solomon, J. B., 314(133), *318*
Solomons, I. A., 409(53), 411, *434*
Solunina, I. A., 426(148), *436*
Somerville, A. R., 118(17), *146*
Sommereyns, G., 105(62), *109*, 501(22), 520(96, 97), *534*, *536*
Song, M. K., 297(55), *316*
Sorkin, E., 501(15), *534*
Šorm, F., 257(46a), 259(55), 280(46a), *282*
Sorma, P. S., 311(118), *318*
South, D., 77(177), *88*
Souto, J., 402(31b), *433*
Specker, H., 506(53), *535*
Spedding, D. J., 100(35), *109*
Spencer, E. Y., 357(11a), *395*
Spetsig, L. O., 454(46), *495*
Spilker, P., 45(64), *85*
Spitzer, A., 471(111), *497*
Sporer, A. H., 325(53, 54), *351*
Spotswood, T. M., 440(7), 441(7), *494*
Spratt, W. E., 101(39), *109*
Sprenger, H.-E., 92(7), *108*
Sprince, H., 284(12), 290(12, 34), 292(12), *315*, *316*
Springler, H., 238(98), *247*
Sreenivasaya, M., 183(54), *198*
Sri-Ram, J., 169(90), *171*
Srivastava, S. P., 118(23), *146*
Stadler-Denis, A., 105(62), *109*, 520(97), *536*
Stadtman, E. R., 178(47), 181(47), 183(47), *197*
Stärk, H., 105(58), *109*
Stahl, E., 35(13), *84*
Stanek, J., 300(82), *317*
Stanishevskaya, E. M., 324, 329(46), *351*
Stanislavová, J., 81(195), *88*, 257(47), *282*
Starikova, N. D., 122(69), *148*
Stark, J. B., 183(50), *198*
Stark, W., 454(40), *495*
Stárka, L., 225(62), 226(63), 229(62, 63), *246*
Stauffer, J. F., 409(45), *434*
Steck, W., 300(73), *317*
Steel, A. E., 522(116), 524(116), *537*
Steelink, C., 312(120), *318*
Steim, J. M., 105(60), *109*
Stein, P. C., 45(46), 51(46), *85*
Steinnes, E., 249(4), 259(4), *281*
Steinegger, E., 466(97), 469(97), *497*
Steiniger, H., 399(14a), *433*
Stella, C., 158(42), *170*
Stephan, R., 475(134), *498*
Stepushkina, V. V., 419(102), 424(102), *435*
Štěrbíková, J., 143(174), 145(174), 146(180), *151*
Stern, F., 172(4), 194(4), *197*
Stern, J. S., 111(3), *146*
Stevens, B. J., 286(19), *315*
Stevens, F. C., 141(148), *150*
Stevens, H. M., 501(16a), 520(16a), 522(124, 125, 126), 525(124, 125, 126), *534*, *537*
Stewart, G. H., 8, 12(18), 13, 17(43a), *28*
Stich, W., 336(112, 113), 337(112, 113), 340(112, 113), *353*
Stock, R., 54(100), 66(100), *86*
Stoddart, R. W., 48(83), *86*, 154(23), 158(23), *170*
Stoll, A., 320(9), *350*
Storherr, R. W., 384(47), *396*
Stota, S., 300(82), *317*
Stotz, E., 191(68), 193(68), *198*, 340(125, 126), *353*
Stowe, B. B., 283(4), *315*
Strain, H. H., 1(1), 2(6a), *4*, 6(1), 7(11), 9(1, 20), 11(31), 16(1), 22(55), 23(55), 24(55), 25(55), *27*, *28*, *29*, 78(181), *88*, 319(2, 3), 320(7, 8, 10), 321(16), 322(7, 8, 16, 24, 25), 323(7, 31, 33, 34, 37, 38, 39), 324(7, 16, 33, 37, 44), 325(16), 326(24, 58), 327(16, 25, 39, 63, 65, 66, 67, 72), 328(8, 37, 39), 329(7, 8, 31, 85), 330(58, 88), 331(24), 332(24, 58), 333(16, 58), *350*, *351*, *352*
Štrauchová, O., 415(81), *435*
Street, H. V., 32(3a), 45(59, 60), 73(156), *84*, *85*, *87*, 462(73), *496*
Strickland, R. G., 75(167), *88*
Stroh, H. H., 164(67), 167(82, 83), *171*
Strom, R., 423(122, 123), *436*
Stuchlík, M., 439(3), 467(86a), *494*, *496*
Studier, M. H., 254(23), 257(23), *281*
Stürchen-Kirchner, H., 133(117a), *149*
Stupnicka, E., 213(20), *245*
Stupnicki, R., 213(20), *245*
Stutter, E., 399(14a), *433*
Stylianov, M., 213(19), *245*
Su, J.-Ch., 259(58), *282*

Subers, M. H., 485(164), 488, *499*
Sugawara, N., 503(43a), *535*
Sugihara, M., 45(69), *85*
Šulc, M., 42(34), *84*, 217(54), 229(54), *245*
Sullivan, H. R., 409(49), *434*
Sun, M., 525(131), *537*
Sundaram, T. K., 311(118), *318*
Sundt, E., 301(95), *317*
Surak, J. G., 522(108), 524(108), *536*
Sutherland, B. J., 454(42), *495*
Sutherland, J. W., 460(68), *496*
Suzuki, Y., 183(61), *198*
Svec, W. A., 320(7), 322(7), 323(7, 39), 324(7), 327(39, 67), 328(39), 329(7), *350*, *351*, *352*
Svendsen, B. A., 124(82), *148*
Svendson, A. B., 308(106), 313(106), 314(106), *317*
Svoboda, W. A., 378(40), *396*
Svobodová-Leblová, S., 423(119), *436*
Swaminathan, M., 64(124), 65(124), *87*
Swanson, J. L., 220(57), 229(57), *245*
Swartwout, J. R., 155(36), *170*
Sweeney, J. T., 177(38), 183(38), *197*
Swell, L., 175, *197*
Syalitskaya, N. M., 161(54), *170*
Sylvester, J. C., 409(52), *434*
Symons, R. A., 259(54), *282*
Synge, R. L. M., 2(8a, 8b, 8c), 8, *27*
Syper, L., 444(18), *494*
Szabó, G., 399(15), *433*
Szabo, M., 533(163), *538*
Szarfoldi, J., 431(152), *436*
Szczepaniak, S., 65(129), *87*, 132(115), *149*
Szekerke, M., 142(165), 144(165), *150*
Szente, A., 485(166), *499*
Szent-Györgi, A., 77(176), *88*
Szilágyi, I., 398(8), *433*
Szotyori, K. S., 342(135), *353*, 432(159), *437*
Szumilo, H., 467(80), *496*

Tait, J. F., 216(30), *245*
Takeda, K., 232(79), 234(79), *246*
Talbot, G., 419(105), *435*
Tambiah, M. S., 286(14), *315*
Tamura, Z., 527(147), 530(147, 149), *537*
Tanaka, K., 232(79), 234(79), *246*
Tanguay, R., 138(133), *150*
Tanimukai, H., 285, 291, *315*
Tanos, B., 297(52), *316*
Tantivatana, P., 323(35), *351*
Tanzer, J. M., 35(9), *84*
Tapper, B. A., 160(52), *170*
Taruno, C., 425(133), *436*
Taylor, A. O., 345(163), 348(163), 349(163), 350(163), *354*
Taylor, D. A. H., 241(102), 244(102), *247*
Taylor, H., 520(98), *536*
Taylor, H. D., 410(56), *434*
Taylor, J. D., 424(125), *436*
Taylor, K. B., 472(115), *497*
Taylor, K. W., 145(178), *151*
Taylor, M. W., 194(100), *199*
Tazoe, K., 422(118), *436*
Teague, R. T., Jr., 440(6), *494*
Tegeler, G., 232(73), *246*
Tetrault, P. A., 416(83), *435*
Teoule, R., 250(10a), *281*
Terbojevich, M., 127(103), *149*
Teuchy, H., 301(94), *317*
Teutloff, A., 431(153), 432(153), *436*
Tewari, S. N., 461(70), *496*
Thilo, E., 490(170), 491(170), *499*
Thimann, K. V., 283(4), *315*
Thirkell, D., 42(31), *84*
Thoma, J. A., 7(3, 4), 17(4), *27*, 67(139), *87*
Thomas, H., 93(11), *108*
Thomas, J. A., 259(60), *282*
Thomas, M. R., 327(65, 66, 72), *352*
Thomas, R., 407(39), *434*
Thompson, A. C., 300(74), *317*
Thompson, A. R., 178(44), 181(44), 183(44), *197*
Thompson, C. R., 231(38), *246*
Thompson, E. O. P., 141(143, 145), *150*
Thompson, D. E., Jr., 442(10), *494*
Thompson, J. F., 68(141), *87*
Thompson, R. H. S., 185(71), *198*
Thomson, J., 379(41), 380(41), 381(41), *396*
Thorn, G. D., 387(55), *396*
Thornber, J. P., 161(59), *171*
Thornburg, W. W., 391(64), *396*, 420(110), *435*
Thorp, R. H., 216(31), *245*
Thorpe, W. V., 297(50), 300(50), 307(137), 312(50, 122), 313(50), *316*, *318*
Tichá, M., 160(49), *170*
Tigane, E., 121(63), 127(11, 63), *148*, *149*

Tipper, C. F. H., 454(50), 455(50), 456(50), *495*
Tiselius, A., 177(34), *197*
Tocksteinova, D., 520(102), *536*
Todd, Lord, 418(92), *435*, 490(168), 491(168), *499*
Toelgyessy, J., 104(52), *109*
Toennis, G., 25(56), *29*
Tokushigo, A. M., 174(19), 194(19), *197*
Tomingas, R., 441(8a), *494*
Tominska, M., 520(101), *536*
Tomisek, A. J., 51(94), *86*, 99(30), *108*
Tomita, Y., 323(29), 326(29), *351*
Tonchev, T. A., 154(19), *169*
Toporova, E. G., 399(13), *433*
Torracca, E., 48(76), *85*
Toschi, G., 185(69), *198*
Toth, M., 152(4), *169*
Toth, S., 120(56), *147*
Toth, T., 507(56), *535*
Touchstone, J. C., 13(36), *28*
Tovey, K. C., 257(47b), *282*
Towers, G. H. N., 298(58), *316*
Toyoda, K., 329(81), *352*
Toyozawa, I., 118(20), *146*
Trabal, F., 172(4), 194(4), *197*
Trafford, D. J. H., 213(18), 217(39), *245*
Treadwell, C. R., 175, *197*
Trevarthen, J., 290(27), *316*
Trischmann, H., 164(72), *171*
Trofimova, N. I., 478(141), *498*
Trojna, M., *318*
Truter, E. V., 3, *5*
Trzaska, J., 126(90), 129(90), 131(90), *148*
Tsarenko, N. Ya., 468(91), *496*
T'Siobbel, J. B., 478(144), *498*
Tsukamoto, H., 168(87), *171*
Tsukiura, H., 409(62), *434*
Tswett, M., 2, *5*, 319(1), 321(1), *350*
Tucker, T. C., 57(108), *86*
Tulus, R., 478(142), *498*
Tunin, D., 408(43), *434*
Turba, F., 238(91), 243(91), *246*
Turina, S., 17(41a), *28*
Turkovic, I., 462(73a), *496*
Tušl, J., 469(103a), *497*
Tyburczyk, W., 120(24), *147*
Tyson, B., 81(192), *88*
Tyler, S. A., 15(38), *28*

Udalova, T. P., 447(22), *495*
Ueno, Y., 173(10), 178(10), 181(10), 195(10), 196(10), *197*
Ulbrich, V., 452(38), 453, *495*
Ullah, S., 154(18), *169*
Ullmans, S., 416(83), *435*
Umbarger, H. E., 173(14), 191(66), 193(66), 196(14), *197*, *198*
Undenfriend, S., 103(48), *109*, 286(17), *315*
Unterman, H. W., 409(51), 410(51), *434*
Ursick, J. A., 531(153a), *537*
Urx, M., 413, 414(73), *434*
Usui, T., 232(81), 233(81), 234(81), *246*

Vacek, Z., 300(82), *317*
Vahidi, A., 111(9a), *146*
Vályi-Nagy, T., 398(8), *433*
Vaňásek, V., 444(18a), *494*
Van Dalen, E., 105(66), *109*, 504(46), *535*
Vanderhaeghe, H., 418(91), *435*
Vanderheiden, B. S., 250(10), 257(10), 270(10), *281*, 493(175), *499*
Van der Venne, M. T., 478(144), *498*
Van Duuren, B. L., 55(102), *86*
Vanecek, S., 468(95), *497*
van Erkelens, P. C., 104(50), *109*
van Heijenoort, J., 141(159), *150*
van Kley, H., 142(164), 143(164), *150*
van Raaphorst, J. G., 504(45a), *535*
van Rheenen, D. L., 115(10b), *146*
Van Sumere, C. F., 301(94), *317*
van Wyk, C. J., 122(70), *148*
Varner, E. L., 104(54), *109*
Varshney, K. G., 520(92), 522(92), 524(92), *536*
Vasatko, J., 454(43), *495*
Vasic, J., 478(143a), *498*
Vassort, P., 45(58), *85*
Večeřa, M., 111(7, 8), 118(8), *146*, 450(32), *495*
Večerková, J., 67(140), 81(195), *87*, *88*
Veech, R. L., 338(124), 340(124), *353*
Vega, F. A., 37(17), *84*
Veiga, L. A., 160(51), *170*
Venkatasubramanian, T. A., 63(119), *87*, 132(113), *149*
Venter, P. J., 122(70), *148*
Verachtert, H., 257(34), 258, *281*

Verma, M. R., 301(91), 313(126a), *317*, *318*, 532(155a), *537*
Veronese, F. M., 120(25), *147*
Vetlugina, L. A., 402(30), *433*
Vetterli, W. A., 120(48), *147*, 469(103), 470(103), *497*
Viemer, S., 122(65), *148*
Viil, J., 122(65), *148*
Villaneuva, V. R., 127(108), *149*
Virgin, H. I., 322(21), *350*
Virtanen, A. I., 185(75), *198*
Vischer, E., 251(12), 258, *281*, *282*, 416(85), *435*
Visser, B. J., 35(10), *84*
Viswanathan, C. V., 177(33), 185(33), 190(33), 196(33, 103), *197*, *199*
Vitul'skaya, N. V., 322(23), *350*
Vlitos, A. J., 283(2), 287(2), 290(2), 292(2), *315*
Voegelein, W., 120(48), *147*, 469(103), 470(103), *497*
Vogt, G., 126(88), 132(88), *148*
Voinova, Z. E., 419(102), 424(102), *435*
Volcani, B. E., 424(130), *436*
Volodarskii, I. L., 482(156), 483(156), *498*
Vomhoff, D. W., 57(108), *86*
von Arx, E., 31(2), *84*, 216(26), 226(44), *245*
Vondrávek, M., 397(1, 4), 398(1), 399(4), 400(1), 401(1), 402(24), 407(4), 408(1, 4), 410(1, 4), 413(4), 414(4), 415(1), 416(1), *433*
Vondráčková, J., 413(73), 414(73), 415(81), *434*, *435*
Vora, V. C., 416(88), *435*
Voser, W., 416(85), *435*
Vrba, Z., 482(154), *498*
Vrtiskova, A., 402(27), *433*

Wachtmeister, C. A., 342(136), *353*
Wade, E. H. M., 127(110), 127(111), *149*
Wade, R., 142(165), 144(165), *150*
Wagman, G. H., 411(63, 64, 65, 66), 412, 413(67), *434*
Wagner, G., 300(67), *316*
Wagner, H., Jr., 141(158), *150*
Wagner, H., 185(84), 187(84), *198*
Wagner, J., 120(51), *147*
Wakae, M., 409(62), *434*
Waksmundzki, A., 67(134), *87*
Waldi, D., 461(72), *496*
Waldon-Edward, D. M., 111(2), *146*
Walker, D. G., 185(70), *198*
Walker, P. G., 26(62), *29*
Walker, W. R., 522(110), 524(110), *536*
Walker, J. B., 409(60), *434*
Walker, M. S., 409(60), *434*
Walker, R., 470(109), *497*
Wall, H. M., 164(75), 166(75), *171*
Wall, P. E., 296(49), *315*
Wallenfelds, K., 164(72), *171*
Wallhaeusser, K. H., 416(86), *435*
Walter, A. D., 185(73), *198*
Walter, H. P., 490(170), 491(170), *499*
Walton, H. F., 503(33), *534*
Wantier, G., 105(62), *109*, 501(22), 520(97), *534*, *536*
Wapnir, R. A., 127(102), 127(102), *149*
Warren, F. L., 185(70), *198*
Warren, J. C., 220(57), 229(57), *245*
Washington, C. L., 256(31), 258(31), *281*
Washington, R. J., 136(124), *149*
Watanabe, N., 232(76) *246*
Waterfield, W. R., 101(38), *109*
Waterhouse, C., 216(38), 229(38), *245*
Watson, M. A., 154(20), *169*
Watt, P. R., 311(116), *318*
Watts, R. R., 365(30), 371(30), *396*
Weaver, V. C., 1(3), 2(3), 3, *4*, 10(22), 11, 12(22), 18(22), 20(22), *28*, 36(15), 40(15), 41(15), 42(15, 35), 43(37), 44(15, 37), 49(15), 51(15), 67(15), *84*, 92(5), 100(5), *108*
Webb, R. A., 501(16a), 520(16a), *534*
Weber, E., 104(56), *109*
Webster, C., 361(19), *395*
Weg, M. W., 57(108), *86*
Wei, R., 421(117), 422(117), 423(117), *436*
Weidner, G., 77(175), *88*
Weil-Malherbe, H., 338(124), 340(124), *353*
Weinhold, A. R., 26(60), *29*
Weinstein, M. J., 411(64, 65), 412(65), *434*
Weiss, A., 516(85), *536*
Weiss, F., 77(175), *88*, 241(101), 243(101), 244(101), *247*
Weissbach, H., 286(17), *315*
Weisz, H., 73(160), *88*
Weitkamp, H., 80(190), *88*
Welford, G. A., 104(55), *109*, 533(162), *538*

Weller, L. E., 283(3), *315*
Wells, R. A., 507(59), 522(59), 524(59), *535*
Weltzien, H. C., 386(54), 387(54), *396*
Wender, S. H., 299, 300(73), *316*, *317*, 345(156, 167), 348(156, 159, 167), *354*
Werum, L. N., 420(110), *435*
West, G. B., 26(58), *29*
Westall, R. G., 18(46), *28*
Westenskov, G., 216(34), *245*
Westman, A. E. R., 525(140), *537*
Westphal, V., 177(32), 185(82), 187(82), *197*, *198*
Wettstein, A., 416(85), *435*
Whaley, H. A., 416(87), *435*
Wheat, R. W., 152(5), *169*
Wheeler, T. E., 143(173), *151*
White, I. G., 423(124), *436*
White, K., 297(50), 300(50, 81), 312(50), 313(50), *316*, *317*
Whitehead, D. F., 410(56), *434*
Whitehead, J. K., 139(136), *150*
Whitfield, A. E., 191(96), 193(96), *198*
Whitman, A. L., 409(52), *434*
Wichtl, M., 238(88), *246*
Wickberg, B., 165(79), *171*
Wieker, W., 490(170), 491(170), *499*
Wieland, T., 41(23), *84*, 191(65), 193(65), *198*
Wierzchowski, P., 120(52), *147*
Wiesner, H. J., 98(28), *108*
Wiesner, I., 494(181), *499*
Wikoff, H. M., 335, *352*
Wild, G., 408(43), *434*
Wilder, J., 257(34), 258(34), *281*
Wilkinson, J. H., 308(110), *317*
Williams, A. H., 308(111, 112), 313(112), 314(112), *317*
Williams, C. B., 138(132), *150*
Williams, D. H., 418(92), *435*
Williams, N. J., 142, *150*
Williams, N. R., 155(33), 164(75), 166(75), *170*, *171*
Williams, R. T., 296(48), 307(136), 312(136), 314(48, 136), *316*, *318*, 478(141), *498*
Williamson, D. E., 460(68), *496*
Willoughby, M., 202(13), *244*
Willstätter, R., 320(9), *350*
Wilson, A. T., 100(33, 35), *109*
Wilson, C. M., 161(58), *171*
Wilson, D. V., 127(98), 129(98), *148*
Wilson, J. R., 297(55), *316*
Wilson, O. S., 120(40), *147*
Wingo, W. J., 74(165), *88*, 102(42), *109*
Winkler, H., 507(62), *535*
Winsten, W. A., 253(22), *281*, 409(54), 411, *435*
Winteringham, F. P. W., 102(43), 104(49), 105(49, 57), *109*
Wise, C. S., 67(135), *87*
Wiśniewski, W., 468(91a), *496*
Wisse, J. H., 35(10), *84*
With, T. K., 335(106), 336(106), 337(106), 340(106), *353*
Witkop, B., 120(29), *147*, 418(98), *435*
Witkowski, H., 467(82), *496*
Wittmann, H., 328(75), *352*
Wittwer, S. H., 283(3), *315*
Wohnlich, J., 154(27), *170*
Wojciak, Z., 467(83), *496*
Wolf, F., 439(2), *494*
Wolf, F. T., 324(45), *351*
Wolf, M., 41(20), *84*
Wolff, J. B., 323(28), 330(28), *351*
Wolman, Y., 141(157), *150*
Wolstenholme, W. A., 141(159), *150*
Wong, E., 345(163), 348(163), 349(163), 350(163), *354*
Wong, J. T., 127(111), *149*
Wood, D. J., 313(125), *318*, 342(140, 141), *353*
Wood, H. C. S., 424(127), *436*
Wood, T., 485(162, 162a, 163), 486(162, 162a), 489, 490(162, 162a), *498*, *499*
Woolf, L. I., 34(6), *84*
Wooten, V., 232(82), *246*
Wren, J. J., 420(106), *435*
Wright, H. E., Jr., 300(68), 310(68), *316*
Wright, W. W., 413, 414(69), *434*
Wroński, M., 474(131), *498*
Wuthier, R. E., 42(31), *84*
Wyborny, L., 27(66), *29*

Yablokova, O. G., 122(69), *148*
Yakobson, L. M., 407(34), *433*
Yakubov, G. Z., 397(3), 400(3), 401(3), 402(3, 30), 415(3), *433*
Yamada, A., 238(97), *246*
Yamaguchi, K., 329(79), *352*, 525(132), *537*

Yamamoto, A., 168(87), *171*
Yamamoto, Y., 323(29), 326(29), *351*
Yamatodani, S., 408(41), *434*
Yamasaki, K., 232(81), 233(81), 234(81), *246*
Yamazaki, M., 431(154, 155), 432(155), *436*
Yaneva, A., 467(86), *496*
Yang, C. H., 299, *316*
Yang, T.-I., 259(58), *282*
Yao, K.-Y., 71, *87*
Yarnall, F. E., 356(8), *395*
Yasin, R., 391(63), *396*
Yauger, W. L., Jr., 297(56), *316*
Yeh, K.-H., 493(176), *499*
Yip, G., 379(42), 380(42), 381(42), *396*
Yoe, J. H., 507(65), 519(65), *535*
Yoneda, H., 533(157), *537*
Yoshida, T., 418(98), *435*
Yoshino, T., 45(69), *85*, 525(137), *537*
Young, H. Y., 384(50), *396*
Yuan, C.-Y., 493(176), *499*

Zähner, H., 416(85), *435*
Zaffaroni, A., 52(96), *86*, 191(68), 193(68), *198*, 216(37), 227(49), *245*
Zahradnik, R., 185(81), 187(81), *198*
Zajić, J., 447(21), *495*
Zak, B., 96(21), *108*
Zalucky, T. B., 312(119), *318*
Zamfir, J., 533(163), *538*
Zamojski, A., 419(105), *435*
Zamorani, A., 127(100), 129(100), *149*
Zelenina, L. M., 162(63), *171*
Zeller, N., 517(87), 522(87), *536*
Zelnik, R., 238(89), 240(89), *246*
Zenk, M. H., 127(108a), *149*
Zentner, H., 159(47), *170*
Ziegler, M., 507(62), 512(71), *535*
Zinsser, H. H., 172(4), 194(4), *197*
Zirakishvili, Z. V., 522(107), *536*
Ziti, L. M., 238(89), 240(89), *246*
Zobel, M., 431(153), 432(153), *436*
Zommers, Z., 136(126), *149*
Zsadon, B., 468(93a), *496*
Zubrzycki, Z. J., 104(53), *109*
Züllig, H., 322(20), *350*
Zukriegel, H., 469(104), *497*
Zumoff, B., 56(104), *86*
Zweig, G., 3(12), *5*, 8, 9(17), 26(60), *27*, *29*, 33(4), 35(11), 45(42), 53(4), 64(4), 73(4), 75(169), 83(4), *84*, *85*, *88*, 96(18), 99(31), 100(31), 102(18), 104(18), 107(18), *108*, 120(44), 121(44), 122(44), 123(44), 126(44), 127(44), 132(44), 133(44), 134(44), 138(44), 139(44), *147*, 152(1), 153(1), 158(1), 163(1), *169*, 178(104), 179(104), 184(104), 186(104), *199*, 200(6), 201(6), 202(6), 216(6), 229(6), 236(6), 237(6), 242(6), 243(6), 244(6), *244*, 249(3), 250(3), 251(3), 252(3), 253(3), 255(3), 257(3), 258(3), *281*, 322(18), 323(18), 324(18), 326(18, 56), 327(18), 329(18, 86), 331(18), *350*, *351*, *352*, 355(2, 4, 5), 360(4), 361(4), 362(4), 375(39), 377(39), 378(39), 379(2, 39), 390(2), 395(2), *395*, *396*, 442(9), 443(9), 477(9), *494*, 500(6), 506(54), 518(54), 520(54), *534*, *535*
Zwierz, K., 74(164), *88*

SUBJECT INDEX

A

Absorbents, 42–43
Absorption, humidification and, 51
Absorption spectroscopy, 97
Acacetin, 345
Acetaldehyde, 113
Acetic acid, 12, 41, 46, 52, 113, 116, 126, 132, 139, 143, 145–146, 154–155, 178, 180, 182, 187, 191
  glacial, 41, 136
Acetic anhydride, 41
Acetoacetate, 174
Acetoacetic acid, 186
4-Acetoaminobenzalazine, 451
Acetoguaiacone, 306
Acetophenone, 449
Acetophenone azine, 451
Acetone, 32, 74, 122–123, 130, 132, 139, 176, 186–187
5-Acetylamino-6-amino-3-methyluracil, 265
Acetylation, 53
Acetyl-CoA, 487
N-Acetyl derivatives, 168–169
Acetyldigitoxin-α, 239
Acetyldigitoxin-β, 239
Acetyldigoxin-α, 239
Acetyldigoxin-β, 239
Acetyleburicoic acid, 234
Acetylgitoxin-α, 239
Acetylgitoxin-β, 239
Acetylglucosamine, 163
N-Acetylindoxyl, 295
N-Acetylisatin, 295
N-Acetylneomycins, 409
N-Acetyl-*S*-(2-nitrophenyl)-L-cysteine, 308
N-Acetyl-S-(4-nitrophenyl)-L-cysteine, 308
Acetyl phosphate, 181
Acetyl tryptophan, 295
N-Acetyltryptophan, 287
5-Acetyluracil, 267
Acid hydrolysis, 248–249
Acid salts, preparation of, 172–173
Acids
  aliphatic
    chromatography of, 172–196
    detection reagents for, 191–195
    nonvolatile water-soluble, 179–183
    paper for, 176–177
    quantitative analysis of, 195–196
    sample preparation for, 172–176
    separation of on ion-exchange papers, 47
    solvents for, 177–191
Aconitic acid, 26, 184
Aconitine, 466
Acridine derivatives, 11
Acrylonitrile fibers, 469
Actinomycetes, 417
Actinomycins, 418
Activation analysis, 4
Acumycin, 416
Acyl derivatives, 275
Adenine, 111, 129, 266, 270, 272, 276–277, 279, 487
Adenine-*N*′-oxide, 265
Adenosine, 264, 270, 273, 276–277, 487
Adenosine 3′:5′-cyclic phosphate, 278
Adenosine diphosphate, 276–277, 486
Adenosine 5′-diphosphate, 262, 278
Adenosine monophosphate, 270, 276
Adenosine 2′-monophosphate, 278
Adenosine 3′-monophosphate, 278
Adenosine 5′-monophosphate, 277, 278
Adenosine 3′-phosphate, 486
Adenosine 5′-phosphate, 262, 486
Adenosine tetraphosphate, 487
Adenosine triphosphate (ATP), 277, 486
Adenosine 5′-triphosphate, 262, 278
S-Adenosylmethionine, 262
Adipic acid, 181, 184
Adipylpenicillin, 407
ADP, *see* Adenosine diphosphate

Adrenaline, 26, 115, 117
m-Adrenaline, 115
Adrenosterone, 226
Adsorption, 11, 20
  ion exchange and, 12
  partition and, 11–12
  silica paper and, 42
  of sugars, 152
  tailing and, 22
Adsorption chromatography, 32, 319
Aetioporphyrin, 336
Aglycones, chromatography of, 237–244
Alanine, 123, 128, 130, 135, 139
α-Alanine, 132
Alanyl peptides, 142
Alcohol, 50, 442–447
  aliphatic, 442
    primary, 443
    secondary, 443
    tertiary, 443
  free, 442
  monohydric, 442
  sugar, 443
Alcohol derivatives, 443–445
Aldehydes, aliphatic, 447–448
Aldohexoses, 158–159
Aldonic acids, 160
Aldopentoses, 158–159
Aldoses, 159, 163
Aldosterone, 202, 225–226
Aldrin, 359
Algae, pigments in
  R_F values of, 333
  separation of, 331–333
Aliphatic compounds, *see* specific compounds
Alizarin, 115, 516–517
Alkali metals, detection of, 78
Alkaline earths, 25–26
Alkaline hydrolysis, 104, 249
Alkaloids, 20, 460–469, *see also* specific types
  anabasis, 462
  detection of, 461–462
  Papaveraceae, 464–465
  R_F values for 464–466
  separation of on ion-exchange papers, 47
  solonaceous, 464
  systematic analysis of, 81
Alkoxy acids, 179
1-Alkoxy-2(3)-hydroxy-3(2)-aminocyclohexanes, 120
Alkyl 2,4-dinitrobenzyl ethers, 444–445
Alkyl hydroperoxides, 455
Alkyl phosphates, 490
  R_F values for, 491
Alkyl phosphites, 490
Alkyl phosphoric acids, 493
N'-Alkylstreptomyclamines, 409
Alkyl xanthates, 444
Allantoin, 271
β-Allocortol, 213
Allocryptopine, 465
Allocymarin, 239
Alloemicymarin, 239
Allonolactone, 166
Alloperiplocymarin, 240
Alloperiplogenin acetate, 240
Allopregnane-3α,20β-diol, 213
Allostrophanthidin, 239
N-Allylthiourea, 476
Alumina, 48
Aluminum hydroxide, 3, 42
Altronolactone, 166
Amines, 41–42
  adrenaline type, 47
  aliphatic, 110–120, 133
    3,5-dinitrobenzoyl derivatives of, 116
    R_F values of, 116
    tryptamine derivatives of, 116
  aromatic, 110–120
  densitometry of, 117–118
  detection of, 113–117
  double spots and, 26
  experiments with, 118–120
  fluorescence in, 118
  labile, 110
  paper used with, 111
  quantitative estimation of, 117–118
  R_F values of, 112–113
  secondary, 113, 115
  solvents for, 111–113
  volatile, 110, 120
1-Amino-3-acetaminobenzene, 113
1-Amino-4-acetaminobenzene, 113
4-Aminoacetophenone, 451
4-Aminoacetophenone azine hydrochloride, 451
Amino acid(s), 37, 42, 46, 120–140, 143

acetylated, 127
acidic, 126
allenic, 127
centrifugal chromatography of, 71
chromatographic techniques for, 120–139
coincidence detection and, 81
cytoactive, 144
dansyl, 138
detection of, 132–133
DNP, 133, 138–139
elution of, 134
estimation of, 90
experiments on, 138–140
in human tissue, 138, 142
isotope indication of, 103–104
multidevelopment of, 67
multiple spotting in, 26
paper used for, 123–126
peak broadening in, 13–14
quantitative determination of, 134–138
R_F values of, 128–130, 144
reagent solution for, 91
sample preparation in, 121–123
separation of, 2, 39–40, 53, 83, 110
serum, 131
solvents for, 126–127
spot measurement of, 93
straight-chain, 131
sulfur, 127
two-dimensional chromatography of, 70
ultraviolet light detection of, 134, 139
visual comparison of, 134
Amino acid chromatograms, 80–81
Amino acid derivatives, 138–139
Aminoadipoyl-7-aminocephalosporanic acid, 408
Amino alcohols, 139
R_F values of, 114
p-Aminobenzoic acid, 160, 303, 422
p-Aminobenzylpenicillin, 407
5-Amino-6-carboxyuracil, 263
7-Aminocephalosporanic acid derivatives, 409
2-Amino-3-chlorophenol, 307
2-Amino-4-chlorophenol, 112, 308
2-Amino-5-chlorophenol, 208, 308
2-Amino-6-chlorophenol, 308
3-Amino-2-chlorophenol, 307
3-Amino-4-chlorophenol, 307
3-Amino-5-chlorophenol, 308
4-Amino-2-chlorophenol, 308
4-Amino-3-chlorophenol, 307
5-Amino-2-chlorophenol, 308
6-Amino-2,8-dihydroxypurine, 262
6-Amino-1,3-dimethyl-5-formylaminouracil, 267
2-Amino-4,6-dimethylpyrimidine, 269
6-Amino-2,4-dimethylpyrimidine, 269
6-Amino-1,3-dimethyluracil, 269
2-Aminodiphenylamine, 111
4-Aminodiphenylamine, 111
p-Aminodiphenylamine, 118
2-Amino-4,6-deoxy-5-methylpyrimidine, 264
2-Amino-4,6-deoxypyrimidine, 264
2-Amino-4,6-dinitrophenol, 112
Aminoethylcellulose paper, 44–45
6-Amino-5-formylamino-4-hydroxypyrimidine, 263
6-Amino-5-formylamino-3-methyl-4-oxypyrimidine, 265
6-Amino-5-formylamino-1-methyl-4-oxypyrimidine, 264
6-Amino-5-formylamino-1-methyluracil, 263
6-Amino-5-formylamino-3-methyluracil, 264
6-Amino-5-formylaminouracil, 262
2-Amino-3-hydroxy acids, aliphatic, 127
2-Amino-4-hydroxy-6-methylpyrimidine, 267
2-Amino-8-hydroxypurine, 262
6-Amino-2-hydroxypurine, 263
6-Amino-8-hydroxypurine, 263
2-Amino-4-hydroxypyrimidine, 266
4-Amino-5-imidazolecarboxamide, 266, 272
2-Amino-5-imidazolecarboxamide ribonucleoside, 264
4-Amino-5-imidazolecarboxamidine, 272
Amino-*J* acid, 313
6-Amino-3-methyl-5-(*N*-formylmethylamino)uracil, 267
2-Amino-5-methyl-4-oxypyrimidine, 268
2-Amino-6-methylpurine, 265
2-Amino-4-methylpyrimidine, 268
5-Amino-6-methyluracil, 264

6-Amino-1-methyluracil, 266
1-Amino-2-naphthol-*N*,4-disulfonic acid, 482
o-Aminonaphthosulfonic acid, 483
2-Amino-1-naphthol-*N*-sulfonic acid, 482
2-Amino-1-naphthol-*O*-sulfonic acid, 482
2-Amino-5-nitrophenol, 112
6-Aminopenicillanic acid, 407
2-Aminophenol, 113
3-Aminophenol, 113
4-Aminophenol, 113
m-Aminophenol, 119, 305, 308
o-Aminophenol, 119, 157, 307
p-Aminophenol, 119, 159, 305, 308
2-Aminophenol-5-sulfonamide, 113
m-Aminophenyl glucoside, 305
m-Aminophenyl sulfuric acid, 305
p-Aminophenyl sulfuric acid, 305
3-Aminopropanol, 114
2-Aminopurine, 264
2-Aminopyrimidine, 268
Amino salts, multiple zoning and, 26
2-Aminostilbene, 112
Amino sugars, 26, 132, 160, 168–169
α-Amino-*n*-tetradecanoic acid, 126
6-Amino-2-thiouracil, 266
3-Aminotriazole (ATA), 383–384
5-Aminouracil, 263
6-Aminouracil, 264, 273
5-Aminouridine, 262
Ammonia, 39, 113, 116
Ammoniacal solvents, multiple spotting and, 26
Ammonium *n*-butyl *n*-propyl phosphate, 491
Ammonium dibenzyl hydrogen phosphate, 491
Ammonium diethyl phosphate, 491
Ammonium ethyl *n*-propyl phosphate, 491
Ammonium molybdate, 141
Ammonium molybdophosphate, 47
AMP, *see* Adenosine monophosphate
Amyl acetate, 41
t-Amyl alcohol–dioxane–H_2O, 131
n-Amylamine, 116
β-Amylase, 154
tert-Amyl hydroperoxide, 456
Anagyrine, 464
$\Delta^{1,4}$-Androstadien-10-ol-3,17-dione, 211
$\Delta^{4,6}$-Androstadien-17β-ol-3-one, 226
$\Delta^{4,6}$-Androstadiene-3,17-dione, 226
Androstane derivatives, 222–223
Androstane-3α,17β-diol, 213
Androstane-3β,17β-diol, 213
Androstane-3,11β-diol-17-one, 225
Androstane-3,17-dione, 225–226
5β-Androstane-3,11,17-trione, 219
Androstan-3β-ol-17-one, 225
5β-Androstan-3,11,17-one, 204
Δ^4-Androstene-3,17-dione, 211, 226
4-Androstene-3,17-dione, 204, 219
Δ^4-Androstene-3,11,17-trione, 225
Δ^{16}-Androsten-3α-ol, 213
Δ^{16}-Androsten-3β-ol, 213
Δ^4-Androsten-1α-ol-3,17-dione, 211
Δ^4-Androsten-11β-ol-3,17-dione, 225
Δ^5-Androsten-3β-ol-17-one, 225
Androsterone, 204, 219, 226
Androsterone benzoate, 204
Angolamycin, 416
"α"-Anhydrodigitoxigenin, 241
"β"-Anhydrodigitoxigenin, 241
"β"-Anhydrodigoxigenin, 241
"β"-Anhydrodigoxigenone, 241
Anhydrodemethylchlortetracycline, 414
"β"-Anhydrosarmentogenin, 241
Anhydrotetracyclin, 414–415
"β"-Anhydrouzarigenin, 241
Anhydrovitamin A, 429
Anhydrovitamin A_2, 429
Aniline, 73, 112, 159
Aniline hydrogen phthalate, 158–159, 164
Aniline xylose, 193
Anion(s)
  chromatography of, 525–530
  detection of, 527–530
  paper for, 525
  sample preparation for, 525
  solvents for, 526–527
Anion-exchange chromatography, 44, 155, 254
Anisaldazine, 451
Anisaldehyde, 449
p-Anisidine, 119
p-Anisidine-periodate, 160
Anthocyanides, R_F values for, 343
Anthocyanins, 340–351
  R_F values for, 343–344
Anthoxanthins, 341–350

Anthracene, 100, 195, 439
Anthranilic acid, 287, 289, 303
Anthranol, 195
Anthraquinone dyes, 469, 473
Anthraquinones, 450
Antibiotics
  amphoteric, 404
  analysis, classification, and identification of, 401–406
  antibacterial, 399
  antifungal, 399
  antiviral, 399
  bioautography of, 49, 398–399
  chromatographic spectra of, 401–402
  chromatographic techniques for, 400–401
  chromatography of, 397–419
  classes of, 406–419
  cytostatic, 399
  detection of, 398–400
  macrolides, 415–417
  pH value of, 404–406
  polypeptide, 417–418
  R_F values of, 403–404
  sample preparation for, 397
  separation of on ion-exchange papers, 47
  spot measurement of, 93
  systematic analysis of, 81
  ultraviolet densitometry of, 399–400
Antimony chloride, 203
Antimony trichloride, 313
Antioxidants, 391–393
Apigenin, 345
Apigeninidin, 343
Apiin, 346
Apoatropine, 464
Aporheine, 465
Arabinose, 156, 159
L-Arabinose, 166–167
Arachidic acid, 182, 188
Aralkyl hydrazones, 167
Arginine, 123, 128, 130, 141
Armepavine, 465
Aromatic acids, 296–315
  chromatography of, 296–315
  detection of, 311–315
  R_F values of, 310
  sample preparation in, 296–300
  solvents for, 300–310
Arsenate, 529
Arsenite, 529
Arsenoorganic compounds, 394–395
Arylamines, 120
Ascending chromatography, 2, 53, 57–61, 93, 131, 217, 256
Ascending strip chromatography, 64
Ascorbic acid, 184, 419, 423, 431–432
Asclepiadaceae, 155
Ashbya gossypii, 423
Asparagine, 126, 128, 130, 133
Aspartic acid, 26, 61–62, 128, 130, 133, 135
Association, trailing zones and, 21
Astralagin, 345
ATP, *see* Adenosine triphosphate
Atropine, 462
Auranetin, 345
Aureusidin, 347
Aureusin, 347
Aurone aglycones, 347
Autography, sublimation, 100
Automobile exhaust, phenols in, 299
Autoradiography, 78, 98–101, 103
  of purines and pyrimidines, 253
  ultraviolet absorption photography and, 81
Avicularin, 345
5-Azacytosine, 259
Azaleatin, 345
6-Azathymine, 268
6-Azauridine, 264
Azelaic acid, 181, 184, 188
Azines, 167–168, 451–452
Azo dye, 110, 384, 469, 472–473

B

Bacillus subtilis, 398
Bacitracins, 417
Baicalein, 345
Barbituric acid, 265
Barium, 104
Barium hydroxide, 121
Beer's Law, 96–97, 162
Behenic acid, 182, 188
Behenolic acid, 189
Belladonna alkaloids, 467–468
Belladonnine, 464

Benzalacetophenone, 449
Benzalazine, 451
Benzaldehyde, 449
1:2-Benzanthracene, 441
Benzelpenicillin, 404
Benzel phosphate, 492
Benzene, 11, 50, 52, 100, 116, 173, 175, 177, 219, 325
Benzene–chloroform, 219
Benzene-dimethyl sulfoxide, 155
Benzhydrol, 443
Benzidine, 111–112, 119, 159
o-Benzidine, 111
2:3-Benzofluorene, 439
Benzoic acid, 304
Benzoin, 443
Benzo[*c*]phenanthrene, 441
3:4-Benzophenanthrene, 440
Benzophenone, 449
Benzophenone azine, 452
3:4-Benzopyrene, 439–441
Benzoyl peroxide, 456
5-Benzyloxyindole-3-acetic acid, 288
Benzyl-α-pyridyl phosphate, 492
Berberine, 464
Beryllium, 104
BHC, 359
Bile acids, 232–237
  chromatography of, 232–237
  conjugated, 236
  detection of, 237
  free, 236
  quantitative estimation of, 237
  sample preparation for, 232–233
Bioautography, 78
  of antibiotics, 398–399
  of purines and pyrimidines, 253
Biochanin A, 346
Biotin, 419, 422
4-*N,N*-Bis(2-chloroethyl)aminobenzaldehyde, 292
4-*N,N*-Bis(2-chloroethyl)amino-2-tolualdehyde, 292
Bis-α-hydroxystreptomycyclamine, 409–410
Blood, 138
  isolation of pentachlorophenol from, 297–298
    of phosphoric esters from, 484–485
  peptides in, 142
  sugar in, 154
Blood oxytocin, bioassay of, 143
Blood serum, isolation of indoles from, 285–286
Borate, 529
Borax, 155
Brain tissue, 138
  peptides in, 142
Brassidic acid, 189
Brestan, 386
2-Bromanthraquinone, 450
Bromate, 529
Bromide, 529
Bromine, 78, 371
Bromine vapor, 474–475
Bromoacetic acid, 178
Bromoacetone, 175
4-Bromoaniline, 112
Bromobutyric acid, 178
1-Bromo-2-chloranthraquinone, 450
Bromocresol green, 192–193
Bromocresol purple (BCP), 15, 192, 380
5-Bromodeoxyuridine, 268
Bromodiols, 447
1-Bromonaphthalene, 116
o-Bromophenol, 309
Bromophenol blue, 145, 191–193
Bromopropionic acid, 178
N-Bromosuccinimide, 365
Bromothymol blue, 192
5-Bromouridine, 266, 280
5-Bromouridine 2′,3′-cyclic phosphate, 280
5-Bromouridine 3′-phosphate, 280
5-Bromouridine 2′(3′)-phosphate, 280
5-Bromouridylyl-(3′→5′)-adenosine, 280
5-Bromouridylyl-(3′→5′)-uridine, 280
Bromthymol blue, 371–372
Brown potentiometer, 101
Brucine, 462, 466
Bufadienolides, chromatography of, 237–244
Buffer, tailing and, 21
Bufotenin, 288
Bufotenine, 116
Butein, 346
Butin, 346
n-Butoxyacetic acid, 179
5-Butoxymethyluracil, 269
sec-Butyl alcohol, 135

n-Butylamine, 114, 116
sec-Butylamine, 116
tert-Butylamine, 116
n-Butyl, *sec*-butyl, isobutyl hydroperoxide, 456
t-Butylchromate, 212–213
n-Butyl ether, 177
tert-Butyl ethyl hydrogen phosphate, 491
tert-Butyl hydroperoxide, 456
tert-Butyl perbenzoate, 456
tert-Butyl phosphate, 491
Butyric acid, 180, 182, 195
n-Butyric acid, 128, 178
Butyrophenone, 449
Butyrophenone azine, 452
Butyryl phosphate, 181

C

Cadmium acetate, 135–136
Caffeic acid, 184, 303, 310
Caffeine, 273, 465
Caffeylglucose, 167
Calciferol, 210
Calcium, 104
Calycotomine, 464
Cannabis alkaloids, 467
Cannabiscitrin, 345–346
Capersinidin, 343
Capillary analysis, 2, 53, 63, 319
Capric acid, 182, 188
n-Caproic acid, 178, 180, 182, 188, 195
Caproyl phosphate, 181
Caprylic acid, 180, 182, 188
Captan, 386
Carbamate insecticides, *see* Insecticides
Carbohydrates, 442
  chromatography of, 152–169
  detecting reagents for, 158–161
  elution of, 161–162
  hydrolysis of, 154, 163
  hydroxyl groups, 152
  paper used for, 155
  quantitative analysis of, 161–163
  sample preparation in, 153–154
  solvents for, 157–158
  titration of, 162
  ultraviolet light detection of, 158
Carbomycin, 416
Carbon, 121
Carbon disulfide, 50
Carbon tetrachloride, 50
Carbonate, 529
Carbonyl compounds, 18
4-Carboxybenzalazine, 451
Carboxyl groups, 12
Carboxylic acid, 170
Carboxymethylcellulose, 44–45, 145
6-Carboxypurine, 263
Cardenolides, chromatography of, 237–244
Cardiac aglycones, 237–244
Cardiac glycosides, 237–244
  chromatography of, 237–244
  mobility data for, 244
  paper for, 238
  solvents for, 238–242
  spot detection in, 243–244
β-Carotene, 427, 429
Carotenes, 333
Catechol, 304
Catecholamine derivatives, 118
Catecholamines, 111, 120
  R_F values of, 115
Catenulin, 409–410
Cation exchange, 124
Cation paper chromatography, 10
Cation retention, 12
Cations, 41
  chromatographic techniques for, 512–515
  detection of, 515–518
  elution of, 519–520
  inorganic analysis of, 500–525
  paper for, 501–506
  sample preparation for, 500–501
  solvents for, 506–512
Cellophane sheets, in chromatogram evaluation, 79
Cellulose, elution and, 95
α-Cellulose, 37, 45
Cellulose chromogram paper, 323, 326
Cellulose fibers, 469
Cellulose paper, 10, 323, 326
  copolymerized, 41
  ion-exchange, 124
  modified, 41–49
  pores of, 21
  pure, 36–41

Cellulose phosphate paper, 43–45
Cellulose slurry, 40
Cellulose thin-layer chromatography, 3
Cephalosporin, 408–409
Cephalosporin C, 407–409
Cephalosporin N, 407–408
Cephalosporins P_1-P_5, 409
Cephalothin, oral, 409
Cerenox, 386
Ceresan, 386
Ceric ammonium nitrate, 313
Cerium(IV) phosphate, 48
Cernuoside, 347
Cerotic acid, 189
Cetylpyridinium chloride, 167
Cevadine, 466
Chalcone aglycones, 346
Chelerythrine, 465
Chelidonine, 465
Chelilutine, 465
Chelirubine, 465
Chemical detection, 73–77
Chenodeoxycholic acid, 236
Chitin, 163
Chloramphenicol, 400, 404
Chlorate, 529
Chlordane, 359
Chloride, 529
Chlorin a_2 (methyl ester), 336
Chlorinated insecticides, *see* Insecticides
Chlorine, 20, 78, 105, 164, 192
1-Chloro-2-aminoethane, 114
3β-Chloro-Δ^5-androsten-17-one, 225–226
2-Chloraniline, 112
3-Chloraniline, 112
4-Chloraniline, 112
m-Chloraniline, 308
o-Chloraniline, 64, 307
p-Chloraniline, 64, 308
2-Chloranthraquinone, 450
1-Chloroanthraquinone, 450
Chloroacetic acid, 178
Chlorobenzilate, 359
1-Chloro-2-bromoanthraquinone, 450
1-Chloro-5-bromoanthraquinone, 450
2-Chloro-3-bromoanthraquinone, 450
3-Chlorocatechol, 307
4-Chlorocatechol, 307
4-Chloro-*m*-cresol, 309
4-Chloro-*o*-cresol, 309
6-Chloro-*m*-cresol, 309
6-Chloro-*o*-cresol, 309
4-Chloro-2,3-dimethylphenol, 309
4-Chloro-2,5-dimethylphenol, 309
4-Chloro-3,5-dimethylphenol, 309
6-Chloro-3,4-dimethylphenol, 309
4-Chloro-5-ethyl-3-methylphenol, 309
1-Chloro-2-fluoroanthraquinone, 450
1-Chloro-5-fluoroanthraquinone, 450
2-Chloro-3-fluoroanthraquinone, 450
Chloroform, 50, 175, 219
Chlorogenic acid, 184, 299
3-Chloro-2-hydroxypropyl ethers, 447
1-Chloro-2-iodoanthraquinone, 450
1-Chloro-5-iodoanthraquinone, 450
2-Chloro-3-iodoanthraquinone, 450
3-Chloro-4-methoxyaniline, 112
3-Chloro-6-methoxyaniline, 112
3-Chloro-2-methylaniline, 112
3-Chloro-6-methylaniline, 112
4-Chloro-2-methylaniline, 112
2-Chloro-3-nitrophenol, 307
2-Chloro-4-nitrophenol, 308
2-Chloro-5-nitrophenol, 308
2-Chloro-6-nitrophenol, 308
3-Chloro-2-nitrophenol, 307
3-Chloro-4-nitrophenol, 307
3-Chloro-5-nitrophenol, 308
4-Chloro-2-nitrophenol, 308
4-Chloro-3-nitrophenol, 307
5-Chloro-2-nitrophenol, 308
o-Chlorophenol, 306, 309
p-Chlorophenol, 309
L-2-Chlorophenylmercapturic acid, 307
L-3-Chlorophenylmercapturic acid, 307
L-4-Chlorophenylmercapturic acid, 307
4-Chlorophenyl phosphate, 492
Chlorophenol red, 192
Chlorophyll, 195, 333
  R_F values for, 427
Chlorophyll a, 333
Chlorophyll b, 333
Chlorophyll c, 333
Chloroplast pigments, 42
  anomalous zones in, 333–334
  chromatography of, 319–334
  detection and identification of, 326–328
  experiments with, 329–333
  fat-soluble, 319–334
  paper for, 322–323

quantitative methods for, 328–329
sample preparation for, 319–322
solvents for, 323–325
2-Chloropropionic acid, 178
3-Chloropropionic acid, 178
5-Chloro-2,3,6-(1*H*)-pyridinetrione, 367
Chloroquinol, 307
4-Chlororesorcinol, 307
Chlorotetracycline, 400, 404, 414–415
2-Chlorothiophenol, 307
3-Chlorothiophenol, 307
4-Chlorothiophenol, 307
β-Cholestanol, 218
Cholestan-3-one, 210
Cholestenone, 222
Δ^4-Cholesten-3-one, 210
Cholesterol, 190, 205–206, 208, 210, 218, 443
Cholesterol acetate, 218
Cholesterol ester, 175–176, 190
Cholic acid, 211, 234–236
Choline, 422
Chondroitin sulfate, 155
Chromate, 529
Chromatogram
defined, 1
overloading of, 25
pH type, 404–406
preservation and storage of, 78–80
qualitative evaluation of, 78–80, 90–108
summarized, 401–402
Chromatographic data, publication of, 18
Chromatographic mobility, 15
Chromatographic peak, 12–13
Chromatographic spectra, 17
Chromatographic spot width, 13
Chromatographic system, defined, 6
Chromatography
adsorption, 319
of agylcones, 237–244
of aliphatic acids, 172–196
of amines, 110–120
of amino acids, 120–139
anion-exchange column, 254
of anions, 525–530
of antibiotics, 397–419
of aromatic acids, 296–315
ascending, 2, 53, 57–61, 93, 131, 217, 256
of bile acids, 232–237
of carbohydrates, 152–169
of cardiac glycosides, 237–244
centrifugal, 13, 32, 71–72, 93–94, 124–126, 156, 216
circular, 2, 14–15, 63, 72, 156, 216, 253
column-disk combination, 73
cone-stripe, 156
defined, 1
descending, 2, 53–57, 82, 93, 217, 254, 256
development, 3, 16, 23, 53–67
diagonal, 81
Durchlauf, 54
equipment for, 31
of flavonoids, 341–350
history of, 1–3
horizontal, 61–63, 65, 73
of indoles, 283–296
of inorganic ions, 500–533
ion-exchange column, 248
laboratory for, 30–31
linear, 14
with longitudinal barriers, 73
of naturally occurring pigments, 319–350
of organic compounds, 438–494
of peptides, 140–143
of pesticides, 355–395
of phenols, 296–315
of porphyrins, 334–341
precipitation, 32, 146, 153
preparative, 82–83
progress in, 3–4
of proteins, 143–146
of purines, 248–280
purity of, 20
of pyrimidines, 248–280
quantitative, 90–108
rear-phase, 73
reversed-phase, 19, 41–42, 52–53, 73, 111, 200, 228–229, 512
salting-out, 401–402
sources of error in, 105–108
steroid, 200–232
of steroid acids, 232–237
techniques in, 30–83
theory and fundamentals of, 6–27
thermal paper, 73, 132

thin-layer, 3, 15, 17, 31, 40–41, 82, 107, 120, 254
two-dimensional, 14, 26, 68–72, 102–103, 126–127, 131, 139, 216
Chromic–sulfuric acid charring solutions, 48
Chromium, 532
Chrysene, 439
Chrysin, 345
Chrysoeriol, 345
Cigarette smoke, phenolic acids in, 299
Cinchonine, 466
Cinnamaldehyde, 449
Cinnamic acid, 304
Cinnamoylformic acid, 186
Cinnamyl alcohol, 443
Circular chromatography, 2, 14–15, 63, 72, 156, 216, 253
Citric acid, 26, 135, 181, 184
CMP, *see* Cytidine monophosphate
Cobalt, 47, 519–520, 532
Cocaine, 466
Co-chromatography, 15–16
radioactive spot identification with, 99
spot identification and, 80
Coenzyme Q, 419
Coenzyme Q_6–Coenzyme Q_{10}, 430
Colchicine, 466
Collagen, 143
Collidine, 31, 126
2,4,6-Collidine, 132–133
Color tests
AACS, 203–204
Carr–Price, 203–205
Column chromatography, 3, 15, 120
Column-disk combination chromatography, 73
Comets, 26
Concentration gradient, molecular diffusion and, 13
Cone-stripe chromatography, 156
Conical paper chromatography, 64
Congo red, 371
Copper, 37, 47, 519–520
Copper nitrate, 133
Copper oxychloride, 386
Copper sulfate, 314
Coproporphyrin, 336
Coproporphyrin I, 337
Coproporphyrin III, 337
Coprostanol, 210
Coptisine, 465
Co-Ral, 368
Coreopsin, 347
Corticosteroids, 225
Corticosterone, 205, 225–226
Cortisol, 205, 225–226
Δ^1-Cortisol, 205
Cortisol metabolites, 202
Cortisone, 205, 225–226
Δ^1-Cortisone, 205
Cortisone acetate, 225
Corysamine, 465
Cosmetin, 346
o-Coumaric acid, 304
7-Coumarinylglucoside, 305
7-Coumarinyl sulfuric acid, 305
Countercurrent distribution, 2, 8, 73
Craig countercurrent distribution apparatus, 8
Creatine, 47
Creatine phosphate, 487
Creep-back, 105–106
m-Cresol, 123, 304, 306, 309
o-Cresol, 304, 306, 309
p-Cresol, 304, 306
p-Cresolmonomethylol, 304
Cryogenine, 400
Crystallinity, cellulose paper strength and, 10
Crystal violet, 54, 471
CTP, *see* Cytidine triphosphate
2-*p*-Cumaroylglucose, 167
3-*p*-Cumaroylglucose, 167
6-*p*-Cumaroylglucose, 167
1-*p*-Cumaroyl-β-D-glucose, 167
4-*p*-Cumaroylrhamnose, 167
Cumene hydroperoxide, 456
Cumyl, 2-hydroxy-2-phenylethyl peroxide, 456
Cupric acetate, 193
Curare alkaloids, 467
Cyanide, 160, 529
Cyanidin, 343
Cynanidinglycosides, 344
Cyanocobalamin, 425
Cyanoglycosides, 160
m-Cyanophenol, 309
p-Cyanophenol, 309
3,5-Cycloandrostan-6β-ol-17-one, 225

Cyclohexane, 50, 52, 111, 118
Cyclohexane hydroperoxide, 456
Cyclohexene hydroperoxide, 456
Cyclohexylammonium benzyl *n*-butyl phosphate, 491
Cyclohexylammonium isopropyl *n*-propyl phosphate, 491
Cyclohexylammonium isopropyl *n*-propyl pyrophosphate, 491
Cymarin, 239–242
Cymarin acetate, 240
Cymarol, 239–242
Cystathionine, 129
Cysteic acid, 129–130
Cysteine, 25, 121, 128
Cysteine methyl ester, 129
Cystine, 25, 128, 130
Cytidine, 264, 271, 277
Cytidine 5′-diphosphate, 278
Cytidine 2′-monophosphate, 271, 278
Cytidine 2′:3′-monophosphate, 278
Cytidine 3′-monophosphate, 278
Cytidine 5′-monophosphate, 277, 278
Cytidine 5′-phosphate, 262
Cytidine triphosphate, 277, 486
Cytidine 5′-triphosphate, 278
Cytochrome, 145
Cytosine, 266, 271–272, 275, 277, 279, 464
Cytosine 5-carboxylic acid, 262

D

Dactylin, 346
Daidzein, 346
Damarenolic acid, 234
Dansyl derivatives, 139
Datura alkaloids, 461
DDD, 361
DDE, 359
DDT, 105, 355, 359, 361
Deacetyldigilanid A, B, and C, 239
DEAE paper, 124–126
Decanoic acid, 180
Decarbamoylnovobiocin, 419
9-Decenoic acid, 189
n-Decylamine, 116
Deguelin, 390
Dehydrochlortetracycline, 414
7-Dehydrocholesterol, 206, 210, 218
Dehydrocholic acid, 234–235
Dehydrocorydaline, 464
Dehydrodeguelin, 390
Dehydroepiandrosterone, 204, 211, 219, 226
6-Dehydroestrone, 208
16-Dehydrosteroids, 224
Deionization, 154
Delnav, 368
Delphinidin, 343
Delphinidin glycosides, 344
Demecolcine, 466
Demethylchlortetracycline, 414
Demethyltetracycline, 414
Densitometry, 95–99
Deoxyadenosine, 265, 271, 273
Deoxyadenosine 5′-diphosphate, 278
Deoxyadenosine 5′-monophosphate, 271, 278
Deoxyadenosine 5′-triphosphate, 278
Deoxycholic acid, 234–236
Deoxycorticosterone, 205, 225
11-Deoxycorticosterone, 225
11-Deoxycorticosterone acetate, 225
11-Deoxycortisone, 225–226
11-Deoxycortisone acetate, 225
Deoxycytidine, 266, 271
Deoxycytidine 5′-phosphate, 262
Deoxycytidine 5′-monophosphate, 271, 278
Deoxycytidylic acid, 273
Deoxyglucose, 159
Deoxyglucose monophosphate, 271
2-Deoxyglucose 6-phosphate, 487
Deoxyguanosine, 264, 271, 273
Deoxyguanosine 5′-diphosphate, 278
Deoxyguanosine 5′-monophosphate, 278
Deoxyguanosine 5′-triphosphate, 278
Deoxyinosine, 265, 271
Deoxymonovinylhydroxyethyldeuteroporphyrin, 336
Deoxynucleosides, 279
Deoxyprotoporphyrin, 336
Deoxyribonucleic acids, 259
Deoxyribonucleotides, 270
Deoxyribo-5′-nucleotides, 275
Deoxyribosides, 274
2-Deoxy sugars, 160

6-Deoxy-L-talose, 166
Deoxyuridine, 266, 271, 273
Deoxyuridine 5′-phosphate, 273
Desalting, 16, 46, 121–122, 127, 153
antibiotics, 400–401
electrolytic, 32
Descending chromatography, 2, 53–57, 82, 93, 217, 254, 256
Deuteroporphyrin, 336
Deuteroporphyrin IX, 337
De Vilbiss atomizers, 75
1,2-Diacetyloleandomycin, 416
1,3-Diacetyloleandomycin, 416
2,3-Diacetyloleandomycin, 416
Diacetyl peroxide, 456
Diagonal chromatography, 81
Dialkyl hydrazones, 167
Diamines, 113
Diaminobenzene derivatives, 120
2,5-Diamino-4,6-deoxypyrimidine, 262
4,5-Diamino-2,6-dihydroxypyrimidine, 272
2,6-Diamino-5-formylamino-4-hydroxypyrimidine, 262
2,6-Diamino-5-formylamino-4-methoxypyrimidine, 269
4,6-Diamino-5-formylamino-2-methylpyrimidine, 264
3,4-Diamino-6-hydroxypyrimidine, 272
4,5-Diamino-2-hydroxypyrimidine, 272
2,6-Diamino-7-methylpurine, 263
4,5-Diamino-6-methylpyrimidine, 268
1,2-Diamino-4-nitrobenzene, 113, 174
1,3-Diamino-4-nitrobenzene, 113
2,4-Diamino-6-oxypyrimidine, 265
2,6-Diaminopurine hemisulfate, 264
4,6-Diamino-2-oxypyrimidine, 265
4,6-Diaminopyrimidine, 267
4,6-Diamino-2-thiopyrimidine, 268
Diammonium P^1,P^2-di-*n*-butyl pyrophosphate, 491
Diammonium P^1,P^2-diethyl pyrophosphate, 471
o-Dianisidine, 159
Diatoxanthin, 333
Diazinon, 368
Diazomethane, 173
1:2,5:6-Dibenzathracene, 439
1:2;3:4-Dibenzopyrene, 439
1:2;4:5-Dibenzopyrene, 439
1:2;6:7-Dibenzopyrene, 439
3:4;9:10-Dibenzopyrene, 439
Dibrom, 368
Dibromoacetic acid, 178
2,6-Dibromo-*N*-chloro-*p*-quinoneimine (DCQ), 370
2,3-Dibromopropionic acid, 178
Di-*n*-butylamine, 116
2,5-Di-*t*-butylhydroquinone, 392
Di-*n*-butyl peroxide, 455
2,4-Dichloroaniline, 112
3,3′-Dichlorobenzidine, 112
4,6-Dichloro-*m*-cresol, 309
4,6-Dichloro-*o*-cresol, 309
3,4-Dichlorocatechol, 307
3,5-Dichlorocatechol, 307
4-5-Dichlorocatechol, 307
Dichlorodifluoromethane, 75
5,5′-Dichloro-3,3′-dimethoxybenzidine, 112
2,4-Dichloro-3,5-dimethylphenol, 309
Dichloroethane, 50
2,4-Dichloro-5-ethyl-3-methylphenol, 309
2,3-Dichlorophenol, 307
2,4-Dichlorophenol, 306–307
2,5-Dichlorophenol, 307
2,6-Dichlorophenol, 307
3,4-Dichlorophenol, 307
3,5-Dichlorophenol, 307
2,6-Dichlorophenolindophenol, 398
2,4-Dichlorophenoxyacetic acid esters, 381
2,3-Dichlorophenylmercapturic acid, 307
L-2,4-Dichlorophenylmercapturic acid, 307
L-2,6-Dichlorophenylmercapturic acid, 307
3,4-Dichlorophenylmercapturic acid, 307
L-3,5-Dichlorophenylmercapturic acid, 307
Di(*p*-Chlorophenyl)pyrophosphate, 492
2,3-Dichloroquinol, 307
2,5-Dichloroquinol, 307
2,6-Dichloroquinol, 307
2,4-Dichloroquinonechlorimide, 425
2,6-Dichloroquinonechlorimide, 312
2,4-Dichlorothiophenol, 307
2,6-Dichlorothiophenol, 307
3,5-Dichlorothiophenol, 307
Dichromate, 529

Dicyclohexyl peroxide, 455
Dieldrin, 359
Diethanolamine, 114, 116
2′,5′-Di-*O*-(1-ethoxyethyl)-5-bromouridine-3′-phosphate, 280
2′,5′-Di-*O*-(1-ethoxyethyl)-5-iodouridine-3′-phosphate, 280
Diethylamine, 116
Diethylaminoethylcellulose paper, 44–45
Diethylene glycol, 443, 446
Diethylene glycol–acetic acid–Tetralin, 187
Diethylenetriamine, 114
Diethyl ether, 50, 120, 122, 154
N,N-Diethylethylenediamine, 114
Diethylstilbestrol, 208
N,N′-Diethylthiourea, 476
Diethyl-3,5,6-trichloro-2-pyridyl phosphate, 367
O,O-Diethyl-*O*-3,5,6-trichloro-2-pyridylphosphorothioate (Dursban), 367
N-Diethyltryptamine, 289
Differential charring, 77
Differential migration, 9–10
Diffuse zones, 21
Diffusion, 21
Digilanid A, 239
Digilanid B, 239
Digilanid C, 239
Diginin, 241
Digitalinum verum, 239, 242
Digitoxigenin, 239, 241–242
Digitoxigenin acetate, 241–242
Digitoxigenin glycoside, 242
Digitoxin, 239, 241–242
Digoxigenin, 241
Digoxigenin diacetate, 241
Digoxigenone, 241
Digozin, 239
Di-*n*-hexyl peroxide, 455
Dihydrodeguelin, 390
Dihydroeugenol, 306
Dihydrostreptomycin, 409–410
Dihydrovitamin D_2, 420
Dihydroxyacetone phosphate, 486
2,4-Dihydroxyacetophenone azine, 451
2,5-Dihydroxyacetophenone azine, 451
2,6-Dihydroxyacetophenone azine, 451
3α,17α-Dihydroxy-21acetoxy-5β-pregnane-11,20-dione, 205
3α,11β-Dihydroxy-5α-androstan-17-one, 219
3β,11β-Dihydroxy-5α-androstan-17-one, 219
3β,11β-Dihydroxy-5β-androstan-17-one, 219
3,5-Dihydroxybenzoic acid, 310
2,4-Dihydroxybenzophenone azine, 452
2,2′-Dihydroxydiphenyl, 309
Di-2-hydroxyethyl peroxide, 456
Dihydroxyindole, 295
3,12-Dihydroxy-7-ketocholanic acid, 235
11β,17α-Dihydroxy-3-keto-Δ^4-etienic acid, 234
Dihydroxymethyl peroxide, 456
6,8-Dihydroxy-2-methylpurine, 263
3,4-Dihydroxyphenylacetic acid, 310, 315
p-(2,4-Dihydroxyphenyl)azophenol, 305
p-(2,4-Dihydroxyphenyl)azophenyl sulfuric acid, 305
3-(3,4-Dihydroxyphenyl)propionic acid, 310
17α,21-Dihydroxy-5α-pregnane-3,11,20-trione, 205
17α,21-Dihydroxy-5β-pregnane-3,11,20-trione, 205
3β,21-Dihydroxy-5α-pregnan-20-one, 205
2,6-Dihydroxypurine (xanthine), 272
2,8-Dihydroxypurine, 272
6,8-Dihydroxypurine, 263, 268
2,4-Dihydroxypyrimidine-6-methylsulfone, 267
Diiodotyrosine, 132, 142
3,5-Diido-L-tyrosylglycine, 142
Diisopropanolamine, 114
Diisopropyl ether, 155
3,12-Diketocholanic acid, 235
2,3-Dimethoxybenzalazine, 451
3,3′-Dimethoxybenzidine, 112
2,4-Dimethoxy-5-methylpyrimidine, 269
Dimethoxyphenylpenicillin, 407
5,6-Dimethoxytryptamine, 287
N,N-Dimethyl, 116
2,4-Dimethylacetophenone, 449
2,5-Dimethylacetophenone, 449
3,4-Dimethylacetophenone, 449
2,4-Dimethylacetophenone azine, 451
2,5-Dimethylacetophenone azine, 451
3,4-Dimethylacetophenone azine, 451
Dimethylamine, 116

4-Dimethylaminobenzalazine, 451
p-Dimethylaminobenzaldehyde, 111, 116, 118, 193, 290, 292, 311–312
p-Dimethylaminocinnamaldehyde (MCA), 290–292
2-Dimethylamino-6-hydroxypurine, 267, 274
2-Dimethylamino-6-hydroxypurine riboside, 274
5-Dimethylamino-1-napthalene sulfonyl chloride, 139
6-Dimethylaminopurine, 268
6-Dimethylaminopurine ribonucleoside, 267
2,3-Dimethylaniline, 112
2,4-Dimethylaniline, 112
2,5-Dimethylaniline, 112
2,6-Dimethylaniline, 112
N,N-Dimethylaniline, 112
2,3-Dimethylarabinose, 165
2,4-Dimethylarabinose, 165
2,5-Dimethylarabinose, 165
DL-β,β-Dimethylaspartic acid, 129
Dimethylformamide, 12, 52, 116
2,3-Dimethylgalactose, 165
2,4-Dimethylgalactose, 165
2,5-Dimethylgalactose, 165
3,4-Dimethylgalactose, 165
Dimethylglyoxime, 80
2,5-Dimethylhexane-2,5-dihydroperoxide, 436
1,7-Dimethylhypoxanthine, 268
1,7-Dimethylguanine, 263
Dimethylmannose, 165
2,4-Dimethylol-*o*-cresol, 304
2,6-Dimethylol-*p*-cresol, 304
2,4-Dimethylolphenol, 304
2,6-Dimethylolphenol, 304
N,N-Dimethyl oxide, 116
2,3-Dimethylphenol, 309
2,4-Dimethylphenol, 306
2,5-Dimethylphenol, 306, 309
2,6-Dimethylphenol, 306
3,4-Dimethylphenol, 306, 309
3,5-Dimethylphenol, 306, 309
Dimethyl-*p*-phenylenediamine, 206
4,6-Dimethylpyrimidine, 269
Dimethylpyruvic acid, 186
3,4-Dimethylrhamnose, 165
Dimethylsulfoxide, 133, 155, 164
α,α,-Dimethyltryptamine, 287
N-Dimethyltryptamine, 289
N,N-Dimethyltryptamine, 287
1,3-Dimethyluracil, 269, 273
5,6-Dimethyluracil, 268
1,3-Dimethyluric acid, 266
1,17-Dimethyluric acid, 266
3,7-Dimethyluric acid, 264
1,3-Dimethylxanthine, 268
1,7-Dimethylxanthine, 268
3,7-Dimethylxanthine, 267
3,8-Dimethylxanthine, 267
2,3-Dimethylxylose, 165
2,4-Dinitro-5-aminophenyl, 139
3,5-Dinitrobenzoate, 442–444
3,5-Dinitrobenzoate esters, 443
3,5-Dinitrobenzoyl chloride, 314
3,5-Dinitrobenzoyl derivatives, 120
2,4-Dinitro-5-fluoroaniline, 139
2,4-Dinitrofluorobenzene (DNFB), 122
2,4-Dinitrophenyl, 110–111
2,4-Dinitrophenylamino acids, 123
Dinitrophenylation, 122
Dinitrophenyl (DNP) derivatives, 122, 173
2,4-Dinitrophenylhydrazines, 173–174, 213–214
2,4-Dinitrophenylhydrazones, 18, 25, 175–176, 447, 452–454
Dinoxanthin, 333
Dinucleoside phosphates, 280
Dinucleotide, 423
Diolefins, 438
Diols, 447
Diosmetin, 345
Di-*n*-pentyl peroxide, 455
Dipeptides
  diastereoisomeric, 143
  stereoisomeric, 143
Diphenates, 442
Diphenylamine, 119, 160
Diphenylcarbazone, 193, 518
N,N'-Diphenyl-*p*-phenylenediamine, 392–393
Diphenyl pyrophosphate, 492
Diphosphate, 531
1,3-Diphosphoglyceric acid, 487
2,3-Diphosphoglyceric acid, 486
Diphosphopyridine nucleotide, 486

Dipicrylamine, 518
Dipole-dipole forces, 11
Dipping, application of reagents by, 74, 91, 107
Di-*n*-propylamine, 116
P^1,P^2-Di-*n*-propyl pyrophosphate, 491
Dipropylenetriamine, 114
Dipterex, 368, 370
Di-α-pyridyl phosphate, 492
Disaccharides, 155, 157
1,4-Disaccharides, 152
Dissociation, trailing zones and, 21
Di-Syston, 368
Dithioammelide, 478
Dithiocarbamate fungicides, 386
Dithiocarbamates, 478-479
2,2′-Dithiobis(5-nitropyridine), 475
Dithiouracil, 269, 273
Ditolyl pyrophosphate, 492
DNA, *see* Deoxyribonucleic acid
DNFB, 139
DNP-amino acid, 127
2-Dodecenoic acid, 189
9-Dodecenoic acid, 189
n-Dodecylamine, 116
Donnan permeable membrane theory, 12
DOPA, 115, 117
Dopamine, 115, 117
Double tailing, 22, 24, 35
Dowex-50 resins, 121, 122
Dragendorff reagent, 461, 467–468
Driving force, 9–10
Dry-start chromatography, 44
Dulcitol, 443
Dursban, 367
Dyes, 41–42, 145–146, *see also* specific types
  centrifugal chromatography of, 71
  cosmetic, 471
  dichloroquinoxaline reactive, 473
  direct, 470
  elatography of, 73
  fast base, 472
  food, 470
  group identification of, 473
  membrane chromatography of, 473
  metal, 473
  oil-soluble, 473
  pharmaceutical, 470–471
  photosensitizing, 471–472
  solvent front marking with, 54
  spot testing of, 1–2
  synthetic, 469–473
  systematic analysis of, 81
  two-dimensional chromatography of, 72
  vinylsulfone reactive, 472

E

Eburicoic acid, 234
Ecteola cellulose paper, 44–45
Eddy diffusion, 13
Edeine, 418
EDTA, 37, 117
Ehrlich's reagent, 290, 292
Elaidic acid, 188, 195
Elatography, 72–73
Electrolysis, 121
π-Electron acceptors, 74
Electrophoresis, 73, 120–121, 124, 140–142, 146, 248, 254
β-Eleostearic acid, 189
Elliptone, 390
Elution
  accuracy of, 98
  of radioactive substances, 99–101
  simultaneous, 93
  in spot estimation, 93–95
Emetine, 466
Emicymarin, 239–240, 242
Emicymarin acetate, 240
Enanthic acid, 188, 195
Endrin, 359
Enzymes, separation of on ion-exchange papers, 47
Ephedrine, 466
Epiandrosterone, 204, 219, 226
Epianhydrodemethylchlortetracycline, 414
Epianhydrotetracycline, 414
Epichloretetracycline, 414
Epidemethylchlortetracycline, 414
Epidemethyltetracycline, 414
Epilupinine, 464
Epinephrine, 111, 117
Epioxytetracycline, 414
Epitetracycline, 414
EPN, 368
Epoxy resins, 494

Equilibrium, 12–14
  conditions of, 16, 21
  distribution of, 18
  resistance to, 14
Equilenin, 208
Equilin, 208
Ergosterol, 206, 210, 218
Ergot alkaloids, 132, 462–463
Eriodictyl, 346
Erucic acid, 188, 195
Erythromycin, 404, 415–416
Erythrose 4-phosphate, 486
Escherichia coli, 399
Eserine, 466
Estradiol, 225
Estradiol-17α, 208
Estradiol-17β, 208
Estriol-3,16α,17β, 208
Estriol-3,6β,17β, 208
Estrogens, 225
  detection of, 208–209
  R_F values for, 219, 225
Estrone, 208, 222, 226
Ethanol, 32, 113, 115, 118, 122, 132–133, 135–136, 139, 143, 153–154, 159, 161, 163–164, 173
17α-Ethinylestradiol, 208
Ethion, 368
Ethoxyacetic acid, 179
2-Ethoxylaniline, 112
2′,3′-*O*-Ethoxymethylene-5-bromouridine, 280
2′,3′-*O*-Ethoxymethylene-5-iodouridine, 280
2′,3′-*O*-Ethyxymethylene-3-methyl-5-bromouridine, 280
2′,3′-*O*-Ethoxymethylene-3-methyl-5-iodouridine, 280
2′,3′-*O*-Ethoxymethyleneuridine, 280
5-Ethoxymethyl-6-methyluracil, 269
5-Ethoxymethyluracil, 269
5-Ethoxymethyluridine, 267
3-Ethoxypropionic acid, 179
Ethyl acetate, 32, 52, 54, 123, 160, 174
Ethylamine, 114, 116
2-Ethylamino-6-hydroxypurine, 274
Ethyl benzoate, 48
S-Ethylcysteine, 129
Ethylene diamine, 114
Ethylenediaminetetraacetic acid, *see* EDTA
Ethylene glycol, 443, 446
p-Ethylguaiacol, 306
2-Ethyl-*n*-hexanoic acid, 179
Ethyl hydroperoxide, 456
5-Ethyl-3-methylphenol, 309
17α-Ethyl-19-nortestosterone, 204
m-Ethylphenol, 306
o-Ethylphenol, 306
p-Ethylphenol, 306
Ethyl *n*-propyl hydrogen phosphate, 491
N-Ethylthiourea, 476
Ethyl-3,5,6-trichloro-2-pyridyl phosphate, 367
O-Ethyl-*O*-3,5,6-trichloro-2-pyridyl phosphate, 367
α-Ethyltryptamine, 287
N-Ethyltryptamine, 289
17α-Ethynyltestosterone, 221
Etiocholane-3α,17β-diol, 213
Etiocholan-3α-ol-11,17-dione, 225
Etiocholanolone, 226
Etiocholan-3α-ol-17-one, 225
Δ⁹-Etiocholanolone, 211
Eugenol, 306

F

Fast Red Salt GG, 312
Fatty acids, 62
  chloresteryl esters of, 190–191
  2,4-dinitrophenylhydrazides of, 179,
  higher, 175–176, 185–191, 193
  hydroxamate derivatives of, 178–179
  lower, 177–178, 192
  radiochromatography of, 104
  separation on ion-exchange papers, 47
Feed additives, 391–395
Ferbam, 386
Ferric ammonium sulfate, 161
Ferric chloride, 191–192, 370
Ferric dimethyldithiocarbamate, 386
Ferricyanide, 161
Ferrocyanide, 130, 529
Ferrous sulfate, 1
Ferrous thiocyanate, 455
Ferulic acid, 115, 167, 310
2-Feruloylglucose, 167
3-Feruloylglucose, 167

4-Feruloylglucose, 167
6-Feruloylglucose, 167
1-Feruloyl-β-D-glucose, 167
Filter paper, 2, 32, 37, 41
Fingerprinting, 140–142
Fisetin, 345
Flavinadenine, 423
Flavinadenine dinucleotide, 423
Flavin mononucleotide, 423
Flavins, 423
  R_F values of, 423
Flavone aglycones, 345, 346
Flavone glycosides, 346
Flavonoid compounds, 110, 341–350
  chromatography of, 341–350
  colors and color reactions of, 349
  detection of, 347–348
  experiments with, 348–350
  paper for, 342
  R_F values for, 345–347
  sample preparation of, 341–342
  separation of on ion-exchange papers, 47
  solvents for, 342
  techniques for, 342–347
Flavonol aglycones, 345
Flavonol glycosides, 345
Flowering plants, separation of pigments in, 329–331
Flow rate, peak broadening and, 13–14
5-Fluoracil, 267
Fluoranthene, 439
1-Fluoranthraquinone, 450
Fluorescein, 77, 471
Fluorescence, in amines, 118
Fluorescent spot measurement, 96
Fluoride, 529
2-Fluoroanthraquinone, 450
1-Fluoro-2-chloranthraquinone, 450
5-Fluorodeoxyuridine, 267
Fluorometers, 96
Fluorometric scanner, 96
Fluorometry, 98, 120, 138–139
Folic acid, 422, 424
Folin phenol reagent, 111
Food additives, 391–395
Food dyes, 469
Formaldehyde, 293
Formamide, 111, 116
Formate, 501
Formazan, 161
Formic acid, 121, 178, 180
Formonetin, 346
5-Formylamino-3-methyl-6-methylaminouracil, 266
5-Formylcytosine, 265
5-Formyldeoxyuridine, 265
5-Formyluracil, 265
5-Formyluridine, 263
Foromacidin A, 416
Foromacidin B, 416
Foromacidin C, 416
Frontal analysis, 12
Frontal separation, 23–24
Fructose, 156, 159, 487
D-Fructose, 160, 168
Fructose 1,6-diphosphate, 487
Fructose oligosaccharides, 159
Fructose 1-phosphate, 437
Fructose 6-phosphate, 487
Fructuronic acid, 166
Fruit juices, sugar in, 154
L-Fucose, 166
Fucoxanthin, 333
Fumaric acid, 184
Fungicides, 386–388, *see also* specific kinds
  chromatography of, 386–388
  detection of, 387–388
  paper for, 386
  R_F values for, 386
  solvents for, 387
Furfural, 449
6-Furfurylaminopurine, 269

G

Galactitol, 155
Galactonolactone, 166
Galactosamine, 155, 169
Galactose, 155–156, 155
D-Galactose, 160, 168
Galacturonic acid, 166
Gallic acid, 303
Galuteolin, 346
Gas chromatography, 1, 355
GDP, *see* Guanosine diphosphate
Geiger counter, 101
Geiger-Müller counter, 101, 103, 105
Genista alkaloids, 464

Genisten, 346
Gentamicin complex components, 411–413
Gentistic acid, 303, 310
Gibberellic acid, 104
Girard's hydrazones *T*, 234
Gitaloxin, 242
Gitoxigenin, 239, 241–242
Gitoxigenin diacetate, 241
Gitoxigenin glycoside, 242
Gitoxin, 239–240, 242
Glass fiber paper, 3, 48–49, 111, 155, 236, 287
Glass fibers, borosilicated, 48–49
Glass paper, 123, 234–236
Glaucamine, 465
D-Glucital, 155
Glucogitaloxin, 242
Gluconolactone, 166
Glucosamine, 155, 169
Glucose, 155–156, 164, 167, 487
D-Glucose, 160, 168
Glucose disulfates, 167
Glucose oxidase reagent, 159
Glucose 1-phosphate, 487
Glucose 6-phosphate, 487
Glucose 3-sulfate, 167
Glucose 6-sulfate, 167
Glucosides, 164, 347
  R_F values for, 166
Glucuronic acid, 155, 166
Glucuronolactone, 166
Glutamic acids, 26, 61–62, 128, 130
Glutamine, 126, 128, 130
Glutaric acid, 181, 184
Glutathione, 129
Glyceraldehyde-3-phosphate, 486
Glycerate acid, 184
Glyceric acid, 184
Glycerol, 443, 446–447
Glycerol 1-phosphate, 487
Glycerol 2-phosphate, 487
Glycine, 51, 118, 126, 128, 130, 135, 139
Glycochenodeoxycholic acid, 236
Glycocholic acid, 236
Glycodeoxycholic acid, 236
Glycolate acid, 184
Glycolic acid, 184
Glycolithocholic acid, 236
Glycols, 446–447
Glycoproteins, 161
α-Glycosidase, 154
Glycyl-L-3,5-diiodotyrosine, 142
Glycyl-3,5-diiodo-L-tyrosylglycine, 142
Glycyl-L-thyroxine, 142
Glycyl-L-thyroxylglycine, 142
Glycyl-L-tyrosine, 142
Glycyl-L-tyrosylglycine, 142
Glyoxal, 161
Glyoxal-bis(2-hydroxyanil), 517
Glyoxylic acid, 186
GMP, *see* Guanosine monophosphate
Gold chloride, 293
Gossypetin, 345
Gossypin, 346
Gossypitrin, 346
Gradient development, 73
Gramine, 116, 287, 295
Grote reagent, 387
GTP, *see* Guanosine triphosphate
Guaicol, 306
Guanidine, 113
  derivatives of, 120
Guanine, 263, 271–272, 274, 276–277, 279
Guanosine, 263, 271, 274, 276–277
Guanosine diphosphate (GDP), 277, 486
Guanosine 5′-diphosphate, 278
Guanosine monophosphate, 271, 276
Guanosine 2′-monophosphate, 278
Guanosine 3′-monophosphate, 278
Guanosine 5′-monophosphate, 277, 278
Guanosine 3′-phosphate, 486
Guanosine 5′-phosphate, 262, 486
Guanosine triphosphate, 277, 486
Guanosine 5′-triphosphate, 278
Gulonolactone, 166
Guthion, 368, 370

H

Hager's solvent, 329
Hair dyes, phenols in, 298–299
Hamycin, 416
Hematoporphyrin IX, 337
Hemotoporphyrin IX dimethyl ester, 337
Hemin, 336
Hemoglobin, 47
Heparin, 155
Heparin sulfate, 155

Heptachlor, 359
Heptachlor epoxide, 359
Heptadecanoic acid, 182
Heptaene, 416
n-Heptane 3,3′-dihydroperoxide, 456
Heptanoic acid, 129, 180, 182
n-Heptanoic acid, 178
Heptaphosphate, 531
n-Heptyl hydroperoxide, 3-methylhexyl-3-hydroperoxide, 456
n-Heptyl, 2-hydroxyethyl peroxide, 456
D-*n*-Heptyl peroxide, 456
Herbicides, 379–385
  detection of, 380
  organochlorine, 379–383
  paper for, 380
  phenoxy acid, 381
  symmetrical triazines, 383
Heroine, 466
Hesperitin, 346
HETP (height equivalent to theoretical plate), 8, 13, 21
Hexacarboxylic porphyrin, 336
2-Hexadecenoic acid, 189
Hexamethylenediamine, 114
Hexane, 52, 116
n-Hexanoic acid, 179
n-Hexanol, 126
Hexanophenone, 448
Hexaphosphate, 531
Hexasaccharide, 163
Hexosamines, 156, 160
Hexoses, 160
n-Hexylamine, 116
High-frequency measurement, 105
Hippuric acid, 303
Hirsuitidin, 343
Histamine, 26, 123
Histidine, 128, 130, 132, 136
Histones, 146
Homocysteic acid, 129
Homocystine, 129
Homocystine thiolacetone, 129
Homoeriodictyol, 346
Homogentistic acid, 303
Homoserine, 129
Homovanillic acid, 115, 117
Hordenine, 466
Humidification, by preequilibration, 51
Hyaluronic acid, 155
Hydrastinine, 466
(—)-β-Hydrastine, 465
Hydrastinine, 465
Hydrazides, 139
Hydrazine, 115
Hydrocarbons
  aromatic, 439–441
  chromatography of, 438–441
  polycyclic, 440
Hydrochloric acid, 154–155
Hydrogen bonding, 10–11, 21, 50
Hydrogen di(2-ethylhexyl) orthophosphoric acid, 126
Hydrogen peroxide, 274, 314, 456
Hydrogen selenide, 474
Hydrogen sulfide, 516
Hydrolysis, enzymatic, 160
trans-4-Hydroperoxy-5-biomothymine, 274
cis-4-Hydroperoxy-5-hydroxythymine, 274
trans-4-Hydroperoxy-5-hydroxythymine, 274
Hydrophilic solvents, 11, 52
Hydrophobic solvents, 42, 52–53
Hydroquinoline, 155
Hydroquinone, 304
Hydrous oxides, 47
Hydroxamates, preparation of, 173–175
Hydroxamic acids, 234
2-Hydroxyacetophenone, 449
3-Hydroxyacetephenone, 449
4-Hydroxyacetophenone, 449
2-Hydroxyacetophenone azine, 451
4-Hydroxyacetophenone azine, 451
Hydroxy acids, 442
n-Hydroxy acids, 302, *see also* specific acids
o-Hydroxy acids, 302, *see also* specific acids
p-Hydroxy acids, 302, 303, *see also* specific acids
3α-Hydroxy-5β-androstane-11,17-dione, 219
11β-Hydroxy-5α-androstane-3,17-dione, 219
3α-Hydroxy-5α-androstan-17-one, 219
3α-Hydroxy-5β-androstan-17-one, 219
3α-Hydroxy-5β-androstan-17-one, 219
3β-Hydroxy-5β-androstan-17-one, 219
17β-Hydroxy-5β-androstan-3-one, 204

3α-Hydroxy-5β-androst-9-en-17-one, 219
3β-Hydroxy-5-androsten-17-one, 219
11β-Hydroxyandrosterone, 226
Hydroxyanisole, butylated, 392
3-Hydroxyanthranilic acid, 287, 289, 303
5-Hydroxyanthranilic acid, 287
2-Hydroxybenzalazine, 451
m-Hydroxybenzaldehyde, 449
o-Hydroxybenzaldehyde, 306
p-Hydroxybenzaldehyde, 449
m-Hydroxybenzoic acid, 309–310
o-Hydroxybenzoic acid, 309
p-Hydroxybenzoic acid, 309–310
m-Hydroxybenzophenone, 449
o-Hydroxybenzophenone, 449
p-Hydroxybenzophenone, 449
4-Hydroxybenzophenone azine, 452
m-Hydroxybenzyl alcohol, 304
o-Hydroxybenzyl alcohol, 304
p-Hydroxybenzyl alcohol, 304
6-Hydroxycatecholamines, 120
o-Hydroxycinnamic acid, 310
7-Hydroxycoumarin, 305
5-Hydroxydeoxyuridine, 264
3α-Hydroxy-7,12-diketocholanic acid, 235
o-Hydroxydiphenyl, 309
10-Hydroxy-$\Delta^{1,4}$-estradiene-3,17-dione, 208
2-Hydroxyestradiol-17β, 208
1α-Hydroxyestrone, 208
N-(2-Hydroxyethyl)ethylenediamine, 114
2-Hydroxyethyl hydroperoxide, 456
11β-Hydroxyetiocholanolone, 226
cis-4-Hydroxy-5-hydroperoxythymine, 274
trans-4-Hydroxy-5-hydroperoxythymine, 274
5-Hydroxyindoleacetic acid, 287–288, 303
7-Hydroxyindoleacetic acid, 287
3α-Hydroxy-12-ketocholanic acid, 235
17α-Hydroxy-3-keto-Δ^4-etienic acid, 234
3-Hydroxykynurenine, 289
D-3-Hydroxykynurenine, 287
L-3-Hydroxykynurenine, 287
D-5-Hydroxykynurenine, 287
L-5-Hydroxykynurenine, 287
Hydroxylamine, 26
2-Hydroxylignoceric acid, 189
Hydroxylupanine, 464
2-Hydroxy-3-methoxybenzalazine, 451
2-Hydroxy-3-methoxybenzaldehyde, 449
2-Hydroxy-4-methoxybenzophenone azine, 452
2-Hydroxy-4-methoxybenzophenone hydrazone, 452
3-Hydroxy-4-methoxycinnamic acid, 115
3-Hydroxy-4-methoxymandelic acid, 115
6-Hydroxy-2-methylaminopurine, 266, 274
2-Hydroxy-5-methylbenzophenone, 449
2-Hydroxy-5-methylbenzophenone azine, 452
7-Hydroxy-4-methylcoumarin, 305
5-Hydroxymethylcytosine, 265
5-Hydroxymethyldeoxycytidine, 263
5-Hydroxymethyldeoxyuridine, 263–264
8-Hydroxy-7-methylguanine, 263
Hydroxymethyl hydroperoxide, 456
4-Hydroxy-2-methylmercaptopyrimidine, 269
5-Hydroxymethylorotic acid, 263, 266
2-Hydroxy-6-methylpurine, 264
2-Hydroxy-4-methylpyrimidine, 267
4-Hydroxy-6-methylpyrimidine, 268
5-Hydroxymethyluracil, 264–265
2-Hydroxymyristic acid, 189
2-Hydroxypalmitic acid, 189
o-Hydroxyphenylacetic acid, 115
p-Hydroxyphenylacetic acid, 310
3-(*p*-Hydroxyphenyl)propionic acid, 310
o-Hydroxyphenylpyruvic acid, 186
p-Hydroxyphenylpyruvic acid, 186
m-Hydroxyphenylsulfamic acid, 305
N-Hydroxyphthalonimide, 133
21-Hydroxy-4-pregnane-3,11,20-trione, 205
3β-Hydroxy-5α-pregnan-20-one, 204
11α-Hydroxyprogesterone, 205
17α-Hydroxyprogesterone, 226
Hydroxyproline, 123, 128, 134
2-Hydroxypurine, 272
6-Hydroxypurine(hypoxanthine), 272
8-Hydroxypurine, 265, 272
Hydroxypurines, R_F values for, 272
4-Hydroxypyrimidine, 267
Hydroxyquinol, 304
3-Hydroxyquinoline, 305
6-Hydroxyquinoline, 305
7-Hydroxyquinoline, 305
8-Hydroxyquinoline, 305, 513, 517

Hydroxysparteine, 464
2-Hydroxystearic acid, 189
Hydroxystreptomycin, 409–410
Hydroxytoluene, butylated, 392
5-Hydroxytryptamine, 26, 257
6-Hydroxytryptamine, 289
5-Hydroxytryptophan, 287–288
Hyperin, 345
Hyoscamine, 462, 464
Hyposulfite, 162
Hypoxanthane, 276
Hypoxanthine, 265, 271–273, 277, 279, 487

I

Idonolactone, 166
IDP, *see* Inosine diphosphate
Imidan, 376–377
Imidazolepropionic acid, 27
Imidazoles, 272
Imino acids, 77
IMP, *see* Inosine monophosphate
Indican, 287, 295
Indican glucoside, 295
Indigo disulfonate, 295
Indigo tetrasulfonate, 295
Indigotin, 295
Indium, 47
Indole-3-acetamide, 294
Indole-3-acetic acid, 287, 294
Indole-3-acetonitrile, 288, 295
Indole-3-acrylic acid, 287–288
Indole-3-aldehyde, 287–288, 294
Indoleamines, 288
Indole-3-butyric acid, 287, 294
Indole-3-butyronitrile, 287, 295
Indole-3-carboxylic acid, 287–288, 294
Indole-3-lactic acid, 287
Indole-3-propionic acid, 287, 294
β-Indolepyruvic acid, 186
Indoles
  chromatography of, 283–296
  detection reagents for, 290–291
  quantitative methods for, 291–296
  R_F values for, 287
  solvents for, 286–296
Indolylacetonitrile, 287
Indolyl-3-pyruvic acid, 287
Indoxylacetate, 295
Indoxyl sulfate, 287
Infrared spectra, spot detection and, 80
Inks, 14
Inorganic acids and bases, as solvents, 50
Inorganic analysis, 500–533
Inorganic phosphates, 104
Inosine, 263, 271, 276–277, 487
Inosine diphosphate (IDP), 277, 486
Inosine 5′-diphosphate, 278
Inosine monophosphate (IMP), 271, 276, 486
Inosine 5′-monophosphate, 277, 278
Inosine triphosphate (ITP), 277, 486
Inosine 5′-triphosphate, 278
Inositol, 162, 422, 446
Inositol phosphates, 490–492
Insecticides, 42, 391, *see also* specific kinds
  carbamate, 375–379
  chlorinated, 357–363
  hydrolyzed, 47
  organophosphate, 363–375
  syngerists and, 390
Insect repellant, 391
Insulin, 141, 145
Invertase, 154
Iodate, 47, 529
Iodide, 47, 529
Iodine, 105, 115, 143
Iodine vapor, as detection reagent, 74
Iodoacetic acid, 178
1-Iodoanthraquinone, 450
2-Iodoanthraquinone, 450
Iodobutyric acid, 178
I-Iodo-2-chloranthraquinone, 450
o-Iodophenol, 309
p-Iodophenol, 309
p-Iodophenylsulfonyl chloride, ^{131}I-labeled, 103
Iodopropionic acid, 178
5-Iodouridine, 280
5-Iodouridine 2′,3′-cyclic phosphate, 280
5-Iodouridine 2′(3′)-phosphite, 280
5-Iodouridylyl-(3′→5′)-uridine, 280
Ion exchange chromatography, 11–12, 20, 32, 41, 44, 47–48, 81, 120, 124, 126, 139–140, 153–154, 248, 400, 513
Ion exchange papers, 43, 145, 257
Ion-exchange resins, 32, 42, 45–48, 120–121

Ion-exchange retention, 12
Ionic solutions, 32
Ionization, trailing zones and, 21
Ions
  inorganic, 45, 47–48
  organic, 45–47
Iridin, 346
Irigenin, 346
Irigenol, 346
Iron, 37
Isatin, 77, 295
Isoamylamine, 116
Isoanhydrovitamin A, 429
Isobarbituric acid, 264
Isochlortetracycline, 414
Isochroman hydroperoxide, 456
Isochroman peroxide, 456
Isocitric acid, 184
Isocytosine, 273
Isoflavone aglycones, 346
Isoflavone glucosides, 346
Isoleucine, 54, 121, 123, 126, 128, 130, 139
Isonovobiocin, 419
Isopropanol, 52, 122
Isopropanolamine, 114
Isopropanol-*n*-butanol–isoamyl alcohol–formic acid, 130
Isopropanol–formic acid, 138
Isopropyl alcohol, 132
N-Isopropylthiourea, 476
Isoptone, 390
Isoquercitrin, 345
Isoquinoline alkaloids, 467
Isorhamnetin, 345
l-α-Isosparteine, 464
Isotrophanthidin, 239
Isosystox, 370
Isotetracycline, 414
Isotherm linearity, 21
Isotope indicators, 103–104
Isotripiperidine, 35
Isovaleric acid, 180
Itaconic acid, 184
ITP, *see* Inosine triphosphate

J

Jatrorrhizine, 464

K

Kagi-Miescher solution, 204–205
Kampferol, 345
Kanamycins, 409–411
Kapacidin, 416
Kelthane, 359
Kerosene, 52, 337–338
α-Keto acid 2,4-dinitrophenylhydrazones, 183–185
α-Keto acid 4-nitro-2-alkylquinoxalinols, 185
α-Keto acids, 25, 173–174, 176, 183, 193, 196
  R_F values for, 186
α-Keto-ε-aminocaprioic acid, 186
11-Ketoandrosterone, 226
α-Ketobutyric acid, 186
7-Ketocholesterol, 210
7-Ketodehydroepiandrosterone, 226
3-Keto-$\Delta^{8,24(28)}$-ergostadien-21-oic acid, 234
6-Ketoestradiol, 208
7-Ketoestrone, 208
3-Keto-Δ^4-etienic acid, 234
11-Ketoetiocholanolone, 226
α-Ketoglutaric acid, 184, 186
α-Ketoisocaproic acid, 186
α-Ketoisovaleric acid, 186
α-Keto-γ-methiolbutyric acid, 186
D-α-Keto-β-methylvaleric acid, 186
Ketones, 442
  aliphatic, 447–448
  uncombined, 447–448
Ketoses, 158–159, 163
Ketosteroids, 225
  R_F values for, 225
17-Ketosteroids, 219
  relative mobilities of, 219
Kidney, peptides in, 142
Kieselguhr-loaded papers, 42
Kilner jar, 118
Kinetic transport, 13
Korlan (Trolene), 368
Kynuramine, 289
Kynurenic acid, 289, 303
Kynurenine, 289
D-Kynurenine, 287
L-Kynurenine, 287

L

Lactic acid, 184
Lactobacillus leichmannii, 425
Lactones, 166
Lactose, 156, 160, 164
Lanosterol, 206, 210
Lateral spreading, 23
Laurate, 190
Lauric acid, 175–176, 182, 188, 190
Lauryl alcohol, 111
Lead, 47, 104, 154
Lecithin, 236, 493
Leptosidin, 347
Leucine, 54, 123, 126, 128, 130, 139
Levulinic acid, 184
Lignoceric acid, 189
Linarin, 346
Lindane, 359
Linear flow, R_F value and, 14
Line oven technique, 73
Linoleate, 190
Linoleic acid, 175–176, 189–190, 195
Linolenate, 190
Linolenic acid, 175–176, 188, 190
Lipids, 32, 41–42, 121
Lipophilic compounds, 52
Liquid, immiscible, in partition chromatography, 12
Liquid chromatography, 1, 4
Liquid scintillation, 101, 103
Lithocholic acid, 235–236
Liver, peptides in, 142
Lobeline, 466
Lumichrome, 423
Luminflavin, 423
Lumisterol, 206, 218
Lupanine, 464
Lupinine, 464
Lutein, 333
Luteolin, 345
Luteolinidin, 343
Lutidine, 31, 126, 336
Lysine, 128, 130
  α and ε peptides of, 143

M

Macarpine, 465
Magnamycin, 415
Malathion, 368
Malethion, 370
Maleic acid, 184
Maleic hydrazide, 385
Malic acid, 184
Malonic acid, 159, 180, 184
Maltose, 156, 160, 168
Malvidin, 343
Malvidin glycosides, 344
Mandelic acid, 26, 303
Maneb fungicide, 386
Mannitol, 443
D-Mannitol, 155
Mannonolactone, 166
Mannose, 159, 164
D-Mannose, 160, 168
Mannosidodihydrostreptomycin, 410
Mannosidostreptomycin, 409–410
Marein, 347
Maritimein, 347
Maritimetin, 347
Mass spectrometry, 82, 141
Mass transfer, 13–14
Mecambrine, 465
Melatonin, 289
Membrane chromatography, 473
Menadione, 422
Menazon, 368
2-Mercapto-5-anilino-1,3,4-thiadiazole, 518
8-Mercapto-6-methylpurine, 268
6-Mercaptopurine, 273
β-Mercaptopyruvate, 174
β-Mercaptopyruvic acid, 186
Mercaptotriazines, 478
Mercuric acetate, 154
Mercury adducts, 444
Mercury compounds, preparation of, 175
Mesochlorin, 336
Mesoporphyrin, 336
Mesoporphyrin IX, 337
Metal complexes, 531–533
Metal ions
  appearance after Oxine-Kojic spray, 516
  direction of, 104
  R_F values for, 508–511
  separations of, 42, 48
Metanephrine, 117
Metanil yellow, 371

Meteloidine, 464
Methanol, 52, 115–117, 126, 136, 146, 160, 169
Methionine, 25, 123, 128, 130
Methionine sulfone, 129
Methionine sulfoxide, 25, 129
Methoxyacetic acid, 179
2-Methoxyaniline, 112
2-Methoxybenzalazine, 451
o-Methoxybenzaldehyde, 449
p-Methoxybenzyl alcohol, 443
4-Methoxy-*n*-butyric acid, 179
Methoxychlor, 359
2-Methoxy-5-chloroaniline, 112
5-Methoxy-*N,N*-dimethyl oxide, 116
5-Methoxy-*N*-dimethyltryptamine, 289
5-Methoxy-*N,N*-dimethyltryptamine, 116
Methoxydoisynolic acid, 208
2-Methoxyestriol, 208
2-Methoxyestrone, 208
Methoxyethylmercuric chloride, 386
3-Methoxy-4-hydroxy acids, 303, *see also* specific acids
1-Methoxy-2(3)-hydroxy-3(2)-bromocyclohexanes, 447
4-Methoxy-3-hydroxymandelic acid, 117
2-Methoxy-5-methylaniline, 112
5-Methoxy-*N*-methyltryptamine, 116
5-Methoxymethyluracil, 268
5-Methoxymethyluridine, 265
6-Methoxypurine, 268
6-Methoxypurine ribonucleoside, 268
5-Methoxytryptamine, 287, 289
6-Methoxytryptamine, 287
4′-Methylacetophenone, 448
1-Methyladenosine, 264, 266
3-Methyladenine, 267
7-Methyladenine, 266
9-Methyladenine, 267
2-Methyladenine sulfate, 267
Methylamines, 114, 116, 118
6-Methylamino-5(*N*-formylmethylamino) uracil, 265
6-Methylaminopurine, 268
6-Methylaminopurine deoxyribonucleoside, 263
6-Methylaminopurine ribonucleoside, 265
5-Methylaminouracil, 264
2-Methylaniline, 113
N-Methylaniline, 112
Methyl-β-L-arabinopyranoside, 116
2-*C*-Methyl-L-arabinose, 155, 166
Methylarsinebisdimethyldithiocarbamate, 386
DL-*threo*-β-Methylaspartic acid, 129
5-Methylbarbituric acid, 268
1-Methyl-1:2-benzanthracene, 441
3-Methyl-1:2-benzanthracene, 441
4-Methyl-1:2-benzanthracene, 441
Methylbenzo-(*c*)-phenanthrene, 441
4-(α-Methylbenzyl)-2-phenylphenyl, 306
3-Methyl-5-bromouridine, 280
Methyl cellosolve, 177
2-Methyl-5-chloraniline, 112
20-Methylcholanthrene, 439
4-Methylcoumarin-7-yl sulfuric acid, 305
N-Methylcysteine, 129
S-Methylcysteine, 129
S-Methylcysteine *S*-oxide, 129
S-Methylcysteine sulfoxide, 129
5-Methylcytidine, 264
Methylcytidine, 464
1-Methylcytosine, 267
5-Methylcytosine, 267
5-Methyldeoxycytidine, 262, 267
5-Methyldeoxycytidine 5′-phosphate, 262
N-Methyldiethanolamine, 114
α-Methyl-DOPA, 115
3,4-Methylenedeoxyphenol synergists, 390
Methyl ethyl ketone, 126
Methyl-α-L-fucopyranoside, 166
5-Methyl-2-furfural, 449
2-Methylgalactose, 168
S-Methylglutathione, 129
p-Methylguaiacol, 306
1-Methylguanine, 264, 274
3-Methylguanine, 264
7-Methylguanine, 264
1-Methylguanine sulfate, 263
1-Methylguanosine, 263, 274
7-Methylguanosine, 263
Methyl hydroperoxide, 456
5-Methyl-5-hydroxybarbituric acid, 265
1-Methylhypoxanthine, 267
2-Methylhypoxanthine, 265
3-Methylhypoxanthine, 265
7-Methylhypoxanthine, 266
8-Methylhypoxanthine, 266
9-Methylhypoxanthine, 266
2-Methylindole, 294

3-Methylindole, 294
1-Methylinosine, 264
7-Methylinosine, 263
3-Methyl-5-iodouridine, 280
4-Methylmannose, 165
6-Methylmercaptopurine, 269
1-Methyl-6-methylamino-5-(*N*-formylmethylamino)uracil, 268
3-Methyl-6-methylamino-5-(*N*-formylmethylamino)uracil, 268
3-Methyl-6-methylaminouracil, 269
Methyl-2-*C*-methyl-β-L-arabinopyranoside, 166
Methyl-2-*C*-methyl-β-L-ribopyranoside, 166
Methyl orange, 371
5-Methylorotic acid, 267
Methylparathion, 368
Methylpentoses, 158
Methylphenyl phosphate, 492
Methylphenyl polysiloxanes, 484
6-Methylpurine, 267
Methyl red, 192, 314
Methyl-α-L-rhamnopyranoside, 166
3-Methylrhamnone, 165
Methyl-β-D-ribopyranoside, 166
2-*C*-Methyl-L-ribose, 155, 166
Methyl salicylate, 48
N-Methylserotonin, 288
Methyltestosterone, 226
17-Methyltestosterone, 225
6-Methylthiouracil, 273
6-Methyl-2-thiouracil, 269
N-Methylthiourea, 476
1-Methylthymine, 269
2-Methyl-4,5,6-triaminopyrimidine sulfate, 266
1-Methyltryptamine, 287
2-Methyltryptamine, 287
α-Methyltryptamine, 287
N-Methyltryptamine, 288
N-Methyltryptophan, 287
1-Methyluracil, 268
3-Methyluracil, 269
6-Methyluracil, 268, 273
1-Methyluric acid, 263
3-Methyluric acid, 263
7-Methyluric acid, 263
3-Methyluridine, 268
5-Methyluridine, 265
Methyl violet, 471
1-Methylxanthine, 267
3-Methylxanthine, 265
7-Methylxanthine, 264
2-Methylxylose, 165
Methymycin, 415–416
Migration, 126
  anomalous, 21
Millon's reagent, 313
Mitox, 359
Mobile phase, 52–53
  peak broadening in, 13–14
  R_M values and, 18–20
Mobile phase velocity, 13, 18
"Moist paper" techniques, 7
Molecular diffusion
  peak broadening and, 13
  of solute zone, 13
Molecular migration, 8
Molecular structure
  chromatographic behavior and, 7
  migration behavior and, 17
Molybdenum, X-ray fluorescence detection of, 78
Molybdophosphoric acid, 216
1-Monoacetyloleandomycin, 416
2-Monoacetyloleandomycin, 416
3-Monoacetyloleandomycin, 416
Monoalkyhydrazones, 167
Mono(*p*-chlorophenyl) phosphate, 492
Monoethanolamine, 114
Monoethyl phosphate, 492
Monoisopropanolamine, 114
Monophenyl phosphate, 492
Monophenyl pyrophosphate, 492
Monophosphate, 531
Monosaccharides, 54, 154, 157, 162–163, 442
Monotolyl phosphate, 492
Monovinylhydroxyethyldeuteroporphyrin, 336
Monovinylmonohydroxyethyldeuteroporphyrin IX, 337
Morgan-Elson reagent, 169
Morin, 345
Morphine, 465
Mouron, 384–385
Mucopolysaccharides, 47, 155
  acidic, 164

Multiple development, 23, 57, 61, 67, 132, 216
Multiple spotting, 21, 24–27, 35, 37
Multiple zoning, 24–27, 117, 169
Muramine, 465
Musts, amino acids in, 122
Myricetin, 345
Myricitrin, 346
Myristate, 190
Myristic acids, 175–176, 182, 188, 190

N

Nabam fungicide, 386
2-Naphthaldehyde, 449
Naphthalenes, 483
α-Naphthol, 141, 306, 309
β-Naphthol, 64, 306, 309, 312
Naphthols, 20, 64, 306, 309, 312
1,2-Naphthoquinones, 183
1-Naphthylamine, 113, 482
α-Naphthylamine, 119
2-Naphthylamine, 113, 120, 482
β-Naphthylamine, 159
1-Naphthylamine-2,4-disulfonic acid, 482
1-Naphthylamine-4,7-disulfonic acid, 482
1-Naphthylamine-7,*N*-disulfonic acid, 482
1-Naphthylamine-4,*N*-disulfonic acid, 482
2-Naphthylamine-1,6-disulfonic acid, 482
2-Naphthylamine-1,*N*-disulfonic acid, 482
2-Naphthylamine-6,*N*-disulfonic acid, 482
2-Naphthylamine-8,*N*-disulfonic acid, 482
1-Naphthylamine-2-sulfonic acid, 482
1-Naphthylamine-4-sulfonic acid, 482
1-Naphthylamine-*N*-sulfonic acid, 482
2-Naphthylamine-*N*-sulfonic acid, 482
β-Naphthylamine–thymol–ethanol-H_3PO_4, 159
1-Naphthylamine-2,4,7-trisulfonic acid, 482
N-(1-Naphthyl)ethylenediamine, 312
1-Naphthylisothiocyanate derivatives, 139
β-Naphthylphosphate, 492
β-Naphthyl-α-pyridyl phosphate, 492
1-(1-Naphthyl)2-thiourea (ANTU), 391
Narbomycin, 416
Narcolides, 415–416
Narcotin, 465
Naringenin, 346
Naringin, 346
Neamine, 409–410
Nemacide (VC-13), 368
Neodiadinoxanthin, 333
Neodinoxanthin, 333
Neofucoxanthin, 333
Neohesperidin, 346
Neomycin A, 410
Neomycin B, 409–410
Neomycin C, 409–410
Neomycins, *N*-acetyl, 399
Neoperidinin, 333
Neoxanthin, 333
Nervifolin, 239
Nessler reagent, 193
Neutron activation analysis, 104–105
Niacin, 419, 422
Niacinamide, 419
Nickel, 47, 80, 519--520, 532–533
Nicotiflorin, 345
Nicotinamide, 422
Ninhydrin, 46, 91, 113, 126, 130, 132–136, 138, 141, 143, 169
p-Nitraniline, 312
Nitrate, 529
Nitric acid, 154
Nitrilotriacetic acid, 117
2-Nitroaniline, 112
3-Nitroaniline, 112
4-Nitroaniline, 112
p-Nitroaniline, 193, 293
  diazotized, 117
Nitrobenzenediazonium fluoborate, 312
p-Nitrobenzenediazonium fluoborate, 378–379
Nitrobenzoic acid, 459–460
Nitrocellulose membranes, 145
2-Nitro-4-chloroaniline, 112
4-Nitro-2-chloroaniline, 112
Nitro compounds, 457–460
Nitrofuran derivatives, 457–458
Nitrofurantoin, 400
2-Nitro-4-methoxyaniline, 112
4-Nitro-2-methyoxyaniline, 112
2-Nitro-4-methylaniline, 112
3-Nitro-6-methylaniline, 112
4-Nitro-2-methylaniline, 112
6-Nitro-2-naphthylamine-8-sulfonic acid, 518
m-Nitrophenol, 308–309
o-Nitrophenol, 307, 309

p-Nitrophenol, 308–309
Nitrophenols, 308–309, 458
4-Nitrophenylhydrazone, 448–449
Nitrophthalates, 442
Nitroquinoxalinol derivatives, 174
1-Nitroso-2-naphthol (N2N), 293
Nitrosophenols, 458
Nitrosotocopherols, 428–429
5-Nitrouracil, 273
Noactine, 416
Nonadecanoic acid, 182
Nonanoic acid, 129, 180, 182
n-Nonanoic acid, 178
Nonequilibrium, 13
Noradrenaline, 26, 115
19-Nor-5α-androstane-3α,17β-diol, 213
19-Nor-5α-androstane-3β,17β-diol, 213
19-Nor-5β-androstane-3α,17β-diol, 213
19-Nor-5β-androstane-3β,17β-diol, 213
19-Norandrost-4-ene-17α-ethyl-17β-ol-3β-propionate, 204
Nordihydroguaiaretic acid, 392
Norepinephrine, 111, 117
Norhyoscyamine, 464
Norleucine, 129
Nortangeretin, 345
Norvaline, 128
Norwogonin, 345
Novobiocin, 419
Nucleic acid, 105, 248–250
Nucleosides
  defined, 248
  R_F values for, 280
Nucleotides
  defined, 248
  R_F values for, 280
  separation of, 3, 45

O

2-Octadecenoic acid, 189
n-Octadecylamine, 116
Octanoic acid, 129, 180
n-Octanoic acid, 178
1-Octanol, 443
Octaphosphate, 531
n-Octylamine, 116
Okanin, 346
Oleandomycin, 415–416
Oleandrigenin, 239
Oleandrin, 239
Oleanolic acid, 234
Oleate, 190
Olefins, 438–439
Oleic acid, 175–176, 188, 190, 195
Oligosaccharides, 154, 157–158, 163–165
  acidic, 158
Opium alkaloids, 463, 468
Orcinol, 159, 304
Organic acids
  radiochromatography of, 104
  R_F values for, 104
  as solvents, 50
  spot measurement of, 93
Organic compounds, chromatography of, 438–494
Organic liquids, volatile, 52
Organic peroxides, 454–457
  R_F values for, 456
Organochlorine herbicides, 379–383
  detection in, 380
  paper for, 380
  quantitative estimation in, 381–383
  sample preparation for, 379
  solvents for, 380
Organophosphate insecticides, 363–375
  chromatography of, 365
  detection of, 365–372
  experiments with, 374–375
  paper for, 364
  quantitative estimation of, 371–372
  sample preparation for, 363–364
  solvents for, 365–372
Organophosphorus insecticides, detection of, 77
Orobol, 346
Orotic acid, 264, 271
Orotidine, 262
Orotidine monophosphate, 262
Oscillometry, 105
Oscine, 464
Oxalacetic acid, 185
Oxalate, 501
Oxalic acid, 180, 184
  salts of, 26
Oxalyl chloride, 176
Oxidation, 41
Oxo compounds, 447–454
17-Oxosparteine, 464

Oxygen, presence of during migration, 24
Oxytetracycline, 414

P

Palmatine, 465
Palmitate, 190
Palmitic acid, 175–176, 182, 188, 190
Panagomic acid, 425
Pancreas, of rat, 145
Pantothenic acid, 419, 422
Pantothenol, 419
Papaveraceae alkaloids, 464–465
Papaverine, 465
Paper, 36–49
  absorbent-powder loaded, 42–43
  acetylated, 111, 440–441
  acid-washed, 37, 501
  adsorbent-loaded, 42–44, 229
  for aliphatic acids, 176–177
  for alkaloids, 462–467
  aluminum hydroxide, 42
  for amines, 111
  for amino acids, 123–126
  aminoethylcellulose, 44–45
  anion-exchange, 155
  for anions, 525
  buffered, 123
  capillary pore structure of, 4
  for carbamate insecticides, 377–378
  for carbohydrates, 155
  carboxymethylcellulose, 44–45
  for cardiac glycosides, 238
  care of, 107
  cation-exchange, 124
  for cations, 501–506
  cellulose phosphate, 43–45
  chemically modified, 41
  chloroplast pigments for, 322–323
  circular, 14–15
  DEAE, 124–126
  diethylaminoethylcellulose, 44–45
  dimethyl sulfoxide-impregnated, 155, 164
  ecetola cellulose, 44–45
  fan-shaped, 118
  ferrocyanide, 130
  for flavonoids, 342
  formamide-impregnated, 111
  for fungicides, 386
  geometry of, 16
  glass, 123, 234–236
  glass fiber, 48–49, 111, 155, 236, 287
  impregnated, 41–42, 111, 155, 164, 512–513
  ion-exchange, 43, 47–48, 257
  kieselguhr, 42
  loaded, 4, 42–48, 124, 229
  modified, 41–49, 501–506
  nature of, 16
  for organochlorine herbicides, 380
  for organophosphate insecticides, 364
  paraffin-oil impregnated, 111
  phenolphthalein, 122
  for porphyrins, 335–336
  pure cellulose, 36–41
  resin-loaded, 124
  role of, 10–11
  selective retention of, 9
  silica, 42
  silica gel, 42
  silicic acid, 71
  silicone-treated, 42, 177
  sodium molybdate impregnated, 155
  stannic phosphate, 503
  for steroids, 229
  synthetic organic ion-exchange resin loaded, 45–47
  wedge-shaped, 13
  zinc oxide, 92
Paper chromatography, *see also* Chromatography, Paper
  cation, 10
  frontal, 12
  future of, 3–4
  history of, 1–3
  quantitative, 90–108
  R_F values in, 14
  sources of error in, 105–108
  techniques of, 30–83
  theory and mechanism of, 6–27
Paper chromatography-electrophoresis, 121–122
Paper electrophoresis, 73, 156, *see also* Electrophoresis
Paper partition chromatography, 2–3, 11
Papyrus, 1
Paraformaldehyde, 118, 293
Paraffin oil, 52, 111
Parathion, 368, 370

Paromomycin, 410
Parthane, 359
Partition, 126
  adsorption, 11–12
  ion exchange and, 513
  of sugars, 152
Partition chromatography, 12, 45
Partition coefficient, 19
Partition equilibrium, 14
Partogrid, 79
Patuletin, 345
PCNB, 359
Peak broadening, 13–15
Pelargonic acid, 181, 188, 195
Pelargonidin, 343–344
Pelargonidin glycosides, 343–344
Penicillins, 400, 406–408
  natural, 407
  pharmaceutical, 407
Penicilloic acids, 407
Penophytin, 427
Pentachlorophenol, 297–298
Pentadecanoic acid, 182
Pentaene, 416
Pentaerythritol, 443
3β,11β,17α,20β-21-Pentahydroxyallopregnane, 211
Pentamethylenediamine, 114
Pentaphosphate, 531
Pentoses, 160, 169
Peonidin, 343
Peonidin glycosides, 344
Peptide hydrolyzates, 121
Peptide mapping, 140–142
Peptides, 37, 120–121, 127, 132, 139–143
  cyclic, 141
  chromatography of, 140–143
  dinitroaminophenyl, 139
  higher, 141
  hydrolysis of, 142
  R_F values of, 144
  scanning of, 141
  ultraviolet light detection in, 141
Peracetic acid, 456
Peracid treatment, 175
Perchloric acid, 41, 76
Percrotonic acid, 456
Per-*n*-hexoic acid, 456
Peridinin, 333
Periplocymarin, 240
Periplogenin, 240–241
Periplogenin acetate, 240
Periodate, 49, 475
Per-*n*,per-isobutyric acid, 456
Permeable membrane theory, 12
Peroxides
  aliphatic, 454–455
  organic, 454–457
Peroxybenzoic acids, 455–457
  R_b values for, 457
Perpropionic acid, 456
Per-*n*-valeric acid, 456
Perylene, 440
Pesticides, chromatography of, 41, 355–395
Petrolatum, 52
Petroleum ether, 21, 31, 50, 52, 121, 324–325, 388
Petunidin, 343
Petunidin glycosides, 344
Phenolphthalein paper, 122
Phenols, 42, 54, 68–69, 163, 296–315, 442
  chlorinated, 307–308
  chromatographic techniques for, 310–311
  conjugated, 305
  detection of, 311–315
  chromatography of, 296–315
  sample preparation of, 296–300
  solvents for, 300–310
  water-saturated, 52, 70
Phenol violet, 191
Phenoxy acid herbicides, R_F values for, 381
2-Phenoxyethanol, 443
Phenylacetic acid, 304
Phenylalanine, 128, 130, 139
Phenylazobenzenesulfonic acid dyes, R_F values of, 306
Phenylbenzyl phosphate, 492
1-Phenyl-1,3-butanedione, 448
trans-4-Phenyl-3-buten-2-one, 448
o-Phenylenediamine, 119, 193
p-Phenylenediamine, 119, 455
L-Phenylephrine, 115
Phenylethylamine, phenolic derivative of, 120
β-Phenylethylamine derivatives, 120
Phenylhydrazine, 26

m-Phenylphenol, 306
o-Phenylphenol, 306
p-Phenylphenol, 306
3-Phenyl-1-propanol, 443
Phenyl-2-propanone, 448
Phenylpropionic acid, 304
Phenyl-α-pyridyl phosphate, 492
Phenylpyruvic acid, 186
Phenyl-2-thienylketone, 448
Phenylthiohydantoin amino acids, 123
Phenylthiohydantoins, 139
Phloroglucinol, 304, 310, 313
Phosdrin, 368, 370
Phosphate, 486, 529–530
  inorganic, 492, 530–531
Phosphate buffer, 126
Phosphate esters, aromatic, 492
Phosphatidylethanolamine, 493
Phosphatidylserine, 493
Phosphenic acid, 493
Phosphoenol pyruvate, 486
6-Phosphogluconic acid, 487
3-Phosphoglycerate, 486–487
Phosphoglycolaldehyde, 487
Phosphoglycolic acid, 487
Phosphohydroxypyruvic acid, 487
Phospholactic acid, 487
Phospholipids, 493–494
Phosphomolybdic acid, 467-468
Phosphomolybdotungstic acid, 117
Phosphonic acid, 493
Phosphonous acid, 493
Phosphopeptides, 141
Phosphorescence spectra, spot identification and, 80
Phosphoribonic acid, 487
Phosphoribulose pyrophosphates, 486
Phosphoric acid, 155, 203
Phosphoric esters, 484–493
Phosphorous acids, 493
Phosphorous compounds, 484–494
Phosphorylated carbohydrates, 46
Phosphotungstic acid, 203
Photocell detector, 96
Photography
  in chromatogram recording, 79–80
  in spot estimation, 92
Photometric scanner, 96
Photometric spot measurement, 95–98
Photometry, 136
Physical detection methods, 77–78
Physostigmine, 466
Phytosterol, 218
Picric acid, 26
Picromycin, 415–416
Pigments
  chromatographic technique for, 325–326
  detection and identification of, 326–328
  double tailing zones and, 22–24
  extraction of, 320–322
  flavonoid, 341–350
  naturally occurring, 319–350
  quantitative methods for, 328–339
Pilocarpine, 466
Pimaricin, 416
Pimelic acid, 181
Pinacryptol yellow, 469
Pinocembrin, 346
Piperonal, 449
Piperonal azine, 451
Piperonyl butoxide, 390
Piperonyl cyclonene, 390
Piperonyl sulfoxide, 390
Placenta, peptides in, 142
Planimetry, 92, 104
Plant material, 121–122
  amino acids in, 135
  isolation of chlorogenic acid from, 299
    of indoles from, 283–284
    of phenolic acids from, 298
    preparation of sample in, 319–320
  sugar in, 154
Plant pigments, 319–350
  R_F values for, 427
  two-dimensional development of, 71
Plasma, 133
Plastid pigment chromatography, 11–12
Plastid pigment zones, 22
Plastoquinone A, 427
Plate concept theory, 8
Plate height, 8–9
Platinum oxide, 175
Podospicatin, 346
Polar compounds, rapid ascending separation of, 42
Polar liquids, 41
Polarography, 105, 162
Polyamide fibers, 469
Polyamines, 113

aliphatic, 114
Polyaminocarboxylic acids, 117–118
Polyenes, 416–417
Polyethylenethiuram disulfide, 386
Polymyxins, 417–418
Polyols, 160
Polypeptides, 417–418
Polyporenic acid A, 234
Polyram, 386
Polysaccharides, 154, 163–164
Porphyrin esters, 336
Porphyrin methyl esters
preparation of, 335
R_F values of, 337
Porphyrins, 334–341
chromatographic techniques for, 334–341
detection of, 148, 340
experiments with, 340–341
extraction of, 334–335
paper for, 335–336
quantitative analysis of, 340
R_F values for, 336–337
sample preparation for, 334–335
solvents for, 336–340
techniques for, 340
Potassium acetate, 143
Potassium ferricyanide, 113, 126, 161, 194, 518
Potassium hydroxide, 192
Potassium iodide, 115, 517
Potassium iodoplatinate, 461, 467–468
Potassium permanganate, 388–389
Potassium persulfate, 118
Potassium phosphate, 124
Potentiometer, sliding-wire, 78
Pratensein, 346
Precipitation chromatography, 32, 146, 153
Prednisolone, 211
Preequilibrium, 51
saturation and, 54
Pregnane derivatives, 223
Pregnane-3α,20α-diol, 213
5α-Pregnane-3β,20α-diol, 204
5α-Pregnane-3β,20β-diol, 204
Pregnane-3α,20α-diol diacetate, 213
Pregnane-3,20-dione, 225–226
5β-Pregnane-3,20-dione, 205
Pregnane-3α,17α,20α-triol, 213
Pregnanolone, 211, 213
Δ^4-Pregnene-2α,17α,21-triol-20-one, 211
Pregnenelone, 204
Procaine, 423
Processed foods, isolation of phenols and aromatic acids from, 298
Progesterone, 205, 211, 220, 225–226
Proline, 128, 134
Propanediol phosphate, 487
Propionic acid, 26, 178, 180
Propionyl phosphate, 181
Propiophenone, 449
Propiophenone azine, 451
n-Propoxyacetic acid, 179
n-Propylamine, 114, 116
n-Propyldihydrogen phosphate, 491
Propylene glycol, 52, 446
Propylphenyl phosphate, 492
n-Propylgallate, 392
Propyl-2-thiouracil, 269
Protamines, 145–146
Protein hydrolyzates, 70, 120, 127, 132, 143
Proteins, 121, 132, 143–146
chromatography of, 143–146
elimination of, 121–122
fingerprinting of, 140
precipitation of, 154
serum, 146
Protocatechuic acid, 303, 310
Protopine, 465
Protoporphyrin, 336–337
Protoporphyrin III, 337
Protoporphyrin IX, 337
Provitamins, separation of, 218
Prunetin, 346
Psilocin, 289
Pteridines, 424
Pteroic acid, 424
Pulchellidin, 343
Pulse height analyzer, 105
Pulse height discrimination, 103
Purine, 267, 272–273
Purine alkaloids, 467
Purines, 248–280
chromatographic applications in, 259
chromatographic techniques for, 248–280
coincidence detection and, 81
detection of, 77, 250

enzymatic, 253
quantitative methods in, 257–259
sample preparation in, 248–250
solvents for, 255–257
Purpurea glycoside A, 241
Purpurea glycoside B, 241
Pyrene, 439–440
Pyrethrins, 388–389
Pyridine, 50, 113, 126, 139, 142, 153–154
Pyridinium barbiturate, 161
Pyridoxal, 425
Pyridoxal phosphates, 424
Pyridoxal 5-phosphate, 425
Pyridoxamine, 425
Pyridoxamine 5-phosphate, 425
Pyridoxic acid, 425
Pyridoxines, 422, 424–426
Pyridoxol, 425
Pyridoxol 5-phosphate, 425
Pyrimidines, 248–280
chromatographic applications of, 259
chromatographic systems in, 255–257
chromatographic techniques for, 248–280
coincidence detection and, 81
detection of, 77, 250–254
enzymatic, 253
quantitative methods for, 257–259
sample preparation for, 248–250
solvents for, 255–257
Pyrogallol, 304, 395
Pyrogallol carboxylic acid, 303
Pyrophosphate, 486
inorganic, 492
Pyrrochlorin, 336
Pyrrole pigments, 334–341
Pyrrolizidine alkaloids, 461
Pyrroporphyrin, 336
Pyruvate acid, 184
Pyruvic acid, 161, 181, 186

Q

Quantification, 3
Quenching, 103
Quercetagetin, 345
Quercetin, 345, 516, 518
Quercetrin, 345
Quercimeritrin, 345
Quinalizarin, 517
Quinidine, 466
Quinine, 195, 466
Quinolinic acid, 303
8-Quinolyl glucoside, 305
2-Quinolyl sulfuric acid, 305
6-Quinolyl sulfuric acid, 305
7-Quinolyl sulfuric acid, 305
8-Quinolyl sulfuric acid, 305
Quinoneoximebenzoyl hydrazone, 386
Quinones, 426, 450
R_F values for, 427
Quinoxalines, 417

R

R_b value, defined, 457
R_{BCP}, 15
R_F value, 2, 7–8, 12–19
of aglycones, 241
of aliphatic amines, 116
of aliphatic polyamines, 114
of alkaloids, 464–466
of amino acids, 123, 126–131, 135
of amino alcohols, 114
of anthocyanidins, 343
of anthocyanins, 343–344
of antibiotics, 401–404
of aromatic acids, 302–305, 310
of aromatic amines, 111–113
of aromatic hydrocarbons, 439
in carbohydrates, 155–156
of cardiac glycosides and aglycones, 239–241
of catecholamines, 115, 117
of cholesteryl esters of higher fatty acids, 190
of chlorinated pesticides, 359
of chlorinated phenols, 307–308
of corticosteroids, estrogens, and ketosteroids, 225
critical solvent volume and, 107
of cytoactive amino acids and peptides, 144
defined, 14
of 16-dehydrosteroids, 224
of deoxyribonucleotides and ribonucleotides, 270
after descending development, 54
of 2,4-dinitrophenylhydrazides of fatty acids, 182

of estrogens, 219
factors influencing, 15–17
of flavins, 423
of flavonoid compounds, 345–347
of fungicides, 386
of glucosides, 166
of higher fatty acids, 188, 190
of hydroperoxidic thymine derivatives, 274
of hydroxamic acid derivatives, 180
of hydroxypurines, aminohydroxypyrimidines, and imidazoles, 272
of imines, 112–113, 120
of indoleamines, 288
of indoles, 287
influence of temperature and humidity on, 52
of inorganic phosphates, 531
of α-ketoacids, 186
of ketosteroids, 221
of lactones, 166
of lower fatty acids, 178
of marine algea pigments, 333
for metal ions, 508–511
of methylated sugars, 165
multiple zoning and, 26
of nitro compounds, 460
of 4-nitrophenylhydrazones of aromatic carbonyl compounds, 449
of nucletoides, nucleosides, and dinucleoside phosphates, 280
of organic acids, 184
of organic compounds, 456
of peptides, 142–143
of phenols, 302–305, 309–310
of phenylazobenzenesulfonic acid dyes, 306
of porphyrinmethyl esters, 337
of porphyrins, 336
of progesterone analogs, 220
for quinones, fat-soluble vitamins, and plant pigments, 427
rapid measurement of, 78–79
reproducibility of, 17–18
of semivolatile alkoxy acids, 179
spot identification and, 80–81
stereochromatography and, 82
of steroids and steroid esters, 222–223, 226
of streptomycins, neomycins, and kanamycins, 410
of sugar acids, 166
of sugar esters, 167
of sugars, 152
of tetracycline antibiotics, 414
of tocopherols and vitamin K, 430
transfer coefficients and, 73
of tryptamine derivatives, 116
in two-solvent systems, 51
for uncombined ketones, 448
unidimensional multiple development and, 67
for vitamins, 420, 422, 430
R_M values, 7
mathematical derivation of, 18–20
R_P values, of progesterone analogs, 220
R_{Pi} values, of deoxyribonucleotides and ribonucleotides, 270
R_{PG} value, defined, 392
R_R value, defined, 14
R_S values, stereochromatography and, 82
Radial chromatography, 14–15, 53, 62–67, 145, 319
β-Radiation, 99, 101
Radioactive ink, 99
Radioactive substances, 99, 533
chromatographic techniques for, 98–105
detection of, 78
sample spotting and, 35
Radioactivity, measurement of, 90
Radioassay, of monosaccharides, 162–163
Radioautography, 133
of carbohydrates, 169
Radiochromatography, double decomposition, 104
Radioisotope techniques, 98–105
Radiometric detection methods, 101–105
Radiometric precipitation, 104
Raffinose, 156
Random walk model, 8
Rauwolfia alkaloids, 462–463
Reaction time, 91
Rear-phase chromatography, 73
Reflection densitometry, 97
Regression lines, slopes of, 106–107
Reichstein S steroid, 205
Reichstein's Substance A, 211
Relomycins, 416

Reproducibility, 17–18, 36
in descending development, 56–57
on impregnated papers, 52
pretreatment and, 51
Resin-loaded papers, 45–47
Resistive action, 9–11
Resolution, 8, 36
in ascending development, 57
humidity and, 51
maximum, 13
sample spotting and, 34
in slow-moving substances, 54
Resorcinol, 304
α-Resorcylic acid, 303
β-Resorcylic acid, 303, 310
γ-Resorcylic acid, 303
Retamine, 464
Reversed-phase chromatography, 19, 41–42, 52–53, 73, 111, 200, 228–229, 512
Reversed-phase partition, 176
Reynoutrin, 345
Rhamnetin, 345
Rhamnose, 156, 164–165
L-Rhamnose, 166, 168
Rhizopterin, 424
Rhodamine, 54
Rhodium, 47
Rhodochlorin, 336
Rhodoporphyrin, 336
Rhoeadine, 465
Rhoeagenine, 465
Rhoifolin, 346
Rhothane, 359
Riboflavin, 422–424
Riboflavin phosphate, 423
Ribolumazines, 423
Ribonucleic acids (RNA), 146, 259
Ribonucleosides, 279
Ribonucleotides, R_F and R_{pi} values for, 270
Ribose, 487
D-Ribose, 166–167
Ribose-5-phosphate, 487
5-Ribosyluracil, 263
Ribulose 1,5-diphosphate, 487
Ribulose 5-phosphate, 486
Ricinoleic acid, 189
Rifomycins, 418–419
Ring oven technique, 173
RNA, *see* Ribonucleic acids
Robinetin, 345
Robinin, 345
Rodenticides, 390
Roemeridine, 465
Rogor (Dimethoate), 368
Rogor-oxon, 368
Ronnel, 368, 370
Rosidinin, 343
Rotenone, 389, 390
Rubeanic acid, 517–518
Rutin, 345, 423

S

Sakuranetin, 346
Salicylaldazine, 451
Salicylaldehyde, 449
Salicylic acid, 302, 310
Salicylsulfonic acid, 370–371
Salts
inorganic, 121
multiple spotting and, 25–26
Sample application, 3
Sample preparation, 32
Sample spotting, 32–36
Sanguilutine, 465
Sanguinarine, 465
Sanguirubine, 465
Santoquin(6-ethoxy-2,2,4-trimethyl-1,2-dihydroquinoline), 392
Saponaretin, 345–346
Saponin, 241
Sarmentocymarin, 240, 242
Sarmentogenin, 240–242
Sarmentogenin acetate, 240
Sarmentogenin diacetate, 241
Sarmentoside A, 240
Sarnovide mono(?)acetate, 241
Sarverogenin, 241–242
Sarverosid, 242
Saturation, in descending development, 54
Scandium, 47
Scanners, 101–103
gas-flow, 101
two-dimensional, 102
Scanning
in carbohydrates, 169
direct, 95–98
of radioactive substances, 101–103

β-Scintillation, 104
Scintillation counting, 48
Scintillation spectrometry, 163
Scopolamine, 464
Seawater, amino acids in, 122
Sebacic acid, 181, 184, 188
Sedoheptulose 7-phosphate, 487
Selective retention, 9
Semicarbazide, 193
Separation
  degree of sorption and, 10
  extent of, 20
  factors determining, 12–14
  mechanism of, 6
  prediction of, 6–8
Separation index, 20
Separation zones, 48
Sephadex G-25, 141
Sepiapterin, 424
Serine, 128, 130
Serotonin, 288
Serum, 133
Sesamin, 390
Sesamolin, 390
Siberian silkworms, sugar in, 154
Silica gel, 2–3, 48–49
Silica gel paper, 42
Silicic acid, 3, 49
Silicic acid paper, 71
Silicone oil, 111
Silicone paper, 41, 177
Silicon oil, 52
Silicotungstic acid, 206
Silver nitrate, 192, 359–361, 370, 380, 394, 529
Sinapic acid, 310
Sincurarine, 26
Sincurarine salts, R_F values and, 26
Siphonaxanthin, 333
Siphonein, 333
β-Sitosterol, 206, 218
Skatole, 287–288
Skellysolve B-benzene, 219
Sodium-acetylkynurenine, 289
Sodium azide, 387
Sodium bicarbonate, 367
Sodium chloride, 367
Sodium citrate, 142
Sodium dithionite, 517
Sodium lauryl sulfate, 161
Sodium molybdate, 155
Sodium nitrate, 117
Sodium nitroprusside, 113–115
Sodium orthophosphate, 12
Sodium periodate, 160
Sodium rhodizonate, 517
Sodium tetraphenylboron, 518
Sodium versenate, 159
Solubilization chromatography, 45
Solute detection, 4
Solute quantification, 4
Solute(s)
  determination of, 90
  ionic forms of, 25
  oxidation of, 25
  polar, 49
  polymerization of, 25
  precipitation of, 25
  remaining at origin, 24
  sorptive equilibrium of, 12
Solute spots, separation of, 12
Solute zone, movement and distance traveled by, 9, 15
Solution chromatography, 1
Solvent(s)
  aging of, 16
  with alcohols, 126
  for aliphatic acids, 177–191
  alkaline, 126
  for alkaloids, 462–467
  for amines, 111–113
  for amino acids, 126–127
  for anions, 526–527
  for aromatic acids, 300–310
  for bile acids, 233–236
  for carbamate insecticides, 377–378
  for carbohydrates, 157–158
  for cardiac glycosides, 238–242
  for cations, 506–512
  change in composition of, 16
  for chlorinated insecticides, 358–359
  for chloroplast pigments, 323–325
  composition of, 16
  critical volume of, 107
  defined, 1
  direct phase, 49
  extraction, 32
  for fat-soluble chloroplast pigments, 323–325
  for flavonoids, 342

for fungicides, 387
immiscible, 2
impurities in, 16
for indoles, 286–290
with ketones, 126
for metal ions, 524–525
odor of, 31
for organochlorine herbicides, 380
for organophosphate insecticides, 365–372
phenol-based, 126
for phenols, 300–310
polarity of, 50
for porphyrins, 336–340
for purines, 255–257
for pyrethrins, 388
for pyrimidines, 255–257
role of, 9–10
sample spotting and, 32–36
selection of, 6, 49–51
stationary polar, 52
for steroids, 227–229
techniques with, 49–53
two-phase, 49
volume of in reservoir, 16
water-miscible, 49
for water-soluble vitamins, 420–421

Solvent flow, 9, 14–20, 49–53
ascending, 57–61
automatic stoppage of, 56
descending, 53–57
distorted, 23
rule of, 123
unidimensional multiple development of, 67

Solvent front
distance traveled by, 14
dyes used to mark, 54
projections from, 22
radial distance to, 14

Solvent front detector, 61
Solvent reservoir, 16
Solvent systems, types of, 51–53
Solvent velocities, 13
Somogyi's reagent, 154, 162

Sorbent
HETP of, 8
selection of, 6
theoretical state concept of, 8

Sorbitol, 443, 446
L-Sorbose, 168
Sorption, irreversible, 24
Sorptive equilibrium, 12
Spark chambers, 101
Sparteine, 464, 466
Spectrometry, liquid scintillating, 103
Spectrophotofluorometer, 118
Spectrophotometry, 90, 93, 126, 139, 159
Spectroscopy, 82
Spermatophytes, separation of pigments in, 329–331
Sphingomyelin, 493
Spinal fluid, isolation of indoles from, 286
Spiraeoside, 345
Spongothymidine, 267
Sporidesmolides, 417

Spot
area of, 92–93, 134
bearding of, 126
color density of, 134
errors in producing, 105–106
estimation of, 91–105
ideal, 21
identification of, 80–82
initial, 90, 105
migration values for, 14–20
multiple, 21, 24–27, 35, 37
photometric measurement of, 95–98
remaining at origin, 24
spreading of, 13
total color of, 134
transmittance, 98
visual estimation of, 73–78, 92

"Spot Tests" (Feigl), 74

Spotting
multiple, 35
procedures for, 33–35
sample, 32–36
saturation and, 54

Spraying, of reagents, 74–76, 91, 107
Squalene, 210
Stains, 22, *see also* Spot
Stannous chloride, 517
Staphylococcus aureus, 398
Starch, 164
Stationary nonpolar solvents, 52–53

Stationary phase, 13
aqueous, 51–52
longitudinal changes in, 23

peak broadening in, 13–14
R_M values in, 18–20
tailing and, 21
Stationary polar organic solvents, 52
Stearate, 190
Stearic acid, 175–176, 182, 188, 190, 195
Stereochromatography, 81–82
Steroids, 41–42, 81, 200–232, 408
chromatographic applications for, 232
chromatographic data on, 279
chromatographic techniques for, 202, 217–227, 232–237
color tests for, 203–216, *see also* Color tests
vic-glycol, 214–216
identification of, 216–217
α-ketol, 214–216
mobility-structure relationships for, 231
quantitative analysis for, 231
papers for, 229
physical detection of, 216
reaction of, 204–206
sample preparation and application for, 200–202
Sterols, 42, 210–211
Steryl esters, 42
Stigmasterol, 206, 218
Stilbene, 440
Stillopsillin, 346
Stillopsin, 347
Streaked zones, 21–22
Streaking, 106, 110
in carbohydrates, 169
in ion-exchange resin-loaded papers, 46
procedures for, 35–36
Streptococcus faecalis, 424
Streptomycin monoaldehyde ammonium, 410
Streptomycins, 409–411
Streptomycin sulfate, 410
Streptothricin, 410
Streptothricin I, 410
Streptothricin II, 410
Streptothricin III, 410
Strobane, 359
Strontium, 104
Strophanthidin, 239, 241–242
Strophanthidin acetate, 240
Strophanthidol, 239–242
Strophanthidol acetate, 240
Stropeside, 239, 242
Strychnine, 462, 466
Stylophylline, 465
Stylopine, 465
Suberic acid, 184
Sublimation autography, 100
Succinic acid, 180, 184
6-Succinoaminopurine, 266
Succinyl peroxide, 456
Sucrose, 156, 158–162, 314
Sudan IV, 193
Sugar acids, 160, 164
R_F values for, 166
Sugar alcohols, 165
Sugar esters, 164–167
R_F values for, 167
Sugar hydrazones, 167–168
Sugar phosphates, 104, 485
Sugars, 42
adsorption of, 152
aldohexose, 152
aldomethylose, 152
amino, 154
C-labeled, 103
coincidence techniques and detection, 81
elimination of, 121–122
estimation of, 90
isomeric, 152
methylated, 164–165
oligosaccharide, 152
partition of, 152
ring form of, 152
R_F values for, 165–168
spot measurement of, 93
urinary, 154
Sugar sulfates, 167
m-Sulfamidobenzoic acid, 310
o-Sulfamidobenzoic acid, 310
p-Sulfamidobenzoic acid, 310
Sulfanilic acid, 304
diazotized, 312
Sulfate, 104, 529
Sulfathiazole, 400
Sulfenamides, aromatic, 477
Sulfide, 529–530
Sulfinomides, aromatic, 477
Sulfite, 529
Sulfonamides, 477–478
Sulfonates, aliphatic, 479

Sulfonic acids, 479–483
β-Sulfopyruvic acid, 186
Sulfur, X-ray fluorescence detection of, 78
Sulfur amino acids, 127
Sulfur compounds, 473–483
  aniline derivatives of, 475
Sulfuric acid, 121, 154
Sulphurein, 347
Sulphuretin, 347
Sumatrol, 390
Suprasterol II, 218
Swerved zones, 23
Syringic acid, 184, 303, 310
Systox (Demeton), 368, 370
Systox-*S*-methylsulfone, 368
Systox-*S*-methylsulfoxide, 368

T

Tachysterol, 218
Tagaturonic acid, 166
Tailing, 21–22
  double, 24, 35
Tailing zones, 21–22
Talonolactone, 166
Tannic acid, 303, 517
Tartaric acid, 181, 184
Tartrate, 501
Taurine, 129
Taurochenodeoxycholic acid, 236
Taurocholic acid, 236
Taurodeoxycholic acid, 236
Taurolithocholic acid, 236
Tectiridin, 346
Tectirigenin, 346
Tectochrysin, 345
Temperature
  control of in laboratory, 31
  descending development and, 54
  effect of, 16, 44, 91
  glass fiber papers and, 48
  gradients in, 4
  influence of, 16, 44, 91
  multiple spots and, 24
  reproducibility and, 17
  selection of, 6
Tephrosin, 390
TEPP, 368
Terephthalaldehyde, 292
Terephthaladazine, 452
Teropterin, 424
Terpenes, 42
Testosterone, 204, 225–226
Testosterone benzoate, 204
Testosterone propionate, 204
7,7,8,8-Tetracyanoquinodimethan, 133
Tetracycline, 400, 413–4–5
Tetracycline antibiotics, 414
Tetracycline hydrochloride, 414
Tetracyclinemethylenelysine, 414
9-Tetradecenoic acid, 189
n-Tetradecyclamine, 116
Tetraene, 416
Tetrahydroberberine, 465
Tetrahydrofuran hydroperoxide, 456
2′-*O*-Tetrahydropyranyluridylyl-(3′→5′)-2′,3′-*O*-ethoxymethylene-5-bromouridine, 280
2′-*O*-Tetrahydropyranyluridylyl-(3′→5′)-2,3-*O*-ethoxymethylene-5-iodouridine, 280
2′-*O*-Tetrahydropyranyluridylyl-(3′→5′)-2′,3′-*O*-ethoxymethylene-3-methyl-5-bromouridine, 280
2′-*O*-Tetrahydropyranyluridylyl-(3′→5′)-2′,3′-*O*-ethoxymethylene-3-methyl-5-iodouridine, 280
Tetralin hydroperoxide, 456
Tetrametaphosphate, 531
Tetramethylenediamine, 114
Tetramethylgalactose, 165
Tetramethylmannose, 165
Tetramethylthiuram disulfide, 386
1,3,7,9-Tetramethyluric acid, 269
Tetraphenylborate, 415
Tetraphosphate, 531
Thebaine, 465
Theobromine, 273
Theophylline, 273, 465
Theophylline-7-acetic acid, 269
Thermal detection, 159
Thermal paper chromatography, 73, 132
Therobromine, 465
Thiamine, 421–423
Thiamine hydrochloride, 421
Thiamine monophosphate, 421
Thiamine phosphate, 421
Thiazine, 472
Thimet, 368, 370

Thin-layer chromatography, 3, 15, 17, 31, 40–41, 82, 107, 120, 254
Thioammeline, 478
2-Thiobarbituric acid, 158, 267
Thiocyanates, 479, 529
2-Thiocytosine, 267
Thiodon, 359
Thiohydantoins, 139
5′-Thiomethyladensoine, 269
5′-Thiomethylinosine, 266
Thionyl chloride, 173, 176
2-Thioorotic acid, 266
Thiosulfate, 104, 529–530
2-Thiothymine, 269
Thiouracil, 273
Thioureas, 120, 390, 475–477
Thiram, 386
Thiurams, 478–479
Threonine, 128, 130
Thryonines, 126
Thymidine, 268, 271, 273
Thymidine 5′-diphosphate, 278
Thymidine monophosphate (TMP), 271
Thymidine 5′-monophosphate, 278
Thymidine 5′-phosphate, 263
Thymidine 5′-triphosphate, 278
Thymine, 268, 271, 273–274, 279
  R_F values and, 274
Thymol, 306
Thymol blue, 191
Thymyl phosphate, 492
Thymyl-α-pyridyl phosphate, 492
Thyroxine, R_F values for, 142
L-Thyroxylglycine, 142
Tigloidine, 464
Titanium, 47
Titrated compounds, radioisotope techniques for, 99–100
Titration, 105
TMP, *see* Thymidine monophosphate
Tobacco alkaloids, 462
Tocol, 430
α-Tocopherol, 218, 392, 420, 422, 427, 430
β-Tocopherol, 429–430
γ-Tocopherol, 429–430
δ-Tocopherol, 430
ε-Tocopherol, 429–430
η-Tocopherol, 429
σ-Tocopherol, 429
DL-α-Tocopherol, 420
Tocopherols, 428-430
  R_F values for, 430
α-Tocopheryl acetate, 420
α-Tocopherylquinone, 427, 430
o-Tolidine, 113, 455
Toluene, 324–325
Toluidine, 111
o-Toluidine, 119
p-Toluidine, 119
  elatography of, 73
Tongues, detection reagent and, 22
Toxaphene, 359
Toxicarol, 390
Trailing zones, 21–22, 37
Transfer coefficients, 73
Transmission densitometry, 48, 97
Transuranium, 47
Triacetyloleandomycin, 416
2,4,5-Triaminopyrimidine, 272
4,5,6-Triaminopyrimidine, 272
4,5,6-Triaminopyrimidine sulfate, 264
Triazines, 120
Tribromoacetic acid, 178
Tricarballylic acid, 184
Tricetinidin, 343
Trichloroacetic acid, 115, 122, 145, 158, 178, 203
2,4,5-Trichloroaniline, 112
2,4,6-Trichloro-*m*-cresol, 309
Trichloroethylene, 340
N-Trichloromethylthiotetrahydrophthalimide, 386
3,5,6-Trichloro-2-pyridinol, 367
3,5,6-Trichloro-2-pyridyl phosphate, 367
Trichomycin, 416
Tricin, 345
Tridecanoic acid, 182
2-Tridecanone, 448
Triethanolamine, 114
Triethylamine, 163
Triethylene glycol, 443
Triethylenetetramine, 114
Trifluoroacetic acid, 121
3α,6α,7β-Trihydroxycholanic acid, 235
3α,6β,7α-Trihydroxycholanic acid, 235
3α,6β,7β-Trihydroxycholanic acid, 235
11β,17α,21-Trihydroxy-5β-pregnane-3,20-dione, 205

3β,17α,21-Trihydroxy-5-pregnen-20-one, 205
2,6,8-Trihydroxypurine (uric acid), 272
Triisopropanolamine, 114
Triketohydrindenehydrate, 132
Trimetaphosphate, 492, 531
3,4,5-Trimethoxybenzalazine, 451
3,4,6-Trimethylamannose, 165
2,3,5-Trimethylarabinose, 165
2,3,4-Trimethylarabinose, 165
Trimethylenediamine, 114
2,3,4-Trimethylgalactose, 165
2,3,5-Trimethylgalactose, 165
2,3,6-Trimethylgalactose, 165
2,4,6-Trimethylgalactose, 165
2,3,4-Trimethylglucose, 165
2,3,5-Trimethylphenol, 306
2,4,6-Trimethylolphenol, 304
1,3,7-Trimethylxanthine, 269
1,3,8-Trimethylxanthine, 269
2,3,4-Trimethylxylose, 165
2,4,6-Trinitrophenol, 309
Triphenylmethanol, 443
2,3,5-Triphenyltetrazolium chloride, 160–161, 398
2,3,5-Triphenyl-2*H*-tetrazolium chloride, 163
Triphenyltin acetate, 386
Triphosphate, 531
Triphosphopyridine nucleotide, 487
Tripeptides, 141
α-Tripiperidine, 35
Trishydrochloride, 145
Trisodium citrate, 45
Trithion, 368
Tritium, 99–100, 103
Tropine alkaloids, 64–65, 463
Tropine, 464
Tryptamine, 111, 115–117, 287–288, 295
Tryptamine derivatives, 118
Tryptophan, 51, 71, 121, 127–128, 130, 287–288, 295
Tryptophan metabolites, 120
Two-dimensional chromatography, 14, 26, 68–72, 102–103, 126–127, 131, 139, 216
Tylosin, 416
Tyramine, 115, 117
Tyrisine, 130
Tyrosine, 121, 123, 126, 128, 130
  R_F values for, 142
L-Tyrosylglycine, 142

U

UDP, *see* Uridine diphosphate
UDP-apiose, 259
Ultrafiltration, 121
Ultraviolet absorption photography, 81
Ultraviolet densitometry, 399–400
Ultraviolet light detection, 77–81, 133
  organophosphate insecticides and, 365–370
  of phenols and aromatic acids, 314–315
  of purines and pyrimidines, 250
  of pyrethrins, 389
  spot identification and, 80–81
Ultraviolet strip scanner, 96
Umbelliferone, 305
UMP, *see* Uridine monophosphate
Undecenoic acid, 182
10-Undecenoic acid, 189
Undecylenic acid, 195
Undecylic acid, 188
Uracil, 266, 271, 273–274, 276–277, 279
Uracil-*t*-acetic acid, 265
Uracil-5-carboxylic acid, 265
Uranium, 47
Urbazit, 386
Urea, 115, 159, 206
  isomeric, 120
Uric acid, 262, 271–273, 276
Urdine, 264, 271, 273–274, 277, 280
Uridine 2′,3′-cyclic phosphate, 280
Uridine diphosphate (UDP), 277
Uridine 5′-diphosphate, 262, 278
Uridine diphosphate glucose, 262, 486
Uridine monophosphate, 271
Uridine 2′-monophosphate, 279
Uridine 2′:3′-monophosphate, 279
Uridine 3′-monophosphate, 279
Uridine 5′-monophosphate, 279
Uridine 2′(3′)-phosphate, 280
Uridine 5′-phosphate, 262
Uridine triphosphate (UTP), 277
Uridine 5′-triphosphate, 262, 278
Uridylyl-(3′ →5′)-5-bromouridine, 280
Uridylyl-(3′ →5′)-iodouridine, 280
Uridylyl-(3′ →5′)-3-methyl-5-bromouridine, 280

Uridylyl-(3′→5′)-3-methyl-5-iodouridine, 280
Uridylyl-(3′→5′)-uridine, 280
Urinary 17-ketosteroids, 201
Urine, 120–123, 126, 130, 133, 136, 138, 154, 174, 201–202
  ascorbic acid in, 431–432
  extraction of phenols and aromatic acids from, 296–297
  isolation of indoles from, 284–285
  two-dimensional development of, 71
Uronic acids, 154, 158–161, 164
Uroporphyrin, 336
Uroporphyrin I, 337
Uroporphyrin III, 337
UTP, *see* Uridine triphosphate
Uzarigenin, 241

V

Valeric acid, 180, 188, 195
n-Valeric acid, 178
Valeroidine, 464
Valeryl phosphate, 181
Valine, 123, 126, 128, 130
  radioactive, 124
Valylhydroxamate, 124
Vamidothion, 368
Vamidothionsulfoxide, 368
Vanadium, 47
Vancomycin, 410
Van Deemter equation, for gas chromatography, 9
Van der Waals forces, 11
Vanilla bean, 298
Vanillic acid, 115, 303, 310, 316
Vanillin, 449
Vanillin azine, 451
Vanilmandelic acid, 115
Veratraldazine, 451
Veratraldehyde, 449
Veratridine, 466
Verodoxin, 242
Vincamine, 466
Violaxanthin, 333
Viomycin, 410
Vitamin A, 218, 419–420, 422, 426–427
  acetate, 215, 320, 429
  acid, 426
  alcohol, 420, 426
  aldehyde, 419, 426
  esters, 426
  Palmitate, 420, 429
Vitamin A_1, 429
  acid, 429
  alcohol, 429
  aldehyde, 429
Vitamin A_2, 426, 429
  acetate, 429
  alcohol, 429
  aldehyde, 429
Vitamin B_1, 419
Vitamin B_2, 419, 423
Vitamin B_6, 419, 424–425
Vitamin B_{12}, 419, 422, 425
Vitamin B_{15}, 426
Vitamin C, 422, 431–432
Vitamin D, 419–420
Vitamin D_2, 206, 210, 218, 420, 426–427
Vitamin D_3, 210, 213, 420
Vitamin E, 419–420, 428–430
Vitamin K, 419–420, 427–430
Vitamin K_1, 430
Vitamin K_2, 430
Vitamin K_3, 419
Vitamins, 41
  chromatography of, 419–432
  detection of, 103
  fat-soluble, 420, 426–430
  R_F values for, 420, 422, 427–430
  separation of, 47, 218
  water-soluble, 420–426
Vitexin, 345
Volume fractions, 19

W

Warfarin, 391
Wash liquid, defined, 1
Water, 11
  as solvent, 50
Water-cellulose complex, 11
Wedge chromatogram, 64
Wedge strip chromatography, 64, 66
Wesson Oil, 12
Wet-start chromatography, 44, 124
Whatman papers, 12, 37–45, 48, 66, 82–83, 111, 114–115, 118, 121, 123–127, 130, 132–133, 138–139, 141, 143, 145, 153–155, 163–164, 176–177,

238, 310–311, 322, 335–336, 340, 342, 380
Wines, amino acids in, 122

X

Xanthates, 442, 444
Xanthine, 263, 271–273, 276, 279
Xanthophyll lutein, 24
Xanthophylls, 427
Xanthosine, 262, 271, 276
Xanthurenic acid, 289, 303
Xanthurenic acid 8-methyl ether, 289
Xanthydrol, 115–117, 293
X-ray film, in detection, 99
X-ray fluorescence, 78
Xylan, 164
Xylose, 156, 159
D-Xylose, 160, 167
Xylulose 5-phosphate, 486

Y

Yeast
  sugar in, 154
  two-dimensional chromatography of, 71
Yellow RFS, 371

Z

Zaffaroni systems, 52
Zimmerman reagent, 212–213
Zinc, 47, 519–520
Zinc dimethyldithiocarbamate, 386
Zinc ethylenebisdithiocarbamate, 386
Zinc oxide paper, 92
Zinc uranyl acetate, 518
Zineb, 386
Ziram, 386
Zirconium, 104
Zirconium phosphate, 47
Zones
  anomalous, 21–27
  arrowhead-shaped, 23
  boundaries of, 6, 14
  circular, 62–63
  detection of, 3
  diffuse, 21
  displacement of centers in, 12
  distance traveled by, 14
  double-tailing, 22
  elution of, 48
  factors determining separation of, 12–14
  flattened, 23
  heart-shaped, 23
  initial, 8, 13, 24, 26, 35–36, 54, 107, 153
  irregular, 21–24
  lateral diffusion of, 107
  lateral spreading of, 23
  lost, 24
  measurement of, 90, 98
  molecular diffusion of, 13
  movement of, 9
  multiple, 12, 24–27, 117
  narrowed, 23
  overlapping, 98
  pattern of, 17
  plastid pigment, 22
  radial distance to, 14
  rapid-fading, 91
  shape of, 18
  spreading, 8, 21
  streaked, 21–22
  swerved, 23
  tailing, 21–22, 37
  trailing, 21–22, 37
  velocity of, 18
  V-shaped, 23
Zymosterol, 206, 210, 218